CRC Handbook of Laser Science and Technology

Volume III
Optical Materials
Part 1: Nonlinear Optical Properties/Radiation Damage

Editor

Marvin J. Weber, Ph.D.
Lawrence Livermore National Laboratory
University of California
Livermore, California

CRC Press, Inc.
Boca Raton, Florida

Library of Congress Cataloging-in-Publication Data
Main entry under title:

Optical materials.

(CRC handbook of laser science and technology ; v. 3)
Bibliography: p.
Contents: pt. 1. Nonlinear optical properties
radiation damage.
1. Optical materials. 2. Optical materials--
Effect of radiation on. 3. Nonlinear optics.
I. Weber, Marvin J., 1932- . II. Series.
QC374.068 1986 621.36 86-983
ISBN 0-8493-3503-5

Direct all inquiries to CRC Press, Inc., 2000 Corporate Blvd., N.W., Boca Raton, Florida, 33431.

International Standard Book Number 0-8493-3503-5

Library of Congress Card Number 86-983
Printed in the United States

PREFACE

Laser action has been observed in all forms of matter and spans a spectrum ranging from radiowaves to X-rays. The object of the *CRC HANDBOOK OF LASER SCIENCE AND TECHNOLOGY* is to provide a concise, readily accessible source of critically evaluated data for workers in all areas of laser research and development. The emphasis is on the presentation of tabular and graphical data compiled by recognized authorities. Definitions of properties and references to the original data sources and to supplementary reviews and surveys are also provided, as appropriate. The previous two volumes in this series dealt with laser action in all media and contained extensive tables of laser transitions and references. Volumes III, IV, and V are devoted to the physical and chemical properties of optical materials used in laser systems and applications.

The earlier *CRC Handbook of Lasers with Selected Data on Optical Technology* contained several sections on optical materials. These sections have been updated and expanded and many new sections added to form Volumes III to V. The materials covered are almost exclusively condensed matter. Because many properties are dependent in varying degrees on preparation methods, materials imperfections and measurement techniques, several sections included discussions and descriptions of these specific characteristics and of materials compositions.

Optical materials for laser systems encompass an extremely wide range of special property requirements and operating wavelengths and environments. Of necessity, the topics covered in these volumes are selective. In some sections it was possible to be exhaustive; in others a more general survey is provided because extensive data tabulations already exist elsewhere. The applications of optical materials are continually expanding. Therefore an attempt has been made throughout to include not only currently useful materials but also representative examples of broader classes of materials of possible future interest. One can frequently use observed trends in materials properties with composition to select and tailor new materials having specific operating characteristics.

Data on optical materials can be presented from different points of view — by material, by properties, or by application. For laser materials no single approach seemed fully appropriate, therefore several formats have been utilized. A number of properties may be relevant to a given application. As an example, for transmitting materials one may be interested in optical, thermal, and mechanical properties, thus these properties are grouped together within a single section. However, not all properties are covered within a particular section. Because of their special character, properties such as optical nonlinearities, radiation damage, and fabrication are discussed separately. Characteristics of specific classes of materials such as glasses may also be covered in several sections depending upon its use as a transmitting material, a filter material, an optical waveguide material, or a laser host material. Indices at the end of each volume list individual materials and where data on specific properties are located.

With the advent of lasers, nonlinear optical phenomena have become important and have been the subject of intense study and application. The properties of materials for harmonic generation and two-photon absorption, nonlinear refractive index, and stimulated Raman scattering properties of various opticals materials are included in Volume III. Data on radiation damage of optical crystals and glasses are also surveyed in this volume. Volume IV covers materials for fundamental uses: transmission (laser windows and lenses), filtering, reflection, and polarization. Materials for more specialized uses involving linear electrooptic, magnetooptic, elastooptic, and photorefractive effects and liquid crystals are also covered in this volume. Volume V presents data on properties of materials for optical waveguides, optical storage and recording, phase conjugation, lasers, and quantum counter applications. Other sections cover optical coatings and thin films. The final section describes fabrication techniques and procedures for all types of optical materials.

Laser-induced damage to optical components is an extremely important consideration for many laser applications. Although it was originally planned to include a section on laser damage, this topic will be covered separately elsewhere. In this regard, I welcome comments about the contents and presentation in the present volumes and suggestions for materials and properties to be included in future editions.

A handbook can never be completely current with the journal literature. Because of the very nature of the preparation and publication process, one must accept the fact that a handbook becomes out-of-date at the time of the final type setting. Although all sections for Volumes III to V were solicited concurrently, there were delays in the receipt of some manuscripts. Some sections were updated, but variations in timeliness, as evident from the reference dates, remain.

These volumes are the result of the efforts and talents of many people to whom I am indebted. I thank especially the contributors for the time devoted to preparing these compilations and texts and the Advisory Board and contributors for their numerous helpful comments and suggestions regarding the content and format. The staff of CRC Press, and Senior Editor Marsha Baker in particular, have my thanks and appreciation for the preparation of these volumes. Finally, I am grateful to my wife Pauline for her generous support of this project.

Marvin J. Weber
Danville, California
February 1984

THE EDITOR

Marvin J. Weber received his education at the University of California, Berkeley, and was awarded the A.B., M.A., and Ph.D. degrees in physics in 1954, 1956, and 1959. After graduation Dr. Weber joined the Research Division of the Raytheon Company where he was a Principal Scientist. As Manager of Solid State Lasers, his group developed many new rare earth laser materials. While at Raytheon, he also discovered luminescence in bismuth germanate, a scintillator crystal widely used for the detection of high energy particles and radiation.

During 1966-67 Dr. Weber was a Visiting Research Associate in the Department of Physics, Stanford University.

In 1973 Dr. Weber joined the Laser Fusion Program at the Lawrence Livermore National Laboratory. As Head of Basic Materials Research, he had the responsibility for the physics and characterization of optical materials for high-power laser systems. His work on laser glass resulted in an Industrial Research IR-100 Award in 1979 for research and development of fluorophosphate laser glasses. In 1983 Dr. Weber was the recipient of the George W. Morey Award of the American Ceramics Society for his basic studies of fluorescence and stimulated emission in glasses and for the insight that research has provided into glass structure.

Dr. Weber has published numerous scientific papers and review articles in the areas of lasers, luminescence, optical spectroscopy, and magnetic resonance in solids and has been granted several patents on solid-state laser materials. He is a Fellow of the American Physical Society, a Fellow of the Optical Society of America, and a member of the American Ceramics Society, the Materials Research Society, and the American Association for Crystal Growth.

Among other activities, Dr. Weber has been a consultant for the National Science Foundation's Division of Materials Research, a member of the National Academy of Science - National Research Council Evaluation Panel for the National Bureau of Standard's Inorganic Materials Division, and a participant in various advisory panels. He is currently an Associate Editor of the *Journal of Luminescence* and a member of the Editorial Advisory Board of the *Journal of Non-Crystalline Solids*.

In 1984 Dr. Weber began a temporary transfer assignment with the Division of Materials Sciences, Office of Basics Energy Sciences, of the U.S. Department of Energy in Washington, D.C., where he has been involved with planning for advanced synchrotron radiation facilities and applications of computers for materials simulations.

CONTRIBUTORS

E. Joseph Friebele, Ph.D.
Research Physicist
Naval Research Laboratory
Washington, D.C.

Shobha Singh, Ph.D.
Distinguished Member, Technical Staff
AT&T Bell Laboratories
Murray Hill, New Jersey

Fred Milanovich, Ph.D.
Physicist
Lawrence Livermore National Laboratory
Livermore, California

W. Lee Smith, Ph.D.
Physicist
Therma-Wave Corporation
Fremont, California

Richard T. Williams, Ph.D.
Reynolds Professor of Physics
Department of Physics
Wake Forest University
Winston-Salem, North Carolina

HANDBOOK OF LASER SCIENCE AND TECHNOLOGY

VOLUME I: LASERS AND MASERS

VOLUME II: GAS LASERS

HANDBOOK OF LASER SCIENCE AND TECHNOLOGY

VOLUME IV: OPTICAL MATERIALS
PART 2: PROPERTIES

VOLUME V: OPTICAL MATERIALS
PART 3: APPLICATIONS, COATING, AND FABRICATION

HANDBOOK OF LASER SCIENCE AND TECHNOLOGY

VOLUME III: OPTICAL MATERIALS
PART 1: NONLINEAR OPTICAL PROPERTIES/RADIATION DAMAGE

TABLE OF CONTENTS

Section 1
Nonlinear Optical Properties

1. NONLINEAR OPTICAL PROPERTIES

1.1. NONLINEAR OPTICAL MATERIALS

S. Singh

INTRODUCTION

When a material substance is subjected to electromagnetic radiation, the electrons of the medium tend to be polarized. In the electric dipole approximation, this effect is characterized by an induced polarization, $\mathbf{P}_j^{(1)}$, or the electric dipole moment per unit volume of the medium phenomenologically by the linear relation:

$$\mathbf{P}_j^{(1)}\omega = \epsilon_o \chi_{jk}^{(1)}(\omega) E_k(\omega) \qquad (C \cdot \text{m}^{-2} \text{ in MKS units}) \tag{1}$$

where E_k is a component of the electric field strength associated with the incident radiation, ϵ_o is the permittivity of free space having the value of 8.854×10^{-12} C · V^{-1} · m^{-1} in MKS units or $(^1/_4\pi)$ stat$C$·statV^{-1}·cm^{-1} in CGS (esu) units, and $\chi_{jk}^{(1)}$ (ω), a second-rank tensor for anisotropic media, is a complex quantity which is responsible for the familiar optical phenomena of absorption, emission, reflection, and refraction. Thus,

$$\chi_{jk}^{(1)}(\omega) = \chi_{jk}^{'(1)}(\omega) + i\chi_{jk}^{''(1)}(\omega) \tag{2}$$

The index of refraction, $n(\omega)$, and dielectric constant, $\epsilon(\omega)$, of a medium are related to the real part of the susceptibility by:

$$n_{jk}^2(\omega) = \epsilon_{jk}^{(\omega)}/\epsilon_o = 1 + \chi_{jk}^{'(1)}(\omega) \quad \text{(MKS units)} \tag{3}$$

$$= \epsilon_{jk}(\omega) = 1 + 4\pi\chi_{jk}^{'(1)}(\omega) \quad \text{(CGS units)} \tag{4}$$

The absorption constant, $\alpha(\omega)$, is related to the imaginary part of the susceptibility as:

$$\alpha(\omega) = (\omega/c)\chi_{jk}^{''(1)}(\omega) \qquad (\text{m}^{-1},\text{MKS}) \tag{5}$$

$$= (4\pi\omega/c)\chi_{jk}^{''(1)}(\omega) \ (\text{cm}^{-1},\text{CGS}) \tag{6}$$

The number of independent components of the dielectric tensor, ϵ_{jk} (or the linear susceptibility tensor $\chi_{jk}^{(1)}$), for different crystal classes and for isotropic media are given in Table 1.1.1.

For large values of the optical fields, such as those associated with intense radiation from lasers, the induced polarization in a medium is no longer a linear function of E. In order to describe the nonlinear optical effects which can occur under such conditions, it is convenient to expand **P** as a power series in the electric field, E, and magnetic field, H, and their time and space gradients present in the nonlinear medium.[1-3] Thus, for a lossless nonmagnetic medium one can write qualitatively:

Table 1.1.1
FORM OF THE DIELECTRIC TENSOR ϵ_{jk} (OR FIRST-ORDER SUSCEPTIBILITY TENSOR $\chi_{jk}^{(1)}$ FOR VARIOUS CRYSTAL CLASSES

Cubic and isotropic system	$\begin{bmatrix} \epsilon_{11} & 0 & 0 \\ 0 & \epsilon_{11} & 0 \\ 0 & 0 & \epsilon_{11} \end{bmatrix}$
Tetragonal, trigonal, and hexagonal systems	$\begin{bmatrix} \epsilon_{11} & 0 & 0 \\ 0 & \epsilon_{11} & 0 \\ 0 & 0 & \epsilon_{33} \end{bmatrix}$
Orthorhombic system	$\begin{bmatrix} \epsilon_{11} & 0 & 0 \\ 0 & \epsilon_{22} & 0 \\ 0 & 0 & \epsilon_{33} \end{bmatrix}$
Monoclinic system (unique axis oy)	$\begin{bmatrix} \epsilon_{11} & 0 & \epsilon_{13} \\ 0 & \epsilon_{22} & 0 \\ \epsilon_{13} & 0 & \epsilon_{33} \end{bmatrix}$
Triclinic system	$\begin{bmatrix} \epsilon_{11} & \epsilon_{12} & \epsilon_{13} \\ \epsilon_{21} & \epsilon_{22} & \epsilon_{23} \\ \epsilon_{31} & \epsilon_{32} & \epsilon_{33} \end{bmatrix}$

$$\begin{aligned} \mathbf{P}_j(\omega_\sigma) = \epsilon_o[&\chi_{jk}(\omega_\sigma)E_k(\omega_\sigma) + \chi_{jkl}(\omega_\sigma)K_k(\omega_\sigma)E_l(\omega_\sigma) \\ &+ g_1\chi_{jkl}(-\omega_\sigma;\omega_1,\omega_2)E_k(\omega_1)E_l(\omega_2) - g_2 i\chi_{jkl}^{H}(-\omega_\sigma;\omega_1,\omega_2)E_k(\omega_1)H_l(\omega_2) \\ &+ g_3 i\chi_{jklm}(-\omega_\sigma;\omega_1,\omega_2)E_k(\omega_1)K_l(\omega_2)E_m(\omega_2) \\ &+ g_4\chi_{jklm}(-\omega_\sigma;\omega_1,\omega_2,\omega_3)E_k(\omega_1)E_l(\omega_2)E_m(\omega_3) \\ &+ g_5\chi_{jklm}^{HH}(-\omega_\sigma;\omega_1,\omega_2,\omega_3)E_k(\omega_1)E_l(\omega_2)H_m(\omega_3) \\ &+ g_6\chi_{jklm}^{H}(-\omega_\sigma;\omega_1,\omega_2,\omega_3)E_k(\omega_1)E_l(\omega_2)H_m(\omega_3) + \ldots] \end{aligned} \tag{7}$$

where j, k, l, m . . . are Cartesian subscripts obtained by the Einstein summation convention of repeated indexes j, k, l, m . . . $= x, y, z$, and g_1, g_2, g_3 . . . g_6 are degeneracy factors arising from intrinsic permutation symmetry. The frequencies (having both positive and negative values) satisfy the relation:

$$\omega_\sigma = \omega_1 + \omega_2 + \omega_3 + \ldots \tag{8}$$

In Equation 7, $\mathbf{P}(\omega)$, $E(\omega)$, and $H(\omega)$ are the Fourier complex amplitudes at ω. Thus,

$$\mathbf{P}(\mathrm{r,t}) = (1/2)[\mathbf{P}(\omega)\exp i(\mathrm{k}\cdot\mathrm{r} - \omega t) + c\cdot c] \tag{9}$$

$$E(\mathrm{r,t}) = (1/2)[E(\omega)\exp i(\mathrm{k}\cdot\mathrm{r} - \omega t) + c\cdot c] \tag{10}$$

$$H(\mathrm{r,t}) = (1/2)[H(\omega)\exp i(\mathrm{k}\cdot\mathrm{r} - \omega t) + c\cdot c] \tag{11}$$

Also,

Table 1.1.2
NONLINEAR OPTICAL PROCESSES[9,11]

Susceptibility	Optical process	Ref.
$\chi_{jkl}(-2\omega;\omega,\omega)$	Second harmonic generation (SHG) (g = 1/2)	12
$\chi_{jkl}(0;\omega,-\omega)$	Optical rectification (OR) (g = 1/2)	13
$\chi_{jkl}(-\omega;0,\omega)$	Linear electrooptic or Pockel's effect (EO) (g = 2)	14
$\chi_{jkl}(-\omega_p \pm \omega_s;\omega_p,\omega_s)$ $\omega_r+\omega_p=\omega_s$	Parametric (sum) or difference mixing (PG) (g = 1)	15—23
$\chi_{jkl}(\omega_i)$	Optical activity and frequency mixing	24, 25
$\chi_{jklm}(-3\omega;\omega,\omega,\omega)$	Third-harmonic generation (THG) (g = 1/4)	26—28
$\chi_{jklm}(-2\omega;0,\omega,\omega)$	Electric-field-induced second-harmonic generation (FISHG) (g = 3/4)	26, 28
$\chi_{jklm}(-\omega_4;\omega_1,\omega_2,\omega_3,\omega_4)$	Three-wave sum mixing (g = 3/2)	
$\chi_{jklm}(-\omega;\omega,0,0)$	Quadratic electrooptic or DC kerr effect (g = 3/4)	50, 51
$\chi_{jklm}(-\omega;\omega,\omega,-\omega)$	Optic-field induced birefringence or self-focusing (g = 3/4)	29, 30
$\chi_{jklm}(\omega_s;\omega_p,-\omega_p,\omega_s)$	Optical kerr effect (g = 3/2)	48
$\chi_{jklm}(\omega_2 - 2\omega_1;\omega_1,\omega_1,-\omega_2)$	Three wave mixing (TWM) (g = 3/4)	28, 52
$\mathrm{Im}\chi_{jklm}(-\omega_s;\omega_p,-\omega_p,\omega_s)$	Stimulated Raman, Brillouin, and electronic Raman scattering (SRS, SBS, SERS) (g = 3/2)	39—47
$\mathrm{Im}\chi_{jklm}(-\omega;\omega,\omega_p,-\omega_p)$	Two-photon absorption and inverse Raman effect (g = 3/2)	35, 37, 38
$\mathrm{Im}\chi_{jklm}(-\omega;\omega,\omega,-\omega)$	Two-photon absorption (single frequency) (g = 3/4)	31—34, 36
$\mathrm{Im}\chi_{jklmno}(-\omega;\omega,\omega,\omega,-\omega,-\omega)$	Three consecutive photon absorption (g = 15/4)	49

$$\left.\begin{aligned} E(-\omega) &= E^*(\omega) && \text{(a)} \\ \omega_{-j} &= -\omega_j && \text{(b)} \\ K_{-j} &= -K_j^* && \text{(c)} \\ K &= n\omega/c && \text{(d), } c \text{ being the velocity of light in vacuum} \end{aligned}\right\} \quad (12)$$

The nonlinear optical susceptibilities, $\chi_{jk\,l}$, $\chi_{jk\,lm}$, and higher-order ones give rise to a large variety of nonlinear optical phenomena. A number of these susceptibility tensors and the types of optical processes associated with them are listed in Table 1.1.2. Also included in the table are the relevant references and the degeneracy factors. Review articles and introductory texts on nonlinear optical effects can be found in References 3 to 11.

More than a decade has elapsed since the last publication of the author's chapter on nonlinear optical materials in Reference 54. Since then, the number of nonlinear materials measured has more than doubled. In the present revised chapter this compilation is brought up to date through 1982. In addition, at the suggestion of the editor, Dr. M. Weber, data on third-order optical susceptibilities are also included.

SECOND-ORDER POLARIZATION AND ITS SYMMETRY PROPERTIES

For the general case of monochromatic waves of angular frequencies ω_1, ω_2 . . . ω_r

incident upon a medium, the *j*th Cartesian component of the induced polarization density, $\mathbf{P}_j^{(r)}$, at the frequency $\omega_\sigma = \omega_1 + \omega_2 + \ldots + \omega_r$ due to the *r*th order nonlinear susceptibility, $\chi^{(r)}$ can be written as:

$$\mathbf{P}_j^{(r)}(\omega_\sigma) = \epsilon_o\left[\sum_{\alpha 1r \cdot \alpha r} g(-\omega_\sigma)\chi_{j\alpha_r \cdot \cdot \alpha_r}\{-\omega_\sigma;\omega_1,\omega_2\cdots\omega_r E_{\alpha 1}(\omega_1)E_{\alpha 2}(\omega_2)\cdots E_{\sigma_r}(\omega_r)\}\right] \tag{13}$$

where g is a degeneracy factor arising from the number of distinguishable permutations of the frequencies. The summation is carried over all of the distinct sets of ω_1, $\omega_2 \ldots \omega_r$.

For the specific case of SHG, $\omega_1 = \omega_2 = \omega$ and $g = {}^1/_2$, the second-order polarization at the harmonic frequency 2ω is given by:

$$\mathbf{P}_j^{(2)}(2\omega) = \epsilon_o \sum_{k,l}(1/2)\chi_{jkl}(-2\omega;\omega,\omega)E_k(\omega)E_l(\omega) \tag{14}$$

There is a considerable amount of confusion in the literature about the relation between induced nonlinear polarization and nonlinear susceptibility. The following convention is adopted here to avoid this confusion. The second-order nonlinear coefficient is denoted by $d_{jkl}^{(2)}$ and is related to the second-order nonlinear susceptibility tensor $\chi_{jkl}^{(2)}$ by:

$$\chi_{jkl}^{(2)} = 2\, d_{jkl}^{(2)} \tag{15}$$

In order to express the induced nonlinear polarization due to various second-order nonlinear processes, $\chi_{jkl}^{(2)}$ in Equation 13 is replaced by $2\, d_{jkl}^{(2)}$. Thus, in the case of SHG, the induced polarization at the harmonic frequency 2ω given by Equation 14 is expressed as:

$$\mathbf{P}_j^{(2)}(2\omega) = \epsilon_o \sum_{k,l} d_{jkl}^{(2)}(-2\omega;\omega,\omega)E_k(\omega)E_l(\omega) \tag{16}$$

Similarly, the induced polarizations associated with other second-order processes are expressed in terms of *d* by using the proper g values from Table 1.1.2.

The SHG coefficient d_{jkl} is a $3 \times 3 \times 3$ third-rank tensor. Since the order in which the electric field components are written in the right hand side of Equation 16 is of no significance, the *d*s satisfy the permutation symmetry:

$$d_{jkl}(-2\omega,\omega,\omega) = d_{jlk}(-2\omega;\omega,\omega) \tag{17}$$

This property is similar to that of the piezoelectric tensor (by virtue of which an electric polarization is produced in a medium when it is subjected to a stress). By analogy with the piezoelectric tensor, for convenience it is possible to write the nonlinear optical tensor, $d_{jkl}^{(2)}$, in a contracted form which reduces the maximum number of independent SHG tensor elements to 18. In the contracted form, the symmetric suffixes *k* and *l* are replaced by a single suffix, *m*, that takes the values 1 to 6. The relation between the contracted and uncontracted elements can be expressed by:

$$d_{jm} = d_{jkl}\ ;\ m = \begin{cases} k & \text{if } k = l \\ g - (k + l) & \text{if } k \neq l \end{cases} \tag{18}$$

Thus, the relation between *m* and *kl* is

$$\begin{array}{rl} kl & m \\ 11 \leftrightarrow & 1 \\ 22 \leftrightarrow & 2 \\ 33 \leftrightarrow & 3 \\ 23 \text{ or } 32 \leftrightarrow & 4 \\ 31 \text{ or } 13 \leftrightarrow & 5 \\ 12 \text{ or } 21 \leftrightarrow & 6 \end{array} \tag{19}$$

In the contracted form, the 18 elements of d_{jm} and the components of the second-order polarization at the harmonic frequency 2ω can be written in the matrix form:

$$\begin{vmatrix} P_x \\ P_y \\ P_z \end{vmatrix} = \epsilon_o \begin{vmatrix} d_{11} & d_{12} & d_{13} & d_{14} & d_{15} & d_{16} \\ d_{21} & d_{22} & d_{23} & d_{24} & d_{25} & d_{26} \\ d_{31} & d_{32} & d_{33} & d_{34} & d_{35} & d_{36} \end{vmatrix} \begin{vmatrix} E_x^2 \\ E_y^2 \\ E_z^2 \\ 2\,E_yE_z \\ 2\,E_xE_z \\ 2\,E_xE_y \end{vmatrix} \tag{20}$$

Care should be exercised in using Equation 20 in parametric generation. For example, the z component of the polarization induced by two waves, $E_x(\omega_1)$ and $E_y(\omega_2)$, polarized in the x and y axes, respectively,

$$\mathbf{P}_z(\omega_1 + \omega_2) = \epsilon_o d_{36} E_x(\omega_1) E_y(\omega_2) \tag{20A}$$

The factor 2 does not apply here because $E_y(\omega_1) = E_x(\omega_2) = 0$.

The number of nonvanishing, independent elements of the second-harmonic tensor depends upon the point-group symmetry of the medium. Since the nonlinear tensor, d_{jkl}, is a polar tensor of odd rank, its elements are identical to zero for any medium possessing a center of inversion. This has the implication that all liquids, gases, and centrosymmetric solids have zero susceptibility tensors of ranks 3, 5, 7, and so on. For the 21 out of 32 crystal classes that lack a center of inversion, the contracted form of the second-harmonic tensor is given in Table 1.1.3. For the procedure to obtain the number of nonvanishing elements of a third-rank tensor subject to the point group symmetry of a crystal, see References 55 and 56. We follow the IRE convention[57] in choosing the nonlinear optic axes X, Y, and Z.

In certain cases, the number of nonvanishing elements of the nonlinear optical tensor is further reduced by an additional symmetry condition[58] if it is assumed that the optical frequencies involved lie far enough removed from the characteristic UV and IR absorption bands of the medium. The dispersive effects can then be neglected and the nonlinear optical tensor becomes independent of the frequencies and should become symmetric for the interchange of any two suffixes in d_{jkl}. As a result of this symmetry the following equalities are obtained:

$$\left.\begin{array}{l} d_{12} = d_{26}\,;\; d_{13} = d_{35}\,;\; d_{14} = d_{25} = d_{36} \\ d_{15} = d_{31}\,;\; d_{16} = d_{21}\,;\; d_{23} = d_{24} \text{ and } d_{24} = d_{32} \end{array}\right\} \tag{21}$$

Table 1.1.3
FORM OF THE SHG-TENSOR FOR THE VARIOUS CRYSTAL CLASSES

TRICLINIC SYSTEM

Class 1 − C_1

$$\begin{bmatrix} d_{11} & d_{12} & d_{13} & d_{14} & d_{15} & d_{16} \\ d_{21} & d_{22} & d_{23} & d_{24} & d_{25} & d_{26} \\ d_{31} & d_{32} & d_{33} & d_{34} & d_{35} & d_{36} \end{bmatrix}$$

MONOCLINIC SYSTEM

Class *m* − C_2

$$\begin{bmatrix} d_{11} & d_{12} & d_{13} & 0 & 0 & d_{16} \\ d_{21} & d_{22} & d_{23} & 0 & 0 & d_{26} \\ 0 & 0 & 0 & d_{34} & d_{35} & 0 \end{bmatrix} \quad m \perp Z$$

Class *m* − C_s

$$\begin{bmatrix} d_{11} & d_{12} & d_{13} & 0 & d_{15} & 0 \\ 0 & 0 & 0 & d_{24} & 0 & d_{26} \\ d_{31} & d_{32} & d_{33} & 0 & d_{35} & 0 \end{bmatrix} \quad m \perp Y$$

Class 2 − C_2

$$\begin{bmatrix} 0 & 0 & 0 & d_{14} & d_{15} & 0 \\ 0 & 0 & 0 & d_{24} & d_{25} & 0 \\ d_{31} & d_{32} & d_{33} & 0 & 0 & d_{36} \end{bmatrix} \quad 2 \parallel Z$$

Class 2 − C_2

$$\begin{bmatrix} 0 & 0 & 0 & d_{14} & 0 & d_{16} \\ d_{21} & d_{22} & d_{23} & 0 & d_{25} & 0 \\ 0 & 0 & 0 & d_{34} & 0 & d_{36} \end{bmatrix} \quad 2 \parallel Y$$

(IRE-convention)

ORTHORHOMBIC SYSTEM

Class *mm*2 − C_{2v}

$$\begin{bmatrix} 0 & 0 & 0 & 0 & d_{15} & 0 \\ 0 & 0 & 0 & d_{24} & 0 & 0 \\ d_{31} & d_{32} & d_{33} & 0 & 0 & 0 \end{bmatrix}$$

Class 222 − D_2

$$\begin{bmatrix} 0 & 0 & 0 & d_{14} & 0 & 0 \\ 0 & 0 & 0 & 0 & d_{25} & 0 \\ 0 & 0 & 0 & 0 & 0 & d_{36} \end{bmatrix}$$

TETRAGONAL SYSTEM

Class 4 − C_4

$$\begin{bmatrix} 0 & 0 & 0 & d_{14} & d_{15} & 0 \\ 0 & 0 & 0 & d_{15} & -d_{14} & 0 \\ d_{31} & d_{31} & d_{33} & 0 & 0 & 0 \end{bmatrix}$$

Class $\bar{4}$ − S_4

$$\begin{bmatrix} 0 & 0 & 0 & d_{14} & d_{15} & 0 \\ 0 & 0 & 0 & -d_{15} & d_{14} & 0 \\ d_{31} & -d_{31} & 0 & 0 & 0 & d_{36} \end{bmatrix}$$

Class 4*mm* − C_{4v}

$$\begin{bmatrix} 0 & 0 & 0 & 0 & d_{15} & 0 \\ 0 & 0 & 0 & d_{15} & 0 & 0 \\ 0 & 0 & 0 & 0 & 0 & d_{36} \end{bmatrix}$$

Class $\bar{4}$2 *m* − D_{2d}

$$\begin{bmatrix} 0 & 0 & 0 & d_{14} & 0 & 0 \\ 0 & 0 & 0 & 0 & d_{14} & 0 \\ 0 & 0 & 0 & 0 & 0 & d_{36} \end{bmatrix}$$

Class 422 − D_4

$$\begin{bmatrix} 0 & 0 & 0 & d_{14} & 0 & 0 \\ 0 & 0 & 0 & 0 & -d_{14} & 0 \\ 0 & 0 & 0 & 0 & 0 & 0 \end{bmatrix}$$

TRIGONAL SYSTEM

Class 3 − C_3

$$\begin{bmatrix} d_{11} & -d_{11} & 0 & d_{14} & d_{15} & -d_{22} \\ -d_{22} & d_{22} & 0 & d_{15} & -d_{14} & -d_{11} \\ d_{31} & d_{31} & d_{33} & 0 & 0 & 0 \end{bmatrix}$$

Class 3*m* − C_{3v}

$$\begin{bmatrix} 0 & 0 & 0 & 0 & d_{15} & -d_{22} \\ -d_{22} & d_{22} & 0 & d_{15} & 0 & 0 \\ d_{31} & d_{31} & d_{33} & 0 & 0 & 0 \end{bmatrix} \quad m \perp X$$

(IRE-convention)

Table 1.1.3 (continued)
FORM OF THE SHG-TENSOR FOR THE VARIOUS CRYSTAL CLASSES

Class 3*m* – C_{3v}

$$\begin{bmatrix} d_{11} & -d_{11} & 0 & 0 & d_{15} & 0 \\ 0 & 0 & 0 & d_{15} & 0 & -d_{11} \\ d_{31} & d_{31} & d_{33} & 0 & 0 & 0 \end{bmatrix} \quad m \perp Y$$

Class 32 – D_3

$$\begin{bmatrix} d_{11} & -d_{11} & 0 & d_{14} & 0 & 0 \\ 0 & 0 & 0 & 0 & -d_{14} & -d_{11} \\ 0 & 0 & 0 & 0 & 0 & 0 \end{bmatrix}$$

HEXAGONAL SYSTEM

Class $\bar{6}$ – C_{3h}

$$\begin{bmatrix} d_{11} & -d_{11} & 0 & 0 & 0 & -d_{22} \\ -d_{22} & d_{22} & 0 & 0 & 0 & -d_{11} \\ 0 & 0 & 0 & 0 & 0 & 0 \end{bmatrix}$$

Class 6 – C_6

$$\begin{bmatrix} 0 & 0 & 0 & d_{14} & d_{15} & 0 \\ 0 & 0 & 0 & d_{15} & -d_{14} & 0 \\ d_{31} & d_{31} & d_{33} & 0 & 0 & 0 \end{bmatrix}$$

Same as Class 4 – C_4

Class $\bar{6}m2$ – D_{3h}

$$\begin{bmatrix} 0 & 0 & 0 & 0 & 0 & -d_{22} \\ -d_{22} & d_{22} & 0 & 0 & 0 & 0 \\ 0 & 0 & 0 & 0 & 0 & 0 \end{bmatrix} \quad m \perp X \text{ (IRE-convention)}$$

Class $\bar{6}m2$ – D_{3h}

$$\begin{bmatrix} d_{11} & -d_{11} & 0 & 0 & 0 & 0 \\ 0 & 0 & 0 & 0 & 0 & -d_{11} \\ 0 & 0 & 0 & 0 & 0 & 0 \end{bmatrix} \quad m \perp Y$$

Class 6*mm* – C_{6v}

$$\begin{bmatrix} 0 & 0 & 0 & 0 & d_{15} & 0 \\ 0 & 0 & 0 & d_{15} & 0 & 0 \\ d_{31} & d_{31} & d_{33} & 0 & 0 & 0 \end{bmatrix}$$

Same as Class 4*mm* – D_{4v}

HEXAGONAL SYSTEM

Class 622 – D_6

$$\begin{bmatrix} 0 & 0 & 0 & d_{14} & 0 & 0 \\ 0 & 0 & 0 & 0 & -d_{14} & 0 \\ 0 & 0 & 0 & 0 & 0 & 0 \end{bmatrix}$$

Same as Class 422 – D_4

CUBIC SYSTEM

Class 23 – *T*

$$\begin{bmatrix} 0 & 0 & 0 & d_{14} & 0 & 0 \\ 0 & 0 & 0 & 0 & d_{14} & 0 \\ 0 & 0 & 0 & 0 & 0 & d_{14} \end{bmatrix}$$

Class $\bar{4}3m$ – T_d

$$\begin{bmatrix} 0 & 0 & 0 & d_{14} & 0 & 0 \\ 0 & 0 & 0 & 0 & d_{14} & 0 \\ 0 & 0 & 0 & 0 & 0 & d_{14} \end{bmatrix}$$

Class 432 – 0

All elements vanish

SECOND-HARMONIC POWER AND COHERENCE LENGTH

The second-harmonic power, $\mathbf{P}_{2\omega}$, generated by a single-mode Gaussian beam of power $\mathbf{P}_{\omega}$, incident on a plane parallel slab of thickness L of a nonlinear crystal is given by:

$$\mathbf{P}_{2\omega}^{(MKS)} = \frac{2\omega^2 \cdot L^2 \cdot \mathbf{P}_{\omega}^2 \cdot d_{jm}^2 \exp\left\{-L\left(\alpha_{\omega} + \frac{1}{2}\alpha_{2\omega}\right)\right\}}{\pi w_1^2 \cdot \epsilon_o c^3 n_{\omega}^2 \cdot n_{2\omega}} \left[\frac{\mathrm{Sin}^2 L \cdot \Delta K/2}{(L \cdot \Delta K/2)^2}\right] \quad (22A)$$

$$\mathbf{P}_{2\omega}^{(CGS)} = \frac{128\pi^2\omega^2 \cdot L^2 \cdot \mathbf{P}_{\omega}^2 \cdot d_{jm}^2 \exp\left\{-L\left(\alpha_{\omega} + \frac{1}{2}\alpha_{2\omega}\right)\right\}}{c^3 w_1^2 n_{\omega}^2 n_{2\omega}} \left[\frac{\mathrm{Sin}^2 L \cdot \Delta K/2}{(L \cdot \Delta K/2)^2}\right] \quad (22B)$$

where

ω = angular frequency of the fundamental wave
w_1 = spot radius of the fundamental Gaussian beam and is defined as the radius at which the electric field in the beam is 1/e of its value on the beam axis
d_{jm} = pertinent SHG coefficient
n_{ω} = principal index of refraction of the crystal at the fundamental wavelength
$n_{2\omega}$ = principal index of refraction of the crystal at the second harmonic wavelength
ϵ_o = permittivity of free space
c = velocity of light in vacuum
ΔK = $2K_{\omega} - K_{2\omega}$, the wave vector mismatch between the fundamental and second harmonic waves
α_{ω} = absorption coefficient of the crystal at the fundamental wavelength, and
$\alpha_{2\omega}$ = absorption coefficient of the crystal at the second harmonic wavelength.

The powers $\mathbf{P}_{\omega}$ and $\mathbf{P}_{2\omega}$ are those internal to the crystal.

In general, because of dispersion in the medium, $\Delta K \neq 0$ and the bracketed terms in the right of Equations 22A and 22B indicate that the second-harmonic power undergoes periodic oscillations, known as Maker[59] oscillations, as a function of thickness, the period of oscillation being given by $L_{coh} = \pi/\Delta K$, which is known as the "coherence length". For the case when the fundamental beam propagates along a principal axis of the crystal (which is assumed in Equation 22A and 22B), the coherence length is given by:

$$L_{coh} = \frac{\pi c}{2\omega(n_{2\omega} - n_{\omega})} = \frac{\lambda_1}{\Delta(n_{2\omega} - n_{\omega})} \quad (23)$$

where λ_1 is the free-space wavelength of the fundamental wave. In an isotropic medium, since n_{ω} and $n_{2\omega}$ depend upon the polarization direction of the fundamental and second-harmonic waves, respectively, it follows from Equation 23 that in such media, each nonlinear optical coefficient has a coherence length associated with it.

By substituting $L = L_{coh} = \pi/\Delta K$ in Equations 22A and 22B and using the power reflection coefficients to convert from internal to external powers, the following expressions are obtained:

$$\mathbf{P}_{2\omega}^{ext} = \left[\frac{64}{(1 + n_{\omega})^4(1 + n_{2\omega})^2}\right]\left[\frac{8\omega^2}{\pi^3 w_0^2 \epsilon_o c^3}\right] L_{coh}^2 d_{jm}^2 \mathbf{P}_{\omega}^{2ext}$$

$$\cdot \exp\left[-L\left(\alpha_{\omega} + \frac{\alpha_2\omega}{2}\right)\right] \quad [\mathrm{MKS}] \quad (24A)$$

$$\mathbf{P}^{ext}_{2\omega} = \left[\frac{64}{(1 + n_\omega)^4(1 + n_{2\omega})^2}\right]\left[\frac{512\ \omega^2}{w_0^2c^3}\right] L^2_{coh}d^2_{jm}\mathbf{P}^{2ext}_{\omega}$$

$$\cdot \exp\left[-L\left(\alpha_\omega + \frac{\alpha_2}{2}\omega\right)\right] \text{[CGS]} \qquad (24B)$$

Equations 24A and 24B are valid only for low values of the indexes of refraction ($n < 2$, $n_{2\omega} - n_\omega < 0.1$). A more rigorous procedure[60] yields the following equations:

$$\mathbf{P}^{ext}_{2\omega} = \left[\frac{128 n_{2\omega}}{(1 + n_\omega)^3(1 + n_{2\omega})^3(n_\omega + n_{2\omega})}\right]\left[\frac{8\omega^2}{\pi^3 w_0^2 \epsilon_o c^3}\right] L^2_{coh}d^2_{jm}\mathbf{P}^{2ext}_{w}$$

$$\exp\left[-L\left(\alpha_\omega + \frac{\alpha_2\omega}{2}\right)\right] \qquad \text{(MKS units)} \qquad (25A)$$

$$\mathbf{P}^{ext}_{2\omega} = \left[\frac{128\ n_{2\omega}}{(1 + n_\omega)^3(1 + n_{2\omega})^3(n_\omega + n_{2\omega})}\right]\left[\frac{512\ \omega^2}{w_0^2c^3}\right] L^2_{coh}d^2_{jm}\mathbf{P}^{2ext}_{\omega}$$

$$\exp\left[-L\left(\alpha_\omega + \frac{\alpha_2\omega}{2}\right)\right] \qquad \text{(CGS units)} \qquad (25B)$$

PHASE-MATCHED SECOND-HARMONIC GENERATION

It follows from Equations 22A and 22B that for efficient second-harmonic conversion, the wavevector mismatch, ΔK, between the polarization wave and the generated wave should be zero. This is termed "phase matching". Because of the normal dispersion of the refractive index between the fundamental and second-harmonic wavelengths, phase matching is difficult to achieve in an isotropic crystal. However, it was shown by Giordmaine[61] and Maker et al.[59] independently that, in an anisotropic crystal, it is often possible to obtain $\Delta K = 0$ by a suitable choice of direction of propagation and polarization.

For the general case of mixing of two waves of frequencies ω_1 and ω_2 and wavevectors K_1 and K_2, the energy and momentum conservations are written as:

$$\omega_3 = \omega_1 + \omega_2 \qquad (26)$$

$$K_3 = K_1 + K_2 + \Delta K \qquad (27)$$

Phase-matching is achieved whenever $\Delta K = 0$. Using Equation 12d in Equation 27 for $\Delta K = 0$ yields the condition:

$$n_3 = \frac{\omega_1 n_1}{\omega_1 + \omega_2} + \frac{\omega_2 n_2}{\omega_1 + \omega_2} \qquad (28)$$

Two types of phase-matching can occur in principle in birefringent crystals with both uni- and biaxial characteristics.

Type I. ω_1 and ω_2 are of the same polarization (i.e., both either ordinary (o) rays or both extraordinary (e) rays).

Type II. ω_1 and ω_2 are of orthogonal polarizations (i.e., one ray is an e ray, the other is an o ray).

SHG — PHASE-MATCHING IN UNIAXIAL CRYSTALS

Details on phase-matching in uniaxial crystals can be found in References 62 to 64. Whether a given frequency is propagated as an e ray or as an o ray depends on the crystal being positively or negatively uniaxial. Thus,

$$\left.\begin{array}{l}\omega_1 \equiv \text{e wave}\ ,\ \omega_2 \equiv \text{e wave}\ ,\ \omega_3 \equiv \text{o wave[Type I]}\\ \omega_1 \equiv \text{o wave}\ ,\ \omega_2 \equiv \text{e wave}\ ,\ \omega_3 \equiv \text{o wave[Type II]}\end{array}\right\}\ \text{For the positive uniaxial}$$

$$\left.\begin{array}{l}\omega_1 \equiv \text{o wave}\ ,\ \omega_2 \equiv \text{o wave}\ ,\ \omega_3 \equiv \text{e wave [Type I]}\\ \omega_1 \equiv \text{e wave}\ ,\ \omega_2 \equiv \text{o wave}\ ,\ \omega_3 \equiv \text{e wave [Type II]}\end{array}\right\}\ \text{For the negative uniaxial}$$

For the case of two waves which can propagate in a uniaxial crystal with a common wave normal making an angle, θ_m, with the optic axis, the refractive indexes are given by:

$$n^{\text{ord}} = n^{\text{o}}\ ;\ \frac{1}{[n^{\text{e}}(\theta_m)]^2} = \frac{\text{Cos}^2\theta_m}{(n^{\text{o}})^2} + \frac{\text{Sin}^2\theta_m}{(n^{\text{e}})^2} \qquad (29)$$

where n^{o} and n^{e} are the ordinary and extraordinary refractive indexes, respectively. If n^{o} and n^{e} are known at each of the particular frequencies ω_1, ω_2, and ω_3, then the phase-matching angle, θ_m, may be calculated for Type I or Type II phase-matching conditions using Equation 29. Thus, for SHG, in a negative ($n^{\text{e}} < n^{\text{o}}$) uniaxial crystal the index-matching conditions are

$$\text{for Type I:}\quad n^{\text{e}}_{2\omega}(\theta_m) = n^{\text{o}}_{\omega} \qquad (30)$$

$$\text{for Type II:}\quad n^{\text{e}}_{2\omega}(\theta_m) = \frac{1}{2}\,[\text{n}^{\text{e}}_{\omega}(\theta_m) + \text{n}^{\text{o}}_{\omega}] \qquad (31)$$

For positive ($n^{\text{e}} > n^{\text{o}}$) uniaxial crystals, index matching requires:

$$\text{for Type I:}\quad n^{\text{o}}_{2\omega} = \text{n}^{\text{e}}_{\omega}(\theta_m) \qquad (32)$$

$$\text{for Type II:}\quad n^{\text{o}}_{2\omega} = \frac{1}{2}\,[\text{n}^{\text{e}}_{\omega}(\theta_m) + \text{n}^{\text{o}}_{\omega}] \qquad (33)$$

By combining Equations 32 and 33, each with Equation 29, the phase-matching angles θ_m(I) and θ_m(II) are given by:

$$\text{Sin}^2\theta_m\ \text{(I)} = \frac{(n^{\text{o}}_{2\omega})^{-2} - (n^{\text{o}}_{\omega})^{-2}}{(n^{\text{e}}_{\omega})^{-2} - (n^{\text{o}}_{\omega})^{-2}} \qquad (34)$$

$$\text{Sin}^2\theta_m\ \text{(II)} = \frac{[\text{n}^{\text{o}}_{\omega}/(2n^{\text{o}}_{2\omega} - n^{\text{o}}_{\omega})]^2 - 1}{(n^{\text{o}}_{\omega}/n^{\text{e}}_{\omega})^2 - 1} \qquad (35)$$

for positive uniaxial. The phase-matching angles for a negative uniaxial crystal can be similarly obtained by combining Equations 30 and 31, each with Equation 29.

It should be kept in mind that phase-matching is not possible in all uniaxial crystals. Crystals with large birefringence and small dispersion are more likely to show phase-matching, particularly of the Type II. The phase-matching angle θ_m(II) is always greater than θ_m(I), and if the Type II process is possible, then so is the Type I process.

Table 1.1.4
EFFECTIVE TENSOR ELEMENTS d_{eff} FOR PHASE-MATCHED SHG IN POSITIVE AND NEGATIVE (−) UNIAXIAL CRYSTALS[11]

Class	For Type I phase matching	For Type II phase matching
$\bar{4}3m(T_d)$	$d^+ = -d_{14}\mathrm{Cos}2\theta_r\mathrm{Sin}2\theta_m$	$d^+ = -d_{14}\mathrm{Sin}2\theta_r\mathrm{Sin}\theta_m$
$23(T)$	$d^- = -d_{14}\mathrm{Sin}2\theta_r\mathrm{Sin}\theta_m$	$d^- = -2d_{14}\mathrm{Cos}2\theta_r\mathrm{Cos}\theta_m\mathrm{Sin}\theta_m$
$\bar{4}2m(D_{2d})$	$d^+ = -d_{14}\mathrm{Cos}2\theta_r\mathrm{Sin}2\theta_m$	$d^+ = -d_{14}\mathrm{Sin}2\theta_r\mathrm{Sin}\theta_m$
	$d^- = -d_{36}\mathrm{Sin}2\theta_r\mathrm{Sin}\theta_m$	$d^- = -(d_{14} + d_{36})\mathrm{Cos}2\theta_r\mathrm{Cos}\theta_m\mathrm{Sin}\theta_m$
$\bar{6}m2(D_{3h})$	$d^+ = -d_{22}\mathrm{Cos}3\theta_r\mathrm{Cos}^2\theta_m$	$d^+ = -d_{22}\mathrm{Sin}3\theta_r\mathrm{Cos}\theta_m$
	$d^- = -d_{22}\mathrm{Sin}3\theta_r\mathrm{Cos}\theta_m$	$d^- = -d_{22}\mathrm{Cos}3\theta_r\mathrm{Cos}^2\theta_m$
$\bar{6}(C_{3h})$	$d^+ = d_{11}\mathrm{Sin}3\theta_r - d_{22}\mathrm{Cos}3\theta_r\mathrm{Cos}^2\theta_m$	$d^+ = -(d_{11}\mathrm{Cos}3\theta_r + d_{22}\mathrm{Sin}3\theta_r)\mathrm{Cos}\theta_m$
	$d^- = -(d_{11}\mathrm{Cos}3\theta_r + d_{22}\mathrm{Sin}3\theta_r)\mathrm{Cos}\theta_m$	$d^- = (d_{11}\mathrm{Sin}3\theta_r - d_{22}\mathrm{Cos}3\theta_r)\mathrm{Cos}^2\theta_m$
$32(D_3)$	$d^+ = d_{11}\mathrm{Sin}3\theta_r\mathrm{Cos}^2\theta_m + d_{14}\mathrm{Sin}2\theta_m$	$d^+ = -d_{11}\mathrm{Cos}3\theta_r\mathrm{Cos}\theta_m$
	$d^- = -d_{11}\mathrm{Cos}3\theta_r\mathrm{Cos}\theta_m$	$d^- = d_{11}\mathrm{Sin}3\theta_r\mathrm{Cos}^2\theta_m - d_{14}\mathrm{Cos}\theta_m\mathrm{Sin}\theta_m$
$4mm(C_{4v})$	$d^+ = 0$	$d^+ = -d_{15}\mathrm{Sin}\theta_m$
$6mm(C_{6v})$	$d^- = -d_{31}\mathrm{Sin}\theta_m$	$d^- = 0$
$6(C_6)$	$d^+ = d_{14}\mathrm{Sin}2\theta_m$	$d^+ = -d_{15}\mathrm{Sin}\theta_m$
$4(C_4)$	$d^- = -d_{31}\mathrm{Sin}\theta_m$	$d^- = -d_{14}\mathrm{Sin}\theta_m\mathrm{Cos}\theta_m$
$3m(C_{3v})$	$d^+ = -d_{22}\mathrm{Cos}3\theta_r\mathrm{Cos}^2\theta_m$	$d^+ = -d_{22}\mathrm{Sin}3\theta_r\mathrm{Cos}\theta_m - d_{15}\mathrm{Sin}\theta_m$
	$d^- = -(d_{31}\mathrm{Sin}\theta_m + d_{22}\mathrm{Sin}3\theta_r\mathrm{Cos}\theta_m)$	$d^- = -d_{22}\mathrm{Cos}3\theta_r\mathrm{Cos}^2\theta_m$
$\bar{4}(S_4)$	$d^+ = -(d_{14}\mathrm{Cos}2\theta_r + d_{15}\mathrm{Sin}2\theta_r)\mathrm{Sin}2\theta_m$	$d^+ = -(d_{14}\mathrm{Sin}2\theta_r - d_{15}\mathrm{Cos}2\theta_r)\mathrm{Sin}\theta_m$
	$d^- = d_{31}\mathrm{Cos}2\theta_r\mathrm{Sin}\theta_m - d_{36}\mathrm{Sin}2\theta_r\mathrm{Sin}\theta_m$	$d^- = -[(d_{14} + d_{36})\mathrm{Cos}2\theta_r + (d_{15} + d_{31})\mathrm{Sin}2\theta_r] \times \mathrm{Sin}\theta_m\mathrm{Cos}\theta_m$
$3(C_3)$	$d^+ = (d_{11}\mathrm{Sin}3\theta_r - d_{22}\mathrm{Cos}3\theta_r)\mathrm{Cos}^2\theta_m + d_{14}\mathrm{Sin}2\theta_m$	$d^+ = -(d_{11}\mathrm{Cos}3\theta_r + d_{22}\mathrm{Sin}3\theta_r)\mathrm{Cos}\theta_m - d_{15}\mathrm{Sin}\theta_m$
	$d^- = -(d_{11}\mathrm{Cos}3\theta_r + d_{22}\mathrm{Sin}3\theta_r)\mathrm{Cos}\theta_m - d_{31}\mathrm{Sin}\theta_m$	$d^- = (d_{11}\mathrm{Sin}3\theta_r - d_{22}\mathrm{Cos}3\theta_r)\mathrm{Cos}^2\theta_m - d_{14}\mathrm{Sin}\theta_m\mathrm{Cos}\theta_m$

When phase-matched SHG is achieved by propagating the fundamental along a direction different from a principal axis of a birefringent crystal, it is termed "critical phase matching" (CPM). Under these conditions, the nonlinear coefficient, d_{jm}, in Equations 22A and 22B must be replaced with an effective nonlinear coefficient, $(d_{jm})_{eff}$. References 11, 64, and 65 have derived the effective SHG tensor elements, d_{eff}, for all the uniaxial crystal classes and are listed in Table 1.1.4.

The effective nonlinear tensor elements, d_{eff}, are defined in terms of the direction cosines relating the principal axes (X, Y, Z) to the laboratory system (x, y, z) shown in Figure 1.1.1. The matrix of direction cosines a_{jm} relating the two systems are obtained by combining a clockwise ($+x$ toward $+z$) rotation, θ, about the laboratory y axis giving (x', y', Z) with a counterclockwise ($+x$ toward y) rotation, θ_r, about the resulting principal Z axis to give (X, Y, Z),

$$\begin{pmatrix} X \\ Y \\ Z \end{pmatrix} = \alpha_{jm}\begin{pmatrix} x \\ y \\ z \end{pmatrix} = \begin{pmatrix} \mathrm{Cos}\theta_r\mathrm{Cos}\theta & \mathrm{Sin}\theta_r & \mathrm{Cos}\theta_r\mathrm{Sin}\theta \\ -\mathrm{Sin}\theta_r\mathrm{Cos}\theta & \mathrm{Cos}\theta_r & -\mathrm{Sin}\theta_r\mathrm{Sin}\theta \\ -\mathrm{Sin}\theta & 0 & \mathrm{Cos}\theta \end{pmatrix}\begin{pmatrix} x \\ y \\ z \end{pmatrix} \tag{36}$$

The inverse of this unitary matrix is its transpose $(a^{-1})_{jm} = a_{mj}$. The matrix a_{jm} can be used to transform the induced nonlinear polarization in Equation 14 from the principal axis system to the laboratory axis system:

$$\mathbf{P}_m(2\omega) = \epsilon_o \sum_{n,0=x,y,z}\left(\sum_{jkl=xyz} \alpha_{jm}\mathrm{d}_{jkl}(-2\omega;\omega,\omega)\alpha_{kn}\alpha_{l0}\right)E_n(\omega)E_0(\omega) \tag{37}$$

where $m = x, y,$ or z. The term in brackets is the effective SHG coefficient for an input

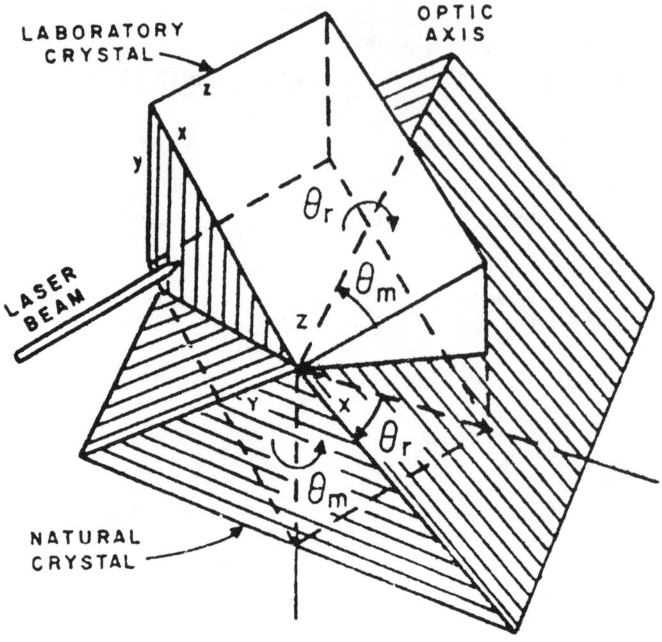

FIGURE 1.1.1. Diagram showing interrelation between principal dielectric axis system X, Y, Z and the laboratory axis system, x, y, z. (From Kurtz, S. K., *Quantum Electronics: A Treatise*, Vol. 1, Robin, H. and Tang, C. L., Eds., Academic Press, New York, 1975, 209. With permission.)

electric field $E_n^{(\omega)}E_0^{(\omega)}$. The values of d_{eff} listed in Table 1.1.4 are obtained by substituting the nonvanishing coefficients, d_{jm} from Table 1.1.3 in Equation 37.[11]

PHASE-MATCHING IN BIAXIAL CRYSTALS

Phase-matching conditions for second-harmonic generation in biaxial crystals are discussed in References 66 and 67. The conditions for both critical (CPM) and noncritical (NCPM) phase-matching in biaxial crystals have been analyzed in Reference 66 and are listed in Tables 1.1.5 and 1.1.6, respectively. It is assumed that the principal indexes of refraction n^X, n^Y, and n^Z obey the relations:

$$n_{2\omega}^Z > n_\omega^Z \ , \ n_{2\omega}^Y > n_\omega^Y \ , \ n_{2\omega}^X > n_\omega^X$$

and

$$n_{2\omega}^Z - n_\omega^Z \simeq n_{2\omega}^Y - n_\omega^Y \simeq n_{2\omega}^X - n_\omega^X << n_\omega^X$$

with the convention:

$$n_{2\omega}^Z > n_{2\omega}^Y > n_{2\omega}^X \text{ and } n_\omega^Z > n_\omega^Y > n_\omega^X$$

For a wave propagating in a biaxial crystal at angles θ and ϕ with respect to the principal axes, Z and X, respectively, the refractive index is given by Fresnel's equation:[68]

$$\frac{\text{Sin}^2\theta\text{Cos}^2\phi}{(n)^{-2} - (n^x)^{-2}} + \frac{\text{Sin}^2\theta\text{Sin}^2\phi}{(n)^{-2} - (n^y)^{-2}} + \frac{\text{Cos}^2\theta}{(n)^{-2} - (n^z)^{-2}} \tag{38}$$

The loci ϕ(θ) for a given type of phase-matching, in the general case, has no simple form.

Table 1.1.5
THE REFRACTIVE-INDEX CONDITIONS FOR CRITICAL PHASE MATCHING IN BIAXIAL CRYSTALS[66]

$$\text{For } n_{2\omega}^{Z} > n_{\omega}^{Z} > n_{2\omega}^{Y} > n_{\omega}^{Y} > n_{2\omega}^{X} > n_{\omega}^{X}: \left[\begin{array}{l} n_{2\omega}^{X} < \frac{1}{2}(n_{\omega}^{X} + n_{\omega}^{Y});\ n_{2\omega}^{Y} < \frac{1}{2}(n_{\omega}^{Y} + n_{\omega}^{Z}) \\ n_{2\omega}^{X} > \frac{1}{2}(n_{\omega}^{X} + n_{\omega}^{Y});\ n_{2\omega}^{Y} < \frac{1}{2}(n_{\omega}^{Y} + n_{\omega}^{Z}) \\ n_{2\omega}^{X} > \frac{1}{2}(n_{\omega}^{X} + n_{\omega}^{Y});\ n_{2\omega}^{Y} > \frac{1}{2}(n_{\omega}^{Y} + n_{\omega}^{Z}) \\ n_{2\omega}^{X} > \frac{1}{2}(n_{\omega}^{X} + n_{\omega}^{Y});\ n_{2\omega}^{Y} > \frac{1}{2}(n_{\omega}^{Y} + n_{\omega}^{Z}) \\ n_{2\omega}^{X} > \frac{1}{2}(n_{\omega}^{X} + n_{\omega}^{Z}) \end{array}\right.$$

$$\text{For } n_{2\omega}^{Z} > n_{\omega}^{Z} > n_{2\omega}^{Y} > n_{2\omega}^{X} > n_{\omega}^{Y} > n_{\omega}^{X}: \left[\begin{array}{l} n_{2\omega}^{X} < \frac{1}{2}(n_{\omega}^{X} + n_{\omega}^{Z});\ n_{2\omega}^{Y} < \frac{1}{2}(n_{\omega}^{Y} + n_{\omega}^{Z}) \\ n_{2\omega}^{X} < \frac{1}{2}(n_{\omega}^{X} + n_{\omega}^{Z});\ n_{2\omega}^{Y} > \frac{1}{2}(n_{\omega}^{Y} + n_{\omega}^{Z}) \\ n_{2\omega}^{X} > \frac{1}{2}(n_{\omega}^{X} + n_{\omega}^{Z});\ n_{2\omega}^{Y} > \frac{1}{2}(n_{\omega}^{Y} + n_{\omega}^{Z}) \end{array}\right.$$

$$\text{For } n_{2\omega}^{Z} > n_{2\omega}^{Y} > n_{\omega}^{Z} > n_{\omega}^{Y} > n_{2\omega}^{X} > n_{\omega}^{X}: \left[\begin{array}{l} n_{2\omega}^{X} < \frac{1}{2}(n_{\omega}^{X} + n_{\omega}^{Y}) \\ n_{2\omega}^{X} > \frac{1}{2}(n_{\omega}^{X} + n_{\omega}^{Y});\ n_{2\omega}^{X} < \frac{1}{2}(n_{\omega}^{X} + n_{\omega}^{Z}) \\ n_{2\omega}^{X} > \frac{1}{2}(n_{\omega}^{X} + n_{\omega}^{Z}) \end{array}\right.$$

$$\text{For } n_{2\omega}^{Z} > n_{2\omega}^{Y} > n_{\omega}^{Z} > n_{2\omega}^{X} > n_{\omega}^{Y} > n_{\omega}^{X}: \left[\begin{array}{l} n_{2\omega}^{X} < \frac{1}{2}(n_{\omega}^{X} + n_{\omega}^{Z}) \\ \\ n_{2\omega}^{X} > \frac{1}{2}(n_{\omega}^{X} + n_{\omega}^{Z}) \end{array}\right.$$

When the wave propagation is confined within one of the principal planes ZX or ZY, in which case $\Phi = 0$ or $90°$, respectively, then Equation 38 is greatly simplified. The phase-matching angle, θ_m, is then easily calculated by combining the appropriate index-matching condition with Equation 38. For example, for a wave propagating in the principal plane, ZX, of a negative biaxial crystal, the phase-matching angles θ_m (I) and θ_m (II) for Types I and II processes, respectively, are given by:

$$\mathrm{Sin}^2\theta_m\ (\mathrm{I}) = \frac{(n_{\omega}^{Y})^{-2} - (n_{2\omega}^{X})^{-2}}{(n_{2\omega}^{z})^{-2} - (n_{2\omega}^{x})^{-2}} \tag{39}$$

$$\mathrm{Sin}^2\theta_m\ (\mathrm{II}) = \frac{[n_{\omega}^{x}/2n_{2\omega}^{y} - n_{\omega}^{y}]^2 - 1}{(n_{\omega}^{x}/n_{\omega}^{z})^2 - 1} \tag{40}$$

In order to obtain the corresponding phase-matching angles for waves propagating in the *ZY* plane, n^x and n^y are interchanged in Equations 39 and 40.

In the general case, when the directions of propagation are not along the crystal principal axes, one must use the effective nonlinear coefficients by taking the projections of the nonlinear optical polarizations generated in the directions of the optical fields.

Table 1.1.6
THE REFRACTIVE-INDEX CONDITIONS AND POLARIZATIONS FOR NONCRITICAL PHASE MATCHING IN BIAXIAL CRYSTALS[66]

Type of $p \cdot m$ (propagation-axis)	Index condition	Fundamental polarization E (ω)	Harmonic polarization $E(2\omega)$
Type I (X)	$n^{X}_{2\omega} > n^{Y}_{\omega}; n^{Y}_{2\omega} = n^{Z}_{\omega}$	$E^{z}(\omega)$	$E^{y}(2\omega)$
Type I (X)	$n^{X}_{2\omega} < n^{Y}_{\omega}; n^{Y}_{2\omega} = n^{Z}_{\omega}$	$E^{z}(\omega)$	$E^{y}(2\omega)$
Type I (Y)	$n^{X}_{2\omega} = n^{Z}_{\omega}$	$E^{z}(\omega)$	$E^{x}(2\omega)$
Type I(Z)	$n^{Y}_{2\omega} < n^{Z}_{\omega}; n^{X}_{2\omega} = n^{Y}_{\omega}$	$E^{y}(\omega)$	$E^{x}(2\omega)$
Type I (Z)	$n^{Y}_{2\omega} > n^{Z}_{\omega}; n^{X}_{2\omega} = n^{Y}_{\omega}$	$E^{y}(\omega)$	$E^{x}(2\omega)$
Type II (X)	$n^{X}_{2\omega} > \frac{1}{2}(n^{X}_{\omega} + n^{Y}_{\omega}); n^{Y}_{2\omega} = \frac{1}{2}(n^{Y}_{\omega} + n^{Z}_{\omega})$	$E^{y}(\omega), E^{z}(\omega)$	$E^{y}(2\omega)$
Type II (X)	$n^{X}_{2\omega} < \frac{1}{2}(n^{X}_{\omega} + n^{Y}_{\omega}); n^{Y}_{2\omega} = \frac{1}{2}(n^{Y}_{\omega} + n^{Z}_{\omega})$	$E^{y}(\omega), E^{z}(\omega)$	$E^{y}(2\omega)$
Type II (Y)	$n^{X}_{2\omega} = \frac{1}{2}(n^{X}_{\omega} + n^{Z}_{\omega})$	$E^{x}(\omega), E^{z}(\omega)$	$E^{x}(2\omega)$
Type II (Z)	$n^{Y}_{2\omega} < \frac{1}{2}(n^{Y}_{\omega} + n^{Z}_{\omega}); n^{X}_{2\omega} = \frac{1}{2}(n^{X}_{\omega} + n^{Y}_{\omega})$	$E^{x}(\omega), E^{y}(\omega)$	$E^{x}(2\omega)$
Type II (Z)	$n^{Y}_{2\omega} > \frac{1}{2}(n^{Y}_{\omega} + n^{Z}_{\omega}); n^{X}_{2\omega} = \frac{1}{2}(n^{X}_{\omega} + n^{Y}_{\omega})$	$E^{x}(\omega), E^{y}(\omega)$	$E^{x}(2\omega)$

The effective second-order nonlinear coefficients, d_{eff}, for biaxial crystals have been derived in Reference 67 and the results for various crystal classes are listed in Table 1.1.7. The angles θ, Φ, and δ that appear in this table are shown in Figure 1.1.2, in which the incident beam is along H, A, and B are two optic axes in the XZ plane, and e_1 and e_2 indicate the two orthogonal polarization directions. The angle δ is related to θ, ϕ, and Ω (angle between the Z axis and the optic axis) by:[69]

$$\mathrm{Cot}2\delta = \frac{\mathrm{Cot}^2\Omega\ \mathrm{Sin}^2\theta - \mathrm{Cos}^2\theta\mathrm{Cos}^2\phi + \mathrm{Sin}^2\phi}{\mathrm{Cos}\theta - \mathrm{Sin}2\phi} \tag{41}$$

SHG CONVERSION EFFICIENCY WITH FOCUSED BEAM — FOR CRITICAL PHASE-MATCHING

When critical phase-matched SHG is used with a focused beam, there is a mismatch, ΔK, of the wavevector for small deviations from the phase-matched direction due to the finite divergence of the beam. By using Equations 22A, 23, 29, and 34, it can be shown that for a negative crystal of length L and Type I phase-matching:

$$\Delta\theta_m = -\left(\frac{1.39\ \lambda_1}{2\pi L}\right)\frac{(n^{o}_{2\omega})^2\tan\theta_m}{n^{o}_{\omega}[(n^{o}_{2\omega})^2 - (n^{o}_{\omega})^2]} \tag{42}$$

Table 1.1.7
EFFECTIVE TENSOR ELEMENTS d_{eff} FOR PHASE-MATCHED SHG IN BIAXIAL CRYSTALS

Class	d_{eff} for Type I phase matching	d_{eff} for Type II phase matching
$222(D_2)$ (Orthorhombic)	$d_{36}[\mathrm{S\ in}\ 2\theta\mathrm{Cos}2\phi\mathrm{Cos}\delta(3\mathrm{Sin}^2\delta - 1)$ $+ \mathrm{Sin}\theta\mathrm{Sin}2\phi\mathrm{Sin}\delta(3\mathrm{Cos}^2\theta\mathrm{Cos}^2\delta + 3\mathrm{Cos}^2\delta - 1)]$	$d_{36}[\mathrm{Sin}2\theta\mathrm{Cos}2\phi\mathrm{Sin}\delta(3\mathrm{Cos}^2\delta - 1)$ $- \mathrm{Sin}\theta\mathrm{Sin}2\phi\mathrm{Cos}\delta(3\mathrm{Cos}^2\theta\mathrm{Sin}^2\delta + 3\mathrm{Sin}^2\delta - 1)]$
$mm2\ (C_{2v})$ (Orthorhombic)	$(d_{32} - d_{31})(3\mathrm{Sin}^2\delta - 1)\mathrm{Sin}\theta\mathrm{Cos}\theta\mathrm{Sin}2\phi\mathrm{Cos}\delta$ $- 3(d_{31}\mathrm{Cos}^2\phi + d_{32}\mathrm{Sin}^2\phi)\mathrm{Sin}\phi\mathrm{Cos}^2\theta\mathrm{Sin}\delta\mathrm{Cos}^2\delta$ $+ (d_{31}\mathrm{Sin}^2\phi + d_{32}\mathrm{Cos}^2\phi)\mathrm{Sin}\theta\mathrm{Sin}\delta(3\mathrm{Sin}^2\delta - 2)$ $+ d_{33}\mathrm{Sin}^3\theta\mathrm{Sin}\delta\mathrm{Cos}^2\delta$	$(d_{32} - d_{31})(3\mathrm{Cos}^2\delta - 1)\mathrm{Sin}\theta\mathrm{Cos}\theta\mathrm{Sin}2\phi\mathrm{Cos}\delta$ $- 3(d_{31}\mathrm{Cos}^2\phi + d_{32}\mathrm{Sin}^2\phi)\mathrm{Sin}\theta\mathrm{Cos}^2\theta\mathrm{Sin}^2\delta\mathrm{cos}\delta$ $- (d_{31}\mathrm{Sin}^2\phi + d_{32}\mathrm{Cos}^2\phi)\mathrm{Sin}\theta\mathrm{Cos}\delta(3\mathrm{Cos}^2\delta - 2)$ $- d_{33}\mathrm{Sin}^3\theta\mathrm{Sin}^2\delta\mathrm{Cos}\delta$
$2(\parallel Y)(C_2)$ (Monoclinic)	$- (d_{22}\mathrm{Sin}^2\phi + 3d_{21}\mathrm{Cos}^2\phi)\mathrm{Cos}^3\theta\mathrm{Sin}\phi\mathrm{Sin}\delta\mathrm{Cos}^2\delta$ $- [d_{22}\mathrm{Cos}^2\phi + d_{21}(3\mathrm{Sin}^2\phi - 2)]\mathrm{Cos}\theta\mathrm{Sin}\phi\mathrm{Sin}^3\delta$ $+ 2[d_{22}\mathrm{Cos}^2\phi - d_{21}(3\mathrm{Cos}^2\phi -$ $1)]\mathrm{Cos}\theta\mathrm{Sin}\theta\mathrm{Sin}\delta\mathrm{Cos}^2\delta$ $- 2[d_{22}\mathrm{Sin}^2\phi - d_{21}(3\mathrm{Sin}^2\phi -$ $1)]\mathrm{Cos}^2\theta\mathrm{Cos}\phi\mathrm{Sin}^2\delta\mathrm{Cos}\delta$ $- [d_{22}\mathrm{Sin}^2\phi + d_{21}(\mathrm{Sin}^2\phi + 1)]\mathrm{Cos}^2\theta\mathrm{Cos}\phi\mathrm{Cos}^3\delta$ $- d_{34}\mathrm{Sin}^2\theta\mathrm{Cos}\delta[3\mathrm{Cos}\theta\mathrm{Sin}\phi\mathrm{Sin}\delta\mathrm{Cos}\delta -$ $\mathrm{Cos}\phi(3\mathrm{Cos}^2\delta - 2)]$ $+ d_{36}[\mathrm{Sin}2\theta\mathrm{Cos}2\phi\mathrm{Cos}\delta(3\mathrm{Sin}^2\delta - 1)$ $+ \mathrm{Sin}\theta\mathrm{Sin}2\phi\mathrm{Sin}\delta(3\mathrm{Cos}^2\theta\mathrm{Cos}^2\delta + 3\mathrm{Cos}^2\delta - 1)]$	$(d_{22}\mathrm{Sin}^2\phi + 3d_{21}\mathrm{Cos}^2\phi)\mathrm{Cos}^3\theta\mathrm{Sin}\phi\mathrm{Sin}^2\delta\mathrm{Cos}\delta$ $+ [d_{22}\mathrm{Cos}^2\phi + d_{21}(3\mathrm{Sin}^2\phi - 2)]\mathrm{Cos}\theta\mathrm{Sin}\phi\mathrm{Cos}^3\delta$ $- 2[d_{22}\mathrm{Cos}^2\phi - d_{21}(3\mathrm{Cos}^2\phi -$ $1)]\mathrm{Cos}\theta\mathrm{Sin}\phi\mathrm{Sin}^2\delta\mathrm{Cos}\delta$ $- 2[d_{22}\mathrm{Sin}^2\phi - d_{21}(3\mathrm{Cos}^2\phi -$ $1)]\mathrm{Cos}^2\theta\mathrm{Cos}^2\theta\mathrm{Cos}\phi\mathrm{Sin}\delta\mathrm{Cos}^2\delta$ $- [d_{22}\mathrm{Sin}^2\phi + d_{21}(\mathrm{Sin}^2\phi + 1)]\mathrm{Cos}^2\theta\mathrm{Cos}\phi\mathrm{Sin}^3\delta$ $+ d_{34}\mathrm{Sin}^2\theta\mathrm{Sin}\delta[3\mathrm{Cos}\theta\mathrm{Sin}\phi\mathrm{Sin}\delta\mathrm{Cos}\delta +$ $\mathrm{Cos}\phi(3\mathrm{Sin}^2\delta - 2)]$ $+ d_{36}[\mathrm{Sin}2\theta\mathrm{Cos}2\phi\mathrm{Sin}\delta(3\mathrm{Cos}^2\delta - 1)$ $- \mathrm{Sin}\theta\mathrm{Sin}2\phi\mathrm{Cos}\delta(3\mathrm{Cos}^2\theta\mathrm{Sin}^2\delta + 3\mathrm{Sin}^2\delta - 1)]$

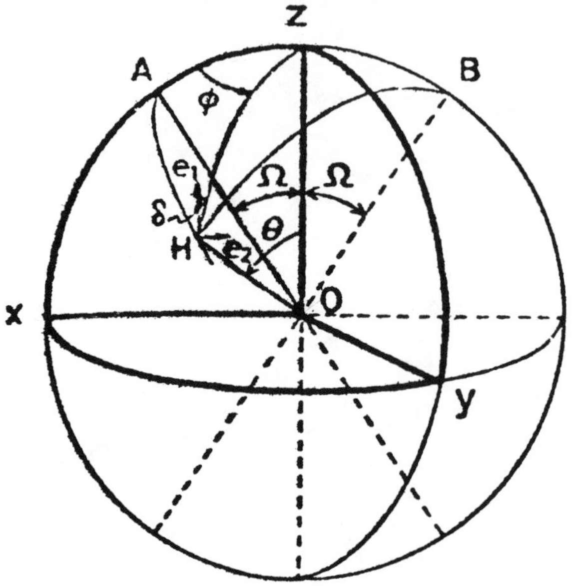

FIGURE 1.1.2. The propagation and polarization directions of optical waves in biaxial crystals. (From Hiromasa, I., Hatsuhiko, N., and Humio, I., *J. Appl. Phys.*, 46, 3992, 1975. With permission.)

For negative crystals and Type I process:

$$\Delta\theta_m = -\left(\frac{1.39\lambda}{2\pi L}\right)\frac{(n^o_\omega)^2\tan\theta_m}{(n^o_{2\omega})[(n^o_\omega)^2 - (n^o_{2\omega})^2]} \tag{43}$$

where λ_1 is the wavelength of the fundamental beam and $\Delta\theta_m$ is the angular half-width of the phase-matched SHG after which it drops to half of its peak value. As a result of this, the second-harmonic conversion efficiency of the crystal is limited.

Another disadvantage of critical phase-matching is that for waves of finite aperture, there is a ‘‘walk-off’’ of the second-harmonic wave from the polarization wave due to double refraction. These limitations can be described[65] in terms of a focusing parameter:

$$\xi = L/\mathrm{b} = L\lambda_1/2\pi \mathrm{n}_\omega \mathrm{w}_0^2 \tag{44}$$

and a double refraction parameter:

$$\mathrm{B} = \rho(\pi L\mathrm{n}_\omega/2\lambda_1)^{1/2} \tag{45}$$

where b $(= 2\pi \mathrm{w}_0^2/\lambda)$ is the confocal parameter of the beam in the crystal and ρ is the double refraction angle. For the beams propagating in a principal plane, ρ can be obtained from the relation:[62,70]

$$\tan\rho = \frac{[(n^o/n^e)^2 - 1]\tan\theta_m}{1 + (n^o/n^e)^2\tan^2\theta_m} \tag{46}$$

For focused beams, the optimum phase-matched second-harmonic power, $P_{2\omega}$, inside a

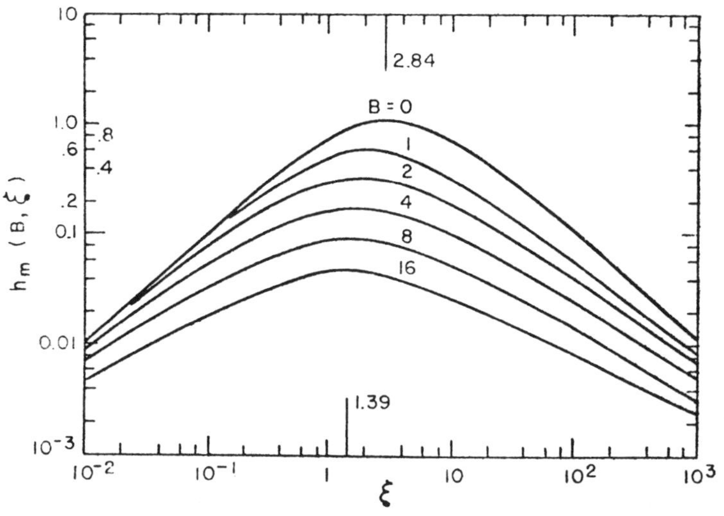

FIGURE 1.1.3. SHG power represented by the function H_m (B,ξ) for optimum phase matching as a function of focusing parameter $\xi = \ell/b$ for several values of double-refraction parameter $B = \rho(\ell k_1)^{1/2}/2$. Vertical lines indicate optimum focusing in the limits of small and large B. (From Boyd, G. D. and Kleinman, D. A., *J. Appl. Phys.*, 39, 3597, 1968. With permission.)

nonlinear medium of thickness L due to its effective nonlinear optical coefficient, $(d_{jm})_{\text{eff}}$, is given by:[65]

$$\mathbf{P}_{2\omega} = A \cdot (\mathbf{P}_\omega)^2 \cdot L \cdot K_\omega \cdot h_m \cdot \exp\left[-L\left(\alpha_\omega + \frac{1}{2}\alpha_{2\omega}\right)\right] \tag{47}$$

where

$$A = \begin{cases} 2\omega^2/\pi\epsilon_o c^3 n_\omega^2 n_{2\omega})(d_{jm})^2_{\text{eff}} & \text{W}^{-1}\text{(MKS)} \quad (48) \\ (128\pi^2\omega^2/c^3 n_\omega^2 n_{2\omega})(d_{jm})^2_{\text{eff}} & \text{(erg/sec)}^{-1}\text{(CGS)} \quad (49) \end{cases}$$

$K_\omega = 2\,\pi/\lambda_1$, wavevector of the fundamental, h_m is a function containing B, and ξ and other quantities are defined in "Second-Harmonic Power and Coherence Length". The powers outside the medium are obtained by using the factors appearing in Equations 24A, 24B, or 25A and 25B. The function h_m is conveniently plotted as a function of the focusing parameter ξ for several values of B in Figure 1.1.3 which is reproduced from Reference 65. For a given beam radius, w_o it is often convenient to define an aperture length:[62]

$$l_a = \frac{w_0\sqrt{\pi}}{\rho} \tag{50}$$

and an effective length of focus:

$$l_f = \pi w_0^2 K_\omega/2 \tag{51}$$

Then, in limiting cases, the asymptotic form of Equation 47, neglecting absorption is[65]

$$\mathbf{P}_{2\omega} = \frac{A\,(\mathbf{P}_\omega)^2}{w_0^2}\begin{cases} L^2 & \text{for } (l_a, l_f >> L) \quad\quad \zeta = << 1 & (52A) \\ Ll_a & \text{for } (l_f >> L >> l_a) & (52B) \\ l_f l_a & \text{for } (L >> l_f >> l_a) & (52C) \\ 4l_f^2 & \text{for } (L >> l_a >> l_f) \quad\quad \zeta >> 1 & (52D) \\ 4.75 l_f^2 & \text{for } (l_a >> L >> L >> l_f) & (52E) \end{cases}$$

SECOND-HARMONIC CONVERSION EFFICIENCY FOR NONCRITICAL PHASE MATCHING

Phase-matched SHG for $\theta_m = 90°$ is termed "noncritical phase matching" (NCPM). In this case, the walk-off angle $\rho = 0$, i.e., the direction of energy flow (Poynting vector) of the fundamental and second harmonic beams, is collinear. Also, the effects of beam divergence vanish. The second harmonic generated by an unfocused beam is then approximated by Equation 52A. For optimizing 90° phase matching with a focused beam, the curve for $B = 0$ in Figure 1.1.3 is used, from which it is found that $\xi = 2.84$ and $h_m = 1.068$. Using these values in Equation 47, the optimum second-harmonic power is given by

$$\mathbf{P}_{2\omega} = 1.068\ A\,(\mathbf{P}_\omega)^2 L \cdot K_\omega \exp\left[-L\left(\alpha_\omega + \frac{1}{2}\alpha_{2\omega}\right)\right] \tag{53}$$

A convenient method of obtaining NCPM consists of adjusting the temperature of the nonlinear crystal to an appropriate value T_m, at which the refractive index of the fundamental equals that of the second harmonic. The phase-matching temperature is given by the equation:

$$T_m = T_o + \frac{n_\omega - n_{2\omega}}{\dfrac{dn_{2\omega}}{dT} - \dfrac{dn_\omega}{dT}} \tag{54}$$

where T_o is the room temperature.

TWO-WAVE MIXING

Sum Frequency Generation

For the case of mixing of two waves at ω_1 and ω_2 to yield $\omega_3 = \omega_1 + \omega_2$, the induced polarization from Equations 13 and 15 is

$$\mathbf{P}_j(\omega_3) = 2\,\epsilon_o \sum_{kl} \chi_{jkl}(-\omega_3;\omega_1,\omega_2)E_k(\omega_1)E_l(\omega_2) \tag{55}$$

The power $\mathbf{P}_{\omega 3}$ generated at the frequency ω_3 by mixing beams of powers $\mathbf{P}_{\omega 1}$ and $\mathbf{P}_{\omega 2}$ at the frequencies ω_1 and ω_2, respectively, in a medium of length L is given by:

$$\mathbf{P}_{\omega 3} = \frac{4\,\omega_3^2 L^2 \mathbf{P}_{\omega 1}\mathbf{P}_{\omega 2} d_{jm}^2\, w_3^2 \exp\left[-\frac{1}{2}(\alpha_{\omega 1} + \alpha_{\omega 2} + \alpha_{\omega 3})L\right]}{\pi\epsilon_o c^3 n_{\omega 1} n_{\omega 2} n_{\omega 3} w_1^2 w_2^2} \tag{56A}$$

$$\left[\frac{\mathrm{Sin}(L\Delta K/2)}{(L\Delta K/2)}\right]^2 \quad \text{[MKS]}$$

$$\mathbf{P}_{\omega 3} = \frac{256\ \pi^2\omega_3^2L^2\mathbf{P}_{\omega 1}\mathbf{P}_{\omega 2}d_{jm}^2\mathrm{w}_3^2\exp\left[-\frac{1}{2}(\alpha_{\omega 1} + \alpha_{\omega 2} + \alpha_{\omega 3})L\right]}{n_{\omega 1}n_{\omega 2}n_{\omega 3}c^3\mathrm{w}_1^2\mathrm{w}_2^2} \tag{56B}$$

$$\left[\frac{\mathrm{Sin}(L\Delta K/2)}{(L\Delta K/2)}\right]^2 \quad [\mathrm{CGS}]$$

where, $n_{\omega 1}$, $n_{\omega 2}$, and $n_{\omega 3}$ are refractive indexes at ω_1, ω_2, and ω_3, respectively, $\alpha_{\omega 1}$, $\alpha_{\omega 2}$, and $\alpha_{\omega 3}$ are absorption coefficients at ω_1, ω_2, and ω_3, respectively, $\Delta K = K_3 - K_2 - K_1$, (the wavevector mismatch) and the other quantities are defined in Equations 22A and 22B.

The optimum power generated at the sum frequency, ω_3, using focused Gaussian beams inside a medium of length L is given in Reference 65:

$$\mathbf{P}_{\omega 3} = \frac{4\ \omega_1\omega_2\omega_3\mathbf{P}_{\omega 1}\mathbf{P}_{\omega 2}L(d_{jm})_{\mathrm{eff}}^2h_m\exp\left[-\frac{1}{2}(\alpha_{\omega 1} + \alpha_{\omega 2} + \alpha_{\omega 3})L\right]}{\pi\epsilon_o n_{\omega 3}^2c^4} \quad [\mathrm{MKS}] \tag{57A}$$

$$\mathbf{P}_{\omega 3} = \frac{256\ \pi^2\omega_1\omega_2\omega_3\mathbf{P}_{\omega 1}\mathbf{P}_{\omega 2}L(d_{jm})_{\mathrm{eff}}^2 \cdot h_m\exp\left[-\frac{1}{2}(\alpha_{\omega 1} + \alpha_{\omega 3} - \alpha_{\omega 2})L\right]}{c^4n_{\omega 3}^2}$$

$$\left[\frac{\mathrm{Sin}(L\Delta K/2)}{(L\Delta K/2)}\right]^2 \quad [\mathrm{CGS}] \tag{57B}$$

where the various quantities have been defined in Equations 22A, 22B, 56A, 56B, and 46.

Difference Frequency Generation

For the difference frequency $\omega_2 = \omega_3 - \omega_1$, the induced polarization is

$$\mathbf{P}_j(\omega_2) = 2\epsilon_o \sum_{k,l} \chi_{jkl}(-\omega_2; \omega_3, -\omega_1)E_k(\omega_3)E_l^*(\omega_1) \tag{58}$$

and the power generated at ω_1 is given by:

$$\mathbf{P}_{\omega 2} = \frac{4\ \omega_2^2L^2\mathbf{P}_{\omega 1}\mathbf{P}_{\omega 3}d_{jm}^2\mathrm{w}_2^2\exp\left[-\frac{1}{2}(\alpha_{\omega 1} + \alpha_{\omega 3} - \alpha_{\omega 2})L\right]}{\pi\epsilon_o c^3n_{\omega 1}n_{\omega 2}n_{\omega 3}\mathrm{w}_1^2\mathrm{w}_3^2} \tag{59A}$$

$$\left[\frac{\mathrm{Sin}(L\Delta K/2)}{(L\Delta K/2)}\right]^2 \quad [\mathrm{MKS}]$$

$$\mathbf{P}_{\omega 2} = \frac{256\ \pi^2\omega_2^2L^2\mathbf{P}_{\omega 1}\mathbf{P}_{\omega 3}\mathrm{d}_{jm}^2\mathrm{w}_2^2\exp\left[-\frac{1}{2}(\alpha_{\omega 1} + \alpha_{\omega 3} - \alpha_{\omega 2})L\right]}{n_{\omega 1}n_{\omega 2}n_{\omega 3}c^3\mathrm{w}_1^2\mathrm{w}_3^2} \tag{59B}$$

$$\left[\frac{\mathrm{Sin}(L\Delta K/2)}{(L\Delta K/2)}\right]^2 \quad [\mathrm{CGS}]$$

For the optimum power generated at the frequency ω_2 using focused Gaussian beams inside a medium of length L, the expressions are[65]

$$\mathbf{P}_{\omega 2} = \frac{4\,\omega_3^2\omega_1 L\mathbf{P}_{\omega 1}\mathbf{P}_{\omega 3}(d_{jm})_{\text{eff}}^2 \cdot h_m \exp\left[-\frac{1}{2}(\alpha_{\omega 1} + \alpha_{\omega 3} - \alpha_{\omega 2})L\right]}{\pi\epsilon_o n_{\omega 2} n_{\omega 3} c^4} \quad \text{[MKS]} \qquad (60A)$$

$$\mathbf{P}_{\omega 2} = \frac{256\,\pi^2\omega_3^2\omega_1 L\mathbf{P}_{\omega 1}\mathbf{P}_{\omega 3}(d_{jm})_{\text{eff}}^2 h_m \exp\left[-\frac{1}{2}(\alpha_{\omega 1} + \alpha_{\omega 3} - \alpha_{\omega 2})L\right]}{n_{\omega 2} n_{\omega 3} c^4} \quad \text{[CGS]} \qquad (60B)$$

THIRD-ORDER POLARIZATION AND ITS SYMMETRY PROPERTIES

Of the many interesting nonlinear optical phenomena that have their origin in the third-order nonlinear susceptibility χ_{jklm}, we will consider only the three-wave mixing (TWM), third-harmonic generation (THG), and self-focusing (SF).

Third-Harmonic Generation

The induced third-order polarization at the harmonic frequency, 3ω, is obtained from Equation 13:

$$\mathbf{P}_j(3\omega) = \epsilon_o \sum_{klm} (1/4)\chi_{jklm}(-3\omega;\omega,\omega,\omega)E_k(\omega)E_l(\omega)E_m(\omega) \qquad (61)$$

Most experimentalists use the third-order nonlinear coefficient, C_{jklm}, which is related to the third-order susceptibility, χ_{jklm}, by:

$$\chi_{jklm}^{(3)} = 4\,C_{jklm}^{(3)} \qquad (62)$$

which reduces Equation 61 to:

$$\mathbf{P}_j(3\omega) = \epsilon_o \sum_{klm} C_{jklm}(-3\omega;\omega,\omega,\omega)E_k(\omega)E_l(\omega)E_m(\omega) \qquad (63)$$

The third-order nonlinear coefficient is a fourth-rank tensor[4,14,72] with $3^{r+1} = 81$ (r is the order of the process) elements. The form of the third-order susceptibility tensor C_{jklm}, $(\omega_4, \omega_1, \omega_2, \omega_3)$, for the 32 crystal classes and isotropic media has been derived in Reference 4. The elements of the 3×27 matrixes are listed in Table 1.1.8.

For the special case of THG for which the matrixes must be invariant under interchange of the indexes k, l, and m ($C_{jklm} = C_{jkml} = C_{jlkm} = C_{jlmk}$), a contracted notation is used which reduces the 3×27 arrays to 3×10 arrays. Thus, $C_{jklm} \rightarrow C_{jn}$, where n runs from 0 to 9 according to:

klm		*n*
$xxx = 111$	⟷	1
$yyy = 222$	⟷	2
$zzz = 333$	⟷	3

(62)

Table 1.1.8
THE FORM OF THE 3 × 27 THIRD-ORDER SUSCEPTIBILITY TENSOR C_{jklm} $(-\omega_4;\omega_1,\omega_2,\omega_3)$ FOR THE 32 CRYSTAL CLASSES AND ISOTROPIC MEDIA

TRICLINIC

CLASSES: 1 (C_1), $\bar{1}(S_2)$
There are 81 independent nonzero elements

MONOCLINIC

CLASSES: $2(C_2)$, $m(C_{1h})$, $2/m(C_{2h})$
There are 41 independent nonzero elements

C_{111}; C_{2222}, C_{3333}
C_{1121}; C_{1122}; C_{1123}; C_{1132}; C_{1131}; C_{1133}; C_{2211}; C_{2212};
C_{2213}; C_{2232}; C_{2231}; C_{2233}; C_{3311}; C_{3312}; C_{3313};
C_{3321}; C_{3322}; C_{3323}; C_{1232}; C_{1223}; C_{1322}; C_{2132}
C_{2123}; C_{2213}; C_{2231}; C_{2312}; C_{2321}; C_{3122}; C_{3212}
C_{3221}; C_{1113}; C_{3111}; C_{1311}; C_{1131}; C_{1333}; C_{3313}
C_{3133}; C_{3331}

ORTHORHOMBIC

CLASSES: 222 (D_2); $mm2$ (C_{2V}); mmm (D_{2h})
There are 21 independent nonzero elements

C_{1111}; C_{2222}; C_{3333}
C_{1121}; C_{1122}; C_{1123}; C_{1132}
C_{1131}; C_{1133}; C_{2211}; C_{2212}
C_{2213}; C_{2232}; C_{2231}; C_{2233}
C_{3311}; C_{3312}; C_{3313}; C_{3321}
C_{3322}; C_{3323}

TETRAGONAL

CLASSES: $4(C_4)$; $\bar{4}(S_4)$, and $4/m$ (C_{4h})
There are 21 independent elements out of a total of 41 nonzero elements

$C_{1111} = C_{2222}$; C_{3333}
$C_{3311} = C_{3322}$; $C_{1233} = -C_{2133}$; $C_{1122} = C_{2211}$; $C_{1112} = -C_{2221}$
$C_{1133} = C_{2233}$; $C_{3312} = -C_{3321}$; $C_{1212} = C_{2121}$; $C_{1121} = -C_{2212}$
$C_{3131} = C_{3232}$; $C_{1323} = -C_{2313}$; $C_{1221} = C_{2112}$; $C_{1233} = -C_{2122}$
$C_{1313} = C_{2323}$; $C_{3132} = -C_{3231}$; $C_{2111} = -C_{1222}$
$C_{3113} = C_{3223}$; $C_{3123} = -C_{3213}$; $C_{1331} = C_{2332}$; $C_{1332} = -C_{2331}$

CLASSES: 422 (D_4), $4mm$ (C_{4V}), $\bar{4}2m$ (D_{2d}), $4/mmm$ (D_{4h})

$C_{1111} = C_{2222}$; C_{3333}
$C_{2233} = C_{3322}$; $C_{3311} = C_{1133}$; $C_{1122} = C_{2211}$
$C_{2323} = C_{3232}$; $C_{3131} = C_{1313}$; $C_{1212} = C_{2121}$
$C_{2332} = C_{3223}$; $C_{3113} = C_{1331}$; $C_{1221} = C_{2112}$

Table 1.1.8 (continued)
THE FORM OF THE 3 × 27 THIRD-ORDER SUSCEPTIBILITY TENSOR C_{jklm} $(-\omega_4;\omega_1,\omega_2,\omega_3)$ FOR THE 32 CRYSTAL CLASSES AND ISOTROPIC MEDIA

TRIGONAL

CLASSES: $3(C_3)$; $\bar{3}(S_6)$
There are 27 independent elements out of a total of 73 nonzero elements

$$C_{3333} \qquad (C_{1122} = C_{2211})$$
$$C_{1111} = C_{2222} = C_{1122} + C_{1221} + C_{1212} \qquad (C_{1221} = C_{2112})$$
$$(C_{1212} = C_{2121})$$
$$C_{2233} = C_{1133};\ C_{1233} = -C_{2133};\ C_{3322} = C_{3311};\ C_{3312} = -C_{3321}$$
$$C_{3223} = C_{3113};\ C_{3123} = -C_{3213};\ C_{2332} = C_{1331};\ C_{1332} = -C_{2331}$$
$$C_{2323} = C_{1313};\ C_{1323} = -C_{2313};\ C_{3232} = C_{3131};\ C_{3132} = -C_{3231}$$
$$(C_{2212} = -C_{1121})$$
$$C_{1122} = -C_{2221} = C_{2212} + C_{2122} + C_{1222} \qquad (C_{2122} = -C_{1211})$$
$$(C_{1222} = -C_{2111})$$
$$C_{2223} = -C_{2113} = -C_{1213} = -C_{1123}$$
$$C_{2232} = -C_{2131} = -C_{1231} = -C_{1132}$$
$$C_{2322} = -C_{2311} = -C_{1321} = -C_{1312}$$
$$C_{3222} = -C_{3211} = -C_{3121} = -C_{3112}$$
$$C_{1113} = -C_{1223} = -C_{2123} = -C_{2213}$$
$$C_{1311} = -C_{1232} = -C_{2132} = -C_{2321}$$
$$C_{3111} = -C_{3122} = -C_{3212} = -C_{3221}$$

CLASSES: $3m\ (C_{3v})$, $\bar{3}m\ (D_{3d})$, and $32\ (D_3)$
There are 14 independent elements out of a total of 37 nonzero elements

$$C_{3333} \qquad (C_{1122} = C_{2211})$$
$$C_{1111} = C_{2222} = C_{1122} + C_{1221} + C_{1212} \qquad (C_{1221} = C_{2112})$$
$$(C_{1212} = C_{2121})$$
$$C_{2233} = C_{1133};\ C_{2223} = -C_{2113} = -C_{1213} = -C_{1123}$$
$$C_{3322} = C_{3311};\ C_{2232} = -C_{2131} = -C_{1231} = -C_{1132}$$
$$C_{3223} = C_{3113};\ C_{2322} = -C_{2311} = -C_{1321} = -C_{1312}$$
$$C_{2332} = C_{1331};\ C_{3222} = -C_{3211} = -C_{3121} = -C_{3112}$$
$$C_{2323} = C_{1313}$$
$$C_{3232} = C_{3131}$$

HEXAGONAL

CLASSES: $6\ (C_6)$, $\bar{6}\ (C_{3h})$, $6/m\ (C_{6h})$
There are 19 independent elements out of a total of 41 nonzero elements

$$C_{3333} \qquad (C_{1122} = C_{2211})$$
$$C_{1111} = C_{2222} = C_{1122} + C_{1221} + C_{1212} \qquad (C_{1221} = C_{2112})$$
$$(C_{1212} = C_{2121})$$
$$C_{2233} = C_{1133};\ C_{1233} = -C_{2133};\ C_{3322} = C_{3311};\ C_{3312} = -C_{3321}$$
$$C_{3223} = C_{3113};\ C_{3123} = -C_{3213};\ C_{2332} = C_{1331};\ C_{1332} = -C_{2331}$$
$$C_{2323} = C_{1313};\ C_{1323} = -C_{2313};\ C_{3232} = C_{3131};\ C_{1332} = -C_{3231}$$
$$(C_{2112} = C_{1121})$$
$$C_{1112} = -C_{2221} = C_{2212} + C_{2121} + C_{1222} \qquad (C_{2122} = -C_{1211})$$
$$(C_{1222} = -C_{2111})$$

Table 1.1.8 (continued)
THE FORM OF THE 3 × 27 THIRD-ORDER SUSCEPTIBILITY TENSOR C_{jklm} $(-\omega_4;\omega_1,\omega_2,\omega_3)$ FOR THE 32 CRYSTAL CLASSES AND ISOTROPIC MEDIA

HEXAGONAL

CLASSES: 622 (D_6), 6*mm* (D_{6v}), $\bar{6}m2$ (D_{3h}), 6/*mmm* (D_{3h})
There are 10 independent elements out of a total of 21 elements

C_{3333}; $(C_{1122} = C_{2211})$
$C_{1111} = C_{2222} = C_{1122} + C_{1221} + C_{1212}$ $(C_{1221} = C_{2112})$
$C_{1111} = C_{2222} = C_{1122} + C_{1221} + C_{1212}$ $(C_{1211} = C_{2121})$
$(C_{1212} = C_{2121})$
$C_{2233} = C_{1133}$; $C_{3322} = C_{3311}$; $C_{3223} = C_{3113}$; $C_{2332} = C_{1331}$
$C_{2323} = C_{1313}$; $C_{3232} = C_{3131}$

CUBIC

CLASSES: 23(*T*), *m*3(T_h)
There are 7 independent elements out of a total of 21 nonzero elements

$C_{1111} = C_{2222} = C_{3333}$
$C_{2233} = C_{3311} = C_{1122}$
$C_{3322} = C_{1133} = C_{2211}$
$C_{2323} = C_{3131} = C_{1212}$
$C_{3232} = C_{1313} = C_{2121}$
$C_{2332} = C_{3113} = C_{1221}$
$C_{3223} = C_{1331} = C_{2112}$

CLASSES: 432(0), $\bar{4}3m$ (T_d), and m3m (O_h)
There are 4 independent elements out of a total of 21 nonzero elements

$C_{1111} = C_{2222} = C_{3333}$
$C_{2233} = C_{3322} = C_{3311} = C_{1133} = C_{1122} = C_{2211}$
$C_{2323} = C_{3232} = C_{3131} = C_{1313} = C_{1212} = C_{2121}$
$C_{2332} = C_{3223} = C_{3113} = C_{1331} = C_{1221} = C_{2112}$

ISOTROPIC MEDIA

There are 3 independent elements out of a total of 21 nonzero elements

$C_{1111} = C_{2222} = C_{3333}$
$C_{1111} = C_{1122} + C_{1212} + C_{1221}$
$C_{2233} = C_{3322} = C_{3311} = C_{1133} = C_{1122} = C_{2211}$
$C_{2323} = C_{3232} = C_{3131} = C_{1313} = C_{1212} = C_{2121}$
$C_{2332} = C_{3223} = C_{3113} = C_{1331} = C_{1221} = C_{2112}$

$$
\begin{array}{lll}
\left.\begin{array}{l} xxy = 112 \\ xyx = 121 \\ yxy = 212 \end{array}\right\} \longleftrightarrow 4 & \left.\begin{array}{l} yyx = 221 \\ yxy = 212 \\ xyy = 122 \end{array}\right\} \longleftrightarrow 8 & \\
\left.\begin{array}{l} yyz = 223 \\ yzy = 232 \\ zyy = 322 \end{array}\right\} \longleftrightarrow 5 & \left.\begin{array}{l} zzy = 332 \\ zyz = 323 \\ yzz = 233 \end{array}\right\} \longleftrightarrow 9 & (62) \\
\left.\begin{array}{l} zzx = 331 \\ zxz = 313 \\ xzz = 133 \end{array}\right\} \longleftrightarrow 6 & \left.\begin{array}{l} xyz = 123 \\ xzy = 132 \\ yzx = 231 \\ yxz = 213 \\ zxy = 312 \\ zyx = 321 \end{array}\right\} \longleftrightarrow 0 & \\
\left.\begin{array}{l} xxz = 113 \\ xzx = 131 \\ zxx = 311 \end{array}\right\} \longleftrightarrow 7 & &
\end{array}
$$

For index j, 1, 2, and 3 represent X, Y, and Z, respectively. In a medium in which the optical frequencies ω and 3ω lie far from the electronic and vibrational regions, the number of independent elements are further reduced to 15 by using the Kleinman conjecture:[74] $C_{jklm} = C_{jklm}$.

The third-harmonic polarizability matrixes, C_{jn}, for the 32 crystal point groups are listed in Table 1.1.9.[73] The axis of highest symmetry is denoted by Z, except for the cubic groups, whose threefold axes are always (111). A twofold axis normal to Z is always denoted by Y. A vertical reflection plane occurring in the absence of a twofold Y axis contains X and Z.

Unlike the second-order susceptibility, which vanishes for centrosymmetric materials, the third-order (and higher odd-order) susceptibility exists for all crystal classes. Consequently, third-order nonlinear processes can occur in solids, liquids, and free atoms and molecules.

In the contracted form the 3 × 10 elements of $C_{jn}\,(-3\omega;\omega,\omega,\omega)$ and the components of the third-order polarization can be written in the form:

$$
\begin{bmatrix} \mathbf{P}_X \\ \mathbf{P}_Y \\ \mathbf{P}_Z \end{bmatrix} = \begin{bmatrix} C_{11} & C_{12} & C_{13} & C_{14} & C_{15} & C_{16} & C_{17} & C_{18} & C_{19} & C_{10} \\ C_{21} & C_{22} & C_{23} & C_{24} & C_{25} & C_{26} & C_{27} & C_{28} & C_{29} & C_{20} \\ C_{31} & C_{32} & C_{33} & C_{34} & C_{35} & C_{36} & C_{37} & C_{38} & C_{39} & C_{30} \end{bmatrix} \begin{bmatrix} E_x^3 \\ E_y^3 \\ E_z^3 \\ 3\,E_yE_z^2 \\ 3\,E_y^2E_z \\ 3\,E_xE_z^2 \\ 3\,E_x^2E_z \\ 3\,E_xE_y^2 \\ 3\,E_x^2E_y \\ 6\,E_xE_yE_z \end{bmatrix} \qquad (63)
$$

When a single-mode Gaussian beam of power, P_ω, is incident on a nonlinear medium of length L, the third harmonic power $\mathbf{P}_{3\omega}^{\mathrm{THG}}$, generated is given by:

$$
\mathbf{P}_{3\omega}^{\mathrm{THG}} = \frac{12\,\omega^2 C_{jn}^2 L^2 \mathbf{P}_\omega^3 \exp\left[-\frac{1}{2}(\alpha_{3\omega} + 3\,\alpha_\omega)L\right]}{\pi^2 \epsilon_o^2 c^4 w_0^4 n_\omega^3 n_{3\omega}} \left[\frac{\mathrm{Sin}(L\Delta K/2)}{(L\Delta K/2)}\right]^2 \quad [\mathrm{MKS}] \qquad (64\mathrm{A})
$$

Table 1.1.9
FORM OF THE THG-TENSOR FOR THE VARIOUS CRYSTAL CLASSES

TRICLINIC

Classes 1(C_1), $\bar{1}$(S_2)

$$\begin{bmatrix} C_{11} & C_{12} & C_{13} & C_{14} & C_{15} & C_{16} & C_{17} & C_{18} & C_{19} & C_{10} \\ C_{21} & C_{22} & C_{23} & C_{24} & C_{25} & C_{26} & C_{27} & C_{28} & C_{29} & C_{20} \\ C_{31} & C_{32} & C_{33} & C_{34} & C_{35} & C_{36} & C_{37} & C_{38} & C_{39} & C_{30} \end{bmatrix}$$

MONOCLINIC

Classes 2(C_2), *m*(C_{1h}), 2/*m*(C_{2h})

$$\begin{bmatrix} C_{11} & C_{12} & 0 & C_{14} & 0 & C_{16} & 0 & C_{18} & C_{19} & 0 \\ C_{21} & C_{22} & 0 & C_{24} & 0 & C_{26} & 0 & C_{28} & C_{29} & 0 \\ 0 & 0 & C_{33} & 0 & C_{35} & 0 & C_{37} & 0 & 0 & C_{30} \end{bmatrix}$$

ORTHORHOMBIC

Classes 222 (D_2), *mm*2 (D_{2v}), *mmm* (D_{2h})

$$\begin{bmatrix} C_{11} & 0 & 0 & 0 & 0 & C_{16} & 0 & C_{18} & 0 & 0 \\ 0 & C_{22} & 0 & C_{24} & 0 & 0 & 0 & 0 & C_{29} & 0 \\ 0 & 0 & C_{33} & 0 & C_{35} & 0 & C_{37} & 0 & 0 & 0 \end{bmatrix}$$

TETRAGONAL

Classes 4(C_4), $\bar{4}$(S_4), 4/*m* (C_{4h})

$$\begin{bmatrix} C_{11} & C_{12} & 0 & C_{14} & 0 & C_{16} & 0 & C_{18} & C_{19} & 0 \\ -C_{12} & C_{11} & 0 & C_{18} & 0 & -C_{19} & 0 & -C_{14} & C_{16} & 0 \\ 0 & 0 & C_{33} & 0 & C_{35} & 0 & C_{35} & 0 & 0 & 0 \end{bmatrix}$$

Classes 422 (D_4), 4*mm* (C_{4v}), $\bar{4}$2*m* (D_{2d}), 4/*mmm* (D_{4h})

$$\begin{bmatrix} C_{11} & 0 & 0 & 0 & 0 & C_{16} & 0 & C_{18} & 0 & 0 \\ 0 & C_{11} & 0 & C_{18} & 0 & 0 & 0 & 0 & C_{16} & 0 \\ 0 & 0 & C_{33} & 0 & C_{35} & 0 & C_{35} & 0 & 0 & 0 \end{bmatrix}$$

TRIGONAL

Classes 3(C_3), $\bar{3}$(S_6)

$$\begin{bmatrix} C_{11} & C_{12} & 0 & C_{12} & C_{15} & C_{16} & -C_{15} & C_{11} & C_{19} & 2 \times C_{25} \\ -C_{12} & C_{11} & 0 & C_{11} & C_{25} & -C_{19} & -C_{25} & -C_{12} & C_{16} & 2 \times C_{15} \\ C_{31} & C_{32} & C_{33} & -3C_{32} & C_{35} & 0 & C_{35} & -3C_{31} & 0 & 0 \end{bmatrix}$$

Classes 32(D_3), 3*m* (C_{3v}, $\bar{3}$*m* (D_{3d})

$$\begin{bmatrix} C_{11} & 0 & 0 & 0 & C_{15} & C_{16} & -C_{15} & C_{11} & 0 & 0 \\ 0 & C_{11} & 0 & C_{11} & 0 & 0 & 0 & 0 & C_{16} & 2C_{15} \\ C_{31} & 0 & C_{33} & 0 & C_{35} & 0 & C_{35} & -3C_{31} & 0 & 0 \end{bmatrix}$$

Table 1.1.9 (continued)
FORM OF THE THG-TENSOR FOR THE VARIOUS CRYSTAL CLASSES

HEXAGONAL

Classes $6(C_6)$, $\bar{6}(C_{3h})$, $6/m$ (C_{6h})

$$\begin{bmatrix} C_{11} & C_{12} & 0 & C_{12} & 0 & C_{16} & 0 & C_{11} & C_{19} & 0 \\ -C_{12} & C_{11} & 0 & C_{11} & 0 & -C_{19} & 0 & -C_{12} & C_{16} & 0 \\ 0 & 0 & C_{33} & 0 & C_{35} & 0 & C_{35} & 0 & 0 & 0 \end{bmatrix}$$

Classes 622 (D_6), $6mm$ (C_{6v}), $\bar{6}m$ $2(D_{3h})$, $6/mmm$ (D_{6h})

$$\begin{bmatrix} C_{11} & 0 & 0 & 0 & 0 & C_{16} & 0 & C_{11} & 0 & 0 \\ 0 & C_{11} & 0 & C_{11} & 0 & 0 & 0 & 0 & C_{16} & 0 \\ 0 & 0 & C_{33} & 0 & C_{35} & 0 & C_{35} & 0 & 0 & 0 \end{bmatrix}$$

CUBIC

Classes $23(T)$, m $3(T_h)$

$$\begin{bmatrix} C_{11} & 0 & 0 & 0 & 0 & C_{16} & 0 & C_{18} & 0 & 0 \\ 0 & C_{11} & 0 & C_{16} & 0 & 0 & 0 & 0 & C_{18} & 0 \\ 0 & 0 & C_{11} & 0 & C_{16} & 0 & C_{18} & 0 & 0 & 0 \end{bmatrix}$$

Classes 432(0), $\bar{4}3m$ (T_d), m $3m$ (O_h)

$$\begin{bmatrix} C_{11} & 0 & 0 & 0 & 0 & C_{16} & 0 & C_{16} & 0 & 0 \\ 0 & C_{11} & 0 & C_{16} & 0 & 0 & 0 & 0 & C_{16} & 0 \\ 0 & 0 & C_{11} & 0 & C_{16} & 0 & C_{16} & 0 & 0 & 0 \end{bmatrix}$$

$$\mathbf{P}_{3\omega}^{\mathrm{THG}} = \frac{(16 \times 192)\pi^2\omega^2 C_{jn}^2 L^2 \mathbf{P}_\omega^3 \exp\left[-\frac{1}{2}(\alpha_{3\omega} + 3\,\alpha_\omega)L\right]}{c^4 w_0^4 n_\omega^3 n_{3\omega}}$$

$$\left[\frac{\mathrm{Sin}(L\Delta K/2)}{(L\Delta K/2)}\right]^2 \quad [\mathrm{CGS}] \qquad (64\mathrm{B})$$

$n_{3\omega}$ = refractive index at 3ω
$\alpha_{3\omega}$ = absorption coefficient at 3ω
ΔK = $K_{3\omega} - 3K_\omega$, wavevector mismatch
C_{jn} = THG coefficient and the other quantities have been defined in Equations 22A and 22B. The powers are those internal to the medium

Equations 64A and 64B are valid only for an unfocused or lightly focused Gaussian beam whose confocal parameter, b, is much greater than the length, L, of the nonlinear medium.

In the case of tight focusing ($b \ll L$) in a negatively dispersing nonlinear medium ($\Delta K < 0$), the third harmonic power is given by

$$\mathbf{P}_{3\omega}^{\mathrm{THG}} = \frac{(16 \times 48)\,\pi^4 b^4 \omega^2 C_{jn}^2 (\Delta K)^2 \mathbf{P}_{\omega}^3 \exp\left[\mathrm{b}\Delta K - \frac{1}{2}(\alpha_{3\omega} + 3\,\alpha_{\omega})L\right]}{\mathrm{n}_{\omega}^3 n_{3\omega} c^4 \mathrm{w}_0^4} \quad \text{[CGS]} \qquad (64C)$$

$$\mathbf{P}_{3\omega}^{\mathrm{THG}} = \frac{3\omega^2 b^4 C_{jn}^2 \mathbf{P}_{\omega}^2 (\Delta K)^2 \exp\left[b\Delta K - \frac{1}{2}(\alpha_{3\omega} + 3\,\alpha_{\omega})L\right]}{n_{\omega}^3 n_{3\omega} c^4 \epsilon_o^2 \mathrm{w}_0^4} \quad \text{[MKS]} \qquad (64D)$$

$$\text{For } \Delta K \geqslant 0, \quad \mathbf{P}_{3\omega}^{\mathrm{THG}} = 0 \qquad (64E)$$

In general, the wavevector mismatch $\Delta K \neq 0$ due to dispersion in the medium and the third-harmonic power will oscillate, reaching a maximum for $L = L_{\mathrm{coh}}$, where the coherence length:

$$L_{\mathrm{coh}} = \frac{\pi}{\Delta K} = \pi c/3\omega(n_{\omega} - n_{3\omega}) = \lambda_1/6(n_{\omega} - \mathrm{n}_{3\omega}) \qquad (65)$$

λ_1 being the free space wavelength of the fundamental wave.

For gaseous nonlinear media, the microscopic nonlinear coefficient C_{jn}^{MIC} (per molecule) is related to the bulk nonlinear coefficient C_{jn} by:

$$C_{jn} = \mathrm{N}C_{jn}^{\mathrm{MIC}} \cdots \qquad (65A)$$

where N is the number of molecules per unit volume.

Three-Wave Mixing (TWM) — Sum Frequency Generation

For the general case of three-wave mixing represented by $\omega_4 = \omega_1 + \omega_2 + \omega_3$, the induced third-order polarization is given by:

$$\mathbf{P}_j(\omega_4) = \epsilon_o\, \Sigma\, 6\, C_{jklm}(-\omega_4;\omega_1,\omega_2,\omega_3)E_k(\omega_1)E_m(\omega_3)E_l(\omega_2) \qquad (66)$$

The power generated, $P_{\omega 4}$, at the sum frequency inside a medium of length L is

$$\mathbf{P}_{\omega4}^{\mathrm{MIX}} = \frac{144\,\omega_4^2 C_{jn}^2 \mathbf{P}_{\omega1}\mathbf{P}_{\omega2}\mathbf{P}_{\omega3}\mathrm{w}_4^2}{\pi^2\epsilon_o^2 n_{\omega1}n_{\omega2}n_{\omega3}n_{\omega4}c^4\,\mathrm{w}_1^2\mathrm{w}_2^2\mathrm{w}_3^2}\left[\frac{e^{(i\Delta\mathrm{K}-\Delta\alpha)\mathrm{L}} - 1}{i\Delta\mathrm{K} - \Delta\alpha}\right]^2 \cdot e^{-\alpha_{\omega4}\cdot L} \quad \text{[MKS]} \qquad (67A)$$

$$\mathbf{P}_{\omega4}^{\mathrm{MIX}} = \frac{(256)(144)\pi^2\omega_4^2 C_{jn}^2 \mathbf{P}_{\omega1}\mathbf{P}_{\omega2}\mathbf{P}_{\omega3}\cdot\mathrm{w}_4^2}{n_{\omega1}n_{\omega2}n_{\omega3}n_{\omega4}c^4\mathrm{w}_1^2\mathrm{w}_2^2\mathrm{w}_3^2}\left[\frac{e^{(i\Delta K-\Delta\alpha)L} - 1}{i\Delta K - \Delta\alpha}\right]^2 e^{-\alpha_{\omega4}\cdot L} \quad \text{[CGS]} \qquad (67B)$$

where

$n_{\omega j}$ = refractive index at the frequency ω_j
$P_{\omega j}$ = power inside the medium at ω_j
w_j = beam spot radius at ω_j
ΔK = $K_4 - K_3 - K_2 - K_1$ = wavevector mismatch
Δ_α = $(^1/_2)\,(\alpha_{\omega1} + \alpha_{\omega2} + \alpha_{\omega3} - \alpha_{\omega4})$; $\alpha_{\omega j}$ being the absorption coefficient at ω_j
ϵ_o = free space permittivity
c = velocity of light in vacuum

The form of the tensor $C_{jklm}\,(-\omega_4;\omega_1,\omega_2,\omega_3)$ for the 32 crystal classes and isotropic media have been computed in Reference 4 and are listed in Table 1.1.9.

Difference Frequency Generation

Optical difference mixing of the form $\omega_3 = \omega_1 + \omega_1 - \omega_2$ has been useful for studying weak nonlinear susceptibilities. An advantage of this process is that materials which are not transparent to third-harmonic or sum frequency can still be studied to measure C_{jn}.

The induced polarization at ω_3 is

$$\mathbf{P}_j(\omega_3) = \epsilon_o \sum_{klm} 3\, C_{jklm}(-\omega_3;\omega_1,\omega_1,-\omega_2)E_k(\omega_1)E_l(\omega_1)E_m^*(\omega_2) \tag{68}$$

and the power generated at ω_3 is given by:

$$\mathbf{P}_{\omega 3}^{\mathrm{MIX}} = \frac{36\ \omega_3^2 C_{jn}^2 \mathbf{P}_{\omega 1}^2 \mathbf{P}_{\omega 2} \mathrm{w}_3^2}{\pi^2 \epsilon_o^2 c^4 n_{\omega 1}^2 n_{\omega 2} n_{\omega 3} \cdot \mathrm{w}_2^2 \mathrm{w}_1^4} \left[\frac{e^{(i\Delta K - \Delta\alpha)L} - 1}{i\Delta K - \Delta\alpha}\right]^2 e^{-\alpha_{\omega 3} \cdot L} \quad \text{[MKS]} \tag{69A}$$

$$\mathbf{P}_{\omega 3}^{\mathrm{MIX}} = \frac{(256)(36)\pi^2 \omega_3^2 C_{jn}^2 \mathbf{P}_{\omega 1}^2 \mathbf{P}_{\omega 2} \cdot \mathrm{w}_3^2}{n_{\omega 1}^2 n_{\omega 2} n_{\omega 3} c^4 \cdot \mathrm{w}_2^2 \mathrm{w}_1^4} \left[\frac{e^{(i\Delta K - \Delta\alpha)\mathrm{L}} - 1}{i\Delta K - \Delta\alpha}\right]^2 e^{-\alpha_{\omega 3} \cdot L} \quad \text{[CGS]} \tag{69B}$$

where the various quantities have been defined in Equations 67A and 67B.

Self-Induced Refractive Index Change

An intense monochromatic beam of frequency ω will experience a changed index of refraction due to the third-order nonlinear polarization induced by it in a medium. The polarization component at frequency ω in terms of the optical electric fields can be written as:

$$\mathbf{P}_j(\omega) = \epsilon_o \chi_{jk}^{(1)}(-\omega;\omega)E_k(\omega) + 3\epsilon_o C_{jklm}(-\omega;\omega,-\omega,\omega)E_j(\omega)E_k(\omega)E_k(\omega)E_l^*(\omega) \tag{70A}$$

In terms of the incident power, $\mathbf{P}_\omega$, and the linear index of refraction n_o, the change of refractive index $\delta n\ (\omega)$ is

$$\delta n(\omega) = \frac{6C_{jm}(-\omega;\omega,-\omega,\omega) \cdot \mathbf{P}_\omega}{\pi \mathrm{w}^2 \epsilon_o c n_0^2} \tag{70B}$$

where w is the beam spot radius. If we denote:

$$n(\omega) = n_0 + n_2|E(\omega)|^2 \tag{70C}$$

then the nonlinear index of refraction coefficient $n_2\ (\omega)$ is given by:

$$n_2(\omega) = \frac{(3/2)C_{jm}(-\omega;\omega,-\omega,\omega)}{\mathrm{n}_0} \quad \text{[MKS]} \tag{70D}$$

$$= \frac{12\pi C_{jm}(-\omega;\omega,-\omega,\omega)}{n_0} \quad \text{[CGS]} \tag{70E}$$

For a nonlinear medium of cubic symmetry, the coefficient of nonlinear index is given by References 130a and 403.

For a beam linearly polarized along an (1, 0, 0) axis:

$$n_2 = \frac{(3/2)}{n_0} C_{1111} \quad \text{[MKS]} \tag{70F}$$

$$n_2 = \frac{12\pi}{n_0} C_{1111} \quad [\text{CGS}] \tag{70G}$$

For a beam circularly polarized in the plane perpendicular to the (1, 0, 0) axis:

$$n_2 = \frac{(3/4)}{n_0} (2C_{1122} - C_{1221} + C_{111}) \quad [\text{MKS}] \tag{70H}$$

$$n_2 = \frac{6\pi}{n_0} (2C_{1122} - C_{1221} + C_{1111}) \quad [\text{CGS}] \tag{70I}$$

For a beam circularly polarized in the plane perpendicular to the (111) axis:

$$n_2 = \frac{(1/2)}{n_0} (4C_{1122} - C_{1221} + C_{1111}) \quad [\text{MKS}] \tag{70J}$$

$$n_2 = \frac{4\pi}{n_0} (4C_{1122} - C_{1221} + C_{1111}) \quad [\text{CGS}] \tag{70K}$$

In an isotropic medium, for a beam of linear polarization:

$$n_2 = \frac{(3/2)}{n_0} (2C_{1122} + C_{1221}) \quad [\text{MKS}] \tag{70L}$$

$$n_2 = \frac{12\pi}{n_0} (2C_{1122} + C_{1221}) \quad [\text{CGS}] \tag{70M}$$

For a circularly polarized beam:

$$n_2 = \frac{3C_{1122}}{n_0} \quad [\text{MKS}] \tag{70N}$$

$$n_2 = \frac{24\pi}{n_0} (C_{1122}) \quad [\text{CGS}] \tag{70P}$$

The critical power, $\mathbf{P}_{cr}$, at which the self-focusing overcomes the diffraction spread is given by:[29]

$$\mathbf{P}_{cr} = \frac{(1 \cdot 22\lambda)^2 c\pi\epsilon_o}{64 n_2} \quad [\text{MKS}] \tag{70Q}$$

$$\mathbf{P}_{cr} = \frac{5 \cdot 763\lambda^2 C}{16\pi^3 n_2} \quad [\text{CGS}] \tag{70R}$$

PHASE-MATCHED PROPAGATION OF THIRD-ORDER NONLINEAR OPTICAL POLARIZATION

The phase-matching techniques of Giordmaine[61] and Maker et al.[59] have been extended by Midwinter and Warner[75] to three-wave interactions involving the third-order polarizability tensor, C_{jklm}, in solids.

For the general case of mixing of waves of frequencies ω_1, ω_2, and ω_3 and wavevectors K_1, K_2, and K_3, the energy and momentum conservation require that:

$$\omega_4 = \omega_1 + \omega_2 + \omega_3 \tag{71A}$$

$$K_4 = K_1 + K_2 + K_3 \tag{71B}$$

Phase matching is achieved when $\Delta K = K_4 - (K_1 + K_2 + K_3) = 0$. Using Equation 12d and Equation 71B yields:

$$n_4 = \frac{\omega_3 n_3}{(\omega_3 + \omega_2 + \omega_1)} + \frac{\omega_2 n_2}{(\omega_3 + \omega_2 + \omega_1)} + \frac{\omega_1 n_1}{(\omega_3 + \omega_2 + \omega_1)} \tag{72}$$

Three types of phase matching is possible in uniaxial crystals.

Type I: ω_3, ω_2, and ω_1 are either all o rays or all e rays resulting in ω_4 as an e ray or an o ray, respectively.
Type II: ω_4 and one of the waves ω_3, ω_2 or ω_1 are either o rays or e rays, making the remaining two either e rays or o rays, respectively.
Type III: ω_4 and two of the waves ω_3, ω_2, or ω_1 are either o rays or e rays, making the remaining wave either an e ray or an o ray, respectively.

The refractive indexes of the o and e rays propagating in a direction, θ, measured from the optic axis of a uniaxial crystal are given by Equation 29.

If n^o and n^e are known for each of the frequencies ω_4, ω_3, ω_2, and ω_1, the phase matching angle, θ, will be generated in uniaxial crystals as a cone while the azimuthal angle, ϕ, of the wave normal direction varies from 0 to 2 π. The power output generated at ω_4 will in general depend upon the chosen value of ϕ (measured from the principal x axis), the phase matching angle, θ, and the symmetry of the C_{jklm} tensor relevant to the crystal under consideration.

For critical phase matching, the nonlinear coefficient C_{jn} in Equation 64A and 64B must be replaced with an effective nonlinear coefficient $(C_{jn})_{eff}$. The effective third-order nonlinear tensor elements, C_{eff}, for all the uniaxial and isotropic crystal classes have been derived in Reference 75 and are listed here in Table 1.1.10.

While in a crystalline medium the birefringence can be used to match the phase velocities of fundamental and harmonic radiation by compensating the material dispersion, it is not possible to do the same in liquids and gases. The possibility of using anomalous dispersion to achieve phase matching was suggested by a number of early nonlinearoptics workers[1,14,24,27] and has been demonstrated in liquids and gases.[76-78]

The principle of phase-matched THG using anomalous dispersion is based on the introduction of a localized absorption intermediate between the fundamental and harmonic wavelength, resulting in a matching of the refractive indexes at these wavelengths.

An attractive system[77] for phase-matched THG in gases consists of using alkali metal vapors as nonlinear media which are anomalously dispersive if the driving frequency is chosen below and its harmonic chosen above the principle resonance line. A normally dispersive buffer gas, such as xenon, is added such that the refractive indexes of the mixture are equal at the fundamental and harmonic wavelengths. To calculate the required ratio of pressures of buffer gas to metal vapor, it is necessary to know the index of refraction vs. wavelength for the constituents. The refractive index is calculated from the Sellmeier equation:

Table 1.1.10
EFFECTIVE NONLINEAR COEFFICIENT C_{eff} FOR PHASE MATCHED THG IN POSITIVE (+) AND NEGATIVE (−) UNIAXIAL AND ISOTROPIC CRYSTALS

Class	$C_{eff}^{\pm}$ for Type I P.M.	$C_{eff}^{\pm}$ for Type II P.M.	$C_{eff}^{\pm}$ for Type III P.M.
Isotropic	$C^{\pm} = 0$	$C^{\pm} = (1/3)\, C_{11}$	$C^{-} = 0$
23(T) $\bar{4}3m(T_4)$ 432.(0)	$C^{+} = -{}^1/_4(C_{11} - 3C_{16}\mathrm{Cos}^3\theta\mathrm{Sin}4\phi)$ $C^{-} = {}^1/_4(C_{11} - 3C_{16}\mathrm{Cos}\theta\mathrm{Sin}4\phi)$ $C^{-} = {}^1/_4(C_{11} - 3C_{16}\mathrm{Cos}\theta\mathrm{Sin}4\phi)$	$C^{\pm} = {}^1/_2(C_{11} - 3C_{16})\mathrm{Cos}^2\theta\mathrm{Sin}^2 2\phi + C_{16}$	$C^{+} = -{}^1/_4(C_{11} - 3C_{16}\mathrm{Cos}\theta\mathrm{Sin}4\phi)$ $C^{-} = -{}^1/_4(C_{11} - 3C_{16}\mathrm{Cos}\theta\mathrm{Sin}4\phi)$
$\bar{6}(C_{3h})$ $6(C_6)$ $\bar{6}m2(D_{3h})$ $6mm\ (C_{6v})$ $622\ (D_6)$	$C^{+} = 0$ $C^{-} = 0$	$C^{\pm} = {}^1/_3C_{11}\mathrm{Cos}^2\theta + C_{16}\mathrm{Sin}^2\theta$	$C^{-} = 0$ $C^{+} = 0$
$3m(C_{3v})$ $32(D_3)$ $\bar{3}m(D_{3d})$	$C^{+} = -{}^3/_2C_{10}\mathrm{Cos}\theta\mathrm{Sin}2\theta\mathrm{Cos}3\phi$ $C^{-} = C_{10}\mathrm{Sin}\theta\mathrm{Cos}3\phi$	$C^{\pm} = {}^1/_3C_{11}\mathrm{Cos}^2\theta + C_{16}\mathrm{Sin}^2\theta + C_{10}\mathrm{Sin}2\theta\mathrm{Sin}3\phi$	$C^{-} = -{}^3/_2C_{10}\mathrm{Cos}\theta\mathrm{Sin}2\theta\mathrm{Cos}3\phi$ $C^{+} = C_{10}\mathrm{Sin}\theta\mathrm{Cos}3\phi$
$3(C_3)$ $\bar{3}(S_6)$	$C^{+} = -{}^3/_2(C_{10}\mathrm{Cos}3\phi + C_{15}\mathrm{Sin}3\phi)\mathrm{Cos}\theta\mathrm{Sin}2\theta$ $C^{-} = (C_{10}\mathrm{Cos}3\phi + C_{15}\mathrm{Sin}3\phi)\mathrm{Sin}\theta$	$C^{\pm} = (C_{10}\mathrm{Sin}3\phi - C_{15}\mathrm{Cos}3\phi)\mathrm{Sin}2\theta + {}^1/_3C_{11}\mathrm{Cos}^2\theta + C_{16}\mathrm{Sin}^2\theta$	$C^{-} = -{}^3/_2(C_{10}\mathrm{Cos}3\phi + C_{15}\mathrm{Sin}3\phi)\mathrm{Cos}\theta\mathrm{Sin}2\theta$ $C^{+} = (C_{10}\mathrm{Cos}3\phi + C_{15}\mathrm{Sin}3\phi)\mathrm{Sin}\theta$
$4(C_4)$ $\bar{4}(S_4)$ $4/m(C_{4h})$	$C^{+} = -\mathrm{Cos}^3\theta[{}^1/_4(C_{11} - 3C_{18})\mathrm{Sin}4\phi - C_{21}\mathrm{Cos}4\phi]$ $C^{-} = \mathrm{Cos}\theta[{}^1/_4(C_{11} - 3C_{18})\mathrm{Sin}4\phi - C_{21}\mathrm{Cos}4\phi]$	$C^{\pm} = \mathrm{Cos}^2\theta[{}^1/_2(C_{11} - 3C_{18})\mathrm{Sin}^2 2\phi - C_{21}\mathrm{Sin}4\phi] + (C_{16}\mathrm{Sin}^2\theta + C_{18}\mathrm{Cos}^2\theta)$	$C^{-} = -\mathrm{Cos}^3\theta[{}^1/_4(C_{11} - 3C_{18})\mathrm{Sin}4\phi - C_{21}\mathrm{Cos}4\phi]$ $C^{+} = \mathrm{Cos}\theta[{}^1/_4(C_{11} - 3C_{18})\mathrm{Sin}4\phi - C_{21}\mathrm{Cos}4\phi]$
$4mm(C_{4v})$ $\bar{4}2m(D_{2d})$ $422(D_4)$ $4/mmm(D_{4h})$	$C^{+} = -{}^1/_4(C_{11} - 3C_{18})\mathrm{Cos}^3\theta\mathrm{Sin}4\phi$ $C^{-} = {}^1/_4(C_{11} - 3C_{18})\mathrm{Cos}\theta\mathrm{Sin}4\phi$	$C^{\pm} = {}^1/_2(C_{11} - 3C_{18})\mathrm{Cos}^2\theta\mathrm{Sin}^2 2\phi + (C_{16}\mathrm{Sin}^2\theta + C_{18}\mathrm{Cos}^2\theta)$	$C^{-} = -{}^1/_4(C_{11} - 3\ C_{18})\mathrm{Cos}^3\theta\mathrm{Sin}4\phi$ $C^{+} = {}^1/_4(C_{11} - 3C_{18})\mathrm{Cos}\theta\mathrm{Sin}4\phi$

$$n(\lambda) - 1 = \frac{Ne^2}{8\pi^2 mc^2\epsilon_o}\sum_{j,k}\frac{\rho(j)f_{jk}}{\left(\frac{1}{\lambda_{jk}^2} - \frac{1}{\lambda^2}\right)} = \frac{Nr_e}{2\pi}\sum_{j,k}\frac{\rho(j)f_{jk}}{\left(\frac{1}{\lambda_{jk}^2} - \frac{1}{\lambda^2}\right)} \tag{73}$$

where

N	=	number density of atoms
r_e	=	classical electron radius = $2.818 \times 10^{-5}\ m$
f_{jk}	=	oscillator strength of the transition of wavelength λ_{jk}
$\rho(j)$	=	fractional population of level j

The density, N, is related to the vapor pressure p (torr), by

$$N = 9.66084 \times \frac{10^{24}p(\text{torr})}{\lambda} \quad \text{atoms/m}^3 \tag{74}$$

where T = temperature in K. Data on vapor pressures of a wide range of materials are given in References 79 and 80.

For the alkalis, at pressures $\simeq$ 1 torr, p is approximated by:

$$p(\text{torr}) \simeq \exp(-A/\text{T} + D) \tag{75}$$

where A and D are constants given in Reference 81. Also included in Reference 80 are the data on f_{jk} for a number of wavelengths of interest in various alkali vapors. For the inert gases, data on refractive index have been given in Reference 82.

As the temperature and thus the density of alkali atoms increases, the temperature tolerance necessary to maintain phase matching decreases. A maximum variation of $\delta(\Delta K) = \pi/L$ yields a maximum allowed temperature variation of:[82]

$$\delta(\text{T}) = \frac{L_c}{L}\text{T}\left[\frac{A}{\text{T}} - 1\right]^{-1} \tag{76}$$

In some cases, such as shorter wavelengths of $\simeq$ 100 nm, the dispersion of the inert gas may itself become negative, e.g., when the third-harmonic or sum frequency lies above and close to a transition of high f_{jk}. It should be possible to use phase-matched mixtures of inert gases over parts of the region 146.9 to 50 nm.

MEASUREMENT OF NONLINEAR COEFFICIENTS

Second-Order Susceptibility

The techniques for measuring the SHG coefficients have been considered in detail in Reference 11. The most commonly employed methods are briefly described below.

Phase-Matched Method

When phase matching is possible in a crystal, the effective interaction length is increased from the normal range of a few microns to distances of several centimeters, as a result of which the second harmonic powers can reach detectable levels.

This method has been used for absolute measurements of SHG coefficients in a number of materials both in the visible [83-85] and infrared.[86,87] The mathematical relations required to calculate the magnitude of d_{eff} are given in the section "SHG Conversion Efficiency with Focused Beam — for Critical Phase-Matching".

Since this method is limited to phase-matchable coefficients only, it can not always be used to determine all of the nonvanishing elements of SHG tensor of a given crystal class.

Maker Fringe Method

This technique is based upon the fact that when a plane parallel plate of a nonlinear material is rotated about a principal axis perpendicular to the laser beam, the second-harmonic intensity goes through periodic maxima and minima (termed as Maker-fringe[59]). These oscillations result from the angular variation of the wavevector mismatch, ΔK, between the fundamental and the harmonic waves:

$$|\Delta K| = |2K_\omega - K_{2\omega}| = (4\pi/\lambda)|n_\omega \text{Cos } \theta'_\omega - n_{2\omega}\theta'_{2\omega}| \quad (77)$$

where θ'_ω and $\theta'_{2\omega}$ are the angles of refraction of the fundamental and second-harmonic waves, respectively. For nonnormal incidence of the fundamental beam on a sample of thickness L, the second-harmonic power, $\mathbf{P}_{2\omega}$, varies from Equations 22A and 22B as

$$\mathbf{P}_{2\omega} \propto \text{Sin}^2\psi/\psi^2 \quad (78)$$

$$\psi = (\pi/2)[L/L_{\text{coh}}(\theta)] \quad (78A)$$

$$L_{\text{coh}}(\theta) = \lambda/4(n_\omega \text{Cos } \theta_\omega - n_{2\omega}\text{Cos } \theta_{2\omega}) \quad (78B)$$

where θ is the angle of incidence of the laser beam.

From Equation 78A, the condition for the mth minimum is

$$\psi/(\pi/2) = L/L_{\text{coh}}(\theta_m) = 2m \quad (79)$$

By carefully fitting[60] of the minima of the fringes, θ_m, to Equation 79, it is possible to determine the coherence length at normal incidence (given in Equation 23) to within $\pm 1\%$.

Alternatively, the coherence length at normal incidence can be approximated for a uniaxial crystal as

$$L_{\text{coh}}(\theta) = L(\text{Sin}^2\theta_{m+1} - \text{Sin}^2\theta_m)/4n_\omega n_{2\omega} \quad (80)$$

where θ_{m+1} and θ_m are the angular separation of $(m + 1)$ and mth minima. From the coherence length thus determined, together with the value of the second harmonic power at normal incidence obtained from the locus of the Maker fringe peaks, the nonlinear coefficient can be calculated by using Equations 25A and 25B.

The approximation given by Equation 80 can also be used for biaxial crystals with appropriate modifications. For example, the nonlinear coefficient, d_{31}, of a crystal of symmetry $mm2$ can be determined by rotating the crystal about its principal z axis. The coherence length, L_c^{31}, can be determined from the Maker fringes by using the relation:

$$L_{\text{coh}}^{31} = \frac{L[(n_\omega^Y)^2 - n_\omega^X n_{2\omega}^Z]}{4(n_\omega^X - n_{2\omega}^Z)n_{2\omega}^Z(n_\omega^Y)^2}[\text{Sin}^2\theta_{m+1} - \text{Sin}^2\theta_m] \quad (80A)$$

Similarly, L_e^{32} is obtained by interchanging X and Y in Equation 80A.

In order to observe the Maker fringes due to the nonlinear coefficient, d_{14}, of a biaxial crystal of symmetry 222, a sample with its surface normal making a 45° angle with y and z axes is rotated about the x axis. In that case, the coherence length, L_{coh}^{14}, can be obtained from the Maker fringe minima using:

$$L_{coh}^{14} = \frac{[L(n_\omega^Z)^2 - n_\omega^Y n_{2\omega}^X][\mathrm{Sin}^2\theta_{m+1} - \mathrm{Sin}^2\theta_m]}{4(n_\omega^Z)^2 n_{2\omega}^X \left[n_{2\omega}^X - \dfrac{\sqrt{2}\, n_\omega^y n_\omega^z}{\{(n_\omega^y)^2 + (n_\omega^z)^2\}^{1/2}} \right]} \quad (80B)$$

Expressions similar to Equation 80B can be obtained for the coherence lengths L_{coh}^{25} and L_{coh}^{36}. In order to obtain L_{coh}^{14} for a uniaxial crystal, use $n_\omega^x = n_\omega^y = n_\omega^o$.

Although, in principle, a CW laser can be used, only pulsed and Q-switched lasers have been employed in using the Maker fringe method. A majority of SHG coefficients reported for materials transparent in the visible have been obtained by this technique.

Wedge Technique

The Maker fringe technique presents difficulties in IR-transmitting materials. The coherence lengths for most materials in the IR are large due to smaller dispersion. Consequently, the number of Maker fringes obtained by rotating a plane parallel slab of the material is insufficient (in many instances less than one) to determine L_{coh}. Because of the high Fresnel reflection coefficient for materials with high indexes of refraction, significant changes in the fundamental intensity distribution are caused by the standing-wave pattern inside the plane parallel slab. Furthermore, for samples which are many coherence lengths thick, the absorption losses can be significant.

The above-mentioned difficulties are overcome by using the wedge technique,[88] which is closely related to the Maker fringe method. In this technique, the plane parallel slab is replaced with a prism-shaped wedge which is translated perpendicular to the fundamental beam. The wedge has an apex angle of about 2 to 3° and the thickness at the base is about 5 to 10 times the coherence length.

A fundamental beam incident in the sample generates second harmonic, and is multiply reflected and transmitted. Because of the wedge in the sample, each successive transmitted beam travels in a different direction. A focusing optics together with a detector can be placed so as to collect only the initial transmitted beam and its second harmonic. Thus, the Fabri-Perot and multiple reflection interference effects, which can be troublesome in high refractive-index materials, are eliminated. When the wedge is translated, the light path length in the sample may be varied and L_{coh} determined from the fringes.

The measurements are made on thin samples placed in the near field (no diffraction) of a laser beam in the TEM_{oo} Gaussian mode. The SHG power, $\mathbf{P}_{2\omega}$, produced by the fundamental beam, $\mathbf{P}_\omega$, traversing the wedge of a lossless medium is[89]

$$\mathbf{P}_{2\omega} = A\, \mathbf{P}_\omega^2 \frac{(2L_{coh}/\pi)^2}{w_1^2} \cdot C(\eta\psi) \quad (81)$$

where the constant A is given by Equations 48 and 49, w_1 is the spot radius of the fundamental beam:

$$\eta = w_1 \tan\theta / L_{coh}, \text{ for wedge angle } \theta \quad (82)$$

Ψ = phase mismatch for the translating wedge and is given for normal incidence by

$$\psi = \frac{\pi}{2}\left(\frac{L}{L_{coh}(0)}\right) + \frac{\pi}{2}\left(\frac{x \tan\theta}{L_{coh}(0)}\right) = \frac{\pi}{2}\frac{L(x)}{L_{coh}(0)} \quad (83)$$

L = thickness of the wedge at the beam center ($x = 0$), x being measured along a direction perpendicular to the incident beam.

$L_{coh}^{(0)} = \frac{\Delta x}{2} \cdot \tan\theta$, Δx being the distance between SHG minima, and

$$C(\eta\psi) = [1 - \operatorname{Cos} 2\psi \exp(-\pi^2\eta^2/16)]/2 \tag{84}$$

If the fundamental beamwidth is negligible ($\eta < 0.1$), then for a lossy medium,

$$\mathbf{P}_{2\omega} = A\,\mathbf{P}_{\omega}^2(L^2/w_i^2)F \tag{85}$$

where

$$F = \frac{e^{-\alpha_2 L}(1 - e^{-\alpha L})^2 + 4e^{-\alpha L}\operatorname{Sin}^2\psi}{(\alpha L)^2 + 4\psi^2} \tag{86}$$

$$\alpha = \alpha_1 - \alpha_2/2 \tag{86A}$$

α_1 and α_2 being the absorption coefficients at the fundamental and second-harmonic wavelengths, respectively.

In Equations 81 and 85 the powers are those inside the medium.

In terms of SHG power, $P_{2\omega}'$, measured outside the crystal, the relative magnitude of an unknown d_{eff} to that of a standard d_{eff}^{std} is given by:

$$\frac{d_{eff}}{d_{eff}^{std}} = \left[\frac{\mathbf{P}_{2\omega}'}{\mathbf{P}_{2\omega}'^{nd}}\right]^{1/2} \cdot \frac{L_{coh}^{std}}{L_{coh}} \cdot \frac{(n_\omega + 1)^2(n_{2\omega} + 1)}{[(n_\omega + 1)^2(n_{2\omega} + 1)]^{std}} \cdot \frac{C^{std}(\eta)}{C(\eta)} \tag{87}$$

Parametric Fluorescence Method

This is a useful technique for accurate absolute measurements of d. It is based upon the observation of spontaneous parametric emission[90-92] resulting from a two-wave interaction satisfying the momentum matching condition $K_p = K_g + K_i$ at three frequencies which satisfy the relation $\hbar\omega_p = \hbar\omega_g + \hbar\omega_i$, where ω_p, ω_g and ω_i are the pump, signal, and idler frequencies, respectively.

The power, $\mathbf{P}_s$, scattered from the incident pump power, $\mathbf{P}_p$, into the signal beam integrated over angle and frequency is given by:[93]

$$\mathbf{P}_s = (\beta L \mathrm{P}_p/b)\Omega \tag{89}$$

where

Ω = collection solid angle of the signal detection system, and is given by $\pi\theta^2$, θ being the acceptance angle within the nonlinear crystal. For a stop of radius r placed at the focus of a lens of focal length, F:

$$\Omega = \pi r^2/n_g^2 F^2 \tag{90}$$

n_s being the refractive index of the nonlinear crystal at the signal frequency.

L = length of the crystal:

$$b = \omega_s\left(\frac{\partial n_s}{\partial \omega_s}\right) - \omega_i\left(\frac{\partial n_i}{\partial \omega_i}\right) = \frac{1}{c}\left[n_s - n_i - \lambda_z\left(\frac{dn_s}{d\lambda_s}\right) + \lambda_i\left(\frac{dn_i}{d\lambda_i}\right)\right] \tag{91}$$

where n_g and n_i are the refractive indexes at the signal and idler wavelengths of λ_g and λ_i, respectively, and

$$\beta = \begin{cases} 2\hbar\omega_z^4 d_{eff}^2 n_s/(2\pi)^2\epsilon_o c^4 n_i n_p, & m^{-1}[\text{MKS}] \quad (92) \\ 32\pi\hbar\omega_z^4\omega_i d_{eff}^2 n_s/c^4 n_i n_p, & \text{cm}^{-1}[\text{CGS}] \quad (93) \end{cases}$$

where n_p is the refractive index at the pump wavelength.

Raman Scattering Method

In noncentrosymmetric crystals with simultaneously Raman and IR-active modes of given symmetry type, measurement of the absolute scattering efficiencies for longitudinal and transverse modes S_L and S_T and the corresponding frequencies ω_L and ω_T is sufficient to determine the nonlinear coefficient d_{jm}. For crystals with zinc-blend structure (for which it is assumed that the plasma frequency is much less than the T_O frequency and the phonon wave vector is much greater than the T_O frequency divided by the velocity of light) the Raman efficiencies inside the medium are given by:

$$S_{L,T} = \sigma_{L,T}\rho^{-1}|d\alpha_{12}/dQ_3|_{L,T}^2 \quad (94)$$

with

$$\sigma_{L,T} = \frac{\hbar\omega_s^4(\bar{n}_{L,T} + 1)l d\Omega}{32\ \pi^2\epsilon_o^2 c^4\omega_{L,T}} \quad (95)$$

where ρ is the reduced mass density, ω_T and ω_L are the transverse and longitudinal mode frequencies, ω_s is the Stokes frequency, $\bar{n}$ is the Bose factor, l is the scattering length, and $d\Omega$ is the scattering solid angle.

The nonlinear coefficients in terms of measured quantities are given by:

$$D^{-1} = a[1 \pm (\sigma_T S_L/\sigma_L S_T)^{1/2}] \quad (96)$$

$$r_{jkl}^2 = S_T n_\infty^2(1 + D^{-1})^2/\epsilon_o n^8 a\omega_T^2\sigma_T \quad (97)$$

$$\xi_{jkl} = \epsilon_o n^4 r_{jkl}/(1 + D) = 4\epsilon_o d_{lkj}(2\omega;\omega,\omega) \quad (98)$$

where

$$a = \omega_T^2(\omega_L^2 - \omega_T^2)^{-1} \quad (99)$$

n_∞ is the high-frequency refractive index, n is the refractive index at the optical frequency, and r_{jkl} is the electrooptic coefficient $(d_{jkl}(-\omega_L;\omega_L,0)$.

For polar crystals possessing orthorhombic symmetry or higher, the method of oblique phonon scattering can be used, the details of which are found in Reference 95.

Third-Order Susceptibility

The techniques for measuring the third-order nonlinear coefficients C_{jn} are similar to those for d_{jm}. The most commonly employed methods are described below.

Phase-Matched Method

In the situation where the confocal parameter $b \gg L$, optimum harmonic efficiency is obtained if $\Delta K = 0$, in which case the harmonic power increases by $(\pi L/2L_{coh})$. Under these conditions the third-harmonic power $\mathbf{P}_{3\omega}^{THG}$ generated by a fundamental beam of power $\mathbf{P}_\omega$ inside a medium of length L is obtained from Equations 64A and 64B as:

$$\mathbf{P}_{3\omega}^{THG} = \frac{12\,\omega^2 C_{eff}^2 L^2 \mathbf{P}_\omega^3 \exp\left[-\frac{1}{2}(\alpha_{3\omega} + 3\alpha_\omega)L\right]}{\pi^2 \epsilon_o^2 c^4 w_0^4 n_\omega^3 n_{3\omega}} \quad [MKS] \qquad (100A)$$

$$\mathbf{P}_{3\omega}^{THG} = \frac{(16 \times 192)\pi^2 \omega^2 C_{eff}^2 L^2 \mathbf{P}_\omega^3 \exp\left[-\frac{1}{2}(\alpha_{3\omega} + 3\alpha_\omega)L\right]}{c^4 w_0^4 n_\omega^3 n_{3\omega}} \quad [CGS] \qquad (100B)$$

where C_{eff} is the effective nonlinear coefficient and the other quantities have been defined in Equations 64A and 64B.

On the other hand, for tight focusing ($b \ll L$) in the center of a negatively dispersive nonlinear medium ($\Delta K = K_{3\omega} - 3K_\omega < 0$), the third-harmonic power is given by Equations 64C and 64D.[96] The conversion efficiency is maximum for $b\Delta K = -4$. For a normally dispersive medium (positive dispersion $\Delta K > 0$) and also for a nominally phase-matched medium ($\Delta K = 0$) the harmonic power generated in an infinite medium is zero.

For confocal focusing ($b = L$) at the center of a negatively dispersive medium, the harmonic power is maximized when $L\Delta K = -3.5$, and is given by:

$$\mathbf{P}_{3\omega}^{THG} = \frac{(16)(48)(2 \cdot 46)\pi^2 L^2 C_{jn}^2 \mathbf{P}_\omega^3}{c^4 n_\omega^3 n_{3\omega} \cdot w_0^4} \quad [CGS] \qquad (101A)$$

$$\mathbf{P}_{3\omega}^{THG} = \frac{(3)(2 \cdot 46)\omega^2 L^2 C_{jn}^2 \mathbf{P}_\omega^3}{n_\omega^3 n_{3\omega} c^4 \epsilon_o^2 \pi^2 w_0^4} \quad [MKS] \qquad (101B)$$

Maker Fringe Method

This technique is based upon the measurement of the coherence length, L_{coh}, of the medium by observing the Maker oscillations of the third-harmonic power. These oscillations can be obtained in a plane parallel slab of the material by rotating it about a vertical axis or in a wedge of the sample by translating it perpendicular to the fundamental beam.

For a lightly focused fundamental Gaussian beam with a confocal parameter $b \gg L$, the third-harmonic power transmitted through one coherence length of the nonlinear medium is obtained from Equations 64A and 64B as:

$$\mathbf{P}_{3\omega}^{THG} = \frac{(48)\omega^2 C_{jn}^2 L_{coh}^2 P_\omega^3 \exp\left[-\frac{1}{2}(\alpha_{3\omega} + 3\alpha_\omega)L\right]}{\pi^4 w_0^4 \epsilon^2 n_\omega^3 n_{3\omega} \cdot c^4} \quad [MKS] \qquad (102A)$$

$$\mathbf{P}_{3\omega}^{THG} = \frac{(64 \times 192)\omega^2 C_{jn}^2 L_{coh}^2 P_\omega^3 \exp\left[-\frac{1}{2}(\alpha_{3\omega} + 3\alpha_\omega)L\right]}{c^4 w_0^4 n_\omega^3 n_{3\omega}} \quad [CGS] \qquad (102B)$$

where the various quantities have been defined in Equations 100A and 100B. By using the value of L_{coh} determined from the Maker-fringe minima and the third-harmonic power at normal incidence obtained from the locus of the Maker fringe peaks, together with the appropriate transmission factors for the fundamental and harmonic waves, the nonlinear coefficient can be estimated.

Third-harmonic generation can also be observed in reflection.[97] This method is particularly useful for strongly absorbing media such as metals.[98] For a fundamental wave incident at an angle θ_i, the coherence length is given by:

$$(1/2)(3\omega/c)[(\epsilon_{\omega}\text{Sin}^2\theta_1)^{1/2} - (\epsilon_{3\omega} - \text{Sin}^2\theta_1)^{1/2}]L_{coh} = \pi \tag{103}$$

where ϵ_{ω} and $\epsilon_{3\omega}$ are the complex dielectric constants at the fundamental and third-harmonic frequency, respectively.

When the fundamental and third-harmonic fields are both polarized parallel to the plane of incidence, the Fresnel factor is given by:

$$F_{11} = 1024\,\pi^2 \left[\frac{\epsilon_{\omega}}{\epsilon_{\omega} - \epsilon_{3\omega}}\right]^2$$

$$\frac{\text{Cos}^6\theta_1[\epsilon_{3\omega}(\epsilon_{\omega} - \text{Sin}^2\theta_1)^{1/2} - \epsilon_{\omega}(\epsilon_{3\omega} - \text{Sin}^2\theta_1)^{1/2}]^2}{[(\epsilon_{\omega} - \text{Sin}^2\theta_1)^{1/2} + \epsilon_{\omega}\text{Cos}\theta_1]^6[\epsilon_{3\omega}\text{Cos}\theta_1 - \epsilon_{\omega}(\epsilon_{3\omega} - \text{Sin}^2\theta_i)^{1/2}]^2} \tag{104}$$

The Fresnel factor for the fundamental and third-harmonic waves polarized normal to the plane of incidence is

$$F_{\P} = \frac{1024\,\pi^2\text{Cos}^6\theta_i}{[\text{Cos}\theta_1 + (\epsilon_{\omega} - \text{Sin}^2\theta_1)^{1/2}]^6[\text{Cos}\theta_1 + (\epsilon_{3\omega} - \text{Sin}^2\theta_1)^{1/2}]^2 \; [(\epsilon_{\omega} - \text{Sin}^2\theta_1)^{1/2} + (\epsilon_3 - \text{Sin}^2\theta_i)^{1/2}]^2} \tag{105}$$

Three-Wave Mixing Method

Optical difference mixing[28,52] of the form $\omega_3 + 2\omega_1 - \omega_2$ or $\omega_4 = 2\omega_2 - \omega_1$, where the frequencies ω_1 and ω_2 are obtained from a laser and the ω_3 or ω_4 radiation is created by the mixing, has been widely used for studying weak nonlinear susceptibilities. This technique permits the use of near phase-matched conditions so that detectable harmonic levels are achieved with nonlinear coefficients too small to be observed by THG or sum-frequency mixing where phase matching is, in general, not possible. Materials which are not transparent to third harmonic can still be studied using this technique.

In order to minimize the interference multiple reflection problems, the sample is slightly wedged and tilted away from normal incidence or antireflection coated. For a focused Gaussian beam of confocal parameter much greater ($b \geqslant 10L$) than the thickness of the nonlinear medium, the total power at the mixing frequency is obtained from Equations 69A and 69B. In order to estimate the power outside the medium, each of these equations must be multiplied by

$$R = \frac{(1 - R_1)^2\,(1 - R_2)}{1 + R_3} \tag{106}$$

where R_1, R_2, and R_3 are the reflection coefficients at ω_1, ω_2, and ω_3, respectively.

Change of Refractive Index Method

The third-order nonlinear optical coefficient $C_{jn}\,(-\omega;\omega,-\omega,\omega)$ can be determined from the coefficient of nonlinear refractive index, n_2 (Equations 70F to 70P). A particularly attractive technique of accurate and direct measurement of n_2 is to determine the nonlinear change of optical phase by interferometric comparisons with an unperturbed beam. This technique has been used to determine $C_{jn}(-\omega;\omega,-\omega,\omega)$ of a large number of cubic solids and isotropic liquids.

Some other processes involving self-induced polarization changes (SIPC) such as the intensity-dependent rotation of the vibrational ellipse[30] and intensity-induced birefringence[47] have proven useful in determining $C_{jn}(-\omega;\omega,-\omega,\omega)$.

FIELD-INDUCED SECOND HARMONIC GENERATION METHOD

When the frequency, ω_2, in the process of three-wave mixing (see Section "Difference Frequency Generation") is allowed to approach zero, the mixed frequency $2\omega_1 - \omega_2$ approaches 2 ω_1, the second harmonic. This limit is responsible for the phenomenon of (electrical) field-induced second-harmonic generation (FISHG)[26,28] which has been observed in solids, liquids, and gases. The symmetry properties of the nonlinear coefficients $C_{jn}(-2\omega;\omega,\omega,\rho)$ are readily derived by symmetrizing the coefficients of Table 1.1.8 between their middle two indexes. The induced polarization density may be written as:

$$\mathbf{P}_j(2\omega) = \epsilon_o \sum_{klm} 3C_{jklm}(-2\omega;0,\omega,\omega)E_k(0)E_l(\omega)E_m(\omega) \qquad (106A)$$

The nonlinear coefficient $C_{jn}(-2\omega;\omega,\omega,0)$ can be determined by measuring the second-harmonic radiation when a DC electric field is applied to the medium. The third-order process of FISHG, of course, occurs both in noncentrosymmetric as well as centrosymmetric media.

Other methods by which the third-order nonlinear coefficients have been determined include the observation of phenomena such as stimulated Raman scattering (SRS), DC (or low frequency) Kerr effect, AC Kerr effect (including the Raman-induced Kerr effect), electric-field induced absorption (FIA), and resonant DC Kerr effect. These phenomena have been discussed in Reference 403.

MEASUREMENT OF THE SIGN OF NONLINEAR OPTICAL COEFFICIENTS

The SHG coefficients in two piezoelectric antipodes, i.e., two crystals which are inversion images of each other, will have opposite signs. The absolute sign of a second-harmonic coefficient will depend upon the definition of the + or − direction of the polar axis. We use the standard piezoelectric convention[57] according to which an electrode at the plus end of the polar axis becomes negative upon compression (negative stress) along that axis.

In the case of THG, the fourth-rank susceptibility lensor does not change sign upon inversion so there is no need to differentiate between the + and − direction. Two methods have been extensively used for determining the signs of nonlinear optical coefficients:

1. The relative signs of nonlinearities in two samples may be determined by the interference method.[99,100] This method is based upon the fact that when harmonic or mixed frequency is generated by the same fundamental beam (or beams) in two consecutive plane parallel slabs (or wedges) A and B of nonlinear material, a definite phase relationship exists between the electric field at the harmonic frequency created by the two samples. If E_A and E_B are the field amplitudes generated in A and B, respectively, A and B in series generate total amplitude $E_A \pm E_B$, the sign being + for χ of the same sign and − for χ of opposite sign. The interference method has also been used for simultaneous determination of the relative magnitude and sign of nonlinear optical coefficients.[102]
2. The other method involves the observation of Maker fringes due to the combined effect of two nonlinear coefficients in a nonlinear material.[103] The fundamental beam initially polarized at 45° to the vertical (∥ z-crystal axis), passes through a polarizer whose electric-field transmission direction makes an angle, β, with respect to the vertical. Following the sample is an analyzer for the harmonic whose transmission direction with respect to the vertical is given by the angle α. A properly oriented platelet or wedge of the sample is mounted between the polarizers and the SHG powers $\mathbf{P}_{2\omega}(\alpha,\beta)$ due to individual coefficients separately as well as simultaneously are ob-

served. The relative signs of the nonlinear coefficients are readily concluded from the amplitudes of the maxima of the second-harmonic powers thus observed.

A list of experimentally determined signs of second-harmonic coefficients for a number of materials is given in Table 1.1.11.

THEORETICAL ESTIMATE OF THE NONLINEAR OPTICAL COEFFICIENTS

Optical nonlinearities have an electronic origin. The creation of the rth-order harmonic component can be visualized as a step-wise process in which a strong electric field interacts with ground state (valence) electrons and produces an induced dipole moment at the incident frequency, ω. This induced dipole moment interacts with the field to create a 2ω variation of excited and mixed-state populations. These populations in turn interact with the field to produce a third-harmonic component of the dipole moment. Multiple interactions of this type occur simultaneously without involving any transitions and resulting in the rth harmonic of the dipole moment.

Formal quantum mechanical expressions for the nonlinear susceptibilities for localized systems have been derived by Armstrong et al.[1] and by Butcher[4] using standard time-dependent perturbation theory. Following Reference 4, the theoretic expression for the rth order nonlinear susceptibility is given by:

$$\chi^{(r)}_{j_0 j_1 \cdots j_r}(-\omega_\sigma;\omega_1,\omega_2,\ldots\omega_r) = \frac{S_r(-1)^r N}{r\,\hbar\,\epsilon_o} \times \sum_{\nu_0\nu_1\ldots\nu_r} \rho^0_{\nu_0\nu_0}$$

$$\frac{Q^{j_0}_{\nu_0\nu_1}\cdot Q^{j_1}_{\nu_1\nu_2}\cdots Q_{\nu_r\nu_0}}{(\Omega_{\nu_1\nu_0}-\omega_1-\omega_2\ldots\omega_r)(\Omega_{\nu_2\nu_0}-\omega_2-\ldots\omega_r)\ldots(\Omega_{\nu_r\nu_0}-\omega_r)} \quad \text{[MKS]} \qquad (107)$$

where

O^j_{lm}	is the (lm)th matrix element of the electric-dipole-moment operator O_j between stationary states $\|l>$, $\|m>$ of the unperturbed atom.
Ω_{lm}	$= (E_l - E_m)/\hbar$, the transition frequencies between energy levels E_l, E_m.
$\hbar$	$= h/2\pi$, h being Planck's constant.
$\rho^0_{v0'0}$	is the probability of occupancy of the ground level v_0.
N	is the number density of atoms.
S_T	is the total symmetrization operation in the γth order, it implies that the expression which follows it is to be summed over all (γ + 1) permutations of the pairs $j_0\omega_\sigma$, $j_1\omega_1 \ldots j_r\omega_r$.

and the summation $\sum_{v_0 \cdot v_r}$ are over all of the atomic states, which include those for which $\Omega_{\nu_m\nu_0} = 0$, and also include integrations over the continuum state.

Thus, for the special case of SHG, the nonlinear susceptibility is given by:

$$\chi_{jkl}(-2\omega;\omega,\omega) = \frac{N}{\hbar^2\epsilon_o}\sum_{abc}\rho^0_{aa}\left[\frac{Q^{(j)}_{ab}Q^{(k)}_{bc}Q^{(l)}_{ca}}{(\Omega_{ab}-2\omega)(\Omega_{dc}-\omega)}\right.$$

$$\left.+\frac{Q^{(k)}_{ab}Q^{(j)}_{bc}Q^{(l)}_{ca}}{(\Omega_{ab}+\omega)(\Omega_{ac}-\omega)}+\frac{Q^{(k)}_{ab}Q^{(l)}_{bc}Q^{(j)}_{ca}}{(\Omega_{ab}+\omega)(\Omega_{ac}+2\omega)}\right] \qquad (107A)$$

Table 1.1.11
SIGNS OF NONLINEAR OPTICAL COEFFICIENTS

Material	Coefficients and their absolute signs						Ref.
1 Barium sodium niobate $Ba_2NaNb_5O_{15}$	$d_{31}(-)$	$d_{32}(-)$	$d_{33}(-)$	$d_{15}(-)$	$d_{24}(-)$		188
2 Barium titanate $BaTiO_3$	$d_{31}(-)$		$d_{33}(-)$	$d_{15}(-)$			227
3 Cadmium germanium arsenide $CdGeAs_2$					$C_{16}(-)$	$C_{18}(-)$	122b
4 Cadmium selenide CdSe	$d_{31}(-)$		$d_{33}(+)$				188
5 Cadmium sulfide CdS	$d_{31}(-)$		$d_{33}(+)$				227
6 Cadmium telluride CdTe				$d_{14}(+)$			188
7 Cuprous bromide CuBr				$d_{14}(-)$			351
8 Cuprous chloride				$d_{14}(-)$			340
9 Cuprous iodide CuI	$d_{14}(-)$						
10 Gadolinium molybdate $Gd_2(MoO_4)_3$	$d_{31}(-)$	$d_{32}(+)$	$d_{33}(-)$	$d_{15}(-)$	$d_{24}(+)$		188
11 Gallium antimonide GaSb	$d_{14}(+)$						88
12 Gallium arsenide GaAs	$d_{14}(+)$	$C_{18}(+)$					88, 101
13 Gallium phosphide GaP	$d_{14}(+)$						357
14 Germanium Ge						$C_{11}(+)$	101
15 Guanidine aluminum sulfate hexahydrate (GASH) $(CN_3H_6)As(SO_4)_2{\cdot}6H_2O$	$d_{22}(-)$	$d_{31}(+)$	$d_{33}(+)$				302
16 Indium arsenide InAs	$d_{14}(+)$						88
17 Lead germanate $PbGe_3O_{11}$	$d_{11}(-)$	$d_{22}(-)$	$d_{31}(+)$	$d_{33}(-)$			275
18 Lead niobate $PbNb_4O_{11}$	$d_{31}(+)$	$d_{32}(-)$	$d_{33}(-)$	$d_{15}(+)$	$d_{24}(-)$		190
19 Lithium formate monohydrate $LiCOOH{\cdot}H_2O$	$d_{31}(+)$	$d_{32}(-)$	$d_{33}(+)$				188
20 Lithium gallate $LiGaO_2$	$d_{31}(-)$	$d_{32}(+)$	$d_{33}(-)$				188, 227
21 Lithium iodate $LiIO_3$	$d_{31}(-)$		$d_{33}(-)$				188
22 Lithium niobate $LiNbO_3$	$d_{31}(-)$		$d_{33}(-)$	$d_{22}(+)$			227, 306
23 Lithium sodium formate monohydrate $Li_{0.9}Na_{0.1}COOH{\cdot}H_2O$		$d_{32}(-)$					197
24 Lithium tantalate $LiTaO_3$	$d_{31}(-)$		$d_{33}(-)$	$d_{22}(+)$			188, 227
25 Methyl-(2,4-dinitro phenyl)-amino propanoate (MAP) $C_{10}H_{11}N_3O_6$				$d_{25}(-)$			166
26 Potassium dihydrogen phosphate KH_2PO_4			$d_{36}(+)$				154
27 Potassium niobate $KNbO_3$	$d_{31}(-)$	$d_{32}(+)$	$d_{33}(-)$	$d_{15}(-)$	$d_{24}(+)$		193

Table 1.1.11 (continued)
SIGNS OF NONLINEAR OPTICAL COEFFICIENTS

Material	Coefficients and their absolute signs					Ref.
28 Quartz α-SiO_2	$d_{11}(+)$					188
29 Silicon Si					$C_{18}(+)$	101, 133
30 Silicon carbide SiC	$d_{31}(+)$		$d_{33}(-)$	$d_{15}(+)$		337
31 Silver gallium sulfide $AgGaS_2$				$d_{36}(+)$		89
32 Silver iodide AgI	$d_{31}(+)$		$d_{33}(-)$			338
33 Sodium formate NaCOOH	$d_{31}(-)$					197
34 Terbium molybdate $Tb_2(MoO_4)_3$	$d_{31}(-)$	$d_{32}(+)$	$d_{33}(-)$	$d_{15}(-)$	$d_{24}(+)$	188
35 Zinc oxide ZnO	$d_{31}(+)$		$d_{33}(-)$	$d_{15}(+)$		227
36 Zinc selenide ZnSe	$d_{14}(+)$			$d_{36}(+)$		188
37 Zinc sulfide (Z) β-ZnS	$d_{14}(+)$			$d_{36}(+)$		188, 340
38 Zinc sulfide (h) α-ZnS	$d_{31}(-)$		$d_{33}(+)$			340
39 Zinc telluride ZnTe	$d_{14}(+)$					188

Similarly, the expression for the THG nonlinear susceptibility is

$$\chi_{jklm}(-3\omega;\omega,\omega,\omega) = \frac{N}{\hbar^3\epsilon_o}\sum_{abcd}\rho^0_{aa}\left[\frac{Q^{(j)}_{ab}Q^{(k)}_{bc}Q^{(l)}_{cd}q^{(m)}_{da}}{(\Omega_{ba}-3\omega)(\Omega_{ca}-2\omega)(\Omega_{da}-\omega)}\right.$$
$$+\frac{Q^{(k)}_{ba}Q^{(j)}_{bc}Q^{(l)}_{cd}Q^{(m)}_{da}}{(\Omega_{ba}+\omega)(\Omega_{ca}-2\omega)(\Omega_{da}-\omega)}+\frac{Q^{(k)}_{ab}Q^{(l)}_{bc}Q^{(j)}_{cd}Q^{(m)}_{da}}{(\Omega_{ba}+\omega)(\Omega_{ca}+2\omega)(\Omega_{da}-\omega)}$$
$$\left.+\frac{Q^{(k)}_{ab}Q^{(l)}_{bc}Q^{(m)}_{cd}Q^{(j)}_{da}}{(\Omega_{ba}+\omega)(\Omega_{ca}+2\omega)(\Omega_{da}+3\omega)}\right] \tag{107B}$$

When the fundamental frequency is not in close resonance with particular levels, then it is necessary to carry out a summation over all intermediate states, and both the magnitudes and signs of the matrix elements must be known. While in principle it is possible to calculate the magnitude of the nonlinear optical coefficients from Equation 107, the complicated appearance of this expression prevents any attempt to extract accurate numerical values from them. This approach necessitates the use of extremely accurate wavefunctions — a condition which is rarely met.

The results of the quantal treatment have been used in calculating the third-order susceptibility of free atoms and molecules of alkali vapors[81,104] and rare-gases,[96,105] for which the relevant matrix elements are obtained from measured absorption f-numbers and/or approximate radial wavefunctions.

In the case of solids, the application of the quantum mechanical approach has been limited to simple low-ionicity crystals of group IV and III-V semiconductors in the frequency range above the lattice resonances but below the electronic transitions.[107,108] In Reference 107, use has been made of tetrahedral sp^3 bonding orbitals for the ground-state wavefunction, with p orbitals pointing in the crystallographic[111] directions. The authors of Reference 108 considered the crystal as a huge molecule with each bond effectively localized, and for the ground-state wavefunctions they used the approximate linear combination of atomic orbitals — molecular orbitals (LCAO-MO).

An empirical, but more practical approach to estimating the second-order coefficients was proposed by Miller,[109] who suggested that if electric polarizations instead of electric fields are used as the independent variables to describe SHG, then the SHG-coefficient $d_{jkl}^{(2\omega)}$ can be expressed in terms of the linear optical susceptibility $\chi_{jk}^{(\omega)}$ (defined in Equation 3)and a new third-rank tensor $\delta_{jkl}^{(2\omega)}$ (called Miller-δ) by the relation:

$$d_{jkl}^{(2\omega)} = \epsilon_o \sum_{r,s,t} \chi_{jr}^{(2\omega)} \chi_{kj}^{(\omega)} \chi_{lt}^{(\omega)} \cdot \delta_{jkl}^{(2\omega)}, \qquad m \cdot V^{-1} \quad [\text{MKS}] \tag{108}$$

For crystals with symmetry higher than triclinic or monoclinic, Equation 108 can be rewritten in the principal axis coordinate system of the $\chi_j^{(2\omega)}$-tensor as:

$$d_{jkl}^{2\omega} = \epsilon_o \chi_{jj}^{(2\omega)} \chi_{kk}^{(\omega)} \chi_{ll}^{(\omega)} \cdot \delta_{jkl}^{(2\omega)}, \qquad m \cdot V^{-1} \tag{108A}$$

Data on SHG coefficients show that the variation in the elements of $\delta_{jkl}^{(2\omega)}$ from one material to another generally ranges over two orders of magnitude smaller interval than that exhibited by the corresponding d'_{jkls}. Thus, if in the empirical relation, Equation 108A, $\delta_{jkl}^{(2\omega)}$ is taken to be a constant ($\simeq$ 0.07 m^2/C) for all materials, then an order of magnitude estimate of $d_{jkl}^{(2\omega)}$ for any material in its transparency region can be obtained from its refractive index data. In terms of the indexes of refraction, the Miller-δ is

$$\delta_{jkl}^{(2\omega)} = d_{jkl}^{(2\omega)} / \epsilon_o (n_{2\omega}^2 - 1)(n_{\omega}^2 - 1), \qquad m^2 \cdot C^{-1} [\text{MKS}] \tag{108B}$$

It should be pointed out that Miller's rule cannot be used to predict the dispersion of the nonlinear susceptibilities, especially when one of the frequencies is near a lattice resonance or electronic absorption. Therefore, care must be exercised in extrapolating experimental values of $d_{jkl}^{(2\omega)}$ or $\delta_{jkl}^{(2\omega)}$ from one frequency region to another.

A generalization of Miller's rule to third-order nonlinear susceptibilities would yield:

$$C_{jkl}^{(2\omega)} = \epsilon_o \chi_{jj}^{(3\omega)} \cdot \chi_{kk}^{(\omega)} \chi_{ll}^{(\omega)} \cdot \delta_{jklm}^{(3\omega)} \tag{109}$$

The near constancy of Miller-δ for second-order magnitude nonlinear susceptibilities has been explained in terms of an anharmonic oscillator model using both classical[110,111] and quantum mechanical[5,12] treatments. For a nonlinear material containing N optical electrons per unit volume moving in an anharmonic potential V_{jkl} (units of $C^2 \cdot m^{-4}$), the Miller-δ is given by:

$$\delta_{jkl}^{(2\omega)} = \frac{-3V_{\text{jkl}}}{N^2 e^3} \qquad (m^2 \cdot C^{-1}) \tag{110}$$

which is independent of the frequencies of the applied fields.

A number of models[113-123] based on the dielectric theory of Phillips[124] and Van Vechtan[125] have been successfully used in describing the nonlinear susceptibility of various solids. Semiempirical calculations of second-order nonlinear susceptibilities have been made for ferroelectric materials containing the BO_6-type octahedra.[126,127]

In Tables 1.1.12 and 1.1.13a are listed the theoretically estimated values of second- and third-order nonlinear coefficients, respectively, for a number of materials by various models.

Table 1.1.12
CALCULATED VALUES OF SECOND-ORDER NONLINEAR OPTICAL COEFFICIENT, d_{jm}, BY VARIOUS MODELS

All Values = $10^{12} \times d_{jm}$ in $m \cdot V^{-1}$

Material Ref.	d_{jm}	107	108	113	114	115	116	117	118	119	120	121	122a	123	127
AgI					20										
AlAs	d_{14}				297			135							
AlN	d_{31}				−11										
	d_{33}				28		−17								
AlSb	d_{14}		147		232			177							
AlP	d_{14}							104							
$Ba_{0.27}Sr_{0.75}Nb_2O_{5.78}$	d_{31}														−2.4
	d_{33}														−6.0
$Ba_2NaNb_5O_{15}$	d_{31}														−7.6
	d_{32}														−8.8
	d_{33}														−17
BAs	d_{14}						366								
BeO	d_{31}				−1.9										
	d_{33}				4.6		−0.5								
BN	d_{14}				21			11.4							
BP	d_{14}		0	—		15		12							
$CdGeAs_2$	d_{14}										410.6				
$CdGeP_2$	d_{14}										268				
$CdSiAs_2$	d_{14}										318				
$CdSiP_2$	d_{14}										222				
$CdSnAs_2$	d_{14}										222				
$CdSnAs_2$	d_{14}										587				
											251				
$CdSnP_2$	d_{14}														
CdS	d_{31}				−10	25									
	d_{33}				25	52							78		
CdSe	d_{31}				−12	54									
	d_{33}				30	109							112		
CdTe	d_{14}			195	57	147		125	149	276.5		281	176	69.1	
CuBr	d_{14}				−5.4										
	d_{33}				13										

CuCl	d_{14}					7.7	13		23					23	
CuI	d_{14}				29			64.5					92		
GaAs	d_{14}	−181	398	398	256	239		169	209.5	178		226	128	79.6	
GaN	d_{31}				−17										
	d_{33}				43		−26								
GaP	d_{14}		293	147	178	142		95	130	117		157	86	50.2	
GaSb	d_{14}	−3266	335	754	492	503		241	754	325		404	226	146.6	
InAs	d_{14}	−635	859	670	329	402		201	503	235		358	209.5	134	
InN	d_{33}						−49								
InP			587	356	222			142				136			
InSb	d_{14}	−1088	1361	1403	591	796		337	838	415		599	383	230.4	
$K_3Li_2Nb_5O_{15}$	d_{31}														−6.4
	d_{33}														−9.9
$LiNbO_3$	d_{31}														−6.57
	d_{33}														−36.9
	d_{22}														+3.4
$LiTaO_3$	d_{31}														−1.47
	d_{33}														−17
											251				
	d_{22}														+2.3
SiC(z)	d_{14}				70		−72	30							
SiC(w)	d_{31}														
	d_{33}							−83						19.5	
SiO_2	d_{11}					0.4	+0.5								
	d_{14}					0									
$ZnGeAs_2$	d_{14}										180				
$ZnGeP_2$	d_{14}										201				
ZnO	d_{31}				−4.8	5									
	d_{33}				12	10	−12						18.4		
ZnS(z)	d_{14}			52	34	36		39	35.6	73		90		20.9	
ZnSe	d_{14}			90	39	59		57	63	111		132	71	31.4	
$ZnSiAs_2$	d_{14}										83.8				
$ZnSiP_2$	d_{14}										142				
$ZnSnAs_2$	d_{14}										323				
$ZnSnP_2$	d_{14}										264				
ZnTe	d_{14}			132	111	126	+126	106	153	188		235	134	60.75	
ZnS(w)	d_{31}			−14	20										
	d_{33}				34	40							55		

Table 1.1.13a
CALCULATED VALUES OF THIRD-ORDER NONLINEAR OPTICAL COEFFICIENT, C_{jn}, FOR SOME SOLIDS BY VARIOUS MODELS

All Values = $C_{jn} \times 10^{20}\ m^2 \cdot V^{-2}$

Material	Coeff.	λ (μm)	Ref. 107	Ref. 115	Ref. 122b	Ref. 128	Ref. 129	Ref. 130	Ref. 130
AlSb	C_{11}						3.15		
CaF_2	C_{11}							0.0014	0.0014
$CdGeAs_2$	C_{11}				6.3				
	C_{16}				4.9				
	C_{18}				4.55				
CdS	C_{11}			0.875				1.05	0.42
GaAs	C_{11}				3.32				
	C_{18}				2.45				
	$1/3\ (C_{11} + C_{18})$	0.4	−17.5						
	C_{11}					2.73	8.4	3.85	10.5
GaP	C_{11}						3.5		
	C_{11}						25.9		
	$(1/3)(C_{11} + C_{18})$	0.4	−2800						
Ge	C_{11}			56	14		42.	14.3	73.5
	$1/3\ (C_{11} + C_{18})$	1.0	−122.5						
InAs	C_{11}						21.7		
	$1/3\ (C_{11} + C_{18})$	0.2	−210						
	C_{18}					129.5			
InSb	C_{11}						21.7		
	C_{18}					1505	5.25		
	$(1/3)(C_{11} + C_{18})$	0.4	−1750						
InP	C_{11}						10.5		
KBr	C_{11}							0.013	0.0046
KCl	C_{11}							0.0067	0.0032
KI	C_{11}							0.031	0.011
LiF	C_{11}			0.0008				0.0011	0.00098
MgO	C_{11}							0.014	0.02

NaCl	C_{11}				0.007	0.0028
	C_{18}		0.0098			
NaF	C_{11}				0.00105	0.00046
PbS	C_{11}		980			
Si	C_{11}	3.15	1.78	6.12	3.36	21.0
SiO_2	C_{11}				0.0039	0.0049

Table 1.1.13b

CALCULATED VALUES OF THIRD-ORDER NONLINEAR OPTICAL COEFFICIENT, C_{11}, OF SOME GASEOUS ATOMS AND MOLECULES

All Values = $C_{11} \times 10^{50}\ m^5 \cdot V^{-2}$

Material	λ (μm)	Ref. 81	Ref. 96	Ref. 104	Ref. 105	Ref. 106	Ref. 130	Ref. 146
Ar					0.043		0.0098	
CO	9.33							6.6
CO_2							0.021	
Cs	1.064	−4095						
	0.6943	1138						
H_2							0.007	
He	0.6943		0.0042		0.003	0.0014	0.0004	
K	1.064	91.4						
	0.6943	21.6					0.023	
Kr	0.6943				0.1106			
Li	1.064	43.8						
	0.6943	105						
Na	1.064	−138						
	0.6943	111						
N_2							0.01	
Ne					0.0027		0.0007	
Rb	1.064	2152		3752				
	0.6943	288						
SF_6	10.6							1.1×10^4
Xe					0.345			

In Table 1.1.13(b), the nonlinear coefficient C_{jn}^{mic} represents C_{jn}/N, the nonlinear coefficient per atom or molecule, where N is the number density of atoms or molecules in the medium.

UNITS OF NONLINEAR OPTICAL COEFFICIENTS

In the literature, both MKSQ/SI and CGS/esu system of units are widely used. These are summarized below.

Quantity	MKSQ/Si-units	CGS/esu-units
Polarization P	Cm^{-2}	stat $C \cdot cm^{-2}$
Linear susceptibility $\chi^{(1)}$ (or index of refraction)	Dimensionless	Dimensionless
Second-order nonlinear optical susceptibility $\chi^{(2)}$ (or second-order nonlinear coefficient $d_{jm}^{(2)}$	$m \cdot V^{-1}$ $\equiv m^{3/2} \cdot J^{-1/4}$	$cm \cdot statV^{-1}$ $\equiv cm^{3/2} \cdot erg^{-1/4}$
Second-order Miller — $\delta^{(2)}$	$m^2 \cdot C^{-1}$ $\equiv m^{5/2} \cdot J^{1/4} \cdot C^{-2}$	$cm^2 \cdot Stat\ C^{-1}$ $\equiv cm^{5/2} \cdot erg^{1/4} \cdot Stat\ C^2$
Third-order nonlinear optical susceptibility $\chi^{(3)}$ (or third-order nonlinear coefficient $C_{jn}^{(3)}$)	$m^2 \cdot V^{-2}$ $\equiv m^3 \cdot J^{-1}$	$cm^2 \cdot statV^{-2}$ $\equiv cm^3 \cdot erg^{-1}$
Third-order microscopic nonlinear optical susceptibility $\chi^{(3)\ mic}$ (or third-order microscopic nonlinear coefficient $C_{jn}^{(3)\ mic}$)	$m^5 \cdot V^{-2}$ $\equiv m^6 \cdot J^{-1}$	$cm^5 \cdot statV^{-2}$ $\equiv cm^6 \cdot erg^{-1}$
nth-order nonlinear optical susceptibility $\chi^{(n)}$ (or nth order nonlinear coefficient)	$(m/V)^{n-1}$ $\equiv (m^{3/2} \cdot J^{-1/4})^{n-1}$	$(cm/statV)^{n-1}$ $\equiv (cm^3/erg)^{\frac{n-1}{2}}$
nth order microscopic nonlinear optical susceptibility $\chi^{(n)mic}$ (or nth order microscopic nonlinear coefficient)	$m^3 \cdot (m/V)^{n-1}$	$cm^3 \cdot (cm^3/erg)^{\frac{n-1}{2}}$

Some workers[146,147,151] have used an alternative MKS definition of the nonlinear optical susceptibility (or coefficient) by omitting ϵ_0 from Equation 7. In this case, $\chi^{(n)} = \epsilon_\theta \chi^{(n)}$ [MKSQ/SI] and $\chi^{(3)}$, e.g., has the units $A \cdot s \cdot m \cdot V^{-3}$ or $J \cdot m \cdot V^{-4}$.

In order to convert the nonlinear optical coefficents from MKSQ/SI to CGS/esu system of units, the following relations are used.

$$d_{jm}^{(2)}[\text{MKSQ/SI}] = [4\pi/(3 \times 10^4)]d_{jm}^{(2)} \quad [\text{CGS/esu}] \tag{111}$$

$$\delta_{jm}^{(2)}[\text{MKSQ/SI}] = [(3/4\pi) \times 10^5]\delta_{jm}^{(2)} \quad [\text{CGS/esu}] \tag{112}$$

$$C_{jn}^{(3)}[\text{MKSQ/SI}] = [4\pi/(3 \times 10^4)^2]C_{jn}^{(3)} \quad [\text{CGS/esu}] \tag{113}$$

$$C_{jn}^{(3)mic}[\text{MKSQ/SI}] = [4\pi/10^6(3 \times 10^4)^2] \cdot C_{jn}^{(3)mic} \quad [\text{CGS/esu}] \tag{114}$$

For an nth order nonlinear process, the susceptibility in MKSQ/SI units is related to that in CGS/esu units by

$$\chi^{(n)}[\text{MKSQ/SI}] = [4\pi/(3 \times 10^4)^{n-1}] \cdot \chi^{(n)} \quad [\text{CGS/esu}] \tag{115}$$

$$\chi^{(n)mic}[\text{MKSQ/SI}] = [4\pi/10^6(3 \times 10^4)^{n-1}]\chi^{(n)mic} \quad [\text{CGSj/esu}] \tag{116}$$

ABSOLUTE STANDARDS OF NONLINEAR OPTICAL COEFFICIENTS

A knowledge of the absolute magnitude of the nonlinear optical susceptibilities is important for an understanding of the origin of the nonlinearities and also for designing various nonlinear optical devices. Most of the measurements of nonlinear optical coefficients available in the literature have been made against some "known" standard materials such as α-SiO_2 $NH_4H_2PO_4$(ADP), KH_2PO_4(KDP), LiF, and He, which are transparent in the visible and UV, and GaAs and Ge for the IR-transmitting materials.

In order to convert from relative to absolute values we have used the values of second- and third-order nonlinear standard coefficients listed in Table 1.1.14. Also included in this table are the SHG standard coefficients recommended by Kurtz et al.[154] and Levine and Bethea.[155] While our standard values of SHG coefficients of α-SiO_2, $NH_4H_2PO_4$ and KH_2PO_4 are essentially in agreement with those of Reference 155, they are substantially lower (by 30% or more) than those of Reference 154.

The above difference in the two scales of absolute values may be attributable to the fact that while the scale adopted by us is based upon the absolute measurements of the SHG-coefficient $d_{36}^{(2\omega)}$ of ADP by Francois[84] the scale used in Reference 154 is derived from the absolute measurements of $d_{31}^{(2\omega)}$ of $LiIO_3$. The ADP crystals have the advantage that they are readily available in large size of reproducible optical and dielectric quality. They are highly transparent at the fundamental and second-harmonic wavelengths of 0.6328 and 0.3164 μm, respectively, used in Reference 84. The absolute values of the SHG coefficients of α-SiO_2, KDP, and α-$LiIO_3$ were in turn derived from the careful relative measurements of Jerphagnon and Kurtz[156] and Jerphagnon.[157] On the other hand, single crystals of $LiIO_3$ of reproducible optical quality are not easy to grow.[158,159] The pump wavelengths of 0.4880 and 0.5145 μm used by Choy and Byer[160] to measure the absolute value of $d_{31}^{2\omega}$ ($LiIO_3$) are in close proximity of the band edge which could result in some resonant contribution to the nonlinearity. By making relative measurements of d_{36} of KDP with respect to d_{31} of $LiIO_3$ at the fundamental wavelength of 1.318 μm, Choy and Byer arrived at an absolute value for d_{36}-KDP, which is higher than that used by us. Here we would like to point out that KDP crystals[161] are strongly absorbing at 1.318 μm, which makes such measurements suspicious. For this reason we have preferred the Levine-Bethea scale.

TABLES OF NONLINEAR OPTICAL COEFFICIENTS

The experimentally observed values of the third-order nonlinear coefficients, C_{jn} of a number of solids and liquids are listed in Table 1.1.15a. The values of C_{jn}^{min}, which represents the third-order nonlinear coefficient per atom or molecule for gaseous species, are given in Table 1.1.15b. Also included in these tables are the fundamental wavelength λ_1, and the relevant nonlinear optical process used.

Table 1.1.16 gives the experimental values of SHG coefficients d_{jm} and δ_{jm} for a variety of crystals. Also included in the table are the fundamental wavelength λ_1, the pertinent refractive index n_ω and $n_{2\omega}$, and the reference SHG coefficient (if any) relative to which the measurements were made. If the measurement was absolute rather than relative, it was denoted by "Abs".

Data have been compiled from the literature through 1983. In addition, many unpublished results of the author and many colleagues have been included.

With regard to the convention for the absolute signs[188] of SHG coefficients for the III-V and II-VI crystals the outward normal from the metal A face (element II or III face) is defined as +[111] and +z for the zinc-blende and wurzite forms, respectively. For the general class of polar crystals, the +z direction is taken as the outward normal from the z-cut face that develops a negative charge upon compression along z. For the ferroelectric crystals, +z is also the direction of the spontaneous polarization $\mathbf{P}_s$. The definitions for polar crystals are consistent with +z defined as the outward normal from the z-cut face that develops a positive charge on cooling the crystal when $d\mathbf{P}s/dT < 0$, and a negative charge on cooling when $d\mathbf{P}s/dT$ is positive.

Table 1.1.14
REFERENCE VALUES USED FOR CONVERSION FROM RELATIVE TO ABSOLUTE MAGNITUDES

Material	Nonlinear coefficient	λ (μm)	Levine-Bethea scale[155]		Kurtz, Jerphagnon-Choy[154]		These tables	
$NH_4H_2PO_4$	d_{36}	0.6328					(0.57) × 10^{12}	$m{\cdot}V^{-1}$
	d_{36}	1.064	0.53 × 10^{-12}	$m{\cdot}V^{-1}$	0.76 × 10^{-12}	$m{\cdot}V^{-1}$	0.50 × 10^{-12}	$m{\cdot}V^{-1}$
	δ_{36}	0.6328					2.92 × 10^{-2}	$m^2{\cdot}C^{-1}$
	δ_{36}	1.064	3.1 × 10^{-2}	$m^2{\cdot}C^{-1}$	4.5 × 10^{-2}	$m^2{\cdot}C^{-1}$	2.92 × 10^{-2}	$m^2{\cdot}C^{-1}$
KH_2PO_4	d_{36}	1.064	0.44 × 10^{-12}	$m{\cdot}V^{-1}$	0.63 × 10^{-12}	$m{\cdot}V^{-1}$	0.41 × 10^{-12}	$m{\cdot}V^{-1}$
	δ_{36}	1.064	2.8 × 10^{-2}	$m^2{\cdot}C^{-1}$	4.0 × 10^{-2}	$m^2{\cdot}C^{-1}$	2.62 × 10^{-2}	$m^2{\cdot}C^{-1}$
α-SiO_2	d_{11}	1.064	0.33 × $10^{-}{}_{12}$	$m{\cdot}V^{-1}$	0.5 × −12	$m{\cdot}V^{-1}$	0.32 × 10^{-12}	$m{\cdot}V^{-1}$
	δ_{11}	1.064	1.5 × 10^{-2}	$m^2{\cdot}C^{-1}$	2.2 × 10^{-2}	$m^2{\cdot}C^{-1}$	1.42 × 10^{-2}	$m^2{\cdot}C^{-1}$
	C_{11}	1.89					0.0059 × 10^{-20}	$m^2{\cdot}V^{-2}$
$LiTO_3$	d_{31}	1.064	5.53 × 10^{-12}	$m{\cdot}V^{-1}$	7.11 × 10^{-12}	$m{\cdot}V^{-1}$	4.96 × 10^{-12}	$m{\cdot}V^{-1}$
	δ_{31}	1.064	5.0 × 10^{-2}	$m^2{\cdot}C^{-1}$	6.54 × 10^{-2}	$m^2{\cdot}C^{-1}$	4.63 × 10^{-2}	$m^2{\cdot}C^{-1}$
GaAs	d_{14}	10.6	90 × 10^{-12}	$m{\cdot}V^{-1}$	134 × 10^{-12}	$m{\cdot}V^{-1}$	134 × 10^{-12}	$m{\cdot}V^{-1}$
	δ_{14}	10.6	1.1 × 10^{-2}	$m^2{\cdot}C^{-1}$	1.7 × 10^{-2}	$m^2{\cdot}C^{-1}$	1.24 × 10^{-2}	$m^2{\cdot}C^{-1}$
	C_{11}	10.6					16.8 × 10^{-20}	$m^2{\cdot}V^{-2}$
Ge	C_{11}	10.6					140 × 10^{-20}	$m^2{\cdot}V^{-2}$
LiF	C_{11}	1.89					0.0014 × 10^{-20}	$m^2{\cdot}V^{-2}$
He	C_{11}^{mic}	0.6943					0.00245 × 10^{-50}	$m^5{\cdot}V^{-2}$

In Table 1.1.17 are listed the refractive index of some of the nonlinear materials. For uniaxial crystals, n^e and n^o represent the extraordinary and ordinary indexes, respectively. For biaxial crystals we have followed the IRE convention[57] in determining the relations among the principal optic axes x, y, and z, the crystallographic axes a, b, and c, and the piezoelectric axes X, Y, and Z. These relations are summarized in Table 1.1.18.

Since many authors do not follow the IRE convention, we have rearranged, if necessary, the principal optic axes wherever the relation between the crystallographic and principal axes is specified by the authors. As a result of this, some of the SHG tensor components given in Table 1.1.16 may be different from those listed in the original publications. Also included in Table 1.1.17 are dispersion equations for the refractive indexes, the measured coherence lengths l_c, phase-matching angles θ_m, phase-matching temperature T_m, range of transparency (or transmission spectra), and damage threshold, wherever such data are available.

A large number of solids, most of which are organic molecular compounds, have been studied by second-harmonic generation in powders. In these studies the SHG intensity is compared to that of a reference material and no determination of d_{jm} and δ_{jm} is reported. The results on such materials are summarized in Table 1.1.19, together with the fundamental wavelength, λ_1, and the relative second-harmonic intensity, $I_{2\omega}$. The particle size in most cases ranges between 50 and 150 μm.

A list of major symbols and abbreviations used in this monograph is given in Table 1.1.20.

ACKNOWLEDGMENTS

This monograph is dedicated to the memory of my father Het Ram, who passed away while I was busy writing the manuscript. I am thankful to my sons Satish and Sushil, and my wife Kamlesh for help in preparing the tables.

Table 1.1.15a
THIRD-ORDER NONLINEAR OPTICAL COEFFICIENT C_{jn}

Material	NLO-process	Coefficient $C_{jn} \times 10^{20} m^2 \cdot V^{-2}$	λ (μm)	Ref.
Acetone	$(-\omega;\omega,\omega,-\omega)$	$C_{11} = 0.252 \pm 0.056$	0.6943	405
CH_3COCH_3				
Aluminum (film)	$(-3\omega;\omega,\omega,\omega)$	$C_{jn} = 56.3 \pm 29$	1.06	132
Al				
Aluminum oxide	$(-2\omega_1 +$	$C_{11} = 0.00399 \pm 0.0005$	0.5250	152
Al_2O_3	$\omega_2;\omega_1,\omega_1,-\omega_2)$	$C_{11} = 0.0159 \pm 0.002$	0.5250	152
Sapphire	$(-\omega;\omega,\omega,\omega)$	$C_{11} \leq 0.28$	0.6943	405
Ruby				
4-Aminopyridine + I_2	$(-2\omega;\omega,\omega,0)$	$C_{11} = 0.02835 \pm 15\%$	1.318	410g
complex		$C_{11} = 0.035 \pm 15\%$	1.318	410g
$(NH_2C_5H_4N + I_2)$				
Conc. 0.49×10^{20} cm^{-3}				
Conc. 0.65×10^{20} cm^{-3}				
Ammonium dihydrogen	$(-3\omega;\omega,\omega,\omega)$	$C_{11} = 0.0104$	1.06	143
phosphate (ADP)		$C_{18} = 0.0098$	1.06	143
$NH_4H_2PO_4$				
Aniline	$(-2\omega;\omega,\omega,0)$	$C_{11} = 0.1141 \pm 15\%$	1.06	410b
$C_6H_5NH_2$				
Barium fluoride	$(-2\omega_1 +$	$C_{11} = 0.00387 \pm 0.00042$	0.5750	411
BaF_2	$\omega_2;\omega_1,\omega_1,-\omega_2)$	$C_{18} = 0.00159 \pm 0.00014$	0.5750	411
		$C_{11} + 3C_{18} = 0.00859 \pm 0.00065$	0.5750	411
Benzoyl chloride	$(-\omega;\omega,\omega,-\omega)$	$C_{18} = 0.42$	0.5000	47
C_6H_5COCl				
Benzene	$(-\omega;\omega,\omega,-\omega)$	$C_{11} = 0.518 \pm 0.07$	0.6943	405
C_6H_6		$C_{18} = 0.098$	0.5000	47
		$C_{18} = 0.0782$	0.6943	403
	$(-\omega;\omega,\omega,-\omega)$	$C_{18} = 0.091$	0.6940	130a
	$(-\omega;\omega,\omega,-\omega)$	$C_{18} = 0.042$	0.6940	130a
	$-(\omega_1,\omega_1,\omega_2,-\omega_2)$	$C_{18} = 0.028$	0.4880, 0.6940	130a
	$(-2\omega_1 +$	$C_{11} = 0.56$	0.6943	28
	$\omega_2,\omega_1,\omega_1,-\omega_2)$			
		$C_{11} = 0.0242 \pm 0.0024$	0.5250	152
		$C_{11} = 0.00859 \pm 0.00037$	0.545	411
		$C_{18} = 0.00252 \pm 0.00014$	0.545	411
		$C_{18} = 0.00303 \pm 0.00019$	0.545	411
	$(-3\omega;\omega,\omega,\omega)$	$C_{11} = 0.0184 \pm 0.0042$	1.89	134
	$(-2\omega;\omega,\omega,0)$	$C_{11} = 0.02215 \pm 15\%$	1.06	410b
		$C_{11} = 0.0196 \pm 15\%$	1.318	410d
90% Benzene +	$(-2\omega;\omega,\omega,0)$	$C_{11} = 0.02121 \pm 15\%$	1.318	410d
10% pyridine				
$C_6H_6 + C_5H_5N$				
Beryllium	$(-3\omega;\omega,\omega,\omega)$	$\bar{C}_{jn} \leq 1.16$	1.06	132
(polished)				
Be				
Borosilicate glass	$(-2\omega_1 +$	$C_{11} = 0.0018$	0.6943	28
	$\omega_2;\omega_1,\omega_1,-\omega_2)$			
BSG		$C_{11} = 0.0129 \pm 0.002$	0.5250	152
Bromobenzene	$(-2\omega_1 +$	$C_{11} = 0.084$	0.6943	28
C_6H_5Br	$\omega_2;\omega_1,\omega_1,-\omega_2)$	$C_{11} = 0.02913 \pm 15\%$	1.06	410b
	$(-2\omega;\omega,\omega,0)$			
Bromoform	$(-\omega;\omega,\omega,-\omega)$	$C_{18} = 0.0098$	0.6943	28
(Tribromomethane)				
$CHBr_3$				

Table 1.1.15a (continued)
THIRD-ORDER NONLINEAR OPTICAL COEFFICIENT C_{jn}

Material	NLO-process	Coefficient $C_{jn} \times 10^{20} m^2 \cdot V^{-2}$	λ (μm)	Ref.
BK-7 glass	$(-\omega;\omega,\omega,-\omega)$	$C_{18} = 0.00257$	0.6943	408
Cadmium fluoride	$(-2\omega_1 +$	$C_{11} = 0.0068 \pm 0.0010$	0.5750	411
CdF_2	$\omega_2;\omega_1,\omega_1,-\omega_2)$	$C_{18} = 0.0022 \pm 0.0003$	0.5750	411
		$C_{11} + 3C_{18} = 0.0133 \pm 0.0014$	0.5750	411
Cadmium germanium	$(-3\omega;\omega,\omega,\omega)$	$C_{11} = 182 \pm 84$	10.6	122b
arsenide		$C_{16} = 175$	10.6	122b
$CdGeAs_2$		$C_{18} = -35$	10.6	122b
p-type: 5×10^{16} cm^{-3}				
Cadmium sulfide	$(-2\omega_1 +$	$C_{11} = 2.24$	0.6943	28
CdS	$\omega_2;\omega_1,\omega_1,-\omega_2)$	$C_{11} = 1050$	0.53	144
Calcite	$(-2\omega_1 +$	$C_{11} = 0.0196 \pm 0.003$	0.5250	152
$CaCO_3$	$\omega_2;\omega_1,\omega_1,-\omega_2)$	$C_{11} = 0.0084 \pm 0.00037$	0.530	411
		$C_{11} = 0.0078 \pm 0.00033$	0.556	411
		$C_{33} = 0.0047 \pm 0.0009$	0.530	411
Calcium fluoride	$(-2\omega_1 +$	$C_{11} + 3C_{18} = 0.0154 \pm 0.0022$	0.5250	152
CaF_2	$\omega_2;\omega_1,\omega_1,-\omega_2)$	$C_{11} = 0.002 \pm 0.0006$	0.575	411
		$C_{18} = 0.00089 \pm 0.00023$	0.575	411
		$C_{11} + 3C_{18} = 0.0046 \pm 0.00047$	0.575	411
		$C_{11} = 0.005$	0.6943	28
		$C_{18} = 0.0025$	0.6943	28
	$(-3\omega\omega,\omega,\omega)$	$C_{18}/C_{11} = 0.5067$	1.06	132
Carbon disulfide	$(-\omega;\omega,\omega,-\omega)$	$C_{18} = 0.476$	0.6940	130a
CS_2		$C_{18} = 0.219$	0.6940	130a
		$C_{18} = 0.5 \pm 0.0146$	0.6940	403
		$C_{18} = 0.063$	0.6940	28
		$C_{18} = 0.560$	0.5000	47
	$(-\omega_1;\omega_1,\omega_2,-\omega_2)$	$C_{18} = 0.266$	0.4880, 0.6940	130a
	$-(2\omega,\omega,\omega,0)$	$C_{11} = 0.06233 \pm 15\%$	1.06	410b
Carbon tetrachloride	$(-\omega;\omega,\omega,-\omega)$	$C_{18} = 0.00168 \pm 30\%$	0.6940	28
CCl_4		$C_{18} = 0.007$	0.6940	406
		$C_{18} = 0.0126$	0.6940	30a
		$C_{18} = 0.0056$	0.6940	30a
		$C_{18} = 0.0035$	0.4880,	30a
	$(-\omega_1;\omega_1,\omega_2-\omega_2)$		0.6940	
	$(-2\omega;\omega,\omega,0)$	$C_{11} = 0.0182 \pm 15\%$	1.06	410b
β-Carotene (glass)	$(-3\omega;\omega,\omega,\omega)$	$C_{11} = 0.263 \pm 0.08$	1.89	140
$C_{40}H_{56}$				
Chloroform	$(-\omega;\omega,\omega,-\omega)$	$C_{18} = 0.007 \pm 30\%$	0.6940	28
(trichloromethane)				
$CHCl_3$				
Chlorobenzene	$(-2\omega;\omega,\omega,0)$	$C_{11} = 0.00875 \pm 15\%$	1.06	410b
C_6H_5Cl				
Copper	$(-3\omega,\omega,\omega,-\omega)$	$\bar{C}_{jn} = 5.88 \pm 3.3$	1.06	132
Cu		$\bar{C}_{jn} = 6.7 \pm 3.3$	1.06	132
Polished				
Evaporated				
Diamond	$(-3\omega;\omega,\omega,\omega)$	$C_{11} + 3C_{18} = 0.1456 \pm 10\%$	1.06	132
	$-(2\omega_1 +$	$C_{11} + 3C_{18} = 0.163 \pm 0.046$	1.06	136
	$\omega_2;\omega_1,\omega_1,-\omega_2)$	$C_{11} + 3C_{18} = 0.07383 \pm 0.0019$	0.407	411
		$C_{18} = 0.01218 \pm 0.0003$	0.407	411
		$C_{11} + 3C_{18} = 0.0569 \pm 0.0019$	0.450	411
		$C_{18} = 0.01022 \pm 0.00019$	0.450	411
		$C_{11} + 3C_{18} = 0.05133 \pm 0.0019$	0.500	411
		$C_{11} + 3C_{18} = 0.04396 \pm 0.0015$	0.545	411

Table 1.1.15a (continued) THIRD-ORDER NONLINEAR OPTICAL COEFFICIENT C_{jn}

Material	NLO-process	Coefficient $C_{jn} \times 10^{20} m^2 \cdot V^{-2}$	λ (μm)	Ref.
		$C_{18} = 0.00803 \pm 0.00019$	0.545	411
		$C_{11} = 0.02147 \pm 0.0009$	0.545	411
		$C_{18} = 0.00803 \pm 0.0003$	0.545	411
		$C_{18} = 0.00803 \pm 0.0003$	0.545	411
		$C_{18} = 0.00859 \pm 0.0004$	0.565	411
		$C_{18} = 0.00737 \pm 0.0003$	0.565	411
		$C_{11} + 3C_{18} = 0.03948 \pm 0.0019$		
1,2 Dichloroethane 1,2 $C_2H_4Cl_2$	$(-2\omega;\omega,\omega,0)$	$C_{11} = 0.042 \pm 15\%$	1.06	410b
o-Diiodobenzene *o*-$C_6H_4I_2$	$(-2\omega;\omega,\omega,0)$	$C_{11} = 0.0847 \pm 15\%$	1.06	410b
p-Dioxane O-$(CH_2)_2$-O-$(CH_2)_2$	$(-2\omega;\omega,\omega,0)$	$C_{11} = 0.01351 \pm 15\%$	1.06	410e
Dodecane $C_{12}H_{26}$	$(-2\omega;\omega,\omega,0)$	$C_{11} = 0.0146 \pm 15\%$	1.06	410b
trans-4-Dimethylamino-4-nitrostilbene (DANS) @ 152°C	$(-2\omega,\omega,\omega,0)$	$C_{11} = 0.4585 \pm 15\%$	1.318	410h
ED-4 glass	$(-2\omega_1 + \omega_2;\omega_1,\omega_1,-\omega_2)$	$C_{11} = 0.01498 \pm 0.0011$	0.525	152
Ethyl alcohol	$(-\omega;\omega,\omega,-\omega)$	$C_{18} = 0.00315$	0.6940	407
C_2H_5OH		$C_{11} = 0.196 \pm 0.042$	0.6943	405
Fluorobenzene C_6H_5F	$(-2\omega;\omega,\omega,0)$	$C_{11} = -0.009135 \pm 15\%$	1.06	410b
o-Fluoronitrobenzene *o*-$C_6H_4NO_2F$	$(-2\omega;\omega,\omega,\omega,0)$	$C_{11} = 0.5915 \pm 15\%$	1.06	410b
m-Fluoronitrobenzene *m*-$C_6H_4NO_2F$	$(-2\omega;\omega,\omega,0)$	$C_{11} = 0.3745 \pm 15\%$	1.06	410b
p-Fluoronitrobenzene *p*-$C_6H_4NO_2F$	$(-2\omega;\omega,\omega,0)$	$C_{11} = 0.364 \pm 15\%$	1.06	410b
Gallium arsenide	$(-2\omega_1 +$	$C_{11} = 16.80 \pm 10\%$	10.6	101
GaAs	$\omega_2;\omega_1,\omega_1,-\omega_2)$	$C_{18} = 4.2 \pm 0.168$	10.6	101
$n_e = 1.5 \times 10^6$ cm^{-3}		$C_{11} = 2.45$	10.6	52
$n_e = 1.5 \times 10^{16}$ cm^{-3}		$C_{11} = 2.45$	9.6	52
High resistivity		$C_{18} = 1.78 \pm 10\%$	10.6	137
High resistivity		$C_{11} = 3.39 \pm 10\%$	10.6	137
	$(-3\omega;\omega,\omega,-\omega)$	$C_{11} = 62 \pm 50\%$	1.06	132
Germanium	$(-2\omega_1 +$	$C_{11} = +140 \pm 50\%$	10.6	101
Ge	$\omega_2;\omega_1,\omega_1,-\omega_2)$	$C_{18} = 85.4 \pm 2.8$	10.6	101
		$C_{11} = 140 \pm 50\%$	10.6	133
		$C_{18} = +85.4$	10.6	133
		$C_{18}/C_{11} = 0.52 \pm 2\%$	10.6	137
	$(-3\omega;\omega,\omega,-\omega)$	$C_{11} = 42.8 \pm 80\%$	1.06	132
		$C_{11} = 12 \pm 3.6$	1.06	141
		$C_{11} = 23.5 \pm 50\%$	1.06	132
Glycerine $CH_2OHCHOH$ -CH_2OH	$(-\omega;\omega,\omega,-\omega)$	$C_{11} = 0.196 \pm 0.07$	0.6943	405
Gold	$(-3\omega;\omega,\omega,-\omega)$	$C_{11} = 16.4 \pm 7.9$	1.06	141
Au		$\bar{C}_{jn} = 7.1 \pm 2.5$	1.06	132
Evaporated		$\bar{C}_{jn} = 7.2 \pm 2.5$	1.06	132
Evaporated				
Polished				
n-Hexane	$(-\omega;\omega,\omega,-\omega)$	$C_{18} = 0.00126 \pm 30\%$	0.6940	28
$CH_3(CH_2)_4CH_3$	$(-2\omega;\omega,\omega,0)$	$C_{11} = 0.0092 \pm 15\%$	1.06	410b
Heptane C_7H_{16}	$(-2\omega;\omega,\omega,0)$	$C_{11} = 0.0101 \pm 15\%$	1.06	410b

Table 1.1.15a (continued)
THIRD-ORDER NONLINEAR OPTICAL COEFFICIENT C_{jn}

Material	NLO-process	Coefficient $C_{jn} \times 10^{20} m^2 \cdot V^{-2}$	λ (μm)	Ref.
Indium arsenide	$(-2\omega_1 +$	$C_{11} = 63$	10.6	52
InAs	$\omega_2;\omega_1,\omega_1,-\omega_2)$	$C_{11} = 63$	9.6	52
$n_e = 2.6 \times 10^{16}$ cm^{-3}				
Iodobenzene	$(-2\omega;\omega,\omega,0)$	$C_{11} = +0.0609 \pm 15\%$	1.06	410b
C_6H_5I				
K-8 glass	$(-\omega;\omega,\omega,-\omega)$	$C_{11} = 0.21 \pm 0.042$	0.6943	405
LaSF-7 glass	$(-\omega;\omega,\omega,-\omega)$	$C_{18} = 0.014$	0.6940	409
Lithium fluoride	$(-2\omega_1 +$	$C_{11} = 0.0048 \pm 0.0008$	0.5250	152
LiF	$\omega_2;\omega_1,\omega_1,-\omega_2)$	$C_{11} = 0.0028$	0.6943	28
		$C_{18} = 0.00126$	0.6943	28
	$(-3\omega;\omega,\omega,-\omega)$	$C_{11} = 0.0014 \pm 0.00002$	1.89	134
		$3C_{18} + \times C_{11} = 0.0042$	0.6943	28
		$C_{11} = 0.0042$	1.06	97
		$C_{18}/C_{11} = 0.4533$	1.06	132
Lithium iodate	$(-3\omega;\omega,\omega,-\omega)$	$C_{12} = 0.2285$	1.06	135
$LiIO_3$		$C_{35} = 6.66 \pm 1$	1.06	135
LSO-glass	$(-\omega;\omega,\omega,-\omega)$	$C_{18} = 0.0026$	0.694	407
Magnesium	$(-3\omega;\omega,\omega,-\omega)$	$\bar{C}_{jn} \leqslant 2.15$	1.060	132
Mg (polished)				
Magnesium oxide	$(-2\omega_1 +$	$C_{11} = 0.014$	0.6943	28
MgO	$\omega_2 \cdot \omega_1,\omega_1,-\omega_2)$	$C_{18} = 0.0077$	0.6943	28
	$(-3\omega;\omega,\omega,-\omega)$	$C_{11} = 0.0336$	1.06	97
Mercury cadmium telluride	$(-2\omega_1 +$	$C_{11} = 1.75 \times 10^{10}$	10.6	139
HgCdTe	$\omega_2;\omega_1,\omega_1,-\omega_2)$	$C_{11} = 1.89 \times 10^{6}$	10.6	144
$Hg_{.08}Cd_{0.2}Te$				
Methanol	$(-\omega;\omega,\omega,-\omega)$	$C_{18} = 0.00042 \pm 30\%$	0.6940	28
CH_3OH	$(-2\omega;\omega,\omega,0)$	$C_{11} = 0.0161 \pm 15\%$	1.06	410e
Methyl benzene	$(-2\omega_1 +$	$C_{11} = 0.056$	0.6943	28
$C_6H_5CH_3$	$\omega_2;\omega_1,\omega_1,-\omega_2)$			
Methyl iodide	$(-2\omega;\omega,\omega,0)$	$C_{11} = 0.1166 \pm 15\%$	1.06	410b
CH_3I				
Methylene iodide	$(-2\omega;\omega,\omega,0)$	$C_{11} = +0.1690 \pm 15\%$	1.06	410b
CH_2I_2				
m-Nitroaniline	$(-2\omega;\omega,\omega,0)$	$C_{11} = 1.162 \pm 15\%$	1.318	410f
m-$C_6H_4 \cdot NH_2 \cdot NO_2$				
o-Nitroaniline	$(-2\omega\omega,\omega,0)$	$C_{11} = 1.6905 \pm 15\%$	1.318	410f
o-$C_6H_4 \cdot NH_2 \cdot NO_2$				
p-Nitroaniline	$(-2\omega;\omega,\omega,0)$	$C_{11} = +6.895 \pm 15\%$	1.318	410f
p-$C_6H_4 \cdot NH_2 \cdot NO_2$		$C_{11} = 0.3395 \pm 15\%$	1.318	410h
@ 129°C				
Nitrobenzene	$(-\omega;\omega,\omega,-\omega)$	$C_{18} = 0.42$	0.500	47
$C_6H_5NO_2$		$C_{18} = 0.322$	0.6943	130a
		$C_{18} = 0.1447$	0.6943	130a
	$(-\omega_1;\omega_1,\omega_2,-\omega_2)$	$C_{18} = 0.182$	0.4880, 0.694	130a
	$(-\omega;\omega,\omega,-\omega)$	$C_{11} = 1.148 \pm 0.140$	0.6943	405
	$(-2\omega;\omega,\omega,0)$	$C_{11} = 0.7 \pm 0.107$	1.06	410a
	$(-2\omega;\omega,\omega,0)$	$C_{11} = 0.585 \pm 15\%$	1.06	410b
		$C_{11} = 0.5845 \pm 15\%$	1.318	410f
@ 150°C		$C_{11} = +0.3605 \pm 15\%$	1.06	410b
@ 178°C		$C_{11} = +0.2905 \pm 15\%$	1.06	410
50% Nitrobenzene	$(-2\omega;\omega,\omega,0)$	$C_{11} = 0.35 \pm 0.056$	1.06	410
in benzene				
$C_6H_5NO_2 + C_6H_6$				
25% Nitrobenzene		$C_{11} = 0.173 \pm 0.028$	1.06	410a
in benzene				
$C_6H_5NO_2 + C_6H_6$				

Table 1.1.15a (continued)
THIRD-ORDER NONLINEAR OPTICAL COEFFICIENT C_{jn}

Material	NLO-process	Coefficient $C_{jn} \times 10^{20} m^2 \cdot V^{-2}$	λ (μm)	Ref.
10% Nitrobenzene in benzene $C_6H_5NO_2 + C_6H_6$		$C_{11} = 0.056 \pm 0.009$	1.06	410a
Nitrocyclohexane $C_6H_{11} \cdot NO_2$	$(-2\omega;\omega,\omega,0)$	$C_{11} = 0.03955$	1.06	410b
Nitromethane	$(-2\omega;\omega,\omega,0)$	$C_{11} = 0.042 \pm 0.0093$	1.06	410a
CH_3NO_2		$C_{11} = 0.031425 \pm 15\%$	1.06	410b
1-Nitropropane $C_3H_7 \cdot NO_2$	$(-2\omega;\omega,\omega,0)$	$C_{11} = 0.0289 \pm 15\%$	1.06	410b
2-Nitropropane $CH_3(CH - NO_2)CH_3$	$(-2\omega;\omega,\omega,0)$	$C_{11} = 0.0221 \pm 15\%$	1.06	410b
Poly-γ-benzyl-L-glutamate PBLG	$(-2\omega;\omega,\omega,0)$	$C_{11} = 0.0462$	1.06	410c
Potassium bromide	$(-2\omega_1 +$	$C_{11} = 0.042$	0.6943	28
KBr	$\omega_2;\omega_1,\omega_1,-\omega_2)$	$C_{18} = 0.0154$	0.6943	28
	$(-3\omega;\omega,\omega,-\omega)$	$C_{11} = 0.0392$	1.06	97
		$C_{18}/C_{11} = 0.3667$	1.06	132
Potassium chloride	$-2\omega_1 +$	$C_{11} = 0.0266$	0.6943	28
KCl	$\omega_2;\omega_1,\omega_1,-\omega_2)$	$C_{18} = 0.0081$	0.6943	28
	$(-3\omega;\omega,\omega,-\omega)$	$C_{11} = 0.0168$	1.06	97
		$C_{18}/C_{11} = 0.28$	1.06	132
Potassium dihydrogen phosphate (KDP) KH_2PO_4	$(-3\omega;\omega,\omega,-\omega)$	$C_{11} - 3C_{18} = 0.04$	1.06	135
Potassium iodide	$(-2\omega_1 +$	$C_{11} = 0.0035$	0.6943	28
KI	$\omega_2;\omega_1,\omega_1,-\omega_2)$	$C_{18} = 0.00216$	0.6943	28
Pyridine	$(-\omega;\omega,\omega,-\omega)$	$C_{18} = 0.014$	0.6940	28
C_5H_5N	$(-2\omega;\omega,\omega,0)$	$C_{11} = 0.0721 \pm 15\%$	1.06	410b
Pyridine-iodine complex $(C_5H_5N - I_2)$				
Conc. = $0.47 \times 10^{20} cm^{-3}$	$(-2\omega;\omega,\omega,0)$	$C_{11} = +0.03955 \pm 15\%$	1.318	410d
Conc. = $1.09 \times 10^{20} cm^{-3}$		$C_{11} = +0.07315 \pm 15\%$	1.318	410d
Conc. = $1.7 \times 10^{20} cm^{-3}$		$C_{11} = +0.09905 \pm 15\%$	1.318	410d
Conc. = $1.97 \times 10^{20} cm^{-3}$		$C_{11} = +0.10465 \pm 15\%$	1.318	410d
Pyridine + iodine + chlorine complex $C_5H_5N - ICl$				
Conc. = $0.7 \times 10^{20} cm^{-3}$	$(-2\omega;\omega,\omega,0)$	$C_{11} = 0.0308 \pm 15\%$	1.318	410g
Conc. = $0.79 \times 10^{20} cm^{-3}$		$C_{11} = 0.0308 \pm 15\%$	1.318	410g
Quinoline				
C_6H_4N:CHCH:CH	$(-\omega;\omega,\omega,-\omega)$	$C_{11} = 0.966 \pm 0.098$	0.6943	405
SF-7 glass	$(-\omega;\omega,\omega,-\omega)$	$C_{18} = 0.01108$	0.694	408
Silicon	$(-2\omega_1 + \omega_2;\omega_1,\omega_1,-\omega_2)$	$C_{11} = 8.4 \pm 10\%$	10.6	101
Si		$C_{18} = +4.03 \pm 0.252$	10.6	101
		$C_{11} = 7.0 \pm 10\%$	10.6	133
		$C_{18} = +3.36$	10.6	133
(p-type: $10^{14} cm^{-3}$)		$C_{11} = 3.85 \times 10^4$	1.06	144

Table 1.1.15a (continued)
THIRD-ORDER NONLINEAR OPTICAL COEFFICIENT C_{jn}

Material	NLO-process	Coefficient $C_{jn} \times 10^{20} m^2 \cdot V^{-2}$	λ (μm)	Ref.
	$(-3\omega;\omega,\omega,-\omega)$	$C_{11} = 115.4 \pm 50\%$	1.06	132
		$C_{11} = 82.8 \pm 30\%$	1.06	132
		$C_{11} = 60.7 \pm 9.7$	1.06	141
Silicon dioxide	$(-2\omega_1 +$	$C_{11} = 0.014$	0.6943	28
α-SiO_2	$\omega_2;\omega_1,\omega_1,-\omega_2)$	$C_{11} = 0.0059 \pm 50\%$	1.89	134
	$(-3\omega;\omega,\omega,-\omega)$			
Silica (fused)	$(-2\omega_1 +$	$C_{11} = 0.0098 \pm 0.0017$	0.5250	152
	$\omega_2;\omega_1,\omega_1,-\omega_2)$	$C_{11} = 0.0098$	0.6943	28
	$(-\omega;\omega,\omega,-\omega)$	$C_{18} = 0.0017$	0.6940	407
	$(-\omega;\omega,\omega,-\omega)$	$C_{11} = 0.672 \pm 0.126$	0.6943	405
Silver	$(-3\omega;\omega,\omega,-\omega)$	$C_{11} = 6.68 \pm 2.4$	1.06	141
Ag		$C_{jn} = 6.2 \pm 2.5$	1.06	132
Evaporated		$C_{jn} = 7.78 \pm 2.5$	1.06	132
Evaporated				
Polished				
Sodium chloride	$(-\omega_1 +$	$C_{11} = 0.0238$	0.6943	28
	$\omega_2;\omega_1,\omega_1,-\omega_2)$			
NaCl		$C_{18} = 0.0101$	0.6943	28
	$(-3\omega;\omega,\omega,-\omega)$	$C_{11} = 0.0168$	1.06	97
		$C_{18}/C_{11} = 0.4133$	1.06	132
Sodium fluoride	$(-3\omega;\omega,\omega,-\omega)$	$C_{11} = 0.0035$	1.06	97
NaF				
trans-Stilbene	$(-2\omega;\omega,\omega,0)$	$C_{11} = 0.09695 \pm 15\%$	1.064	410b
trans-$C_6H_5CH =$				
CHC_6H_5				
@ 157°C				
Strontium fluoride	$(-2\omega_1 +$	$C_{11} = 0.00205 \pm 0.0005$	0.575	411
SrF_2	$\omega_2;\omega_1,\omega_1,-\omega_2)$	$C_{18} = 0.0014 \pm 0.00019$	0.575	411
		$C_{11} + 3C_{18} = 0.00616 \pm 0.0005$	0.575	411
Strontium titanate	$(-2\omega_1 +$	$C_{11} = 5.6$	0.6943	28
	$\omega_2;\omega_1\omega_1,-\omega_2)$			
$SrTiO_3$		$C_{18} = 2.63$	0.6943	28
Terbium aluminum	$(-2\omega_1 +$	$C_{11} = (3.1 \pm 0.62) \times 10^6$	4.0	138
garnet	$\omega_2;\omega_1\omega_1,-_2)$	$C_{18} = (0.95 \pm 0.2) \times 10^6$	4.0	138
$Tb_3Al_5O_{12}$				
TF-7 glass	$(-\omega;\omega,\omega,-\omega)$	$C_{11} = 0.42 \pm 0.098$	0.6943	405
Toluene	$(-\omega;\omega,\omega,-\omega)$	$C_{11} = 0.672 \pm 0.07$	0.6943	405
$C_6H_5CH_3$		$C_{18} = 0.05133$	0.6943	130a
	$(-\omega_1;\omega_1,\omega_2,-\omega_2)$	$C_{18} = 0.063$	0.4880, 0.694	130a
	$(-2\omega;\omega,\omega,0)$	$C_{11} = 0.0238 \pm 15\%$	1.06	410b
Water				
H_2O	$(-\omega;\omega,\omega,-\omega)$	$C_{11} = 0.098 \pm 0.042$	0.6943	405
		$C_{18} = 0.00042 \pm 30\%$	0.6940	28
	$(-2\omega;\omega,\omega,0)$	$C_{11} = +0.0616 \pm 15\%$	1.06	410e
Ø-Xylene	$(-\omega;\omega,\omega,-\omega)$	$C_{11} = 0.616 \pm 0.126$	0.6943	405
$C_6H_4(CH_3)_2$				
p-Xylene	$(-2\omega;\omega,\omega,0)$	$C_{11} = +0.02079 \pm 15\%$	1.06	410b
p-$C_6H_4(CH_3)_2$				
Yttrium aluminum	$(-2\omega_1 +$	$C_{11} = 0.03052 \pm 0.0018$	0.5250	152
garnet (YAG)	$\omega_2;\omega_1,\omega_1,-\omega_2)$			
$Y_3Al_5O_{12}$				
	$(-\omega;\omega,\omega,-\omega)$	$C_{18} = 0.0084$	0.6940	407
		$C_{18} = 0.0095$	0.6940	407

Table 1.1.15b
THIRD-ORDER NONLINEAR OPTICAL COEFFICIENT, C_{jn}^{mic}

Material	NLO-Process	Coefficient $C_{jn}^{mic} \times 10^{50}$ m^5·V^{-2}	λ (μm)	Ref.
4-Aminopyridine + I_2	$(-2\omega;\omega,\omega,0)$	C_{12} = +57.05 ± 15%	1.318	410g
NH_2-C_5H_4N + I_2				
Conc.: 0.49 × $10^{20}cm^{-3}$		C_{11} = +54.95 ± 15%	1.318	410g
Conc.: 0.49 × $10^{20}cm^{-3}$		C_{11} = +58.8 ± 15%	1.318	410g
Aniline	$(-2\omega;\omega,\omega,0)$	C_{11} = +2.723 ± 15%	1.318	410f
C_6H_5·NH_2				
Argon	$(-2\omega_1 +$	C_{11} = 0.0217 ± 10%	0.6943	142
Ar	$\omega_2;\omega_1,\omega_1,-\omega_2)$			
	$(-3\omega;\omega,\omega,-\omega)$	C_{11} = 0.0875	0.308	148
		C_{11} = 0.0441 ± 0.007	0.6943	96
	$(-2\omega;0,\omega,\omega)$	C_{22} = 0.0833 ± 0.0027	0.6943	412
Benzene	$(-2\omega;\omega,\omega,0)$	C_{11} = 0.728 + 15%	1.318	410d
C_6H_6	$(-2\omega;\omega,\omega,0)$	C_{11} = +0.819 ± 15%	1.06	410b
Benzene (90% + pyridine (10%)	$(-2\omega;\omega,\omega,0)$	C_{11} = 0.8085 ± 15%	1.318	410d
C_6H_6 + C_5H_5N				
Bromotrifluoromethane	$(-2\omega;0,\omega,\omega)$	C_{22} = 0.4375 ± 0.03	0.6943	414
$CBrF_3$				
Bromobenzene	$(-2\omega;\omega,\omega,0)$	C_{11} = 0.889 ± 15%	1.06	410b
C_6H_5Br				
Cadmium vapor	$(-3\omega;\omega,\omega,-\omega)$	C_{11} = 700	1.06	149
Cd		C_{11} = 70	0.5320	149
Carbon dioxide	$(-2\omega_1 +$	C_{11} = 0.028 + 10%		
CO_2	$\omega_2;\omega_1,\omega_1,-\omega_2)$			142
	$(-3\omega;\omega,\omega,-\omega)$	C_{11} = 0.054 ± 0.008	0.6943	96
Carbon disulfide	$(-2\omega;\omega,\omega,0)$	C_{11} = +1.116 ± 15%	1.06	410b
CS_2				
Carbon monoxide	$(-2\omega_1 +$	C_{11} = 0.0252 ± 10%	0.6943	142
CO	$\omega_2;\omega_1,\omega_1,-\omega_2)$			
	$(-3\omega;\omega,\omega,-\omega)$	C_{11} = 1.95	9.33	147
Carbon tetrachloride	$(-2\omega_1 +$	C_{jn}^{mic} = 0.98	0.6943	415
CCl_4	$\omega_2;\omega_1,\omega_1,-\omega_2)$			
	$(-2\omega;0,\omega,\omega)$	C_{jn}^{mic} = 0.98	0.6943	415
Cesium vapor	$(-3\omega;\omega,\omega,-\omega)$	C_{11} = 3.5 × 10^5	0.6943	150
Cs				
Chlorobenzene	$(-2\omega;\omega,\omega,0)$	C_{11} = +0.28 ± 15%	1.06	410b
C_6H_5Cl				
Chlorotrifluoromethane	$(-2\omega;0\omega,\omega)$	C_{22} = 0.2142 ± 0.009	0.6943	414
$CClF_3$				
Deuterium	$(-2\omega_1 +$	C_{11} = 0.0182 ± 10%	0.6943	142
D_2	$\omega_2;\omega_1,\omega_1,-\omega_2)$			
1,2,Dichloroethane	$(-2\omega;\omega,\omega,0)$	C_{11} = +1.1865 ± 15%	1.06	410b
1,2,$C_2H_4Cl_2$				
Difluoromethane	$(-2\omega;\omega,\omega,0)$	C_{22} = 0.108 ± 0.008	0.6943	414
CH_2F_2				
o-Diiodobenzene	$(-2\omega;\omega,\omega,0)$	C_{11} = +2.142 ± 15%	1.06	410b
$C_6H_4I_2$				
m-Dinitrobenzene	$(-2\omega;\omega,\omega,0)$	C_{11} = +11.375 ± 15%	1.32	410e
m-$C_6H_4(NO_2)_2$				
@ 140°C				
m-Dinitrobenzene (13%) + benzene	$(-2\omega;\omega,\omega,0)$	C_{11} = +6.65 ± 15%	1.06	410e
m-$C_6H_4(NO_2)_2$ + C_6H_6				
p-Dioxane	$(-2\omega;\omega,\omega,0)$	C_{11} = 0.581 ± 15%	1.06	410e
O–$(CH_2)_2$–O–$(CH_2)_2$				

Table 1.1.15b (continued)
THIRD-ORDER NONLINEAR OPTICAL COEFFICIENT, C_{jn}^{mic}

Material	NLO-Process	Coefficient $C_{jn}^{mic} \times 10^{50}$ $m^5 \cdot V^{-2}$	λ (μm)	Ref.
Dodecane @ 23°C $C_{12}H_{26}$	$(-2\omega;\omega,\omega,0)$	$C_{11} = +1.708 \pm 15\%$	1.06	410b
@ 131°C		$C_{11} = +1.8445 \pm 15\%$	1.06	410b
@ 161°C		$C_{11} = +1.8585 \pm 15\%$	1.06	410b
trans-4-Dimethylamino-4′-nitrostilbene (DANS) @ 152°C	$(-2\omega;\omega,\omega,0)$	$C_{11} = 346.5 \pm 15\%$	1.318	410h
Ethane C_2H_6	$(-2\omega_1 + \omega_2;\omega_1,\omega_1,-\omega_2)$	$C_{11} = 0.0868 \pm 10\%$	0.6943	142
Fluorobenzene C_6H_5F	$(-2\omega;\omega,\omega,0)$	$C_{11} = -0.315 \pm 15\%$	1.06	410b
Fluoromethane CH_3F	$(-2\omega;\omega,\omega,0)$	$C_{22} = 0.1675 \pm 0.021$	0.6943	414
m-Fluoronitrobenzene *m*-$C_6H_4 \cdot F \cdot NO_2$	$(-2\omega;\omega,\omega,0)$	$C_{11} = +10.99 \pm 15\%$	1.06	410b
o-Fluoronitrobenzene	$(-2\omega;\omega,\omega,0)$	$C_{11} = +15.855 \pm 15\%$	1.06	410b
p-Fluoronitrobenzene *p*-$C_6H_4F \cdot NO_2$	$(-2\omega;\omega,\omega,0)$	$C_{11} = +10.78 \pm 15\%$	1.06	410b
Helium	$(-2\omega_1 + \omega_2;\omega_1,\omega_1 - \omega_2$	$C_{11} = 0.00245$	0.6943	96
He	$(-3\omega;\omega,\omega,-\omega)$	$C_{11} = 0.00122$	0.6943	142
	$(-2\omega;0,\omega,\omega)$	$C_{22} = 0.0027$	0.6943	412
Heptane C_7H_{16}	$(-2\omega;\omega,\omega,0)$	$C_{11} = +0.84 \pm 15\%$	1.06	410b
Hexane $C_{16}H_{14}$	$(-2\omega;\omega,\omega,0)$	$C_{11} = +0.7105 \pm 15\%$	1.06	410b
Hydrogen H_2	$(-2\omega_1 + \omega_2;\omega_1,\omega_1,-\omega_2)$	$C_{11} = 294 \pm 10\%$	0.6943	142
	$(-3\omega;\omega,\omega,-\omega)$	$C_{11} = 0.028 \pm 0.0042$	0.6943	96
Iodobenzene C_6H_5I	$(-2\omega;\omega,\omega,0)$	$C_{11} = +1.7465 \pm 15\%$	1.06	410b
Krypton	$(-3\omega;\omega,\omega,-\omega)$	$C_{11} = 0.1351 \pm 0.0262$	0.6943	96
Kr	$(-2\omega;0,\omega,\omega)$	$C_{22} = 0.2037 \pm 0.0098$	0.6943	412
Methane CH_4	$(-2\omega_1 + \omega_2;\omega_1,\omega_1,-\omega_2)$	$\bar{C}_{jn}^{mic} = 0.196$	0.6943	415
	$(-2\omega;0,\omega,\omega)$	$\bar{C}_{jn}^{mic} = 0.182$	0.6943	415
	$(-\omega;0,0,\omega)$	$C_{22} = 0.1806 \pm 0.013$	0.6943	414
	$(-2\omega;0,\omega,\omega)$	$C_{jn}^{mic} = 0.1708 \pm 0.0084$	0.6943	416
	$(-2\omega_1 + \omega_2;\omega_1,\omega_1,-\omega_2)$	$C_{22} = 0.1925 \pm 0.0161$	0.6943	413
	$(-2\omega;\omega,\omega,0)$	$C_{11} = 0.0413 \pm 10\%$	0.6943	142
		$C_{11} = +0.147 \pm 15\%$	1.06	410b
Methanol $CH_3 \cdot OH$	$(-2\omega;\omega,\omega,0)$	$C_{11} = +0.301 \pm 15\%$	1.06	410e
Methyl iodide CH_3I	$(-2\omega;\omega,\omega,0)$	$C_{11} = +2.156 \pm 15\%$	1.06	410b
Methylene iodide CH_2I_2	$(-2\omega;\omega,\omega,0)$	$C_{11} = +2.562 \pm 15\%$	1.06	410b
Neon	$(-2\omega;0,\omega,\omega,)$	$C_{22} = 0.00735 \pm 0.00024$	0.6943	412
Ne	$(-3\omega;\omega,\omega,-\omega)$	$C_{11} = 0.00312 \pm 0.00053$	0.6943	96
Nitric oxide NO	$(-2\omega_1 + \omega_2;\omega_1,\omega_1,-\omega_2)$	$C_{11} = 0.0588 \pm 10\%$	0.6943	142
m-Nitroaniline *m*-$C_6H_4 \cdot NH_2 \cdot NO_2$	$(-2\omega;\omega,\omega,0)$	$C_{11} = 29.75 \pm 15\%$	1.318	410f

Table 1.1.15b (continued)
THIRD-ORDER NONLINEAR OPTICAL COEFFICIENT, C_{jn}^{mic}

Material	NLO-Process	Coefficient $C_{jn}^{mic} \times 10^{50}$ $m^5 \cdot V^{-2}$	λ (μm)	Ref.
O-Nitroaniline $O\text{-}C_6H_4 \cdot NH_2 \cdot NO_2$	$(-2\omega;\omega,\omega,0)$	$C_{11} = 43.05 \pm 15\%$	1.318	410f
p-Nitroaniline @ 129°C	$(-2\omega;\omega,\omega,0)$	$C_{11} = 58.45 \pm 15\%$	1.318	410h
p-Nitroaniline $p\text{-}C_6H_4 \cdot NH_2 \cdot NO_2$	$(-2\omega;\omega,\omega,0)$	$C_{11} = +173.6 \pm 15\%$	1.318	410f
p-Nitroaniline (1%) in methanol	$(-2\omega;\omega,\omega,0)$	$C_{11} = +273.0 \pm 15\%$	1.06	410e
p-Nitroaniline (2—3%) in methanol	$(-2\omega;\omega,\omega,0)$	$C_{11} = +301 \pm 15\%$	1.06	410e
p-Nitroaniline (4 to 7%) in methanol	$(-2\omega,\omega,\omega,o)$	$C_{11} = +238 \pm 15\%$	1.06	410e
Nitrobenzene @ 23°C $C_6H_5NO_2$	$(-2\omega;\omega,\omega,o)$	$C_{11} = +15.155 \pm 15\%$	1.318	410f
@ 150°C		$C_{11} = +12.25 \pm 15\%$	1.06	410b
@ 178°C		$C_{11} = +11.2 \pm 15\%$	1.06	410b
15% Nitrobenzene in *p*-xylene		$C_{11} = +9.1 \pm 15\%$	1.06	410e
11% Nitrobenzene in *p*-dioxane	$(-2\omega;\omega,\omega,o)$	$C_{11} = +10.045 \pm 15\%$	1.06	410e
17% Nitrobenzene in methanol	$(-2\omega;\omega,\omega,o)$	$C_{11} = +18.2 \pm 15\%$	1.06	410e
50% Nitrobenzene in aniline	$(-2\omega;\omega,\omega,o)$	$C_{11} = +11.725 \pm 15\%$	1.06	410e
Nitrocyclohexane $C_6H_{11} \cdot NO_2$	$(-2\omega;\omega,\omega,o)$	$C_{11} = +1.575$	1.06	410b
Nitrogen N_2	$(-2\omega_1 + \omega_2;\omega_1,\omega_1,-\omega_2)$	$C_{11} = 0.0189 \pm 10\%$	0.6943	142
	$(-3\omega;\omega,\omega,-\omega)$	$C_{11} = 0.03745 \pm 0.006$	0.6943	96
Nitromethane $CH_3 \cdot NO_2$	$(-2\omega;\omega,\omega,o)$	$C_{11} = +0.672 \pm 15\%$	1.06	410b
1-Nitropropane $1,C_3H_7-NO_2$	$(-2\omega;\omega,\omega,o)$	$C_{11} = +1.0045 \pm 15\%$	1.06	410b
2-Nitropropane $CH_3(CH-NO_2)CH_3$	$(-2\omega;\omega,\omega,o)$	$C_{11} = 0.763 \pm 15\%$	1.06	410b
Oxygen O_2	$(-2\omega_1 + \omega_2;\omega_1,\omega_1,-\omega_2)$	$C_{11} = 0.0182 \pm 10\%$	0.6943	142
Poly-γ-benzyl-L-glutamate (PBLG)	$(-2\omega;\omega,\omega,o)$	$C_{11} = +5.2 \times 10^6 \pm 50\%$	1.06	410c
Pyridine-iodine complex $(C_6H_5N + I_2)$	$(-2\omega;\omega,\omega,o)$	$C_{11} = +60.90 \pm 15\%$	1.318	410d
Conc.: $0.47 \times 10^{20} cm^{-3}$		$C_{11} = 76.3 \pm 15\%$	1.318	410d
Conc.: $1.09 \times 10^{20} cm^{-3}$				
Conc.: $1.7 \times 10^{20} cm^{-3}$		$C_{11} = +72.1 \pm 15\%$	1.318	410d
Conc.: $1.97 \times 10^{20} cm^{-3}$		$C_{11} = +66.15 \pm 15\%$	1.318	410d
Pyridine + iodine + chlorine complex				
Conc.: $0.70 \times 10^{20} cm^{-3}$	$(-2\omega;\omega,\omega,o)$	$C_{11} = +21.0 \pm 15\%$	1.318	410g
Conc.: $0.79 \times 10^{20} cm^{-3}$		$C_{11} = +18.55 \pm 15\%$	1.318	410g
Rubidium vapor Rb	$(-3\omega;\omega,\omega,-\omega)$	$C_{11} = 4900$	1.064	151
Silicon tetrafluoride SiF_4	$(-2\omega_1 + \omega_2;\omega_1,\omega_1,-\omega_2)$	$C_{11} = 0.0315 \pm 10\%$	0.6943	142
trans-Stilbene	$(-2\omega;\omega,\omega,o)$	$C_{11} = 5.39 \pm 15\%$	1.064	410h
Sulfur hexafluoride SF_6	$(-2\omega_1 + \omega_2;\omega_1,\omega_1,-\omega_2)$	$C_{11} = 0.035 \pm 10\%$	0.6943	142
	$(-3\omega;\omega,\omega,-\omega)$	$C_{11} = 5862$	10.6	146

Table 1.1.15b (continued)
THIRD-ORDER NONLINEAR OPTICAL COEFFICIENT, C_{jn}^{mic}

Material	NLO-Process	Coefficient $C_{jn}^{mic} \times 10^{50}$ m^5·V^{-2}	λ (μm)	Ref.
Tetrafluoromethane	$(-2\omega;o,\omega,\omega)$	$C_{22} = 0.0637 \pm 0.0021$	0.6943	414
CF_4	$(-\omega;o,o,\omega)$	$C_{jn}^{mic} = 0.0875 \pm 0.005$	0.6943	416
Toluene $C_6H_5CH_3$	$(-2\omega;\omega,\omega,o)$	$C_{11} = +1.0465 \pm 15\%$	1.06	410b
Trifluoromethane CHF_3	$(-2\omega_1 + \omega_2;\omega_1,\omega_1,-\omega_2)$	$C_{jn}^{mic} = 0.1071$	0.6943	415
	$(-2\omega;o,\omega,\omega)$	$C_{22} = 0.0952 \pm 0.0042$	0.6943	414
Water H_2O	$(-2\omega;\omega,\omega,o)$	$C_{11} = +0.504 \pm 15\%$	1.06	410e
Xenon	$(-2\omega;o,\omega,\omega)$	$C_{22} = 0.5635 \pm 0.0392$	0.6943	412
Xe		$C_{11} = 0.3426 \pm 0.0655$	0.6943	96
p-Xylene *p*-$C_6H_4(CH_3)_2$	$(-2\omega;\omega,\omega,o)$	$C_{11} = +1.0465 \pm 15\%$	1.06	410b

Table 1.1.16
SECOND-HARMONIC COEFFICIENTS

System — Monoclinic

Material	Symmetry class	$d_{jm} \times 10^{12}$ (*m/V*)	$\delta_{jm} \times 10^2$ (m^2/C)	λ_1 (μm)	n_ω	$n_{2\omega}$	Measured relative to	Ref.
Ammonium tartrate	2-C_2	$d_{21} = 0.24 \pm 0.04$	$\delta_{21} = 0.86 \pm 0.14$	1.153	1.5667	1.5781	d_{36}^{KDP}	171
$(NH_4)_2C_4H_4O_6$		$d_{36} = 0.25 \pm 0.04$	$\delta_{36} = 1.01 \pm 0.16$	1.153	1.5628	1.5303	d_{36}^{KDP}	171
		$0.835d_{34} \pm 0.165\ d_{16}$ $-0.733d_{14} = 0.37 \pm 0.5$		1.153			d_{36}^{KDP}	171
Benzanthracene	2-C_2	$d_{21} = 0.064$		0.6943			$d_{11}^{\alpha\text{-}SiO_2}$	167
$C_{18}H_{12}$		$d_{22} = 0.0096$		0.6943			$d_{11}^{\alpha\text{-}SiO_2}$	167
		$d_{16} = 0.003$		0.6943			$d_{11}^{\alpha\text{-}SiO_2}$	167
7-Diethylamino-4-	2-C_2	$d_{21} = 4.1$	$\delta_{21} = 4.52$	1.06	1.871	1.667	d_{36}^{KDP}	165
methylcoumarin (DMC)		$d_{22} = 1.6$	$\delta_{22} = 4.26$	1.06	1.616	1.667	d_{36}^{KDP}	165
$C_{14}H_{17}NO_2$		$d_{23} = 0.53$	$\delta_{23} = 2.27$	1.06	1.506	1.667	d_{36}^{KDP}	165
		$d_{25} = 0.61$	$\delta_{25} = 1.16$	1.06	1.743	1.524	d_{36}^{KDP}	165
Dipotassium tartrate	2-C_2	$d_{21} = 0.12$	$\delta_{21} = 0.65$	0.6943	1.4893	1.5693	d_{36}^{KDP}	163
hemihydrate		$d_{22} = 4.1$	$\delta_{22} = 18.3$	0.6943	1.5194	1.5693	d_{36}^{KDP}	163
$K_2C_4H_4O_6 - {}^1/_2\ H_2O$		$d_{23} = 0$	$\delta_{23} = 0$	0.6943	1.5294	1.5693	d_{36}^{KDP}	163
		$d_{25} = 0.18$	$\delta_{25} = 0.87$	0.6943	1.5043	1.5693	d_{36}^{KDP}	163
Lithium sulfate	2-C_2	$d_{22} = 0.40 \pm 0.06$	$\delta_{22} = 2.8 \pm 0.4$	1.064	1.4657	1.4868	$d_{11}^{\alpha\text{-}SiO_2}$	162
monohydrate		$d_{23} = 0.29 \pm 0.04$	$\delta_{23} = 1.96 \pm 0.27$	1.064	1.4752	1.4868	$d_{11}^{\alpha\text{-}SiO_2}$	162
$Li_2SO_4 \cdot H_2O$		$d_{34} = 0.25 \pm 0.04$	$\delta_{34} = 1.77 \pm 0.28$	1.064	1.4704	1.4769	$d_{11}^{\beta\text{-}SiO_2}$	162
Methyl-(2,4-dinitro	2-C_2	$d_{23} = 10.67 \pm 1.33$	$\delta_{23} = 18.3 \pm 2.3$	1.064	—	1.7100	$d_{11}^{\alpha\text{-}SiO_2}$	166
phenyl)-amino-		$d_{22} = 11.7 \pm 1.3$	$\delta_{22} = 28.3 \pm 3.1$	1.064	1.5991	1.7100	$d_{11}^{\alpha\text{-}SiO_2}$	166
propanoate (MAP)		$d_{21} = 2.35 \pm 0.5$	$\delta_{21} = 5.5 \pm 1.25$	1.064	—	1.7100	$d_{11}^{\alpha\text{-}SiO_2}$	166
$C_{10}H_{12}N_3O_6$		$d_{25} = -0.35 \pm 0.3$	$\delta_{25} = -0.69 \pm 0.59$	1.064	—	1.7100	$d_{11}^{\alpha\text{-}SiO_2}$	166
Phenanthrene	2-C_2	$d_{21} = 0.013$		0.6943			$d_{11}^{\alpha\text{-}SiO_2}$	167
$C_{14}H_{10}$		$d_{22} = 0.026$		0.6943			$d_{11}^{\alpha\text{-}SiO_2}$	167
		$d_{16} = 0.003$		0.6943			$d_{11}^{\alpha\text{-}SiO_2}$	167

Sucrose								
$C_{12}H_{22}O_{11}$	2-C_2	$d << d_{36}^{ADP}$					d_{36}^{ADP}	168
Triglycine sulfate (TGS)								
$(NH_2CH_2COOH)_3 \cdot H_2SO_4$	2-C_2	$d_{23} = 0.32$	$\delta_{23} = 1.09$	0.6943	1.567	1.618	d_{36}^{KDP}	164
Ammonium malate								
NH_4–OOC–CHOH–CH_2–COOH–H_2O	m-C_s	$d_{32} \simeq 0.45$		0.60			d_{36}^{KDP}	274
2-Methyl-4-nitro-aniline (MNA)	m-C_s	$d_{11} = 160 \pm 40$	$\delta_{11} = 94 \pm 23$	1.064	1.8	2.2	$d_{11}^{\alpha\text{-}SiO_2}$	169
CH_3–NH_2–NO_2–C_6H_4		$d_{12} = 24 \pm 6$	$\delta_{12} = 24 \pm 6$	1.064	1.8	1.8	$d_{11}^{\alpha\text{-}SO_2}$	169
Potassium malate	m-C_s	$d_{11} = 6.62 \pm 20\%$		0.630			d_{36}^{KDP}	170
COOK–-CHOH–$CH_2 \cdot 1.5H_2O$		$d_{12} = 0.6 \pm 20\%$		0.630			d_{36}^{KDP}	170
		$d_{13} = 0.66 \pm 20\%$		0.630			d_{36}^{KDP}	170
		$d_{15} = 2.52 \pm 20\%$		0.630			d_{36}^{KDP}	170
		$d_{24} = 1.39 \pm 20\%$		0.630			d_{36}^{KDP}	170
		$d_{26} = 0.53 \pm 20\%$		0.630			d_{36}^{KDP}	170
		$d_{31} = 0.66 \pm 20\%$		0.630			d_{36}^{KDP}	170
		$d_{32} = 2.52 \pm 20\%$		0.630			d_{36}^{KDP}	170
		$d_{33} = 3.97 \pm 20\%$		0.630			d_{36}^{KDP}	170
		$d_{35} = 0.26 \pm 20\%$		0.630			d_{36}^{KDP}	170
System — Orthorhombic								
Ammonium oxalate	222-D_2	$d_{14} = 0.37$	$\delta_{14} = 1.89$	1.06	1.4802	1.5996	d_{36}^{KDP}	173
monohydrate		$(^1/_2)(d_{25} + d_{36}) = 0.68 \pm 0.06$					d_{36}^{KDP}	173
$(NH_4)_2C_2O_4 \cdot H_2O$								
Barium formate	222-D_2	$d_{14} = 0.11 \pm 11\%$	$\delta_{14} = 0.34 \pm 11\%$	1.064	1.6017	1.5773	d_{36}^{KDP}	178
$Ba(COOH)_2$		$d_{25} = 0.11 \pm 14\%$	$\delta_{25} = 0.34 \pm 14\%$	1.064	1.5900	1.6019	d_{36}^{KDP}	178
		$d_{36} = 0.11 \pm 17\%$	$\delta_{36} = 0.34 \pm 17\%$	1.064	1.5702	1.6407	d_{36}^{KDP}	178
		$d_{36} = 0.13 \pm 11\%$		1.064				
Benzophenone								
$(C_6H_5)_2$-CO	222-D_2	$d_{36} = 0.35$		0.6943			$d_{11}^{\alpha\text{-}SiO_2}$	167
1,8-Dinitronaphthalene	222-D_2	$d_{25} = 0.35$		0.6943			$d_{11}^{\alpha\text{-}SiO_2}$	167
$C_{10}H_6O_4N_2$		$d_{36} = 0.35$		0.6943			$d_{11}^{\alpha\text{-}SiO_2}$	167

Table 1.1.16 (continued)
SECOND-HARMONIC COEFFICIENTS

System — Orthorhombic

Material	Symmetry class	$d_{jm} \times 10^{12}$ (m/V)	$\delta_{jm} \times 10^{2}$ (m^2/C)	λ_1 (μm)	n_ω	$n_{2\omega}$	Measured relative to	Ref.
D-Glucose	222-D_2	$d_{14} \simeq 0.32$		1.064			$d_{11}^{\alpha\text{-}SiO_2}$	172
Glutamic acid hydrochloride $C_5H_{10}O_4NCl$	222-D_2	$d_{36} = 0.17$		1.064			d_{36}^{ADP}	179
Hippuric acid $C_6H_5CO–NH(CH_2CO_2H)$	222-D_2	$d_{36} = 2.5$	$\delta_{14} = 6.3$	0.6943	1.561	1.78	d_{36}^{ADP}	191
α-Iodic acid α-HIO_3	222-D_2	$d_{36} = 6.4 \pm 1.64$	$\delta_{36} = 3.89 \pm 0.96$	1.064	1.9391	1.8547	$d_{11}^{\alpha\text{-}SiO_2}$	174
							$d_{31}^{LiNbO_3}$	174
		$d_{36} = 5.15 \pm 0.16$	$\delta_{36} = 3.13 \pm 0.1$	1.064	1.9391	1.8547	$d_{33}^{\alpha\text{-}SiO_2}$ $d_{33}^{LiIO_3}$	175
		$d_{36} = 4.18 \pm 0.78$	$\delta_{36} = 2.54 \pm 0.47$	1.1526	1.935	1.845	d_{36}^{KDP}	176
		$d_{36} = 4.95 \pm 0.74$	$\delta_{36} = 3.01 \pm 0.43$	1.1526			d_{36}^{ADP}	176
3-Methyl-4-nitropyridine-1-oxide (POM) $NO_2·CH_3NOC_5H_4$	222-D_{2h}	$d_{36} = 6.4 \pm 15\%$	$\delta_{36} = 9.85 \pm 15\%$	1.0642	1.745	1.660	$d_{11}^{\alpha\text{-}SiO_2}$	181
		$d_{36} = 5.53 \pm 15\%$	$\delta_{36} = 8.53 \pm 15\%$				and d_{36}^{KDP}	181
5-Nitrouracil $C_4N_3H_3O_4$	222-D_2	$d_{36} = 5.15 \pm 1.03$	$\delta_{36} = 5.66 \pm 1.13$	1.064	1.82	1.71	$d_{33}^{LiIO_3}$	180
Strontium formate $Sr(COOH)_2$	222-D_2	$d_{14} = 0.51$	$\delta_{14} = 2.1$	1.064	1.553	1.545	d_{36}^{KDP}	192
Strontium formate dihydrate $Sr(COOH)_2·2H_2O$	222-D_2	$d_{14} = 0.33$	$\delta_{14} = 1.83$	1.064	1.517	1.488	d_{36}^{KDP}	192
D-Thronine	222-D_2	$d_{14} = 0.4 \pm 0.06$	$\delta_{14} = 1.51 \pm 0.24$	1.064	1.5821	1.5243	$d_{11}^{\alpha\text{-}SiO_2}$	177
		$d_{25} = 0.43 \pm 0.06$	$\delta_{25} = 1.63 \pm 0.24$	1.064	1.5471	1.5965	$d_{11}^{\alpha\text{-}SiO_2}$	177
		$d_{36} = 0.40 \pm 0.06$	$\delta_{36} = 1.53 \pm 0.24$	1.064	1.5440	1.6043	$d_{11}^{\alpha\text{-}SiO_2}$	177
Acenaphthene $C_{10}H_6CH_2CH_2$	$mm2$-C_{2v}	$d_{31} = 0.14 \pm 15\%$		0.694	1.4065		$d_{11}^{\alpha\text{-}SiO_2}$	167
		$d_{33} = 1.21 \pm 15\%$		0.6943	1.6201		$d_{11}^{\alpha\text{-}SiO}$	167

m-Aminophenol (AP) OH -C_6H_4-NH_2	*mm*2-C_{2v}	d_{31} = 1.03 ± 15%	δ_{31} = 2.88 ± 15%	1.064			d_{36}^{KDP}	203
		d_{32} = 1.44 ± 15%	δ_{32} = 2.62 ± 15%	1.064			d_{36}^{KDP}	203
		d_{33} = 1.48 ± 15%	δ_{33} = 5.5 ± 15%	1.064			d_{36}^{KDP}	203
		d_{15} = 0.98 ± 15%	δ_{15} = 2.62 ± 15%	1.064			d_{36}^{KDP}	203
		d_{24} = 1.27 ± 15%	δ_{24} = 2.36 ± 15%	1.064			d_{36}^{KDP}	203
		d_{31} = 0.48 ± 0.08	δ_{31} = 1.28	1.064	1.639	1.589	$d_{11}^{\alpha\text{-}SiO_2}$	205
		d_{32} = 1.52 ± 0.26	δ_{32} = 2.75	1.064	1.736	1.589	$d_{11}^{\alpha\text{-}SiO_2}$	205
		d_{33} = 2.15 ± 0.38	δ_{33} = 7.64	1.064	1.562	1.589	$d_{11}^{\alpha\text{-}SiO_2}$	205
Antimony niobate $SbNbO_4$	*mm*2-C_{2v}	d_{32} = 4.72 ± 0.82	—	1.058			d_{36}^{KDP}	211
Antimony tantalate $SbTaO_4$	*mm*2-C_{2v}	d_{32} = 4.1 ± 0.82	—	1.058			d_{36}^{KDP}	211
Barium magnesium fluoride $BaMgF_4$	*mm*2-C_{2v}	d_{31} = 0.023 ± 20%	δ_{31} = 0.17 ± 20%	1.064	1.467	1.467	$d_{11}^{\alpha\text{-}SiO_2}$	216
		d_{32} = 0.042 ± 20%	δ_{32} = 0.36 ± 20%	1.064	1.439	1.467	$d_{11}^{\alpha\text{-}SiO_2}$	216
		d_{33} = 0.016 ± 20%	δ_{33} = 0.12 ± 20%	1.064	1.458	1.467	$d_{11}^{\alpha\text{-}SiO_2}$	216
		d_{15} = 0.023 ± 20%	δ_{15} = 0.17 ± 20%	1.064	1.4625	1.473	$d_{11}^{\alpha\text{-}SiO_2}$	216
		d_{24} = 0.023 ± 20%	δ_{24} = 0.19 ± 20%	1.064	1.4485	1.450	$d_{11}^{\alpha\text{-}SiO_2}$	216
		d_{31} = 0.023 ± 23%	δ_{31} = 0.17 ± 23%	1.064	1.4674	1.4678	d_{36}^{KDP}	178
		d_{32} = 0.035 ± 12%	δ_{32} = 0.29 ± 12%	1.064	1.4436	1.4678	d_{36}^{KDP}	178
		d_{33} = 0.0094 ± 14%	δ_{33} = 0.07 ± 14%	1.064	1.4604	1.4678	d_{36}^{KDP}	178
		d_{24} = 0.025 ± 17%	δ_{24} = 0.22 ± 17%	1.064	1.4436	1.4508	d_{36}^{KDP}	178
Barium sodium niobate	*mm*2-C_{2v}	d_{31} = −12.8 ± 1.28	δ_{31} = −2.13 ± 0.14	1.0642	2.2570	2.2502	$d_{11}^{\alpha\text{-}SiO_2}$	185, 188
$Ba_2NaNb_5O_{15}$		d_{32} = −12.8 ± 0.64	δ_{32} = −2.10 ± 0.11	1.0642	2.2584	2.2502	$d_{11}^{\alpha\text{-}SiO_2}$	185, 188
		d_{33} = −17.6 ± 1.28	δ_{33} = −3.59 ± 0.28	1.0642	2.1700	2.2502	$d_{11}^{\alpha\text{-}SiO_2}$	185, 188
		d_{15} = −12.8 ± 0.64	δ_{15} = −1.97 ± 0.1	1.0642	2.2135	2.3656	$d_{11}^{\alpha\text{-}SiO_2}$	185, 188
		d_{24} = −12.8 ± 0.64	δ_{24} = −2.10 ± 0.11	1.0642	2.2142	2.3673	$d_{11}^{\alpha\text{-}SiO_2}$	185, 188
Barium zinc fluoride	*mm*2-C_{2v}	d_{31} = 0.008 ± 20%	δ_{31} = 0.042 ± 20%	1.06	1.514	1.517	$d_{11}^{\alpha\text{-}SiO_2}$	216

Table 1.1.16 (continued)
SECOND-HARMONIC COEFFICIENTS

System — Orthorhombic

Material	Symmetry class	$d_{jm} \times 10^{12}$ (m/V)	$\delta_{jm} \times 10^{2}$ (m^2/C)	λ_1 (μm)	n_ω	$n_{2\omega}$	Measured relative to	Ref.
$BaZnF_4$		$d_{32} = 0.08 \pm 20\%$	$\delta_{32} = 0.45 \pm 20\%$	1.06	1.499	1.517	$d_{11}^{\alpha\text{-}SiO_2}$	216
		$d_{33} = 0.035 \pm 20\%$	$\delta_{33} = 0.19 \pm 20\%$	1.06	1.507	1.517	$d_{11}^{\alpha\text{-}SiO_2}$	216
		$d_{15} = 0.011 \pm 20\%$	$\delta_{15} = 0.06 \pm 20\%$	1.06	1.5105	1.524	$d_{11}^{\alpha\text{-}SiO_2}$	216
2-Bromo-4-nitro-	$mm2$-C_{2v}	$d_{31} = 6.15 \pm 30\%$	$\delta_{31} = 13.1 \pm 30\%$	1.064			d_{36}^{KDP}	203
aniline		$d_{32} = 14.35 \pm 30\%$	$d_{32} = 18.34 \pm 30\%$	1.064			d_{36}^{KDP}	
$Br(C_6H_3NO_2NH_2)$		$d_{33} = 2.79 \pm 30\%$	$\delta_{33} = 11.0 \pm 30\%$	1.064			d_{36}^{KDP}	
		$d_{15} = 4.51 \pm 30\%$	$\delta_{15} = 9.17 \pm 30\%$	1.064			d_{36}^{KDP}	
		$d_{24} = 11.48 \pm 30\%$	$\delta_{24} = 13.36 \pm 30\%$	1.064			d_{36}^{KDP}	
m-Bromonitrobenzene	$mm2$-C_{2v}	$d_{31} = 2.56 \pm 0.48$	$\delta_{31} = 4.74$	1.064	1.683	1.695	$d_{11}^{\alpha\text{-}SiO_2}$	205
(BNB)		$d_{32} = 2.94 \pm 0.8$	$\delta_{32} = 4.89$	1.064	1.705	1.695	$d_{11}^{\alpha\text{-}SiO_2}$	205
$Br(C_6H_4NO_2)$		$d_{33} = 5.12 \pm 0.64$	$\delta_{33} = 10.24$	1.064	1.656	1.695	$d_{11}^{\alpha\text{-}SiO_2}$	205
Calcium iodate	$mm2$-C_{2v}	$d_{31} = 0.72$	$\delta_{31} = 1.95$	1.064	1.582	1.688	$d_{33}^{LiIO_3}$	213
hexahydrate		$d_{32} = 0.81$	$\delta_{32} = 1.84$	1.064	1.625	1.688	$d_{33}^{LiIO_3}$	213
$Ca(IO_3)_2 6H_2O$		$d_{33} = 2.1$	$\delta_{33} = 4.27$	1.064	1.653	1.688	$d_{33}^{LiIO_3}$	213
		$d_{15} = 0.57$	$\delta_{15} = 1.50$	1.064	1.6175	1.625	$d_{33}^{LiIO_3}$	213
		$d_{24} = 0.036$	$\delta_{24} = 0.08$	1.064	1.6390	1.654	$d_{33}^{LiIO_3}$	213
2-Chloro-4-nitroaniline	$mm2$-C_{2v}	$d_{31} = 6.15 \pm 15\%$	$\delta_{31} = 13.6 \pm 15\%$	1.064			d_{36}^{KDP}	203
$Cl(C_6H_3NO_2NH_2)$		$d_{32} = 14.35 \pm 15\%$	$\delta_{32} = 18.6 \pm 15\%$	1.064			d_{36}^{KDP}	203
		$d_{33} = 2.67 \pm 15\%$	$\delta_{33} = 13.1 \pm 15\%$	1.064			d_{36}^{KDP}	203
		$d_{15} = 5.33 \pm 15\%$	$\delta_{15} = 11.0 \pm 15\%$	1.064			d_{36}^{KDP}	203
		$d_{24} = 14.35 \pm 15\%$	$\delta_{24} = 16.8 \pm 15\%$	1.064			d_{36}^{KDP}	203
m-Chloronitrobenzene		$d_{31} = 2.94 \pm 0.64$	$\delta_{31} = 6.72$	1.064	1.6557	1.663	$d_{11}^{\alpha\text{-}SiO_2}$	205
(CNB)	$mm2$-C_{2v}	$d_{32} = 2.56 \pm 0.8$	$\delta_{32} = 5.35$	1.064	1.6626	1.663	$d_{11}^{\alpha\text{-}SiO_2}$	205
$Cl(C_6H_4NO_2)$		$d_{33} = 4.96 \pm 0.8$	$\delta_{32} = 11.92$	1.064	1.624	1.663	$d_{11}^{\alpha\text{-}SiO_2}$	205
2,4-Diaminitoluene	$mm2$-C_{2v}	$d_{31} = 0.41 \pm 0.08$	$d_{31} = 0.92 \pm 0.18$	1.064	1.5930	1.7676	d_{36}^{KDP}	209

(m-Tolylenediamine)MTD		$d_{32} = 0.94 \pm 0.2$	$\delta_{32} = 1.12 \pm 1.24$	1.064	1.7644	1.7676	d_{36}^{KDP}	209
$CH_3C_6H_3(NH_2)_2$		$d_{33} = 0.66 \pm 0.16$	$\delta_{33} = 0.90 \pm 0.22$	1.064	1.7240	1.7676	d_{36}^{KDP}	209
		$d_{15} = 0.37 \pm 0.08$	$\delta_{15} = 0.84 \pm 0.18$	1.064	1.6585	1.6226	d_{36}^{KDP}	209
		$d_{24} = 1.07 \pm 0.21$	$\delta_{24} = 1.26 \pm 0.25$	1.064	1.7442	1.8189	d_{36}^{KDP}	209
1,3-Diiodobenzene	$mm2$-C_{2v}	$d_{32} = 2.70 \pm 10\%$		0.6943			$d_{11}^{\alpha\text{-}SiO_2}$	167
$C_6H_4I_2$		$d_{33} = 0.83 \pm 10\%$		0.6943			$d_{11}^{\alpha\text{-}SiO_2}$	167
m-Dinitrobenzene	$mm2$-C_{2v}	$d_{31} = 3.39 \pm 30\%$		0.6943	—	—	$d_{11}^{\alpha\text{-}SiO_2}$	167
$C_6H_4(NO_2)_2$		$d_{32} = 2.32 \pm 30\%$		0.6943			$d_{11}^{\alpha\text{-}SiO_2}$	167
		$d_{33} = 3.11 \pm 30\%$		0.6943			$d_{11}^{\alpha\text{-}SiO_2}$	167
		$d_{31} < 1.03 \pm 50\%$	$\delta_{31} < 2.88 \pm 50\%$	1.064			d_{36}^{KDP}	203
		$d_{32} = 1.85 \pm 50\%$	$\delta_{32} = 5.5 \pm 50\%$	1.064			d_{36}^{KDP}	203
		$d_{33} = 0.66 \pm 50\%$	$\delta_{33} = 4.72 \pm 50\%$	1.064			d_{36}^{KDP}	203
		$d_{15} < 0.82 \pm 50\%$	$\delta_{15} < 2.1 \pm 50\%$	1.064			d_{36}^{KDP}	203
		$d_{24} = 1.48 \pm 50\%$	$\delta_{24} = 6.81$	1.064			d_{36}^{KDP}	203
		$d_{31} = 0.48$	$\delta_{31} = 1.22$	1.064	1.706	1.493	$d_{11}^{\alpha\text{-}SiO_2}$	206
		$d_{32} = 1.92$	$\delta_{32} = 5.82$	1.064	1.655	1.493	$d_{11}^{\alpha\text{-}SiO_2}$	206
		$d_{33} = 0.45$	$\delta_{33} = 3.0$	1.064	1.473	1.493	$d_{11}^{\alpha\text{-}SiO_2}$	206
		$d_{31} \leqslant 1.68$	$\delta_{31} = 4.2$	1.064	1.7094	1.4912	d_{36}^{KDP}	207
		$d_{32} = 2.54 \pm 0.5$	$\delta_{32} = 7.79 \pm 1.53$	1.064	1.6538	1.4912	d_{36}^{KDP}	207
		$d_{33} = 0.7 \pm 0.1$	$\delta_{33} = 4.76 \pm 0.68$	1.064	1.4714	1.4912	d_{36}^{KDP}	207
		$d_{32} = 1.24 \pm 0.08$	$\delta_{32} = 3.9 \pm 0.25$	1.153	1.6477	1.4905	d_{36}^{KDP}	208
Gadolinium molybdate	$mm2$-C_{2v}	$d_{31} = -2.49 \pm 0.37$	$\delta_{31} = -2.03 \pm 0.3$	1.0642	1.8146	1.9102	$d_{11}^{\alpha\text{-}SiO_2}$	187—189
$Gd_2(MoO_4)_3$		$d_{32} = +2.42 \pm 0.36$	$\delta_{32} = +1.97 \pm 0.3$	1.0642	1.8142	1.9102	$d_{11}^{\alpha\text{-}SiO_2}$	187—189
		$d_{33} = -0.044 \pm 0.008$	$\delta_{33} = -0.035 \pm 0.006$	1.0642	1.8637	1.9102	$d_{11}^{\alpha\text{-}SiO_2}$	187—189
		$d_{15} = -2.62 \pm 0.4$	$\delta_{15} = -2.14 \pm 0.33$	1.0642	1.8386	1.8549	$d_{11}^{\alpha\text{-}SiO_2}$	187—189
		$d_{24} = +2.58 \pm 0.39$	$\delta_{24} = +2.11 \pm 0.32$	1.0642	1.8384	1.8545	$d_{11}^{\alpha\text{-}SiO_2}$	187—189
Lead niobate	$mm2$-C_{2v}	$d_{31} = +6.5 \pm 0.97$	$\delta_{31} = 0.82 \pm 0.12$	1.0642	2.2979	2.4396	$d_{11}^{\alpha\text{-}SiO_2}$	190
$PbNb_4O_{11}$		$d_{32} = -5.87 \pm 0.88$	$\delta_{32} = -0.73 \pm 0.1$	1.0642	2.3010	2.4396	$d_{11}^{\alpha\text{-}SiO_2}$	190
		$d_{33} = -8.88 \pm 1.32$	$\delta_{33} = -1.05 \pm 0.16$	1.0642	2.3254	2.4396	$d_{11}^{\alpha\text{-}SiO_2}$	190
		$d_{15} = +5.89 \pm 0.88$	$\delta_{15} = +0.74 \pm 0.1$	1.0642	2.3115	2.4113	$d_{11}^{\alpha\text{-}SiO_2}$	190

Table 1.1.16 (continued)
SECOND-HARMONIC COEFFICIENTS

System — Orthorhombic

Material	Symmetry class	$d_{jm} \times 10^{12}$ (m/V)	$\delta_{jm} \times 10^{2}$ (m^2/C)	λ_1 (μm)	n_ω	$n_{2\omega}$	Measured relative to	Ref.
		$d_{24} = -5.42 \pm 0.39$	$\delta_{24} = -0.67 \pm 0.1$	1.0642	2.3131	2.4137	$d_{11}^{\alpha\text{-}SiO_2}$	190
Lithium formate monohydrate	$mm2$-C_{2v}	$d_{31} = d_{15} = +0.096 \pm 0.02$	$\delta_{31} = +1.14$	1.064	1.3593	1.5229	$d_{11}^{\alpha\text{-}SiO_2}$	194, 188
		$d_{32} = d_{24} = -1.12 \pm 0.08$	$\delta_{32} = -7.24$	1.064	1.4673	1.5229	$d_{11}^{\alpha\text{-}SiO_2}$	194, 188
$LiCHO_2 \cdot H_2O$		$d_{33} = +1.63 \pm 0.096$	$\delta_{33} = +8.8$	1.064	1.5035	1.5229	$d_{11}^{\alpha\text{-}SiO_2}$	194, 188
Lithium hydrazine	$mm2$-C_{2v}	$d_{31} = 0.16$	$\delta_{31} = 2.0$	1.064	1.4005	1.4057	$d_{11}^{\alpha\text{-}SiO_2}$	223
fluoroberyllate		$d_{32} = 0.064$	$\delta_{32} = 0.82$		1.3962	1.4057		223
$Li(N_2H_5) \cdot BeF_4$		$d_{33} = 0.32$	$\delta_{33} = 4.06$		1.3983	1.4057		223
Lithium hydrogen	$mm2$-C_{2v}	$d_{31} = 0.03$	$\delta_{31} = 0.11$	1.064	1.571	1.570	$d_{11}^{\alpha\text{-}SiO_2}$	224
phosphite		$d_{32} = 0.16$	$\delta_{32} = 0.60$		1.560	1.570	$d_{11}^{\alpha\text{-}SiO_2}$	224
LiH_2PO_3		$d_{33} = 0.43$	$\delta_{33} = 1.65$		1.555	1.570	$d_{11}^{\alpha\text{-}SiO_2}$	224
		$d_{15} = 0.035$	$\delta_{15} = 0.12$		1.563	1.590	$d_{11}^{\alpha\text{-}SiO_2}$	224
		$d_{24} = 0.17$	$\delta_{24} = 0.64$		1.5575	1.575	$d_{11}^{\alpha\text{-}SiO_2}$	224
		$d_{31} = 0.035$	$\delta_{31} = 0.16$	1.064	1.530	1.540	$d_{11}^{\alpha\text{-}SiO_2}$	225
		$d_{32} = 0.19$	$\delta_{32} = 0.83$		1.541	1.540	$d_{11}^{\alpha\text{-}SiO_2}$	225
		$d_{33} = 0.45$	$\delta_{33} = 2.1$		1.526	1.540	$d_{11}^{\alpha\text{-}SiO_2}$	225
Lithium indium	$mm2$-C_{2v}	$d_{31} = 9.9 \pm 15\%$	$\delta_{31} = 3.04$	10.6	2.0609	2.1181	d_{14}^{GaAs}	212
sulfide		$d_{32} = 8.58 \pm 15\%$	$\delta_{32} = 2.91$	10.6	2.0229	2.1181	d_{14}^{GaAs}	212
$LiInS_2$		$d_{33} = 15.8 \pm 15\%$	$\delta_{33} = 4.89$		2.0577	2.1181	d_{14}^{GaAs}	212
Lithium metagallate	$mm2$-C_{2v}	$d_{31} = d_{15}\ -0.07 \pm 0.01$	$\delta_{31} = \delta_{15} = -0.10 \pm 0.05$	1.064	1.7433	1.7705	$d_{11}^{\alpha\text{-}SiO_2}$	195, 188
$LiGaO_2$		$d_{32} = d_{24} = +0.15 \pm 0.02$	$\delta_{32} = \delta_{24} = +0.23 \pm 0.09$	1.064	1.7131	1.7705	$d_{11}^{\alpha\text{-}SiO_2}$	195, 188
		$d_{33} = -0.59 \pm 0.06$	$\delta_{33} = -0.77 \pm 0.08$	1.064	1.7399	1.7705	$d_{11}^{\alpha\text{-}SiO_2}$	195, 188

Lithium sodium	$mm2\text{-}C_{2v}$	$d_{31} = d_{15} = 0.055$	$\delta_{31} = \delta_{15} = 0.57$	1.064	1.3762	1.5356	$d_{31}^{LiNbO_3}$	197
formate monohydrate		$d_{32} = d_{24} = -0.44 \pm 0.2$	$\delta_{32} = \delta_{24} = -2.95 \pm 1.1$	1.064	1.4524	1.5356	$d_{31}^{LiNbO_3}$	197
$Li_{0.9}Na_{0.1}(COOH){\cdot}H_2O$		$d_{33} = 0.93 \pm 0.27$	$d_{33} = 4.55$	1.064	1.5181	1.5356	$d_{31}^{LiNbO_3}$	197
m-Nitroaniline (NA)	$mm2\text{-}C_{2v}$	$d_{31} = 20.9 \pm 15\%$	$\delta_{31} = 36.7$	1.064			d_{36}^{KDP}	203
$NO_2C_6H_4{\cdot}NH_2$		$d_{32} < 0.41 \pm 15\%$	$\delta_{32} < 0.79$	1.064			d_{36}^{KDP}	203
		$d_{33} = 9.43 \pm 15\%$	$\delta_{33} = 22.3$	1.064			d_{36}^{KDP}	203
		$d_{15} = 22.55 \pm 15\%$	$\delta_{15} = 39.3$	1.064			d_{36}^{KDP}	203
		$d_{31} = 12.48 \pm 1.28$	$\delta_{31} = 19.52 \pm 2.0$	1.064	1.719	1.700	$d_{11}^{\alpha\text{-}SiO_2}$	205
		$d_{32} = 1.02 \pm 0.22$	$\delta_{31} = 1.82 \pm 0.39$	1.064	1.683	1.700	$d_{11}^{\alpha\text{-}SiO_2}$	205
		$d_{33} = 13.12 \pm 1.28$	$\delta_{33} = 28.57 \pm 2.8$	1.064	1.630	1.700	$d_{11}^{\alpha\text{-}SiO_2}$	205
		$d_{31} = 8.64 \pm 1.92$	$\delta_{31} = 14.94 \pm 3.3$	1.319	1.711	1.661	$d_{11}^{\alpha\text{-}SiO_2}$	205
		$d_{33} = 8.0 \pm 1.6$	$\delta_{33} = 19.09 \pm 3.8$	1.319	1.625	1.661	$d_{11}^{\alpha\text{-}SiO_2}$	205
		$d_{31} = 9.6 \pm 35\%$	$\delta_{31} = 13.47 \pm 35\%$	1.064	1.74	1.72	$d_{33}^{LiIO_3}$	204
		$d_{32} = 1.07 \pm 35\%$	$\delta_{32} = 1.67 \pm 35\%$	1.064	1.71	1.72	$d_{33}^{LiIO_3}$	204
		$d_{33} = 11.2 \pm 35\%$	$\delta_{33} = 21.78 \pm 35\%$	1.064	1.65	1.72	$d_{33}^{LiIO_3}$	204
Polyvinylidene fluoride	$mm2\text{-}C_{2v}$	$d_{31} = 0.16$	$\delta_{31} \cong 1.7$	1.064	$n_{av} \cong$	1.42	$d_{11}^{\alpha\text{-}SiO_2}$	202
PVF_2		$d_{32} \cong 0$	$\delta_{32} \cong 0$	1.064			$d_{11}^{\alpha\text{-}SiO_2}$	202
CH_2CF_{2n}		$d_{33} \cong 0.32$	$\delta_{33} \cong 3.4$	1.064			$d_{11}^{\alpha\text{-}SiO_2}$	202
Potassium chloroiodate	$mm2\text{-}C_{2v}$	$d_{31} = 0.049 \pm 0.004$	$\delta_{31} = 0.07 \pm 0.006$	1.064	1.74	1.70	$d_{33}^{LiIO_3}$	215
$K_2H(IO_3)_2Cl$		$d_{32} = 0.86 \pm 0.12$	$\delta_{32} = 0.96 \pm 0.13$		1.82	1.70	$d_{33}^{LiIO_3}$	215
		$d_{33} = 4.51 \pm 0.20$	$\delta_{33}\ 9.86 \pm 0.44$		1.67	1.70	$d_{33}^{LiIO_3}$	215
		$d_{24} = 0.90 \pm 0.16$	$\delta_{24} = 0.96 \pm 0.17$		1.745	1.88	$d_{33}^{LiIO_3}$	215
Potassium dihydrogen	$mm2\text{-}C_{2v}$	$d_{31} = 0.51 \pm 0.04$	$\delta_{31} = 2.99 \pm 0.26$	0.6943			d_{36}^{KDP}	201
phosphate		$d_{32} = 0.34 \pm 0.04$	$\delta_{32} = 2.02 \pm 0.26$	0.6943			d_{36}^{KDP}	
KH_2PO_4		$d_{33} < 0.013$	$\delta_{33} < 0.09$	0.6943			d_{36}^{KDP}	
(phase below −150°)		$d_{15} = 0.50 \pm 0.04$	$\delta_{15} = 2.90 \pm 0.26$	0.6943			d_{36}^{KDP}	
		$d_{24} = 0.34 \pm 0.04$	$\delta_{24} = 1.94 \pm 0.26$	0.6943			d_{36}^{KDP}	
Potassium diphthalate	$mm2\text{-}C_{2v}$	$d_{31} = 0.22$	$\delta_{31} = 0.77$	1.15			d_{36}^{KDP}	210
$KHC_8H_4O_4$		$d_{32} = 0.06$	$\delta_{32} = 0.19$	1.15			d_{36}^{KDP}	210
		$d_{31} = 0.68$	$\delta_{31} = 2.3$	0.63			d_{36}^{KDP}	210
		$d_{32} = 0.16$	$\delta_{32} = 0.38$	0.63			d_{36}^{KDP}	210
Potassium fluoroiodate	$mm2\text{-}C_{2v}$	$d_{31} = \pm 0.57 \pm 25\%$	$\delta_{31} = \pm 1.99 \pm 25\%$	1.064	1.554	1.617	$d_{33}^{LIO_3}$	214
KIO_2F_2		$d_{32} = \pm 0.16 \pm 25\%$	$\delta_{32} = \pm 0.47 \pm 25\%$		1.597	1.617	$d_{33}^{LIO_3}$	214

Table 1.1.16 (continued)
SECOND-HARMONIC COEFFICIENTS

System — Orthorhombic

Material	Symmetry class	$d_{jm} \times 10^{12}$ (*m/V*)	$\delta_{jm} \times 10^2$ (m^2/C)	λ_1 (μm)	n_ω	$n_{2\omega}$	Measured relative to	Ref.
		$d_{33} = \pm 2.79 \pm 25\%$	$\delta_{33} = \pm 8.22 \pm 25\%$		1.594	1.617	$d_{33}^{LIO_3}$	214
		$d_{15} = 0.49 \pm 25\%$	$\delta_{25} = 1.57 \pm 25\%$		1.574	1.569	$d_{33}^{LIO_3}$	214
		$d_{24} = 0.25 \pm 25\%$	$\delta_{24} = 0.73 \pm 25\%$		1.5955	1.621	$d_{33}^{LIO_3}$	214
Potassium niobate	$mm2\text{-}C_{2v}$	$d_{31} = -12.88 \pm 1.03$	$\delta_{31} = -2.25 \pm 0.18$	1.064	2.2574	2.2029	$d_{33}^{LIO_3}$	193
$KNbO_3$		$d_{32} = +11.34 \pm 1.03$	$\delta_{32} = +2.16 \pm 0.2$	1.064	2.2200	2.2029	$d_{33}^{LIO_3}$	193
		$d_{33} = -19.58 \pm 1.03$	$\delta_{33} = -4.7 \pm 0.25$	1.064	2.1196	2.2029	$d_{33}^{LIO_3}$	193
		$d_{15} = -12.36 \pm 2.0$	$\delta_{15} = -2.08 \pm 0.34$	1.064	2.1885	2.3807	$d_{33}^{LIO_3}$	193
		$d_{24} = +11.85 \pm 2.0$	$\delta_{24} = +2.22 \pm 0.37$	1.064	2.1698	2.3224	$d_{33}^{LIO_3}$	193
Potassium penta-borate tetrahydrate $KB_5O_8{\cdot}4H_2O$	$mm2\text{-}C_{2v}$	$d_{31} = 0.046$	$\delta_{31} = 0.27$	0.4342	1.496	1.496	Abs	219
		$d_{32} = 0.003$	$\delta_{32} = 0.025$		1.444	1.496	Abs	219
Potassium rubidium	$mm2\text{-}C_{2v}$	$d_{31} = 6.5$	$\delta_{31} = 5.5$	1.064	1.743	1.891	d_{36}^{KDP}	222
phosphotitanate		$d_{32} = 5.0$	$\delta_{32} = 4.26$	1.064	1.753	1.891	d_{36}^{KDP}	222
$K_xRb_{1-x}TiOPO_4$		$d_{33} = 13.7$	$\delta_{33} = 8.8$	1.064	1.837	1.891	d_{36}^{KDP}	222
		$d_{15} = 6.1$	$\delta_{15} = 5.4$	1.064	1.790	1.781	d_{36}^{KDP}	222
		$d_{24} = 7.6$	$\delta_{24} = 6.39$	1.064	1.795	1.790	d_{36}^{KDP}	222
Potassium thiomolybdate	$mm2\text{-}C_{2v}$	$d_{31} = -+0.96 \pm 20\%$	$\delta_{31} = -+0.69 \pm 20\%$	1.32	1.870	1.875	$d_{11}^{\alpha\text{-}SiO_2}$	218
potassium chloride		$d_{32} = -+1.24 \pm 20\%$	$\delta_{32} = -+0.98 \pm 20\%$	1.32	1.839	1.875	$d_{11}^{\alpha\text{-}SiO_2}$	218
$K_2MoOS_3{\cdot}KCl$		$d_{33} \pm 5.12 \pm 20\%$	$\delta_{33} = \pm 4.37 \pm 20\%$	1.32	1.815	1.875	$d_{11}^{\alpha SiO_2}$	218
		$d_{15} = -+0.93 \pm 20\%$	$d_{15} - +0.66 \pm 20\%$	1.32	1.8425	1.944	$d_{11}^{\alpha\text{-}SiO_2}$	218
		$d_{24} = -+1.12 \pm 20\%$	$d_{24} = -+0.81 \pm 20\%$	1.32	1.8545	1.904	$d_{11}^{\alpha\text{-}SiO_2}$	218
α-Resorcinol	$mm2\text{-}C_{2v}$	$d_{31} = 0.53 \pm 30\%$	$\delta_{31} = 2.1 \pm 30\%$	1.064			d_{36}^{KDP}	203
(1,3-Dihydroxybenzene)		$d_{32} = 0.66 \pm 30\%$	$\delta_{32} = 2.1 \pm 30\%$	1.064			d_{36}^{KDP}	203
$C_6H_4(OH)_2$		$d_{33} = 0.70 \pm 30\%$	$\delta_{33} = 1.83 \pm 30\%$	1.064			d_{36}^{KDP}	203
		$d_{24} = 0.49 \pm 30\%$	$\delta_{24} = 1.57 \pm 30\%$	1.064			d_{36}^{KDP}	203

		$d_{31} = 1.49 \pm 20\%$	$\delta_{31} = 4.97 \pm 20\%$	1.064	1.560	1.627	$d_{33}^{LIO_3}$	204
		$d_{32} = 0.54 \pm 20\%$	$\delta_{32} = 1.49 \pm 20\%$	1.064	1.605	1.627	$d_{33}^{LIO_3}$	204
		$d_{33} = 1.25 \pm 20\%$	$\delta_{33} = 3.57 \pm 20\%$	1.064	1.597	1.627	$d_{33}^{LIO_3}$	204
		$d_{31} = 1.38 \pm 10\%$	—	0.6943	—	—	$d_{11}^{\alpha\text{-SiO}_2}$	167
		$d_{33} = 1.07 \pm 10\%$	—	0.6943	—	—	$d_{31}^{LiNbO_3}$	167
Sodium formate	$mm2\text{-}C_{2v}$	d_{31}	δ_{31}	1.064	1.3814	1.5471	d_{31}	197
Na(COOH)		$d_{15} = -0.22 \pm 0.11$	$\delta_{15} = 02.16 \pm 1.1$					
		$d_{32} = d_{15} \cong 0.022$	$\delta_{32} = \delta_{24} \cong 0.15$	1.064	1.4531	1.5471	$d_{31}^{LNbO_3}$	197
		$d_{33} = 0.33 \pm 0.16$	$\delta_{33} = 1.49 \pm 0.7$	1.064	1.5295	1.5471	$d_{31}^{LNbO_3}$	197
Sodium nitrite	$mm2\text{-}C_{2v}$	$d_{31} = 0.074 \pm 0.008$	$\delta_{31} = 1.37 \pm 0.15$	1.064	1.3353	1.4125	$d_{11}^{\alpha\text{-SiO}_2}$	199
$NaNO_2$		$d_{32} = 1.89 \pm 0.25$	$\delta_{32} = 7.76 \pm 1.03$	1.064	1.6319	1.4125	$d_{11}^{\alpha\text{-SiO}_2}$	199
		$d_{33} = 0.094 \pm 0.008$	$\delta_{33} = 1.14 \pm 0.1$	1.064	1.4029	1.4125	$d_{11}^{\alpha\text{-SiO}_2}$	199
		$d_{15} = 0.074 \pm 0.008$	$\delta_{15} = 1.35 \pm 0.15$	1.064	1.3691	1.3455	$d_{11}^{\alpha\text{-SiO}_2}$	199
		$d_{24} = 1.80 \pm 0.25$	$\delta_{24} = 6.9 \pm 0.96$	1.064	1.5174	1.6547	$d_{11}^{\alpha\text{-SiO}_2}$	199
		$d_{31} = d_{15} = 0.18$	$\delta_{31} = \delta_{15} = 0.77$	1.153	1.629	1.4124	—	198
		$d_{32} = d_{24} = 0.76$	$\delta_{32} = \delta_{24} = 13.7$	1.153	1.3391	1.4124	—	198
Terbium molybdate $Tb_2(MoO_4)_3$	$mm2\text{-}C_{2v}$	$d_{31} = -2.99 \pm 0.35$	$\delta_{31} = -1.79 \pm 0.27$	1.064	1.8226	1.9185	$d_{11}^{\alpha\text{-SiO}_2}$	187—189
		$d_{32} = +2.22 \pm 0.33$	$\delta_{32} = +1.74 \pm 0.26$	1.064	1.8222	1.9185	$d_{11}^{\alpha\text{-SiO}_2}$	187—189
		$d_{33} = -0.11 \pm 0.03$	$\delta_{33} = -0.66 \pm 0.18$	1.064	1.8704	1.9185	$d_{11}^{\alpha\text{-SiO}_2}$	187—189
		$d_{15} = -2.52 \pm 0.38$	$\delta_{15} = -1.98 \pm 0.3$	1.064	1.8459	1.8649	$d_{11}^{\alpha\text{-SiO}_2}$	187—189
		$d_{24} = +2.55 \pm 0.35$	$\delta_{24} = +2.01 \pm 0.3$	1.064	1.8458	1.8645	$d_{11}^{\alpha\text{-SiO}_2}$	187—189
1,3,5-Triphenylbenzene	$mm2\text{-}C_{2v}$	$d_{31} = 0.064 \pm 20\%$	$\delta_{31} = 0.17 \pm 20\%$	1.0642	1.51	1.89	$d_{11}^{\alpha\text{-SiO}_2}$	204
$C_{24}H_{18}$		$d_{32} = 0.13 \pm 20\%$	$\delta_{32} = 0.11 \pm 20\%$	1.0642	1.82	1.89	$d_{11}^{\alpha\text{-SiO}_2}$	204
		$d_{33} = 0.64 \pm 20\%$	$\delta_{33} = 0.53 \pm 20\%$	1.0642	1.82	1.89	$d_{11}^{\alpha\text{-SiO}_2}$	204
		System — Tetragonal						
Dicalcium strontium	$4\text{-}C_4$	$d_{31} = 0.061 \pm 0.006$		1.0642			$d_{11}^{\alpha\text{-SiO}_2}$	244
propionate		$d_{33} = 0.12 \pm 0.01$		1.0642			$d_{11}^{\alpha\text{-SiO}_2}$	244
$Ca_2Sr(CH_3-CH_2CO_2)_6$								

Table 1.1.16 (continued)
SECOND-HARMONIC COEFFICIENTS

System — Tetragonal

Material	Symmetry class	$d_{jm} \times 10^{12}$ (m/V)	$\delta_{jm} \times 10^2$ (m^2/C)	λ_1 (μm)	n_ω	$n_{2\omega}$	Measured relative to	Ref.
Methyl-3-isopropyl-4-phenol	$4\text{-}C_4$	$d_{31} = 0.37$	$\delta_{31} = 1.21$	1.064	1.5839	1.5862	$d_{11}^{\alpha\text{-}SiO_2}$	245
		$d_{33} = 0.29$	$\delta_{33} = 1.01$	1.064	1.5691	1.5862	$d_{11}^{\alpha\text{-}SiO_2}$	245
$OH\text{–}C_6H_4\text{–}CH_3\text{=}CH_3\text{–}CH_3\text{–}CH$								
Cadmium gallium sulfide (thiogallate) $CdGa_2S_4$	$\bar{4}\text{-}S_4$	$d_{36} = 25.6 \pm 3.8$	$\delta_{36} = 2.96 \pm 0.44$	1.064	2.327	2.453	$d_{11}^{\alpha\text{-}SiO_2}$	241
Cadmium mercury thiocyanate $Cd[Hg(SCN)_4]$	$\bar{4}\text{-}S_4$	$d_{31} = 6.54 \pm 1.96$	$\delta_{31} = 4.57 \pm 1.37$	1.06	1.9245	1.792	$d_{31}^{LiNbO_3}$	239
		$d_{31} = 6.61 \pm 0.5$	$\delta_{31} = 4.62 \pm 0.35$	1.06	1.9245	1.792	$d_{33}^{LiNbO_3}$	240
		$d_{36} = 1.52 \pm 0.5$	$\delta_{36} = 1.06 \pm 0.35$	1.06			$d_{33}^{LiNbO_3}$	240
Indium thiophosphide $InPS_4$	$\bar{4}\text{-}S_4$	$d_{31} = 25.6 \pm 2.56$	$\delta_{31} = 3.16 \pm 0.32$	1.064	2.2805	2.4868	$d_{31}^{Ba_2NaNb_3O_{15}}$	243
		$d_{31} = 26.3 \pm 2.58$	$\delta_{31} = 3.25 \pm 0.32$	1.064	2.2805	2.4868	$d_{33}^{LiIO_3}$	243
		$d_{36} = 20.5 \pm 2.56$	$\delta_{36} = 2.53 \pm 0.32$	1.064	2.2805	2.4868	$d_{31}^{BaNaNb_3O_{15}}$	243
		$d_{36} = 20.1 \pm 2.1$	$\delta_{36} = 2.48 \pm 0.26$	1.064	2.2865	2.4868	$d_{33}^{LiIO_3}$	243
Mercury gallium sulfide $HgGa_2S_4$	$\bar{4}\text{-}S_4$	$d_{36} = 25.6 \pm 7.7$	$\delta_{36} = 1.56 \pm 0.47$	1.064	2.56	2.65	$d_{11}^{\alpha\text{-}SiO_2}$	242
Tellurium dioxide TeO_2	$422\text{-}D_4$	$d_{14} = 0.61 \pm 0.06$	$\delta_{14} = 0.093 \pm 0.01$	1.064	2.2718	2.3001	$d_{11}^{\alpha\text{-}SiO_2}$	234
		$d_{14} = 0.41 \pm 0.03$	$\delta_{14} = 0.063 \pm 0.005$	1.064	2.2718	2.2953	$d_{11}^{\alpha\text{-}SiO_2}$	235
		$d_{14} = 0.38 \pm 0.03$	$\delta_{14} = 0.058 \pm 0.005$	1.064	2.2718	2.2953	d_{36}^{KDP}	235
		$d_{14} = 0.34 \pm 0.05$	$\delta_{14} = 0.055 \pm 0.008$	1.318	2.2653	2.2523	$d_{11}^{\alpha\text{-}SiO_2}$	236
		$d_{14} = 0.595 \pm 0.09$	$\delta_{14} = 0.091 \pm 0.014$	1.064	2.2718	2.3001	$d_{11}^{\alpha\text{-}SiO_2}$	236
		$d_{14} = 4.13 \pm 1.03$	$\delta_{14} = 0.36 \pm 0.09$	0.659	2.358	2.68	$d_{11}^{\alpha\text{-}SiO_2}$	236

Barium titanate $BaTiO$	$4mm\text{-}C_{4v}$	$d_{15} = -14.35 \pm 1.23$	$\delta_{15} = -1.79$	1.058			d_{36}^{KDP}	226, 227
		$d_{31} = -15.2 \pm 1.23$	$\delta_{31} = -1.92$	1.058			d_{36}^{KDP}	226, 227
		$d_{33} = -5.7 \pm 0.3$	$\delta_{33} = -0.83$	1.058			d_{36}^{KDP}	226. 227
		$d_{15} = 17.0 \pm 1.8$	$\delta_{15} = 1.96 \pm 0.2$	1.0642	2.3175	2.4760	$d_{11}^{\alpha\text{-}SiO_2}$	228
		$d_{31} = 15.7 \pm 1.8$	$\delta_{31} = 1.85 \pm 0.2$	1.0642	2.3379	2.4128	$d_{11}^{\alpha\text{-}SiO_2}$	228
		$d_{33} = 6.8 \pm 1.0$	$\delta_{33} = 0.88 \pm 0.13$	1.0642	2.2970	2.4128	$d_{11}^{\alpha\text{-}SiO_2}$	228
Barium titanium niobate $Ba_6Ti_2Nb_8O_3$	$4mm\text{-}C_{4v}$	$d_{31} = 9.7 \pm 1.8$	$\delta_{31} = 1.5 \pm 0.18$	1.064	2.2506	2.3195	$d_{33}^{LiNbO_3}$	238
		$d_{33} = 13.2 \pm 1.8$	$\delta_{33} = 2.2 \pm 0.18$	1.064	2.2211	2.3195	$d_{33}^{LiNbO_3}$	238
Lead titanate $PbTiO_3$	$4mm\text{-}C_{4v}$	$d_{15} = 33.3 \pm 5$	$\delta_{15} = 1.84 \pm 0.26$	1.0642	2.5704	2.7398	$d_{11}^{\alpha\text{-}SiO_2}$	229
		$d_{31} = 37.6 \pm 5.6$	$\delta_{31} = 2.1 \pm 0.32$	1.0642	2.5712	2.7260	$d_{11}^{\alpha\text{-}SiO_2}$	229
		$d_{33} = 7.5 \pm 1.2$	$\delta_{33} = 0.42 \pm 0.07$	1.0642	2.5692	2.7260	$d_{11}^{\alpha\text{-}SiO_2}$	229
Potassium lithium niobate	$4mm\text{-}C_{4v}$	$d_{31} = 6.18 \pm 1.28$	$\delta_{31} = 1.22 \pm 0.25$	1.0642	2.2057	2.1980	$d_{11}^{\alpha\text{-}SiO_2}$	230a, b
		$d_{33} = 11.2 \pm 1.6$	$\delta_{33} = 2.76 \pm 0.39$	1.0642	2.1113	2.1980	$d_{11}^{\alpha\text{-}SiO_2}$	230a, b
$K_3Li_2Nb_5O_{15}$		$d_{15} = 5.45 \pm 0.54$	$\delta_{15} = 1.05 \pm 0.18$	1.0642	2.1585	2.3297	$d_{11}^{\alpha\text{-}SiO_2}$	230a, b
Potassium lithium niobate $(K_2O)_{0.3}(Li_2O)_{0.7-x}(Nb_2O_5)_x$ x = 0.525 to 0.535]	$4mm\text{-}C_{4v}$	$d_{31} = 10.36 \pm 1.64$		1.064			$d_{31}^{LiNbO_3}$	233
Potassium sodium barium niobate $K_{0.8}Na_{0.2}Ba_2Nb_5O_{15}$	$4mm\text{-}C_{4v}$	$d_{31} = 13.6 \pm 1.6$	$\delta_{31} = 2.2 \pm 0.3$	1.064	2.2601	2.2636	$d_{31}^{LiNbO_3}$	232
Strontium barium niobate $SrBaNb_5O_{15}$	$4mm\text{-}C_{4v}$	$d_{31} = 4.31 \pm 1.32$	$\delta_{31} = 0.69 \pm 0.2$	1.0642	2.2506	2.3092	$d_{11}^{\alpha\text{-}SiO_2}$	231
		$d_{33} = 11.3 \pm 3.3$	$\beta_{33} = 1.93 \pm 0.57$	1.0642	2.2138	2.3092	$d_{11}^{\alpha\text{-}SiO_2}$	231
		$d_{15} = 5.98 \pm 2$	$\delta_{15} = 0.94 \pm 0.3$	1.0642	2.2322	2.3583	$d_{11}^{\alpha\text{-}SiO_2}$	231
Ammonium dihydrogen phosphate $NH_4H_2PO_4$	$\bar{4}2n\text{-}D_{2d}$	$d_{36} = 0.42 \pm 0.03$	$\delta_{36} = 2.2 \pm 0.16$	0.6943	1.5193	1.5004	d_{36}^{KDP}	226
		$d_{14} = 0.40 \pm 0.02$	$\delta_{14} = 2.1 \pm 0.1$	0.6943	1.4973	1.5498	d_{36}^{KDP}	226
		$d_{36} = 0.41 \pm 0.02$	$\delta_{36} = 2.38 \pm 0.15$	1.058	1.5067	1.4816	d_{36}^{KDP}	226, 109

Table 1.1.16 (continued)
SECOND-HARMONIC COEFFICIENTS

System — Tetragonal

Material	Symmetry class	$d_{jm} \times 10^{12}$ (m/V)	$\delta_{jm} \times 10^2$ (m^2/C)	λ_1 (μm)	n_ω	$n_{2\omega}$	Measured relative to	Ref.
		$d_{14} = 0.40 \pm 0.02$	$\delta_{14} = 2.31 \pm 0.11$	1.058	1.4874	1.5277	d_{36}^{KDP}	226, 109
		$d_{36} = 0.84 \pm 0.21$	$\delta_{36} = 4.31 \pm 1.08$	0.6328	1.5217	1.5075	Abs	246
		$d_{36} = 0.544 \pm 0.14$	$\delta_{36} = 3.25 \pm 0.84$	1.15	1.5036	1.4794	Abs	246
		$d_{36} = 0.57 \pm 0.07$	$\delta_{36} = 2.92 \pm 0.36$	0.6328	1.5217	1.5075	Abs	84
		$d_{36} = 0.58 \pm 0.09$	$\delta_{36} = 2.98 \pm 0.46$	0.6328	1.5217	1.5075	Abs	85
		$d_{36} = 0.41 \pm 0.06$	$\delta_{36} = 2.45 \pm 0.36$	1.15	1.5036	1.4794	d_{36}^{KDP}	176
		$d_{36} = 0.29 \pm 0.17$	$\delta_{36} = 1.54 \pm 0.9$	0.6943	1.5193	1.5004	Abs	28
		$d_{36} = 0.84$	$\delta_{36} = 4.64$	0.8250	1.5145	1.4907	Abs	17
		$d_{14} = 0.40 \pm 0.02$	$\delta_{14} = 2.52 \pm 0.11$	0.6943	1.4973	1.5498	d_{36}^{KDP}	201
		$d_{36} = 0.56 \pm 0.02$	$\delta_{36} = 3.0 \pm 0.11$	0.6943	1.5193	1.5004	d_{36}^{KDP}	201
		$d_{36} = 0.53 \pm 0.08$	$\delta_{36} = 2.8 \pm 0.4$	0.6943	1.5193	1.5004	d_{36}^{KDP}	247
Ammonium dihydrogen phosphate (ADP) $NH_2H_2PO_4$	$\bar{4}2m$-D_{2d}	$d_{36} = 0.53$	$\delta_{36} = 3.11$	1.064	1.5066	1.4815	$d_{11}^{\alpha\text{-}SiO_2}$	155
		$d_{36} = 0.762$	$\delta_{36} = 4.5$	1.064	1.5066	1.4815	d_{36}^{KDP}	154
		$d_{36} = 0.87$		0.6328			d_{36}^{KDP}	154
		$d_{36} = 0.85$		0.6943			d_{36}^{KDP}	154
		$d_{36} = 0.72$		1.318	1.4978	1.4764	d_{36}^{KDP}	154
Ammonium dideuterium phosphate (AD*P) $ND_4D_2PO_4$	$\bar{4}2m$-D_{2d}	$d_{36} = 0.495 \pm 0.07$	$\delta_{36} = 2.74 \pm 0.4$	0.6943	1.5138	1.4926	d_{36}^{KDP}	247
Beryllium sulfate $BeSO_4{\cdot}4H_2O$	$\bar{4}2m$-D_{2d}	$d_{36} = 0.30$	$\delta_{36} = 2.24$	0.6328	1.472	1.452	d_{36}^{KDP}	253
		$d_{36} = 0.29 \pm 0.03$	$\delta_{36} = 2.0 \pm 0.2$	0.5321	1.4749	1.4742	$d_{36}^{KD^*P}$	254

Cadmium germanium arsenide	$\bar{4}2m$-D_{2d}	d_{36} = 455.6 ± 91	δ_{36} = 2.96 ± 0.59	10.6	3.5688	3.6933	d_{14}^{GaAs}	255
CdGaAs$_2$		d_{36} = 351 ± 105	δ_{36} = 2.6 ± 0.78	10.6	3.5049	3.6206	d_{36}^{GaAs}	256
Cadmium germanium phosphide CdGeP$_2$	$\bar{4}2m$-D_{2d}	d_{36} = 162 ± 30%	δ_{36} = 2.5 ± 30%	10.6	3.14224	3.19796	d_{36}^{GaAs}	256
Cesium dihydrogen arsenate	$\bar{4}2m$-D_{2d}	d_{36} = 0.22	δ_{36} = 0.81	0.6943	1.5632	1.5722	d_{36}^{KDP}	247
CsH$_2$AsO$_4$		d_{36} = 0.40 ± 0.05	δ_{36} = 1.62 ± 0.2	1.064	1.5516	1.5514	Abs	258
Cesium dideuterium arsenate CsD$_2$AsO$_4$	$\bar{4}2m$-D_{2d}	d_{36} = 0.40 ± 0.05	δ_{36} = 1.64 ± 0.2	1.064	1.5503	1.5495	Abs	258
Copper gallium selenide CuGaSe$_2$	$\bar{4}2m$-D_{2d}	d_{36} = 44.2 ± 10%	δ_{36} = 1.99 ± 10%	10.6	2.6945	2.7220	d_{36}^{GaAs}	259
Copper gallium sulfide	$\bar{4}2m$-D_{2d}	d_{36} = 14.5 ± 15%	δ_{36} = 1.1 ± 15%	10.6	2.4360	2.4746	d_{36}^{GaAs}	89
Copper indium sulfide CuInS$_2$	$\bar{4}2m$-D_{2d}	d_{36} = 10.6 ± 15%	δ_{36} = 0.74 ± 15%	10.6	2.5311	2.5582	d_{36}^{GaAs}	89
Potassium dihydrogen arsenate KH$_2$AsO$_4$	$\bar{4}2m$-D_{2d}	d_{36} = 0.43 ± 0.025	δ_{36} = 1.85 ± 0.1	1.06	1.531	1.572	d_{36}^{KDP}	109
		d_{14} = 0.46 ± 0.02	δ_{14} = 1.96 ± 0.01	1.06	1.554	1.521	d_{36}^{KDP}	109
		d_{36} = 0.315 ± 15%	δ_{36} = 1.23 ± 15%	0.6943	1.562	1.549	d_{36}^{KDP}	201
		d_{36} = 0.45 ± 0.045	δ_{36} = 1.75 ± 0.18	0.6943	1.562	1.549	d_{36}^{KDP}	201
		d_{14} = 0.39 ± 0.045	δ_{14} = 1.50 ± 0.18	0.6943	1.538	1.606	d_{36}^{KDP}	201
Potassium dihydrogen phosphate (KDP) KH$_2$PO$_4$	$\bar{4}2m$-D_{2d}	d_{36} = 0.88 ± 0.29	δ_{36} = 5.75 ± 1.89	1.1526	1.4913	1.4687	Abs	83, 65
		d_{14} = 0.43 ± 0.03	δ_{14} = 2.47 ± 0.17	0.6943	1.4856	1.5335	d_{36}^{KDP}	226
		d_{14} = 0.42 ± 0.02	δ_{14} = 2.67 ± 0.13	1.0582	1.4751	1.5123	d_{36}^{KDP}	226
		d_{14} = 0.39 ± 0.02	δ_{14} = 2.24 ± 0.11	0.6943	1.4856	1.5335	d_{36}^{KDP}	201
		d_{36} = 0.48 ± 0.08	δ_{36} = 3.1 ± 0.5	1.15	1.4913	1.4687	d_{36}^{ADP}	176
		d_{36} = 0.47 ± 15%	δ_{36} = 2.75 ± 15%	0.6943	1.5058	1.4874	d_{36}^{ADP}	274
		d_{36} = 0.46 ± 0.04	δ_{36} = 2.95 ± 0.26	1.06	1.439	1.4706	Abs	261
		d_{36} = 0.44	δ_{36} = 2.81	1.064	1.4942	1.4708	$d_{11}^{\alpha\text{-}SiO_2}$	155
		d_{14} = 0.44 ± 0.003	δ_{36} = 2.54 ± 0.02	1.064	1.5124	1.4768	d_{36}^{KDP}	102
		d_{36} = +0.60	δ_{36} = +4.0	1.318	1.48607	1.46616	$d_{33}^{LIO_3}$	154

Table 1.1.16 (continued)
SECOND-HARMONIC COEFFICIENTS

System — Tetragonal

Material	Symmetry class	$d_{jm} \times 10^{12}$ (m/V)	$\delta_{jm} \times 10^2$ (m^2/C)	λ_1 (μm)	n_ω	$n_{2\omega}$	Measured relative to	Ref.
		d_{36} = 0.712	—	0.6328	1.5074	1.4934		154
		d_{36} = 070	—	0.6943	1.5050	1.4871		154
		d_{36} = 0.63	—	1.06	1.4938	1.4705		154
		d_{36} = 0.69	—	1.15	1.4913	1.4687		154
		d_{36} = 58.1	—	10.6				160
		d_{36} = 65.5	—	3.39				160
Potassium dideuterium phosphate (KD*P) KD_2PO_4	$\bar{4}2m$-D_{2d}	d_{14} = 0.38 ± 0.004	δ_{14} = 2.23 ± 0.02	0.6943	1.4830	1.5285	d_{36}^{KDP}	201
		d_{36} = 0.34 ± 15%	δ_{36} = 2.07 ± 15%	0.6943	1.5022	1.4855	d_{36}^{KDP}	247
		d_{36} = 0.34 ± 0.01	δ_{36} = 2.07 ± 0.06	0.6943	1.5022	1.4855	d_{36}^{KDP}	226
		d_{14} = 0.342 ± 0.02	δ_{14} = 2.01 ± 0.1	0.6943	1.4830	1.5285	d_{36}^{KDP}	226
		d_{36} = 0.38 ± 0.016	δ_{36} = 2.46 ± 0.1	1.0582	1.4978	1.4689	d_{36}^{KDP}	226
		d_{14} = 0.37 ± 0.012	δ_{14} = 2.39 ± 0.08	1.0582	1.4789	1.5085	d_{36}^{KDP}	226
Rubidium dihydrogen arsenate (RDA) RbH_2AsO_4	$\bar{4}2m$-D_{2d}	d_{36} = 0.29 ± 15%	δ_{36} = 1.16 ± 15%	0.6943	1.5543	1.5531	d_{36}^{KDP}	247
		d_{36} = 0.39 ± 0.04	δ_{36} = 1.56 ± 0.16	0.6943	1.5543	1.5531	Abs	262
		d_{36} = 0.45	δ_{36} = 1.80	0.6943	1.5543	1.5531	Abs	263
		d_{36} = 0.47 ± 0.05	δ_{36} = 1.88 ± 0.20	0.6943	1.5543	1.5531	d_{36}^{KDP}	264
Rubidium dihydrogen phosphate (RDP)	$\bar{4}2m$-D_{2d}	d_{36} = 0.47 ± 15%	δ_{36} = 2.75 ± 15%	0.6943	1.4969	1.5020	d_{36}^{KDP}	2
		d_{36} = 0.414 ± 0.045	δ_{36} = 2.42 ± 0.26	0.6943	1.4969	1.5020	d_{36}^{KDP}	264
		d_{14} = 0.49 ± 0.07	δ_{14} = 3.03 ± 0.43	1.0642	1.4813	1.5106	$d_{11}^{\alpha\text{-}SiO_2}$	266
		d_{36} = 0.38 ± 0.04	δ_{36} = 2.39 ± 0.26	1.0642	1.4926	1.4811	$d_{11}^{\alpha\text{-}SiO_2}$	266
Silver gallium selenide $AgGaSe_2$	$\bar{4}2m$-D_{2d}	d_{36} = 49.3 ± 10%	δ_{36} = 3.01 ± 10%	10.6	2.5912	2.5808	d_{36}^{GaAs}	259
		d_{36} = 3.24 ± 5.0	δ_{36} = 1.98 ± 0.3	10.63	2.5915	2.5808	Abs	268
		d_{36} = 37.4 ± 6.0	δ_{36} = 2.29 ± 0.37	10.6	2.5915	2.5808	Abs and	269

		$d_{36} = 67.7 \pm 13$	$\delta_{36} = 3.51$	2.12	2.63455	2.67887	$d_{31}^{LiIO_3}$	160
		$d_{36} = 57.7$	$\delta_{36} = 3.53$	10.6				160
Silver gallium sulfide $AgGaS_2$	$\bar{4}2m$-D_{2d}	$d_{36} = +17.96 \pm 15\%$	$\delta_{36} = 2.23 \pm 15\%$	10.6	2.34722	2.34082	d_{36}^{GaAs}	89
		$d_{14} = 56.6 \pm 18.8$	$\delta_{14} = 7.18 \pm 2.38$	10.6	2.3162	2.3828	d_{14}^{GaAs}	270
		$d_{36} = 23.36 \pm 3.52$		1.064	2.46848	2.61256	$d_{11}^{\alpha\text{-}SiO_2}$	271
		$d_{36} = 20 \pm 4$	$\delta_{36} = 2.48 \pm 0.5$	10.6	2.34722	2.34082	d_{14}^{GaAs}	271
Silver indium selenide	$\bar{4}2m$-D_{2d}	$d_{36} = 55.9 \pm 10\%$	$\delta_{36} = 3.05 \pm 10\%$	10.6	2.62284	2.64470	d_{36}^{GaAs}	259
Zinc silicon arsenide	$\bar{4}2m$-D_{2d}	$d_{36} = 108.5 \pm 15\%$	$\delta_{36} = 1.57 \pm 15\%$	10.6	3.1749	3.2328	d_{36}^{GaAs}	256
Zinc germanium phosphide	$\bar{4}2m$-D_{2d}	$d_{14} = 111.2 \pm 30\%$	$\delta_{14} = 1.98 \pm 30\%$	10.6	3.0908	3.1138	d_{14}^{GaAs}	273
Urea $(NH_2)_2CO$	$\bar{4}2m$-D_{2d}	$d_{36} = 1.35$		0.6			d_{36}^{KDP}	274
		System — Trigonal						
Lead *m*-germanate $PbGe_3O_{11}$	3-C_3	$d_{11} = 0.96 \pm 0.16$	$\delta_{11} = -0.27 \pm 0.05$	1.064	—	—	$d_{11}^{\alpha\text{-}SiO_2}$	275
		$d_{22} = -2.1 \pm 0.3$	$\delta_{22} = 0.6 \pm 0.09$	1.064			$d_{11}^{\alpha\text{-}SiO_2}$	275
		$d_{31} = +0.51 \pm 0.07$	$\delta_{31} = +0.14 \pm 0.02$	1.064			$d_{11}^{\alpha\text{-}SiO_2}$	275
		$d_{33} = -0.79 \pm 0.12$	$\delta_{33} = -0.20 \pm 0.03$	1.064			$d_{11}^{\alpha\text{-}SiO_2}$	275
Sodium periodate $NaH[IO_3(OH)_3]H_2O$	$3 = C_3$	$d_{11} < 0.041 \pm 30\%$	$\delta_{11} < 0.11 \pm 30\%$	1.064	1.614	1.640	$d_{11}^{\alpha\text{-}SiO_2}$	276
		$d_{22} < 0.041 \pm 30\%$	$\delta_{22} < 0.11 = -30\%$	1.064	1.614	1.640	$d_{11}^{\alpha\text{-}SiO_2}$	276
		$d_{31} = 0.5 \pm 30\%$	$\delta_{31} = 1.38 \pm 30\%$	1.064	1.614	1.676	$d_{11}^{\alpha\text{-}SiO_2}$	276
		$d_{33} = 1.85 \pm 30\%$	$\delta_{33} = 3.99 \pm 30\%$	1.064	1.644	1.676	$d_{11}^{\alpha\text{-}SiO_2}$	276
Aluminum phosphate $AlPO_4$	32-D_3	$d_{11} = 0.35 \pm 0.03$	$d_{11} = 1.76 \pm 0.15$	1.0582	1.5156	1.5275	d_{36}^{KDP}	250
		$d_{14} < 0.008$	$\delta_{14} < 0.04$	1.0582	1.5198	1.5275	d_{36}^{KDP}	250
Benzil $C_6H_5COCOC_6H_5$	32-D_3	$d_{11} = 3.6 \pm 0.5$	$\delta_{11} = 8.3 \pm 1.2$	1.0642	1.6313	1.669	$d_{11}^{\alpha\text{-}SiO_2}$	277
Calcium dithionate tetrahydrate $CaS_2O_64H_2O$	32-D_3	$d_{11} = 0.053 \pm 0.02$	$\delta_{11} = 0.24 \pm 0.09$	0.6943	1.545	1.574	$d_{11}^{\alpha\text{-}SiO_2}$	278
Lead dithionate $PbS_2O_64H_2O$	32-D_3	$d_{11} = 0.096$	$\delta_{11} = 0.38$	1.064			$d_{11}^{\alpha\text{-}SiO_2}$	279
		$d_{11} = 0.15$	$\delta_{11} = 0.41$	0.694			$d_{11}^{\alpha\text{-}SiO_2}$	279

Table 1.1.16 (continued)
SECOND-HARMONIC COEFFICIENTS

System — Trigonal

Material	Symmetry class	$d_{jm} \times 10^{12}$ (m/V)	$\delta_{jm} \times 10^{2}$ (m^2/C)	λ_1 (μm)	n_ω	$n_{2\omega}$	Measured relative to	Ref.
Mercuric sulfide	32-D_3	$d_{11} = 50.3 \pm 17$	$\delta_{11} = 2.93 \pm 1$	10.6	2.596	2.626	Abs	280
HgS								
		$d_{11} = 62.8 \pm 21$	$\delta_{11} = 3.66 \pm 1.2$	10.6	2.5956	2.6257		281
		$d_{11} = 47.2 \pm 4$	$\delta_{11} = 1.86 \pm 0.16$	1.32	2.6792	2.919	d_{36}^{KDP}	282
		$d_{11} = 40.85 \pm 6.3$	$\delta_{11} = 2.38 \pm 0.37$	10.6	2.5956	2.6257	$d_{+}^{Proustite}$	282
Neodymium-yttrium	32-D_3	$d_{11} = d_{12} = 1.36 \pm 0.16$	$\delta_{11} = 1.61 \pm 0.19$	1.32	1.762	1.7780	$d_{11}^{\alpha\text{-}SiO_2}$	284
aluminum borate		$d_{14} = d_{25} < 0.01$	$\delta_{14} = 0.01$		1.7275	1.7780	$d_{11}^{\alpha\text{-}SiO_2}$	284
$Nd_xY_{1-x}Al_3(BO_3)_4]$								
$Nd_{0.2}Y_{0.8}Al_3(BO_3)_4$								
Potassium dithionate	32-D_3	$d_{11} = 0.083 \pm 0.014$	$\delta_{11} = 0.65 \pm 0.11$	0.6943	1.4518	1.4728	$d_{11}^{\alpha\text{-}SiO_2}$	285
$K_2S_2O_6$								
		$d_{11} = 0.088 \pm 0.03$	$\delta_{11} = 0.69 \pm 0.23$	0.6943	1.452	1.473	$d_{11}^{\alpha\text{-}SiO_2}$	278
α-Quartz	32-D_3	$d_{11} = 0.34 \pm 0.016$	$\delta_{11} = 1.54 \pm 0.07$	1.0582	1.532	1.542	d_{36}^{KDP}	250
SiO_2		$d_{14} < 0.008$	$\delta_{14} < 0.036$	1.0582	1.536	1.542	d_{36}^{KDP}	250
		$d_{11} = 0.32 \pm 0.04$	$\delta_{11} = 1.42 \pm 0.18$	1.064	1.5341	1.54702	d_{36}^{ADP}	156
		$d_{11} = 0.335$	$\delta_{11} = 1.49$	1.064	1.53413	1.54702	d_{36}^{ADP}	155
		$d_{14} = 0.0029 \pm 0.0003$	$\delta_{14} = 0.013 \pm 0.0013$	1.064	1.53850	1.54702	$d_{11}^{\alpha\text{-}SiO_2}$	286
		$d_{14} = 0.009 \pm 0.0014$		0.5320	1.55164		$d_{11}^{\alpha\text{-}SiO_2}$	286
Rubidium dithionate	32-D_3	$d_{11} = 0.081 \pm 0.03$	$\delta_{11} = 0.64 \pm 0.24$	0.6943	1.453	1.469	$d_{11}^{\alpha\text{-}SiO_2}$	278
$Rb_2S_2O_6$								
Strontium dithionate	32-D_3	$d_{11} = 0.06 \pm 0.02$	$\delta_{11} = 0.24 \pm 0.08$	0.6943	1.545	1.574	$d_{11}^{\alpha\text{-}SiO_2}$	278
tetrahydrate								
$SrS_2O_6{\cdot}4H_2O$								
Selenium	32-D_3	$d_{11} = 79.6 \pm 42$	$\delta_{11} = 4.2 \pm 2.22$	10.6	2.64	2.645	Abs	288
Se								
		$d_{11} = 209.4 \pm 42$	$\delta_{11} = 11 \pm 2.2$	10.6			Abs	289

		$d_{11} = 97 \pm 25$	$\delta_{11} = 5.2 \pm 1.36$	10.6			Abs	87
		$d_{11} = 1840 \pm 860$	$\delta_{11} = 303.8 \pm 142$	28.0	2.12	2.57	d_{11}^{Te}	290
Tellurium	$32\text{-}D_3$	$d_{11} = 5319.8 \pm 837.7$	$\delta_{11} = 1.07 \pm 0.17$	10.5915	4.792	4.850	Abs	292
		$d_{11} = 4188.8 \pm 2094.4$	$\delta_{11} = 0.85 \pm 0.42$	10.5915			Abs	293
		$d_{11} = 5864.3$	$\delta_{11} = 1.18$	10.6			Abs	294
		$d_{11} = 3351.03$	$\delta_{11} = 0.68$	10.5915			Abs	295
		$d_{11} = 1549.8 \pm 387.5$	$\delta_{11} = 0.31 \pm 0.08$	10.6			d_{11}^{InAs}	88
		$d_{11} = 921.8 \pm 293.3$	$\delta_{11} = 0.96 \pm 0.3$	10.6	4.792	4.850	$d_{33}^{Ag_3SbS_3}$	296
		$d_{11} = 570 \pm 190$	$\delta_{-0.65} \pm 0.20$	28.0	4.716	4.775	Abs	297
		$d_{11} = 649.45$	$\delta_{11} = 0.67$	10.6	4.792	4.850	d_{36}^{GaAs}	155
		$d_{11} = 712$	$\delta_{11} = 0.74$	10.6	4.792	4.850	Abs	298
		$d_{11} = 670.4 \pm 290.5$	$\delta_{11} = 0.70 \pm 0.30$	10.6	4.792	4.850	Abs	299
Guanidinium aluminum sulfate hexahydrate [GASH] $(CN_3H_6)As(SO_4)_2{\cdot}6H_2O$	$3m\text{-}C_{3v}$	$d_{22} = -0.105 \pm 0.017$	$\delta_{22} = -0.65$	1.064			$d_{11}^{\alpha\text{-}SiO_2}$	302
		$d_{31} = +0.008 \pm 0.017$	$\delta_{31} = +0.05$	1.064			$d_{11}^{\alpha\text{-}SiO_2}$	302
		$d_{33} = +0.020 \pm 0.003$	$\delta_{33} = 0.18$	1.064			$d_{11}^{\alpha\text{-}SiO_2}$	302
		$d_{22} = \pm 0.08$		0.6943			d_{36}^{KDP}	302, 303
		$d_{31} = \pm 0.007$		0.6943			d_{36}^{KDP}	302, 303
Lithium niobate $LiNbO_3$	$3m\text{-}C_{3v}$	$d_{22} = +2.58 \pm 0.25$	$\delta_{22} = 0.42 \pm 0.04$	1.0582	2.2322	2.3241	d_{36}^{KDP}	
		$d_{31} = -4.88 \pm 0.7$	$\delta_{31} = -0.87 \pm 0.13$	1.0582	2.2322	2.2325	d_{36}^{KDP}	304, 188
		$d_{33} = -43.9 \pm 8.2$	$\delta_{33} = -9.39 \pm 1.75$	1.0582	2.1544	2.2325	d_{36}^{KDP}	304, 188
		$d_{33} = -34 \pm 8.6$	$\delta_{33} = -7.27 \pm 1.84$	1.0582	2.1544	2.2325	d_{36}^{KDP}	304, 188
		$d_{31} = -4.35 \pm 0.4$	$\delta_{31} = -0.8 \pm 0.07$	1.152	2.2278	2.2153	d_{36}^{KDP}	304, 188
		$d_{22} = +2.1 \pm 0.8$	$\delta_{22} = +0.35 \pm 0.14$	1.152	2.2278	2.3037	d_{36}^{KDP}	304, 188
		$d_{31} = -5.14 \pm 0.49$	$\delta_{31} = -0.95 \pm 0.09$	1.152	2.2278	2.2153	d_{36}^{ADP}	176a, b, 188
		$d_{31} = 6.21 \pm 10\%$		$\lambda_p = 0.488$ μm			Abs	93

Table 1.1.16 (continued)
SECOND-HARMONIC COEFFICIENTS

System — Trigonal

Material	Symmetry class	$d_{jm} \times 10^{12}$ (m/V)	$\delta_{jm} \times 10^{2}$ (m^2/C)	λ_1 (μm)	n_ω	$n_{2\omega}$	Measured relative to	Ref.
		$d_{31} = 7.12 \pm 2.5$	$\delta_{31} = 1.27 \pm 0.45$	1.058	2.2322	2.2325	Abs	261
		$d_{31} = 5.45 \pm 0.3$	$\delta_{15} = 0.96 \pm 0.5$	1.064			d_{36}^{ADP}	155
		$d_{31} = -5.93$	$\delta_{31} = -1.13$	1.318	2.21991	2.19710	$d_{31}^{LiIO_3}$	160, 188
		$d_{33} = -31.8$	$\delta_{33} = -7.23$	1.318	2.14539	2.19710	$d_{31}^{LiO_3}$	160, 188
		$d_{33} = -29.1$	$\delta_{33} = -7.34$	2.12	2.12272	2.15610	$d_{31}^{LiO_3}$	160, 188
		$d_{31} = -5.823$		$\lambda_p = 0.488$ μm			Abs	160, 188
Lithium niobate	$3m$ -C_{3v}	$d_{31} = -5.95$		1.064			$d_{11}^{\alpha\text{-}SiO_2}$	306
$(Li/Nb)_m = 1.083$		$d_{22} = +1.93$		1.064			$d_{11}^{\alpha\text{-}SiO_2}$	306
		$d_{33} = -29.8$		1.064			$d_{11}^{\alpha\text{-}SiO_2}$	306
Lithium niobate	$3m$ -C_{3v}	$d_{31} = -4.76$		1.064			$d_{11}^{\alpha\text{-}SiO_2}$	306
$(Li/Nb)_m = 0.946$		$d_{22} = +2.3$		1.064			$d_{11}^{\alpha\text{-}SiO_2}$	306
(congruent-melt)		$d_{33} = -29.7$		1.064			$d_{11}^{\alpha\text{-}SiO_2}$	306
Lithium niobate	$3m$ -C_{3v}	$d_{31} = -3.45$		1.064			$d_{11}^{\alpha\text{-}SiO_2}$	306
$(Li/Nb)_m = 0.852$		$d_{22} = +2.13$		1.064			$d_{11}^{\alpha SiO_2}$	306
		$d_{33} = -27.5$		1.064			$d_{11}^{\alpha\text{-}SiO_2}$	306
Lithium tantalate	$3m$ -C_{3v}	$d_{22} = +1.76 \pm 0.2$	$\delta_{22} = +0.41 \pm 0.05$	1.0582	2.1366	2.2043	d_{36}^{KDP}	305, 188
$LiTaO_3$		$d_{31} = -1.07 \pm 0.2$	$\delta_{31} = -0.25 \pm 0.05$	1.0582	2.1366	2.2089	d_{36}^{KDP}	305, 188
		$d_{33} = -16.4 \pm 2$	$\delta_{33} = -3.72 \pm 0.45$	1.0582	2.1406	2.2089	d_{36}^{KDP}	305, 188

Silver antimony sulfide (Pyrargyrite) Ag_3SbS_3	$3m$ -C_{3v}	$d_{31} = 12.6 \pm 4$	$\delta_{31} = 0.597 \pm 0.19$	10.6	2.7352	2.6221	Abs	86
		$d_{22} = 13.4 \pm 4$	$\delta_{22} = 0.55 \pm 0.16$	10.6	2.7352	2.758	Abs	
Silver arsenic sulfide (proustite) Ag_3AsS_3	$3m$ -C_{3v}	$d_{31} = 12.0$	$\delta_{31} = 0.38$	1.152	2.8028	2.9200	$d_{31}^{LiNbO_3}$	312
		$d_{22} = 20.0$	$\delta_{22} = 0.56$	1.152	2.8028	3.1073	$d_{31}^{LiNbO_3}$	312
		$d_{eff} = \lvert d_{15}\rvert \sin\theta_m +$		10.6	2.697	2.728		313
		$d_{22} \cos\theta_m = 25$		10.6	2.697	2.728		313
		$d_{22} = 26.8 \pm 4$	$\delta_{22} = 1.2 \pm 0.18$	10.59			d_{14}^{GaAs}	314
		$d_{31} = 16.8 \pm 1.0$	$\delta_{22} = 0.91 \pm 0.05$				$d_{-}^{Ag_3SbS_3}$	
		$d_{eff} = \lvert d_{31}\rvert \sin\theta_m + \lvert d_{22}\rvert$ x $\cos\theta_m = 8.93 \pm 2.3$		9.2			Abs	316
Thallium arsenic selenide Tl_3AsSe_3	$3m$ -C_{3c}	$d_{+} = d_{31} \sin(\theta_m + \rho) + d_{22} \cos(\theta_m + \rho)$		10.6			$d_{+}^{Ag_3SbS_3}$	319
Thallium iodate $TlIO_3$	$3m$ -C_{3c}	$d_{33} = 6.85 + 20\%$	$\delta_{33} = 5.98 \pm 20\%$	1.064	1.81	1.87	$d_{33}^{LiIO_3}$	318
		$d_{31} = 3.36 \pm 20\%$	$\delta_{31} = 1.88 \pm 20\%$	1.064	1.96	1.87	$d_{33}^{LiIO_3}$	318
		$d_{15} = 3.49 \pm 20\%$	$\delta_{15} = 1.91 \pm 20\%$	1.064	1.885	2.04	$d_{33}^{LiIO_3}$	318
		$d_{32} = 3.98 \pm 20\%$	$\delta_{32} = 2.23 \pm 20\%$	1.064	1.96	1.87	$d_{33}^{LiIO_3}$	318
		$d_{24} = 3.85 \pm 20\%$	$\delta_{24} = 2.11 \pm 20\%$	1.064	1.885	2.04	$d_{33}^{LiIO_3}$	318
		$d_{22} = 1.11 \pm 20\%$	$\delta_{24} = 0.49 \pm 20\%$	1.064	1.96	2.04	$d_{33}^{LiIO_3}$	318
Tourmaline $(Na,Ca)(MgFe)_3B_3Al_6Si_6$ $(O,OH,F)_{31}$	$3m$-C_{3v}	$d_{15} = 0.24 \pm 0.04$	$\delta_{15} = 0.62 \pm 0.09$	1.064	1.618	1.6433	$d_{11}^{\alpha\text{-}SiO_2}$	317
		$d_{22} = 0.07 \pm 0.01$	$d_{22} = 0.19 \pm 0.03$	1.064	1.6274	1.6433	$d_{11}^{\alpha\text{-}SiO_2}$	317
		$d_{31} = 0.14 \pm 0.03$	$\delta_{31} = 0.37 \pm 0.07$	1.064	1.6274	1.6231	$d_{11}^{\alpha\text{-}SiO_2}$	317
		$d_{33} = 0.50 \pm 0.06$	$\delta_{33} = 1.39 \pm 0.17$	1.064	1.6088	1.6231	$d_{11}^{\alpha\text{-}SiO_2}$	317

System — Hexagonal

Lithium iodate $LiIO_3$	6 -C_6	$d_{31} = -12.2 \pm 1.9$	$\delta_{31} = -10.96 \pm 1.7$	1.06	1.860	1.750	d_{36}^{KDP}	320, 188
		$d_{31} = -4.5 \pm 0.6$	$\delta_{31} = -4.15 \pm 0.56$	1.0845	1.856	1.748	d_{36}^{KDP}	321, 188
		$d_{33} = -3.6 \pm 30\%$	$\delta_{33} = -5.16 \pm 30\%$	1.06	1.719	1.750	d_{36}^{KDP}	321, 188
		$d_{31} = -7.54 \pm 15\%$		$\lambda_p =$ 0.4765			Abs	322, 188

Table 1.1.16 (continued)
SECOND-HARMONIC COEFFICIENTS

System — Hexagonal

Material	Symmetry class	$d_{jm} \times 10^{12}$ (m/V)	$\delta_{jm} \times 10^{2}$ (m^2/C)	λ_1 (μm)	n_ω	$n_{2\omega}$	Measured relative to	Ref.
				0.4880 0.5145μm				
		$d_{31} = -4.96 \pm 0.26$	$\delta_{31} = -4.63 \pm 0.24$	1.0642	1.8517	1.7475	$d_{11}^{\alpha\text{-SiO}_2}$	157, 188
		$d_{33} = -5.15 \pm 0.32$	$\delta_{33} = -7.47 \pm 0.46$	1.0642	1.7168	1.7475	$d_{11}^{\alpha\text{SiO}_2}$	157, 188
		$d_{31} = \pm 10.17 \pm 2.0$		0.5145			Abs	95
		$d_{31} = -5.53$	$\delta_{31} = -5.1$	1.0642	1.8517	1.7475	d_{36}^{ADP}	155, 188
		$d_{15} = 5 \pm 0.5$	$\delta_{15} = 2.85 \pm 0.29$	0.6943	1.9929	1.8021	d_{36}^{KDP}	264
		$d_{31} = -7.215$	$\delta_{31} = -6.54$	$\lambda_p =$ 0.4880			Abs	160, 188, 154
		$d_{31} = -7.33$		$\lambda_p =$ 0.5145			Abs	160, 188, 154
		$d_{31} = -6.82$		1.318			Abs	160, 188, 154
		$d_{33} = -6.75 \pm 0.95$	$\delta_{33} = -1.02$	1.318			$d_{31}^{\text{LiIO}_3}$	160, 188, 154
		$d_{33} = -5.54 \pm 0.61$		1.318			$d_{31}^{\text{LiIO}_3}$	160, 188, 154
		$d_{14} = 0.22 \pm 0.008$	$\delta_{14} = 0.2 \pm 0.007$	1.0642	1.787	1.898	$d_{31}^{\text{LiIO}_3}$	324
		$d_{14} = 0.35 \pm 0.04$	$\delta_{14} = 0.36 \pm 0.04$	1.0642	1.7869	1.8978	$d_{11}^{\alpha\text{-SiO}_2}$	323

Lithium potassium sulfate $LiKSO_4$	6 -C_6	$d_{31} = 0.38$		0.6943			d_{36}^{KDP}	163
		$d_{33} = 0.71$		0.6943			d_{36}^{KDP}	163
Gallium selenide GaSe	$6m2$ -D_{3h}	$d_{22} = 80.4 \pm 16$	$\delta_{22} = 2.88 \pm 0.57$	10.6	2.7870	2.8075	d_{31}^{CdSe}	327
		$d_{22} = 75.4 \pm 10.8$	$\delta_{22} = 2.7 \pm 0.39$	10.6			$d^{Ag_3SbS_3}$	327
		$d_{16} = 558$	$\delta_{16} = 15$	1.06	2.8768	2.9875	d_{14}^{GaAs}	329
		$d_{16} = 972$		0.6943			d_{14}^{GaAs}	329
Gallium sulfide GaS	$6m2$ -D_{3h}	$d_{16} = 135$		0.6943			d_{14}^{GaAs}	329
Indium selenide InSe	$6m2$ -D_{3h}	$d_{16} = 281$		0.694			d_{14}^{GaAs}	329
		$d_{16} = 124$		1.06			d_{14}^{GaAs}	329
Beryllium oxide BeO	$6mm$ -C_{6v}	$d_{33} = -0.20 \pm 0.01$	$\delta_{33} = -0.29 \pm 0.015$	1.064	1.7204	1.7376	$d_{11}^{\alpha\text{-}SiO_2}$	330, 188
		$d_{31} = -0.15 \pm 0.01$	$\delta_{31} = -0.23 \pm 0.015$	1.064			$d_{11}^{\alpha\text{-}SiO_2}$	330, 188
Cadmium selenide CdSe	$6mm$- C_{6v}	$d_{33} = 66.9 \pm 4.2$	$\delta_{33} = 2.98 \pm 0.19$	1.06	2.560	3.036	d_{33}^{CdS}	332
		$d_{15} = 31.0 \pm 7.5$	$\delta_{15} = 2.8 \pm 0.67$	10.6	2.453	2.447	Abs	288
		$d_{31} = -28.5 \pm 6.3$	$\delta_{31} = -2.56 \pm 0.57$	10.6	2.445	2.465	Abs	288, 188
		$d_{33} = +54.5 \pm 12.6$	$\delta_{33} = 4.73 + -1.09$	10.6	2.462	2.465	Abs	
		$d_{31} = -26.8 \pm 10\%$	$\delta_{31} = -2.41 \pm 0.8$	10.6	2.445	2.465	d_{14}^{GaAs}	273, 188
		$d_{31} = -28.25 \pm 15\%$		$\lambda_p = 1.833$ μm			Abs	333, 188
		$d_{33} = +50.6 \pm 5.95$	$\delta_{33} = +3.85 \pm 0.45$	2.12	2.48582	2.55721	$d_{31}^{LiIO_3}$	160, 188
Cadmium sulfide CdS	$6mm$- C_{6v}	$d_{33} = +25.8 \pm 1.6$		1.0582	2.345	2.654	d_{36}^{KDP}	226, 227
		$d_{31} = -13.1 \pm 0.8$		1.0582	2.327	2.654	d_{36}^{KDP}	226, 227
		$d_{15} = 14.4 \pm 0.8$		1.0582	2.336	2.649	d_{36}^{KDP}	226, 227

Table 1.1.16 (continued)
SECOND-HARMONIC COEFFICIENTS

System — Hexagonal

Material	Symmetry class	$d_{jm} \times 10^{12}$ (m/V)	$\delta_{jm} \times 10^{2}$ (m^2/C)	λ_1 (μm)	n_ω	$n_{2\omega}$	Measured relative to	Ref.
		$d_{33} = +44.0 \pm 12.6$	$\delta_{33} = +7.1 \pm 2.0$	10.6	2.263	2.265	Abs	288, 227
		$d_{31} = -26.4 \pm 6.3$	$\delta_{31} = -4.45 \pm 1.06$	10.6	2.242	2.265	Abs	288, 227
		$d_{15} = 28.9 \pm 7.1$	$\delta_{15} = 4.88 \pm 1.2$	10.6	2.252	2.244	Abs	288, 227
Aluminum nitride	$6mm$- C_{6v}	$d_{33} = 7.42 \pm 35\%$		1.064			$d_{11}^{\alpha\text{-SiO}_2}$	335
AlN		$d_{31} \leq 0.30$		1.064			$d_{11}^{\alpha\text{-SiO}_2}$	335
Lithium perchlorate	$6mm$- C_{6v}	$d_{33} = \pm 0.25 \pm 20\%$	$\delta_{33} = \pm 2.41 \pm 20\%$	1.064	1.43	1.44	$d_{11}^{\alpha\text{-SiO}_2}$	336
trihydrate		$d_{31} = \pm 0.22 \pm 20\%$	$\delta_{31} = \pm 1.72 \pm 20\%$	1.064	1.47	1.44	$d_{11}^{\alpha\text{-SiO}_2}$	336
$LiClO_4 \cdot 3H_2O$		$d_{15} = \pm 0.25 \pm 20\%$	$\delta_{15} = \pm 1.90 \pm 20\%$	1.064	1.45	1.49	$d_{11}^{\alpha\text{-SiO}_2}$	336
Silicon carbide	$6mm$- C_{6v}	$d_{31} = +8.6 \pm 0.9$	$\delta_{31} = +0.47$	1.064	2.5830	2.7167	$d_{11}^{\alpha\text{-SiO}_2}$	337
SiC		$d_{33} = -14.4 \pm 1.3$	$\delta_{33} = -0.72$	1.064	2.6225	2.7167	$d_{11}^{\alpha\text{-SiO}_2}$	337
		$d_{15} = +8.0 \pm 0.9$	$\delta_{15} = +0.44$	1.064	2.6027	2.6689	$d_{11}^{\alpha\text{-SiO}_2}$	337
Silver iodide	$6mm$- C_{6v}	$d_{31} = +8.2 \pm 20\%$	$\delta_{31} = +2.1 \pm 20\%$	1.318	2.104	2.200	$d_{11}^{\alpha\text{-SiO}_2}$	338
AgI		$d_{33} = -16.8 \pm 22\%$	$\delta_{33} = -4.1 \pm 22\%$	1.318	2.115	2.200	$d_{11}^{\alpha\text{-SiO}_2}$	338
Zinc silver indium	$6mm$- C_{6v}	$d_{31} = - +7.2 \pm 20\%$	$\delta_{31} = - +0.84 \pm 20\%$	1.064	2.31	2.50	$d_{11}^{\alpha\text{-SiO}_2}$	339
sulfide		$d_{33} = \pm 15.9 \pm 20\%$	$\delta_{33} = \pm 1.73 \pm 20\%$	1.064	2.34	2.50	$d_{11}^{\alpha\text{-SiO}_2}$	339
Zn_3AgInS_5								
Zn_5AgInS_7	$6mm$- C_{6v}	$d_{31} = - +9.22 \pm 20\%$	$\delta^{31} = - +1.03 \pm 20\%$	1.064	2.32	2.50	$d_{11}^{\alpha\text{-SiO}_2}$	339
		$d_{33} = - +20.95 \pm 20\%$	$\delta_{33} = \pm 2.27 \pm 20\%$	1.064	2.34	2.50	$D_{11}^{\alpha\text{-SiO}_2}$	339
Zinc oxide ZnO	$6mm$- C_{6v}	$d_{31} = +1.76 \pm 0.16$	$\delta_{31} = +0.82 \pm 0.08$	1.058	1.939	2.048	d_{36}^{KDP}	250, 227
		$d_{33} = -5.86 \pm 0.16$	$\delta_{33} = -2.60 \pm 0.07$	1.058	1.955	2.048	d_{36}^{KDP}	250, 227

		$d_{15} = 1.93 \pm 0.16$	$\delta_{15} = 0.90 \pm 0.07$	1.058	1.947	2.031	d_{36}^{KDP}	250, 227
α-Zinc sulfide α-ZnS	$6mm$- C_{6v}	$d_{33} = +11.37 \pm 0.07$	$d_{33} = +1.47 \pm 0.09$	1.058	2.299	2.401	d_{33}^{CdS}	332, 340
		$d_{33} = +37.3 \pm 12.6$	$d_{33} = +6.59 \pm 2.23$	10.6	2.219	2.270	Abs	288, 340
		$d_{31} = -18.9 \pm 6.3$	$d_{31} = -3.39 \pm 1.12$	10.6	2.213	2.270	Abs	288, 340
		$d_{15} = 21.37 \pm 8.4$	$d_{15} = 3.82 \pm 1.5$	10.6	2.216	2.265	Abs	288, 340
		$d_{15} = 6.7 \pm 1.0$	$d_{15} = 0.88 \pm 0.13$	1.064	2.296	2.396	$d_{11}^{\alpha\text{-}SiO_2}$	341, 340
		$d_{31} = -7.6 \pm 1.5$	$d_{31} = -0.99 \pm 0.2$	1.064	2.294	2.401	$d_{11}^{\alpha\text{-}SiO_2}$	141, 340
		$d_{33} = +13.8 \pm 1.7$	$d_{33} = +1.78 \pm 0.22$	1.064	2.299	2.401	$d_{11}^{\alpha\text{-}SiO_2}$	341, 340
		System — Cubic						
Sodium bromate $NaBrO_3$	23-T	$d_{14} = 0.19$	$\delta_{14} = 0.48$	0.6943	1.611	1.661	$d_{11}^{\alpha\text{-}SiO_2}$	100
Sodium chlorate $NaClO_3$	23-T	$d_{14} = 0.46$		0.6943	1.512	1.540	$d_{11}^{\alpha\text{-}SiO_2}$	100
Aluminum antimonide AlSb	$\bar{4}3m$- T_d	$d_{14} = 49 \pm 36$	$\delta_{14} = 0.4 \pm 0.29$	1.058	3.3	3.87	d_{36}^{KDP}	344
Bismuth germanium oxide Bi_4GeO_{12}	$\bar{4}3m$- T_d	$d_{14} = 1.28$	$\delta_{14} = 0.41$	1.0642	2.0443	2.1152	$d_{11}^{\alpha\text{-}SiO_2}$	172
Cadmium telluride CdTe	$\bar{4}3m$- T_d	$d_{14} = +167.6 \pm 63$	$\delta_{14} = +7.68 \pm 2.89$	10.6	2.69	2.71	Abs	288, 188
		$d_{14} = +59.0 \pm 24$	$\delta_{14} = +3.15 \pm 1.28$	28.0	2.612	2.693	d_{11}^{Te}	291, 188
Cuprous bromide CuBr	$\bar{4}3m$- T_d	$d_{14} = 13.5 \pm 6.1$	$\delta_{14} = 4.13 \pm 2.0$	1.064	2.011	2.164	$d_{11}^{\alpha\text{-}SiO_2}$	350
		$d_{14} = 8.04 \pm 30\%$	$\delta_{14} = 3.34 \pm 30\%$	10.6	1.970	1.972	d_{14}^{GaAs}	350
		$d_{14} = 4.38 \pm 20\%$	$\delta_{14} = -1.53 \pm 20\%$	1.318			$d_{11}^{SiO_2}$	351
		$d_{14} = -6.37 \pm 20\%$	$\delta_{14} = -2.08 \pm 20\%$	1.064			$d_{11}^{SiO_2}$	351

Table 1.1.16 (continued)
SECOND-HARMONIC COEFFICIENTS

System — Cubic

Material	Symmetry class	$d_{jm} \times 10^{12}$ (m/V)	$\delta_{jm} \times 10^{2}$ (m^2/C)	λ_1 (μm)	n_{ω}	$n_{2\omega}$	Measured relative to	Ref.
		$d_{14} = -6.85 \pm 20\%$	$\delta_{14} = -2.25 \pm 20\%$	1.064			$d_{11}^{SiO_2}$	351
		$d_{14} = -6.53 \pm 20\%$	$\delta_{14} = -1.94 \pm 20\%$	0.946			$d_{11}^{SiO_2}$	351
Cuprous chloride	$\bar{4}3m$- T_d	$d_{14} = \pm 8 \pm 45\%$	$\delta_{14} = \pm 4.43 \pm 45\%$	1.06	1.923	2.012	$d_{11}^{\alpha\text{-}SiO_2}$	350
CuCl		$d_{14} = 6.7 \pm 30\%$	$\delta_{14} = 4.3 \pm 30\%$	10.6	1.893	1.895	d_{14}^{GaAs}	350
		$d_{14} = -6.04 \pm 20\%$	$\delta_{14} = -3.01 \pm 20\%$	0.946			$d_{11}^{\alpha\text{-}SiO_2}$	351
		$d_{14} = -6.08 \pm 20\%$	$\delta_{14} = -3.32 \pm 20\%$	1.064	1.923	2.012	$d_{11}^{\alpha\text{-}SiO_2}$	351
		$d_{14} = -5.47 \pm 20\%$	$\delta_{14} = -3.22 \pm 20\%$	1.318			$d_{11}^{\alpha\text{-}SiO_2}$	351
		$d_{14} = 4.19$	$\delta_{14} = 2.74$	10.6	1.893	1.895		349
Cuprous iodide	$\bar{4}3m$- T_d	$d_{14} = 26.56 \pm 45\%$	$\delta_{14} = 3.82 \pm 45\%$	1.06	2.223	2.392	$d_{11}^{\alpha\text{-}SiO_2}$	350
CuI		$d_{14} = 8.04 \pm 30\%$	$\delta_{14} = 1.67 \pm 30\%$	10.6	2.176	2.178	d_{14}^{GaAs}	350
		$d_{14} = -3.47 \pm 20\%$	$\delta_{14} = -0.48 \pm 20\%$	0.946			$d_{11}^{\alpha\text{-}SiO_2}$	351
		$d_{14} = -5.02 \pm 20\%$	$\delta_{14} = -0.79 \pm 20\%$	1.064	2.223	2.392	$d_{11}^{\alpha\text{-}SiO_2}$	351
		$d_{14} = -3.97 \pm 20\%$	$\delta_{14} = -0.62 \pm 20\%$	1.064	2.223	2.392	$d_{11}^{\alpha\text{-}SiO_2}$	351
		$d_{14} = -4.0 \pm 20\%$	$\delta_{14} = -0.69 \pm 20\%$	1.318			$d_{11}^{\alpha\text{-}SiO_2}$	351
Gallium antimonide	$\bar{4}3m$- T_d	$d_{14} = +628 \pm 63$	$\delta_{14} = +2.84 \pm 0.28$	10.6	3.8	3.82	d_{14}^{InAs}	88
		$d_{14} = 419$	$\delta_{14} = 1.89$					99
Gallium arsenide	$\bar{4}3m$ -T_d	$d_{14} = 229.6 \pm 57.4$	$\delta_{14} = + 1.17 \pm 0.29$	1.0582	3.479	4.352	d_{36}^{KDP}	250
GaAs								
		$d_{14} = + 188.6 \pm$ (factor of 2)	$\delta_{14} = + 2.29 \pm$ (factor of 2)	10.6	3.27	3.30	d_{14}^{InAs}	88
		$d_{14} = 368.7 \pm 125.7$	$\delta_{14} = 4.48 \pm 1.53$	10.6	3.27	3.30	Abs	288
		$d_{36} = 209.5 \pm 13.3$	$\delta_{36} = 1.07 \pm 0.07$	1.058	3.479	4.352	d_{33}^{OdS}	332
		$d_{14} = 274.3 \pm 37.8$	$\delta_{14} = 1.40 \pm 0.19$	1.058	3.479	4.352		99
		$d_{14} = 134.1 \pm 41.9$	$\delta_{14} = 1.24 \pm 0.41$	10.6			$d_{31}^{Ag_3SbS_3}$	296

		$d_{14} = 140 \pm 10$	$\delta_{14} = 0.72 \pm 0.05$	1.06	3.478	4.346	Abs	94
		$d_{14} = 100.6 \pm 21$	$\delta_{14} = 0.52 \pm 0.11$	1.06	3.478	4.346	Abs	354
		$d_{14} = 191 \pm 64$	$\delta_{14} = 0.98 \pm 0.33$	1.058	3.479	4.352	—	344
		$d_{14} = 137$	$\delta_{14} = 0.31$	0.843—08450	3.60	5.9		355
		$d_{14} = 104 \pm 17$		$\lambda_p =$ 10.6 μm $\lambda_s =$ 53.5 GHz			Abs	356
		$d_{14} = 256.5$		0.694			d_{36}^{KDP}	329
		$d_{14} = 135.3$		1.06			d_{36}^{KDP}	329
		$d_{36} = 90.1 \pm 4\%$	$\delta_{36} = 1.1$	10.6	3.27	3.30	$d_{+}^{Ag_3SbS_3}$	155
		$d_{36} = 133.4 \pm 10.4$	$\delta_{36} = 1.31 \pm 0.1$	2.12	3.34652	3.479	$d_{31}^{LiIO_3}$	160
Gallium phosphide GaP	$\bar{4}3m$ -T_d	$d_{14} = 71.8 \pm 12.3$	$\delta_{14} = 1.01 \pm 0.17$	1.058	3.1065	3.424	d_{36}^{KDP}	250
		$d_{36} = 34.6 \pm 2.2$	$\delta_{36} = 0.49 \pm 0.03$	1.058	3.1065	3.424	d_{33}^{CdS}	332
		$d_{14} = + 70.6$		3.39	3.420		Abs	358, 357
		$d_{14} = + 109 \pm$ (factor of 2)	$\delta_{14} = + 2.1 \pm$ (factor of 2)	10.6	3.05	3.11	d_{14}^{InAs}	88
		$d_{36} = 59.2 \pm 10\%$	$\delta_{36} = 0.96 \pm 10\%$	1.318	3.0727	3.2874	$d_{11}^{\alpha\text{-}SiO_2}$	155
		$d_{36} = 59.5 \pm 5.95$	$\delta_{36} = 0.96 \pm 0.1$	1.318	3.07272	3.2874	$d_{31}^{LiIO_3}$	156
		$\delta_{36} = 60 \pm 8.43$	$\delta_{36} = 1.16 \pm 0.16$	2.12	3.0350	3.1065	$d_{31}^{LiIO_3}$	156
Hexamine $N_4(CH_2)_6$	$\bar{4}3m$ -T_d	$d_{14} = 4.1$	$\delta_{14} = 13.6$	1.06	1.577	1.593	d_{36}^{KDP}	359
Indium antimonide InSb	$\bar{4}m$- T_4	$d_{14} = 520 \pm 47$	$\delta_{14} = 1.8$	1.058			d_{36}^{KDP}	99
		$d_{14} = 1634 \pm 503$	$\delta_{14} = 5.59 \pm 1.7$	10.6	3.955		d_{14}^{GaAs}	361
		$d_{14} = 560 \pm 230$	$\delta_{14} = 2.62 \pm 1.08$	28	3.745	3.904	d_{11}^{Te}	291
Indium arsenide	$\bar{4}3m$ -T_d	$d_{14} = 364 \pm 47$		1.058			d_{36}^{KDP}	99
		$d_{14} = 419 \pm 126$	$\delta_{14} = 3.27 \pm 0.1$	10.6	3.49	3.54		288
		$d_{14} = 249 \pm 25\%$		10.6			d_{11}^{Te}	88
		$d_{14} = 419 \pm$ (factor of 2)		10.6			Abs	88

Table 1.1.16 (continued)
SECOND-HARMONIC COEFFICIENTS

System — Cubic

Material	Symmetry class	djm × 10^{12} (m/V)	δ_{jm} × 10^2 (m^2/C)	λ_1 (μm)	n_ω	$n_{2\omega}$	Measured relative to	Ref.
Indium phosphide InP	$\bar{4}3m$ -T_d	d_{14} = 143.5	δ_{14} = 0.81	1.058	3.44	4.24	d_{36}^{KDP}	344
Zinc selenide ZnSe	$\bar{4}3m$ -T_d	d_{14} = + 78.4 ± 29.3	δ_{14} = + 7.7 ± 2.9	10.6	2.42	2.43	Abs	288, 188
		d_{36} = + 26.6 + 1.7	δ_{14} = + 1.87 ± 0.1	1.058	2.48	2.66	d_{33}^{CdS}	332, 188
Zinc sulfide β-ZnS	$\bar{4}3m$ -T_d	d_{14} = + 30.6 ± 8.4	δ_{14} = + 5.1 + 1.4	10.6	2.25	2.26	Abs	288, 188
		d_{36} = + 20.7 ± 1.3	δ_{36} = + 2.73 ± 0.17	1.058	2.289	2.40	d_{33}^{CdS}	332, 188
Zinc telluride ZnTe	$\bar{4}3m$ -T_d	d_{14} = + 92.2 ± 33.5	δ_{14} = + 4.26 ± 1.6	10.6	2.69	2.70	Abs	288, 188
		d_{14} = + 83.2 ± 8.4	δ_{14} = + 1.3 ± 0.23	1.058	2.772	3.182	d_{36}^{KDP}	99, 188
		d_{36} = + 89.6 ± 5.7	δ_{36} = + 2.48 ± 0.16	1.058	2.772	3.182	d_{33}^{CdS}	332, 188

Table 1.1.17.1
ALUMINUM, COPPER, GOLD, AND SILVER (Al, Cu, Au, Ag) REFRACTIVE INDEX AT ROOM TEMPERATURE[364]

Wavelength (μm)	*n* (Al)	*n* (Cu)	*n* (Au)	*n* (Ag)
0.40	0.40	0.85	1.45	0.075
0.45	0.49	0.87	1.40	0.055
0.50	0.62	0.88	0.84	0.050
0.55	0.76	0.72	0.34	0.055
0.60	0.97	0.17	0.23	0.060
0.65	1.24	0.13	0.19	0.070
0.70	1.55	0.12	0.17	0.075
0.75	1.80	0.12	0.16	0.080
0.80	1.99	0.12	0.16	0.090
0.85	2.08	0.12	0.17	0.100
0.90	1.96	0.13	0.18	0.105
0.95	1.75	0.13	0.19	0.110

Table 1.1.17.1a
GOLD (Au) REFRACTIVE INDEX AT ROOM TEMPERATURE[365]

λ(Å)	*n*	λ(Å)	*n*	λ(Å)	*n*
1087	1.115	2476	1.306	4141	1.602
1118	1.105	2513	1.313	4204	1.594
1150	1.111	2554	1.323	4247	1.590
1184	1.129	2597	1.340	4300	1.573
1215	1.153	2642	1.363	4348	1.557
1254	1.187	2685	1.390	4400	1.545
1293	1.203	2737	1.432	4446	1.527
1335	1.206	2771	1.462	4505	1.501
1375	1.226	2801	1.487	4544	1.481
1416	1.265	2846	1.513	4604	1.442
1453	1.290	2899	1.543	4642	1.417
1493	1.297	2952	1.573	4672	1.389
1532	1.328	2989	1.594	4699	1.364
1572	1.390	3035	1.611	4740	1.322
1608	1.431	3074	1.628	4775	1.280
1651	1.429	3125	1.646	4805	1.238
1689	1.421	3157	1.661	4831	1.200
1731	1.404	3202	1.679	4859	1.154
1776	1.374	3240	1.692	4891	1.099
1817	1.353	3285	1.700	4922	1.041
1855	1.333	3327	1.708	4957	0.972
1895	1.311	3372	1.710	4990	0.905
1934	1.299	3414	1.707	5024	0.838
1979	1.291	3456	1.694	5049	0.786
2026	1.290	3497	1.680		
2064	1.285	3538	1.661		
2104	1.276	3584	1.640		
2144	1.277	3638	1.625		
2179	1.280	3683	1.617		
2218	1.285	3742	1.613		
2264	1.290	3780	1.615		
2304	1.293	3840	1.615		
2344	1.292	3910	1.613		
2394	1.296	4038	1.609		
2436	1.298	4092	1.605		

Table 1.1.17.1b
SILVER (Ag) REFRACTIVE INDEX AT ROOM TEMPERATURE[365]

λ(Å)	n	λ(Å)	n	λ(Å)	n
1137	1.164	2466	1.275	3741	0.083
1173	1.166	2503	1.294	3787	0.082
1208	1.153	2551	1.319	3822	0.080
1241	1.137	2592	1.340	3862	0.078
1276	1.130	2638	1.363	3895	0.080
1307	1.118	2677	1.382	3945	0.081
1339	1.096	2710	1.399	3992	0.082
1373	1.060	2749	1.417	4040	0.084
1408	1.029	2792	1.439	4077	0.083
1438	1.001	2819	1.452	4124	0.083
1473	0.965	2859	1.469	4170	0.082
1507	0.936	2893	1.494	4228	0.084
1544	0.915	2932	1.500	4280	0.085
1576	0.910	2968	1.536	4315	0.084
1612	0.903	2995	1.564	4358	0.084
1647	0.898	3029	1.556	4402	0.085
1689	0.901	3052	1.543	4447	0.084
1733	0.909	3086	1.470	4489	0.087
1774	0.923	3128	1.343	4522	0.088
1816	0.943	3153	1.248	4583	0.090
1856	0.966	3187	1.083	4657	0.092
1886	0.985	3227	0.877	4719	0.093
1925	1.010	3257	0.622	4782	0.094
1955	1.028	3280	0.389	4826	0.095
1991	1.052	3300	0.282	4878	0.097
2029	1.080	3330	0.207	4935	0.099
2061	1.106	3368	0.174	5008	0.103
2108	1.139	3415	0.161	5075	0.106
2158	1.168	3459	0.136		
2201	1.190	3494	0.119		
2252	1.207	3528	0.108		
2284	1.218	3565	0.098		
2320	1.225	3607	0.093		
2373	1.238	3662	0.087		
2426	1.254	3704	0.084		

Table 1.1.17.2
ALUMINUM OXIDE (Al_2O_3) REFRACTIVE INDEX AT 24°C[366]

Wavelength (μm)	n^o	n^e
0.2536	1.844	1.834
0.2600	1.8376	1.8280
0.2800	1.8242	1.8149
0.3000	1.8147	1.8056
0.3300	1.8035	1.7947
0.3650	1.7935	1.7849
0.40466	1.78571	1.77724
0.4358	1.78110	1.77275
0.48613	1.77547	1.76724
0.54607	1.77067	1.76254
0.5893	1.76808	1.75999
0.64385	1.76537	1.75734
0.65628	1.76485	1.75682
0.69072	1.76351	1.75549

$dn^o/dT = +13.6 \times 10^{-6}/°C$
$dn^e/dT = +14.7 \times 10^{-6}/°C$

Table 1.1.17.2a ALUMINUM OXIDE (Al_2O_3) INDEX OF REFRACTION OF SYNTHETIC SAPPHIRE FOR THE ORDINARY RAY AT 24°C[393]

λ(μm)	n^o	λ(μm)	n^o
0.26520	1.83360	1.39506	1.74888
0.28035	1.82427	1.52952	1.74660
0.28936	1.81949	1.6932	1.74368
0.29673	1.81595	1.70913	1.74340
0.30215	1.81351	1.81307	1.74144
0.3130	1.80906	1.9701	1.73833
0.33415	1.80184	2.1526	1.73444
0.34662	1.79815	2.24929	1.73231
0.361051	1.79450	2.32542	1.73057
0.365015	1.79358	2.4374	1.72783
0.39064	1.78826	3.2439	1.70437
0.404656	1.78582	3.2668	1.70356
0.43834	1.78120	3.3026	1.70231
0.546071	1.77078	3.3303	1.70140
0.576960	1.76884	3.422	1.69818
0.579066	1.76871	3.5070	1.69504
0.64385	1.76547	3.7067	1.68746
0.706519	1.76303	4.2553	1.66371
0.85212	1.75885	4.954	1.62665
0.89440	1.75796	5.1456	1.61514
1.01398	1.75547	5.349	1.60202
1.12866	1.75339	5.419	1.59735
1.36728	1.74936	5.577	1.58638

Dispersion equation:

$$n^2 - 1 = \sum_i \frac{A_i\lambda^2}{\lambda^2 - \lambda_i^2}$$

At 24°C:

$\lambda_1^2 = 0.00377588$ $A_1 = 1.023798$

$\lambda_2^2 = 0.0122544$ $A_2 = 1.058264$

$\lambda_3^2 = 3213616$ $A_3 = 5.280792$

Table 1.1.17.3a AMMONIUM DIHYDROGEN PHOSPHATE ($NH_4H_2PO_4$)[248]

Wavelength (μm)	n^o	n^e
0.2138560	1.62598	1.56738
0.2288018	1.60785	1.55138
0.2536519	1.58688	1.53289
0.2967278	1.56462	1.51339
0.3021499	1.56270	1.51163
0.3125663	1.55917	1.50853
0.3131545	1.55897	1.50832
0.3341478	1.55300	1.50313
0.3650146	1.54615	1.49720
0.3654833	1.54608	1.49712
0.3662878	1.54592	1.49698
0.3906410	1.54174	
0.4046561	1.53969	1.49159
0.4077811	1.53925	1.49123
0.4358350	1.53578	1.48831
0.4916036		1.48390
0.5460740	1.52662	1.48079
0.5769590	1.52478	1.47939
0.5790654	1.52466	1.47930
0.6328160	1.52166	1.47685
1.013975	1.50835	1.46895
1.128704	1.50446	1.46704
1.152276	1.50364	1.46666

Note: Estimated accuracy = ± 0.00003.

Dispersion equation:

$$n_o^2 = 2.302484 + 1.117089 \times 10^{10}\nu^2/(1 - \nu^2/7.605372 \times 10^9) + 3.751806 \times 10^6/(2.5 \times 10^5 - \nu^2)$$

$$n_e^2 = 2.163077 + 9.670312 \times 10^{-11}\nu^2/(1 - \nu^2/7.785289 \times 10^9) + 1.451540 \times 10^6/(2.5 \times 10^5 - \nu^2)$$

where $\nu = 1/\lambda$ in cm^{-1}.

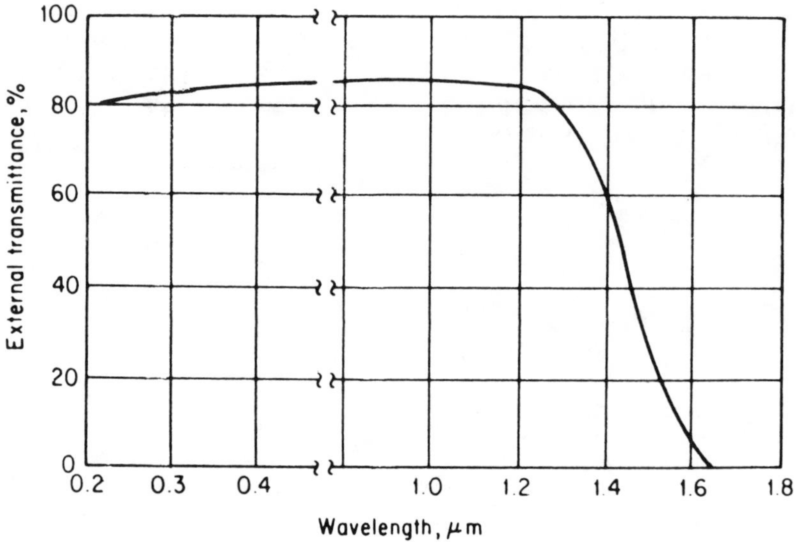

FIGURE 1.1.4. Transmission of ammonium dihydrogen phosphate (ADP) thickness 1.6 mm: (short wavelength portion) and 7.8 mm (long-wavelength portion). (From Deshotels, W. J., *J. Opt. Soc. Am.*, 50, 865, 1960. With permission.)

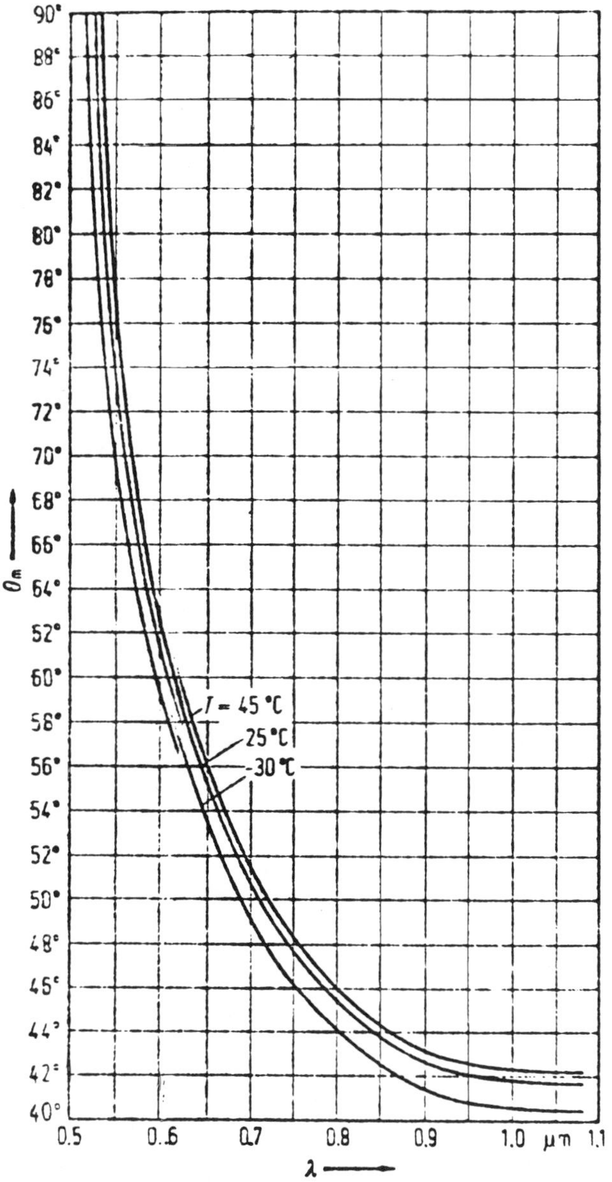

FIGURE 1.1.5. Phase matching angle for $NH_4H_2PO_4$ (ADP) plotted as a function of the fundamental at T = −30°C, T = 25°C, and T = 45°C.[161]

Table 1.1.17.3b
REFRACTIVE INDEX AT ROOM TEMPERATURE[249] AMMONIUM DIHYDROGEN PHOSPHATE (ADP) ($NH_4H_2PO_4$)

Wavelength (μm)	n^o	n^e
0.3653	1.5457	1.4970
0.4047	1.5396	1.4915
0.4078	1.5392	1.4912
0.4358	1.5357	1.4882
0.4916	1.5303	1.4838
0.5461	1.5265	1.4808
0.5779	1.5246	1.4792
0.6234	1.5223	1.4775
0.6907	1.5192	1.4753

Note: Estimated accuracy = ± 0.0001. The change in index of refraction with temperature is given by

$$\Delta n_0 = (n_o^2 - 3.0297n_0 + 2.3004) \times (0.713 \times 10^{-2}) \times (298 - T)$$
$$\Delta n_e = (n_e^3)(0.675 \times 10^{-6}) \times (298 - T),$$

where the temperature, T, is given in K.

Dispersion equation[378]

$$n^2 = A + B\lambda^2/(\lambda^2 - C)$$

	A	B	C
n^o	1.0	1.28196	0.01069
n^e	1.0	1.15607	0.00890

where λ is in μm.

Table 1.1.17.3c
AMMONIUM DIHYDROGEN PHOSPHATE ($NH_4H_2PO_4$)[161]

		Ref.
θ_m	$= 41.9 \pm 1°$ at $\lambda_1 = 1.0582$ μm	250
θ_m	$= 51.9 \pm 1°$ at $\lambda_1 = 0.6943$ μm	250
θ_m	$= 48°$ at $\lambda_1 = 0.6943$ μm	247
θ_m	$= 42°$ at $\lambda_1 = 8250$ μm	17
l_{36}	$= 17.7$ μm, $l_{14} = 6.7$ μm at $\lambda_1 = 0.6943$ μm	226
l_{36}	$= 21.0$ μm, $l_{14} = 31.2$ μm at $\lambda_1 = 1.0582$ μm	226

Table 1.1.17.3d
AMMONIUM DIHYDROGEN PHOSPHATE ($NH_4H_2PO_4$)[251]

Sellmeier equation

$$n^2 = A + B/(1 - C/\lambda^2) + D/(1 - E/\lambda^2)$$, where λ is in μm.

Temp. (K)	Polarization	*A*	*B*	*C*	*D*	*E*
150	o	1.566850	0.758506	0.015017	1.087847	30.0
	e	1.418720	0.745307	0.012965	0.420182	30.0
201	o	1.523685	0.793649	0.014418	1.082612	30.0
	e	1.418169	0.745565	0.012953	0.420131	30.0
300	o	1.446791	0.854884	0.013094	1.071368	30.0
	e	1.417093	0.746071	0.012929	0.420032	30.0
154	o	1.453864	0.820094	0.012771	0.908016	30.0
	e	1.425124	0.717782	0.012175	0.225394	30.0

Temperature-dependent Sellmeier coefficients X

$$X = mT + c$$, where T is in K

Sellmeier coefficients	Fitted constants	n^o	n^e
A	$m \times 10^3$	−0.87835	−0.01089
	c	1.69960	1.42036
B	$m \times 10^3$	0.72007	0.00514
	c	0.64955	0.74453
C	$m \times 10^5$	−1.40526	−0.02471
		0.01723	0.01300
D	$m \times 10^4$	−1.17900	−0.00999
		1.10624	0.42033
E		30.0	30.0

Table 1.1.17.4
CALCIUM FLUORIDE (CaF_2)
REFRACTIVE INDEX AT ROOM TEMPERATURE[367,368]

λ(μm)	n	λ(μm)	n	λ(μm)	n	λ(μm)	n
0.19	1.50500	1.0140	1.42884	2.1608	1.42306	4.000	1.40963
0.20	1.49531	1.08304	1.42843	2.250	1.42258	4.1252	1.40847
0.22	1.48119	1.1000	1.42834	2.3573	1.42198	4.2500	1.40722
0.24	1.47133	1.1786	1.42789	2.450	1.42143	4.4000	1.40568
0.26	1.46397	1.250	1.42752	2.5537	1.42080	4.6000	1.40357
0.28	1.45841	1.3756	1.42689	2.6519	1.42018	4.7146	1.40233
0.30	1.45400	1.4733	1.42642	2.700	1.41988	4.8000	1.40130
0.35	1.44658	1.5715	1.42596	2.750	1.41956	5.000	1.39908
0.40	1.441857	1.650	1.42558	2.800	1.41923	5.3036	1.39522
0.48615	1.43704	1.7680	1.42502	2.880	1.41890	5.8932	1.38712
0.58758	1.43388	1.8400	1.42468	2.9466	1.41823	6.4825	1.37824
0.58932	1.43384	1.8688	1.42454	3.0500	1.41750	7.0718	1.36805
0.65630	1.43249	1.900	1.42439	3.0980	1.41714	7.6612	1.35675
0.6871	1.43200	1.9153	1.42431	3.2413	1.41610	8.2505	1.34440
0.72818	1.43143	1.9644	1.42407	3.4000	1.41487	8.8398	1.33075
0.76653	1.43093	2.0582	1.42360	3.5359	1.41367	9.4291	1.31605
0.88400	1.42980	2.0626	1.42357	3.8306	1.41119		

Dispersion equation:[369]

$$n^2 - 1 = \frac{A_1\lambda^2}{\lambda^2 - \lambda_1^2} + \frac{A_2\lambda^2}{\lambda^2 - \lambda_2^2} + \frac{A_3\lambda^2}{\lambda^2 - \lambda_3^2}$$

where λ is in μm and $A_1 = 0.5675888$, $A_2 = 0.4710914$, $A_3 = 3.8484723$, $\lambda_1^2 = 0.002526430$, $\lambda_2^2 = 0.01007833$, and $\lambda_3^2 = 1200.5560$.

Table 1.1.17.5
CALCITE ($CaCO_3$)
REFRACTIVE INDEXES AT ROOM TEMPERATURE[369]

λ(μm)	n^o	n^e	λ(μm)	n^o	n^e	λ(μm)	n^o	n^e
0.198	—	1.57796	0.410	1.68014	1.49640	1.220	1.63926	1.47870
0.200	1.90284	1.57649	0.434	1.67552	1.49430	1.273	1.63849	
0.204	1.88242	1.57081	0.441	1.67423	1.49373	1.307	1.63789	1.47831
0.208	1.80733	1.56640	0.508	1.66527	1.48956	1.320	1.63767	
0.211	1.85692	1.56327	0.533	1.66277	1.48841	1.369	1.63681	
0.214	1.84558	1.55976	0.560	1.66046	1.48736	1.396	1.63637	1.47780
0.219	1.83075	1.55496	0.589	1.65835	1.48640	1.422	1.63590	
0.226	1.81309	1.54921	0.643	1.65504	1.48490	1.479	1.63490	
0.231	1.80233	1.54541	0.656	1.65437	1.48459	1.497	1.63457	1.47744
0.242	1.78111	1.53782	0.670	1.65367	1.48426	1.541	1.63381	
0.257	1.76038	1.53005	0.706	1.65207	1.48353	1.609	1.63261	
0.263	1.75343	1.52736	0.768	1.61974	1.48259	1.615	—	1.47695
0.267	1.74864	1.52547	0.795	1.64880	1.48216	1.682	1.63127	
0.274	1.74139	1.52261	0.801	1.64869	1.48216	1.749	—	1.47638
0.291	1.72774	1.51705	0.833	1.64772	1.48176	1.761	1.62974	
0.303	1.71959	1.51365	0.867	1.64676	1.48137	1.849	1.62800	
0.312	1.71425	1.51140	0.905	1.64578	1.48098	1.900	—	1.47573
0.330	1.70515	1.50746	0.946	1.64480	1.48060	1.946	1.62602	
0.340	1.70078	1.50562	0.991	1.64380	1.48022	2.053	1.62372	
0.346	1.69833	1.50450	1.042	1.64276	1.47985	2.100	—	1.47492
0.361	1.69317	1.50228	1.097	1.64167	1.47948	2.172	1.62099	
0.394	1.68374	1.49810	1.159	1.64051	1.47910	3.324	—	1.47392

Table 1.1.17.6
CADMIUM SULFIDE (CdS) REFRACTIVE INDEX AT ROOM TEMPERATURE[334]

Wavelength (μm)	n^o	n^e
0.5120		2.751
0.5130		2.743
0.5140		2.737
0.5150	2.743	2.726
0.5160	2.735	2.720
0.5170	2.727	2.714
0.5180	2.718	2.706
0.5190	2.709	2.702
0.5200	2.702	2.698
0.5210	2.700	2.694
0.5220	2.694	2.689
0.5230	2.687	2.685
0.5240	2.681	2.680
0.5250	2.674	2.675
0.5275	2.661	2.665
0.5300	2.649	2.654
0.5325	2.638	2.644
0.5350	2.628	2.637
0.5375	2.617	2.628
0.5400	2.609	2.622
0.5425	2.602	2.612
0.5450	2.594	2.606
0.5475	2.587	2.600
0.5500	2.580	2.593
0.5750	2.528	2.545
0.6000	2.493	2.511
0.6250	2.467	2.484
0.6500	2.446	2.463
0.6750	2.427	2.446
0.7000	2.414	2.432
0.7500	2.390	2.409
0.8000	2.374	2.392
0.8500	2.364	2.378
0.9000	2.359	2.368
0.9500	2.341	2.359
1.0000	2.334	2.352
1.0500	2.328	2.346
1.1000	2.324	2.340
1.1500	2.320	2.336
1.2000	2.316	2.332
1.2500	2.312	2.329
1.3000	2.309	2.326
1.3500	2.306	2.323
1.4000	2.304	2.321

Table 1.1.17.6a
CADMIUM SULFIDE (CdS)

l_{15} = 1.8μ at λ_1 = 1.0582 μm
l_{31} = 1.7 μm at λ_1 = 1.0582 μm — Ref. 226
l_{33} = 1.8 μm at λ_1 = 1.0582 μm
l_{33} = 67 ± 7 μm at λ_1 = 10.6 μm
l_{31} = 50 ± 5 μm at λ_1 = 10.6 μm — Ref. 288
l_{15} = 73 ± 7 μm at λ_1 = 10.6 μm

Dispersion equation:[384]

$$(n^o)^2 = 5.235 + \frac{1.819 \times 10^7}{\lambda^2 - 1.651 \times 10^7}$$

$$(n^e)^2 = 5.239 + \frac{2.076 \times 10^7}{\lambda^2 - 1.651 \times 10^7}$$

where λ is in Å.

Table 1.1.17.7a
CADMIUM GERMANIUM ARSENIDE ($CdGeAs_2$)[256]

Wavelength (μm)	n^o	n^e
2.3000	3.6076	3.7545
2.4000	3.5973	3.7316
2.5000	3.5895	3.7156
2.6000	3.5823	3.7030
2.7000	3.5773	3.6926
2.8000	3.5721	3.6846
2.9000	3.5684	3.6775
3.0000	3.5645	3.6714
3.1000	3.5615	3.6661
3.2000	3.5581	3.6574
3.4000	3.5536	3.6508
3.6000	3.5503	3.6454
3.8000	3.5468	3.6402
4.0000	3.5440	3.6368
4.2000	3.5415	3.6329
4.4000	3.5391	3.6299
4.6000	3.5372	3.6273
4.8000	3.5354	3.6249
5.0000	3.5336	3.6178
5.5000	3.5285	3.6134
6.0000	3.5251	3.6104
6.5000	3.5223	3.6073
7.0000	3.5200	3.6050
7.5000	3.5175	3.6030
8.0000	3.5157	3.6009
8.5000	3.5140	3.5988
9.0000	3.5120	3.5966
9.5000	3.5098	3.5942
10.0000	3.5078	3.5922
10.5000	3.5054	3.5896
11.0000	3.5031	3.5871
11.5000	3.5004	
12.0000	3.4977	
12.5000	3.4950	

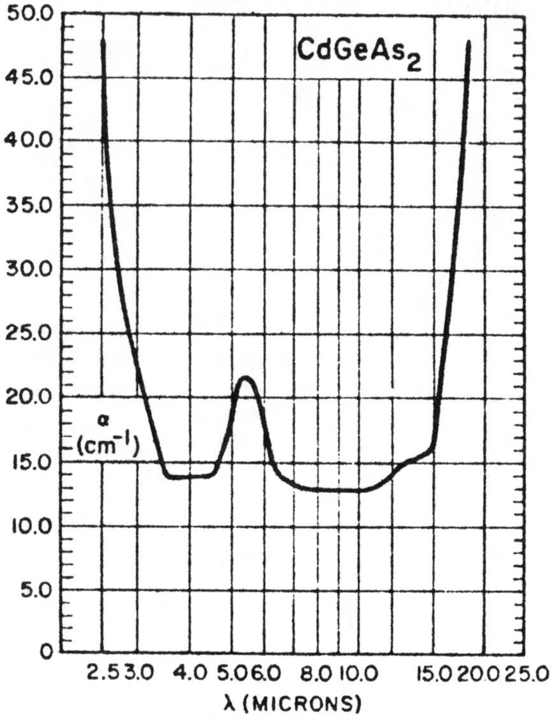

FIGURE 1.1.6. Absorption coefficient of $CdGeAs_2$. (From Boyd, G. D., Buehler, E., Storz, F. G., and Wernick, J. H., *IEEE J. Quantum Electron.*, QE-8, 419, 1972. With permission.)

Note: Phase-matching at λ_1 = 10.6 μm: l_{36} = 21.5 ± 1 μm at λ_1 = 10.6 μm: {Type I; θ_m = 35 ± 1°; Type II; θ_m = 51.6 ± 0.5°}

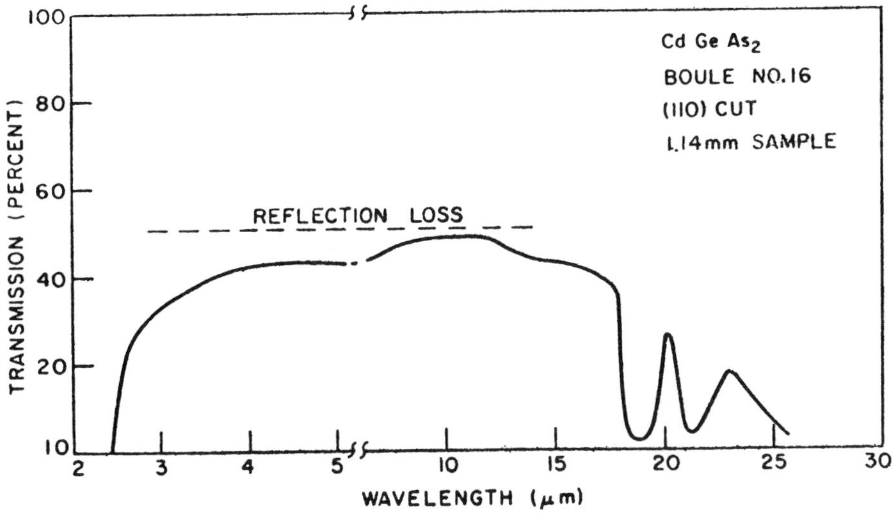

FIGURE 1.1.7. Measured transmission of $CdGeAs_2$. (From Byer, R. L., Kildal, H., and Feigelson, R. S., *Appl. Phys. Lett.*, 19, 237, 1971. With permission.)

Table 1.1.17.7b
CADMIUM GERMANIUM ARSENIDE ($CdGeAs_2$)[257]

Sellmeier equation

$$n^{o^2} = 10.1064 + \frac{2.2988}{(1 - 1.0872/\lambda^2)} + \frac{1.6247}{(1 - 1370/\lambda^2)}$$

$$n^{e^2} = 11.8018 + \frac{1.2152}{(1 - 2.6971/\lambda^2)} + \frac{1.6922}{(1 - 1370/\lambda^2)}$$

where λ is in μm

Table 1.1.17.7c
CADMIUM GERMANIUM ARSENIDE ($CdGeAs_2$)[255]

Room-temperature dispersion equations

$$(n^o)^2 = 4 + \frac{8.891}{1 - (0.5524/\lambda)^2} + \frac{1.886}{1 - (36/\lambda)^2}$$

$$(n^e)^2 = 4 + \frac{9.521}{1 - (0.6847/\lambda)^2} + \frac{1.909}{(1 - (36/\lambda)^2}$$

where λ is expressed in μm. Coherence length measured at λ_1 = 10.6 μm: l_{36} = 22 ± 1 μm; θ_m-Type II measured 52° ± 1.

Table 1.1.17.8
DIAMOND
REFRACTIVE INDEX AT ROOM TEMPERATURE[370]

Wavelength (μm)	n
0.2265	2.7151
0.480	2.4368
0.486	2.4354
0.5358	2.44986?
	2.4237
0.5461	2.4235
	2.42388
0.578	2.41899
0.580	2.4167—2.4238 (Natural)
	2.4142—2.4197 (Synthetic)
0.589	2.4175
0.644	2.4114
0.6563	2.4099
	2.4104

Dispersion equation: $n = A + BL + CL^2 + D^2 + E\lambda^4$, where λ is in μm. For the range 2.5 to 25.0 μm, The constants are

$L = 1/(\lambda^2 - 0.028)$
$A = 2.37553$
$B = 3.36440 \times 10^{-2}$
$C = 8.87524 \times 10^{-2}$
$D = 2.40455 \times 10^{-6}$
$E = 2.21390 \times 10^{-9}$

Table 1.1.17.9a
SILICON (Si)[371]
REFRACTIVE INDEX AT VARIOUS WAVELENGTHS FOR GIVEN TEMPERATURES

λ(μm)	296 K	275 K	202 K	104 K
2.554	3.43681	3.43472	3.42184	3.41172
2.652	3.43529	3.43264	3.42006	3.40896
2.732	3.43367	3.43097	3.41843	3.40754
2.856	3.43224	3.42971	3.41776	3.40611
2.958	3.43102	3.42836	3.41587	3.40475
3.090	3.42987	3.42723	3.41483	3.40365
4.120	3.42304	3.42642	3.40800	3.39695
5.190	3.41974	3.41649	3.40496	3.39388
8.230	3.41629	3.41314	3.40169	3.39064
10.270	3.41551	3.41226	3.40084	3.38989

Table 1.1.17.9b
REFRACTIVE INDEX OF Si AT 26°C[379]

λ(μm)	n	λ(μm)	n	λ(μm)	n	λ(μm)	n
1.3570	3.4975	2.1526	3.4476	4.00	3.4255	7.00	3.4189
1.3673	3.4962	2.3254	3.4430	4.258	3.4242	7.50	3.4186
1.3951	3.4929	2.4373	3.4408	4.50	3.4236	8.00	3.4184
1.3295	3.4795	2.7144	3.4358	5.00	3.4223	8.50	3.4182
1.6606	3.4696	3.00	3.4320	5.50	3.4123	10.00	3.4179
1.7092	3.4664	3.3033	3.4297	6.00	3.4202	10.50	3.4178
1.8134	3.608	3.4188	3.4286	6.50	3.4195	11.04	3.4176
1.9704	3.4537	3.50	3.4284				

Dispersion equation:

$$n = A + BL + CL^2 + D\lambda^2 + E\lambda^4$$

where

$A + 3.41696$ $\quad D = -0.0000209$
$B = 0.138497$ $\quad E = 0.000000148$
$C = 0.013924$ $\quad L = (\lambda^2 - 0.028)^{-1}$

Table 1.1.17.9c SILICON (Si) REFRACTIVE INDEX AT 26°C[380]

Wavelength (μm)	n
2.4373	3.4434
2.50	3.4424
2.7144	3.4393
3.00	3.4361
3.3033	3.4335
3.4188	3.4327
3.50	3.4321
4.00	3.4294
4.258	3.4283
4.50	3.4275
5.00	3.4261
5.50	3.4250
6.00	3.4242
6.50	3.4236
7.00	3.4231
7.50	3.4227
8.00	3.4224
8.50	3.4221
9.00	3.4219
9.50	3.4217
10.00	3.4215
10.50	3.4214
11.00	3.4213
11.04	3.4213
11.50	3.4212
12.00	3.4211
12.50	3.4210
13.00	3.4209
13.50	3.4209
14.00	3.4208
14.50	3.4208
15.00	3.4207
15.50	3.4207
16.00	3.4206
17.00	3.4206
18.00	3.4205
19.00	3.4205
20.00	3.4204
21.00	3.4204
22.00	3.4203
23.00	3.4203
24.00	3.4202
25.00	3.4201

Dispersion equation:[380]

$n = A + BL + CL^2 + D\lambda^2 + E\lambda^4$,
where $L = 1/(\lambda^2 - 0.028)$

$A = 3.41983$ $\quad D = 1.26878 \times 10^{-6}$
$B = 1.59906 \times 10^{-1}$ $\quad E = 1.95104 \times 10^{-9}$
$C = 1.23109 \times 10^{-1}$

where λ is in μm.

Table 1.1.17.10a GERMANIUM (Ge) REFRACTIVE INDEX AT 27°C[372]

Wavelength (μm)	n
2.0581	4.1016
2.1526	4.0919
2.3126	4.0786
2.4374	4.0708
2.577	4.0609
2.7144	4.0552
2.998	4.0452
3.3033	4.0369
3.4188	4.0334
4.258	4.0170
4.866	4.0170
6.238	4.0094
8.66	4.0043
9.72	4.0034
11.04	4.0026
12.20	4.0023
13.02	4.0021

Dispersion equation:

$n = A + BL + CL^2 + D\lambda^2 + E\lambda^4$

where λ is in μm and $L = (\lambda^2 - 0.028)^{-1}$

$A = 3.99931$
$B = 0.391707$
$C = 0.163492$
$D = 0.0000060$
$E = 0.000000053$

Table 1.1.17.10b
GERMANIUM (Ge)[371]
REFRACTIVE INDEX AT VARIOUS WAVELENGTHS FOR GIVEN TEMPERATURES

Wavelength (μm)	297 K	275 K	204 K	94 K
2.554	4.06230	4.05659	4.02528	3.98859
2.652	4.05754	4.05201	4.01955	3.98462
2.732	4.05310	4.04725	4.01511	3.98052
2.856	4.04947	4.04338	4.01139	3.97720
2.958	4.04595	4.03957	4.00796	3.97390
3.090	4.04292	4.03649	4.00485	3.97100
4.120	4.02457	4.01732	3.98662	3.95334
5.190	4.01617	4.00853	3.97820	3.94536
8.230	4.00743	3.99933	3.96934	3.93720
10.270	4.00571	3.99729	3.96745	3.93597
12.360	4.00627	3.99607	3.96625	3.94026

Sellmeier equation:[363]

$n^2 = A + B\lambda^2 (\lambda^2 - C) + D\lambda^2(\lambda^2 - E)$, where λ is in μm.

Temperature-dependent Sellmeier coefficients

$A = -6.040 \times 10^{-3}\, T + 11.05128$
$B = 9.295 \times 10^{-3}\, T + 4.00536$
$C = -5.392 \times 10^{-4}\, T + 0.599034$
$D = 4.151 \times 10^{-4}\, T + 0.09145$
$E = 1.51408\, T + 3426.5$

Table 1.1.17.11
INDIUM ARSENIDE (InAs) REFRACTIVE INDEX AT ROOM TEMPERATURE[373]

λ(μm)	n	λ(μm)	n	λ(μm)	n
0.049	1.139	0.248	1.987	1.38	3.516
0.051	1.135	0.259	2.288	1.68	—
0.054	1.135	0.264	2.617	1.80	—
0.056	1.133	0.269	3.060	2.00	—
0.059	1.131	0.222	3.800	2.07	—
0.062	1.125	0.310	3.678	2.25	—
0.064	1.120	0.335	3.359	2.50	—
0.067	1.110	0.344	3.271	2.76	—
0.070	1.047	0.354	3.227	3.00	—
0.077	0.948	0.387	3.484	3.35	—
0.082	0.894	0.413	3.331	3.40	—
0.089	0.829	0.443	3.817	3.44	—
0.095	0.766	0.451	0.380	3.50	—
0.103	0.745	0.459	4.087	3.65	—
0.108	0.751	0.468	4.119	3.74	3.52
0.112	0.775	0.477	4.192	4.00	3.51
0.123	0.835	0.496	4.489	5.00	3.46
0.136	0.890	0.517	4.558	6.67	3.45
0.153	0.967	0.563	4.320	10.0	3.42
0.172	1.184	0.620	4.101	14.3	3.39
0.180	1.332	0.689	3.934	16.7	3.38
0.188	1.483	0.775	3.800	20.0	3.35
0.195	1.583	0.885	3.696	25.0	3.26
0.211	1.782	1.03	3.613	33.3	2.95
0.225	1.765	1.24	3.548		

Dispersion equation:

$$n^2 = 11.1 + \frac{0.71}{1 - (1/3922\lambda)^2} + \frac{2.75}{1 - (1/218.9\lambda)^2} - 6 \times 10^4\lambda^2$$

Table 1.1.17.12
INDIUM ANTIMONIDE (InSb) REFRACTIVE INDEX AT ROOM TEMPERATURE[373]

λ(μm)	n	λ(μm)	n	λ(μm)	n
0.049	1.15	0.477	3.42	7.00	—
0.052	1.15	0.517	3.82	7.50	—
0.054	1.15	0.539	4.05	7.87	4.001
0.056	1.16	0.564	4.18	8.00	3.995
0.059	1.17	0.590	4.22	8.50	—
0.062	1.17	0.620	4.29	9.00	—
0.065	1.18	0.656	4.71	9.01	3.967
0.069	1.15	0.677	5.08	9.50	—
0.073	1.11	0.689	5.13	10.06	3.953
0.080	1.02	0.708	5.06	11.01	3.937
0.083	0.97	0.775	4.72	12.06	3.920
0.089	0.88	0.886	4.40	12.98	3.912
0.095	0.80	1.03	4.24	13.90	3.902
0.103	0.75	1.24	4.15	15.13	3.881
0.113	0.72	1.55	4.08	15.79	3.873
0.124	0.74	1.60	—	16.96	3.866
0.138	0.80	1.80	—	17.85	3.850
0.155	0.88	2.00	—	18.85	3.843
0.163	0.94	2.07	4.03	19.98	3.826
0.182	1.06			20.0	—
0.207	1.23			21.15	3.814
0.218	1.36			22.20	3.805
0.221	1.38			25.0	3.78
0.234	1.48			26.0	3.74
0.238	1.53			27.0	3.66
0.243	1.57			28.0	3.56
0.248	1.56			29.0	3.50
0.282	1.66			30.0	3.47
0.302	2.19			31.0	3.44
0.310	2.62			32.0	3.39
0.344	3.51			33.0	3.34
0.365	3.51			34.0	3.30
0.413	3.37			35.0	3.25
0.428	3.32			40.0	2.98
0.443	3.32			45.0	2.57

Table 1.1.17.12b
INDIUM ANTIMONIDE ($InSb$)[287]

Wavelength (μm)	*n*
7.87	4.0
8.00	3.99
9.01	3.96
10.06	3.95
11.01	3.93
12.06	3.92
12.98	3.91
13.90	3.90
15.13	3.88
15.79	3.87
16.96	3.86
17.85	3.85
18.85	3.84
19.98	3.82
21.15	3.81
22.20	3.80

l_{14} = 45.2 ± 1.0 μm at λ_1 = 28.0 μm[291]

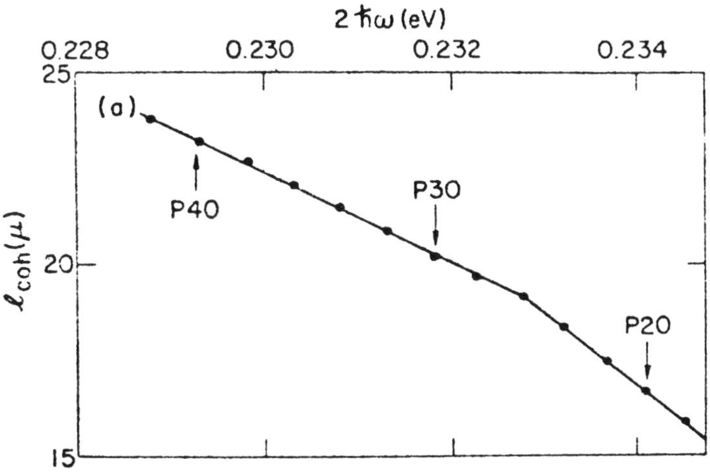

FIGURE 1.1.8. Coherence length ℓ_{14} of InSb as a function of $2\hbar\omega$. (From Wynne, J. J., *Phys. Rev. Lett.*, 27, 17, 1971. With permission.)

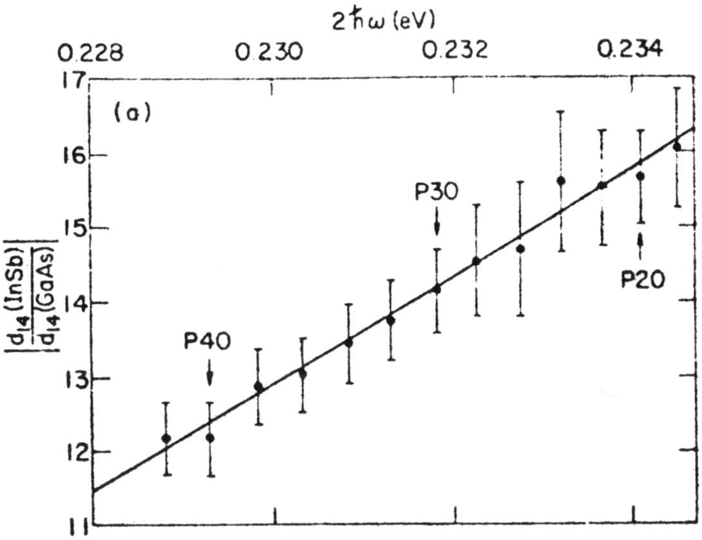

FIGURE 1.1.9. SHG coefficient d_{14} of InSb as a function of $2\hbar\omega$. (From Wynne, J. J., *Phys. Rev. Lett.*, 27, 17, 1971. With permission.)

Table 1.1.17.13
LITHIUM FLUORIDE (LiF) REFRACTIVE INDEX AT ROOM TEMPERATURE[369]

λ(μm)	*n*	λ(μm)	*n*	λ(μm)	*n*	λ(μm)	*n*
0.1935	1.4450	0.254	1.41792	1.50	1.38320	5.50	1.31287
0.1990	1.4413	0.280	1.41188	2.00	1.37875	6.00	1.29745
0.2026	1.4390	0.302	1.40818	2.50	1.37327	6.91	1.260
0.2063	1.4367	0.366	1.40121	3.00	1.36660	7.53	1.239
0.2100	1.4346	0.391	1.39937	3.50	1.35868	8.05	1.215
0.2144	1.4319	0.4861	1.39480	4.00	1.34942	8.60	1.190
0.2194	1.4300	0.50	1.39430	4.50	1.33875	9.18	1.155
0.2265	1.4268	0.80	1.38896	5.00	1.32661	9.79	1.109
0.231	1.4244	1.00	1.38711				

Dispersion equation:[374]

$$n = A + BL + CL^2 + D\lambda^2 + E\lambda^4$$

where $A = 1.38761$, $B = 0.001796$, $C = -0.000041$, $D = -0.0023045$, $E = -0.00000557$, and $L = (\lambda^2 - 0.028)^{-1}$.

Dispersion equation:[375]

$$n^2 - 1 = \frac{a(E_o^2 - E^2)}{(E_o^2 - E^2)^{2+} \gamma^2 E^2} + \frac{b}{E_1^2 - E^2}$$

where $E(ev) = 1239.8521/\lambda(\text{nm})$, $a = 34.76\ (ev)^2$, $b = 236.6\ (ev)^2$, $E_o = 12.632\ ev$, $E_1 = 18.37\ ev$, and $\gamma = 0.33\ ev$.

Table 1.1.17.14a
LITHIUM IODATE ($LiIO_3$)[323] REFRACTIVE INDEX AT ROOM TEMPERATURE

Wavelength (μm)	n^o	n^e
0.4579	1.9186	1.7633
0.4765	1.9124	1.7586
0.4880	1.9089	1.7560
0.4965	1.9065	1.7541
0.5017	1.9051	1.7531
0.5145	1.9018	1.7506
0.5321	1.8978	1.7475
0.6328	1.8815	1.7351
1.0642	1.8517	1.7168

Dispersion equation

$$n_0^2 - 1 = 2.40109\lambda^2/(\lambda^2 - 0.021865)$$
$$n_e^2 - 1 = 1.91359\lambda^2/(\lambda^2 - 0.01940),$$

where λ is in μm.

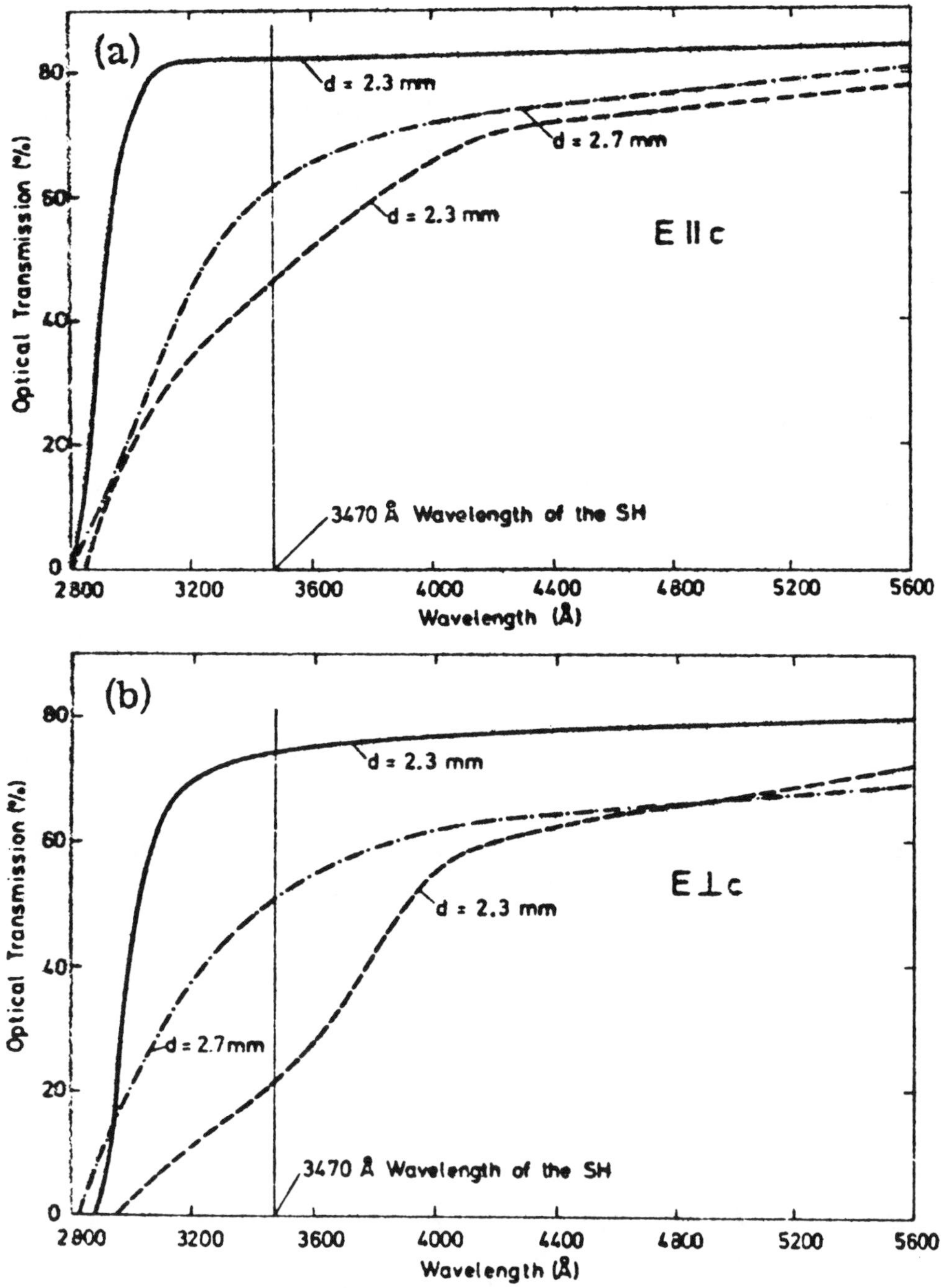

FIGURE 1.1.10. Optical transmission of $LiIO_3$ crystals for the two polarizations (A) E||C and (B) E⊥C. The crystal thickness varied from 2.3 to 2.7 mm. (From Nath, G., Mehmanesch, H., and Gsänger, M., *Appl. Phys. Lett.*, 17, 286, 1970. With permission.)

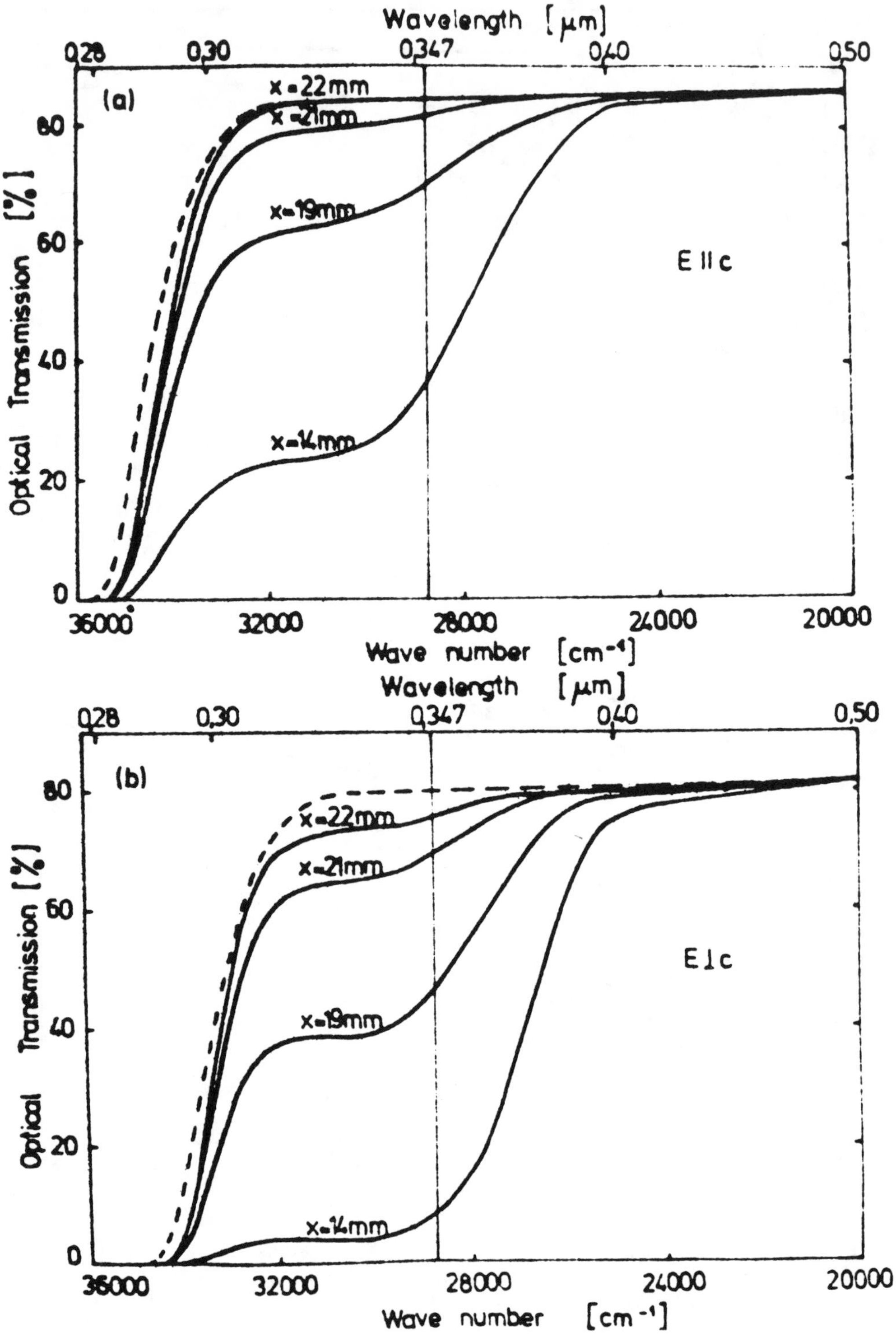

FIGURE 1.1.11. Optical transmission of α-$LiIO_3$ (3.9 mm thick) for different parts of the crystal. x is the distance of the illuminated part of the single crystal from the seed crystal measured along the optic axis of the crystal. (From Feliksinski, T. and Szewczyk, J., *Mater. Res. Bull.*, 16, 1505, 1981. With permission.)

Table 1.1.17.14b LITHIUM IODATE ($LiIO_3$) REFRACTIVE INDEX AT ROOM TEMPERATURE[320]

Wavelength (μm)	n^o	n^e
0.4000	1.948	1.780
0.4360	1.931	1.766
0.5000	1.908	1.754
0.5300	1.901	1.750
0.5780	1.888	1.742
0.6900	1.875	1.731
0.8000	1.868	1.724
1.060	1.860	1.719

θ_m = 30°
ρ = 3°50′ for λ_1 = 1.06 μm — Ref. 320
θ_m = 28.9° for λ_1 = 1.0845 μm
θ_m = 27.2° for λ_1 = 1.1523 μm
ρ = 0.072 rad for λ_1 4li = 1.0845 μm — Ref. 321
l_{31} = 2.1 μm at λ_1 = 1.06 μm
l_{33} = 8.0 μm at λ_1 = 1.06 μm
l_{31} = 2.419 ± 0.003 μm at λ_1 = 1.064 μm
l_{33} = 8.588 ± 0.005 μm at λ_1 = 1.064 μm — Ref. 157

Table 1.1.17.14c LITHIUM IODATE ($LiIO_3$)[325] REFRACTIVE INDEX AT ROOM TEMPERATURE

Wavelength (μm)	n^o	n^e
0.4047	1.9443	1.7826
0.4358	1.9283	1.7706
0.5086	1.9037	1.7521
0.5461	1.8953	1.7457
0.5791	1.8894	1.7413
0.5896	1.8875	1.7406
0.6438	1.8807	1.7346
1.0140	1.8584	1.7176
1.1286	1.8552	1.7152
1.3674	1.8508	1.7122
1.5296	1.8482	1.7101
1.6920	1.8464	1.7089
1.9701	1.8431	1.7072
2.2493	1.8385	1.7050

Table 1.1.17.14d LITHIUM IODATE ($LiIO_3$) REFRACTIVE INDEX AT ROOM TEMPERATURE[326]

Wavelength (μm)	n^o	n^e
0.4545	1.9184	1.7638
0.4579	1.9170	1.7630
0.4658	1.9141	1.7611
0.4727	1.9122	1.7606
0.4765	1.9100	1.7583
0.4880	1.9083	1.7556
0.4965	1.9047	1.7547
0.5017	1.9053	1.7537
0.5145	1.9012	1.7487
0.6328	1.8830	1.7367

Table 1.1.17.14e LITHIUM IODATE ($LiIO_3$)[323] TEMPERATURE- AND WAVELENGTH-DEPENDENT INDEX OF REFRACTION

$$n^2 - 1 = (a_0 + a_1T + a_2T^2)\lambda^2/[\lambda^2 - (b_0 + b_1T + b_2T^2)^2]$$

THE VALUES OF THE SELLMEIER CONSTANTS a_0,a_1,a_2,b_0,b_1,b_2 for $LiIO_3$ BETWEEN 25 AND 215°C

	n^o	n^e
a_0	2.4082035	1.9169995
a_1	$-3.2074499 \times 10^{-4}$	$-2.3603879 \times 10^{-4}$
a_2	5.9441535×10^{-8}	$-7.1344298 \times 10^{-8}$
b_0	1.4777636×10^{-1}	1.3979275×10^{-1}
b_1	1.1592972×10^{-5}	1.2331094×10^{-5}
b_2	3.9918294×10^{-8}	$-4.4550304 \times 10^{-8}$

Table 1.1.17.15 MAGNESIUM OXIDE (MgO) REFRACTIVE INDEX AT 23.3°C[376]

Wavelength (μm)	n
0.36117	1.77318
0.365015	1.77186
1.01398	1.72259
1.12866	1.72059
1.36728	1.71715
1.52952	1.71496
1.6932	1.71281
1.7092	1.71258
1.81307	1.71108
1.97009	1.70885
2.24920	1.70470
2.32542	1.70350
3.3033	1.68526
3.5078	1.68055
4.258	1.66039
5.138	1.63138
5.35	1.62404

Dispersion equation:

$$n^2 = 2.95632 - 0.010162387\,\lambda^2 - 0.0000204968\,\lambda^4 - \frac{0.02195770}{\lambda^2 - 0.01428322}$$

Table 1.1.17.16 POTASSIUM BROMIDE (KBr) REFRACTIVE INDEX AT 22°C[377]

λ(μm)	n	λ(μm)	n	λ(μm)	n	λ(μm)	n
0.404656	1.589752	1.011398	1.54408	6.238	1.53288	17.40	1.50390
0.435835	1.581479	1.12866	1.54258	6.692	1.53225	18.10	1.50076
0.486133	1.571791	1.36728	1.54061	8.662	1.52903	19.01	1.49703
0.508582	1.568475	1.7012	1.53901	9.724	1.52695	19.91	1.49288
0.546074	1.563928	2.44	1.53733	11.035	1.52404	21.18	1.48655
0.587562	1.559965	2.73	1.53693	11.862	1.52200	21.83	1.48311
0.643847	1.555858	3.419	1.53612	14.29	1.51505	23.86	1.47140
0.706520	1.552447	4.258	1.53523	14.98	1.51280	25.14	1.46324

Dispersion equation:

$$n^2 = 2.361323 - 0.00311497\lambda^2 - 0.000000058613\lambda^4 + \frac{0.007676}{\lambda^2} + \frac{0.0156569}{\lambda^2 - 0.0324}$$

Table 1.1.17.17
POTASSIUM CHLORIDE (KCl)
REFRACTIVE INDEX BETWEEN 15° AND 18°C[369]

λ(μm)	n	λ(μm)	n	λ(μm)	n
0.185409	1.82710	0.410185	1.50907	5.3039	1.470013
0.186220	1.81853	0.434066	1.50503	5.8932	1.468804
0.197760	1.73120	0.444587	1.50390	8.2505	1.462726
0.198900	1.72438	0.467832	1.50044	8.8398	1.460858
0.200000	1.71870	0.486149	1.49841	10.0184	1.45672
0.204470	1.69817	0.508606	1.49620	11.786	1.44919
0.208216	1.68308	0.53383	1.49410	12.965	1.44346
0.211078	1.67281	0.54610	1.49319	14.141	1.43722
0.21445	1.66188	0.56070	1.49218	15.912	1.42617
0.21946	1.64745	0.58931	1.49044	17.680	1.41403
0.22400	1.63612	0.58932	1.490443	18.2	1.409
0.23129	1.62043	0.62784	1.48847	18.8	1.401
0.242810	1.60047	0.64388	1.48777	19.7	1.398
0.250833	1.58970	0.656304	1.48727	20.4	1.389
0.257317	1.58125	0.67082	1.48669	21.1	1.379
0.263200	1.57483	0.76824	1.48377	22.2	1.374
0.267610	1.57044	0.78576	1.483282	23.1	1.363
0.274871	1.56386	0.88308	1.481422	24.1	1.352
0.281640	1.55836	0.98220	1.480081	24.9	1.336
0.291368	1.55140	1.1786	1.478311	25.7	1.317
0.308227	1.54136	1.7680	1.475890	26.7	1.300
0.312280	1.53926	2.3573	1.474751	27.2	1.275
0.340358	1.52726	2.9466	1.473834	28.2	1.254
0.358702	1.52115	3.5359	1.473049	28.8	1.226
0.394415	1.51219	4.7146	1.471122		

Dispersion equation:

$$n^2 = \begin{cases} a^2 + \left(\dfrac{M_1}{\lambda^2}\right) + \left(\dfrac{M_2}{\lambda^2} - \lambda_2^2\right) - k\lambda^2 - h\lambda^4 & \text{Ultraviolet} \\ b^2 + \left(\dfrac{M_1}{\lambda^2} - \lambda_1^2\right) + \left(\dfrac{M_2}{\lambda^2} - \lambda_2^2\right) - \left(\dfrac{M_3}{\lambda_3^2} - \lambda^2\right) & \text{Visible} \end{cases}$$

where

$a^2 = 2.174967$
$M_1 = 0.008344206$
$\lambda_1 = 0.0119082$
$M_2 = 0.00698382$
$\lambda^2 = 0.0255550$
$k = 0.000513495$
$h = 0.06167587$
$b^2 = 3.866619$
$M_3 = 5567.715$
$\lambda_3^2 = 3292.47$

Table 1.1.17.18
POTASSIUM IODIDE (KI)
REFRACTIVE INDEX AT ROOM TEMPERATURE[369]

λ(μm)	n	λ(μm)	n	λ(μm)	n	λ(μm)	n
0.248	2.0548	0.436	1.70350	1.083	1.6381	15.91	1.6085
0.254	2.0105	0.486	1.68664	1.18	1.6366	18.10	1.6030
0.265	1.9424	0.546	1.67310	1.77	1.6313	19	1.5997
0.270	1.9221	0.588	1.66654	2.36	1.6295	20	1.5964
0.280	1.8837	0.589	1.66643	3.54	1.6275	21	1.5930
0.289	1.85746	0.656	1.65809	4.13	1.6268	22	1.5895
0.297	1.83967	0.707	1.6537	5.89	1.6252	23	1.5858
0.302	1.82769	0.728	1.6520	7.66	1.6235	24	1.5819
0.313	1.80707	0.768	1.6494	8.84	1.6218	25	1.5775
0.334	1.77664	0.811	1.6471	10.02	1.6201	26	1.5720
0.366	1.74416	0.842	1.6456	11.79	1.6172	27	1.5681
0.391	1.72671	0.912	1.6427	12.97	1.6150	28	1.5629
0.405	1.71843	1.014	1.6396	14.14	1.6127	29	1.5571

Table 1.1.17.19a
POTASSIUM DIHYDROGEN PHOSPHATE (KDP) (KH_2PO_4)[249] REFRACTIVE INDEX AT ROOM TEMPERATURE

Wavelength (μm)	n^o	n^e
0.3653	1.5292	1.4843
0.4047	1.5235	1.4795
0.4078	1.5232	1.4792
0.4358	1.5200	1.4766
0.4916	1.5152	1.4727
0.5461	1.5117	1.4700
0.5791	1.5099	1.4686
0.6234	1.5079	1.4672
0.6907	1.5052	1.4655

Note: Estimated accuracy = ±0.0001. The change in index of refraction with temperature is given by:

$\Delta n_o = (n_o^2 - 1.432) \times (0.402 \times 10^{-4}) \times (298 - T)$

$\Delta n_e = (n_e^2 - 1.105) \times (0.221 \times 10^{-4}) \times (298 - T)$

where the temperature, T, is in K.

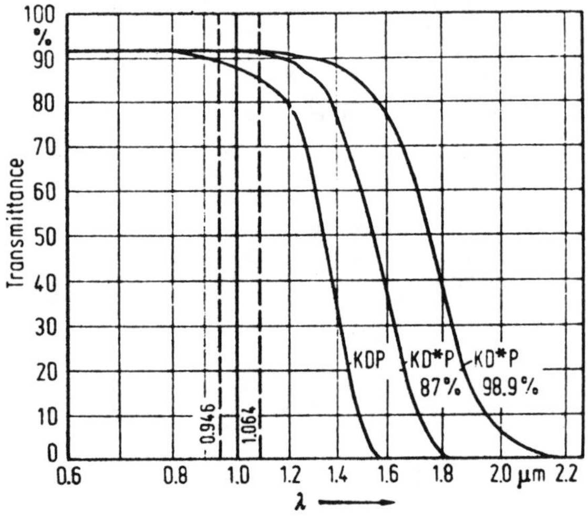

FIGURE 1.1.12. Optical transmission of KH_2PO_4(KDP) and KD_2PO_4(KD*P). Sample thickness 1.225 cm. Data are not corrected for reflection losses.[161]

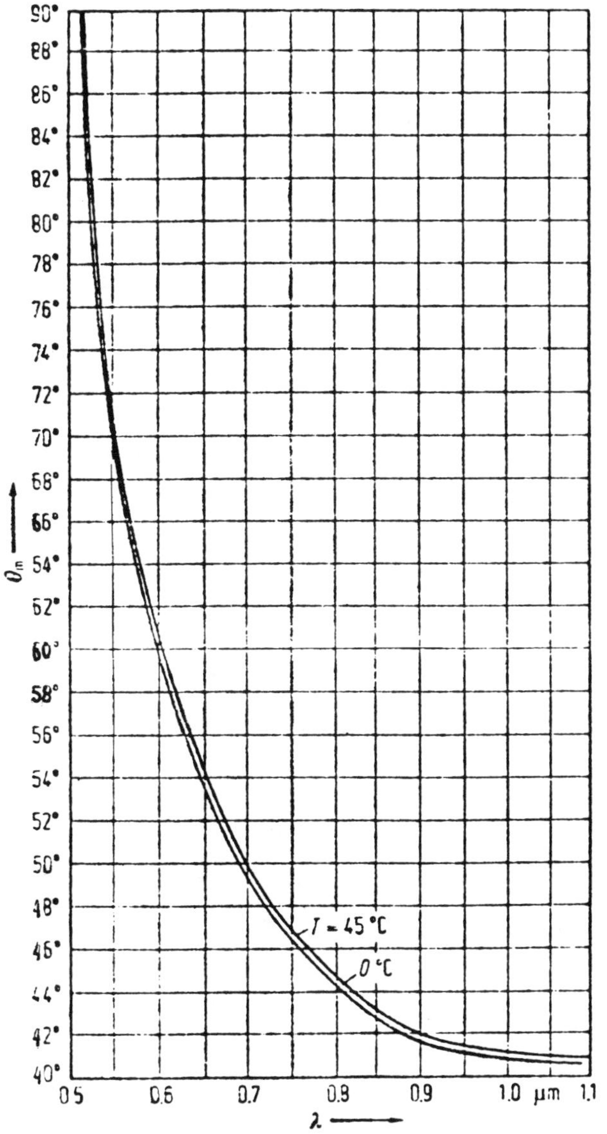

FIGURE 1.1.13. Phase matching angle for KH_2PO_4(KDP) as a function of the fundamental wavelength at T = 0°C and T = 45°C.[161]

Table 1.1.17.19b
POTASSIUM DIHYDROGEN PHOSPHATE (KDP) (KH_2PO_4)[161]

		Ref.
θ_m	= 49° at λ_1 = 0.6943 μm	247
θ_m	= 41° at λ_1 = 1.064 μm	261
θ_m	= 40.3 ± 1° at λ_1 = 1.0582 μm	250
θ_m	= 50.4 ± 1° at λ_1 = 0.6943 μm	250
l_{36}	= 18.5 μm at λ_1 = 0.6943 μm	226
l_{14}	= 7.3 μ at λ_1 = 0.6943 μm	226
l_{36}	= 22.0 μm at λ_1 = 1.0582 μm	226
l_{14}	= 14.6 μm at λ_1 = 1.0582 μm	226

Table 1.1.17.19c
POTASSIUM DIHYDROGEN PHOSPHATE (KDP) (KH_2PO_4)[248]

Wavelength (μm)	n^o	n^e
0.2138560	1.60177	1.54615
0.2288018	1.58546	
0.2446950	1.57228	
0.2464068	1.57105	
0.2536519	1.56631	1.51586
0.2800869	1.55263	1.50416
0.2980628	1.54618	1.49824
0.3021499	1.54433	1.49708
0.3035781		1.49667
0.3125663	1.54117	1.49434
0.3131545	1.54098	1.4919
0.3341478		1.48954
0.3650146	1.52932	1.48432
0.3654833	1.52923	1.48423
0.3662878	1.52909	1.48409
0.3906410		1.48089
0.4046561	1.52341	1.47927
0.4077811	1.52301	1.47898
0.4358350	1.51900	1.47640
0.4916036		1.47254
0.5460740	1.51152	1.46982
0.5769580	1.50987	
0.5790654	1.50977	1.46856
0.6328160	1.50737	1.46685
1.013975	1.49535	1.46401
1.128704	1.49205	1.45917
1.152276	1.49135	1.45893
1.357070	1.48455	
1.523100		1.45521
1.529525		1.45512

Dispersion equation

$$n_0^2 = 2.259276 + 1.008956 \times 10^{-10}\nu^2/(1 - \nu^2/7.726408 \times 10^9) + \frac{3.25130 \times 10^6}{(2.5 \times 10^5 - \nu^2)}$$

$$n_e^2 = 2.132668 + 8.637494 \times 10^{-11}\nu^2/(1 - \nu^2/0.142631 \times 10^9) + 8.069981 \times 10^5/(2.5 \times 10^5 - \nu^2)$$

where $v = 1/\lambda$ in cm^{-1}.

Dispersion equation[278]

$N^2 = A + B\lambda^2/(\lambda^2 - C)$,

where λ is in μm and

	A	*B*	*C*
n^o	1.0	1.24361	0.00959
n^e	1.0	1.12854	0.00841

Table 1.1.17.19d
POTASSIUM DIHYDROGEN PHOSPHATE (KDP) (KH_2PO_4)[251]

Temperature (K)	Polarization	*A*	*B*	*C*	*D*	*E*
201	o	1.455367	0.813422	0.012762	0.908048	30.0
	e	1.424592	0.714882	0.012190	0.225386	30.0
300	o	1.458524	0.799459	0.012741	0.908108	30.0
	e	1.423457	0.708796	0.012221	0.225356	30.0
228	o	1.635778	0.616941	0.015546	0.687676	30.0
	e	1.687221	0.452052	0.016940	0.596194	30.0

Sellmeier equation

$$n^2 = A + B/\left(1 - \frac{C}{\lambda^2}\right) + D/\left(1 - \frac{E}{\lambda^2}\right)$$

where λ is in μm.

Temperature - dependent Sellmeier coefficients (X) of KDP:
($x = mT + c$, T is the temperature in K)

Sellmeier coefficients	Fitted constants	n^o	n^e
A	$m \times 10^3$	0.03185	0.01152
	c	1.44896	1.42691
B	$m \times 10^3$	−0.14114	−0.06139
	c	0.84181	0.72722
C	$m \times 10^5$	−0.02130	0.03104
	c	0.01280	0.01213
D	$m \times 10^4$	0.00575	−0.00198
	c	0.90793	0.22543
E		30.0	30.0

Table 1.1.17.20a
QUARTZ (α-SiO_2
REFRACTIVE INDEX AT ROOM TEMPERATURE[287]

Wavelength (μm)	n^o	n^e
0.185	1.65751	1.68988
0.198	1.65087	1.66394
0.231	1.61395	1.62555
0.340	1.56747	1.57737
0.394	1.55846	1.56805
0.434	1.55396	1.56339
0.508	1.54822	1.55746
0.5893	1.54424	1.55335
0.7680	1.53903	1.54794
0.8325	1.53773	1.54661
0.9914	1.53514	1.54392
1.1592	1.53283	1.54152
1.3070	1.53090	1.53951
1.3958	1.52977	1.53832
1.4792	1.52865	1.53716
1.5414	1.52781	1.53630
1.6815	1.52583	1.53422
1.7614	1.52468	1.53301
1.9457	1.52184	1.53004
2.0531	1.52005	1.52823
2.3000	1.51561	
2.6000	1.50986	
3.0000	1.49953	
3.5000	1.48451	
4.0000	1.46671	
4.2000	1.4569	
5.0000	1.417	
6.4500	1.274	
7.0000	1.167	

Note: l_{11} = 20.65 μm at λ_1 = 1.064 μm.[156]

Table 1.1.17.20b
FUSED SILICA (SiO_2)
REFRACTIVE INDEX AT 20°C[381]

Wavelength (μm)	Computed index	Wavelength (μm)	Computed index
0.213856	1.534307	0.656272	1.456367
0.214438	1.533722	0.667815	1.456067
0.226747	1.522750	0.706519	1.455145
0.230209	1.520081	0.852111	1.452465
0.237833	1.514729	0.894350	1.451835
0.239938	1.513367	1.01398	1.450242
0.248272	1.508398	1.08297	1.449405
0.265204	1.500029	1.12866	1.448869
0.269885	1.498047	1.3622	1.446212
0.275278	1.495913	1.39506	1.445836
0.280347	1.494039	1.4695	1.444975
0.289360	1.490990	1.52952	1.444268
0.296728	1.488734	1.6606	1.442670
0.302150	1.487194	1.681	1.442414
0.330259	1.480539	1.6932	1.442260
0.334148	1.479763	1.70913	1.442057
0.340365	1.478584	1.81307	1.440699
0.346620	1.477468	1.97009	1.438519
0.361051	1.475129	2.0581	1.437224
0.365015	1.474539	2.1526	1.435769
0.404656	1.469618	2.32542	1.432928
0.435835	1.466693	2.4374	1.430954
0.467816	1.464292	3.2439	1.413118
0.486133	1.463126	3.2668	1.412505
0.508582	1.461863	3.3026	1.411535
0.546074	1.460078	3.422	1.408180
0.576959	1.458846	3.5070	1.405676
0.579065	1.458769	3.5564	1.401174
0.587561	1.458464	3.7067	1.399389
0.589262	1.458404		
0.643847	1.456704		

Dispersion equation:

$$n^2 - 1 = \frac{0.6961663\lambda^2}{\lambda^2 - 0.0684043^2} + \frac{0.4079426\lambda^2}{\lambda^2 - 0.1162414^2} + \frac{0.8974794\lambda^2}{\lambda^2 - 9.896161^2}$$

where λ is expressed in μm.

Table 1.1.17.21
SODIUM CHLORIDE (NaCl)
REFRACTIVE INDEX BETWEEN 18 AND 20°C[369]

λ(μm)	*n*	λ(μm)	*n*	λ(μm)	*n*	λ(μm)	*n*
0.19	1.85343	1.1786	1.53031	4.0	1.52190	12.50	1.47568
0.20	1.79073	1.2016	1.53014	4.1230	1.52156	12.9650	1.47160
0.22	1.71591	1.2604	1.52971	4.7120	1.51979	13.0	1.47141
0.24	1.67197	1.3126	1.52937	5.0	1.51890	14.0	1.46189
0.26	1.64294	1.4	1.52888	5.0092	1.51883	14.1436	1.46044
0.28	1.62239	1.4874	1.52845	5.3009	1.51790	14.7330	1.45427
0.30	1.60714	1.5552	1.52815	5.8932	1.51593	15.0	1.45145
0.35	1.58232	1.6	1.52798	6.0	1.51548	15.3223	1.44743
0.40	1.56769	1.6368	1.52781	6.4825	1.51347	15.9116	1.44090
0.50	1.55175	1.6848	1.52764	6.80	1.51200	16.0	1.44001
0.589	1.54427	1.7670	1.52736	7.0	1.51136	17.0	1.42753
0.6400	1.54141	1.8	1.52728	7.0718	1.51093	17.93	1.4149
0.6874	1.53930	2.0	1.52670	7.22	1.51020	18.0	1.41393
0.70	1.53881	2.0736	1.52649	7.59	1.50850	19.0	1.39914
0.7604	1.53682	2.1824	1.52621	7.6611	1.50822	20.0	1.38307
0.7858	1.53607	2.2464	1.52606	7.9558	1.50665	20.57	1.3735
0.80	1.53575	2.3	1.52594	8.0	1.50655	21.0	1.36563
0.8835	1.53395	2.3560	1.52579	8.04	1.5064	21.3	1.352
0.90	1.53366	2.6	1.52525	8.8398	1.50192	22.3	1.3403
0.9033	1.53361	2.6505	1.52512	9.0	1.50105	22.8	1.318
0.9724	1.53253	2.9466	1.52466	9.00	1.50100	23.6	1.299
1.0	1.53216	3.0	1.52434	9.50	1.49980	24.2	1.278
1.0084	1.53206	3.2736	1.52371	10.0	1.49482	25.0	1.254
1.0540	1.53143	3.5	1.52317	10.0184	1.49462	25.8	1.229
1.0810	1.53123	3.5359	1.52312	11.0	1.48783	26.6	1.203
1.1058	1.53098	3.6288	1.52286	11.7864	1.48171	27.3	1.175
1.1420	1.53063	3.8192	1.52238	12.0	1.48004		

Table 1.1.17.22
SODIUM FLUORIDE (NaF)
REFRACTIVE INDEX AT ROOM TEMPERATURE[369]

λ(μm)	n	λ(μm)	n	λ(μm)	n	λ(μm)	n
0.186	1.3930	0.546	1.32640	4.1	1.308	9.8	1.241
0.193	1.3854	0.488	1.32552	4.5	1.305	10.3	1.233
0.199	1.3805	0.589	1.32549	4.7	1.303	10.8	1.222
0.203	1.3772	0.656	1.32436	4.9	1.302	11.3	1.200
0.206	1.3745	0.707	1.32372	5.1	1.301	11.7	1.193
0.210	1.3718	0.720	1.32349	5.3	1.299	12.5	1.180
0.214	1.3691	0.768	1.32307	5.5	1.297	13.2	1.163
0.219	1.3665	0.811	1.32272	5.7	1.295	13.8	1.142
0.227	1.3630	0.842	1.32247	5.9	1.294	14.3	1.118
0.231	1.3606	0.912	1.32198	6.1	1.292	15.1	1.093
0.237	1.3586	1.014	1.32150	6.3	1.290	15.9	1.065
0.240	1.35793	1.083	1.32125	6.5	1.288	16.7	1.034
0.248	1.35500	1.27	1.320	6.7	1.286	17.3	1.000
0.254	1.35325	1.48	1.319	6.9	1.284	18.1	0.963
0.265	1.34999	1.67	1.318	7.1	1.281	18.6	0.924
0.270	1.34881	1.83	1.318	7.3	1.279	19.3	0.881
0.280	1.34645	2.0	1.317	7.5	1.277	19.7	0.838
0.289	1.34462	2.2	1.317	7.7	1.274	20.0	0.82
0.297	1.34328	2.4	1.316	7.9	1.272	20.5	0.75
0.302	1.34232	2.6	1.315	8.1	1.269	21.0	0.70
0.313	1.34062	2.8	1.314	8.3	1.266	21.5	0.65
0.334	1.33795	3.1	1.313	8.5	1.263	22.0	0.55
0.366	1.33482	3.3	1.312	8.7	1.261	22.5	0.45
0.391	1.33290	3.5	1.311	8.9	1.258	23.0	0.33
0.405	1.33194	3.7	1.309	9.1	1.252	23.5	0.25
0.436	1.33025	3.9	1.309	9.4	1.251	24.0	0.24
0.486	1.32818						

Table 1.1.17.24
YTTRIUM ALUMINUM GARNET ($Y_3Al_5O_{12}$) REFRACTIVE INDEX AT ROOM TEMPERATURE[382]

Wavelength (μm)	n
0.4358	1.85291
0.4765	1.84559
0.4880	1.84391
0.4965	1.84271
0.5017	1.84211
0.5145	1.84045
0.5320	1.83815
0.5460	1.83703
0.5790	1.83415
0.5852	1.83317
0.6328	1.82975
0.6402	1.82922
1.0642	1.81523

Table 1.1.17.23
STRONTIUM TITANATE ($SrTiO_3$)
REFRACTIVE INDEX AT 20°C[369]

λ(μm)	n	λ(μm)	n	λ(μm)	n	λ(μm)	n
0.40	2.66386	0.68	2.37135	0.96	2.32035	3.40	2.21370
0.44	2.56007	0.72	2.35998	1.00	2.31633	3.80	2.19428
0.48	2.49751	0.76	2.35055	1.40	2.29073	4.20	2.17265
0.52	2.45553	0.80	2.34260	1.80	2.27498	4.60	2.14865
0.56	2.42544	0.84	2.33583	2.20	2.26109	5.00	2.12212
0.60	2.40285	0.88	2.32997	2.60	2.24676	5.40	2.09290
0.64	2.38532	0.92	2.32486	3.00	2.23111		

Dispersion equation

$$n = A + BL + CL^2 + D\lambda^2 + E\lambda^4$$

where

$A = 2.28355$ $D = -0.0061335$
$B = 0.035906$ $E = -0.0001502$
$C = 0.001666$ $L = (\lambda^2 - 0.028)^{-1}$

λ is in μm.

**Table 1.1.17.25(a)
GALLIUM ARSENIDE (GaAs) REFRACTIVE INDEX AT ROOM TEMPERATURE[352]**

Wavelength (μm)	n
1.127	3.455
1.239	3.425
1.377	3.400
1.550	3.375
1.652	3.366

Sellmeier equation for undoped GaAs:[353]

$$n^2 = A + \frac{B}{\Omega^2 - \gamma^2} + \frac{C}{\gamma_o^2 - \gamma^2}$$

where γ = wave number in cm^{-1}.

$A = 3.5$
$B = 4.5 \times 10^9$
$C(cm^{-2}) = 1.4 \times 10^5$
$\gamma_o(cm^{-1}) = 269$
$\Omega(cm^{-1}) = 24.5 \times 10^3$
$l_{14} = 110 \pm 10$ μm at $\lambda_1 = 10.6$ μm (Reference 288.)
$l_{14} = 104 \pm 7$ μm at $\lambda_1 = 10.6$ μm (Reference 88.)

**Table 1.1.17.25(b)
GALLIUM ARSENIDE (GaAs) REFRACTIVE INDEX AT ROOM TEMPERATURE[369]**

Wavelength λ(μm)	n
0.78 ± 0.01	3.34 ± 0.04
8.0 ± 0.05	3.34 ± 0.04
10.0 ± 0.05	3.135 ± 0.04
11.0 ± 0.05	3.045 ± 0.04
13.0 ± 0.05	2.97 ± 0.04
13.7 ± 0.05	2.895 ± 0.04
14.5 ± 0.05	2.82 ± 0.04
15.0 ± 0.05	2.73 ± 0.04
17.0 ± 0.05	2.59 ± 0.04
19.0 ± 0.05	2.41 ± 0.04
21.9 ± 0.1	2.12 ± 0.04
21.9 ± 0.1	2.12 ± 0.04
21.9 ± 0.1	2.12 ± 0.04

Note: The experimental data appear to be more scattered than the reported experimental errors indicate.

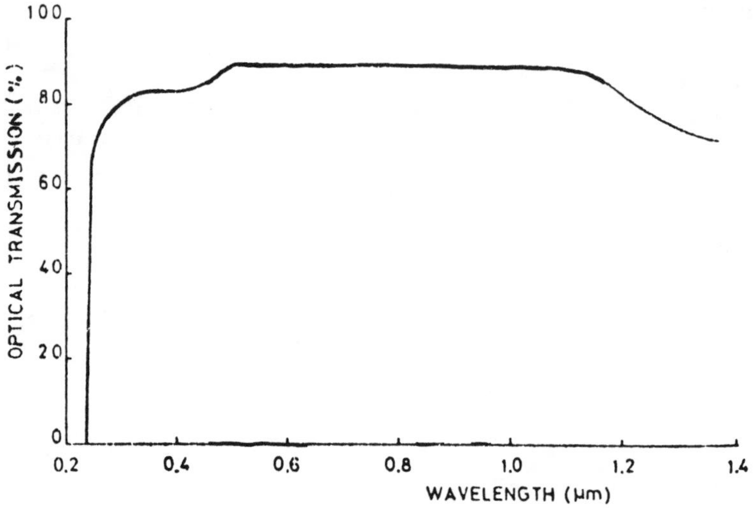

FIGURE 1.1.14. Optical transmission of ammonium tartrate $(NH_4)_2C_4H_4O_6$ at room temperature under normal incidence on (100) plane. Sample thickness = 1.08 mm. (From Mitsui, T., Iio, K., Hamano, K., and Sawada, S., *Opt. Commun.*, 9, 322, 1973. With permission.)

Table 1.1.17.26 AMMONIUM TARTRATE $[(NH_4)_2C_4H_4O_6]$ REFRACTIVE INDEX AT ROOM TEMPERATURE[171]

Wavelength (μm)	$n^{z=c}$	$n^{y=b}$	$n^{x=a}$
0.45	1.5404	1.5899	1.6005
0.50	1.5354	1.5843	1.5916
0.60	1.5292	1.5767	1.5836
0.70	1.5251	1.5720	1.5782
0.80	1.5222	1.5675	1.5754
0.90	1.5200	1.5644	1.5726
1.00	1.5182	1.5618	1.5700
1.10	1.5162	1.5594	1.5677
1.20	1.5146	1.5588	1.5659
1.30	1.5129	1.5578	1.5641
1.40	1.5115	1.5555	1.5620
1.50	1.5097	1.5539	1.5600
1.60	1.5085	1.5505	1.5585

The crystallographic axes have been rearranged to follow the IRE convention: a = 8.808 Å, b = 6.128 Å, c = 7.083 Å. θ_m = 26° 50′ (measured from z-axis) for λ_1 = 1.15 μm; θ_m = 50° 03′ (measured from z-axis) for λ_1 = 1.15 μm; θ_m = 30° 25′ (measured from x-axis) for λ_1 = 1.15 μm.

Table 1.1.17.27
7-DIETHYLAMINO-R-METHYLCOUMARIN (DMC) REFRACTIVE INDEX AT ROOM TEMPERATURE[165]

λ (μm)	n^x	n^y	n^z
0.450		1.715	1.536
0.500	2.097	1.680	1.527
0.530	2.044	1.667	1.524
0.600	1.969	1.649	1.518
0.700		1.636	1.514
0.800		1.627	1.510
0.900		1.621	1.508
1.0			1.506
1.06	1.871	1.616	1.506

Note: θ_m (type I) = 26.8° to x axis; θ_m (type II) = 48.0° to x axis; l_{22} = 5.3 μm; l_{23} = 1.7 μm; d_{23} has sign opposite to d_{21}, d_{22}, and d_{25}.

Table 1.1.17.28
DIPOTASSIUM TARTRATE HEMIHYDRATE ($K_2C_4H_4O_6$ 1/2 H_2O) REFRACTIVE INDEX AT ROOM TEMPERATURE[66]

Wavelength (μm)	n^x	$n^{y=b}$	n^z
0.3650	1.5156	1.5487	1.5630
0.4047	1.5090	1.5409	1.5541
0.4358	1.5049	1.5368	1.5494
0.5461	1.4961	1.5271	1.5384
0.5780	1.4945	1.5253	1.5363
1.014	1.4846	1.5142	1.5238
1.129	1.4832	1.5127	1.5218
1.367	1.4809	1.5102	1.5183

Table 1.1.17.29
LITHIUM SULFATE MONOHYDRATE ($Li_2SO_4H_2O$) REFRACTIVE INDEX AT ROOM TEMPERATURE[66]

Wavelength (μm)	n^x	$n^{y=b}$	$n^{x=c}$
0.3650	1.4771	1.4926	1.5029
0.4047	1.4722	1.4876	1.4980
0.4358	1.4693	1.4849	1.4951
0.4471	1.4686	1.4834	1.4941
0.4713	1.4670		1.4926
0.5016	1.4652	1.4802	1.4905
0.5461	1.4631	1.4782	1.4882
0.5780	1.4619	1.4772	1.4867
0.5876	1.4616	1.4766	1.4866
0.6678	1.4593	1.4743	1.4838
0.7016	1.4585		1.4831
1.014	1.4538	1.4678	1.4777
1.129	1.4525	1.4666	1.4761
1.367	1.4502	1.4636	1.4732
1.530	1.4485		1.4708
1.709	1.4466	1.4588	1.4676

Note: θ_m = 25° at λ_1 = 1.15 μm.

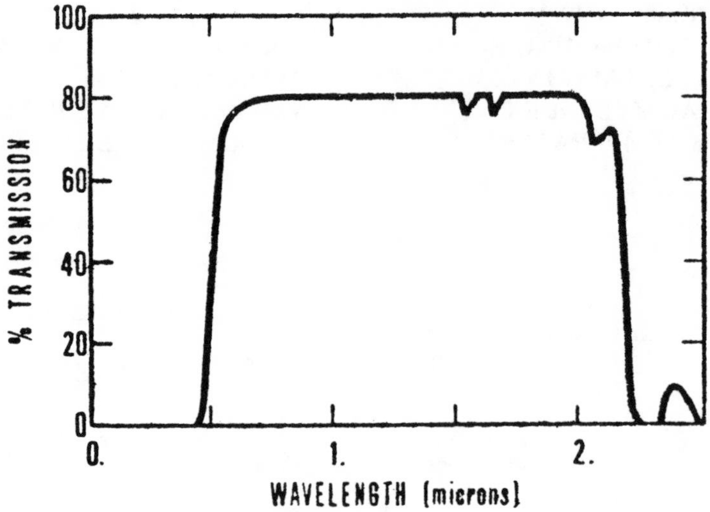

FIGURE 1.1.15. Optical transmission of methyl- (2,4- dinitrophenyl) - aminopropanoate (MAP). $C_{10}H_{12}N_3O_6$ crystal. (From Oudar, J. L. and Hierle, R., *J. Appl. Phys.*, 48, 2699, 1977. With permission.)

Table 1.1.17.30 METHYL-(2,4-DINITRIPHENYL)-AMINOPROPANOATE(MAP) ($C_{10}H_{12}N_3O_6$)[166]

Wavelength (μm)	N^x	$n^{y=b}$	$n^{z=c}$
0.532	2.0353	1.7100	1.5568
1.064	1.8439	1.5991	1.5078

Sellmeier equation

$$n^2 = A + B\lambda^2/(\lambda^2 - C) - D\lambda^2$$

where λ is in μm.

Sellmeier coefficient	n^x	n^y	n^z
A	2.7523	2.3100	2.1713
B	0.6079	0.2258	0.10305
C	0.1606	0.17988	0.16951
D	0.05361	0.01886	0.01667

The crystallographic axes have been rearranged to follow the IRE convention: a = 8.116 Å, b = 11.121 Å, c = 6.829 Å. The angle between a and x is 5.6°; θ_m (type I) = 2.2° ± 1° and −71° ± 1° for λ = 1.06 μm; tan ρ = 0.2; θ_m (type II) = 11° ± 1° for λ_1 = 1.06 μm; tan ρ = 0.042; Damage threshold = 3 × 10^9 w/cm^2.

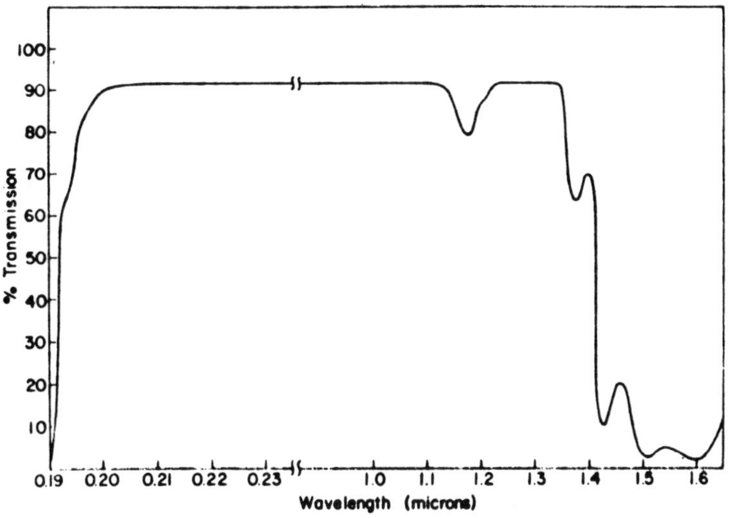

FIGURE 1.1.16. Transmission of sucrose crystal with propagation normal to the b-axis. Note the change of scale on the horizontal axis. (From Halbout, J. M. and Tang, C. L., *IEEE J. Quantum Electron.*, QE-18, 410, 1982. With permission.)

Table 1.1.17.31
SUCROSE ($C_{11}H_{22}O_{11}$)[168]

Sellmeier constant	n^x	n^y	n^z
A	1.8719	1.9703	2.0526
B	0.4660	0.4502	0.3909
C	0.0214	0.0238	0.0252
D	0.0113	0.0101	0.0187

Sellmeier equation:

$$n^2 = A + \frac{B\lambda^2}{\lambda^2 - C} - D\lambda^2$$

with λ in μm.

$\theta_m = 60.5 \pm 1°$ and $16.0 \pm 1°$ with respect to x and z axes, respectively, for $\lambda_1 = 1.06$ μm.

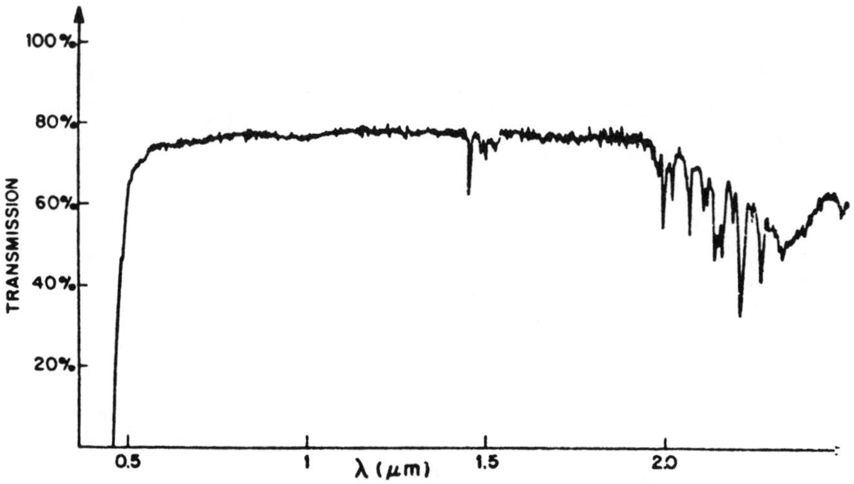

FIGURE 1.1.17. Transmission of a 50-μm plate of 2-methyl-4-nitroaniline. (From Levine, B. F., Bethea, C. G., Thurmond, C. D., Lynch, R. T., and Bernstein, J. L., *J. Appl. Phys.*, 50, 2523, 1979. With permission.)

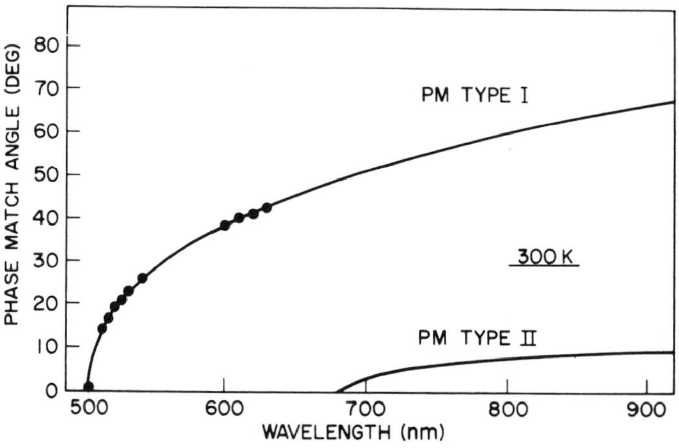

FIGURE 1.1.18. Phase-match angle for second-harmonic generation in potassium malate. ● = experimental values, solid curve calculated from the refractive index. (From Bergman, J. G., Crane, G. R., Levine, B. F., and Bethea, C. G., *Appl. Phys. Lett.*, 20, 21, 1972. With permission.)

Table 1.1.17.32 2-METHYL-4-NITROANILINE (MNA) ($CH_3 \cdot NH_2 \cdot HO_2 \cdot C_6H_4$)[169]

Wavelength (μm)	n^x	n^y
0.5320	2.2	—
0.6328	2.0 ± 0.1	1.6 ± 0.1
1.064	1.8	—

l_{11} = 0.7 ± 0.1 μm for λ_1 = 1.064 μm
θ_m = 55° (external) for d_{11} at λ_1 = 1.064 μm
ρ ~ 0.02 radians at λ_1 = 1.064 μm.

Table 1.1.17.33 POTASSIUM MALATE ($COOK \cdot CHOH \cdot CH_2COOH$ 1.5 H_2O)[170]

Sellmeier constant	n^x	$n^{y=b}$	n^z
A	1.542	1.470	1.339
B	0.8299	0.7473	0.8519
C	0.1419	0.1343	0.1195

Sellmeier equation:

$$n^2 + A + B\lambda^2/(\lambda^2 - C^2)$$

Note: Transparent in the region 0.240 to 1.3 μm.

Table 1.1.17.34
AMMONIUM OXALATE MONOHYDRATE
$[C_2O_4(NH_4)_2 \cdot H_2O]$[66]

Wavelength (μm)	$n^{x=c}$	$n^{y=b}$	$n^{z=a}$
0.4471	1.4460	1.5599	1.6119
0.4713	1.4447	1.5561	1.6084
0.4922	1.4435	1.5544	1.6050
0.5016	1.4426	1.5536	1.6037
0.5461	1.4406	1.5493	1.5993
0.5780	1.4391	1.5470	1.5965
0.5876	1.4388	1.5469	1.5952
0.6678	1.4362	1.5426	1.5892
0.7016	1.4352	1.5408	1.5874
1.014	1.4295	1.5312	1.5763
1.129	1.4276	1.5284	1.5728
1.367	1.4235	1.5222	1.5652

Table 1.1.17.35
BARIUM FORMATE $[Ba(COOH)_2]$[178]

Sellmeier equation

$$n^2 = A + \frac{B}{\lambda^2 - C}$$

where λ is in μm.

Sellmeier constant	$n^{x=a}$	$n^{y=b}$	$n^{z=c}$
A	$2.1238_5 \pm 0.0006$	2.0771 ± 0.0009	$2.1462_9 \pm 0.0007$
B	$0.0086_3 \pm 0.0001$	$0.0076_2 \pm 0.0001$	$0.00736_6 \pm 0.00007$
C	0 ± 0.001	$0.0079_9 \pm 0.007$	$0.0090_8 \pm 0.004$

Note: Range of transmission 0.245 to 2.2 μm. The crystallographic axes given in Reference 178 have been rearranged to follow the IRE convention: C_o = 6.81 Å, a_o = 7.67 Å, and b_o = 8.91 Å.

Coherence lengths measured in μm at λ_1 = 1.064 μm: $l_{14} = 3.24 \pm 2\%$, $l_{25} = 6.13 \pm 2\%$, $l_{36} = 57.8 \pm 9\%$, $l_{36} = 6.03 \pm 2\%$.

Phase matching angles observed:

θ°	φ°
33.3 ± 0.5	0 ± 0.5
71.9 ± 0.5	0 ± 0.5
58.5 ± 1	45 ± 1

$d_{33}/d_{32} < 0$; $d_{33}/d_{31} > 0$.

Table 1.1.17.36
GLUTAMIC ACID HYDROCHLORIDE ($C_5H_{10}O_4NCl$)[179] REFRACTIVE INDEX AT ROOM TEMPERATURE

Wavelength (μm)	n^x	n^y	n^z
0.316	1.606	1.632	1.592
0.633	1.558	1.583	1.545
0.532	1.563	1.587	1.551
1.064	1.545	1.568	1.534

Note: θ_m (type I) = 28° from the y axis for λ_1 = 1.064 μm. Range of transparency: 0.247 to 1.58 μm.

Table 1.1.17.37
HIPPURIC ACID C_6H_5CO-$NH(CH_2CO_2H)$][260] REFRACTIVE INDEX AT ROOM TEMPERATURE

Wavelength (μm)	n^x	n^y	n^z
0.350	1.55	1.61	1.78
0.589	1.5348	1.5921	1.7598
0.700	1.534	1.589	1.755

Table 1.1.17.38
ALPHA-IODIC ACID (α-HIO_3)[174] REFRACTIVE INDEX AT ROOM TEMPERATURE

Wavelength (μm)	$n^{x=a}$	$n^{y=b}$	$n^{z=c}$
0.450	2.0560	2.0184	1.8798
0.500	2.0192	1.9930	1.8621
0.5325	2.0103	1.9829	1.8547
0.550	2.0049	1.9787	1.8497
0.600	1.9922	1.9665	1.8409
0.650	1.9812	1.9571	1.8352
0.700	1.9765	1.9505	1.8308
0.800	1.9672	1.9407	1.8250
0.850	1.9639	1.9378	1.8223
0.900	1.9595	1.9347	1.8206
0.950	1.9564	1.9318	1.8180
1.000	1.9537	1.9292	1.8147
1.065	1.9508	1.9275	1.8123
1.100	1.9484	1.9260	1.8116
1.200	1.9436	1.9230	1.8086

Note: The crystallographic axes have been rearranged to follow the IRE convention: a = 5.8878 Å, b = 7.7333 Å, and c = 5.5379 Å.

Table 1.1.17.38(a)
ALPHA-IODIC ACID (α-HIO_3)[182]

Sellmeier equation:

$$n^2 = A + \frac{B\lambda^2}{\lambda^2 - \lambda_o^2} - C\lambda^2$$

where λ is in μm.

Sellmeier constant	$n^{z=c}$	$n^{y=b}$	$n^{x=a}$
A	2.5761(9)	2.4701(8)	2.6615(8)
B	0.6973(7)	1.2054(8)	1.1316(7)
C	0.0201(5)	0.0152(5)	0.0398(6)
λ_o	0.2356(8)	0.2246(7)	0.2281(7)

Note: At λ = 0.633 μm, the following relations hold between +50°C and −150°C with T in °C:[183]

$n^z = 1.839 - 1.0 \times 10^{-4}T - 0.2 \times 10^{-7}T^2$
$n^y = 1.963 - 1.2 \times 10^{-4}T - 0.7 \times 10^{-7}T^2$
$n^x = 1.988 - 1.5 \times 10^{-4}T - 1.3 \times 10^{-7}T^2$
l_{36} = 3.15 μm for λ_1 = 1.064 μm[175]

Optical rotation of 58.7°/mm and 74.5°/mm at 0.5461 and 0.4360 μm, respectively. Θ_m = 24 and 41° with respect to x axis at λ_1 = 1.065 μm.[174]

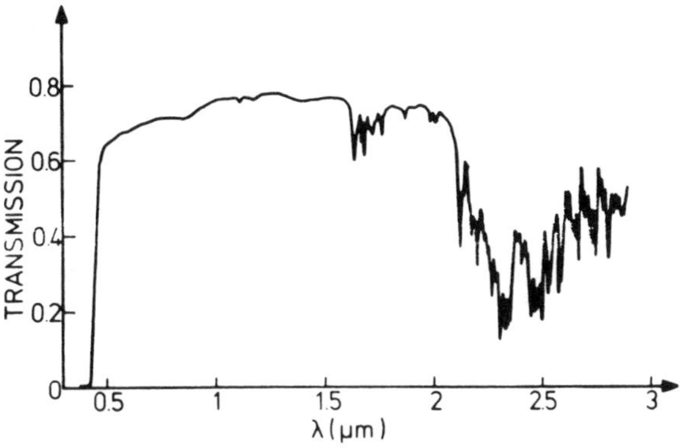

FIGURE 1.1.19. Transmission of a 400 μm-thick plate of 3-methyl-4-nitropyridine-1-oxide (POM). (From Zyss, J., Chemla, D. S., and Nicoud, J. F., *J. Chem. Phys.*, 74, 4800, 1981. With permission.)

Table 1.1.17.39 3-METHYL-4-NITROPYRIDINE-1-OXIDE (POM) ($NO_2{\cdot}CH_3{\cdot}NO{\cdot}C_5H_4$)[181] REFRACTIVE INDEX AT ROOM TEMPERATURE

Wavelength (μm)	$n^{x=a}$	$n^{y=b}$	$n^{z=c}$
0.435	—	—	1.717
0.468	2.114	1.809	1.69
0.480	2.082	1.793	1.682
0.509	2.028	1.766	1.668
0.532	1.997	1.750	1.660
0.546	1.981	1.742	1.656
0.579	1.953	1.728	1.648
0.644	1.915	1.709	1.637
1.0642	1.829	1.663	1.625

Note The crystallographic axes have been rearranged to follow the IRE convention: c = 5.132 Å, a = 6.111 Å, and b = 21.359 Å.

Sellmeier equation:

$$n^2 = A + B/(1 - C/\lambda^2) - D\lambda^2$$

where λ is in μm.

Sellmeier constant	$n^{x=a}$	$n^{y=b}$	$n^{z=c}$
A	2.5521	2.4315	2.4529
B	0.7962	0.3556	0.1641
C	0.1289	0.1276	0.1280
D	0.0941	0.0579	0.0

Note: Θ_m(I) = 54° 18′ 20″, with respect to x axis at λ_1 = 1.064 μm: tanρ = 0.11, Θ_m(II) = 12° 48′ with respect to z axis at λ_1 = 1.064 μm: tanρ = −0.025 ℓc = 1.576 μm at λ_1 = 1.064 μm.

Table 1.1.17.40
5-NITROURACIL
($C_4N_3H_3O_4$)[180]
REFRACTIVE INDEX AT ROOM TEMPERATURE

Wavelength (μm)	n^x	n^y	n^z
0.532	1.57	2.00	1.71
1.064	1.54	1.97	1.67

Θ_m(I) = 36° to x axis at λ_1 = 1.064 μm.
Θ_m(II) = 65° to x axis at λ_1 = 1.064 μm.

Note: Range of transparency from 0.5 to 1.5 μm.

Table 1.1.17.41
STRONTIUM FORMATE
[$Sr(COOH)_2$][192] REFRACTIVE INDEX AT ROOM TEMPERATURE

Wavelength (μm)	n^x	n^y	n^z
0.266	1.613	1.635	1.675
0.355	1.569	1.587	1.612
0.532	1.545	1.560	1.583
1.064	1.528	1.543	1.563

Note: Θ_m(I) = 26° and 72.5° to z-axis at λ_1 = 1.064 μm; Transparency range down to 0.25 μm; Decomposition temperature ~100°C; Damage threshold ~150 MW/cm² at 1.064 μm.

Table 1.1.17.42
STRONTIUM FORMATE DIHYDRATE
[$Sr(COOH)_2 \cdot 2H_2O$][192]
REFRACTIVE INDEX AT ROOM TEMPERATURE

Wavelength (μm)	n^x	n^y	n^z
0.266	1.543	1.598	1.621
0.355	1.509	1.553	1.570
0.5320	1.488	1.526	1.542
1.064	1.477	1.509	1.525

Note: Θ_m(II) = 19° to z axis at λ_1 = 1.064 μm; transparency ranges down to 0.25 μm; decomposition temperature ~100°C; damage threshold ~150 MW/cm² at 1.064 μm.

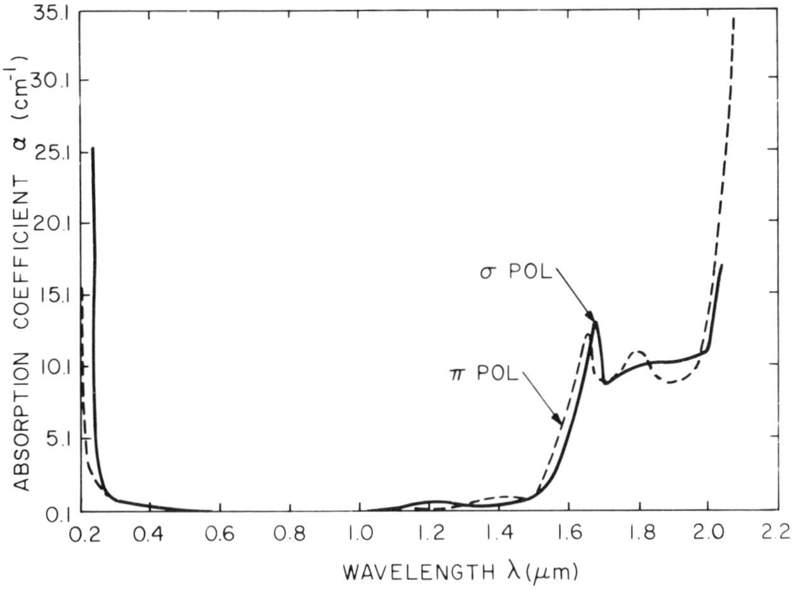

FIGURE 1.1.20. Optical transmission of *d*-threonine single crystal at room temperature. (From Singh, S., Bonner, W. A., Kyle, T., Potopowicz, J. R., and Van Uitert, L. G., *Opt. Commun.*, 5, 131, 1972. With permission.)

Table 1.1.17.43
D-THREONINE ($C_4H_9NO_3$)[117] REFRACTIVE INDEX AT ROOM TEMPERATURE

Wavelength (μm)	$n^{x=z}$	$n^{y=b}$	$n^{z=c}$
0.4579	1.5299	1.6039	1.6125
0.4765	1.5282	1.6017	1.6100
0.4880	1.5272	1.6004	1.6087
0.4965	1.5266	1.5996	1.6077
0.5017	1.5263	1.5991	1.6072
0.5145	1.5254	1.5979	1.6059
0.5321	1.5243	1.5965	1.6043
0.6328	1.5196	1.5898	1.5974
1.0642	1.5114	1.5788	1.5855

Sellmeier equations:

$$(n^x)^2 - 1 = 1.273\ \lambda^2/[\lambda^2 - (0.1032)^2]$$
$$(n^y)^2 - 1 = 1.477\ \lambda^2/[\lambda^2 - (0.1137)^2]$$
$$(n^z)^2 - 1 = 1.497\ \lambda^2/[\lambda^2 - (0.1169)^2].$$

$$\left.\begin{array}{l} l_{14} = 4.63\ \mu\text{m} \\ l_{25} = 5.44\ \mu\text{m} \\ l_{36} = 4.45\ \mu\text{m} \end{array}\right\} \text{at } \lambda_1 = 1.064\ \mu\text{m}.$$

$\Theta_M^{14}(\text{I}) = 24.7°$ and $\Theta_M^{14}(\text{II}) = 40°$ at $\lambda_1 = 1.064\ \mu\text{m}$

Table 1.1.17.44
m-AMINOPHENOL (AP) ($OH\text{-}C_6H_4\text{-}NH_2$)[205]

Sellmeier equation:

$$n^2 = A + B^2/(\lambda^2 - C) - D\lambda^2.$$

Sellmeier constant	$n^{x=c}$	$n^{y=b}$	$n^{z=a}$
A	2.4010	2.6031	2.1591
B	0.2974	0.4205	0.2777
C	0.0750	0.0800	0.0700
D	0.0351	0.0370	0.0137

Note: The crystallographic axes have been rearranged to follow the IRE convention: $c = 6.12$ Å, $a = 8.31$ Å, and b $= 11.23$ Å.

Coherence length measured in μm at $\lambda_1 = 1.024$: $l_{31} = -5.55 \pm 0.1$, $l_{32} = -1.81 \pm 0.04$, and $l_{33} = 9.85 \pm 0.2$.

Phase-match angles measured:

Type I { (0 + 0 = e) 60.4 ± 0.5° ext/y
(e + e = o) 74 ± 0.5° etx/z

Type II { e + o = e 34 ± 0.5° ext/y
e + o = o 40.7 ± 0.5° ext/z

Transparency range: 0.32 to 1.7 μm.

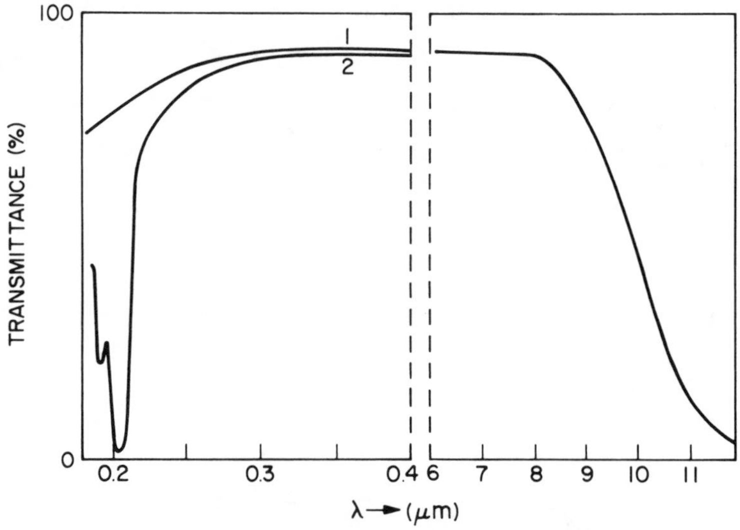

FIGURE 1.1.21. Transmission spectra of $BaMgF_4$. (1) Crystal of optical quality; (2) crystal containing traces of oxygen. (From Recker, K., Wallrafen, F., and Haussühl, S., *J. Cryst. Growth*, 26, 97, 1974. With permission.)

Table 1.1.17.45 BARIUM MAGNESIUM FLUORIDE ($BaMgF_4$)[216] REFRACTIVE INDEX AT ROOM TEMPERATURE

Wavelength (μm)	$n^{x=c}$	$n^{y=b}$	$n^{z=a}$
0.53	1.473	1.450	1.467
1.06	1.467	1.439	1.458

Note: The crystallographic axes given in Reference 216 have been rearranged to follow the IRE convention: $c = 4.125$ Å, $a = 5.81$ Å, and $b = 14.509$ Å.

The coherence lengths measured in μm at $\lambda_1 = 1.064$ μm: $l_{31} \simeq \infty$, $l_{32} = 9.6$, $l_{33} = 29.3$, $l_{15} = 23.3$, $l_{24} = 170$; $d_{33} \cdot d_{32} > 0$. Damage threshold $> 10^9$ W/cm^2.

Table 1.1.17.45(a)
BARIUM MAGNESIUM FLUORIDE [$BaMgF_4$][178] REFRACTIVE INDEX AT ROOM TEMPERATURE

Wavelength (μm)	$n^{x=c}$	$n^{y=b}$	$n^{z=a}$
0.532	1.4742	1.4508	1.4678
1.0642	1.4674	1.4436	1.4604

Note: The crystallographic axes given in Reference 178 have been rearranged to follow the IRE convention: c = 4.125 Å, a = 5.810 Å, and b = 14.509 Å.

Sellmeier equation

$$n^2 = A + B/(\lambda^2 - C)$$

where λ is in μm.

Sellmeier constant	$n^{x=c}$	$n^{y=b}$	$n^{z=a}$
A	$2.1462_9 \pm 0.0007$	$2.007_1 \pm 0.0009$	$2.1238_5 \pm 0.0006$
B	$0.00736_6 \pm 0.00007$	$0.0076_2 \pm 0.0001$	$0.0086_3 \pm 0.0001$
C	$0.0090_8 \pm 0.0004$	0.00799 ± 0.007	0 ± 0.001.

Observed phase-matching angles at $\lambda_1 = 1.064$ μm:

Θ°	Φ°
39.6 ± 0.5	0 ± 0.5
90 ± 0.1	9.2 ± 0.5
18.9 ± 0.5	0 ± 0.5
90 ± 0.5	42.1 ± 1

Table 1.1.17.46
BARIUM SODIUM NIOBATE
$(Ba_2NaNb_5O_{15})$[186]

Wavelength (μm)	$n^{z=c}$	$n^{x=a}$	$n^{y=b}$
0.4579	2.2931	2.4266	2.4286
0.4765	2.2799	2.4077	2.4096
0.4880	2.2726_5	2.3974	2.3992
0.4965	2.2677	2.3904	2.3921
0.5017	2.2648_5	2.3863	2.3880
0.5145	2.2582	2.3769	2.3787
0.5320	2.2501_5	2.3656	2.3673
1.0642	2.1700	2.2570	2.2584

Sellmeier equation:

$$n^2 - 1 = A_0\lambda^2/(\lambda^2 - \lambda_0^2)$$

where λ is in μm.

Sellmeier constant	n^z	n^x	n^y
A_0	3.60287	3.94655	3.95233
λ_0	0.032149	0.040179	0.040252

Table 1.1.17.46(a)
BARIUM SODIUM NIOBATE
$Ba_2NaNb_5O_{15}$ TEMPERATURE- AND WAVELENGTH-DEPENDENT EQUATION FOR THE INDEX OF REFRACTION:

$$n^2(\lambda,T) - 1 = (\alpha_0 + \alpha_1 T + \alpha_2 T^2)\lambda^2/[\lambda^2-(\beta_0 + \beta_1 T + \beta_2 T^2)]$$

For $T > 300°C$

	n^z	n^x	n^y
α_0	3.5989238	3.9490608	3.9624415
α_1	1.9744143×10^{-4}	$-1.6895711 \times 10^{-4}$	-2.450884×10^{-4}
α_2	$-4.2494224 \times 10^{-9}$	1.3340714×10^{-7}	2.8839235×10^{-7}
β_0	3.186415×10^{-2}	4.0053618×10^{-2}	3.9862064×10^{-2}
β_1	5.584291×10^{-6}	1.8781871×10^{-6}	2.9807606×10^{-7}
β_2	1.1992939×10^{-8}	2.0788727×10^{-9}	4.2324985×10^{-9}

$dn^y/dT = -2.5 \times 10^{-15}(°C^{-1})$
$dn^z/dT = 8 \times 10^{-5}(°C^{-1})$

	n^z	$n^{x=y}$
α_0	4.1777819	4.0199195
α_1	$-2.9902804 \times 10^{-3}$	-5.610722×10^{-4}
α_2	4.4549696×10^{-6}	6.9730723×10^{-7}
β_0	1.6503971×10^{-2}	3.7705343×10^{-2}
β_1	7.9803438×10^{-5}	8.9932975×10^{-6}
β_2	$-7.3470132 \times 10^{-8}$	$-4.4328962 \times 10^{-10}$

Coherence lengths measured in μm at $\lambda_1 = 1.0642$ μm:[185] $l_{31} = 40.3$, $l_{32} = 33.7$, $l_{33} = 3.32$, $l_{15} = 1.73$, $l_{24} = 1.72$. $\Theta_m^{31} = 75° 26'$ for $\lambda_1 = 1.064$ $\Theta_m^{32} = 73° 45'$ μm.[185]

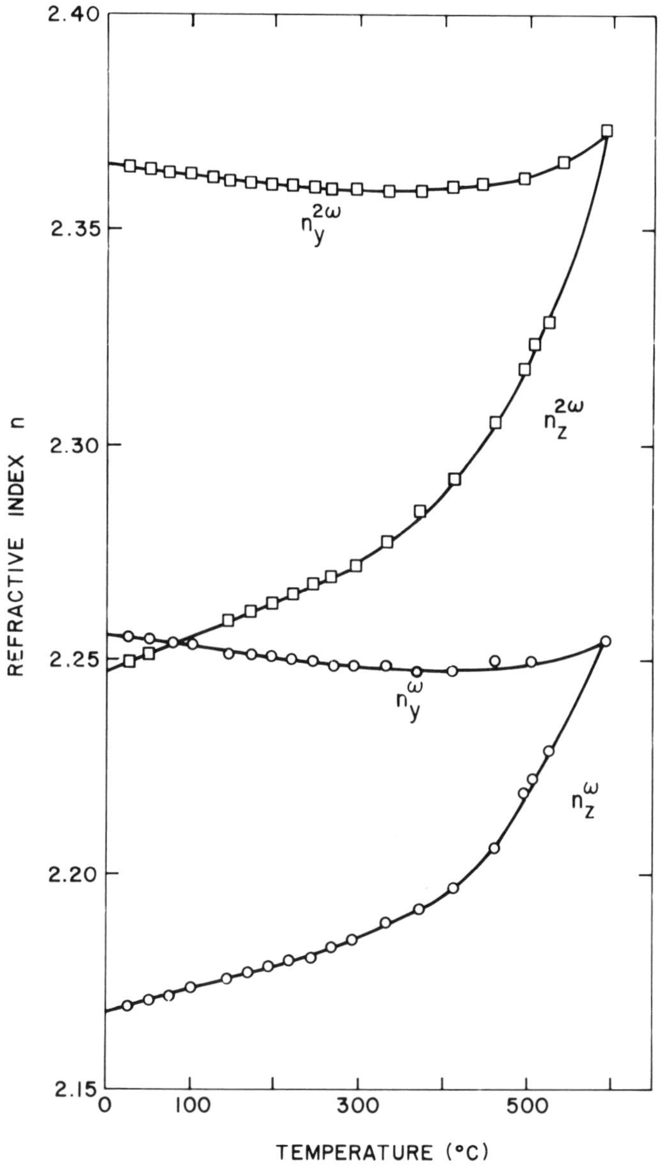

FIGURE 1.1.22. Temperature variation of the indexes of refraction of barium sodium niobate ($Ba_2Na\ Nb_5O_{15}$) at the laser wavelengths of 1.064 and 0.532 μm. (From Singh, S., Draegert, D. A., and Geusic, J. E., *Phys. Rev.*, B2, 2709, 1970. With permission.)

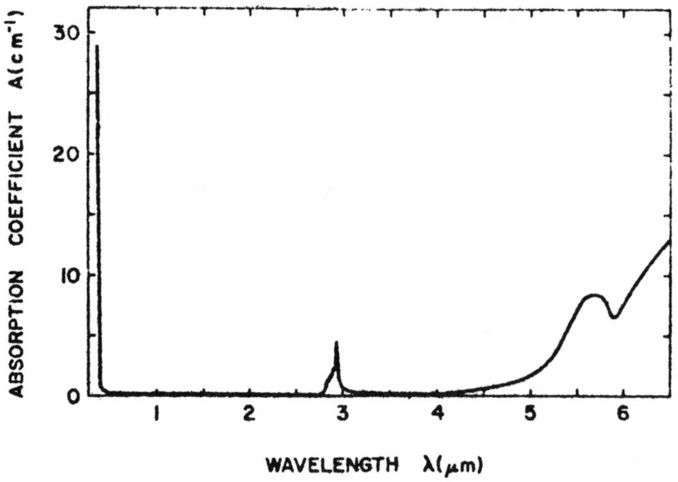

FIGURE 1.1.23. Absorption coefficient of barium sodium niobate at room temperature as a function of wavelength. (From Singh, S., Draegert, D. A., and Geusic, J. E., *Phys. Rev.*, B2, 2709, 1970. With permission.)

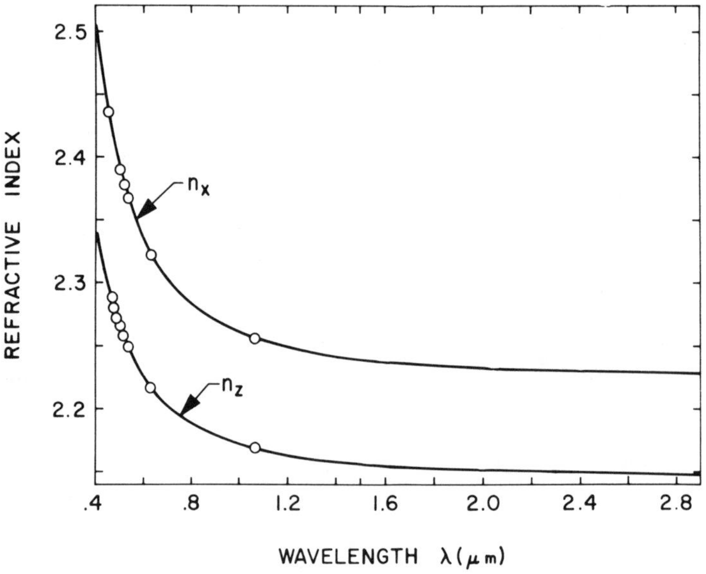

FIGURE 1.1.24. Dispersion of refractive indexes for barium sodium niobate at room temperature. (From Singh, S., Draegert, D. A., and Geusic, J. E., *Phys. Rev.*, B2, 2709, 1970. With permission.)

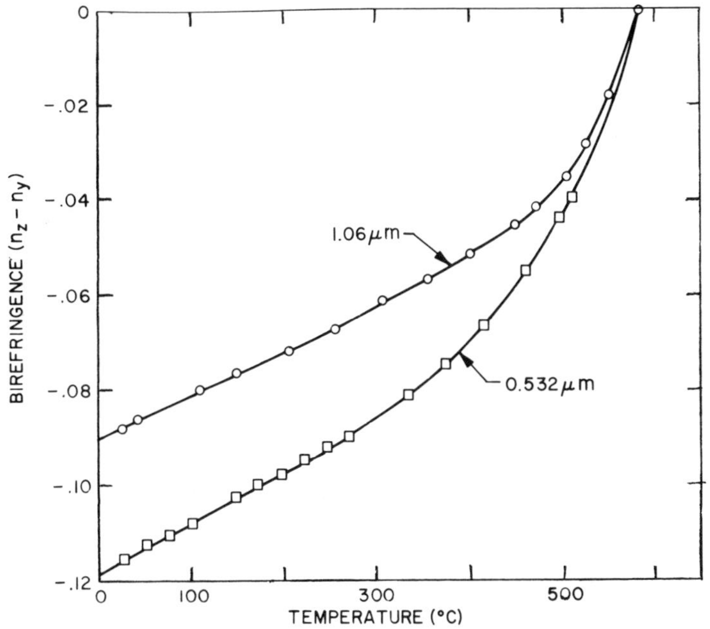

FIGURE 1.1.25. Temperature dependence of the birefringence ($n_z - n_y$) of barium sodium niobate. The curve with circles was obtained at 1.064 μm by the transmission method. The curve with squares was calculated from the index of refraction measurements at 0.532 μm. (From Singh, S., Draegert, D. A., and Geusic, J. E., *Phys. Rev.*, B2, 2709, 1970. With permission.)

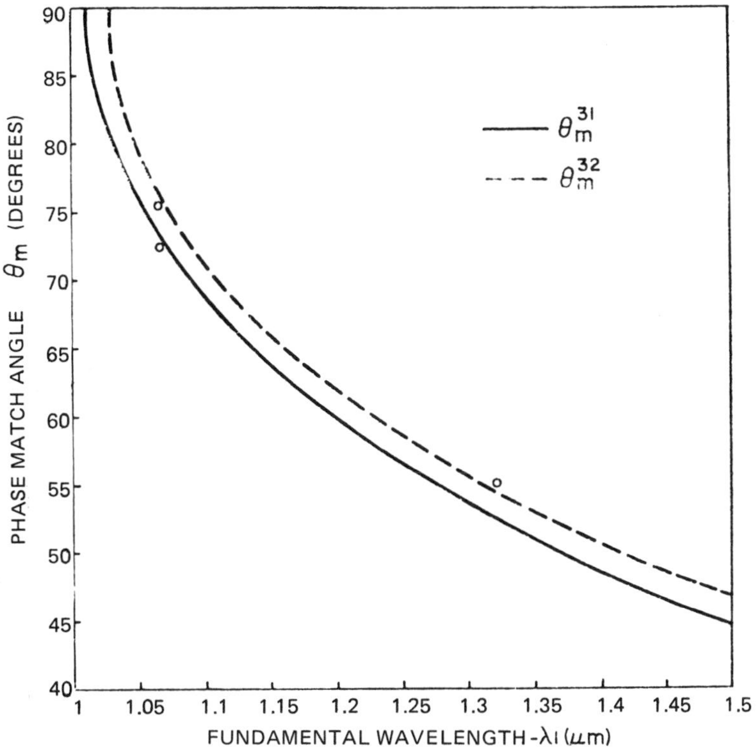

FIGURE 1.1.26. Calculated phase-matching angles θ_m^{31} and θ_m^{32} of barium sodium niobate for fundamental wavelengths between 1.0 and 1.5 μm.[186]

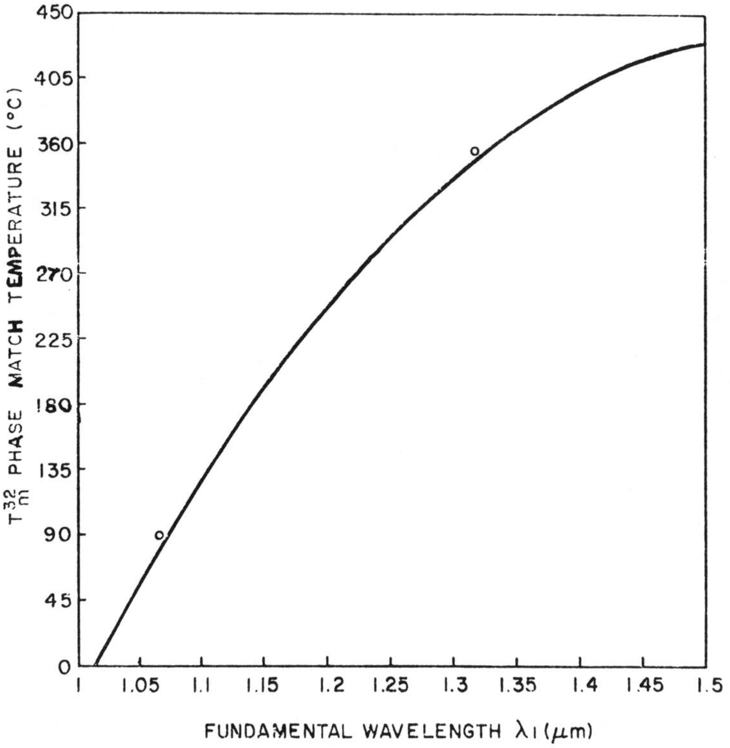

FIGURE 1.1.27. Calculated phase-match temperature T_m^{32} of barium sodium niobate as a function of fundamental wavelengths between 1.0 and 1.5 μm.[186]

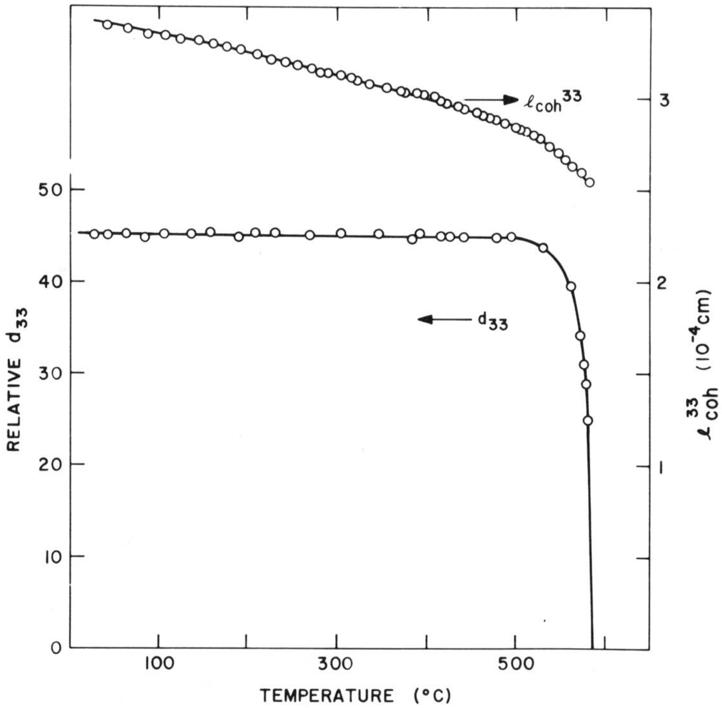

FIGURE 1.1.28. Temperature variation of the coherence length ℓ_{33} and the nonlinear optical coefficient d_{33} of barium sodium niobate. (From Singh, S., Draegert, D. A., and Geusic, J. E., *Phys. Rev.*, 32, 2709, 1970. With permission.)

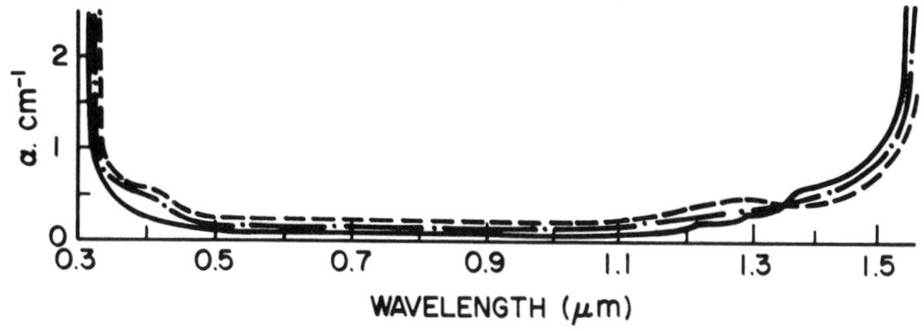

FIGURE 1.1.29. Absorption spectra of *m*-tolylenediamine (MTD) crystals for light polarized X∥a (continuous curve), Y∥b (dashed curve), and Z∥c (chain curve). (From Shigorin, V. D., Shipulo, G. P., Grazhulene, S. S., Musikhin, L. A., and Shekhtman, V. Sh., *Sov. J. Quantum Electron.*, 5, 1393, 1976. With permission.)

Table 1.1.17.47
BARIUM ZINC FLUORIDE ($BaZnF_4$)[216]

Wavelength (μm)	$n^{x = c}$	$n^{y = b}$	$n^{z = a}$
0.53	1.524	1.499	1.517
1.06	1.514	1.490	1.507

Note: The crystallographic axes given in Reference 216 have been rearranged to follow the IRE convention: $c = 4.206$ Å, $a = 5.841$ Å, and $b = 14.563$ Å.

Coherence lengths measured at $\lambda_1 = 1.06$ μm: $l_{33} = 28.0$, $l_{31} = 112$, $l_{32} = 9.82$, $l_{15} = 20.2$, $l_{24} = \infty$; $d_{33}\,d_{31} < 0$, $d_{33}d_{32} > 0$.

Table 1.1.17.48
m-BROMONITROBENZENE (BNB) ($Br\text{-}C_6H_4\text{-}NO_2$)[205]

Sellmeier equation:

$$n^2 = A + B\lambda^2/(\lambda^2 - C) - D\lambda^2$$

Sellmeier constant	$n^{x = a}$	$n^{y = b}$	$n^{z = c}$
A	2.6157	2.7076	2.4608
B	0.2280	0.2150	0.2767
C	0.1000	0.1100	0.0950
D	0.0300	0.0340	0.0186

Coherence lengths measured in μm at $\lambda_1 = 1.064$ μm: $l_{31} = 22.2 \pm 1$, $l_{32} = -26.5 \pm 1.5$, and $l_{33} = 6.8 \pm 0.2$.

Phase-match angles measured:

$$\text{Type I}\begin{cases} o + o = e\ 38.5 \pm 2° \text{ int/x} \\ e + e = o\ 41.5 \pm 2° \text{ int/x} \end{cases}$$

$d_{31}d_{33} > 0$, $d_{32}d_{33} > 0$

Table 1.1.17.49
CALCIUM IODATE HEXAHYDRATE [$Ca(IO_3)_2 \cdot 6H_2O$][213] REFRACTIVE INDEX AT ROOM TEMPERATURE

Wavelength (μm)	$n^{x = a}$	$n^{y = \bullet}$	$n^{z = c}$
0.53	1.625	1.654	1.688
1.06	1.582	1.625	1.653

Note: Coherence length measured in μm at $\lambda_1 = 1.064$ μm: $l_{31} = 2.5$, $l_{32} = 4.2$, $l_{33} = 7.5$, $l_{15} = 33$, $l_{24} = 18$; $d_{31}\,d_{33} < 0$.

Table 1.1.17.50
m-CHLORONITROBENZENE (CNB) ($Cl\text{-}C_6H_4\text{-}NO_2$)[205]

Sellmeier equation:

$$n^2 = A + B\lambda^2/(\lambda^2 - C) - D\lambda^2$$

Sellmeier constant	$n^{x=a}$	$n^{y=b}$	$n^{z=c}$
A	2.4882	2.5411	2.2469
B	0.2384	0.2148	0.3722
C	0.1070	0.1122	0.0810
D	0.0091	0.0135	0.0092

Coherence length measured in μm at λ_1 = 1.064 μm: $l_{31} = 36 \pm 2$, $l_{32} = 665 \pm 50$, $l_{33} = 6.8 \pm 0.3$; $d_{31}d_{33} > 0$, $d_{32}d_{33} > 0$.

Table 1.1.17.51
m-TOLYLENEDIAMINE (MTD) [$CH_3C_6H_3(NH_2)_2$][209] (2,4-DIAMINETOLUENE) REFRACTIVE INDEX AT ROOM TEMPERATURE

Wavelength (μm)	n^x	n^y	n^z
0.436	1.6433	1.8632	1.8019
0.492	1.6296	1.8380	1.7778
0.532	1.6226	1.8189	1.7676
0.546	1.6205	1.8150	1.7644
0.577	1.6163	1.8071	1.7583
0.579	1.6161	1.8069	1.7579
0.589	1.6150	1.8047	1.7564
0.633	1.6108	1.7967	1.7499
1.064	1.5930	1.7644	1.7240
1.153	1.5916	1.7618	1.7220

$l_{32} = 82 \pm 1$ (measured) at λ_1 = 1.064 μm.

Phase-matching angles measured:

$\Theta_m^{31} = 82°$
$\Theta_m = 90°$

Note: Noncritical phase-matching predicted with d_{32} at 1.11 μm fundamental along the x axis. Damage threshold ~300 MW/cm^2.

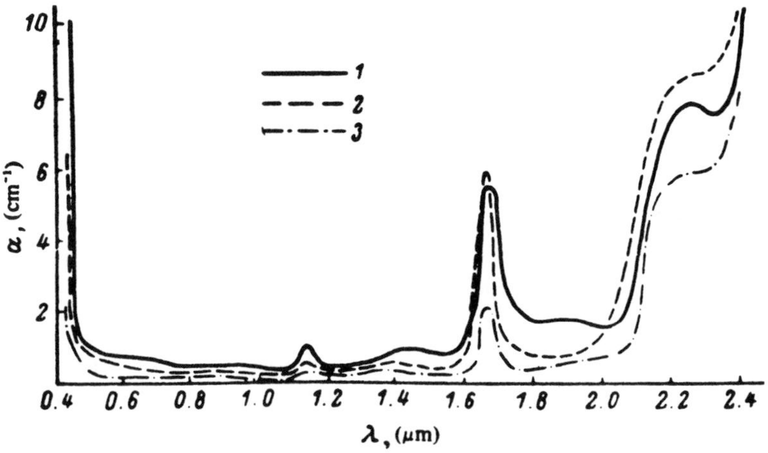

FIGURE 1.1.30. Absorption spectra of meta-dinitrobenzene crystals. Radiation is polarized (1) along X-axis, (2) along Y-axis, (3) along Z-axis. The X, Y, Z, coordinates coincide with a, b, c crystallographic axis, respectively. (From Belikova, G. S., Golovei, M. P., Shigorin, V. D., and Shipulo, G. P., *Opt. Spectrosc. USSR*, 38, 441, 1975. With permission.)

Table 1.1.17.52
META-DINITROBENZENE
$[C_6H_4(NO_2)_2]$[207]

Wavelength (μm)	$n^{x=a}$	$n^{y=b}$	$n^{z=c}$
0.436	1.8027	1.7360	1.5075
0.492	1.7731	1.7104	1.4960
0.532	1.7586	1.6984	1.4912
0.546	1.7554	1.6950	1.4897
0.576	1.7480	1.6889	1.4871
0.579	1.7472	1.6880	1.4868
0.589	1.7454	1.6863	1.4861
0.633	1.7380	1.6800	1.4828
1.064	1.7094	1.6538	1.4714
1.152	1.7072	1.6520	1.4698

$\Theta_m = 35°$ at $\lambda_1 = 1.064$ μm

$l_{31} = 1.22$ μm at $\lambda_1 = 1.064$ μm
$l_{32} = 1.64$ μm
$l_{33} = 13.44$ μm

Table 1.1.17.52(a)
META-DINITROBENZENE
$[C_6H_4(NO_2)_2]$[208]

Wavelength (μm)	$n^{x=a}$	$n^{y=b}$	$n^{z=c}$
0.45	1.7984	1.7225	1.5071
0.50	1.7714	1.7039	1.4996
0.55	1.7570	1.6885	1.4942
0.60	1.7480	1.6781	1.4878
0.65	1.7390	1.6750	1.4846
0.70	1.7336	1.6709	1.4824

The principal indexes have been rearranged so that the polar axis z ≡ c:

$\Theta_m = 34° 45' \pm 30'$ at $\lambda_1 = 1.153$ μm

Coherence lengths measured in μm at $\lambda_1 = 1.064$ μm:[206] $l_{31} = 1.25$, $l_{32} = 1.64$, $l_{33} = 13.1$; $d_{31}d_{32} > 0$, $d_{32}d_{33} > 0$.

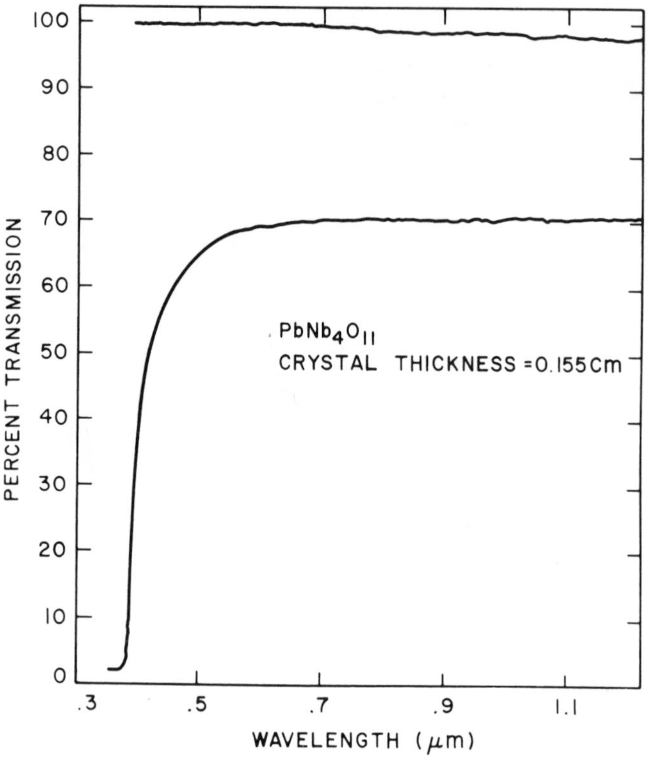

FIGURE 1.1.31. Transmission spectrum of lead niobate $PbNb_4O_{11}$ crystal at room temperature. Spectrum not corrected for reflection.[190]

Table 1.1.17.53
GADOLINIUM MOLYBDATE $[(Gd)_2(MoO_4)_3]$[189]

Wavelength (μm)	$n^{y=b}$	$n^{x=a}$	$n^{z=c}$
0.4579	1.8758	1.8762	1.9342
0.4765	1.8694	1.8699	1.9270
0.4880	1.8659	1.8663	1.9229
0.4965	1.8634	1.8639	1.9201
0.5017	1.8621	1.8625	1.9185
0.5145	1.8588	1.8593	1.9148
0.5321	1.8545	1.8549	1.9102
0.6328	1.8385	1.8390	1.8915
1.064	1.8142	1.8146	1.8637

Dispersion equation:

$(n^x)^2 - 1 = 2.2465\ \lambda^2/(\lambda^2 - 0.0226803)$
$(n^y)^2 - 1 = 2.2450\ \lambda^2/(\lambda^2 - 0.022693)$
$(n^z)^2 - 1 = 2.41957\ \lambda^2/(\lambda^2 - 0.0245458)$

where λ is in μm.

Table 1.1.17.54
LEAD NIOBATE $(PbNb_4O_{11})$[190]

Wavelength (μm)	$n^{x=a}$	$n^{y=b}$	$n^{z=c}$
0.4579	2.4754	2.4766	2.5047
0.4765	2.4554	2.4571	2.4845
0.4880	2.4445	2.4465	2.4735
0.4965	2.4371	2.4392	2.4660
0.5017	2.4329	2.4350	2.4618
0.5145	2.4231	2.4254	2.4518
0.5321	2.4113	2.4137	2.4396
0.6328	2.3644	2.3667	2.3922
1.0642	2.2979	2.3010	2.3254

Dispersion equation:

$(n^x)^2 - 1 = 4.124\ \lambda^2/[\lambda^2 - (0.202)^2]$
$(n^y)^2 - 1 = 4.139\ \lambda^2/[\lambda^2 - (0.2011)^2]$
$(n^z)^2 - 1 = 4.246\ \lambda^2/[\lambda^2 - (0.2014)^2]$

where λ is in μm.

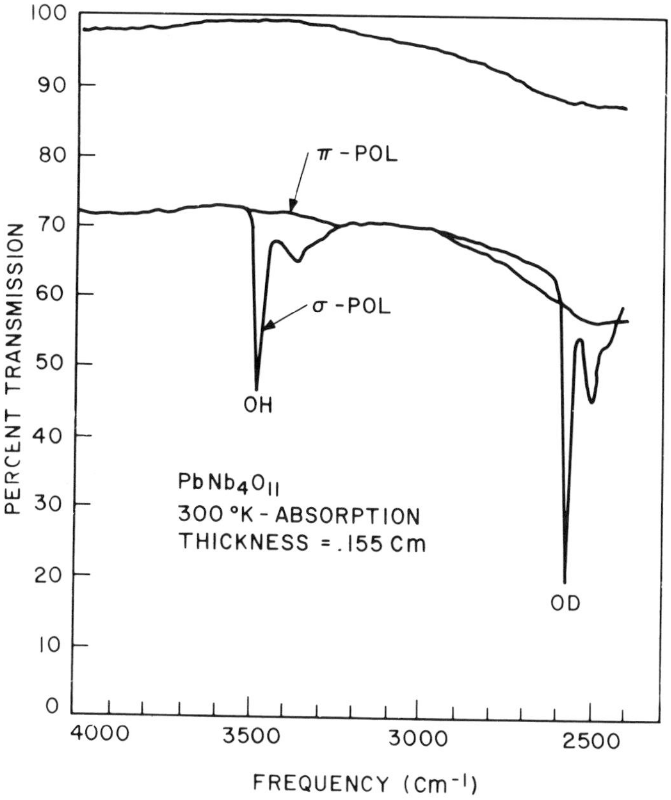

FIGURE 1.1.32. Infrared transmission of lead niobate crystal.[190]

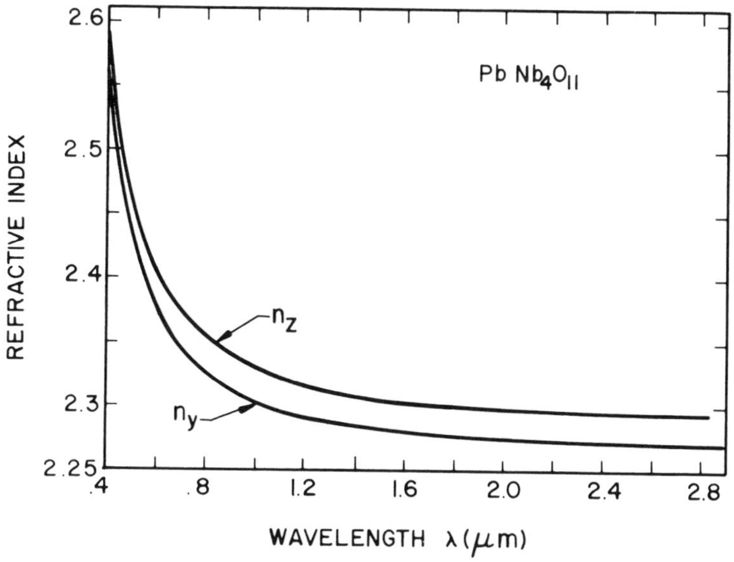

FIGURE 1.1.33. Dispersion of refractive indexes for lead niobate. $PbNb_4O_{11}$ at room temperature as a function of wavelength.[190]

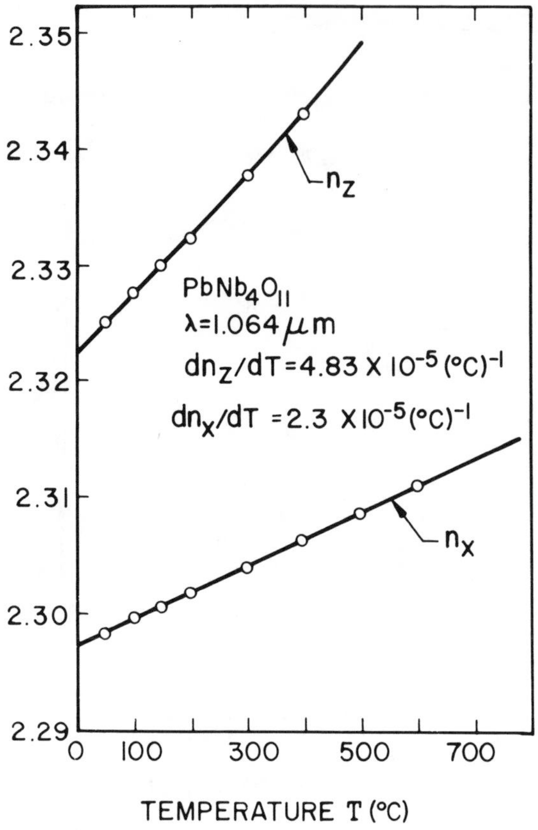

FIGURE 1.1.34. Temperature variation of the indexes of refraction of lead niobate $PbNb_4O_{11}$ at the laser wavelength of 1.064 μm.[190]

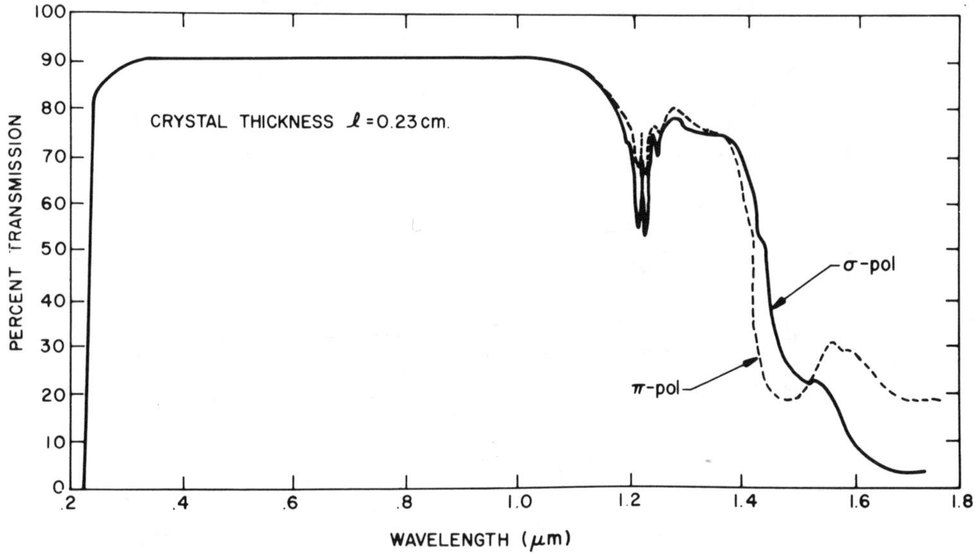

FIGURE 1.1.35. Optical transmission of lithium formate monohydrate (Li $CHO_2 \cdot H_2O$) at room temperature. (From Singh, S., Bonner, W. A., Potopowicz, J. R., and Van Uitert, L. G., *Appl. Phys. Lett.*, 17, 292, 1970. With permission.)

Table 1.1.17.55 LITHIUM FORMATE MONOHYDRATE [$LiCHO_2$-H_2O][194]

Wavelength (μm)	$n^{x=a}$	$n^{y=b}$	$n^{z=c}$
0.4579	1.3708	1.4901	1.5308
0.4765	1.3698	1.4883	1.5286
0.4880	1.3692	1.4873	1.5272
0.4965	1.3688	1.4866	1.5264
0.5017	1.3686	1.4862	1.5258
0.5145	1.3680	1.4851	1.5245
0.5321	1.3675	1.4838	1.5229
0.6328	1.3645	1.4784	1.5163
1.0642	1.3593	1.4673	1.5035

Dispersion equation:

$$(n^x)^2 - 1 = 0.8415\ \lambda^2/[\lambda^2 - (0.0953)^2]$$
$$(n^y)^2 - 1 = 1.14106\ \lambda^2/[\lambda^2 - (0.1183)^2]$$
$$(n^z)^2 - 1 = 1.2454\ \lambda^2/[\lambda^2 - (0.12496)^2]$$

$\Theta_m^{24} = 81.95°$, $\Theta_m^{32} = 55.1°$, $\rho = 0.03$ rad at $\lambda_1 = 1.064$ μm.

Table 1.1.17.57 LITHIUM HYDROGEN PHOSPHITE (LiH_2PO_3)[223]

Wavelength (μm)	$n^{x=c}$	$n^{y=b}$	$n^{z=a}$
0.644	1.533	1.539	1.535
0.579	1.556	1.541	1.537
0.546	1.559	1.544	1.539
0.5090	1.561	1.546	1.542
0.4800	1.563	1.548	1.543
0.4680	1.565	1.549	1.545
0.4350	1.570	1.553	1.548
0.5320			1.540
1.064	1.541	1.530	1.526

The principal axes have been rearranged to follow the IRE convention.

The coherence lengths measured in μm at $\lambda_1 = 1.064$ μm: $l_{31} = 231$, $l_{32} = 26.6$, $l_{33} = 18.3$;

$d_{31}d_{33} < 0$; $d_{32}d_{33} > 0$.
Phase-matching angle $\theta_m = \pm 29.2°$; transparency range: 0.188 to 1.72 μm.

Table 1.1.17.56 LITHIUM HYDRAZINE FLUOROBERYLLATE [$Li(N_2H_5)BeF_4$][224]

The crystallographic axes given in Reference 224 have been rearranged to follow the IRE convention: $c = 5.142$ Å, $a = 8.896$ Å, $b = 9.820$ Å.

Wavelength (μm)	$n^{x=a}$	$n^{y=b}$	$n^{x=c}$
0.4680	1.4114	1.4061	1.4101
0.4800	1.4099	1.404	1.4805
0.5090	1.4077	1.4023	1.4068
0.5460	1.4064	1.4011	1.4053
0.5770	1.4054	1.3999	1.4042
0.6440	1.4038	1.3985	1.4027
1.0600	1.4005	1.3962	1.3983

Sellmeier equation

$$n^2 = A + \frac{B}{1 - C/\lambda^2} - D\lambda^2$$

where λ is in μm.

Sellmeier constant	n^x	n^y	n^z
A	1.95361	1.94103	1.95031
B	0.01303	0.01060	0.01445
C	0.14694	0.15476	0.13966
D	0.00644	0.00357	0.01032

Coherence lengths measured in μm at $\lambda_1 = 1.064$ μm: $l_{31} = 66$, $l_{32} = 27$, $l_{33} = 26$; d_{31} and d_{32} have opposite signs. Transparent between 0.2 and 1.4 μm.

Table 1.1.17.57a LITHIUM HYDROGEN PHOSPHITE (LiH_2PO_3)[224]

The crystallographic axes have been rearranged to follow the IRE convention: $c = 5.06$ Å, $a = 5.17$ Å, and $b = 11.0$ Å.

Wavelength (μm)	$n^{x=c}$	$n^{y=b}$	$n^{z=a}$
0.53	1.590	1.575	1.570
1.06	1.571	1.560	1.555

Coherence lengths measured in μm at $\lambda_1 = 1.064$ μm: $l_{31} = \pm 200$, $l_{32} = +27.3$, $l_{33} = +18$, $l_{15} = \pm 16$, $l_{24} = \pm 15.5$; $d_{33} \cdot d_{31} < 0$.

Table 1.1.17.58
LITHIUM INDIUM SULFIDE
($LiInS_2$)[212]

Wavelength (μm)	$n^{x=a}$	$n^{y=b}$	$n^{z=c}$
0.425	2.4126	2.3472	2.4208
0.450	2.3685	2.3096	2.3766
0.500	2.3095	2.2580	2.3175
0.550	2.2720	2.2244	2.2793
0.600	2.2455	2.2011	2.2536
0.650	2.2265	2.1841	2.2344
0.700	2.2119	2.1712	2.2199
0.750	2.2010	2.1610	2.2085
0.800	2.1918	2.1530	2.1996
0.850	2.1849	2.1465	2.1923
0.900	2.1789	2.1409	2.1863
0.950	2.1737	2.1364	2.1812
1.000	2.1696	2.1325	2.1769
1.100	2.1630	2.1268	2.1706
1.200	2.1579	2.1223	2.1655
1.400	2.1508	2.1158	2.1585
1.600	2.1463	2.1115	2.1538
1.800	2.1430	2.1082	2.1501
2.000	2.1405	2.1057	2.1475
2.200	2.1384	2.1039	2.1454
2.400	2.1367	2.1026	2.1440
2.600	2.1353	2.1012	2.1425
2.800	2.1339	2.0999	2.1411
3.000	2.1325	2.0987	2.1398
3.200	2.1314	2.0976	2.1386
3.400	2.1305	2.0965	2.1372
3.600	2.1291	2.0954	2.1361
3.800	2.1280	2.0941	2.1348
4.000	2.1266	2.0930	2.1335
4.500	2.1237	2.0900	2.1304
5.000	2.1204	2.0867	2.1271
5.500	2.1166	2.0828	2.1229
6.000	2.1128	2.0789	2.1189
6.500	2.1086	2.0750	2.1143
7.000	2.1040	2.0701	2.1096
7.500	2.0990	2.0650	2.1043
8.000	2.0937	2.0595	2.0987
8.500	2.0876	2.0534	2.0924
9.000	2.0816	2.0470	2.0856
9.500	2.0749	2.0398	2.0783
10.000	2.0666	2.0319	2.0703
10.500	2.0585	2.0238	2.0619
11.000	2.0501	2.0146	2.0522

Coherence length measured in μm at $\lambda_1 = 10.6$ μm.

$l_{31} = 40$
$l_{32} = 29.4$
$l_{33} = 42$; $d_{31}d_{33} > 0$; $d_{32}d_{33} > 0$.

For $\Phi = 0$, phase-matched SHG occurs for $\lambda_1 = \lambda_2 = 2.32$ μm, $\lambda_3 = 1.16$ μm, and for $\lambda_1 = \lambda_2 = 5.88$ μm, $\lambda_3 = 2.94$ μm.

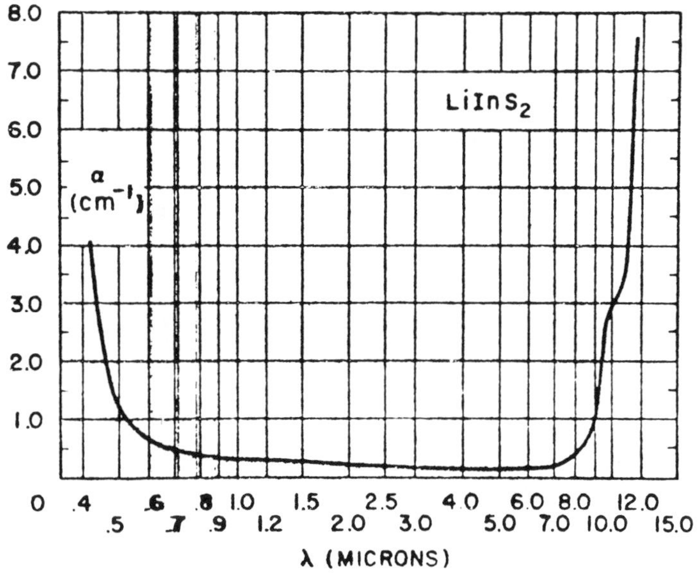

FIGURE 1.1.36. Room temperature absorption coefficient $\alpha(cm^{-1})$ vs. wavelength $\lambda(\mu m)$ for $LiInS_2$, as determined from a sample of thickness 0.524 cm. (From Boyd, G. D., Kasper, H. M., and McFee, J. H., *J. Appl. Phys.*, 44, 2809, 1973. With permission.)

Table 1.1.17.59 LITHIUM METAGALLATE ($LiGaO_2$)[195]

Wavelength (μm)	$n^{x,z}$	n^y
0.4100	1.804	1.7702
0.4500	1.7895	1.757
0.5000	1.7785	1.7466
0.5500	1.7702	1.7395
0.6000	1.7615	1.7343
0.7000	1.7565	1.7268
0.8000	1.7507	1.7218
0.9000	1.7475	1.7185
1.0000	1.7445	1.716
1.2000	1.7405	1.7122
1.4000	1.7372	1.7095
1.6000	1.735	1.707
1.8000	1.7325	1.7045
2.2000	1.7268	1.7005
2.4000	1.7242	1.6978
2.6000	1.7225	1.6955
2.8000	1.720	1.6925

Table 1.1.17.59(a) LITHIUM METAGALLATE ($LiGaO_2$)[196]

Wavelength (μm)	n^x	n^z	n^y
0.4700	1.7852	1.7835	1.7534
0.5000	1.7791	1.7768	1.7477
0.5400	1.7708	1.7683	1.7407
0.5800	1.7653	1.7626	1.7351
0.6200	1.7617	1.7589	1.7311
0.6600	1.7604	1.7578	1.7289

$$\left.\begin{aligned} l_{31} &= 10.5 \pm 0.3\ \mu m \\ l_{32} &= 4.6 \pm 0.1\ \mu m \\ l_{33} &= 9.2 \pm 0.1\ \mu m \end{aligned}\right\} \text{at } \lambda_1 = 1.064\ \mu m.$$[195]

Table 1.1.17.60
LITHIUM SODIUM FORMATE MONOHYDRATE [$Na_{0.1}Li_{0.9}$ (COOH)H_2O] REFRACTIVE INDEX AT ROOM TEMPERATURE)

Sellmeier equation:

$$n^2(\lambda) = A + B^2/(\lambda^2 - \lambda_0^2 - C\lambda^2$$

where λ is in μm.

Sellmeier constant	$n^{x=a}$	$n^{y=b}$	$n^{z=c}$
A	1.1665	1.1942	1.2927
B	0.7210	0.9043	1.0020
C	0.0011	0.0009	0.0050
λ_o	0.1028	1.1211	0.1312

$\theta_m(d^{32})$ = 58.2° and 9.2° at λ_1 = 1.064 μm.

Table 1.1.17.61
m-NITROANILINE (NO_2-C_6H_4-NH_2)[205] REFRACTIVE INDEX AT ROOM TEMPERATURE

Sellmeier equation:

$$n^2 = A + B\lambda^2/(\lambda^2 - C) - D\lambda^2$$

where λ is in μm.

Sellmeier constant	$n^{x=a}$	$n^{y=b}$	$n^{z=c}$
A	2.8102	2.6650	2.4690
B	0.1524	0.1626	0.1864
C	0.1750	0.1719	0.1600
D	0.0294	0.0212	0.0199

Coherence lengths measured in μm at λ_1 = μm: $l_{31} = -13.7 \pm 0.5$, $l_{32} = 16 \pm 0.7$, and $l_{33} = 3.9 \pm 0.2$.
θ_m(int. w.r.t. y axis) = 35 ± 3°, 47 ± 3° at λ_1 = 1.064 μm.
$d_{31}d_{33} > 0$; $d_{32}d_{33} > 0$

Table 1.1.17.61a
m-NITROANILINE (NO_2-C_6H_4-NH_2)[204] REFRACTIVE INDEX AT ROOM TEMPERATURE

Wavelength (μm)	$n^{x=a}$	$n^{y=b}$	$n^{z=c}$
0.532	1.81	1.79	1.72
1.0642	1.74	1.71	1.65

Coherence lengths measured in μm at λ_1 = 1.064 μm: l_{31} = 14.5, l_{32} = 13.6, and l_{33} = 3.9.

Table 1.1 17.62
POTASSIUM CHLOROIODATE [$K_2H(IO_3)_2Cl$][215] REFRACTIVE INDEX AT ROOM TEMPERATURE

Wavelength (μm)	$n^{x=c}$	$n^{y=b}$	$n^{z=a}$
0.53	1.77	1.88	1.70
1.06	1.74	1.82	1.67

Note: The crystallographic axes have been rearranged to follow the IRE convention: *c* = 6.594 Å, *a* = 8.611 Å, and *b* = 15.10 Å.

Coherence lengths measured in μm at λ_1 = 1.064 μm: $l_{31} = -5.98 \pm 0.25$, $l_{32} = -2.19 \pm 0.06$, $l_{33} = +9.25 \pm 0.15$, $l_{24} = +1.97 \pm 0.04$; $d_{32}d_{33} > 0$; $d_{33}d_{31} > 0$.

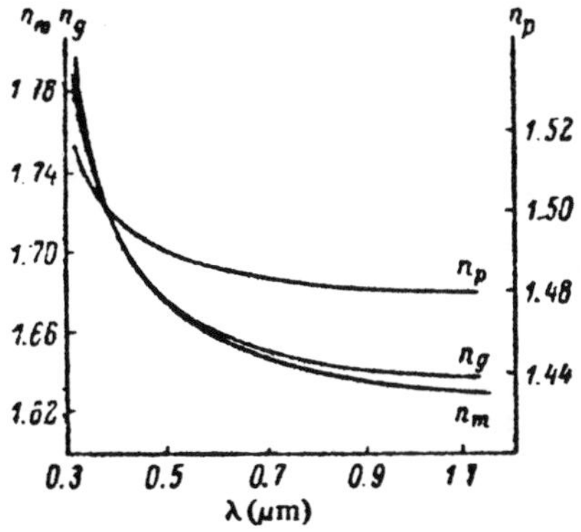

FIGURE 1.1.37. Dispersion of refractive indexes of potassium diphthalate ($KHC_8H_4O_4$). (From Belyaev, G. S., Belikova, G. S., Gilvarg, A. G., Golovei, M. P., Kalinkina, I. N., and Kosowrov, G. I., *Opt. Spectrosc. USSR*, 29, 522, 1970. With permission.)

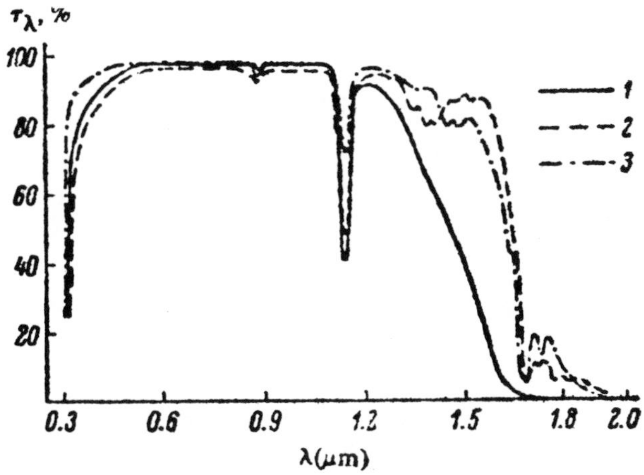

FIGURE 1.1.38. Optical transmission of a potassium diphthalate crystal. Light is polarized (1) along X-axis; (2) along Y-axis; (3) along Z-axis. (From Belyaev, G. S., Belikova, G. S., Gilvarg, A. G., Golovei, M. P., Kalinkina, I. N., and Kosowrov, G. I., *Opt. Spectrosc. USSR*, 29, 522, 1970. With permission.)

Table 1.1 17.63
POTASSIUM DIPHTHALATE ($KHC_8H_4O_4$)[210]

$n^x = 1.63$, $n^y = 1.64$, $n^z = 1.48$ at $\lambda = 1.1$ μm.

$\theta_m = 21°$ and $25°$ at $\lambda_1 = 1.15$ μm
$\theta_m = 46°$ at $\lambda_1 = 0.63$ μm. [210]

Table 1.1 17.64 POTASSIUM FLUOROIODATE (KIO_2F_2)[214] REFRACTIVE INDEX AT ROOM TEMPERATURE

Wavelength (μm)	$n^{x=c}$	$n^{y=a}$	$n^{z=b}$
0.53	1.569	1.621	1.617
1.06	1.554	1.597	1.594

Note: The crystallographic axes have been rearranged to follow the IRE convention: c = 5.97 Å, a = 8.38 Å, and b = 8.41 Å.

Coherence lengths measured in μm at λ_1 = 1.064 μm: l_{31} = +4.21, l_{32} = +13.4, l_{33} = ±11.5, l_{15} = ±55.1, l_{24} = ±10.4; d_{33} $d_{32} > 0$; d_{33} $d_{31} > 0$.

Table 1.1.17.65 POTASSIUM NIOBATE ($KNbO_3$)[193]

Wavelength (μm)	$n^{x=c}$	$n^{y=a}$	$n^{z=b}$
0.4880	2.4190	2.3526	2.2275
0.5145	2.3941	2.3329	2.2116
0.5321	2.3807	2.3224	2.2029
0.6328	2.3291	2.2799	2.1685
1.0642	2.2574	2.2200	2.1196

Note: The crystallographic axes given in Reference 193 have been rearranged to follow the IRE convention: c_0 = 3.971 Å, a_0 = 5.697 Å, and b_0 = 5.722 Å.

Dispersion equation:

$$n^2 - 1 = S_0\lambda_{02}/\ [1 - \lambda_0\lambda]^2$$

where λ is in μm.

Sellmeier constant	$n^{x=c}$	$n^{y=a}$	$n^{z=b}$
$S_0(\mu m^{-2})$	87.67 ± 0.51	97.85 ± 0.40	98.12 ± 0.58
$\lambda_0(\mu m)$	(211.8 ± 0.5) × 10^{-3}	(196.9 ± 0.3 × 10^{-3}	(185.7 ± 0.4) × 10^{-3}

$$[d/dT]\ (n^{y}_{1.06} - n^{z}_{0.532} \simeq \begin{cases} 8.0 \times 10^{-5}\ (^\circ C)^{-1} \text{ between } 25^\circ C \text{ and } 130^\circ C \\ 1.6 \times 10^{-4}\ (^\circ C)^{-1} \text{ at } \sim 180^\circ C \end{cases}$$

Noncritical PM for 1.064 μm fundamental possible for d_{32} at 181 ± 2°C.

$$\left.\begin{array}{l} \theta_m^{31} = 46 \text{ to } 48^\circ \\ \theta_m^{32} = 70 \text{ to } 72^\circ \end{array}\right\} \text{ at } \lambda_1 = 1.064\ \mu m.$$

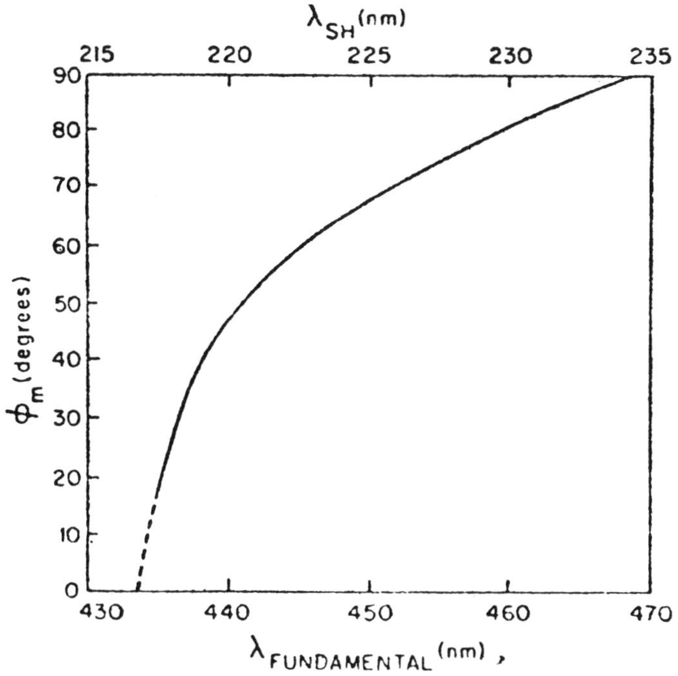

FIGURE 1.1.39. Measured phase-matching angle vs. wavelength for $KB_5O_8{\cdot}4H_2O$. (From Dewey, C. F., Cook, W. R., Hodgson, R. T., and Wynne, J. J., *Appl. Phys. Lett.*, 26, 714, 1975. With permission.)

Table 1.1.17.66
POTASSIUM PENTABORATE TETRAHYDRATE ($KB_5O_8{\cdot}4H_2O$) REFRACTIVE INDEXES AT ROOM TEMPERATURE [220a]

Wavelength (μm)	$n^{x=a}$	$n^{y=b}$	$n^{z=c}$
0.2168			1.4962
0.2345		1.4930[(1)]	
0.390	1.5021	1.4457	1.4327
0.400	1.5005	1.4453	1.4320
0.420	1.4984	1.4438	1.4303
0.450	1.4956	1.4414	1.4280
0.500	1.4917	1.4380	1.4251
0.546	1.4888	1.4357	1.4230
0.600	1.4859	1.4334	1.4211
0.650	1.4839	1.4319	1.4196
0.700	1.4823	1.4306	1.4182
0.730	1.4815	1.4297	1.4176
0.765	1.4813	1.4292	1.4171

Dispersion equation:[220a]

$$n^2 = 1 + 1/(A + B\lambda^2)$$

where λ is in μm.

	n^x	n^y	n^z
A	0.852497	0.972682	1.008157
B	−0.0087588	−0.0087757	−0.0094050

Table 1.1.17.67
POTASSIUM RUBIDIUM PHOSPHOTITANATE (K_xRb_{1-x} $TiOPO_4$) REFRACTIVE INDEX AT ROOM TEMPERATURE

Sellmeier equation for $KTiOPO_4$:

$n^2 = A + B/(1 - C/\lambda^2) + D/(1 - E/\lambda^2,$

where λ is in μm.

Sellmeier constant	n^x	n^y	n^z
A	2.2088890	2.3089380	2.3469479
B	0.7993644	0.72256020	0.9698851
C	0.04706234	0.0540493	0.05793341
D	1.3282970	1.357887	1.647279
E	100	100	100

Note: The relation between crystallographic and principal axes not known. The reference value of d_{36} KDP used in Reference 222 is not known. The SHG-coefficients are reported to be independent of the values of χ.

Walk-off angles for $KTiOPO_4$:

Phase matching	**λ(μm)**	**Walk-off angle**
Type I (in xz-plane)	0.53	3.12°
	1.06	2.67°
Type II (xy-plane)	0.53	0.077°
	1.06	0.057°
Type II (yz-plane)	0.53	0.74°
	1.06	0.64°

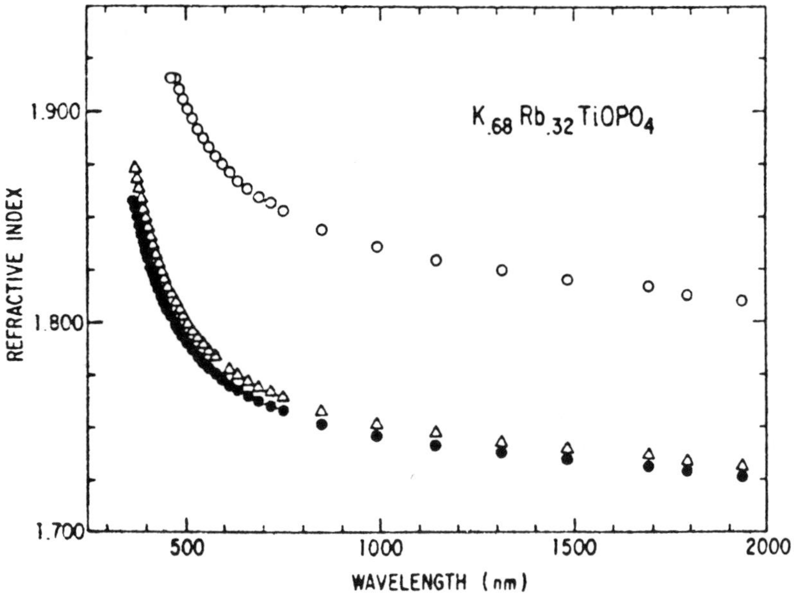

FIGURE 1.1.40. Refractive index vs. wavelength of $K_{0.68}Rb_{0.32}TiOPO_4$. ● — n^X; Δ — n^Y; ○ — n^Z. (From Zumsteg, F. C., Bierlein, J. D., and Gier, T. E., *J. Appl. Phys.*, 47, 4980, 1976. With permission.)

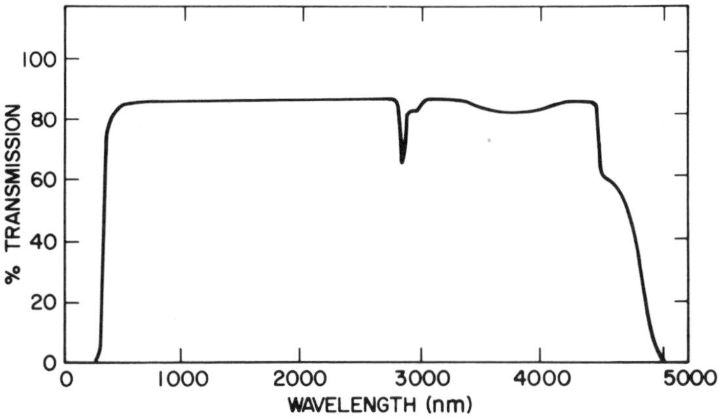

FIGURE 1.1.41. Optical transmission of $RbTiOPO_4$. (From Zumsteg, F. C., Bierlein, J. D., and Gier, T. E., *J. Appl. Phys.*, 47, 4980, 1976. With permission.)

Table 1.1.17.68 POTASSIUM THIOMOLYBDATE POTASSIUM CHLORIDE (K_2MoOS_3KCl)[218] REFRACTIVE INDEX AT ROOM TEMPERATURE

Wavelength (μm)	$n^{x=c}$	$n^{y=b}$	$n^{z=a}$
0.66	1.944	1.904	1.875
1.32	1.870	1.839	1.815

Note: The crystallographic axes given in Reference 218 have been rearranged to follow the IRE convention: c = 6.29 Å, a = 12.36 Å, and b = 12.48 Å.

Coherence lengths measured in μm at λ_1 = 1.32 μm: l_{31} = 62.9, l_{32} = 9.14, l_{33} = 5.49, l_{15} = 3.24, and l_{24} = 4.26; $d_{33}d_{31} < 0$; $d_{33}d_{32} < 0$; Phase matchable for $\lambda_1 \geq 1.37$ μm.

Table 1.1.17.69 1,3-DIHYDROXYBENZENE (α-RESORCINOL) [$C_6H_4(OH)_2$][204]

Wavelength (μm)	$n^{x=a}$	$n^{y=b}$	$n^{z=c}$
0.532	1.589	1.636	1.627
1.064	1.560	1.605	1.597

Coherence length measured in μm for λ_1 = 1.064 μm: l_{31} = 3.7, l_{32} = 9.8, and l_{33} = 8.5.

Table 1.1.17.70 SODIUM FORMATE [Na(COOH)][197]

Sellmeier equation:

$$n^2(\lambda) = A + B\lambda^2/(\lambda^2 - \lambda_0^2) - C\lambda^2$$

where λ is in μm.

Sellmeier constant	$n^{x=c}$	$n^{y=b}$	$n^{z=a}$
A	1.2589	1.2646	1.2515
B	0.8423	0.6381	1.0729
C	0.0005	0.0011	0.0013
λ_o	0.1203	0.1101	0.1314

Note: The crystallographic axes have been rearranged to follow the IRE convention: c_o = 5.05 Å, a_o = 6.32 Å, and b_o = 7.47 Å; $d_{31} \cdot d_{33} < 0$; θ_m^{d31} = 53.0°, 8.1° at 1.064 μm

Table 1.1.17.71 SODIUM NITRITE ($NaNO_2$)[200]

Wavelength (μm)	$n^{x=a}$	$n^{y=c}$	$n^{z=b}$
0.4358	1.3531	1.6900	1.4212
0.4800	1.3500	1.6750	1.4166
0.5086	1.3484	1.6685	1.4158
0.5461	1.3470	1.6620	1.4137
0.5791	1.3458	1.6567	1.4122
0.5889	1.3455	1.6555	1.4120
0.6438	1.3442	1.6510	1.4105

Note: Coherence lengths measured for λ_1 = 1.064 μm:[199] l_{31} = 3.5 ± 0.2 μm, l_{32} = 1.0 ± 0.2 μm, l_{33} = 27 ± 3 μm, l_{15} = 10.5 ± 1.0 μm, and l_{24} = 1.6 ± 0.2 μm.

$$\left.\begin{array}{l} d_{31} \cdot d_{33} < 0 \\ d_{32} \cdot d_{33} > 0 \\ d_{15} \cdot d_{33} < 0 \\ d_{33}\, d_{24} > 0 \end{array}\right\} \text{Ref. 199}$$

θ_m = 56.4° (with x axis)

Table 1.1.17.72
TERBIUM MOLYBDATE
($Tb_2(MoO_4)_3$)[189]

Wavelength (μm)	$n^{y=b}$	$n^{x=a}$	$n^{z=a}$
0.4579	1.8864	1.8867	1.9433
0.4765	1.8797	1.8800	1.9358
0.4880	1.8760	1.8764	1.9315
0.4965	1.8734	1.8739	1.9288
0.5017	1.8720	1.8724	1.9271
0.5145	1.8687	1.8690	1.9232
0.5320	1.8645	1.8649	1.9185
0.6328	1.8476	1.8482	1.8993
1.0642	1.8222	1.8226	1.8704

Dispersion equation:

$$(n^x)^2 - 1 = 2.273955\ \lambda^2/[\lambda^2 - 0.02333].$$
$$(n^y)^2 - 1 = 2.27241\ \lambda^2/[\lambda^2 - 0.023359].$$
$$(n^z)^2 - 1 = 2.443016\ \lambda^2/[\lambda^2 - 0.025133].$$

Table 1.1.17.73
1,3,5-TRIPHENYLBENZENE
($C_{24}H_{18}$)[204]

Wavelength (μm)	$n^{x=c}$	$n^{y=b}$	$n^{z=a}$
0.532	1.53	1.89	1.89
1.064	1.51	1.82	1.82

Note: The crystallographic axes have been rearranged to follow the IRE convention: c = 7.47 Å, a = 11.19 Å, and b = 19.66 Å. Coherence lengths measured in μm at λ_1 = 1.064 μm: l_{31} = 0.7, l_{32} = 4.2, l_{33} = 4.1; $d_{32}d_{33} < 0$.

Table 1.1.17.74
DICALCIUM STRONTIUM PROPIONATE
[$Ca_2Sr(CH_3CH_2CO_2)_6$][244]

Measured coherence lengths in μm at λ_1 = 1.064 μm: l_{31} = 12.3 ± 1.0, l_{33} = 18.2 ± 2.0; $d_{31}d_{33}$ < 0.

Table 1.1.17.75
METHYL-3 ISOPROPYL-4 PHENOL [$OH\text{-}C_6H_4\text{-}CH(CH_3)_3$][245]

Wavelength (μm)	n^o	n^e
0.53	1.6020	1.5862
1.06	1.5839	1.5691

Note: Coherence lengths measured in μm at χ_1 = 1.06 μm: l_{31} = 11.5, l_{33} = 15.5. Not phase-matchable at 1.06 μm.

Table 1.1.17.76
CADMIUM MERCURY THIOCYANATE ($Cd[Hg(SCN_4]^g$)[240]

Wavelength (μm)	n^o	n^e
0.530	2.003_5	1.792
0.633	1.970	1.753
1.060	1.924_5	1.728

Note: Crystals transparent over the range 0.4 to above 2.5 μm with a narrow absorption band at 2.35 μm. Phase matching angle θ_m = 36 ± 2° at λ = 1.06 μm. l_{31} = 1.6 ± 0.2 μm.

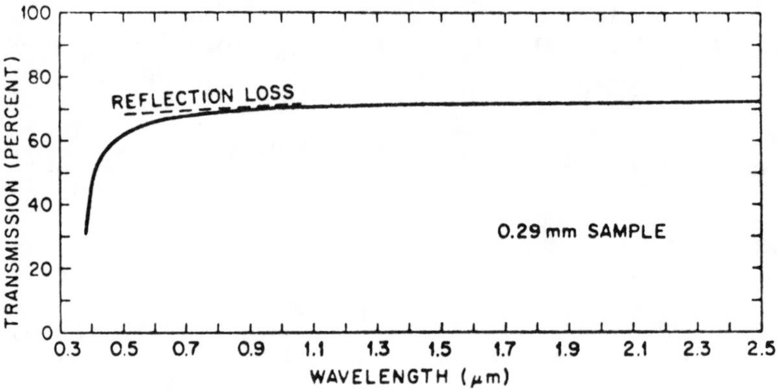

FIGURE 1.1.42. Transmission of $InPS_4$. (From Bridenbaugh, P. M., *Mater. Res. Bull.*, 8, 1055, 1973. With permission.)

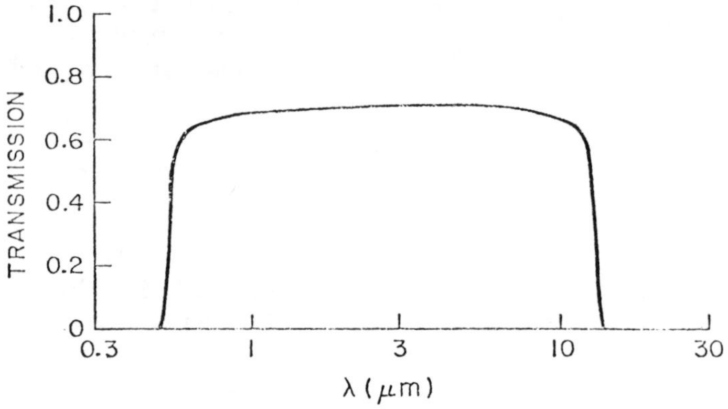

FIGURE 1.1.43. Transmission of mercury gallium sulfide ($HgGa_2S_4$). (From Levine, B. F., Bethea, C. G., Kasper, H. M., and Thiel, F. A., *IEEE Correspondence Ann.*, No. 606QE008, 367, 1976. With permission.)

Table 1.1.17.77
INDIUM THIOPHOSPHIDE ($InPS_4$)[243]

Wavelength (μm)	n^o	n^e
0.5086	2.4253	2.5125
0.5460	2.3963	2.4750
0.5770	2.3784	2.4518
0.6438	2.3495	2.4159
0.5320	2.4056	2.4868 calc.
1.064	2.2805	2.3304 calc.

Note: Coherence lengths measured in μm at λ_1 = 1.064 μm: l_{31} = 1.27 ± 0.05, l_{36} = 1.27 ± 0.05.

Table 1.1.17.78
MERCURY GALLIUM SULFIDE ($HgGa_2S_4$)[242]
REFRACTIVE INDEX AT ROOM TEMPERATURE

Wavelength (μm)	n^o	n^e
0.5145	2.79	2.71
0.5320	—	—
0.6328	2.61	2.55
1.064	2.56	2.50

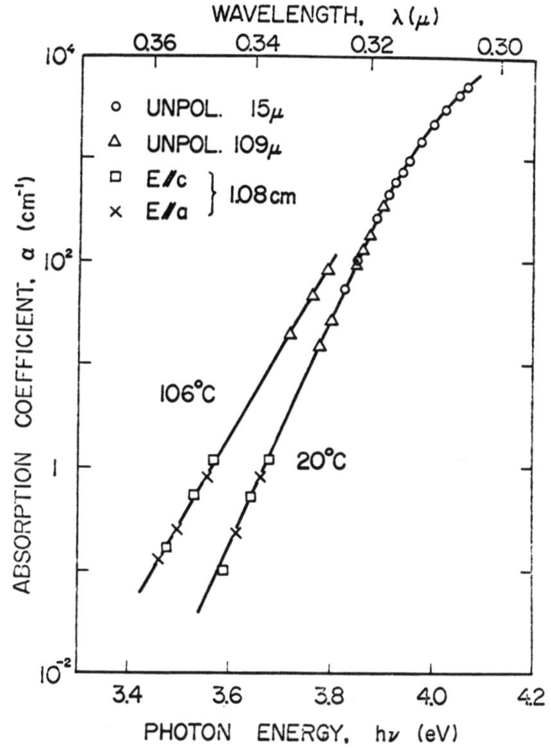

FIGURE 1.1.44. Absorption coefficient of paratellurite (TeO_2) at 20 and 106°C as a function of wavelength. (From Uchida, N., *Phys. Rev.*, B4, 3736, 1971. With permission.)

Table 1.1.17.79
TELLURIUM DIOXIDE (TeO_2)[234]
REFRACTIVE INDEX AT ROOM TEMPERATURE

Wavelength (μm)	n^o	n^e
0.4765	2.3394	2.5064
0.4880	2.3297	2.4951
0.4965	2.3232	2.4874
0.5017	2.3194	2.4829
0.5145	2.3107	2.4727
0.5321	2.3001	2.4602
0.6328	2.2585	2.4112
1.0642	2.2005	2.3431
1.152	2.1955	2.3373

Dispersion equation:

$n^{o2} - 1 = 3.7088\ \lambda^2/(\lambda^2 - 0.038575)$ $n^{e2} - 1 = 4.3252\ \lambda^2/(\lambda^2 - 0.040959)$

Note: Coherence length measured at $\lambda_1 = 1.0642$ μm; $l_{14} = 8.35$. Phase matchable: Type I: $\theta_m(\mathrm{I}) = 58.3°$ at $\lambda_1 = 1.0642$ μm; Type I: $\theta_m(\mathrm{I}) = 42°$ at $\lambda_1 = 1.318$ μm. $l = 0.51$ μm ± 10% at $\lambda_1 = 0.859$ μm.[236] Additional data on refractive indexes and transparency are given in Reference 237.

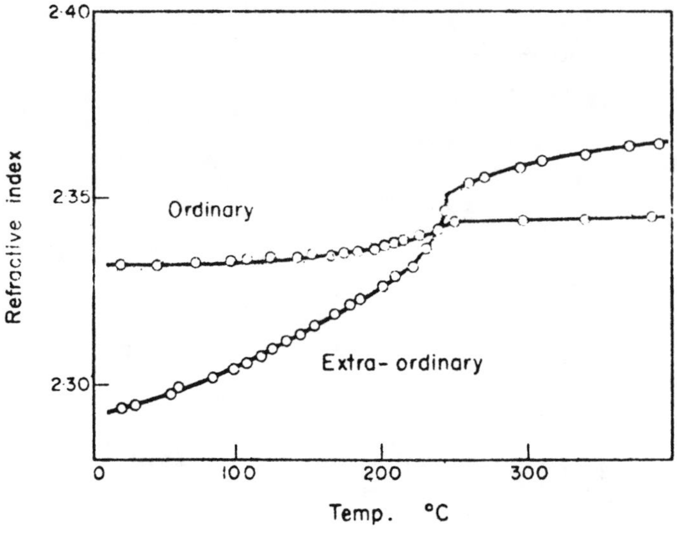

FIGURE 1.1.45. Temperature-variation of refractive indexes (for $\lambda = 0.5893$ μm) of $Ba_6Ti_2Nb_8O_{30}$. (From Itho, Y. and Iwasaki, H., *J. Phys. Chem. Solids*, 34, 1639, 1973. With permission.)

Table 1.1.17.80
BARIUM TITANATE ($BaTiO_3$)[228] AT REFRACTIVE INDEXES AT ROOM TEMPERATURE

Wavelength (μm)	n^o	n^e
0.4579	2.5637	2.4825
0.4765	2.5355	2.4605
0.4880	2.5206	2.4487
0.5145	2.4917	2.4255
0.5321	2.4760	2.4128
0.6328	2.4164	2.3637
1.0642	2.3379	2.2970
2.1284	2.2947	2.2593

Dispersion equation:

$n^{o2} - 1 = 4.239\,\lambda^2/\lambda^2\text{-}(0.2229)^2\ {}^{e2} - 1 = 4.0854\,\lambda^2/[\lambda^2\text{-}(0.2087)^2]$,

where λ is in μm.

Note: Coherence lengths measured in μm at $\lambda_1 = 1.058$ μm:[226] $l_{15} = 1.57$, $l_{31} = 2.90$, $l_{33} = 2.07$.

Table 1.1.17.81
BARIUM TITANIUM NIOBATE ($Ba_6Ti_2Nb_8O_{30}$)[238] REFRACTIVE INDEX AT ROOM TEMPERATURE

Wavelength (μm)	n^o	n^e
0.4358	2.448	2.396
0.4800	2.399	2.351
0.5086	2.376	2.333
0.5461	2.349	2.311
0.5791	2.337	2.297
0.5893	2.332	2.293
0.6438	2.314	2.276

Dispersion equation:

$n^{o2} - 1 = 3.920\,\lambda^2/(\lambda^2 - 0.04040)\ n^{e2} - 1 = 3.804\,\lambda^2/(\lambda^2 - 0.03725)$

where λ is in μm.

d_{31} and d_{33} have the same sign.

Note: Coherence lengths measured in μm at $\lambda_1 = 1.064$ μm: $l_{31} = 3.9$, $l_{33} = 2.7$.

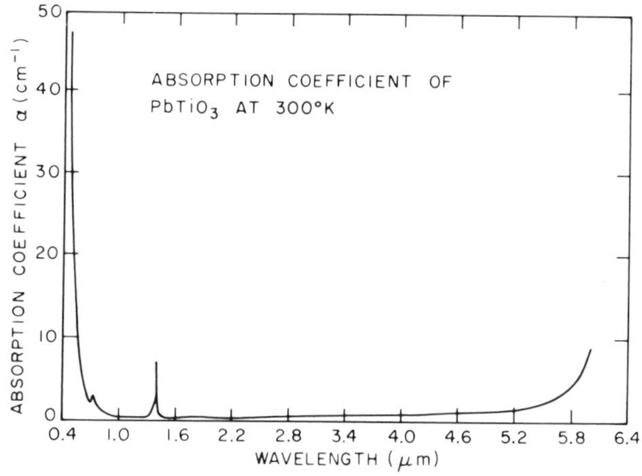

FIGURE 1.1.46. Absorption coefficient of $PbTiO_3$ at 300 K. (From Singh, S., Remeika, J. P., and Potopowicz, J. R., *Appl. Phys. Lett.*, 20, 135, 1972. With permission.)

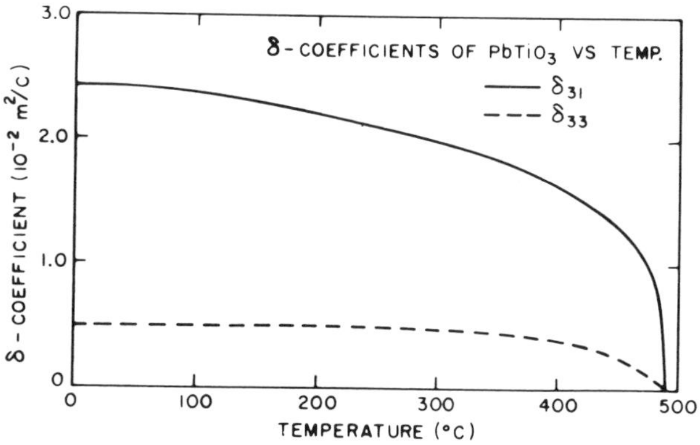

FIGURE 1.1.47. Temperature dependence of Miller-δ for lead titanate ($PbTiO_3$). (From Singh, S., Remeika, J. P., and Potopowicz, J. R., *Appl. Phys. Lett.*, 20, 135, 1972. With permission.)

Table 1.1.17.82
LEAD TITANATE ($PbTiO_3$)[229]
REFRACTIVE INDEX AT ROOM TEMPERATURE

Wavelength (μm)	n^o	n^e
0.4880	2.793	2.7744
0.5017	2.7742	2.7574
0.5145	2.7586	2.7431
0.5321	2.7398	2.7260
0.6328	2.6676	2.6594
1.0642	2.5712	2.5692
1.152	2.5637	2.5623

Dispersion equation:

$n^{o2} - 1 = 5.359\ \lambda^2/[\lambda^2 - (0.224)^2]$ $n^{e2} - 1 = 5.365\ \lambda^2/[\lambda^2 - (0.2170)^2]$

where λ is in μm.

Note: Coherence lengths measured in μm at $\lambda_1 = 1.064$ μm: $l_{15} = 1.53$, $l_{31} = 1.68$, and $l_{33} = 1.65$; $d_{31}/d_{33} < 0$, $d_{15}/d_{31} > 0$.

Table 1.1.17.83
POTASSIUM LITHIUM NIOBATE ($K_3Li_2Nb_5O_{15}$)[230] REFRACTIVE INDEX AT ROOM TEMPERATURE

Wavelength (μm)	n^o	n^e
0.4500	2.4049	2.2512
0.4750	2.3751	2.2315
0.5000	2.3546	2.2144
0.5250	2.3349	2.2010
0.5324	2.3260	2.1975
0.5500	2.3156	2.1900
0.5750	2.3016	2.1801
0.6000	2.2899	2.1720
0.6250	2.2799	2.1645
0.6500	2.2711	2.1586
0.6750	2.2631	2.1529

Dispersion equation:

$n^{o2} - 1 = 3.708\ \lambda^2/(\lambda^2 - 0.04601)\ n^{e2} - 1 = 3.349\ \lambda^2 - 0.03564)$

where λ is in μm.

Note: Phase matchable at $\theta_m = 90°$ for temperatures in the range 60 to 100°C.

Table 1.1.17.84
POTASSIUM LITHIUM NIOBATE $[(K_2O)_{03}(Li_2O)_{0.7-X}(Nb_2O_5(X]$[233]

Wavelength (μm)	n^o	n^e
0.6328	2.278	2.133

Note: For X = 0.52. Phase matchable with θ_m = 90° at 110°C for λ_1 = 0.90 μm and 350°C for λ_1 = 1.06°C for the composition X = 0.525.

Table 1.1.17.85
POTASSIUM SODIUM BARIUM NIOBATE ($K_{0.8}Na_{0.2}Ba_2Nb_5O_{15}$)[232]

Dispersion equation:

$$n^2 = A + \frac{B}{C^2 - (1.2394/\lambda)^2}$$

where λ is in μm.

Temp. (°C)	Constant	n^o	n^e
22	*A*	3.6680	2.9198
	B	24.681	46.737
	C	4.3004	5.1605
312	*A*	3.6307	3.2273
	B	25.357	34.842
	C	4.3004	4.7109
437	*A*	3.5791	3.5539
	B	26.534	26.075
	C	4.3004	4.3004

Note: Phase matchable at θ_m = 90° at −130°C.

Table 1.1.17.86
AMMONIUM DIDEUTERIUM PHOSPHATE ($ND_4D_2PO_4$)[252] REFRACTIVE INDEX AT ROOM TEMPERATURE

Wavelength (μm)	n^o	n^e
0.350	1.5414	1.4923
0.530	1.5198	1.4784
0.690	1.5142	1.4737
1.060	1.5088	1.4712

Note: l_{36} = 8.9 μm; θ_m = 47°.

Table 1.1.17.87
BERYLLIUM SULFATE TETRAHYDRATE ($BeSO_4$-$4H_2O$)[254] REFRACTIVE INDEX AT ROOM TEMPERATURE

Wavelength (μm)	n^o	n^e
0.4154	1.4847	1.4431
0.4825	1.4782	1.4379
0.5321	1.4749	1.4348
0.6328	1.4701	1.4315
0.6471	1.4692	1.4312
0.6764	1.4681	1.4304
0.7525	1.4658	1.4292

Sellmeier equation:

$$n^{o2} - 1 = 1.127884\ \lambda^2/(\lambda^2 - 0.011509)$$

$$n^{e2} - 1 = 1.025413\ \lambda^2/(\lambda^2 - 0.008925)$$

where λ is in μm.

Phase-matching angles θ_m^{36} = 42° at λ_1 = 1.15 μm,[253] = 55° at λ_1 = 0.63 μm,[253] θ_m^{14} = 64° at λ_1 = 1.15 μm,[253] θ_m = 81°, 28' at λ_1 = 0.5321 μm.[254] Transparency range 0.17 to 15 μm.[253]

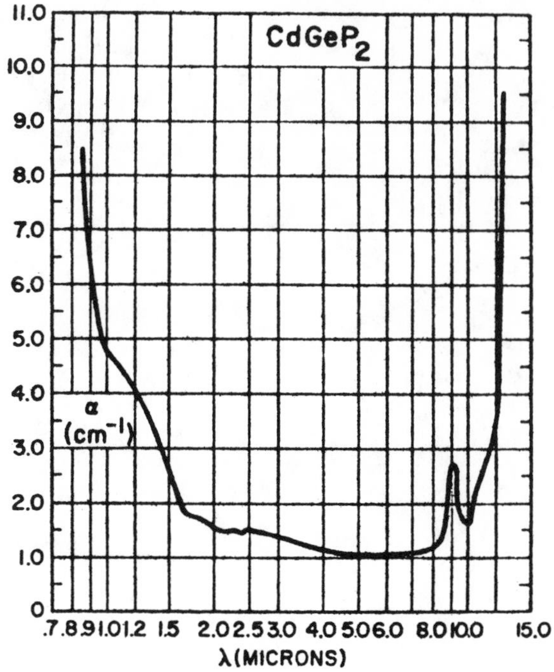

FIGURE 1.1.48. Absorption coefficient of $CdGeP_2$. (From Boyd, G. D., Buehler, E., Storz, F. G., and Wernick, J. H., *IEEE J. Quantum Electron.*, QE-8, 419, 1972. With permission.)

Table 1.1.17.88
CADMIUM GERMANIUM PHOSPHIDE ($CdGeP_2$)[256]

Wavelength (μm)	n^o	n^e	$n^e - n^o$	$10^5 \times dn^o/dT$	$10^5 \times dn^e/dT$	$10^5 \times dB/dT$
0.80	3.4833			35.56		
0.85	3.4403	3.4942	0.0539	29.11	43.35	14.24
0.90	3.4059	3.4505	0.0445	53.72	61.38	7.66
0.95	3.3789	3.4167	0.0378	25.55	26.16	.61
1.00	3.3560	3.3902	0.0342	25.80	28.75	2.95
1.10	3.3232	3.3505	0.0272	23.60	26.37	2.76
1.20	3.2990	3.3222	0.0231	22.25	25.08	2.83
1.30	3.2811	3.3025	0.0214	20.71	22.52	1.81
1.40	3.2669	3.2867	0.0198	20.20	21.42	1.22
1.60	3.2470	3.2643	0.0173	18.68	20.57	1.89
1.80	3.2342	3.2502	0.0161	20.56	21.75	1.19
2.00	3.2255	3.2406	0.2151	18.11	19.31	1.20
2.20	3.2202	3.2340	0.0138	17.12	18.25	1.13
2.40	3.2159	3.2287	0.0128	18.70	19.61	.91
2.60	3.2128	3.2243	0.0115	17.77	19.31	1.55
2.80	3.2096	3.2206	0.0110	17.30	18.96	1.66
3.00	3.2065	3.2175	0.0110	17.69	19.22	1.53
3.20	3.2049	3.2156	0.0106	16.61	18.24	1.63
3.40	3.2026	3.2132	0.0106	16.10	17.65	1.55
3.60	3.2003	3.2108	0.0104	16.39	17.94	1.56
3.80	3.1981	3.2083	0.0102	17.19	18.72	1.53
4.00	3.1963	3.2063	0.0100	17.04	18.61	1.57
4.50	3.1924	3.2026	0.0122	17.44	18.49	1.05
5.00	3.1887	3.1991	0.0103	17.86	18.75	.89

Table 1.1.17.88 (continued)
CADMIUM GERMANIUM PHOSPHIDE ($CdGeP_2$)[256]

Wavelength (μm)	n^o	n^e	$n^e - n^o$	$10^5 \times dn^o/dT$	$10^5 \times dn^e/dT$	$10^5 \times dB/dT$
5.50	3.1830	3.1934	0.0105	17.71	18.51	.80
6.00	3.1800	3.1906	0.0106	18.08	19.03	.95
6.50	3.1768	3.1876	0.0108	18.38	19.56	1.18
7.00	3.1735	3.1846	0.0112	17.69	19.02	1.33
7.50	3.1703	3.1816	0.0113	17.12	18.52	1.40
8.00	3.1669	3.1785	0.0116	15.51	16.73	1.23
8.50	3.1624	3.1743	0.0119	16.58	17.92	1.34
9.00	3.1585	3.1707	0.0122	16.20	17.28	1.08
9.50	3.1547	3.1672	0.0125	14.75	16.69	1.94
10.00	3.1508	3.1636	0.0128	12.88	14.72	1.84
10.50	3.1435	3.1574	0.0139	16.14	16.98	.84
11.00	3.1372	3.1517	0.0145	16.59	17.62	1.03
11.50	3.1321	3.1470	0.0149	15.93	17.02	1.09
12.00	3.1241	3.1402	0.0151	17.14	17.85	0.70
12.50	3.1165	3.1332	0.0167	15.48	16.11	.63

Note: $l = 48.5 \pm 1$ μm at $\lambda_1 = 10.6$ μm.

Sellmeier equation:[251]

$$n^2 = A + B/(1 - C/\lambda^2) + D/(1 - E/\lambda^2)$$

where λ is in μm.

Temp. (°C)	Polarization	*A*	*B*	*C*	*D*	*E*
20	o	5.9677	4.2286	0.2021	1.6351	671.33
	e	6.1573	4.0970	0.2330	1.4925	671.33
118	o	6.3737	3.9281	0.2269	1.6686	671.33
	e	6.9280	3.4442	0.2803	1.5515	671.33

Note: Temperature-dependent Sellmeier coefficients: $X = mT + c$, where $X = A, B, C$, and D. Temperature range from 20 to 118°C.

Coefficients	Constants	n^o	n^e
A	$m \times 10^3$	4.132	7.831
	c	5.900	6.036
B	$m \times 10^3$	−3.056	−6.628
	c	4.275	4.194
C	$m \times 10^5$	25.357	48.144
	c	0.1974	0.2244
D	$m \times 10^4$	3.413	6.026
	c	1.629	1.481
E		671.33	671.33

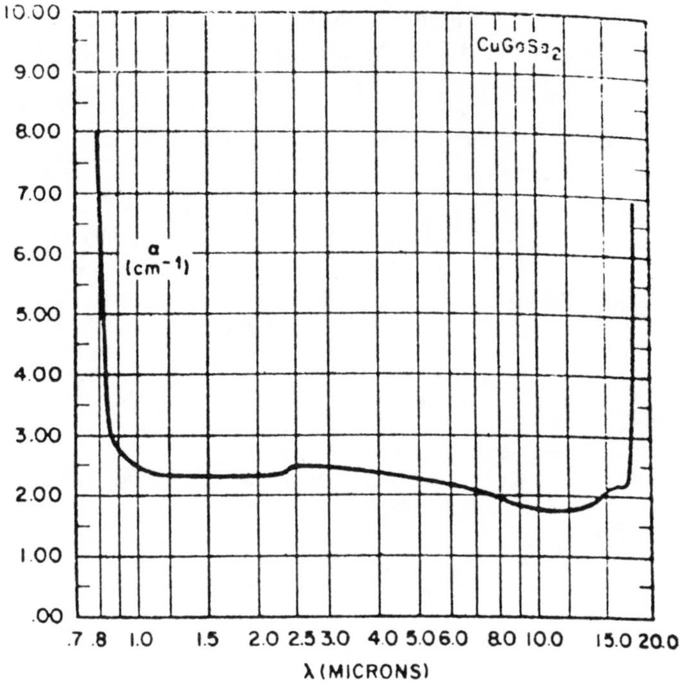

FIGURE 1.1.49. Room-temperature absorption coefficient $\alpha(cm^{-1})$ vs. wavelength (λ) for $CuGaSe_2$, determined from a sample of 0.341 cm thickness. Minimum value of 1.7 cm^{-1} at λ = 11.0 μm is not characteristic. (From Boyd, G. D., Kasper, H. M., McFee, J. H., and Storz, F. G., *IEEE J. Quantum Electron.*, QE-8, 900, 1972. With permission.)

Table 17.89
CESIUM DIHYDROGEN ARSENATE (CsH_2AsO_4)[258] REFRACTIVE INDEX AT ROOM TEMPERATURE

Wavelength (μm)	n^o	n^e
0.3472	1.6027	1.5722
0.5321	1.5733	1.5514
0.6943	1.5632	1.5429
1.0642	1.5516	1.5330

$d(n^e_{2\omega} - n^o_{\omega})/dT = (8.0 \pm 0.2) \times 10^{-6}$°$C^{-1}$ θ_m = 84° 9′ at λ_1 = 1.064 μm; T_m = 48°C at λ_1 = 1.064 μm.

Dispersion equation:[378]

$$n^{o2} - 1 = 1.40840\ \lambda^2/(\lambda^2 - 0.01299)$$

$$n^{e2} - 1 = 1.34731\ \lambda^2/(\lambda^2 - 0.01185)$$

where λ is in μm.

Table 1.1.17.90
CESIUM DIDEUTERIUM ARSENATE (CsD_2AsO_4)[258] REFRACTIVE INDEX AT ROOM TEMPERATURE

Wavelength (μm)	n^o	n^e
0.3472	1.5895	1.5685
0.5321	1.5681	1.5495
0.6943	1.5596	1.5418
1.0642	1.5503	1.5326

$d(n^3_{2\omega} - n^o_{\omega})/dT = (7.8 \pm 0.2) \times 10^{-6}$°$C^{-1}$. θ_m = 79° 21′ at λ_1 = 1.064 μm; T_m = 112.3°C at λ_1 = 1.064 μm.

Dispersion equation:[378]

$$n^{o2} - 1 = 1.39961\ \lambda^2/(\lambda^2 - 0.01156)$$

$$n^{e2} - 1 = 1.34417\ \lambda^2/(\lambda^2 - 0.01155)$$

where λ is in μm.

Table 1.1.17.91
COPPER GALLIUM SELENIDE ($CuGaSe_2$)[259] REFRACTIVE INDEX AT ROOM TEMPERATURE

Wavelength (μm)	n^o	n^e
0.780	2.9580	3.0093
0.800	2.9365	2.9759
0.850	2.8984	2.9197
0.900	2.8716	2.8925
0.950	2.8513	2.8690
1.000	2.8358	2.8513
1.100	2.8115	2.8245
1.200	2.7951	2.8066
1.300	2.7823	2.7928
1.400	2.7725	2.7825
1.600	2.7587	2.7677
1.800	2.7496	2.7579
2.000	2.7430	2.7510
2.200	2.7377	2.7456
2.400	2.7344	2.7419
2.600	2.7315	2.7388
2.800	2.7293	2.7363
3.000	2.7273	2.7344
3.200	2.7257	2.7328
3.400	2.7242	2.7311
3.600	2.7232	2.7300
3.800	2.7220	2.7287
4.000	2.7211	2.7276
4.500	2.7188	2.7252
5.000	2.7170	2.7232
5.500	2.7152	2.7212
6.000	2.7133	2.7192
6.500	2.7116	2.7174
7.000	2.7101	2.7158
7.500	2.7082	2.7136
8.000	2.7060	2.7111
8.500	2.7042	2.7089
9.000	2.7021	2.7065
10.000	2.6974	2.7014
11.000	2.6926	2.6961
12.000	2.6872	2.6898

Note: Coherence length measured for $d_{eff} = \left[\frac{2}{\sqrt{3}}\right] [2d_{44} + d_{36}]/3$; $l_c = 98.6$ μm.

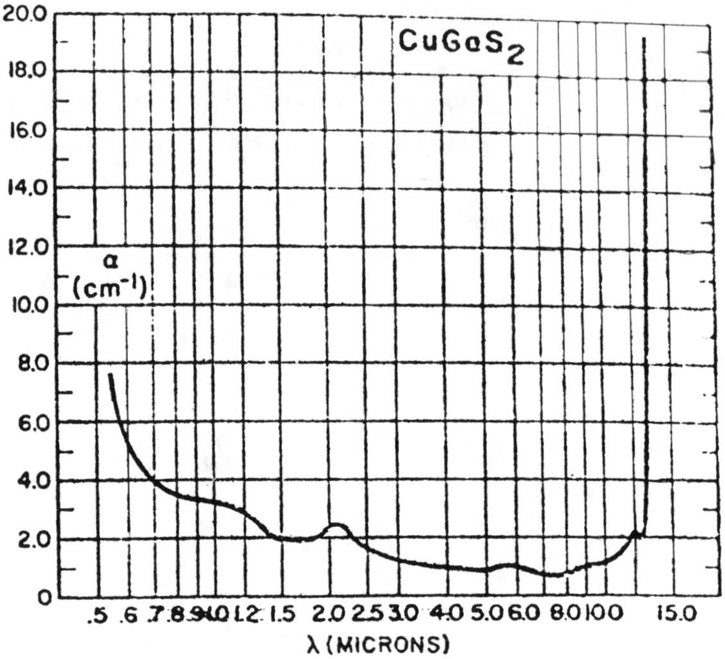

FIGURE 1.1.50. Room temperature absorption coefficient vs. wavelength for $CuGaS_2$. (From Boyd, G. D., Kasper, H., and McFee, J. H., *IEEE J. Quantum Electron.*, 7, 563, 1971. With permission.)

Table 1.1.17.92
COPPER GALLIUM SULFIDE ($CuGaS_2$)[89]

Wavelength (μm)	n^o	n^e	$n^e - n^o$	$10^5 \times dn^o/dT$	$10^5 \times dn^e/dT$
2.7630	2.7813	0.0183	12.95	17.32	4.37
2.7091	2.7141	0.0050			
2.6983	2.7019	0.0036	9.72	10.57	0.84
2.6890	2.6914	0.0024			
2.6804	2.6818	0.0014			
2.6724	2.6731	0.0007			
2.6649	2.6649	−0.0001			
2.6577	2.6570	−0.0007	8.08	8.61	0.53
2.6293	2.6266	−0.0027	6.37	7.64	1.27
2.6100	2.6056	−0.0044	4.64	5.89	1.26
2.5925	2.5886	−0.0040	5.26	6.22	0.96
2.5786	2.5737	−0.0049	6.06	6.51	0.45
2.5681	2.5630	−0.0051	5.64	5.94	0.29
2.5591	2.5538	−0.0053	5.87	6.08	0.21
2.5517	2.5464	−0.0053	5.86	5.98	0.13
2.5406	2.5349	−0.0058	5.75	6.03	0.28
2.5322	2.5265	−0.0058	5.44	5.57	0.13
2.5259	2.5199	−0.0060	5.48	5.74	0.27
2.5209	2.5148	−0.0061	5.44	5.75	0.31
2.5135	2.5073	−0.0062	5.32	5.62	0.30
2.5091	2.5028	−0.0063	4.49	4.82	0.34
2.5051	2.4991	−0.0061	4.52	4.73	0.21
2.5019	2.4956	−0.0063	4.95	5.20	0.24
2.4993	2.4931	−0.0062	5.29	5.64	0.35
2.4975	2.4913	−0.0063	5.08	5.41	0.33

Table 1.1.17.92 (continued)
COPPER GALLIUM SULFIDE ($CuGaS_2$)[89]

Wavelength (μm)	n^o	n^e	$n^e - n^o$	$10^5 \times dn^o/dT$	$10^5 \times dn^e/dT$
2.4959	2.4894	−0.0065	5.15	5.73	0.58
2.4945	2.4880	−0.0065	4.99	5.44	0.44
2.4932	2.4871	−0.0061	5.16	5.35	0.19
2.4922	2.4853	−0.0069	5.09	5.66	0.57
2.4910	2.4841	−0.0069	5.63	6.00	0.37
2.4897	2.4831	−0.0066	5.33	5.56	0.22
2.4884	2.4816	−0.0068	5.76	6.29	0.53
2.4869	2.4801	−0.0068	4.53	4.99	0.46
2.4843	2.4772	−0.0071	4.39	5.12	0.73
2.4803	2.4728	−0.0075	5.68	6.12	0.44
2.4774	2.4694	−0.0080	5.62	6.08	0.46
2.4744	2.4657	−0.0087	5.44	6.24	0.80
2.4715	2.4621	−0.0094	5.11	5.52	0.40
2.4674	2.4582	−0.0092	6.67	6.95	0.28
2.4639	2.4539	−0.0100	4.91	5.52	0.62
2.4589	2.4490	−0.0099	4.70	4.94	0.25
2.4539	2.4436	−0.0103	5.68	6.21	0.54
2.4491	2.4376	−0.0115	4.93	5.44	0.51
2.4429	2.4311	−0.0119	5.19	5.50	0.31
2.4372	2.4234	−0.0138	5.66	6.73	1.07
2.4311	2.4179	−0.0132	5.61	5.72	0.10
2.4275	2.4121	−0.0154	1.26	2.58	1.32
2.4171			4.81		
2.4094			4.80		
2.3999			3.49		

Note: l = 57.7 μm at λ_1 = 10.6 μm; for $d_{eff} = (2d_{14} + d_{36})/3$.

Sellmeier equation:[251]

$$n^2 = A + \frac{B}{1 - C/\lambda^2} + \frac{D}{1 - E/\lambda^2}$$

where λ is in μm.

Temp. (°C)	Polarization	A	B	C	D	E
20	o	3.9064	2.3065	0.1149	1.5479	738.43
	e	4.3165	1.8692	0.1364	1.7575	738.43
120	o	4.0984	2.1419	0.1225	1.5755	738.43
	e	4.4834	1.7316	0.1453	1.7785	738.43

Note: Temperature-dependent Sellmeier constants:[251] X = mT + c, where X = A,B,C, and D. Temperature range 20 to 120°C.

Sellmeier coefficient	Constant	n^o	n^e
A	$m \times 10^3$	1.919	1.668
	c	3.871	4.285
B	$m \times 10^3$	−1.645	−1.376
	c	2.337	1.894
C	$m \times 10^5$	7.607	8.928
	c	0.1134	0.1346
D	$m \times 10^4$	2.756	2.096
		1.542	1.753
E		738.43	738.43

Table 1.1.17.93 COPPER INDIUM SULFIDE ($CuInS_2$)[89] REFRACTIVE INDEX AT ROOM TEMPERATURE

Wavelength (μm)	n^o	n^e
0.90	2.7907	2.7713
0.92	2.7718	2.7536
0.94	2.7567	2.7393
0.96	2.7437	2.7268
0.98	2.7324	2.7162
1.00	2.7225	2.7067
1.10	2.6861	2.6727
1.20	2.6638	2.6510
1.30	2.6478	2.6357
1.40	2.6359	2.6243
1.50	2.6267	2.6156
1.60	2.6195	2.6087
1.80	2.6089	2.5985
2.00	2.6020	2.5918
2.20	2.5961	2.5860
2.40	2.5915	2.5821
2.60	2.5886	2.5789
2.80	2.5860	2.5765
3.00	2.5838	2.5741
3.50	2.5802	2.5707
4.00	2.5760	2.5663
4.50	2.5729	2.5630
5.00	2.5699	2.5598
5.50	2.5673	2.5571
6.00	2.5645	2.5539
7.00	2.5587	2.5474
8.00	2.5522	2.5401
9.00	2.5448	2.5311
10.00	2.5366	2.5225
11.00	2.5274	2.5112
12.00	2.5166	2.4987
12.50	2.5108	

Note: l_c = 70.3 μm at λ_1 = 10.6 μm; for $d_{eff} = (2d_{14} + d_{36})/3$.

Table 1.1.17.94 POTASSIUM DIHYDROGEN ARSENATE (KH_2AsO_4)[260] REFRACTIVE INDEX AT ROOM TEMPERATURE

Wavelength (μm)	n^o	n^e
0.4861	1.5762	1.5252
0.5460	1.5707	1.5206
0.5893	1.5674	1.5179
0.6563	1.5632	1.5146

Note: $\theta_m = 59°$ at $\lambda_1 = 0.6943$.

Dispersion equation:[378]

$$n^{o2} - 1 = 1.41427\ \lambda^2/(\lambda^2 - 0.00990)$$

$$n^{e2} - 1 = 1.25953\ \lambda^2/(\lambda^2 - 0.01108)$$

where λ is in μm.

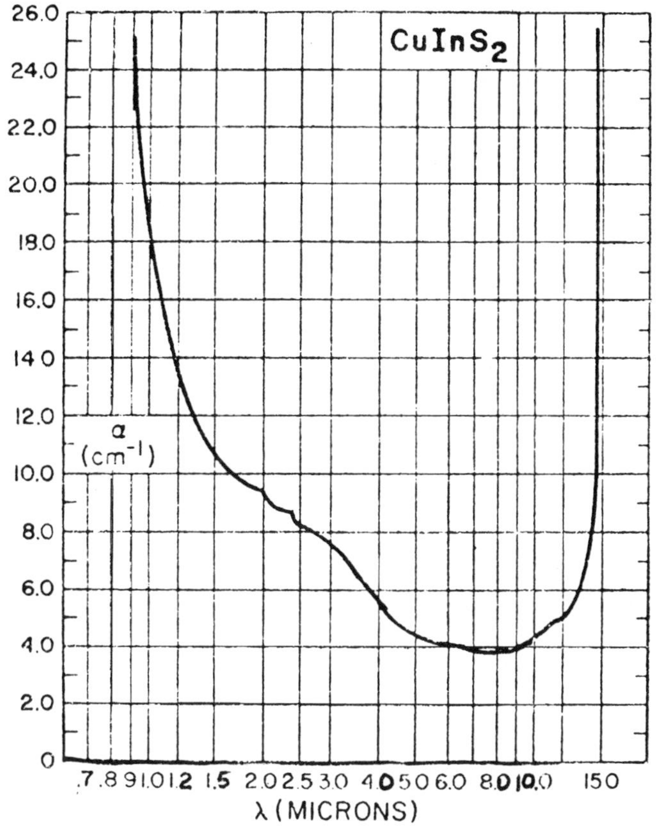

FIGURE 1.1.51. Room temperature absorption coefficient $\alpha(cm^{-1})$ vs. wavelength (λ) for $CuInS_2$. Sample thickness 0.053 cm. Minimum value of $\alpha = 4.0\ cm^{-1}$ at $\lambda = 7.5\ \mu m$ is somewhat unreliable because of the presence of cracks and voids. (From Boyd, G. D., Kasper, H., and McFee, J. H., *IEEE J. Quantum Electron.*, 7, 563, 1971. With permission.)

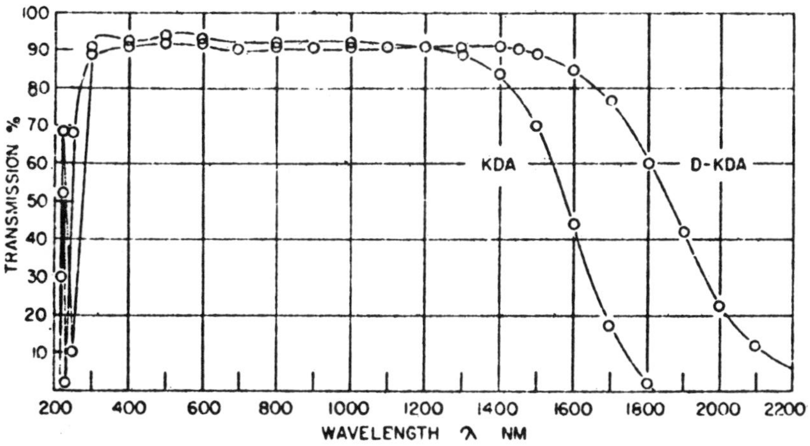

FIGURE 1.1.52. Transmission of KH_2AsO_4 (KDA) and KD_2AsO_4 (KD*A) single crystals. (From Adhav, R. S., *Br. J. Appl. Phys. (J. Phys. D)*, 2, 1969. With permission.)

Table 1.1.17.95
POTASSIUM DIDEUTERIUM PHOSPHATE (KD*P)(KD_2PO_4)[249]
REFRACTIVE INDEX AT ROOM TEMPERATURE

Wavelength (μm)	n^o	n^e
0.4047	1.5189	1.4776
0.4078	1.5185	1.4772
0.4358	1.5155	1.4747
0.4916	1.5111	1.4710
0.5461	1.5079	1.4683
0.5779	1.5063	1.4670
0.6234	1.5044	1.4656
0.6907	1.5022	1.4639
1.000	1.4700	1.4400

Note: Estimated accuracy = ±0.0001; The change in index of refraction with temperature is given by:

$$\Delta n_o = (n_o^2 - 1.047) \times (0.228 \times 10^{-4}) (298\text{-}T)$$
$$\Delta n_e = (n_e^2) \times (0.955 \times 10^{-5}) (298\text{-}T),$$

where T is in K. $\theta = 52°$ at $\lambda_1 = 0.6943$ μm;[247] $l_{36} = 20.6$ μm and $l_{14} = 7.7$ μm at $\lambda_1 = 0.6943$ μm;[226] $l = 21.2$ μm and $l_{14} = 15.8$ μm at $\lambda_1 = 1.0582$ μm.[226]

Sellmeier equation:[251]

$$n^2 = A + B/(1-C/\lambda^2) + D/(1-E/\lambda^2)$$

where λ is in μm.

Room temperature

	A	*B*	*C*	*D*	*E*
n^o	1.661824	0.585337	0.016017	0.691221	30.0
n^e	1.687522	0.447488	0.017039	0.596216	30.0

Dispersion equation:[378]

$$n^2 = A + B\lambda^2/(\lambda^2 - C)$$

	A	*B*	*C*
n^o	1.0	1.23318	0.00931
n^e	1.0	1.12354	0.00828

where λ is in μm.

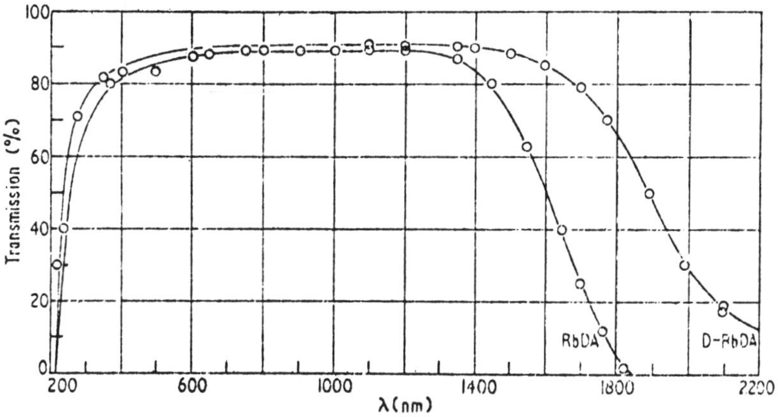

FIGURE 1.1.53. Transmission of RbH_2AsO_4 (RDA) and RbD_2AsO_4 (RD*A) single crystals. (From Adhav, R. S., *J. Appl. Phys.*, 39, 4095, 1968. With permission.)

Table 1.1.17.95(a) POTASSIUM DIDEUTERIUM PHOSPHATE (KD*P). KD_2PO_4.[251]

Sellmeier coefficient	Constant	n^o	n^e
A	$m \times 10^3$	0.33935	0.00343
	c	1.55934	1.68647
B	$m \times 10^3$	−0.41655	−0.06260
	c	0.71098	0.46629
C	$m \times 10^5$	0.64904	0.13626
	c	0.01407	0.01663
D	$m \times 10^4$	0.48281	0.00241
	c	0.67671	0.59614
E		30.0	30.0

Note: Temperature-dependent Sellmeier coefficients (x) of KD*P: $x = mT + c$, where T is in K.

Table 1.1.17.96 POTASSIUM DIPHTHALATE ($KHC_8H_4O_4$)[210]

Crystals transparent from 0.3 to 1.7 μm except for a narrow intense absorption band for all three polarizations at 1.14 μm.[210]

Phase matchable	Type I and II
θ_m = 21°, 25°	for λ_1 = 1.15 μm
= 46°	for λ_1 = 0.63 μm

The relation between the crystallographic and principal axes not given.

Table 1.1.17.97 RUBIDIUM DIHYDROGEN ARSENATE (RDA) (RbH_2AsO_4)

	Ref.
θ_m = 80° at 300 K for λ_1 = 0.6943 μm	247
ℓ_{36} = 158 μm at 0.6943 μm	247
θ_m = 90° at 97.4°C for λ_1 = 0.6943 μm	264
θ_m = 90° at 97°C for λ_1 = 0.6943 μm	263
θ_m = 80°, 19' at 20°C at λ_1 = 0.6943 μm	262
$d(n^e_{2\omega} - n^o_{\omega})/dT = (9.30 \pm 0.4) \times 10^{-6}$°C^{-1}	262

Dispersion equation:[378]

$$n^{o2} - 1 = 1.37723\,\lambda^2/(\lambda^2 - 0.01301)$$
$$n^{e2} - 1 = 1.27283\,\lambda^2/(\lambda^2 - 0.01157)$$

where λ is in μm.

Table 1.1.17.98
RUBIDIUM DIHYDROGEN PHOSPHATE (RDP) (RbH_2PO_4)[266] REFRACTIVE INDEX AT ROOM TEMPERATURE

Wavelength (μm)	n^o	n^e
0.4765	1.514	1.4861
0.4880	1.5132	1.4832
0.4965	1.5126	1.4827
0.5017	1.5121	1.4825
0.5145	1.5116	1.4820
0.5321	1.5106	1.4811
0.6328	1.4976	1.4775
1.0642	1.4926	1.4700

Dispersion equation:

$$n^{o2} - 1 = 1.2068\ \lambda^2/(\lambda^2 - 0.01539)$$
$$n^{e2} - 1 = 1.15123/(\lambda^2 - 0.010048)$$

1.1.17.98a
RDP ROOM TEMPERATURE[267]

Wavelength (μm)	n^o	n^e
0.3472	1.5284	1.4969
0.4358	1.5165	1.4857
0.5468	1.5082	1.4790
0.5893	1.5053	1.4765
0.6943	1.5020	1.4735

	Ref.
ℓ_{36} = 34 μm at λ_1 = 0.6943	247
θ_m = 66° at λ_1 = 0.6943	267
θ_m = 67° at λ_1 = 0.6943	264

Dispersion equation:[378]

$$n^{o2} - 1 = 1.23639\ \lambda^2/(\lambda^2 - 0.01059)$$
$$n^{e2} - 1 = 1.15567\ \lambda^2/(\lambda^2 - 0.00962)$$

Table 1.1.17.99
SILVER GALLIUM SELENIDE ($AgGaSe_2$)[259]

μ	n^o	n^e
0.7250	2.8452	2.8932
0.7500	2.8191	2.8415
0.8000	2.7849	2.7866
0.8500	2.7598	2.7522
9.0000	2.7406	2.7275
0.9500	2.7252	2.7085
1.0000	2.7132	2.6934
1.1000	2.6942	2.6712
1.2000	2.6806	2.6554
1.3000	2.6705	2.6438
1.4000	2.6624	2.6347
1.6000	2.6516	2.6224
1.8000	2.6432	2.6131
2.0000	2.6376	2.6071
2.2000	2.6336	2.6027
2.4000	2.6304	2.5992
2.6000	2.6286	2.5968
2.8000	2.6261	2.5943
3.0000	2.6245	2.5925
3.2000	2.6231	2.5912
3.4000	2.6221	2.5899
3.6000	2.6213	2.5889
3.8000	2.6200	2.5876
4.0000	2.6189	2.5863
4.5000	2.6166	2.5840
5.0000	2.6144	2.5819
5.5000	2.6128	2.5800
6.0000	2.6113	2.5784
6.5000	2.6094	2.5765
7.0000	2.6070	2.5743
7.5000	2.6049	2.5723
8.0000	2.6032	2.5704
8.5000	2.6009	2.5681
9.0000	2.5988	2.5659
9.5000	2.5964	2.5635
10.0000	2.5939	2.5608
10.5000	2.5917	2.5585
11.0000	2.5890	2.5555
11.5000	2.5868	2.5536
12.0000	2.5837	2.5505
12.5000	2.5805	2.5473
13.0000	2.5771	2.5439

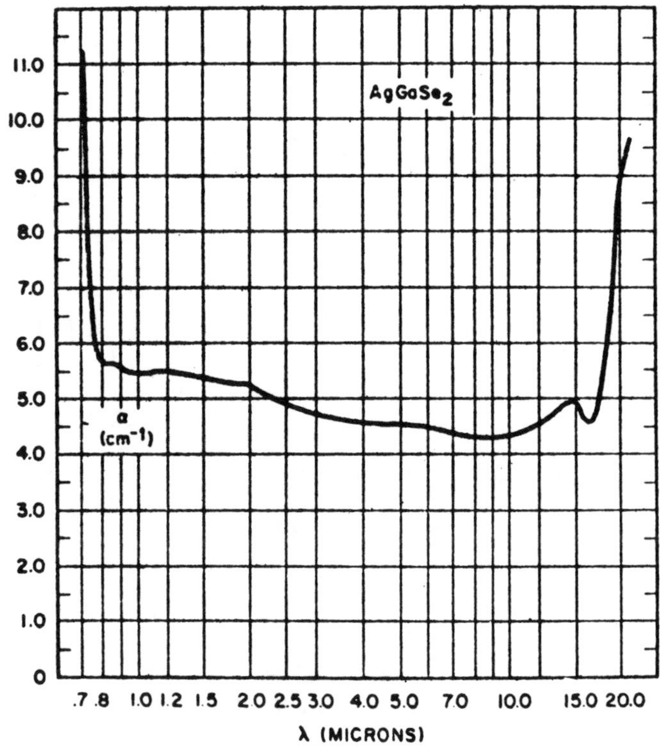

FIGURE 1.1.54. Room temperature absorption coefficient $\alpha(cm^{-1})$ vs. wavelength (λ) for $AgGaSe_2$, as determined from a sample of 0.203 cm thickness. (From Boyd, G. D., Kasper, H. M., McFee, J. H., and Storz, F. G., *IEEE J. Quantum Electron.*, QE-8, 900, 1972. With permission.)

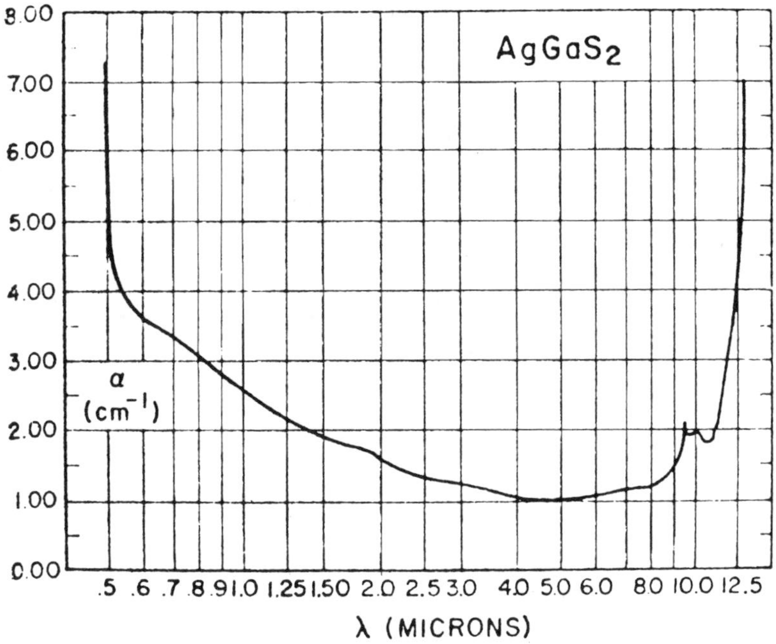

FIGURE 1.1.55. Room-temperature absorption coefficient $\alpha(cm^{-1})$ vs. wavelength (λ) for $AgGaS_2$. Sample thickness 0.226 cm. (From Boyd, G. D., Kasper, H., and McFee, J. H., *IEEE J. Quantum Electron.*, 7, 563, 1971. With permission.)

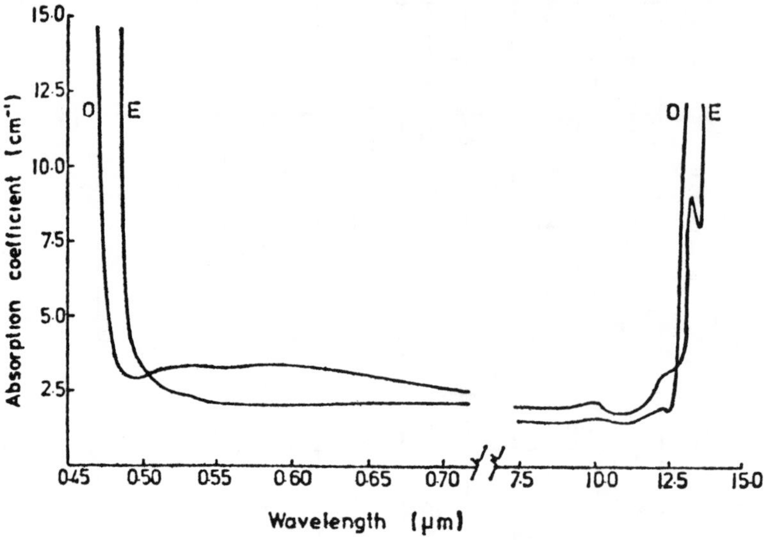

FIGURE 1.1.56. Absorption coefficient for two polarizations in $AgGaS_2$. (From Bhar, G. C. and Smith, R. C., *IEEE J. Quantum Electron.*, QE-10, 546, 1974. With permission.)

Table 1.1.17.99a
SILVER GALLIUM SELENIDE ($AgGaSe_2$)[268]

Sellmeier equation:

$$n^2 = A + \frac{B}{[1 - (D/\lambda)^2]} + \frac{C}{[1 - (E/\lambda)^2}$$

where λ is in μm.

Room temperature

	A	*B*	*C*	*D*(μm)	*E*(μm)
n^o	3.9362	2.9113	1.7954	0.38821	40
n^e	3.3132	3.3616	1.7677	0.38201	40

	Ref.
θ_m = 55.9° at λ_1 = 10.6 μm	268
θ_m = 42.7° for sum-mixing of λ_1 = 10.6 μm λ_2 = 5.33 μm and λ_3 = 3.55 μm	268
θ_m = 61.0° for difference frequency mixing of 10.6, 5.67, and 12.2 μm	268
l_c = 117.3 μm λ_1 = 10.6 μm. For $d_{eff} = (2d_{14} + d_{36})/3.E_\omega$	259
θ_m = 57.5° ± 0.5° for λ_1 = 10.6 μm	269
l_{36} = 237 ± 15 μm at λ_1 = 10.6 μm	268
l_{36} = 12.1 ± 0.7 μm at λ_1 = 2.12 μm	160

Table 1.1.17.99b[257]
SILVER GALLIUM SELENIDE
($AgGaSe_2$)[257]

Sellmeier equation

$$n^2 = A + B/(1 - C/\lambda^2) + D/(1 - E/\lambda^2)$$

where λ is in μm.

	A	*B*	*C*	*D*	*E*
n^o	4.6453	2.2057	0.1879	1.8377	1600
n^e	5.2912	1.3970	0.2845	1.9282	1600

Table 1.1.17.100
SILVER GALLIUM SULFIDE
($AgGaS_2$)[89]

Wavelength (μm)	n^o	n^e
0.4900	2.7148	2.7287
0.5000	2.6916	2.6867
0.5250	2.6303	2.6239
0.5500	2.6190	2.5834
0.5750	2.5944	2.5537
0.6000	2.5748	2.5303
0.6250	2.5577	2.5116
0.6500	2.5437	2.4961
0.6750	2.5310	2.4824
0.7000	2.5205	2.4706
0.7500	2.5049	2.4540
0.8000	2.4909	2.4395
0.8500	2.4802	2.4279
0.9000	2.4716	2.4192
0.9500	2.4644	2.4118
1.0000	2.4582	2.4053
1.1000	2.4486	2.3954
1.2000	2.4414	2.3881
1.3000	2.4359	2.3819
1.4000	2.4315	2.3781
1.5000	2.4280	2.3745
1.6000	2.4252	2.3716
1.8000	2.4206	2.3670
2.0000	2.4164	2.3637
2.2000	2.4142	2.3604
2.4000	2.4119	2.3583
2.6000	2.4102	2.3567
2.8000	3.4094	2.3559
3.0000	2.4080	2.3545
3.2000	2.4068	2.3534
3.4000	2.4062	2.3522
3.6000	2.4046	2.3511
3.8000	2.4024	2.3491
4.0000	2.4024	2.3488
4.5000	2.4003	2.3461
5.0000	2.3955	2.3419
5.5000	2.3938	2.3401
6.0000	2.3908	2.3369
6.3000	2.3874	2.3334
7.0000	2.3827	2.3291
7.5000	2.3787	2.3252
8.0000	2.3757	2.3219
8.5000	2.3699	2.3163
9.0000	2.3663	2.3121
9.5000	2.3606	2.3064
10.0000	2.3548	2.3012
10.5000	2.3486	2.2948
11.0000	2.3417	2.2880
11.5000	2.3329	2.2789
12.0000	2.3266	2.2716
12.5000	2.3177	
13.0000	2.3076	

Table 1.1.17.100a
SILVER THIOGALLATE ($AgGaS_2$)[270]

Dispersion equations:

$$n^{o^2} = 5.728 + \frac{0.2410}{\lambda^2 0.0870} - 0.00210\ \lambda^2$$

$$n^{e^2} = 5.497 + \frac{0.2026}{\lambda^2 - 0.1307} - 0.00233\ \lambda^2$$

where λ is in μm.

	Ref.
$\theta_m = 57.5° \pm 2°$ at $\lambda_1 = 10.6\ \mu m$	270,271
$l_{36} = 1.68 \pm 0.02\ \mu m$ at $\lambda_1 = 10.6\ \mu m$	271
$l_{36} = 365\ \mu m \pm 20\ \mu m$ at $\lambda_1 = 10.6\ \mu m$	271
$\theta_m = 70.8° \pm 0.5°$ at $10.6\ \mu m$	89

Sellmeier equation:[272,257]

$$n^2 = A + B/(1 - C/\lambda^2) + D/(1 - E/\lambda^2)$$

where λ is in μm.

Wavelength range (μm)	Polarization	*A*	*B*	*C*	*D*	*E*
0.485—0.700	o	4.6187	1.3758	0.1205	199.28	950
	e	5.5375	0.5685	0.1725	636.85	950
0.7—13.0	o	2.6149	3.1769	0.0739	2.1328	950
	e	3.0398	2.4973	0.0912	2.1040	950
0.49—12.0	o	3.6280	2.1686	0.1003	2.1753	950
	e	4.0172	1.5274	0.1310	2.1699	950

Table 1.1.17.101
SILVER INDIUM SELENIDE ($AgInSe_2$)[259] REFRACTIVE INDEX AT ROOM TEMPERATURE

λ (μm)	n^o	n^e
1.0500	2.8265	
1.1000	2.7971	2.8453
1.2000	2.7608	2.7906
1.3000	2.7376	2.7595
1.4000	2.7211	2.7385
1.6000	2.6992	2.7116
1.8000	2.6856	2.6949
2.0000	2.6761	2.6838
2.2000	2.6694	2.6758
2.4000	2.6643	2.6702
2.6000	2.6601	2.6658
2.8000	2.6568	2.6623
3.0000	2.6542	2.6592
3.2000	2.6521	2.6570
3.4000	2.6503	2.6551
3.6000	2.6488	2.6533
3.8000	2.6474	2.6516
4.0000	2.6463	2.6504
4.5000	2.6436	2.6475
5.0000	2.6416	2.6451
5.5000	2.6399	2.6431
6.0000	2.6381	2.6414
6.5000	2.6366	2.6398
7.0000	2.6352	2.6379
7.5000	2.6335	2.6361
8.0000	2.6318	2.6343
8.5000	2.6302	2.6327
9.0000	2.6286	2.6310
9.5000	2.6266	2.6290
10.0000	2.6251	2.6274
10.5000	2.6233	2.6254
11.0000	2.6210	2.6229
11.5000	2.6187	2.6204
12.0000	2.6167	2.6183

Note: l_c = 122.3 μm at λ_1 = 10.6 μm for $d_{eff} = (2/\sqrt{3})(2d_{14} + d_{36})/3$.

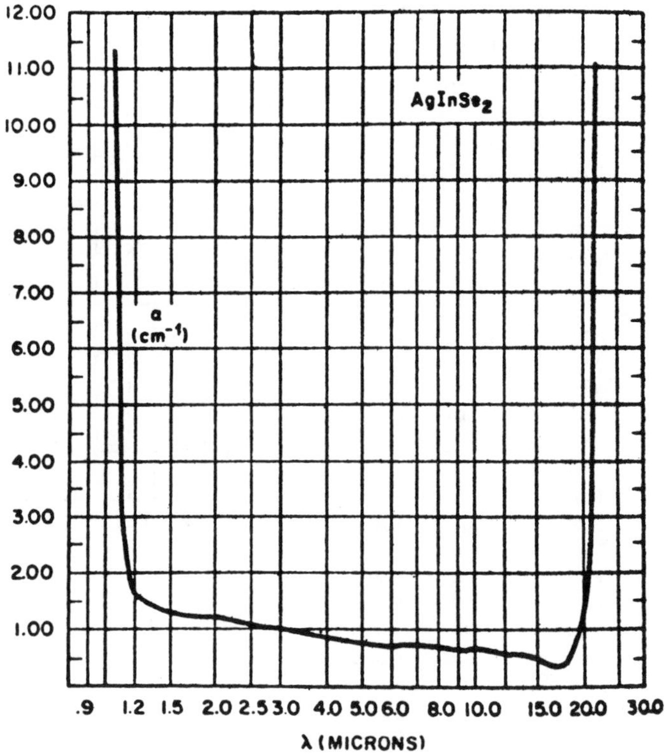

FIGURE 1.1.57. Room-temperature absorption coefficient $\alpha(cm^{-1})$ vs. wavelength (λ) for phosphorus-doped $AgInSe_2$, as determined from a sample of 0.105 cm thickness. (From Boyd, G. D., Kasper, H., McFee, J. H., and Storz, F. G., *IEEE J. Quantum Electron.*, QE-8, 900, 1972. With permission.)

Table 1.1.17.102
ZINC SILICON ARSENIDE ($ZnSiAs_2$)[256]
REFRACTIVE INDEX AT ROOM TEMPERATURE

Wavelength (μm)	n^o	n^e
0.70	3.5579	3.6201
0.75	3.5002	3.5539
0.80	3.4570	3.5050
0.85	3.4210	3.4655
0.90	3.3946	3.4362
0.95	3.3722	3.4116
1.00	3.3551	3.3928
1.10	3.3266	3.3618
1.20	3.3061	3.3394
1.30	3.2901	3.3221
1.40	3.2782	3.3093
1.60	3.2593	3.2889
1.80	3.2485	3.2771
2.00	3.2405	3.2692
2.20	3.2346	3.2620
2.40	3.2296	3.2572
2.60	3.2268	3.2539
2.80	3.2233	3.2506
3.00	3.2210	3.2481
3.20	3.2197	3.2464
3.40	3.2178	3.2447
3.60	3.2162	3.2426
3.80	3.2146	3.2413
4.00	3.2133	3.2402
4.50	3.2106	3.2372
5.00	3.2081	3.2345
5.50	3.2053	3.2317
6.00	3.2025	3.2287
6.50	3.2002	3.2263
7.00	3.1979	3.2241
7.50	3.1956	3.2220
8.00	3.1930	3.2195
8.50	3.1905	3.2168
9.00	3.1874	3.2138
9.50	3.1842	3.2106
10.00	3.1810	3.2074
10.50	3.1772	3.2037
11.00	3.1733	3.1996
11.50	3.1685	3.1953
12.00	3.1626	

Note: l_{36} = 47.2 + 1 μm at λ_1 = 10.6 μm.

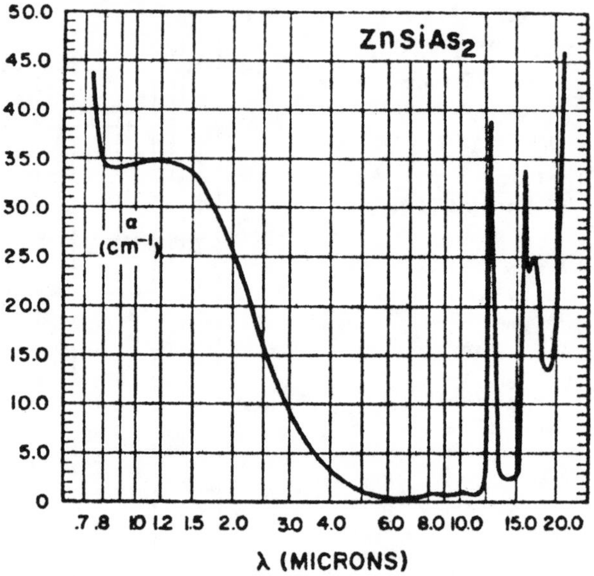

FIGURE 1.1.58. Absorption coefficient of $ZnSiAs_2$. (From Boyd, G. D., Buehler, E., Storz, F. G., and Wernick, J. H., *IEEE J. Quantum Electron.*, QE-8, 419, 1972. With permission.)

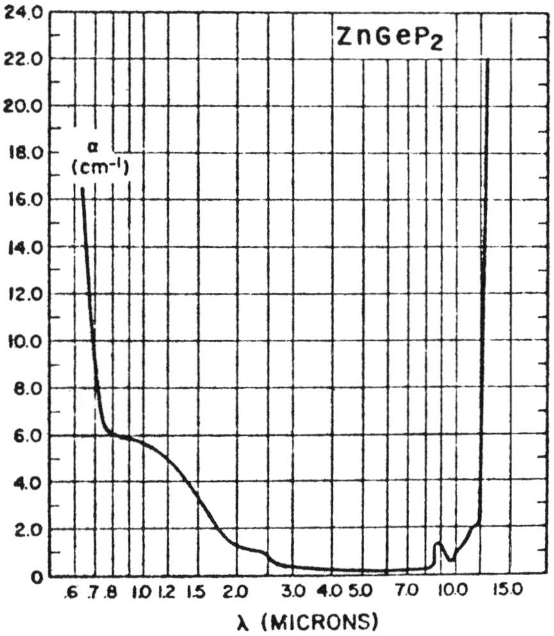

FIGURE 1.1.59. Absorption coefficient $\alpha(\text{cm}^{-1})$ vs. wavelength $\lambda(\mu\text{m})$ in $ZnGeP_2$. (From Boyd, G. D., Buehler, E., Storz, F. G., and Wernick, J. H., *IEEE J. Quantum Electron.*, QE-8, 419, 1972. With permission.)

Table 1.1.17.103
ZINC GERMANIUM PHOSPHIDE ($ZnGeP_2$)[251]

Sellmeier equation:

$$n^2 = A + \frac{B}{1 - C/\lambda^2} + \frac{D}{1 - E/\lambda^2}$$

where λ is in μm.

Temperature (°C)	Polarization					
20	o	4.4733	5.2658	0.1338	1.4909	662.55
	e	4.6332	5.3422	0.1426	1.4580	662.55
70	o	4.4492	5.3338	0.1353	1.4736	662.55
	e	4.5717	5.4513	0.1435	1.4262	662.55

Temperature-dependent Sellmeier coefficients: X = m*T* + c, where X = *A*,*B*,*C*, and *D*. *T* can vary from 20 to 70°C.

Coefficients	Constants	n^o	n^e
A	$m \times 10^3$	−0.4812	−1.230
	c	4.483	4.658
B	$m \times 10^3$	1.361	2.184
	c	5.238	5.298
C	$m \times 10^5$	2.994	1.855
	c	0.1332	0.1422
D	$m \times 10^4$	−3.459	−6.353
	c	1.498	1.471
E		662.55	662.55

Note: $l = 66.6 \pm 1$ μm at $\lambda_1 = 10.6$ μm for $d_{eff} = (2d_{14} + d_{36})/3$.

Table 1.1.17.103a
ZINC GERMANIUM PHOSPHIDE ($ZnGeP_2$)[273]

λ (μm)	n^o	n^e	$n^e - n^o$	$10^5 \times dn^o/dT$	$10^5 \times dn^e/dT$	$10^5 \times dB/dT$
0.64	3.5052	3.5802	0.0750	35.94	37.58	3.54
0.66	3.4756	3.5467	0.0710	31.23	37.34	3.32
0.68	3.4477	3.5160	0.0684	29.52	32.53	3.13
0.70	3.4233	3.4885	0.0652	28.63	31.82	2.97
0.75	3.3730	3.4324	0.0595	26.22	28.26	2.66
0.80	3.3357	3.3915	0.0558	24.69	26.43	2.44
0.85	3.3063	3.3593	0.0530	24.12	25.39	2.27
0.90	3.2830	3.3336	0.0506	22.34	24.61	2.14
0.95	3.2638	3.3124	0.0486	21.32	24.26	2.04
1.00	3.2478	3.2954	0.0476	21.18	23.01	1.95
1.10	3.2232	3.2688	0.0456	20.11	22.08	1.82
1.20	3.2054	3.2493	0.0438	18.63	20.51	1.73
1.30	3.1924	3.2346	0.0423	16.84	20.12	1.66
1.40	3.1820	3.2244	0.0423	15.34	16.55	1.60
1.60	3.1666	3.2077	0.0411	15.10	16.75	1.53
1.80	3.1566	3.1965	0.0403	13.20	14.40	1.47
2.00	3.1490	3.1889	0.0399	14.19	15.29	1.44
2.20	3.1433	3.1829	0.0396	14.60	15.28	1.41
2.40	3.1388	3.1780	0.0391	14.14	15.49	1.39
2.60	3.1357	3.1745	0.0388	15.13	16.80	1.37
2.80	3.1327	3.1717	0.0390	15.48	16.05	1.36
3.00	3.1304	3.1693	0.0388	13.26	13.96	1.35
3.20	3.1284	3.1671	0.0386	14.94	16.28	1.34
3.40	3.1263	3.1647	0.0384	14.40	15.46	1.34
3.60	3.1257	3.1632	0.0376	15.58	16.29	1.33
3.80	3.1237	3.1616	0.0380	14.48	16.53	1.33
4.00	3.1223	3.1608	0.0386	14.26	15.02	1.33
4.20	3.1209	3.1595	0.0386	13.57	15.14	1.33
4.50	3.1186	3.1561	0.0374	15.31	16.60	1.32
4.70	3.1174	3.1549	0.0375	15.51	16.71	1.32
5.00	3.1149	3.1533	0.0383	15.05	16.43	1.32
5.50	3.1131	3.1518	0.0387	14.49	15.42	1.32
6.00	3.1101	3.1480	0.0379	14.58	16.30	1.32
6.50	3.1057	3.1445	0.0387	15.60	16.13	1.32
7.00	3.1040	3.1420	0.0380	12.85	15.01	1.33
7.50	3.0994	3.1378	0.0384	18.15	18.59	1.33
8.00	3.0961	3.1350	0.0389	16.10	17.43	1.33
8.50	3.0919	3.1311	0.0392	15.16	17.37	1.33
9.00	3.0880	3.1272	0.0392	15.56	17.50	1.34
9.50	3.0836	3.1231	0.0395	16.27	17.11	1.34
10.00	3.0788	3.1183	0.0396	16.53	18.41	1.34
10.50	3.0738	3.1137	0.0399	15.40	16.84	1.34
11.00	3.0689	3.1087	0.0398	15.25	16.34	1.34
11.50	3.0623	3.1008	0.0386	14.74	18.32	1.34
12.00	3.0552	3.0949	0.0397	14.24	16.59	1.34

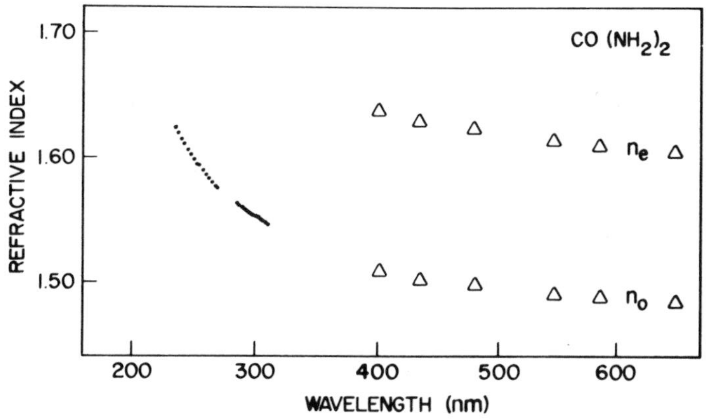

FIGURE 1.1.60. Refractive index vs. wavelength for urea at room temperature. (From Betzler, K., Hess, H., and Loose, P., *J. Mol. Struct.*, 47, 393, 1978. With permission.)

Table 1.1.17.104
UREA [$CO(NH_2)_2$][274]

Sellmeier equation:

$n^2 = A + B/(\lambda^2 - C)$

$A = 2.17$, $B = 0.000$, and $C = 0.028$, where λ is in μm.

Note: $\theta_m = 45°$ for $\lambda_1 = 0.6$ μm.

Table 1.1.17.105
SODIUM PERIODATE ($NaH[IO_3(OH)_3]\ H_2O$)[276] REFRACTIVE INDEX AT ROOM TEMPERATURE

Wavelength (μm)	n^o	n^e
0.53	1.640	1.676
1.06	1.614	1.644

Note: $l_{33} = 8.18$ μm at $\lambda_1 = 1.06$ μm; $l_{31} = 4.30$ μm at $\lambda_1 = 1.06$ μm; $d_{31}\ d_{33} > 0$.

Table 1.1.17.107
LEAD DITHIONATE (PbS_2O_6 — $4H_2O$)[279]

$l_{11} = 2.6$ μm at $\lambda_1 = 0.694$ μm
$l_{11} = 10.4$ μm at $\lambda_1 = 1.064$ μm

Table 1.1.17.106
ALUMINUM PHOSPHATE ($AlPO_4$)[310] REFRACTIVE INDEX AT ROOM TEMPERATURE

Wavelength (μm)	n^o	n^e
0.4	1.5369	1.5465
0.5	1.5287	1.5385
0.6	1.5243	1.5334
0.7	1.5215	1.5301
0.8	1.5192	1.5281
1.0	1.5161	1.5245
1.2	1.5136	1.5223
1.4	1.5112	1.5198
1.6	1.5088	1.5174
1.8	1.5062	1.5145
2.0	1.5034	1.5116
2.2	1.5001	1.5083
2.4	1.4969	1.5048
2.6	1.4928	1.5048

Table 1.1.17.108
MERCURY SULFIDE (α-HgS)[283] REFRACTIVE INDEX AT ROOM TEMPERATURE

Wavelength (μm)	n^o	n^e
0.62	2.9028	3.2560
0.65	2.8655	3.2064
0.68	2.8384	3.1703
0.70	2.8224	3.1489
0.80	2.7704	3.0743
0.90	2.7383	3.0340
1.00	2.7120	3.0050
1.20	2.6884	2.9680
1.40	2.6730	2.9475
1.60	2.6633	2.9344
1.80	2.6567	2.9258
2.00	2.6518	2.9194
2.20	2.6483	2.9146
2.40	2.6455	2.9108
2.60	2.6433	2.9079
2.80	2.6414	2.9052
3.00	2.6401	2.0936
3.20	2.6387	2.9017
3.40	2.6375	2.9001
3.60	2.6358	2.8987
3.80	2.6353	2.8971
4.00	2.6348	2.8963
5.00	2.6267	2.8863
6.00	2.6233	2.8799
7.00	2.6156	2.8741
8.00	2.6112	2.8674
9.00	2.6066	2.8608
10.00	2.6018	2.8522
11.00	2.5914	2.8434

Sellmeier equation:[282]

$$n^2 = A + B/(\lambda^2 - C) - D\lambda^2$$

where λ is in μm.

	A	*B*	*C*	*D*
n^o	6.9443	0.3665	0.1351	0.0019
n^e	8.3917	0.5405	0.1380	0.0027

Sellmeier equation:[257]

$$n^2 = A + B/(1 - C/\lambda^2) + D/(1 - E/\lambda^2)$$

where λ is in μm.

	A	*B*	*C*	*D*	*E*
n^o	4.1506	2.7896	0.1328	1.1378	705
n^e	4.0101	4.3736	0.1284	1.5604	705

Table 1.1.17.108a
MERCURY SULFIDE (α-HgS)[282] REFRACTIVE INDEX AT ROOM TEMPERATURE

Wavelength (μm)	Phase-matching angle θ^m
10.6	20.75° ± 0.75
10.3	20.20° ± 0.75
9.6	18.50° ± 0.75
9.3	17.80° ± 0.75
5.1	14.80° ± 0.75
5.62	14.60° ± 0.75
5.90	14.50° ± 0.75
1.32	54.60° ± 0.50

Note: l_{11} = 100 μm at λ_1 = 10.6 μm.[281]

Table 1.1.17.109
POTASSIUM DITHIONATE ($K_2S_2O_6$)[285] REFRACTIVE INDEX AT ROOM TEMPERATURE

Wavelength (μm)	n^o	n^e
0.313	1.480	1.568
0.334	1.475	1.557
0.365	1.470	1.546
0.405	1.465	1.537
0.436	1.463	1.530
0.546	1.456	1.518
0.578	1.455	1.516
1.014	1.448	1.503
1.367	1.446	1.498
1.709	1.444	1.498
2.93	1.436	1.489
3.39	1.430	1.485

Note: Transmission range 0.215 to 3.9 μm with absorption coefficient ≤ 1 cm^{-1}.

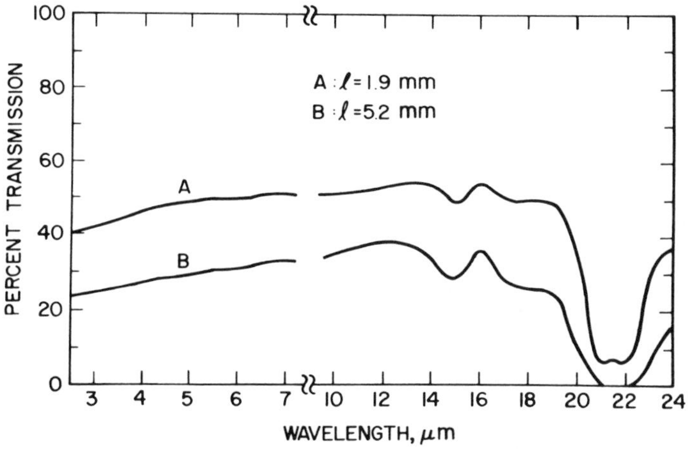

FIGURE 1.1.61. Transmission of trigonal selenium (parallel to the c axis) for two sample lengths. From these data, the absorption coefficients were 1.09 ± 0.02 and 1.40 ± 0.05 cm^{-1} for the wavelengths 10.6 and 5.3 μm, respectively. (From Sherman, G. H. and Coleman, P. D., *J. Appl. Phys.*, 44, 238, 1973. With permission.)

Table 1.1.17.110
SELENIUM (Se)[291]

Wavelength (μm)	n^o	n^e
1.06	2.790 ± 0.008	3.608 ± 0.008
1.15	2.737 ± 0.008	3.573 ± 0.008
3.39	2.65 ± 0.01	3.46 ± 0.01
10.6	2.64 ± 0.01	3.41 ± 0.01

	Ref.
$\theta_m = 10°$ at $\lambda_1 = 10.6$ μm	288
$\theta_m = 6.5°$ at $\lambda_1 = 10.6$ μm	289
$\theta_m = 5.5° \pm 0.3°$ at $\lambda_1 = 10.6$ μm	290

Note: Specific rotation at 10.6 μm = 2.5°/mm and at 3.39 μm = 4.8°/mm.[290] $\lambda_{11} \simeq 400$ μm at $\lambda_1 = 10.6$ μm.[289]

Table 1.1.17.111
TELLURIUM (Te)[297]
REFRACTIVE INDEX AT ROOM TEMPERATURE

Wavelength (μm)	n^o	n^e
8.5	4.801	6.260
8.8	4.799	6.258
9.3	4.798	6.255
9.7	4.795	6.252
10.2	4.793	6.249
10.6	4.792	6.247
10.8	4.791	6.246
11.4	4.789	6.243
12.0	4.785	6.240
12.8	4.781	6.235
13.7	4.776	6.231
14.0	4.775	6.230
14.7	4.772	6.227
15.9	4.767	6.222
17.2	4.761	6.216
18.9	4.753	6.210
20.8	4.744	6.203
23.4	4.734	6.196
26.3	4.722	6.188
28.0	4.716	6.183
30.3	4.706	6.180

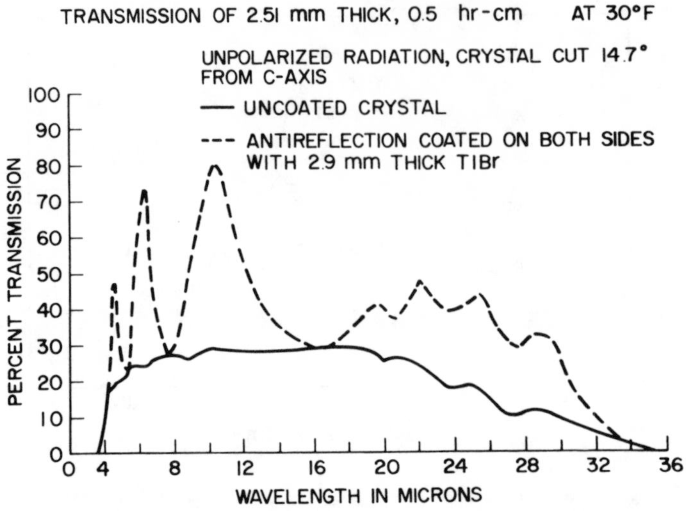

FIGURE 1.1.62. Transmission of tellurium. (From Sherman, G. H. and Coleman, P. D., *IEEE J. Quantum Electron.*, QE-9, 403, 1973. With permission.)

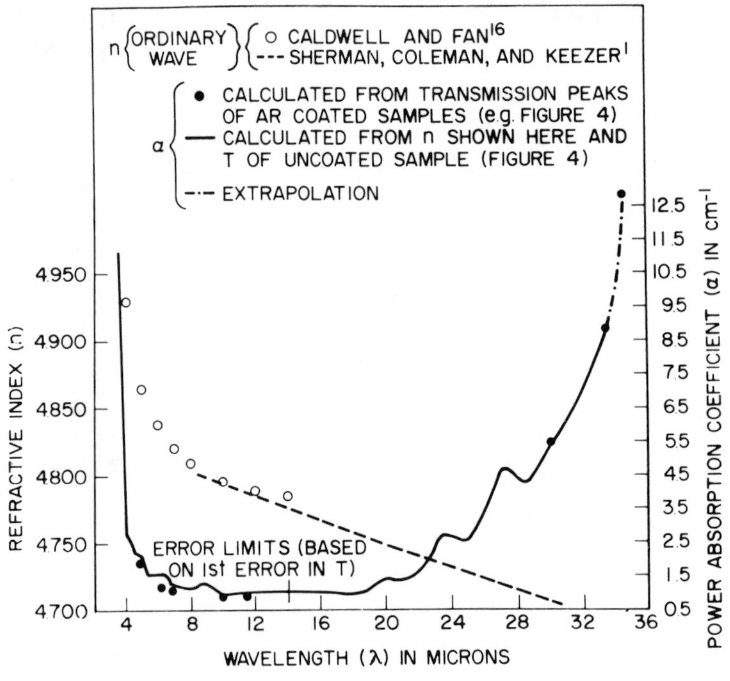

FIGURE 1.1.63. Refractive index and absorption coefficient of tellurium. (From Sherman, G. H. and Coleman, P. D., *IEEE J. Quantum Electron.*, QE-9, 403, 1973. With permission.)

Table 1.1.17.111a[301]
TELLURIUM (Te)

Wavelength (μm)	n^o	n^e
4.0	6.372	4.929
5.0	6.316	4.864
6.0	6.286	4.838
7.0	6.270	4.821
8.0	6.257	4.809
9.0	6.253	4.802
10.0	6.246	4.796
12.0	6.237	4.789
14.0	6.230	4.785

Table 1.1.17.111b
TELLURIUM (Te)[257]

Sellmeier equation:

$$n^2 = A + B/(1 - C/\lambda^2) + D/(1 - E/\lambda^2)$$

where λ is in μm.

Wavelength range (μm)	Polarization	*A*	*B*	*C*	*D*	*E*
4.0—14.0	o	18.5346	4.3289	3.9810	3.7800	11813
	e	29.5222	9.3068	2.5766	9.2350	13521
8.5—30	o	4.0164	18.8133	1.1572	7.3729	10000
	e	1.9041	36.8133	1.0803	6.2456	10000

$\theta_m = 14.17° \pm 0.5$ at $\lambda_1 - 10.6$ μm
$\theta_m = 12.19° \pm 0.5$ at $\lambda_1 = 23.4$ μm
$\theta_m = 13.33 \pm 0.5$ at $\lambda_1 = 26.6$ μm
$\theta_m = 14.07° \pm 0.5$ at $\lambda_1 = 28.0$ μm
} Ref. 300

$\theta_m = 14°4'$ at $\lambda_1 = 27.97$ μm
$\rho = 5°5'$ at $\lambda_1 = 27.97$ μm
} Ref.297

Table 1.1.17.112
GUANIDINIUM ALUMINUM SULFATE HEXAHYDRATE (GASH)
$[(CN_3H_6)As(SO_4)_2 6H_2O]$[302]

$l_{22} = 13.2$ μm
$l_{31} = 391$ μm
$l_{33} = 21.3$ μm
} for $\lambda_1 = 1.064$ μm

$\theta_m = 26.9°$
$n^o = 1.5423$, $n^e = 1.454$ at $\lambda = 0.589$ μm Ref.303
$\theta_m = 34.6°$ at $\lambda_1 = 0.6943$ μm

Table 1.1.17.113
LITHIUM NIOBATE ($LiNbO_3$)[308] (CONGRUENTLY MELTING COMPOSITION) = Li/Nb = 0.946

Wavelength (μm)	n^o	n^e
0.40463	2.4317	2.3260
0.43584	2.3928	2.2932
0.46782	2.3634	2.2683
0.47999	2.3541	2.2605
0.50858	2.3356	2.2448
0.54607	2.3165	2.2285
0.57696	2.3040	2.2178
0.57897	2.3032	2.2171
0.58756	2.3002	2.2147
0.64385	2.2835	2.2002
0.66782	2.2778	2.1953
0.70652	2.2699	2.1886
0.80926	2.2541	2.1749
0.87168	2.2471	2.1688
0.93564	2.2412	2.1639
0.95998	2.2393	2.1622
1.0140	2.2351	2.1584
1.09214	2.2304	2.1545
1.15392	2.2271	2.1517
1.15794	2.2269	2.1515
1.28770	2.2211	2.1464
1.43997	2.2151	2.1413
1.63821	2.2083	2.1356
1.91125	2.1994	2.1280
2.18428	2.1912	2.1211
2.39995	2.1840	2.1151
2.61504	2.1765	2.1087
2.73035	2.1724	2.1053
2.89733	2.1657	2.0999
3.05148	2.1594	2.0946

$n^o = 2.2272$ at $\lambda = 1.1523$ μm
$n^e = 2.2179$ at $\lambda = 0.5761$ μm
$\partial[n^o_{2\omega} - n^e_{2\omega}]/\partial T = -5.9 \times 10^{-5}\,°C^{-1}$

Note: Refractive indexes of other compositional ratios are given in Reference 309.

Table 1.1.17.113a
LITHIUM NIOBATE ($LiNbO_3$)[307] ABSOLUTE REFRACTIVE INDEXES FOR CONGRUENTLY MELTING $LiNbO_3$ at 20°C

Wavelength (μm)	n^o	n^e
0.43584	2.39276	2.29278
0.54608	2.31657	2.22816
0.63282	2.28647	2.20240
1.1523	2.2273	2.1515
3.3913	2.1451	2.0822

Sellmeier equation:[307]

$$n^{o^2} = 4.9048 - \frac{0.11768}{0.04750 - \lambda^2} - 0.027169\,\lambda^2$$

$$n^{e^o} = 4.5820 - \frac{0.099169}{0.044432 - \lambda^2} - 0.021950\,\lambda^2$$

where λ is in μm.

Table 1.1.17.113b
LITHIUM NIOBATE ($LiNbO_3$) TEMPERATURE-DEPENDENT REFRACTIVE INDEXES[307]

$n = n^o[1 + a_1T + a_2T^2 + a_3T^3 + a_4T^4]$

where n^o is the refractive index at 0°C.

Wavelength (μm)	Polarization	a_1	a_2	a_3	a_4
0.633	o	1.898802×10^{-6}	1.814362×10^{-8}	$-2.545860 \times 10^{-11}$	1.851642×10^{-14}
	e	1.714651×10^{-5}	5.330995×10^{-8}	$-7.023143 \times 10^{-11}$	5.590826×10^{-14}
3.39	o	1.302416×10^{-7}	1.031300×10^{-8}	$-1.316482 \times 10^{-11}$	8.880433×10^{-15}
	e	1.384239×10^{-5}	4.564947×10^{-8}	$-5.744349 \times 10^{-11}$	4.912069×10^{-14}

Note: Thermal expansion coefficients α and β in the a-b crystallographic plane are 1.54×10^{-5} and 5.3×10^{-9}, respectively.[307]

Table 1.1.17.113c
LITHIUM NIOBATE ($LiNbO_3$)[306]

Melt composition (Li/Nb)	λ_1 (μm)	ℓ_{31} (μm)	ℓ_{22} (μm)	ℓ_{33} (μm)
1.083	1.064	−49	2.85	3.56
0.946	1.064	144	2.91	3.49
0.852	1.064	49	2.89	3.51

$l_{31} = 14.27$ μm at $\lambda_1 = 1.318$ μm, $l_{33} = 6.42$ μm at $\lambda_1 = 1.318$ μm (Ref. 307)

$\theta_m = 68°$ at $\lambda_1 = 1.058$ μm.[261]

Table 1.1.17.114
LITHIUM TANTALATE ($LiTaO_3$)[310]
REFRACTIVE INDEX AT ROOM TEMPERATURE

Wavelength (μm)	n^o	n^e
0.45	2.2420	2.2468
0.50	2.2160	2.2205
0.60	2.1834	2.1878
0.70	2.1652	2.1696
0.80	2.1538	2.1578
0.90	2.1454	2.1493
1.00	2.1391	2.1432
1.20	2.1305	2.1341
1.40	2.1236	2.1273
1.60	2.1174	2.1213
1.80	2.1120	2.1170
2.00	2.1066	2.1115
2.20	2.1009	2.1053
2.40	2.0951	2.0993
2.60	2.0891	2.0936
2.80	2.0825	2.0871
3.00	2.0755	2.0799
3.20	2.0680	2.0727
3.40	2.0601	2.0649
3.60	2.0513	2.0561
3.80	2.0424	2.0473
4.00	2.0335	2.0377

Table 1.1.17.115
SILVER ANTIMONY SULFIDE (PYRARGYRITE) (Ag_3SbS_3)[311]

Sellimeier equations in the region 1.5 to 10.6 μm at room temperature:

$$n^{o^2} = 1 + \frac{6.585\lambda^2}{\lambda^2 - (0.4)^2} + \frac{0.1133\lambda^2}{\lambda^2 - (15)^2}$$

$$n^{e^2} = 1 + \frac{5.845\lambda^2}{\lambda^2 - (0.4)^2} + \frac{0.0202\lambda^2}{\lambda^2 - (15)^2}$$

with λ in μm.

$\theta_m = 29.0 \pm 1.0°$ at λ_1 at 10.6 μm
$\rho_o = 0.042$ radians
} Ref. 86

Table 1.1.17.116
SILVER ARSENIC SULFIDE (PROUSTILE) Ag_2AsS_3)[312]
ABSOLUTE REFRACTIVE INDEXES AT 20°C

Wavelength (μm)	n^o	n^e
0.5876	—	2.7896
0.6328	3.0190	2.7391
0.6678	2.9804	2.7094
1.014	2.8264	2.5901
1.129	2.8067	2.5756
1.367	2.7833	2.5570
1.530	2.7728	2.5485
1.709	2.7654	2.5423
2.50	2.7478	2.5282
3.56	2.7379	2.5213
4.62	2.7318	2.5178

Sellmeier equation:[312]

$$n^{o^2} = 7.483 + 0.474/(\lambda^2 - 0.09) - 0.0019\lambda^2$$
$$n^{e^2} = 6.346 + 0.342/(\lambda^2 - 0.09) - 0.0011\lambda^2$$

where λ is in μm.

Sellmeier equation:[315] for wavelength range 0.6 to 20 μm:

$$n^{o^2} = 9.220 + \frac{0.4454}{\lambda^2 - 0.1264} - \frac{1733}{1000 - \lambda^2}$$

$$n^{e^2} = 7.007 + \frac{0.3230}{\lambda^2 - 0.1192} - \frac{660}{1000 - \lambda^2}$$

where λ is in μm.

$\theta_m = 21.5 \pm 0.5°$ at $\lambda_1 = 10.6$ μm
$l_{22} = (79 \pm 1)$ μm at $\lambda_1 = 10.6$ μm
} Ref. 314

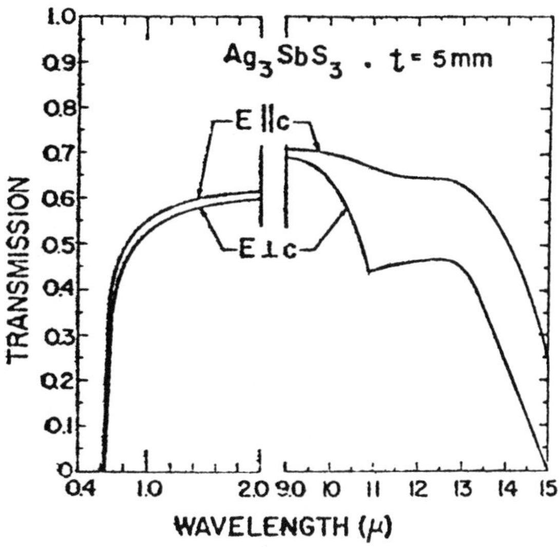

FIGURE 1.1.64. Transmission of a crystal of Ag_3SbS_3. (From Gandrud, W. B., Boyd, G. D., McFee, J. H., and Wehmeier, F. H., *Appl. Phys. Lett.*, 16, 59, 1970. With permission.)

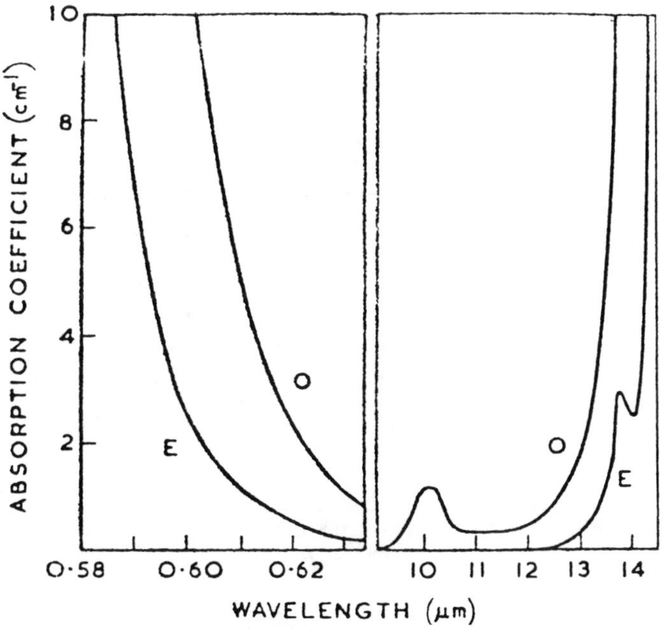

FIGURE 1.1.65. The absorption coefficients of proustile at 20°C. (Hulme, K. F., Jones, O., Davies, P. H., and Hobden, M. V., *Appl. Phys. Lett.*, 10, 133, 1967. With permission.)

Table 1.1.17.117
THALLIUM ARSENIC SELENIDE (Tl_3AsSe_3)[319]
REFRACTIVE INDEX AT ROOM TEMPERATURE

Wavelength (μm)	n^o	n^e
1.553	3.443	3.248
2.66	3.356	3.170
3.29	3.339	3.152
3.365	3.337	3.155
3.38	3.339	3.152
4.35	3.332	3.148
4.46	3.334	3.142
4.55	3.326	3.142
5.26	3.321	3.141
5.3	3.326	3.142

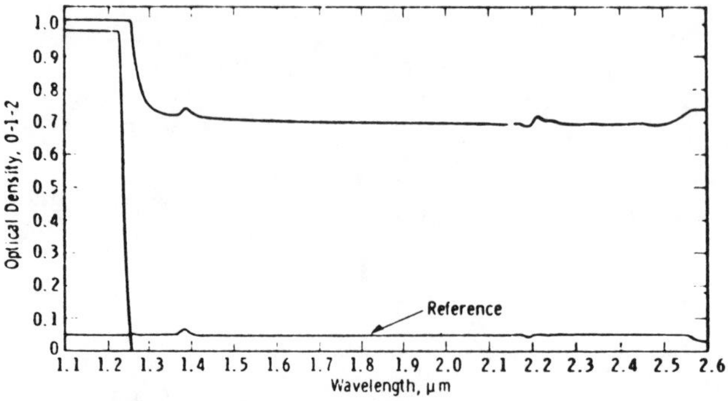

A

Sellmeier equation:[319]

$$n^{o^2} = 1 + (9.977)\frac{\lambda^2}{[\lambda^2 - (0.435)^2]} + (0.067)\frac{\lambda^2}{[\lambda^2 - (20)^2]}$$

$$n^{e^2} = 1 + (8.783)\frac{\lambda^2}{[\lambda^2 - (0.435)^2]} - (0.051)\frac{\lambda^2}{[\lambda^2 - (20)^2]}$$

where λ is in μm.

$\theta_m = 22 \pm 2°$ at $\lambda_1 = 10.6$ μm[319]

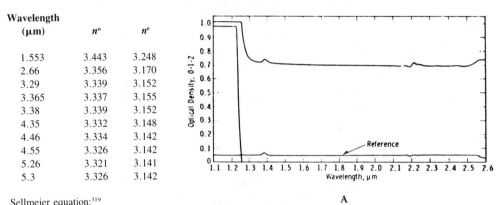

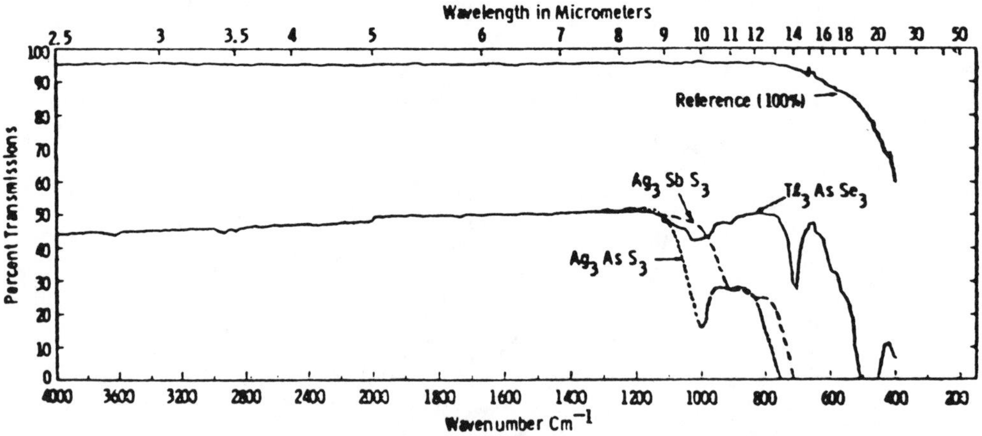

B

FIGURE 66. (A) Optical density of Tl_3AsSe_3 in the 1.1 to 2.6 μm region. Sample thickness of 0.545 cm. Curve not corrected for reflection losses. (B) Comparison of optical transmissions of AG_3AsS_3, $AgSbS_3$, and Tl_3AsSe_3.[319] (From Feichtner, J. D. and Roland, G. W., *Appl. Opt.*, 11, 993, 1972. With permission.)

Table 1.1.17.118
THALLIUM IODATE ($TlIO_3$)[319]
REFRACTIVE INDEX AT ROOM TEMPERATURE

Wavelength (μm)	n^o	n^e
0.532	2.04	1.87
1.064	1.96	1.81

Measured coherence lengths at λ_1 = 1.064 μm:[319] l_{33} = +4.40 μm, l_{31} = −3.21 μm, l_{15} = +1.88 μm, l_{32} = −3.08 μm, l_{24} = +1.8 μm, and l_{22} = +3.34 μm; d_{33} d_{31} > 0; θ_m = 40° at λ_1 = 1.064 μm

Table 1.1.17.119
TOURMALINE
[(Na,Ca)(Mg,Fe)$_3$B$_3$Al$_6$Si$_6$(O,OH,F)$_{31}$][317]

Wavelength (μm)	n^o	n^e
0.4765	1.6474	1.6273
0.4880	1.6465	1.6263
0.4965	1.6457	1.6255
0.5017	1.6454	1.6251
0.5145	1.6446	1.6248
0.5320	1.6433	1.6231
0.6328	1.6378	1.6183
1.0642	1.6274	1.6088

Dispersion equation:

$$n_o^2 - 1 = 1.6346\lambda^2/(\lambda^2 - 0.010734)$$
$$n_e^2 - 1 = 1.57256\lambda^2/(\lambda^2 - 0.011346)$$

where λ is in μm.

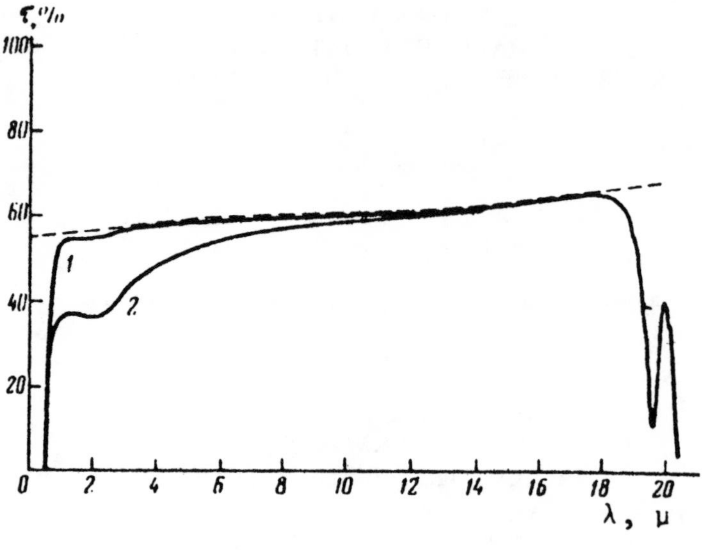

FIGURE 1.1.67. Transmission of GaSe crystal: dashed curve — reflection loss, continuous curve #1 – sample thickness 1.5 mm. Curve #2 for sample thickness = 6.0 mm. (From Abdullaev, G. B., Kulevskii, L. A., Prokhorov, A. M., Savel'ev, A. D., Salaev, E. Yu., and Smirnov, V. V., *JETP Lett.*, 16, 90, 1972. With permission.)

Table 1.1.17.120
GALLIUM SELENIDE (GaSe)[327]

Room temperature dispersion equations:

$$n^{o^2} = \frac{-0.05466}{\lambda^4} + \frac{0.48605}{\lambda^2} + 7.8902 - (0.000824)\lambda^2 - 0.00000273\lambda^4$$

$$n^{e^2} = 6.0476 + \frac{0.3423}{(\lambda^2 - 0.16491)} - 0.001042\,\lambda^2$$

$$\theta_m = 18^\circ\,40' \pm 10' \text{ at } \lambda_1 = 2.36\ \mu\text{m}$$
$$\theta_m = 10^\circ\,10' \pm 20' \text{ at } \lambda_1 = 5.3\ \mu\text{m}$$
$$\theta_m = 12^\circ\,40' \pm 20' \text{ at } \lambda_1 = 10.6\ \mu\text{m}$$

Note: The reference value of $d^{Ag_3AsS_3}$ in Reference 328 = 15.5×10^{-12} m/v.

Table 1.1.17.121
BERYLLIUM OXIDE (BeO)[331]
REFRACTIVE INDEX AT ROOM TEMPERATURE

Wavelength (μm)	n^o	n^e
0.430	1.73039	
0.440	1.72924	1.74556
0.450	1.72820	1.74447
0.460	1.72725	1.74348
0.470	1.72626	1.74251
0.480	1.72542	1.74162
0.490	1.72460	1.74073
0.500	1.72388	1.74002
0.510	1.72308	1.73918
0.520	1.72249	1.73852
0.530	1.72177	1.73779
0.540	1.72121	1.73703
0.550	1.72062	1.73644
0.560	1.72006	1.73588
0.570	1.71950	1.73530
0.580	1.71903	1.73477
0.590	1.71856	1.73423
0.600	1.71795	1.73381
0.610	1.71762	1.73322
0.620	1.71710	1.73279
0.630	1.71668	1.73233
0.640	1.71632	1.73191
0.650	1.71589	1.73156
0.660	1.71554	1.73113
0.670	1.71517	1.73075
0.680	1.71482	1.73041
0.690	1.71450	

Dispersion equation:

$$n_o^2 - 1 = 1.919087\lambda^2/(\lambda^2 - 0.00727575) + 3.972323\lambda^2/(\lambda^2 - 199.31087)$$
$$n_e^2 - 1 = 1.972142\lambda^2/(\lambda^2 - 0.00748564) + 17.5787\lambda^2/(\lambda^2 - 779.49122)$$

where λ is in μm.

$$dn^o/dT = +8.18 \times 10^{-6}/°C$$
$$dn^e/dT = +13.40 \times 10^{-6}/°C$$

$l_{31} = 8.19 \pm 0.03$ μm at $\lambda_1 = 1.064$ μm
$l_{33} = 15.22 \pm 0.06$ μm at $\lambda_1 = 1.064$ μm
} Reference 330

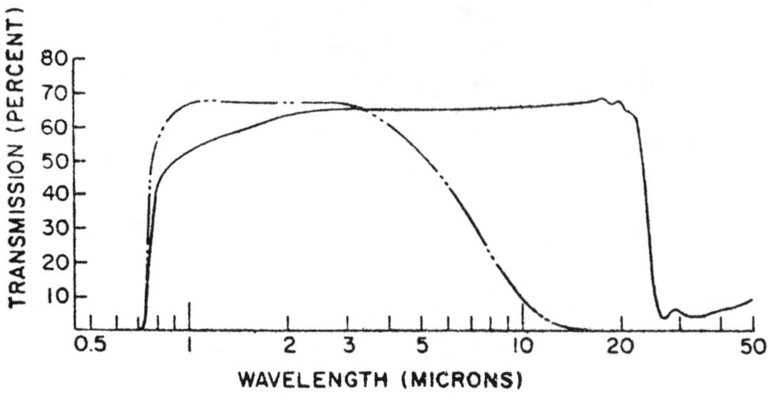

FIGURE 1.1.68. Transmittance of CdSe; dotted curve for as grown crystal, solid curve for selenium compensated crystals. (From Herbst, R. L. and Byer, R. L., *Appl. Phys. Lett.*, 19, 527, 1971. With permission.)

Table 1.1.17.122
CADMIUM SELENIDE (CdSe)[333] REFRACTIVE INDEX AT ROOM TEMPERATURE

Wavelength (μm)	n^o	n^e
1.0139	2.5481	2.5677
1.1287	2.5246	2.5444
1.3673	2.4971	2.5170
1.5295	2.4861	2.5059
1.7109	2.4776	2.4974
2.3253	2.4627	2.4823
3.00	2.4553	2.4748
4.00	2.4500	2.4694
5.00	2.4464	2.4657
6.00	2.4434	2.4625
7.00	2.4398	2.4586
8.00	2.4367	2.4552
9.00	2.4333	2.4514
10.00	2.4294	2.4475
11.00	2.4252	2.4430
12.00	2.4204	2.4379

Sellmeier equation:[257]

$$n^2 = A + B/(1 - C/\lambda^2) + D/(1 - E/\lambda^2)$$

where λ is in μm.

	A	B	C	D	E
n^o	4.2243	1.7680	0.2270	3.1200	3380
n^e	4.2009	1.8875	0.2171	3.6461	3629

Table 1.1.17.122a
CADMIUM SELENIDE REFRACTIVE AT ROOM TEMPERATURE[310]

Wavelength (μm)	n^o	n^e
0.8	2.6448	2.6607
0.9	2.5826	2.6027
1.0	2.5502	2.5696
1.2	2.5132	2.5331
1.4	2.4929	2.5133
1.6	2.4818	2.5008
1.8	2.4732	2.4930
2.0	2.4682	2.4873
2.2	2.4642	2.4840
2.4	2.4612	2.4798
2.6	2.4590	2.4784
2.8	2.4562	2.4757
3.0	2.4542	2.4741
3.2	2.4532	2.4726
3.4	2.4518	2.4714
3.6	2.4509	2.4702
3.8	2.4498	2.4694
4.0	2.4491	2.4685

Table 1.1.17.122b
CADMIUM SELENIDE (CdSe)

l_{33} = 120 ± 10 μm at λ_1 = 10.6 μm
l_{31} = 64 ± 6 μm at λ_1 = 10.6 μm
l_{15} = 152 ± 10 μm at λ_1 = 10.6 μm
} Reference 288

l_{31} = 63.2 ± 1 μm at λ_1 = 10.6 μm[273]

Table 1.1.17.123
LITHIUM PERCHLORATE TRIHYDRATE ($LiClO_4{\cdot}3H_2O$)[336]

$l_{33} = 25.4\ \mu m$ at $\lambda_1 = 1.064\ \mu m$
$l_{31} = -9.7\ \mu m$ at $\lambda_1 = 1.064\ \mu m$
$l_{15} = 8.6\ \mu m$ at $\lambda_1 = 1.064\ \mu m$
$d_{33}d_{31} > 0$
$\theta_m = 86°\ 12'$ at $\lambda_1 = 0.532\ \mu m$
} Reference 336

Table 1.1.17.124
SILICON CARBIDE (SiC)[337]
REFRACTIVE INDEX AT ROOM TEMPERATURE

Wavelength (μm)	n^o	n^e
0.4880	2.6916	2.7423
0.5017	2.6837	2.7337
0.5145	2.6771	2.7261
0.5321	2.6689	2.7167
0.6328	2.6351	2.6794
1.0642	2.5830	2.6225

Dispersion equation:

$n_o^2 - 1 = 5.5515\lambda^2/[\lambda^2 - (0.1625)^2]$
$n_e^2 - 1 = 5.7382\lambda^2/[\lambda^2 - (0.16897)^2]$

where λ is in μm and $l_{15} = 3.84\ \mu m$, $l_{31} = 2.03\ \mu m$, and $l_{33} = 2.79\ \mu m$, all at $\lambda_1 = 1.064\ \mu m$.[337]

Table 1.1.17.125
SILVER IODIDE (AgI)[338]

Wavelength (μm)	n^o	n^e
0.659	2.184	2.200
1.318	2.104	2.115

Note: $l_{31} = 4.12\ \mu m \pm 5\%$ at $\lambda_1 = 1.318\ \mu m$.

Table 1.1.17.126
ZINC SILVER INDIUM SULFIDE (Zn_5AgInS_7)[339]qc
REFRACTIVE INDEX AT ROOM TEMPERATURE

Wavelength (μm)	n^o	n^e
0.532	2.47	2.50
0.639	2.39	2.42
1.064	2.32	2.34

Note: $l_{31} = 1.49\ \mu m$ and $\ell_{33} = 1.63\ \mu m$ at $\lambda_1 = 1.064\ \mu m$.

Table 1.1.17.127
Zn_3AgInS_5[339]
REFRACTIVE INDEX AT ROOM TEMPERATURE

Wavelength (μm)	n^o	n^e
0.532	2.46	2.50
0.639	2.40	2.43
1.064	2.31	2.34

Note: $l_{31} = 1.27\ \mu m$ and $\ell_{33} = 1.46\ \mu m$ at $\lambda_1 = 1.064\ \mu m$.

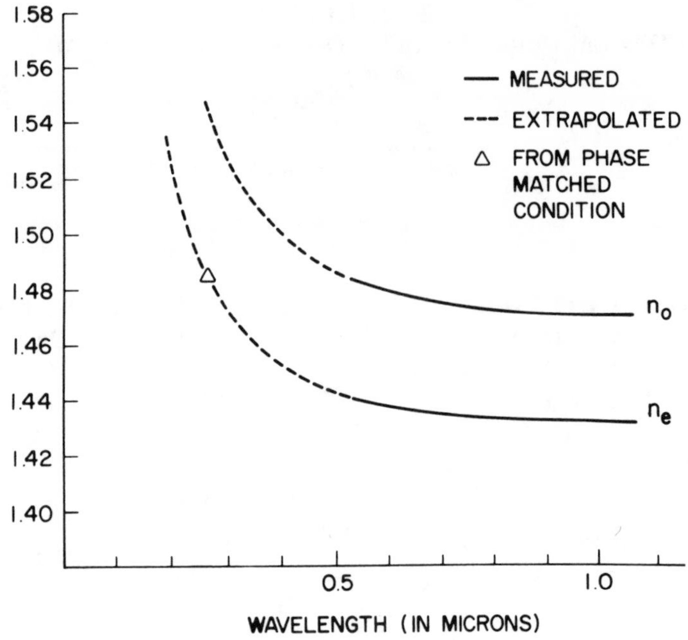

FIGURE 1.1.69. Indexes of refraction vs. wavelength for lithium perchlorate trihydrate ($LiClO_4{\cdot}3H_2O$). (Bergman, J. G., Williams, D., Crane, G. R., and Storey, R. N., *Appl. Phys. Lett.*, 26, 571, 1975. With permission.)

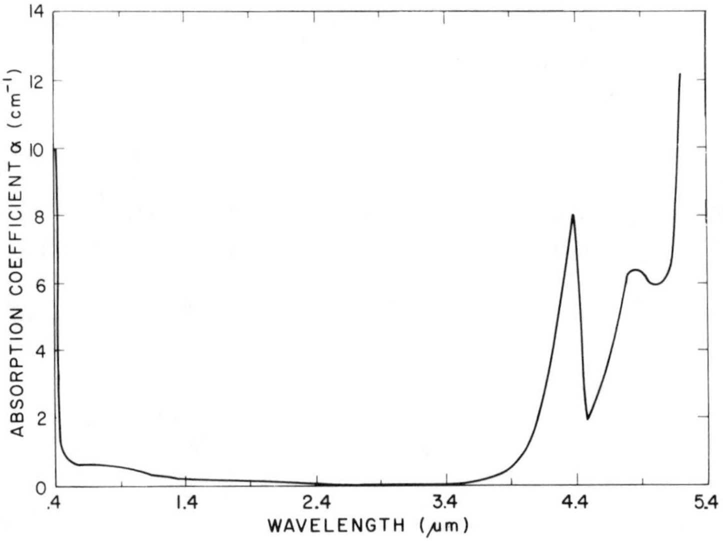

FIGURE 1.1.70. Optical transmission of hexagonal silicon carbide at room temperature. (From Singh, S., Potopowicz, J. R., Van Uitert, L. G., and Wemple, S. H., *Appl. Phys. Lett.*, 19, 53, 1971. With permission.)

Table 1.1.17.128
ZINC OXIDE (ZnO)[310]
REFRACTIVE INDEX AT ROOM TEMPERATURE

Wavelength (μm)	n^o	n^e
0.45	2.1058	2.1231
0.50	2.0511	2.0681
0.60	1.9985	2.0147
0.70	1.9735	1.9897
0.80	1.9597	1.9752
0.90	1.9493	1.9654
1.00	1.9435	1.9589
1.20	1.9354	1.9500
1.40	1.9298	1.9429
1.60	1.9257	1.9402
1.80	1.9226	1.9370
2.00	1.9197	1.9330
2.20	1.9173	1.9313
2.40	1.9152	1.9297
2.60	1.9128	1.9265
2.80	1.9100	1.9251
3.00	1.9075	1.9214
3.20	1.9049	1.9186
3.40	1.9022	1.9160
3.60	1.8994	1.9127
3.80	1.8964	1.9101
4.00	1.8891	1.9068

Table 1.1.17.129
α-ZINC SULFIDE (α-ZnS)[334]
REFRACTIVE INDEX AT ROOM TEMPERATURE

Wavelength (μm)	n^o	n^e
0.3600	2.705	2.709
0.3750	2.637	2.640
0.4000	2.560	2.564
0.4100	2.539	2.544
0.4200	2.522	2.525
0.4250	2.511	2.514
0.4300	2.502	2.505
0.4400	2.486	2.488
0.4500	2.473	2.477
0.4600	2.459	2.463
0.4700	2.448	2.453
0.4750	2.445	2.449
0.4800	2.438	2.443
0.4900	2.428	2.433
0.5000	2.421	2.425
0.5250	2.402	2.407
0.5500	2.386	2.392
0.5750	2.375	2.378
0.6000	2.363	2.368
0.6250	2.354	2.358
0.6500	2.346	2.350
0.6750	2.339	2.343
0.7000	2.332	2.337
0.8000	2.324	2.328
0.9000	2.310	2.315
1.000	2.301	2.303
1.200	2.290	2.294
1.400	2.285	2.288

Note: $l_{33} = 52 \pm 3$ μm, $l_{31} = 51 \pm 3$ μm, and $l_{15} = 53 \pm 3$ μm, all at $\lambda_1 = 10.6$ μm[228]

Table 1.1.17.130
SODIUM BROMATE ($NaBrO_3$)[342]

Dispersion equation at room temperature:

$$n^2 - 1 = 1.3194\lambda^2/[\lambda^2 - (0.09)^2] + 0.2357\lambda^2/[\lambda^2 - (0.2)^2] - 0.0174\lambda^2$$

where λ is in μm.

For an optically active crystal in which the fundamental propagates along [111] direction with circularly polarized light the coherence length is

$$l_c^d = \lambda/4[n_2\omega^d - n_\omega^l]$$

and

$$l_c^l = \lambda/4[n_2\omega^l - n_\omega^d]$$

where *d* and *l* represent dextro and levo, respectively.

optical activity at $\lambda_1 = 0.6943$ μm = 1.4 deg/mm
optical activity at $\lambda_2 = 0.3472$ μm = 11.6 deg/mm.
$\Delta l_c = l_c^d - l_c^l = 2.03 \times 10^{-3}$ μm at $\lambda_1 = 0.6943$ μm

Sign of d_{14} is the same as that of d_{11} α - SiO_2

$n_\omega^d - n_\omega^l = 2.7 \times 10^{-6}$
$n_2\omega^d - n_2\omega^l = 11.2 \times 10^{-6}$

Table 17.131
SODIUM CHLORATE ($NaClO_3$)[343] REFRACTIVE INDEX AT ROOM TEMPERATURE

Wavelength (μm)	n
0.2310	1.616
0.2573	1.585
0.2748	1.572
0.3256	1.549
0.3404	1.544
0.3467	1.542
0.3611	1.539
0.4862	1.522
0.5173	1.519
0.5892	1.515
0.6563	1.513
0.6867	1.512
0.7188	1.511

Dispersion equation:

$$n^2 - 1 = 1.1825\lambda^2/[\lambda^2 - (0.09)^2] + 0.07992\lambda^2/[\lambda^2 - (0.185)^2] - 0.00864\lambda^2$$

where λ is in μm.

Optical activity at $\lambda_1 = 0.6943$ μm = 2.2 deg/mm
Optical activity at $\lambda_2 = 0.3472$ μm = 9.3 deg/mm

$n_\omega^d - n_\omega^\ell = 4.2 \times 10^{-6}$ and $n_{2\omega}^d - n_{2\omega}^\ell = 9 \times 10^{-6}$
where *d* and *l* represent dextro and levo, respectively.
Sign of d_{14} is opposite to that of $d_{11}^{\alpha - SiO_2}$

Table 1.1.17.132
BISMUTH GERMANIUM OXIDE (Bi_4GeO_{12})[345]

Wavelength (μm)	n
0.4765	2.142
0.4880	2.135_7
0.4965	2.131_8
0.5017	2.128_5
0.5145	2.123_7
0.5321	2.115_2
0.6328	2.086_1
1.0642	2.044_3

Dispersion equation:

$$n^2 - 1 = 3.08959\lambda^2/(\lambda^2 - 0.01337)$$

where λ is in μm.

$l_{14} = 3.7$ μm at $\lambda_1 = 1.064$ μm.

Table 1.1.17.133
CADMIUM TELLURIDE CdTe[347]

Wavelength (μm)	n(Temperature 297.5 K)	n(Temperature 83.9 K)
2	2.71772	2.69616
3	2.69933	2.67860
4	2.69258	2.67162
6	2.68570	2.66533
8	2.68044	2.66016
10	2.67513	2.65481
12	2.66932	2.64900
14	2.66280	2.64185
16	2.65451	2.63395
18	2.64492	2.62542
20	2.63428	2.61539
22	2.62265	2.60322
24	2.60928	2.58942
26	2.59466	2.57502
28	2.57724	2.55892
30	2.55916	2.54358

Sellmeier equation:[348]

$$n^2 = A + B\lambda^2/(\lambda^2 - C) + D\lambda^2/(\lambda^2 - E),$$

where λ is in μm.

Temperature-dependent Sellmeier coefficients:[348]

$A = -2.973 \times 10^{-4}T + 3.8466$
$B = 8.057 \times 10^{-4}T + 3.2215$
$C = 0.1866$
$D = -2.160 \times 10^{-2}T + 12.718$
$E = -3.160 \times 10^{1}T + 18753$, where T is the temperature.

Table 1.1.17.133a
CADMIUM TELLURIDE (CdTe)[346] REFRACTIVE INDEX AT ROOM TEMPERATURE

Wavelength (μm)	*n*
0.903	2.91
1.0	2.84
1.1	2.81
1.0—1.3	2.82
7.0—10.0	2.69
10.0	2.69
14.0	2.69

Dispersion equation:

$$n^2 - 1 = 4.68 \pm 1.53\lambda^2/(\lambda^2 - 0.366)$$

where λ is in μm.

$\ell_{14} = 82.0 \pm 2.0$ μm at $\lambda_1 = 28.0$ μm[291]
$\ell_{14} = 186 \pm 10$ μm at $\lambda_1 = 10.6$ μm[348]

Table 1.1.17.134
CUPROUS BROMIDE (CuBr)[350] REFRACTIVE INDEX AT ROOM TEMPERATURE

Wavelength (μm)	*n*
0.4358	$2.336_5 \pm 0.002$
0.4678	$2.229_0 \pm 0.002$
0.4800	$2.207_2 \pm 0.002$
0.5086	$2.171_5 \pm 0.002$
0.5461	$2.241_1 \pm 0.002$
0.5791	$2.122_1 \pm 0.002$
0.5896	$2.117_4 \pm 0.002$
0.6438	$2.096_9 \pm 0.002$
0.7699	$2.069_5 \pm 0.004$

Note: Measured coherence lengths:[350] $\ell_{14} = 2.4$ μm at $\lambda_1 = 1.064$ μm, $\ell_{14} = 353$ μm at $\lambda_1 = 10.6$ μm, $\ell_{14} = 363$ μm at $\lambda_1 = 10.3$ μm, $\ell_{14} = 375$ μm at $\lambda_1 = 9.6$ μm, and $\ell_{14} = 376$ μm at $\lambda_1 = 9.3$ μm.

Measured coherence:[351] $\ell_{14} = 6.4$ μm at $\lambda_1 = 1.318$ μm, $\ell_{14} = 2.4 - 2.6$ μm at $\lambda_1 = 1.064$ μm, and $\ell_{14} = 1.5$ μm at $\lambda_1 = 0.946$ μm

Table 1.1.17.135
CUPROUS CHLORIDE ($CuCl$)[350] REFRACTIVE INDEX AT ROOM TEMPERATURE

Wavelength (μm)	n
0.4047	$2.153_5 \pm 0.001$
0.4078	$2.141_0 \pm 0.001$
0.4358	$2.072_0 \pm 0.001$
0.4678	$2.033_6 \pm 0.001$
0.4800	$2.023_4 \pm 0.001$
0.5086	$2.004_2 \pm 0.001$
0.5461	$1.987_0 \pm 0.001$
0.5791	$1.976_0 \pm 0.001$
0.5896	$1.972_6 \pm 0.001$
0.6438	$1.958_4 \pm 0.001$
0.7699	1.941 ± 0.002

Measured coherence lengths:[350] l_{14} = 4.5 μm at λ_1 = 1.064 μ, l_{14} = 181 μm at λ_1 = 9.3 μm, l_{14} = 176 μm at λ_1 = 9.6 μm, l_{14} = 165 μm at λ_1 = 10.3 μm, l_{14} = 163 μm at λ_1 = 10.6 μm.

Measured coherence lengths:[351] l_{14} = 2.5 μm at λ_1 = 0.946 μm, l_{14} = 4.0 μm at λ_1 = 1.064 μm, l_{14} = 9.2 μm at λ_1 = 1.318 μm.

Dispersion equation:[394]

$$n^2 = 3.580 + \frac{0.03162\lambda^2}{\lambda^2 - (0.4052)^2} + \frac{0.09288}{\lambda^2}$$

λ being in μm.

Table 1.1.17.136
CUPROUS IODIDE (CuI)[350] REFRACTIVE INDEX AT ROOM TEMPERATURE

Wavelength (μm)	n
0.4358	$2.562_1 \pm 0.002$
0.4678	$2.461_7 \pm 0.002$
0.4800	$2.448_5 \pm 0.002$
0.5086	$2.411_0 \pm 0.002$
0.5461	$2.372_6 \pm 0.002$
0.5791	$2.347_5 \pm 0.002$
0.5896	$2.342_8 \pm 0.002$
0.6438	$2.315_6 \pm 0.002$
0.7699	$2.280_2 \pm 0.004$

Coherence lengths:[350] ℓ_{14} = 2.2 μm at λ_1 = 1.064 μm, ℓ_{14} = 437 μm at λ_1 = 9.3 μm, ℓ_{14} = 445 μm at λ_1 = 9.6 μm, ℓ_{14} = 453 μm at λ_1 = 10.3 μm, ℓ_{14} = 461 μm at λ_1 = 10.6 μm.

Coherence lengths:[351] ℓ_{14} = 1.12 μm at λ_1 = 0.946 μm, ℓ_{14} = 1.9 μm at λ_1 = 1.064 μm, ℓ_{14} = 4.4 μm at λ_1 = 1.318 μm.

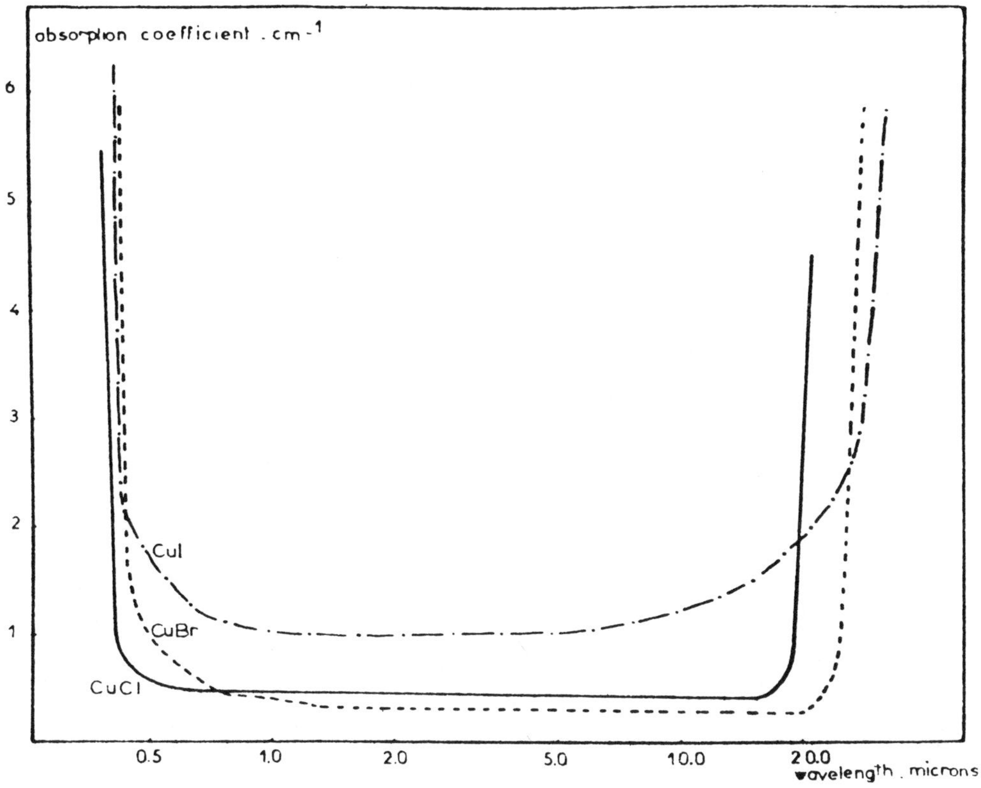

FIGURE 1.1.71. Absorption coefficients of CuCl, CuBr, and CuI vs. wavelength. (From Chemla, D., Kupecek, P., Schwartz, C., Schwab, C., and Goltzene, A., *IEEE J. Quantum Electron.*, QE-7, 126, 1971. With permission.)

Table 1.1.17.137 GALLIUM ANTIMONIDE (GaSb) REFRACTIVE INDEX AT ROOM TEMPERATURE[373]

Wavelength (μm)	n
1.80	3.820
1.90	3.802
2.0	3.789
2.1	3.780
2.2	3.764
2.3	3.758
2.4	3.755
2.5	3.749
3.0	3.898
3.5	3.861
4.0	3.833
5.0	3.824
6.0	3.824
7.0	3.843
8.0	3.843
9.0	3.843
10.0	3.843
12.0	3.843
14.0	3.861
14.9	3.880

Table 1.1.17.138 GALLIUM PHOSPHIDE (GaP)[310]

Wavelength (μm)	n
0.5	3.4595
0.6	3.3495
0.7	3.2442
0.8	3.1830
0.9	3.1430
1.0	3.1192
1.1	3.0981
1.2	3.0844
1.4	3.0646
1.6	3.0509
1.8	3.0439
2.0	3.0379
2.2	3.0331
2.4	3.0296
2.6	3.0271
2.8	3.0236
3.0	3.0215
3.2	3.0197
3.4	3.0181
3.6	3.0166
3.8	3.0159
4.0	3.0137

Table 1.1.17.138a GALLIUM PHOSPHIDE (GaP) REFRACTIVE INDEX AT 24.5°C[357]

Wavelength (μm)	n
0.545	3.4522
0.550	3.4411
0.560	3.4203
0.570	3.4012
0.580	3.3837
0.590	3.3675
0.600	3.3524
0.610	3.3384
0.620	3.3254
0.630	3.3132
0.640	3.3018
0.650	3.2912
0.660	3.2811
0.670	3.2716
0.680	3.2626
0.690	3.2541
0.700	3.2462

Note: Estimated accuracy = ± 0.0012.

Table 1.1.17.138b GALLIUM PHOSPHIDE(GaP)

The nonlinear coefficient in Reference 357 was obtained from the well-known relation Reference 94. $d = -1/4n_\omega^4 r\omega/(1 + C\omega)$, where $n\omega$ is the index at frequency ω, r_ω is the electrooptic coefficient measured at frequency ω, and Cω is the Faust-Henry[358] coefficient determined from the ratio of the Raman scattering efficiencies for the longitudinal (L0) and transverse ω and Cω is the Faust-Henry[358] coefficient (T0) optic modes. Using C = 0.53, $r_{41} = 0.97 \times 10^{-12}$m/v, λ = 3.39 μm, and n = 3.420:

$$d_{14} = 70{\cdot}6 \times 10^{-12} \text{ m/v}$$

	Ref.
ℓ_{14} = 46 ± 3 μm at λ_1 = 10.6 μm	88
ℓ_{36} = 1.61 μm at λ_1 = 1.318 μm	155
ℓ_{36} = 7.5 ± 0.2 μm at λ_1 = 2.12 μm	156
ℓ_{36} = 1.53 ± 0.07 μm at λ_1 = 1.318 μm	156

Table 1.1.17.139 HEXAMINE [$N_4(CH_2)_6$][360] REFRACTIVE INDEX AT ROOM TEMPERATURE

Wavelength (μm)	n
0.4861	1.5984
0.5016	1.5953
0.5461	1.5917
0.5780	1.5899
0.5876	1.5893
0.6676	1.5856

Table 1.1.17.140
INDIUM PHOSPHIDE (InP) REFRACTIVE INDEX AT ROOM TEMPERATURE[373]

Wavelength (μm)	*n*	Wavelength (μm)	*n*
0.302	3.162	0.925	3.396
0.310	3.105	0.930	3.390
0.318	3.054	0.935	3.385
0.326	3.027	0.940	3.379
0.335	3.024	0.945	3.374
0.344	3.047	0.950	3.369
0.354	3.082	0.955	3.364
0.364	3.192	0.960	3.359
0.375	3.441	0.965	3.355
0.387	3.835	0.970	3.351
0.399	4.100	0.975	3.346
0.413	4.083	1.00	3.327
0.427	3.982	1.10	3.268
0.443	3.833	1.20	3.231
0.459	3.754	1.30	3.205
0.477	3.675	1.40	3.186
0.496	3.621	1.50	3.172
0.517	3.567	1.60	3.161
0.539	3.521	1.70	3.152
0.563	3.472	1.80	3.145
0.590	3.450	1.90	3.139
0.620	3.430	2.00	3.134
0.652	3.410	5.00	3.08
		6.00	3.07
		7.00	3.07
		8.00	3.06
		9.00	3.06
		10.0	3.05
		12.0	3.05
		14.0	3.04
		14.85	3.03

Table 1.1.17.141
ZINC SELENIDE (ZnSe)[346] REFRACTIVE INDEX AT ROOM TEMPERATURE

Wavelength (μm)	*n*
0.589	2.61
1.0	2.48
1.5	2.45
2.0	2.44

Dispersion equation:

$$n^2 = 3.855 + 2.045\lambda^2/(\lambda^2 - 0.109)$$

where λ is in μm.

Table 1.1.17.142
ZINC SULFIDE (β-ZnS)[310] REFRACTIVE INDEX AT ROOM TEMPERATURE

Wavelength (μm)	*n*
0.45	2.4709
0.50	2.4208
0.60	2.3640
0.70	2.3333
0.80	2.3146
0.90	2.3026
1.00	2.2932
1.20	2.2822
1.40	2.2762
1.60	2.2716
1.80	2.2680
2.00	2.2653
2.20	2.2637
2.40	2.2604

$\ell_4 = 53 \pm 5$ μm at $\lambda_1 = 10.6$ μm[288]

Dispersion equation:[384]

$$n^2 = 5.131 + \frac{1.275 \times 10^7}{\lambda^2 - 0.732 \times 10^7}$$

Table 1.1.17.143
ZINC TELLURIDE
($ZnTe$)[346]
REFRACTIVE INDEX AT ROOM TEMPERATURE

Wavelength (μm)	n
0.589	3.06
0.620	3.00
0.830	2.84
1.240	2.76
2.06	2.71

Dispersion equation:

$n^2 - 1 = 3.27 + 3.01\lambda^2/(\lambda^2 - 0.142)$

where λ is in μm.

Table 1.1.17.143a
ZINC TELLURIDE
($ZnTe$)[362]
REFRACTIVE INDEX AT ROOM TEMPERATURE

Wavelength (μm)	n
0.569	3.111
0.577	3.085
0.579	3.079
0.589	3.054
0.600	3.035
0.616	3.005
0.650	2.962
0.700	2.913
0.725	2.893
0.750	2.879
0.760	2.871
0.770	2.866
0.800	2.853
1.000	2.790
1.200	2.758
1.300	2.748
1.400	2.741
1.500	2.734
1.515	2.734

$\ell_{14} = 290 \pm 50$ μm at $\lambda_1 = 10.6$ μm[288]

Table 1.1.17.144
AIR

Dispersion equation: (15°C, 760 torr)[395]

$$(n - 1) \times 10^8 = 8342.13 + \frac{2406030}{130 - \sigma^2} + \frac{15997}{38.9 - \sigma^2}$$

(0.033% content of CO_2)

Dispersion equation: (15°C, 760 torr)[396]

$$(n - 1) \times 10^8 = \frac{5791817}{238.0185 - \sigma^2} + \frac{167909}{57.362 - \sigma^2}$$

(Range of validity 0.23 to 1.6945 μm) σ is the vacuum wave number in reciprocal micrometers.

Table 1.1.17.145
ARGON (Ar)

Dispersion equation: (15°C, 760 torr)[397]

$$(n - 1) \times 10^7 = 643.2135 + \frac{286060.21}{144 - \sigma^2} \quad (0°C, 760\ torr)$$

Dispersion equation:

$$(n - 1) \times 10^7 = 678.6711 + \frac{301829.43}{144 - \sigma^2}$$

where σ is the vacuum wave number in reciprocal microns. Range of measurements: 0.4679 to 2.0586 μm.

Table 1.1.17.146
CARBON DIOXIDE (CO_2) REFRACTIVITIES AT STANDARD TEMPERATURE AND PRESSURE[398]

Vacuum wavelength (μm)	$(n - 1)\ 10^7$
1.8172285	4362.699
1.6945208	4373.838
1.5293544	4387.443
1.4756503	4391.562
1.3722327	4399.414
1.3507884	4400.854
1.2960201	4405.031
1.0142573	4427.044
0.9660434	4431.376
0.9227030	4435.584
0.9125471	4436.630
0.8931145	4438.662
0.8777160	4440.321
0.8266794	4446.330
0.7440946	4457.856
0.7247163	4460.979
0.5462258	4505.000
0.5087242	4520.165
0.4801260	4534.159

Dispersion equation:[398]

$$(n - 1)10^5 = \frac{0.154489}{0.0584738 - \sigma^2} + \frac{8309.1927}{210.92417 - \sigma^2} + \frac{287.64190}{60.122959 - \sigma^2}$$

where σ is the wave number in reciprocal micrometers vacuum wavelength. The formula for the compressibility factor was found to be:

$$\frac{PV}{RT} = 1 - P(9.0143 - 0.048529T + 6.980 \times 10^{-5}T^2) \times 10^{-5}$$

where P is the pressure, V the volume, R the universal gas constant, and T the temperature.

Table 1.1.17.147
HELIUM (He) MEASURED REFRACTIVITIES OF HELIUM AT 760 torr, 0°C[399]

Vacuum wavelength (Å)	Refractivity $(n - 1) \times 10^9$
4801.260	35032.25
5087.243	34993.39
5462.258	34950.00
7247.163	34829.74
8266.793	34792.75
9125.471	34771.67
9227.030	34769.72
9660.434	34759.27
10142.573	34752.62
13722.327	34718.03
14756.503	34711.73
15293.544	34709.52
15300.153	34704.58
16945.209	34702.62
20586.910	34693.50

Dispersion equation: (0°C, 760 torr).

$$(n - 1) \times 10^5 = \frac{1467.196}{423.15 - \sigma^2} + \frac{0.00002125}{0.71530 - \sigma^2}$$

where σ is the wave number in units of μm^{-1}.

Table 1.1.17.147a HELIUM (He) REFRACTIVITY AT NPT[400]

Wavelength (Å)	(n − 1) Experimental
920	4.85×10^{-5}
1000	4.53
1100	4.26
1200	4.07
1300	3.96
1400	3.89
1500	3.83
1600	3.78
1700	3.74
1800	3.73
1900	3.77

Table 1.1.17.148 HYDROGEN (H_2) MEASURED REFRACTIVITIES AT 0°C, 760 torr[40]

Vacuum wavelength (μm)	$10^6(n-1)$
1.6945211	136.099
1.3722328	136.289
1.3626384	136.303
1.3507884	136.304
1.0142573	136.747
0.9660435	136.854
0.9227031	136.960
0.9125471	136.987
0.8523783	137.176
0.8266794	137.268
0.7247163	137.746
0.5462252	139.304
0.5087242	139.871
0.4801260	140.398
0.4679465	140.652
0.4359553	141.439
0.4047563	142.407

Dispersion equation:

$$(n - 1)10^6 = 21.113 + \frac{12723.2}{111 - \sigma^2}$$

vacuum wave number in μm^{-1}. The compressibility, Z, was obtained from:

$$Z - 1 = (0.255/T - 23.9/T^2)p,$$

where T is in K, and p is in atmospheres. Also:

$$(n - 1)10^6 = [14895.6/(180.7 - \sigma^2)] + [4903.7/(92 - \sigma^2)].$$

Table 1.1.17.149 NITROGEN (N_2)

Dispersion equation:

$$(n - 1) = \frac{6.3622 \times 10^4}{45.989 \times 10^8 - \nu^2} + \frac{32.453 \times 10^6}{1065.11 \times 10^8 - \nu^2}$$

where ν is the frequency.

Table 1.1.17.150
XENON (Xe)

Dispersion equation:[383]

$$(n - 1) \approx \left[\frac{393235}{46.3012 - 10^{-8}/\lambda^2} + \frac{393235}{59.5779 - 10^{-8}/\lambda^2} + \frac{7366100}{139.8310 - 10^{-8}/\lambda^2}\right] \times 10^{-8}$$

where λ is in centimeters.

Table 1.1.17.151
REFRACTIVE INDEX OF ALKALI METAL VAPORS[81]

The refractive index of the metal vapors is calculated from the Sellmeier equation:

$$n - 1 = \frac{Nr_e}{2\pi} \sum_i \frac{f_i}{[(1/\lambda_i^2) - (1/\lambda^2)]}$$

$r_e = 2.818 \times 10^{-13}$ cm; f_i = oscillator strength of the i-th transition; λ_i = wavelength of the i-th transition in centimeters.

Values of f_i and λ_i for alkali-metal vapors:

Element	Wavelength (μm)	*f*
Li	0.6708	0.744
	0.3234	0.00428
	0.2742	0.00398
Na	0.5892	0.982
	0.3303	0.0142
	0.2853	0.0022
K	0.7076	1.04
	0.4045	0.0154
	0.3447	0.00277
Rb	0.7948	0.395
	0.7800	0.805
	0.4216	0.00532
	0.4202	0.01068
	0.3592	0.000979
	0.3587	0.00196
Cs	0.8944	0.394
	0.8521	0.814
	0.4595	0.00284
	0.4555	0.0174
	0.3890	0.000317
	0.3876	0.00349

Table 1.1.18
RELATION BETWEEN CRYSTALLOGRAPHIC AND REFERENCE AXES FOR NONLINEAR OPTICAL POLARIZABILITY TENSOR[57]

	Point group		Axis identification		
Crystal system	**International**	**Schoenflies**	**Crystallographic**	**Rectangular**	**+/− Axes**
Triclinic	1	C_1		Y ⊥ (010)	Z X
$c_o < a_o < b_o$, α and β > 90°	$\bar{1}$	$C_1(S_2)$		Y ⊥ (010)	
Monoclinic	2	C_2	$b \equiv 2$	X ⊥ (100), Y ≡ b, Z ≡ c	Y
$c_o < a_o$	m	$C_5(C_{ih})$	$b \equiv /m$	X ⊥ (100), Y ≡ b, Z ≡ c	Z X
α = γ = 90°, β > 90°	$2/m$	C_{2h}		X ⊥ (100), Y ≡ b, Z ≡ c	
Orthorhombic	222	$D_2(V)$	$c \equiv 2$, $a \equiv 2$, $b \equiv 2$	X ≡ a, Y ≡ b, Z ≡ c	
$c_o < a_o < b_o$	$mm2$	C_{2v}	z ≡ polar axis ≡ a,b, or c	X is the smaller of the nonpolar axis	Z
α = β = γ = 90°	mmm	$D_{2h}(V_h)$	$c \equiv 2$, $a \equiv 2$, $b \equiv 2$	X ≡ a, Y ≡ b, Z ≡ c	
Tetragonal	4	C_4	$c \equiv 4$	X ≡ a_1, Y ≡ a_2, Z ≡ c	Z
$a_o = b_o$	$\bar{4}$	S_4	$c \equiv 4$	X = a_1, Y ≡ a_2, Z ≡ c	Z
α = β = γ = 90°	$4/m$	C_{4h}	$c \equiv 4$	X ≡ a_1, Y ≡ a_2, Z ≡ c	
	422	D_4	$c \equiv 4$, $a_1 \equiv 2$, $a_2 \equiv 2$	X ≡ a_1, Y ≡ a_2, Z ≡ c	
	$4mm$	C_{4v}	$c \equiv 4$, $a_1 \equiv /m, a_2 \equiv /m$	X ≡ a_1, Y ≡ a_2, Z ≡ c	Z
	$\bar{4}2m$	$D_{2d}(V_d)$	$c \equiv \bar{4}$, $a_1 \equiv 2$, $a_2 \equiv 2$	X ≡ a_1, Y ≡ a_2, Z ≡ c	Z
	$4/mmm$	D_{4h}	$c \equiv 4$, $a_1 = 2$, $a_2 \equiv 2$	X ≡ a_1, Y ≡ a_2, Z ≡ c	

Trigonal	3	C_3	$c \equiv \overline{3}$	$X \equiv a_1$	$Z \equiv c$	Z X
$(a_o)_1) = (a_o)_2 = (a_o)_3$	$\overline{3}$	$C_{3i}(S_6)$	$c \equiv 3$	$X \equiv a_1$	$Z \equiv c$	
	32	D_3	$c \equiv 3,\ a_1 \equiv 2,\ a_2 \equiv 2,\ a_3 \equiv 2$	$X \equiv a_1$	$Z \equiv c$	X
	$3m$	C_{3V}	$c \equiv 3,\ a_1 \equiv /m, a_2 \equiv /m, a_3 \equiv /m$ $a_3 \equiv /m$	$X \equiv a_1$	$Z \equiv c$	Y
	$\overline{3}m$	D_{3d}	$c \equiv \overline{3},\ a_1 \equiv 2,\ a_2 \equiv 2,\ a_3 \equiv 2$	$X \equiv a_1$	$Z \equiv c$	
Hexagonal	6	C_6	$c \equiv \overline{6}$	$X \equiv a_1$	$Z \equiv c$	Z
$(a_o)_1 = (a_o)_2 = (a_o)_3$	$\overline{6}$	C_{3h}	$c \equiv 6$	$X \equiv a_1$	$Z \equiv c$	X Y
	$6/m$	C_{6h}	$c \equiv 6$	$X \equiv a_1$	$Z \equiv c$	
	622	D_6	$c \equiv 6,\ a_1 \equiv 2,\ a_2 \equiv 2,\ a_3 \equiv 2$	$X \equiv a_1$	$Z \equiv c$	
	$6mm$	C_{6V}	$C \equiv 6,\ a_1 \equiv /m, a_2 \equiv /m, a_3 \equiv 2$	$X \equiv a_1$	$Z \equiv c$	Z
	$6m2$	D_{3h}	$c \equiv \overline{6},\ a_1 \equiv 2,\ a_2 \equiv 2,\ a_3 \equiv 2$	$X \equiv a_1$	$Z \equiv c$	X
	$6/mmm$	D_{6h}	$C \equiv 6,\ a_1 \equiv 2,\ a_2 \equiv 2,\ a_3 \equiv 2$	$X \equiv a_1$	$Z \equiv c$	
Cubic	23	T	$a_1 \equiv 2,\ a_2 \equiv 2,\ a_3 \equiv 2$	$X \equiv a_1,\ Y \equiv a_2,\ Z \equiv a_3$		Z
$a_o = b_o = c_o$	$m3$	T_h	$a_1 \equiv 2,\ a_2 \equiv 2,\ a_3 \equiv 2$	$X \equiv a_1,\ Y \equiv a_2,\ Z \equiv a_3$		
$\alpha = \beta = \gamma = 90°$	432	0	$a_1 \equiv 4,\ a_2 \equiv 4,\ a_3 \equiv 4$	$X \equiv a_1,\ Y \equiv a_2,\ Z \equiv a_3$		
	$\overline{4}3m$	T_d	$a_1 \equiv \overline{4},\ a_2 \equiv 4,\ a_3 \equiv 4$	$X \equiv a_1,\ Y \equiv a_2,\ Z \equiv a_3$		Z
	$m3m$	O_h	$a_1 \equiv 4,\ a_2 \equiv 4,\ a_3 \equiv 4$	$X \equiv a_1,\ Y \equiv a_2,\ Z \equiv a_3$		

Table 1.1.19
SHG-POWDER DATA ON NONLINEAR OPTICAL MATERIALS

Material	λ_1 (μm)	Reference material	$I_{2\omega}$	Ref.
m-Nitroaniline (*m*-NA) NO_2–$C_6H_4NH_2$	1.064	ADP	100	386
p-Nitrophenylhydrazione (*p*-NPH) $NO_2C_6H_4NHNH_2$	1.064	ADP	100	386
Dithiocarbamate NCSS–NH_2COO	1.064	ADP	25	386
7-Hydroxycoumarin $C_9H_6O_2$	1.064	ADP	20	386
Coumarin C_6H_4OCOCH:CH	1.064	ADP	10	386
α-Bromo-*p*-nitrotoluene $NO_2C_6H_4CH_2Br$	1.064	ADP	15	386
Isonicotil hydrazide (C_5H_4NCO) $NHNH_2$	1.064	ADP	6	386
p-Nitrobenzaldehyde $NO_2C_6H_4$–CHO	1.064	ADP	5	386
Phenylurea $C_6H_5NHCONH_2$	1.064	ADP	4	386
2-6Dinitrophenol $(NO_2)_2C_6H_3OH$	1.064	ADP	3	386
Magnesium chloroboracite (*mm*2-phase) $Mg_3B\lesseqgtr_{-+}O_{13}Cl$	1.06	α-SiO_2	0.025	387
Chromium chloroboracite ($\bar{4}3m$-phase) $Cr_3B_7O_{13}Cl$	1.06	α-SiO_2	0.01	387
Chromium chloroboracite (*mm*2-phase) $Cr_3B_7O_{13}Cl$	1.06	α-SiO_2	0.02	387
Copper chloroboracite (*mm*2-phase) $Cu_3B_7O_{13}Cl$	1.06	α-SiO_2	0.018	387
Copper chloroboracite ($\bar{4}3m$-phase) $Cu_3B_7O_{13}Cl$	1.06	α-SiO_2	0.01	387
Iron iodoboracite (*mm*2-phase) $Fe_3B_7O_{13}Cl$	1.06	α-SiO_2	0.023	387
Iron iodoboracite ($\bar{4}3m$-phase) $Fe_3B_7O_{13}Cl$	1.06	α-SiO_2	0.01	387
Nickel bromoboracite (*mm*2-phase) $Ni_3B_7O_{13}Cl$	1.06	α-SiO_2	0.03	387
Nickel bromoboracite ($\bar{4}3m$-phase) $Ni_3B_7O_{13}Cl$	1.06	α-SiO_2	0.015	387
Zinc bromoboracite (*mm*2-phase) $ZnB_7O_{13}Br$	1.06	α-SiO_2	0.025	387
*N-N'*Dimethylurea $(CH_3)_2NCONH_2$	1.06	Urea	1.6	388
Di-(*p*-Nitrophenyl)-urea(DNPU) NO_2–C_6H_4–NH–$CONHC_6H_4$–NO_2	1.064	Urea	8.8	389
1-(2,4-Dinitrophenyl)-semicarbazide (DNP-SC) NO_2–C_6H_4–NO_2–NH–NHCO–NH_2	1.064	Urea	8.8	389
3-Dimethylamino-4-nitroacrylo-phenone (DMA NAP) CH_3–CH_3–N–$CHCHCOC_6H_4NO_2$	1.064	Urea	7.5	389
Methyl-(2,4 dinitrophenyl)-aminopropanoate (MAP) CH_3–CH–NH–$COOCH_3NO_2C_6H_4$–NO_2	1.064	Urea	7.5	389
Cholesterol $C_{27}H_{45}OH$	1.064	α-SiO_2	1	172
Picric acid $C_6H_2(NO_2)_3OH$	1.064	α-SiO_2	60	172

Table 1.1.19 (continued)
SHG-POWDER DATA ON NONLINEAR OPTICAL MATERIALS

Material	λ_1 (μm)	Reference material	$I_{2\omega}$	Ref.
P-Anisalazine	1.064	α-SiO_2	30	172
$C_{16}H_{16}O_2N_2$				
ℓ-Histidine	1.064	α-SiO_2	10	172
$C_3H_3N_2CH_2CH(NH_2)COOH$				
4-Aminopyridine	1.064	α-SiO_2	20	172
$NH_2C_5H_4N$				
ℓ-Glutamic acid	1.064	α-SiO_2	8	172
$COOH(CH_2)_2CH–NH_2COOH$				
Rubidium chlorate	1.064	α-SiO_2	20	172
$RbClO_3$				
d-Ribose	1.064	α-SiO_2	1	172
$CH_2CH(CHOH)_3CHO$				
ℓ-Menthol	1.064	α-SiO_2	1.5	172
$C_{10}H_{20}O$				
Urea	1.064	α-SiO_2	400	172
$NO–CH_2–CH_2$				
Tartaric acid	1.064	α-SiO_2	15	172
$C_4H_6O_6$				
MNA	1.064	Urea	22	390
$O_2N–C_6H_4–NH_2$				
MAP	1.064	Urea	10	390
$O_2N–C_6H_4–NO_2–NH–COOCH_3$				
$O_2N–C_6H_4–NH–COOCH_3$	1.064	Urea	1	390
POM	1.064	Urea	13	390
$O_2N–C_5H_4–NO$				
$O_2NC_5H_4N–NO_2–Cl$	1.064	Urea	5	390
$O_2NC_5H_4–N–NO_2–OH$	1.064	Urea	5	390
$O_2N–C_5H_4–N–NO_2–O–C_6H_4$	1.064	Urea	7	390
$NO_2–C_5H_4–N–NO_2–NH–C_6H_4$	1.064	Urea	4	390
$O_2N–C_5H_4–N–NO_2–NH–COOCH_3$	1.064	Urea	3	390
$O_2N–C_5H_4–N–NO_2–NH–C_6H_4$	1.064	Urea	7	390
$O_2N–C_5H_4–N–NO_2–NH–C_{10}H_8$	1.064	Urea	16	390
$O_2N–C_5H_4–N–H–NH–C_6H_4$	1.064	Urea	25	390
$O_2N–C_5H_4–N–H–NHCOOCH_3$	1.064	Urea	1	390
$O_2N–C_5H_4–N–H–O–C_{10}H_8$	1.064	Urea	5	390
$O_2N–C_5H_4–N–H–N–H–C_{10}H_8$	1.064	Urea	2	390
$O_2N–C_5H_4N–NH–C_6H_4–NO_2$	1.064	Urea	4	390
$O_2N–C_5H_4–N{\cdot}H{\cdot}H{\cdot}MeN{\cdot}H{\cdot}Ph{\cdot}OH{\cdot}H{\cdot}CH_3$	1.064	Urea	3	390
$O_2N–C_5H_4–N–CH_3–NH{\cdot}COOCH_3$	1.064	Urea	1	390
$O_2N–C_5H_4–N–CH_3–NH–C_6H_8$	1.064	Urea	25	390
$O_2N–C_5H_4N–CH_3–NHC_{10}H_4$	1.064	Urea	7	390
$O_2N–C_5H_4–N–CH_3–NH–C_{10}H_8$	1.064	Urea	8	390
Tertiary butyl urea	1.064	Urea	<<1	390
$CH_3–CH_3–CH_3–C–NH–CO–NH_2$				
Formyl urea	1.064	Urea	0.15	391
$CHO–NH–C–O–NH_2$				
1,1′-Methylene diurea	1.064	Urea	0.35	391
$NH_2{\cdot}C–O–NH–CH_2–NH–CO–NH_2$				
p-Tolyl urea	1.064	Urea	<<1	391
$CH_3–C_6H_4–NH–C–O–NH_2$				
3,5-Dihydroxylbenzanilide	1.064	Urea	<<1	391
$OH–C_6H_4–OH–C–O–NH_2$				
p-Chlorophenyl-urea	1.064	Urea	0.35	391
$Cl–C_6H_4–NH–C–O–NH_2$				

Table 1.1.19 (continued)
SHG-POWDER DATA ON NONLINEAR OPTICAL MATERIALS

Material	λ_1 (μm)	Reference material	$I_{2\omega}$	Ref.
1-(2,4,Dinitrophenyl) semicarbazide (DNP-SC) NO_2–C_6H_4–NO_2–NH–NH–C–O–NH_2	1.064	Urea	8.8	391
1-Isonicotinoyl-3-thiosemicarbazide NC_5H_4–C–O–NH–NH–C–S–NH_2	1.064	Urea	0.35	391
1-(*p*-Nitrophenyl-sulfonyl)-guanidine NH_2–C–NH–NH–S–O–O–C_6H_4–NO_2	1.064	Urea	~1	391
1-Naphthylurea $C_{10}H_8$–NH–C–O–NH_2	1.064	Urea	<<1	391
1-(2-Naphthyl)urea $C_{10}H_8$–NH–C–O–NH_2	1.064	Urea	<<1	391
1,1-Diphenyl urea C_6H_4–NH–C_6H_4–C–O–NH_2	1.064	Urea	0.25	391
Symmetric diphenyl urea C_6H_4–NH–C–O–NH–C_6H_4	1.064	Urea	<<1	391
N-3-nitrophenyl-*N*′-phenylurea NO_2–C_6H_4–NH–C–O–NH–C_6H_4	1.064	Urea	<<1	391
N-4-Nitrophenyl-*N*′-phenylurea NO_2–C_6H_4–NH–C–O–NH–C_6H_4	1.064	Urea	0.35	391
Di-(*p*-nitrophenyl)urea (DNPU) NO_2–C_6H_4–NH–C–O–NH–C_6H_4–NO_2	1.064	Urea	8.8	391
2-Hydroxy-4-nitrocarbanilide	1.064	Urea	<<1	391
1-Benzoyl-3-(*p*-nitro-phenyl)-2-thiourea NO_2–C_6H_4–NH–C–S–NH–C–O–C_6H_4	1.064	Urea	<<1	391
3-Amino-4-methoxy-benzanilide CH_3O–C_6H_4–NH_2–C–O–NH–C_6H_4	1.064	Urea	0.1	391
m-Nitroaniline NH_2–C_6H_4–NO_2	1.064	Urea	>1	391
2,4-Dinitroaniline NH_2–C_6H_4–NO_2–NO_2	1.064	Urea	<1	391
2,5-Dimethyl-4-nitroaniline NH_2–C_6H_4–CH_3–CH_3–NO_2	1.064	Urea	<1	391
4(-*p*-Aminophenylazo)-benzoic acid NH_2–C_6H_4–N–N–C_6H_4–COOH	1.064	Urea	~1	391
Methyl-(*p*-nitrophenyl)-amino-propanoate CH_3–CH–NH–$COOCH_3$–C_6H_4–NO_2	1.064	Urea	~1	391
3-Dimethylamino-4′-nitroacrylo-phenone (DMA-NAP) CH_3–CH_3–N–CH–CH–C–O–C_6H_4–NO_2	1.064	Urea	7.5	391
Methyl-(2,4 dinitrophenyl)-aminopropanoate (MAP) CH_3–CH–NH–$COOCH_3$–C_6H_4–NO_2–NO_2	1.064	Urea	7.5	391
Lanthanide formate crystal hydrates $Ln(HCOO)_3$–nH_2O–(Ln = Pr; Nd)	1.064	KDP	1	392
Lanthanide formate crystal hydrate with 1,10-phenanthroline $Ln(HCOO)_3$–phen–nH_2O (Ln = Er,Eu,Gd,Tb,Sm)	1.064	KDP	1	392
2,5-Dinitrofluorene	1.064	KDP	6	392
2,4,5,7-Tetranitrofluorenone	1.064	KDP	2	392
N-Nitrourea	1.064	KDP	4	392
N-*N*′-Dimethylurea	1.064	KDP	2	392
Urea oxalate	1.064	KDP	0.2	392
Carbamylurea monohydrate	1.064	KDP	1	392
1-Cyanoguanidine	1.064	KDP	5	392
n-Iodoanisole	1.064	KDP	3	392
2-Methyl-*n*-quinone	1.064	KDP	2	392
4′-Methoxybenzalacetophenone	1.064	KDP	6	392

Table 1.1.19 (continued)
SHG-POWDER DATA ON NONLINEAR OPTICAL MATERIALS

Material	λ_1 (μm)	Reference material	$I_{2\omega}$	Ref.
Tribenzoylmethane	1.064	KDP	7	392
Anhydrous sulfanilic acid	1.064	KDP	5	392
Potassium *n*-aminobenzene sulfonate	1.064	KDP	1	392
4-Methylphenyl-β-styrene-sulfonate	1.064	KDP	2	392
Phenyl-*m*-nitrobenzene sulfonate	1.064	KDP	3	392
Sodium-*m*-nitrobenzene sulfonate	1.064	KDP	1	392
Ethylsulfone-*m*-nitroanilide	1.064	KDP	2	392
N-(2,4,dinitrophenyl)-*n*-toluidine	1.064	KDP	2	392
N-(2,4-dinitrophenyl-*m*-toluidine	1.064	KDP	3	392
N-(2,4-dinitrophenyl)-hydrazine	1.064	KDP	3	392
N-Butyl-2-4-dinitroaniline	1.064	KDP	6	392
4-Diethylaminoazobenzene	1.064	KDP	2	392
N-(2,4 Dinitrophenyl)-*N'*-tosyln-phenylenediamine	1.064	KDP	10	393
3-(*N*-tosyl),4-(*N*-acetyl)-diamino-nitrobenzene	1.064	KDP	3	392
2-(*N*-mesyl),5-(*N*-acetyl)-diamino-nitrobenzene	1.064	KDP	1	392
2-(*N*-acetyl)-amino-4,5-dinitro-aniline	1.064	KDP	20	392
2-Methyl-4-nitro-*N*-methylaniline	1.064	KDP	10	392
2-Methoxy-4-nitro-*N*-methylaniline	1.064	KDP	15	392
n-Nitroaniline with *n*-nitrophenol	1.064	KDP	6	392
2-(β-*n*-Nitrostyryl)-benzoxazole,1,3	1.064	KDP	8	392
1-Hydroxy-2-methyl-5-nitrobenzimidazole	1.064	KDP	20	392
1-Tosyl-2-methyl-5-nitrobenzimidazole	1.064	KDP	8	392
Benzothiazole	1.064	KDP	1	392
2-Pyridone	1.064	KDP	1-2	392
Phenyl-2-thienylketone	1.064	KDP	1	392
1(3-Indolyl)*N*-(2methyl,3bromobutyryl4)isoquinoline perchlorate	1.064	KDP	8	392

Table 1.1.20
LIST OF SYMBOLS AND ABBREVIATIONS

Abs		Absolute measurements
b	m	Confocal parameter for a gaussian beam
c	$m{\cdot}s^{-1}$	Velocity of light
$c{\cdot}c$		Complex conjugate
C_{jn}^{3}, C_{jn}	$m^2{\cdot}V^{-2}$	Third-harmonic generation coefficient
CPM		Critical phase-matching
d_{eff}	$V^{-1}{\cdot}m$	Effective coefficient for SHG
d_{jm}	$V^{-1}{\cdot}m$	SHG coefficient
e	C	Electronic charge
$E(rt)$	$V{\cdot}m^{-1}$	Total electric field
$E(r,w)$	$V{\cdot}m^{-1}{\cdot}s$	Fourier transform of total electric field
E_j	$V{\cdot}m^{-1}$	Amplitude of monochromatic electric field component
EO		Electrooptic effect
F	m	Focal length
f_{jk}		Oscillator strength of $j \rightarrow k$ transition
g		Degeneracy factor
h	J.S.	Planck's constant
$\hbar = h/2\pi$	J.S.	Dirac constant
H	$A\text{-}m^{-1}$	Magnetic field vector
i		Imaginary number $\sqrt{-1}$
I_j	$W{\cdot}m^{-2}$	Intensity of monochromatic wave
I_m		Imaginary part
K,K_ω	m^{-1}	Magnitude of wave vector $(n_\omega{\cdot}\omega/C)$
L	m	Interaction length of the nonlinear medium
L_c,ℓ_c	m	Coherence length
N	m^{-3}	Number density of medium
n^e		Extra-ordinary refractive index
n^o		Ordinary refractive index
n_ω		Refractive index at ω
OR		Optical rectification
p	torr	Pressure
P	W	Power
$\boldsymbol{P}$	$C{\cdot}m^{-2}$	Polarization
$\boldsymbol{P}_j^r(\omega_\sigma)$	$C{\cdot}m^{-2}$	Monochromatic r-th order polarization at ω_σ
$\boldsymbol{P}(r,t)$	$C{\cdot}m^{-2}$	Macroscopic polarization
$\boldsymbol{P}(r,\omega)$	$C{\cdot}m^{-2}{\cdot}s$	Fourier transform of polarization
PG		Parametric generation
PM		Phase matching
$Q{\cdot}Q_{jm}$	$C{\cdot}m$	Electric dipole moment operator and its matrix element
r	m	Spatial position
r_e	m	Classical electron radius
R		Reflection coefficient
SFG		Sum frequency generation
SERS		Stimulated electronic Raman scattering
SRS		Stimulated Raman scattering
SHG		Second-harmonic generation
T	K	Absolute temperature
THG		Third-harmonic generation
TWM,ThWM		Three-wave mixing
ω_0,ω_{10}	m	Gaussian beam-waist radius
Z		Charge number
α	m^{-1}	Single-photon absorption coefficient
ΔK	m^{-1}	Phase-mismatch
ω	$rad{\cdot}S^{-1}$	Frequency (angular)
ω_p,ω_s	$rad{\cdot}S^{-1}$	Pump and signal frequencies
ω_σ	$rad{\cdot}S^{-1}$	$\omega_1 + \omega_2 + \ldots + \omega_\gamma$
δ_{jm}	$m^2{\cdot}C^{-1}$	Miller δ-coefficient

Table 1.1.20 (continued)
LIST OF SYMBOLS AND ABBREVIATIONS

Symbol	Unit	Meaning
ϵ_φ		Dielectric constant at ω
ϵ_0	8.854×10^{-12} $C{,}V\cdot^{-1}m^{-1}$	Dielectric permittivity of free space
χ		Linear susceptibility
$\chi(-\omega_\sigma,\omega_1,\omega_2\ldots\omega_r)$	$(V\cdot m^{-1})^{1-r}$	r-th order macroscopic susceptibility
λ	m	Wave length
ν	Hz	Frequency
χ_{THG}	$m^2\cdot V^{-2}$	Third-order susceptibility
ρ	o	Walk-off angle
θ_m	o	Phase-matching angle
$\tilde{\nu}$	m^{-1}	Wave number

REFERENCES

1. **Armstrong, J. A., Bloembergen, N., Ducuing, J., and Pershan, P. S.,** Interactions between light waves in a nonlinear dielectric, *Phys. Rev.*, 127, 1918, 1962.
2. **Pershan, P. S.,** Nonlinear optical properties of solids: energy considerations, *Phys. Rev.*, 130, 919, 1963.
3. **Terhune, R. W. and Maker, P. D.,** Nonlinear optics, in *Lasers: A Series of Advances*, Vol. 2, Levine, A. K., Ed., Marcel Dekker, New York, 1968, 295.
4. **Butcher, P. N.,** Nonlinear Optical Phenomena, Bulletin 200, Engineering Experiment Station, Ohio State University, Columbus, 1965.
5. **Bloembergen, N.,** *Nonlinear Optics*, W. A. Benjamin, New York, 1965.
6. **Ovander, L. N.,** Nonlinear optical effects in crystals, *Sov. Phys. Uspekhi*, 8, 337, 1965.
7. **Minck, R. W., Terhune, R. W., and Wang, C. C.,** Nonlinear optics, *Appl. Opt.*, 5, 1595, 1966.
8. **Ducuing, J.,** Nonlinear optical processes, Proc. Int. School Phys. Enrico Fermi, in *Quantum Optics*, Glauber, R. J., Ed., Academic Press, New York, 1969, 421.
9. **Hanna, D. C., Yuratich, M. A., and Cotter, D.,** *Nonlinear Optics of Free Atoms and Molecules*, Springer-Verlag, New York, 1979.
10. **Zernike, F. and Midwinter, J.,** *Applied Nonlinear Optics*, John Wiley & Sons, New York, 1973.
11. **Kurtz, S. K.,** *Quantum Electronics: A Treatise*, Vol. 1, *Nonlinear Optics*, Robin, H. and Tang, C. L., Eds., Academic Press, New York, 1975, 209.
12. **Franken, P. A., Hill, A. E., Peters, C. W., and Weinreich, G.,** Generation of optical harmonics, *Phys. Rev. Lett.*, 7, 118, 1961.
13. **Bass, M., Franken, P. A., Ward, J. F., and Weinreich, G.,** Optical rectification, *Phys. Rev. Lett.*, 9, 446, 448, 1962.
14. **Franken, P. A. and Ward, J. F.,** Optical harmonics and nonlinear phenomena, *Rev. Mod. Phys.*, 35, 23, 1963.
15. **Giordmaine, J. A. and Miller, R. C.,** Tunable coherent parametric oscillations in $LiNbO_3$ at optical frequencies, *Phys. Rev. Lett.*, 14, 973, 1965.
16. **Zernike, F., Jr. and Berman, P. R.,** Generation of far-infrared as a difference frequency, *Phys. Rev. Lett.*, 15, 999, 1965.
17. **McMahon, D. H. and Frankin, A. R.,** Detection of nonlinear optical sum generation in ADP using incoherent light, *J. Appl. Phys.*, 36, 2807, 1965.

18. **Wang, C. C. and Racette, G. W.,** Measurement of parametric gain accompanying optical difference frequency generation, *Appl. Phys. Lett.*, 6, 169, 1965.
19. **Martin, M. D. and Thomas, E. L.,** The generation of molecular vibrational frequencies by optical mixing, *Phys. Lett.*, 19, 651, 1966.
20. **Akhmanov, S. A., Kovrigin, A. I., Kolosov, V. A., Piskarskas, A. S., Fadeev, V. V., and Khokhlov, R. V.,** Tunable parametric light generator with KDP crystal, *JETP Lett.*, 3, 241, 1966.
21. **Akhmanov, S. A., Kovrigin, A. I., Piskaraskas, A. S., Fadeev, V. V., and Khokhlov, R. V.,** Observation of parametric amplification in the optical range, *JETP Lett.*, 2, 191, 1965.
22. **Midwinter, J. E. and Warner, J.,** Up-conversion of near infrared to visible radiation in lithium-metaniobate, *J. Appl. Phys.*, 38, 519, 1967.
23. **Warner, J.,** Photomultiplier detection of 10.6μ radiation using optical up-conversion in proustite, *Appl. Phys. Lett.*, 12, 222, 1968.
24. **Giordmaine, J. A.,** Nonlinear optical properties of liquids, *Phys. Rev.*, 138A, 1599, 1965.
25. **Rentzepis, P. M., Giordmaine, J. A., and Wecht, K. W.,** Coherent optical mixing in optically active liquids, *Phys. Rev. Lett.*, 16, 792, 1966.
26. **Terhune, R. W., Maker, P. D., and Savage, C. M.,** Optical harmonic generation in calcite, *Phys. Rev. Lett.*, 8, 404, 1962.
27. **Maker, P. D., Terhune, R. W., and Savage, C. M.,** *Quantum Electronics III*, Grivet, P. and Bloembergen, N., Eds., Columbia University Press, New York, 1964, 1559.
28. **Maker, P. D. and Terhune, R. W.,** Study of optical effects due to an induced polarization third order in the electric field strength, *Phys. Rev.*, 137, A801, 1965.
29. **Chiao, R. Y., Garmire, E., and Townes, C. H.,** Self trapping of optical beams, *Phys. Rev. Lett.*, 13, 479, 1964.
30. **Maker, P. D., Terhune, R. W., and Savage, C. M.,** Intensity dependent changes in the refractive index of liquids, *Phys. Rev. Lett.*, 12, 507, 1964.
31. **Kaiser, W. and Garrett, C. G. B.,** Two photon excitation in CaF_2:E_u^{2+}, *Phys. Rev. Lett.*, 7, 229, 1961.
32. **Abella, I. D.,** *Phys. Rev. Lett.*, 9, 453, 1962.
33. **Peticolas, W. L., Goldsborough, J. P., and Rieckhoff, K. E.,** *Phys. Rev. Lett.*, 10, 43, 1963.
34. **Singh, S. and Stoicheff, B. P.,** Double photon excitation of fluorescence in anthracene single crystals, *J. Chem. Phys.*, 38, 2032, 1963.
35. **Hopfield, J. J., Worlock, J. M., and Kwangjai, P.,** Two quantum absorption spectrum of KI, *Phys. Rev. Lett.*, 11, 404, 1963.
36. **Giordmaine, J. A. and Howe, J. A.,** Intensity induced optical absorption cross section in CS_2, *Phys. Rev. Lett.*, 11, 404, 1963.
37. **Hall, J. L., Jennings, D. A., and McClintock, R. M.,** *Phys. Rev. Lett.*, 11, 364, 1964.
38. **Jones, W. J. and Stoicheff, B. P.,** Inverse Raman spectra: induced absorption at optical frequencies, *Phys. Rev. Lett.*, 13, 657, 1964.
39. **Hougen, J. T. and Singh, S.,** Electronic Raman effect in Pr^{3+} ions in single crystals of $PrCl_3$, *Phys. Rev. Lett.*, 10, 406, 1963.
40. **Elliot, R. J. and Loudon, R.,** The possible observation of electronic Raman transitions in crystals, *Phys. Lett.*, 3, 189, 1963.
41. **Koningstein, J. A. and Sonnich Mortenson, O.,** *Phys. Rev.*, 168, 75, 1968.
42. **Sorokin, P. P., Shiren, N. S., Lankard, J. R., Hammond, E. C., and Kazyaka, T. G.,** *Appl. Phys. Lett.*, 10, 44, 1967.
43. **Rokni, M. and Yatsiv, S.,** *Phys. Lett.*, 24A, 277, 1967.
44. **Eckhardt, G., Hellwarth, R. W., McClung, F. H., Schwarz, S. E., Weiner, D., and Woodbury, E. J.,** *Phys. Rev. Lett.*, 9, 455, 1962.
45. **Chiao, R. Y. and Townes, C. H.,** Stimulated Brillouin scattering and coherent generation of intense hypersonic waves in liquids, *Appl. Phys. Lett.*, 5, 84, 1964.
46. **Hagenlocker, E. E. and Rado, W. G.,** Stimulated Brillouin and Raman scattering in gases, *Appl. Phys. Lett.*, 7, 236, 1965.
47. **Garmire, E. and Townes, C. H.,** Stimulated Brillouin scattering in liquids, *Appl. Phys. Lett.*, 5, 84, 1964.
48. **Mayer, G. and Gires, F.,** Action of an intense light beam on the refractive index of liquids, *Compt. Rend.*, 258, 2039, 1964.
49. **Singh, S. and Bradley, L. T.,** Three photon absorption in naphlhalene crystals by laser excitation, *Phys. Rev. Lett.*, 12, 612, 1964.
50. **Boyle, L. L., Buckingham, A. D., Disch, R. L., and Dummar, D. A.,** *J. Chem. Phys.*, 45, 1318, 1966.
51. **Anastassakis, Iwasa, S., and Burstein, E.,** *Phys. Rev. Lett.*, 17, 1051, 1966.
52. **Patel, C. K. N., Slusher, R. F., and Fleury, P. A.,** Optical nonlinearities due to mobile carriers in semiconductors, *Phys. Rev. Lett.*, 17, 1011, 1966.
53. **Wynne, J. J.,** *Phys. Rev. Lett.*, 29, 650, 1972.

54. **Singh, S.,** Nonlinear Optical Materials, in *Handbook of Lasers,* Presley, R. J., Ed., CRC Press, Boca Raton, 1971, 489.
55. **Bhagvantam, S.,** *Crystal Symmetry and Physical Properties,* Academic Press, New York, 1966.
56. **Nye, J. F.,** *Physical Properties of Crystals,* Oxford University Press, London, 1964.
57. Standards on piezoelectric crystals, *Proc. IRE,* 37, 1378, 1949.
58. **Kleinman, D. A.,** Nonlinear dielectric polarization in optical media, *Phys. Rev.,* 126, 1977, 1962.
59. **Maker, P. D., Terhune, R. W., Nisenoff, M., and Savage, C. M.,** Effects of dispersion and focusing in the production of optical harmonics, *Phys. Rev. Lett.,* 8, 21, 1962.
60. **Jerphagnon, J. and Kurtz, S. K.,** Maker fringes: a detailed comparison of theory and experiment for isotropic and unaxial crystals, *J. Appl. Phys.,* 41, 1667, 1970.
61. **Giordmaine, J. A.,** Mixing of light beams in crystals, *Phys. Rev. Lett.,* 8, 19, 1962.
62. **Boyd, G. D., Ashkin, A., Dziedzic, J. M., and Kleinman, D. A.,** SHG of light with double refraction, *Phys. Rev.,* 137, A1305, 1965.
63. **Kleinman, D. A.,** *Phys. Rev.,* 128, 1761, 1962.
64. **Midwinter, J. E. and Warner, J.,** The effects of phase matching method and of unaxial symmetry on the polar distribution of second order non-linear optical polarization, *Br. J. Appl. Phys.,* 16, 1135, 1965.
65. **Boyd, G. D. and Kleinman, D. A.,** Parametric interaction of focused gaussian light beams, *J. Appl. Phys.,* 39, 3597, 1968.
66. **Hobden, M. V.,** Phase-matched second harmonic generation in biaxial crystals, *J. Appl. Phys.,* 38, 4365, 1967.
67. **Hiromasa, I., Hatsuhiko, N., and Humio, I.,** Generalized study on angular dependence of induced second-order nonlinear optical polarizations and phase matching in biaxial crystals, *J. Appl. Phys.,* 46, 3992, 1975.
68. **Born, M., and Wolfe, E.,** *Principles of Optics,* Pergamon Press, Oxford, 1965.
69. **Tsuboi, S.,** *Henko Kenbikyo* (Polarizing Microscope), Iwanami, Tokyo, 1965.
70. **Nelson, T. J.,** Digital light deflection, *Bell Syst. Tech. J.,* 43, 821, 1964.
71. **Kleinman, D. A., Ashkin, A., and Boyd, G. D.,** Second harmonic generation of light by focused laser beams, *Phys. Rev.,* 145, 338, 1966.
72. **Butcher, P. N. and McLean, T. P.,** The nonlinear constitutive relation in solids at optical frequencies, *Proc. Phys. Soc.,* 81, 219, 1963.
73. **Maker, P. D., Terhune, R. W., and Savage, C. M.,** Optical third harmonic generation, in *Quantum Electronics III,* Grivet, P. and Bloembergen, N., Eds., Columbia University Press, New York, 1964, 1576.
74. **Kleinman, D. A.,** Nonlinear dielectric polarization in optical media, *Phys. Rev.,* 126, 1977, 1962.
75. **Midwinter, J. E. and Warner, J.,** The effects of phase matching method and of crystal symmetry on the polar dependence of third order non-linear optical polarization, Br. *J. Appl. Phys.,* 16, 1667, 1965.
76. **Bey, P. P., Giuliani, J. F., and Rabin, H.,** Generation of a phase-matched optical third harmonic by introduction of anomalous dispersion into a liquid medium, *Phys. Rev. Lett.,* 19, 819, 1967.
77. **Harris, S. E. and Miles, R. B.,** Proposed third harmonic generation in phase matched metal vapors, *Appl. Phys. Lett.,* 19, 385, 1971.
78. **Kung, A. H., Young, J. F., and Harris, S. E.,** *Appl. Phys. Lett.,* 22, 301, 1973.
79. **Honig, R. E. and Kramer, D. A.,** *RCA Rev.,* 30, 285, 1969.
80. **Schins, H. E. J., Van Wijk, R. W. M., and Dorpema, B.,** *Z. Metallkd.,* 62, 330, 1971.
81. **Miles, R. B. and Harris, S. E.,** Optical third harmonic generation in alkali metal vapors, *IEEE J. Quantum Electron.,* 9, 470, 1973.
82. **Leonard, P. J.,** *At. Data Nucl. Data Tables,* 14, 21, 1974.
83. **Ashkin, A., Boyd, G. D., and Dziedzic, J. M.,** Observation of continuous optical harmonic generation with gas masers, *Phys. Rev. Lett.,* 11, 14, 1963.
84. **Francois,** CW measurement of the optical nonlinearity of ammonium dihydrogen phosphate, *Phys. Rev.,* 143, 597, 1966.
85. **Bjorkholm, J. E. and Siegman, A. E.,** Accurate CW measurements of optical second harmonic generation in ammonium dihydrogen phosphate and calcite, *Phys. Rev.,* 154, 851, 1967.
86. **Gandrud, W. G., Boyd, G. D., McFee, J. H., and Wehmeier, F. H.,** Nonlinear optical properties of Ag_3SbS_3, *Appl. Phys. Lett.,* 16, 59, 1970.
87. **Day, G. W.,** Linear and nonlinear optical properties of trigonal selenium, *Appl. Phys. Lett.,* 18, 347, 1971.
88. **Wynne, J. J. and Bloembergen, N.,** Measurement of the lowest order nonlinear susceptibility in III-V semiconductors by second harmonic generation with a CO_2 laser, *Phys. Rev.,* 188, 1211, 1969.
89. **Boyd, G. D., Kasper, H., and McFee, J. H.,** Linear and non-linear optical properties of $AgGaS_2$, $CuGaS_2$ and $CuInS_2$, and theory of the wedge technique for the measurement of nonlinear coefficients, *IEEE J. Quantum Electron.,* 7, 563, 1971.
90. **Khokhlov, R. V.,** *Zh. Eksperim i Teor. Fiz Pisma v Redaktsiy,* 6, 575, 1967; *ov. Phys. JETP Lett.,* 85, 1967.
91. **Harris, S. E., Oshman, M. K., and Byer, R. L.,** *Phys. Rev. Lett.,* 18, 732, 1967.

92. **Smith, R. G., Skinner, J. G., Nilsen, J. W., and Geusic, J. E.,** Solid State Device Research Conference, Santa Barbara, Calif., 1967.
93. **Byer, R. L. and Harris, S. E.,** Power and bandwidth of spontaneous parametric emmission, *Phys. Rev.*, 168, 1064, 1968.
94. **Johnston, W. D. and Kaminow, I. P.,** Contribution to optical nonlinearity in GaAs as determined from Raman scattering efficiencies, *Phys. Rev.*, 188, 1209, 1969.
95. **Otagurs, W. S., Wiemer-Avnear, E., and Porto, S. P. S.,** Determination of the second harmonic generation coefficient and the linear electro optic coefficient in $LiIO_3$ through oblique Raman phonon measurements, *Appl. Phys. Lett.*, 18, 499, 1971.
96. **Ward, J. F. and New, G. H. C.,** Optical third harmonic generation in gases by a focused laser beam, *Phys. Rev.*, 185, 57, 1969.
97. **Wang, C. C. and Baardsen, E. L.,** Study of optical third harmonic generation in reflection, *Phys. Rev.*, 185, 1079, 1969.
98. **Bloembergen, N., Burns, W. K., and Matsuoka, M.,** Reflected third harmonic generated by picosecond laser pulses, *Opt. Commun.*, 1, 195, 1969.
99. **Chang, R. K., Ducuing, J., and Bloembergen, N.,** Dispersion of the optical nonlinearity in semiconductors, *Phys. Rev. Lett.*, 15, 415, 1965.
100. **Simon, H. J. and Bloembergen, N.,** Second harmonic light generation in crystals with natural activity, *Phys. Rev.*, 171, 1104, 1968.
101. **Wynn, J. J.,** Optical third order mixing in GaAs, Ge, Si and InAs, *Phys. Rev.*, 178, 1295, 1969.
102. **Okada, M., Takizawa, K., and Ieiri, S.,** Interference method of accurate determination of the relative magnitudes of nonlinear optical coefficients, *J. Appl. Phys.*, 48, 205, 1977.
103. **Miller, R. C. and Nordland, W. A.,** Relative signs of nonlinear optical coefficients of polar crystals, *Appl. Phys. Lett.*, 16, 174, 1970.
104. **Eicher, H.,** Third-order susceptibility of alkali metal vapors, *IEEE J. Quantum Electron.*, QE-11, 121, 1975.
105. **Dawes, E. L.,** Optical third harmonic coefficients for the inert gases, *Phys. Rev.*, 169, 47, 1968.
106. **Sitz, P. and Yaris, R.,** Frequency dependence of the higher susceptibilities, *J. Chem. Phys.*, 49, 3546, 1968.
107. **Jha, S. S. and Bloembergen, N.,** Nonlinear optical susceptibilities in group IV and III-V semiconductors, *Phys. Rev.*, 171, 891, 1968; **Jha, S. S. and Bloembergen, N.,** Nonlinear optical coefficients in group IV and III-V semiconductors, *IEEE J. Quantum Electron.*, QE-4, 670, 1968.
108. **Flytzanis, C. and Ducuing, J.,** Second order optical susceptibility of III-V compounds, *Phys. Lett.*, 26A, 315, 1968: **Flytzanis, C. and Ducuing, J.,** Second order optical susceptibilities of III-V semiconductors, *Phys. Rev.*, 178, 1218, 1969.
109. **Miller, R. C.,** Optical second harmonic generation in piezoelectric crystals, *Appl. Phys. Lett.*, 15, 17, 1964.
110. **Garrett, C. G. B. and Robinson, F. N. G.,** Miller's phenomenological rule for computing nonlinear susceptibilities, *IEEE J. Quantum Electron.*, (correspondence), QE-2, 328, 1966.
111. **Robinson, F. N. H.,** Nonlinear optical coefficients, *Bell Syst. Tech. J.*, 46, 913, 1967.
112. **Garrett, C. G. B.,** Nonlinear optics, anharmonic oscillators and pyroelectricity, *IEEE J. Quantum Electron.*, QE-4, 70, 1968.
113. **Flytzanis, C.,** Study of second order susceptibilities of III-V and II-VI compounds using phillips model, *Compt. Rend.*, 267B, 555, 1968.
114. **Phillips, J. C. and VanVechtan, J. A.,** Nonlinear optical susceptibilities of covalent crystals, *Phys. Rev.*, 183, 709, 1969.
115. **Levine, B. F.,** Electrodynamical bond-charge calculation of nonlinear optical susceptibilities, *Phys. Rev. Lett.*, 22, 787, 1969.
116. **Levine, B. F.,** A new contribution to the nonlinear optical susceptibility arising from unequal atomic radii, *Phys. Rev. Lett.*, 25, 440, 1970.
117. **Bell, M. I.,** Frequency dependence of miller's rule for nonlinear susceptibilities, *Phys. Rev.*, B6, 516, 1971.
118. **Tang, C. L. and Flytzanis, C.,** Charge-transfer model of the nonlinear susceptibilities of polar semiconductors, *Phys. Rev.*, B4, 2520, 1971.
119. **Tang, C. L.,** A simple molecular orbital theory of the nonlinear optical properties of group III-V and II-VI compounds, *IEEE J. Quantum Electron.*, QE-9, 755, 1973.
120. **Scholl, F. and Tang, C. L.,** Nonlinear optical properties of II-IV-V_2 semiconductors, *Phys. Rev.*, B8, 4607, 1973.
121. **Kleinman, D. A.,** Nonlinear optical susceptibilities of covalent crystals, *Phys. Rev.*, B2, 3139, 1970.

122. (a) **Choy, M. M., Ciraci, S., and Byer, R. L.,** Bond-orbital model for second-order susceptibilities, *IEEE J. Quantum Electron.*, QE-11, 40, 1975; (b) **Chemla, D. S., Begley, R. F., and Byer, R. L.,** Experimental and theoretical studies of third-harmonic generation in the chalcopyrite $CdGeAs_2$, *IEEE J. Quantum Electron.*, QE-10, 71, 1974.
123. **Levine, B. F.,** Bond charge calculation of nonlinear optical susceptibilities of various crystal structures, *Phys. Rev.*, B7, 2600, 1973.
124. (a) **Phillips, J. C.,** A posteriori theory of covalent bonds, *Phys. Rev. Lett.*, 19, 415, 1967; (b) **Phillips, J. C.,** Covalent bands in crystals. I. element of the theory, *Phys. Rev.*, 166, 832, 1968; (c) **Phillips, J. C.,** Covalent bands in crystals. II. Partially icnic binding, *Phys. Rev.*, 168, 905, 1968; (d) **Phillips, J.C.,** Tonicity of the chemical bond in crystals, *Rev. Mod. Phys.*, 42, 317, 1970.
125. (a) **Van Vechtan, J. A.,** Quantum dielectric theory of electronegativity in covalent systems. I. Electronic dielectric constant, *Phys. Rev.*, 182, 891, 1969; (b) **Van Vechtan, J. A.,** Quantum dielectric theory of electro-negativity in covalent systems. II. Ionization potentials and interband transitions, *Phys. Rev.*, 187, 1007, 1969
126. **Wemple, S. H. and DiDomenico, M.,** *Applied Solid-State Science*, Vol. 2, Wolfe, R., Ed., Academic Press, New York, 1972.
127. **Jeggo, C. R. and Boyd, G. D.,** Nonlinear optical polarizability of the niobium — oxygen bond, *J. Appl. Phys.*, 41, 2741, 1970.
128. **Van Vechtan, J. A., Cardona, M., Aspnes, D. E., and Martin, R. M.,** *Proc. 10th Int. Cong. Phys. Semicond.*, 82, 1970.
129. **Flytzanis, C.,** Third order optical susceptibilities in IV-IV and III-V semiconductors, *Phys. Lett.*, 31A, 273, 1970.
130. **Wang, C. C.,** Empirical relation between the linear and the third order nonlinear optical susceptibilities, *Phys. Rev.*, B-2, 2045, 1970; (a) **Wang, C. C.,** Nonlinear susceptibility constants and self-focusing of optical beams in liquids, *Phys. Rev.*, 152, 149, 1966.
131. **Kung, A. H., Young, J. F., Bjorklund, G. S., and Harris, S. E.,** Generation of vacuum ultraviolet radiation in phase-matched Cd vapor, *Phys. Rev. Lett.*, 29, 985, 1972.
132. **Burns, W. K. and Bloembergen, N.,** Third-harmonic generation in absorbing media of cubic or isotropic symmetry, *Phys. Rev.*, B4, 3437, 1971.
133. **Wynne, J. J. and Boyd, G. D.,** Study of optical difference mixing in Ge and Si using a CO_2 gas laser, *Appl. Phys. Lett.*, 12, 191, 1968.
134. **Hermann, J. P.,** Absolute measurements of third order susceptibilities, *Opt. Commun.*, 9, 74, 1973.
135. **Masakatsu, O.,** Third order nonlinear optical coefficients of $LiIO_3$, *Appl. Phys. Lett.*, 18, 451, 1971.
136. **Levenson, M. D., Flytzanis, C., and Bloembergen, N.,** Interference of resonant and nonresonant three-wave mixing in diamond, *Phys. Rev.*, B6, 3962, 1972.
137. **Yablonovitch, E., Flytzanis, C., and Bloembergen, N.,** Anisotropic interference of three-wave and double two-wave frequency mixing in GaAs, *Phys. Rev. Lett.*, 29, 865, 1972.
138. **Shand, M. L.,** Enhancement of third order susceptibility due to electronic transilions in rare-earth ions, *J. Appl. Phys.*, 52, 1470, 1981.
139. **Jain, R. K. and Steel, D. G.,** Large optical nonlinearities and CW degenerate four-wave mixing in HgCdTe, *Opt. Commun.*, 43, 72, 1982.
140. **Hermann, J. P., Ricard, D., and Ducuing, J.,** Optical nonlinearities in conjugated systems: β-carotene, *Appl. Phys. Lett.*, 23, 178, 1973.
141. **Bloembergen, N., Burns, W. K., and Matsuoka, M.,** Reflected third harmonic generated by picosecond laser pulses, *Opt. Commun.*, 1, 195, 1969.
142. **Rado, W. G.,** The nonlinear third order dielectric susceptibility coefficients of gases and optical third harmonic generation, *Appl. Phys. Lett.*, 11, 123, 1967.
143. **Wang, C. C. and Baardsen, E. L.,** Optical third harmonic generation using mode-locked and nonmode-locked lasers, *Appl. Phys. Lett.*, 15, 396, 1969.
144. **Jain, R. K. and Steel, D. G.,** Degenerate four-wave mixing of 10.6 μm radiation in $Hg_{1-x}Cd_xTe$, *Appl. Phys. Lett.*, 37, 1, 1980.
145. **Jain, R. K. and Klein, M. B.,** Degenerate four-wave mixing near the band gap of semiconductors, *Appl. Phys. Lett.*, 35, 454, 1979.
146. **Kildal, H. and Deutsch, T. F.,** Infrared third harmonic generation in molecular gases, *IEEE J. Quantum Electron.*, QE-12, 429, 1976.
147. **Kildal, H.,** Infrared third harmonic generation in phase-matched CO-gas, *IEEE J. Quantum Electron.*, QE-13, 109, 1977.
148. **Reintjes, J.,** Third harmonic coversion of XeCl-laser radiation, *Opt. Lett.*, 4, 242, 1979.
149. **Kung, A. H., Young, J. F., Bjorklund, G. C., and Harris, S. E.,** Generation of vacuum ultraviolet radiation in phase-matched Cd vapor, *Phys. Rev. Lett.*, 29, 985, 1972.
150. **Leung, K. M., Ward, J. F., and Orr, B. J.,** Two-photon resonant optical third harmonic generation in cesium vapor, *Phys. Rev.*, A9, 2440, 1974.

151. **Young, J. F., Bjorklund, G. C., Kung, A. H., Miles, R. B., and Harris, S. E.,** Third harmonic generation in phase-matched Rb vapor, *Phys. Rev. Lett.*, 27, 1551, 1971.
152. **Levenson, M. D.,** Feasibility of measuring the nonlinear index of refraction by third order frequency mixing, *IEEE J. Quantum Electron.*, QE-10, 110, 1974.
153. **Aspnes, D. E.,** Energy-band theory of the second order nonlinear optical susceptibility of crystals of zinc-blend symmetry, *Phys. Rev.*, B6, 4648, 1972.
154. **Kurtz, S. K., Jerphagnon, J., and Choy, M. M.,** Nonlinear dielectric susceptibilities, *Landölt and Bornstein Data Ser.*, 11, 671, 1979.
155. **Levine, B. F. and Bethea, C. G.,** Nonlinear susceptibility of GaP: relative measurements and use of measured values to determine a better absolute value, *Appl. Phys. Lett.*, 20, 272, 1972.
156. **Jerphagnon, J. and Kurtz, S. K.,** Optical nonlinear susceptibilities: accurate relative values for quartz, ammonium dihydrogen phosphate, and potassium dihydrogen phosphate, *Phys. Rev.*, B1, 1738, 1970.
157. **Jerphagnon, J.,** Optical nonlinear susceptibilities of lithium iodate, *Appl. Phys. Lett.*, 16, 298, 1970.
158. **Nath, G., Mehmanesch, H., and Gsänger, M.,** Efficient conversion of a ruby laser to 0.347 μ in low loss lithium iodate, *Appl. Phys. Lett.*, 17, 286, 1970.
159. **Feliksinski, T. and Szewczyk, J.,** Distribution of impurities in lithium iodate crystals, *Mater. Res. Bull.*, 16, 1505, 1981.
160. **Choy, M. M. and Byer, R. L.,** Accurate second order susceptibility measurements of visible and infrared nonlinear crystals, *Phys. Rev.*, B14, 1693, 1976.
161. Cleveland Crystal Information Sheet: Electro-optic Properties of KH_2PO_4 and Ismorphs, unpublished data, 1976.
162. **Singh, H.,** unpublished data.
163. **Sonin, A., Filiminov, A. A., and Suvorov, V. S.,** Generation of the second harmonic of ruby laser radiation in some crystals, *Sov. Phys. Solid State.*, 10, 1481, 1968.
164. **Sonin, A. S. and Suvorov, V. S.,** Nonlinear optical properties of triglycine sulfate single crystals, *Sov. Phys. Solid State*, 9, 1437, 1967.
165. **Southgate, P. D. and Hall, D. S.,** Nonlinear optical susceptibility of a crystalline coumarin dye, *J. Appl. Phys.*, 42, 4480, 1971.
166. **Oudar, J. L. and Hierle, R.,** An efficient organic crystal for nonlinear optics: methyl-(2,4-nitrophenyl)-amino-propanoate, *J. Appl. Phys.*, 48, 2699, 1977.
167. **Gott, J. R.,** Effect of molecular structure on optical second harmonic generation from organic crystals, *J. Phys. B*, 4, 116, 1971.
168. **Halbout, J. M. and Tang, C. L.,** Phase-matched second harmonic generation in sucrose, *IEEE J. Quantum Electron.*, QE-18, 410, 1982.
169. **Levine, B. F., Bethea, C. G., Thurmond, C. D., Lynch, R. T., and Bernstein, J. L.,** An organic crystal with an exceptionally large optical second harmonic coefficient: 2-methyl-4-nitroaniline, *J. Appl. Phys.*, 50, 2523, 1979.
170. **Schuler, L., Betzler, K., Hesse, H., and Kapphan, S.,** Phase-matched second harmonic generation in potassium malate, *Opt. Commun.*, 43, 157, 1982.
171. **Mitsui, T., Iio, K., Hamano, K., and Sawada, S.,** Optical second harmonic generation in ammonium tartrate, *Opt. Commun.*, 9, 322, 1973.
172. **Kurtz, S. K. and Perry, T. T.,** A powder-technique for the evalution of nonlinear optical materials, *J. Appl. Phys.*, 39, 3798, 1968.
173. **Izrailenko, A. N., Yu Orlov, R., and Koptsik, V. A.,** Ammonium oxalate — a new nonlinear optical material, *Sov. Phys. Crystallogr.*, 13, 136, 1968.
174. **Kurtz, S. K. and Perry, T. T.,** Alpha-iodic acid: a solution-grown crystal for nonlinear optical studies and applications, *Appl. Phys. Lett.*, 12, 186, 1968.
175. **Crane, G. R.,** Nonlinear polarizability of alpha-iodic acid, *J. Chem. Phys.*, 62, 3571, 1975.
176. (a) **Bjorkholm, J. E.,** Relative measurements of the optical nonlinearities of KDP, ADP, $LiNbO_3$ and α-HIO_3, *IEEE J. Quantum Electron.*, QE-4, 970, 1968; (b) **Bjorkholm, J. E.,** Correction to relative measurement of the optical nonlinearities of KDP, ADP $LiNbO_3$ and α-HIO_3, *IEEE J. Quantum Electron.*, QE-5, 260, 1969.
177. **Singh, S., Bonner, W. A., Kyle, T., Potopowicz, J. R., and Van Uitert, L. G.,** Second harmonic generation in D-threonine, *Opt. Commun.*, 5, 131, 1972.
178. **Bechthold, P. S. and Haussühl, S.,** Nonlinear optical properties of orthorhombic barium formate and magnesium barium fluoride, *Appl. Phys.*, 14, 403, 1977.
179. **Nelson, M. and Delfino, M.,** Nonlinear dielectric susceptibilities, *Landölt and Bornstein Data Ser.*, 11, 671, 1979.
180. **Bergman, J. G., Crane, G. R., Levine, B. F., and Bethea, C. G.,** Nonlinear optical susceptibility of 5-nitrouracil, *Appl. Phys. Lett.*, 20, 21, 1972.

181. **Zyss, J., Chemla, D. S., and Nicoud, J. F.,** Demonstration of efficient nonlinear optical crystals with vanishing molecular dipole moment: second harmonic generation in 3-methyl-4-nitropyridine-1-oxide, *J. Chem. Phys.*, 74, 4800, 1981.
182. **Naito, H. and Inaba, H.,** *Opto-Electron.*, 4, 335, 1972.
183. **Adukov, A. D., Dobrghanskii, G. F., Kulevskii, L. A., and Polivanov, Y. N.,** *Kratkie Soobshch. Fiz.* No. 10, 40, 1972; *Chem. Abstr.*, 79, 25146W, 1973.
184. (a) **Geusic, J. E., Levinstein, H. J., Rubin, J. J., Singh, S., and Van Uitert, L. G.,** The nonlinear optical properties of $Ba_2NaNb_5O_{15}$, *Appl. Phys. Lett.*, 11, 269, 1967; (b) **Geusic, J. E.,** Erratum to the nonlinear optical properties of $Ba_2NaNb_5O_{15}$, *Appl. Phys. Lett.*, 12, 224, 1968.
185. **Singh, S., Draegert, D. A., and Geusic, J. E.,** Optical and ferroelectric properties of barium sodium niobate, *Phys. Rev.*, B2, 2709, 1970.
186. **Potopowicz, J. R.,** Temperature dependence of the refractive indeces of barium sodium niobate, unpublished.
187. **Miller, R. C., Nordland, W. A., and Nassau, K.,** Nonlinear optical properties of Gd_2 $(M_0O_4)_3$ and Tb_2 $(M_0O_4)_3$, *Ferroelectrics*, 2, 97, 1971.
188. **Miller, R. C. and Nordland, W. A.,** Absolute signs of second harmonic generation coefficients of piezo electric crystals, *Phys. Rev.*, B2, 4896, 1970.
189. **Singh, S., Potopowicz, J. R., Bonner, W. A., and Van Uitert, L. G.,** unpublished.
190. **Singh, S., Bonner, W. A., Potopowicz, J. R., and Van Uitert, L. G.,** Nonlinear optical properties of lead niobate, Electrochem. Soc. Meet., Los Angeles, May, 1970, R. C. Miller, private communication.
191. **Orlov, R. Y.,** Hippuric acid as a source of second harmonics in the optical range, *Sov. Phys. Crystallogr.*, 11, 410, 1966.
192. **Deserno, V. and Haussühl, S.,** Phase-matchable optical nonlinearity in strontium formate and strontium formate dihydrate, *IEEE J. Quantum Electron.*, QE-8, 608, 1972.
193. **Uematsu, Y.,** Nonlinear optical properties of $KNbO_3$ single crystal in the orthorhombic phase, *Jpn. J. Appl. Phys.*, 13, 1362, 1974.
194. **Singh, S., Bonner, W. A., Potopowicz, J. R., and Van Uitert, L. G.,** Nonlinear optical susceptibility of lithium formate monohydrate, *Appl. Phys. Lett.*, 17, 292, 1970.
195. **Miller, R. C., Nordland, W. A., Kolb, E. D., and Bond, W. L.,** Nonlinear optical properties of lithium gallium oxide, *J. Appl. Phys.*, 41, 3008, 1970.
196. **Lenzo, P. V., Spencer, E. G., and Remeika, J. P.,** *Appl. Opt.*, 4, 1036, 1965.
197. **Ito, H., Naito, H., and Inaba, H.,** New phase-matchable nonlinear optical crystals of the formate family, *IEEE J. Quantum Electron.*, QE-10, 247, 1974.
198. **Yandji, T., Iio, K., Hanadate, H., and Sawada, S.,** Study of phase transition of $NaNo_2$ by the second harmonic generation, *J. Phys. Soc. Jpn.*, Suppl. 28, 1970; Proc. Second Int. Meet. Ferroelectricity, 1969.
199. **Inoue, K. and Ishidate, T.,** Temperature dependence of optical nonlinear susceptibility of $NaNO_2$, *Ferroelectrics*, 7, 105, 1974.
200. **Hirotsu, S., Yanagi, T., and Sawada, S.,** Refractive indices of $NaNO_2$ and anisotropic polarizability of NO_2^-, *J. Phys. Soc. Jpn.*, 25, 799, 1968.
201. **Vander Ziel, J. P. and Bloembergen, N.,** Temperature dependence of optical harmonic generation in KH_2PO_4 ferroelectrics, *Phys. Rev.*, 35, A1622, 1964.
202. **Bergman, J. G., McFee, J. H., and Crane, G. R.,** Pyroelectricity and optical second harmonic generation in polyvinylidene fluoride films, *Appl. Phys. Lett.*, 18, 203, 1971.
203. **Southgate, P. D. and Hall, D. S.,** Second harmonic generation and Miller's delta parameter in a series of benzene derivatives, *J. Appl. Phys.*, 43, 2765, 1972.
204. **Bergman, J. G. and Crane, G. R.,** Nonlinear optical susceptibilities of triphenylbenzene, resorcinol, and meta nitroaniline, *J. Chem. Phys.*, 66, 3803, 1977.
205. **Carenco, A., Jerphagnon, J., and Perigaud, A.,** Nonlinear optical properties of some m-disubstituted benzene, *J. Chem. Phys.*, 66, 3806, 1977.
206. **Oudar, J. L. and Perigaud, A.,** Proprieter optiques non lineaires du metanitrobenzene, Nonlinear dielectric susceptibilities, *Landölt and Bornstein Data Ser.*, 11, 671, 1979.
207. **Belikova, G. S., Golovei, M. P., Shigorin, V. D., and Shipulo, G. P.,** Optical second harmonic generation in meta-dinitrogen crystals, *Opt. Spectrosc. USSR*, (English transl.) 38, 441, 1975.
208. **Ito, K., Kusuhara, Y., Hamano, K., and Sawado, S.,** Phase matchable optical nonlinearity in meta-dinitrobenzene, *Jpn. J. Appl. Phys.*, 13, 1299, 1974.
209. **Shigorin, V. D., Shipulo, G. P., Grazhulene, S. S., Musikhin, L. A., and Shekhtman, V. Sh.,** Nonlinear optical properties of molecular crystals of metatolylenediamine, *Sov. J. Quantum Electron.*, 5, 1393, 1976.
210. **Belyaev, G. S., Belikova, G. S., Gilvarg, A. G., Golovei, M. P., Kalinkina, I. N., and Kosowrov, G. I.,** Nonlinear optical properties of potassium diphthalate crystal, *Opt. Spectrosc. USSR*, (English transl.), 29, 522, 1970.

211. **Popolitov, V. I., Ivanova, L. A., Stephanovitch, S. Yu., Chetchkin, V. V., Lobachev, A. N., and Venevtsev, Yu. N.,** Ferroelectrics ABO_4: synthesis of single crystals and ceramics; dielectric and nonlinear optical properties, *Ferroelectrics,* 8, 519, 1974.
212. **Boyd, G. D., Kasper, H. M., and McFee, J. H.,** Linear and nonlinear optical properties of $LiInS_2$, *J. Appl. Phys.,* 44, 2809, 1973.
213. **Morosin, B., Bergman, J. G., and Crane, G. R.,** Crystal structure, linear and nonlinear optical properties of $Ca(IO_3)_2{\cdot}GH_2O$, *Acta Crystallogr.,* B29, 1067, 1973.
214. **Bergman, J. G. and Crane, G. R.,** Structural aspects of nonlinear optics: optical properties of KIO_2F_2 and its related iodates, *J. Chem. Phys.,* 60, 2470, 1974.
215. **Tofield, B., Crane, G. R., and Bergman, J. G.,** Structural aspects of nonlinear optics: optical properties of $K_2H\ (IO_3)_2Cl$ and related compounds, *J. Chem. Soc. Faraday Trans. II,* 70, 1488, 1974.
216. **Bergman, J. G., Crane, G. R., and Guggenheim, H.,** Linear and nonlinear optical properties of ferroelectric $BaMgF_4$ and $BaZnF_4$, *J. Appl. Phys.,* 46, 4645, 1975.
217. **Recker, K., Wallrafen, F., and Haussühl, S.,** Single crystal growth and optical, elastic and piezoelectric properties of polar magnesium barium fluoride, *J. Cryst. Growth,* 26, 97, 1974.
218. **Bergman, J. G. and Crane, G. R.,** Linear and nonlinear optical properties of $K_2M_0OS_3{\cdot}KCl$; *J. Appl. Phys.,* 46, 3530, 1975.
219. **Dewey, H. J.,** Second harmonic generation in $KB_5O_8{\cdot}4H_2O$ from 217.1 to 315.0 nm, *IEEE J. Quantum Electron.,* QE-12, 303, 1976.
220. **Dewey, C. F., Cook, W. R., Hodgson, R. T., and Wynne, J. J.,** Frequency doubling in $KB_5O_8{\cdot}4H_2O$ and $NH_4B_5O_8{\cdot}4H_2O$ to 217.3 nm, *Appl. Phys. Lett.,* 26, 714, 1975; (a) **Cook, W. R. and Hubby, L. M.,** Indices of refraction of potassium pentaborate, *J. Opt. Soc. Am.,* 66, 72, 1976.
221. **Dunning, F. B. and Stickel, R. E., Jr.,** Sun frequency mixing in potassium pentaborate as a source of tunable coherent radiation at wavelength below 217 nm, *Appl. Opt.,* 15, 313, 1976.
222. **Zumsteg, F. C., Bierlein, J. D., and Gier, T. E.,** $K_xRb_{1-x}T_iOPO_4$: a new nonlinear optical material, *J. Appl. Phys.,* 47, 4980, 1976.
223. **Zyss, J., Chemla, D. S., Vilminot, S., Cot, L., and Maurin, M.,** Etudes des propriétés optiques lineaáres et non lineáres du fluoroberyllate de lithuim hydrazine, *Rev. Phys. Appl.,* 12, 1767, 1977.
224. **Bergman, J. G., Ginsberg, A. P., and Maurine, M.,** Polarity of Chemical bonds via Hyperpolarizabilities: Determination of the Direction of the P-H Bond dipole in $L_iH_2PO_3$ from Measurements of Nonlinear Optical Coefficients, BTL TM 79-1535-34, Murray Hill, N.J., 1979.
225. **Chemla, D. S., Beys, L., and Hillaire, P.,** Nonlinear optical properties of lithium hydrogen phosphite, *Opt. Commun.,* 32, 187, 1980.
226. **Miller, R. C., Kleinman, D. A., and Savage, A.,** Quantitative studies of optical harmonic generation in CdS, $BaTiO_3$ and KH_2PO_4 type crystals, *Phys. Rev. Lett.,* 11, 146, 1963.
227. **Miller, R. C. and Nordland, W. A.,** Absolute signs of nonlinear optical coefficients of polar crystals, *Opt. Commun.,* 1, 400, 1970.
228. **Singh, S., Potopowicz, J. R., Camlibel, I., and Wemple, S. H.,** Nonlinear optical properties of melt grown $BaTiO_3$, unpublished.
229. **Singh, S., Remeika, J. P., and Potopowicz, J. R.,** Nonlinear optical susceptibilities of ferroelectric lead titanate, *Appl. Phys. Lett.,* 20, 135, 1972.
230. (a) **Van Uitert, L. G., Singh, S., Levinstein, H. J., Geusic, J. E., and Bonner, W. A.,** A new and stable nonlinear optical material, *Appl. Phys. Lett.,* 11, 161, 1967; (b) **Van Uitert, L. G., et al.,** Errata, A new and stable nonlinear Optical material, *Appl. Phys. Lett.,* 12, 224, 1968.
231. **Singh, S. and Ballman, A. A.,** unpublished data.
232. **Smith, A. W., Burns, G., and O'Kane, D. F.,** Optical and ferroelectric properties of $K_xNa_{1-x}Ba_2Nb_5O_{15}$, *J. Appl. Phys.,* 42, 250, 1971.
233. **Smith, A. W., Burns, G., and Scott, B. A.,** Nonlinear optical properties of potassium — lithium niobates, *J. Appl. Phys.,* 42, 684, 1971.
234. **Singh, S., Bonner, W. A., and Van Uitert, L. G.,** Violation of Kleinman's symmetry condition in paratellurite, *Phys. Lett.,* 38A1, 407, 1972.
235. **Chemla, D. S. and Jerphagnon, J.,** Optical second harmonic generation in paratellurite and Kleinman's symmetry relations, *Appl. Phys. Lett.,* 20, 222, 1972.
236. **Levine, B. F.,** Magnitude and dispersion of Kleinman forbidden nonlinear optical coefficients, *IEEE J. Quantum Electron.,* QE-9, 946, 1973.
237. **Uchida, N.,** Optical properties of single crystal paratellurite (TeO_2), *Phys. Rev.,* B4, 3736, 1971.
238. **Itho, Y. and Iwasaki, H.,** Ferroelectric and optical properties of $Ba_6Ti_2Nb_8O_{30}$ single crystals, *J. Phys. Chem. Solids.,* 34, 1639, 1973.
239. **Sturmer, W. and Deserno, U.,** Mercury thiocyanate-complexes: efficient phase-matchable optical SHG in crystal class 10, *Phys. Lett.,* 32A, 539, 1970.
240. **Bergman, J. G., McFee, J. H., and Crane, G. R.,** Nonlinear optical properties of $CdHg(SCN)_4$ and $ZnHg(SCN)_4$, *Mater. Res. Bull.,* 5, 913, 1970.

241. **Levine, B. F., Bethea, C. G., and Kasper, H. M.,** Nonlinear optical susceptibility of thiogallate $CdGaS_4$, *IEEE J. Quantum Electron.*, QE-10, 904, 1974.
242. **Levine, B. F., Bethea, C. G., Kasper, H. M., and Thiel, F. A.,** Nonlinear optical susceptibility of $HgGa_2S_4$, *IEEE Correspondence Ann.*, No. 606QE008, 367, 1976.
243. **Bridenbaugh, P. M.,** Nonlinear optical properties of $InPS_4$, *Mater. Res. Bull.*, 8, 1055, 1973.
244. **Ishidate, T., Inoue, K., Kameyama, H., and Ishibashi, Y.,** Temperature dependence of second harmonic generation in $Ca_2Sr\ (CH_3CH_2CO_2)_6$, *J. Phys. Soc. Jpn.*, 37, 1176, 1974.
245. **Bergman, J. G. and Jerphagnon, J.,** Linear and nonlinear optical properties of methyl-3-isopropyl-4 phenol, *Chem. Phys. Lett.*, 49, 324, 1977.
246. **McMahon, D. H. and Franklin, A. R.,** Laser focusing effects on second harmonic generation in ADP, *Appl. Phys. Lett.*, 6, 14, 1965.
247. **Suvorov, V. S., Sonin, A. S., and Rez, I. S.,** Some nonlinear optical properties of crystals of the KDP group, *Soc. Phys. JETP*, 26, 33, 1968.
248. **Zernike, F.,** Refractive indices of ammonium dihydrogen phosphate and potassium dihydrogen phosphate between 2000Å and 1.5μ, *J. Opt. Soc. Am.*, 54, 1215, 1964.
249. **Phillips, R. A.,** Temperature variation of the index of refraction of ADP, KDP and deuterated KDP, *J. Opt. Soc. Am.*, 56, 629, 1966.
250. **Miller, R. C. and Savage, A.,** Harmonic generation and mixing of $CaWO_4:Nd^{3+}$ and ruby pulsed laser beams in piezoelectric crystals, *Phys. Rev.*, 128, 2175, 1962.
251. **Ghosh, G. C. and Bhar, G. C.,** Temperature dispersion in ADP, KDP and KD*P for nonlinear devices, *IEEE J. Quantum Electron.*, QE-18, 143, 1982.
252. **Suvorov, V. S. and Sonin, A. S.,** Nonlinear optical materials, *Sov. Phys. Crystallogr.*, 11, 711, 1967.
253. **Golovei, M. P., Dobrzhanskii, G. F., Kosourov, G. I., Kalinkina, I. N., Kortukova, E. I., Likhacheva, Yu. S., and Ogadzhanova, V. V.,** Growth of $BeSO_4{\cdot}4H_2O$ crystals and a study of some of their physical properties, *Sov. Physics Crystallogr.*, 15, 651, 1971.
254. **Kato, K.,** Second harmonic generation at 2660Å in $BeSO_4{\cdot}4H_2O$, *Appl. Phys. Lett.*, 33, 413, 1978.
255. **Byer, R. L., Kildal, H., and Feigelson, R. S.,** $CdGeAs_2$ — a new nonlinear crystal phase matchable at 10.6μm, *Appl. Phys. Lett.*, 19, 237, 1971.
256. **Boyd, G. D., Buehler, E., Storz, F. G., and Wernick, J. H.,** Linear and nonlinear optical properties of ternary $A^{II}B^{IV}C_2^{V}$ chalcopyrite semiconductors, *IEEE J. Quantum Electron.*, QE-8, 419, 1972.
257. **Bhar, G. C.,** Refractive index interpolation in phase-matching, *Appl. Opt.*, 15, 305, 1976.
258. **Kato, K.,** Second harmonic generation in CDA and CD'A, *IEEE J. Quantum Electron.*, QE-10, 616, 1974.
259. **Boyd, G. D., Kasper, H. M., McFee, J. H., and Storz, F. G.,** Linear and nonlinear optical properties of some ternary selenides, *IEEE J. Quantum Electron.*, QE-8, 900, 1972.
260. **Winchell, A. N. and Winchell, H.,** *The Microscopic Characteristics of Artificial Inorganic Solid Substances,* Academic Press, New York, 1964.
261. **Hagen, W. F. and Magnante, P. C.,** Efficient second harmonic generation with diffraction limited and high-spectral-radiance Nd-glass lasers, *J. Appl. Phys.*, 40, 219, 1969.
262. **Kato, K.,** High efficient UV generation at 3472Å in RDA, *IEEE J. Quantum Electron.*, QE-10, 622, 1974.
263. **Sullivan, S. and Thomas, E. L.,** A comparison of ruby second harmonic generation in RDA and other KDP isomorphs, *Opt. Commun.*, 25, 125, 1978.
264. **Pearson, J. E., Evans, G. A., and Yariv, A.,** Measurement of the relative nonlinear coefficients of KDP, RDP, RDA and $LiIO_3$, *Opt. Commun.*, 4, 366, 1972.
265. (a) **Adhav, R. S.,** Some physical properties of single crystals of normal and deuterated rubidium dihydrogen arsenate. II. Electro-optic and dielectric properties, *Br. J. Appl. Phys. (J. Phys. D)*, 2, 1969; (b) **Adhav, R. S.,** Some physical properties of single crystals of normal and deuterated potassium dihydrogen arsenate. II. Electro-optic and dielectric properties, *J. Appl. Phys.*, 39, 4095, 1968.
266. **Singh, S., Potopowicz, J-R., Bonner, W. A., and Van Uitert, L. G.,** Nonlinear optical properties of rubidium dihydrogen phosphate, unpublished.
267. **Vasilevskaya, A. S., Koldobskaya, M. F., Lamova, C. G., Popova, V. P., Regulskaya, T. A., Rez, I. S., Sobesskii, Yu. P., Sonin, A. S., and Suvorov, V. S.,** Some physical properties of rubidium dihydrogen phosphate single crystals, *Sov. Phys. Crystallogr.*, 12, 383, 1967.
268. **Kildal, H. and Mikkelsen, J.,** The nonlinear optical coefficient, phase matching and optical damage in the chalcopyrite $AgGaSe_2$, *Opt. Commun.*, 9, 315, 1973.
269. **Byer, R. L., Choy, M. M., Herbst, R. L., Chemla, D. S., and Feigelson, R. S.,** Second harmonic generation and infrared mixing in $AgGaSe_2$, *Appl. Phys. Lett.*, 14, 65, 1974.
270. **Chemla, D. S., Kupecek, P. J., Robertson, D. S., and Smith, R. C.,** Silver thiogallate. A new material with potential for infrared devices, *Opt. Commun.*, 3, 29, 1971.
271. **Kupecek, P. J., Schwartz, Ch. A., and Chemla, D. S.,** Silver thiogallate ($AgGaS_2$). I. Nonlinear optical properties, *IEEE J. Quantum Electron.*, QE-10, 540, 1974.
272. **Bhar, G. C. and Smith, R. D.,** Silver thiogallate ($AgGaS_2$). II. Linear optical properties, *IEEE J. Quantum Electron.*, QE-10, 546, 1974.

273. **Boyd, G. D., Buehler, E., and Storz, F. G.,** Linear and nonlinear optical properties of $ZnGeP_2$ and CdSe, *Appl Phys. Lett.*, 18, 301, 1971.
274. **Betzler, K., Hess, H., and Loose, P.,** Optical second harmonic generation in organic crystals: urea and ammonium malate, *J. Mol. Struct.*, 47, 393, 1978.
275. **Miller, R. C., Norland, W. A., and Ballman, A. A.,** Nonlinear optical properties of ferroelectric $Pb_5Ge_3O_{11}$, *Opt. Commun.*, 6, 210, 1972.
276. **Crane, G. R. and Bergman, J. G.,** The Iodine Coordination Polyhedron in Hydrated Sodium Periodate Determined from Second Harmonic Generation, BTL TM 77-1527-9, Bell Laboratories, Murray Hill, N.J., 1977.
277. **Jerphagnon, J.,** Optical second harmonic generation in isocyclic and heterocyclic organic compounds, *IEEE J. Quantum Electron.*, QE-7, 42, 1971.
278. **Kovalchuck, V. M. and Perekalina, Z. B.,** Second harmonic generation in crystals of dithionales, *Sov. Phys. Cryst.*, 17, 355, 1972.
279. **Kisel, V. A., Klimova, A. Yu., Kovalchuck, V. M., and Perekalina, Z. B.,** Second harmonic generation in lead dithionate, *Sov. Phys. Solid State*, 15, 625, 1973.
280. **Boyd, G. D., Bridges, T. J., and Burkhardt, E. G.,** Up conversion of 10.6μ radiation to the visible and second harmonic generation in HgS, *IEEE J. Quantum Electron.*, QE-4, 515, 1968.
281. **Jerphagnon, J., Batifol, E., Tsoucaris, G., and Sourbe, M.,** Generation de second harmonique dans le cinabre, *C. R. Acad. Sci. Paris Ser. B*, 265, k495, 1967.
282. **Kupecek, P. J., Chemla, D. S., and LePerson, H.,** Etude des proprietes et de la qualite' de cristaux de cinabre synthetique en vue d'applications en optique non lineaire, *Rev. Phys. Appl.*, 11, 285, 1976.
283. **Bond, W. L., Boyd, G. D., and Carter, H. L., Jr.,** Refractive indices of Hgs (cinnabar) between 0.62 and 11μ, *J. Appl. Phys.*, 38, 4090, 1967.
284. **Dorozhkin, L. M., Kuratev, I. I., Leonyuk, N. I., Timchenko, T. I., and Shestakov, A. V.,** Optical second harmonic generation in a new nonlinear active medium: neodymium-yttrium-aluminum borate crystals, *Sov. Tech. Phys. Lett.*, 7(11), 555, 1982.
285. **Hobden, M. V., Robertson, D. S., Davies, P. H., Hulme, K. F., Warner, J., and Midwinter, J.,** Properties of a phase-matchable nonlinear optical crystal: Potassium dithionate, *Phys. Lett.*, 22, 65, 1966.
286. **Crane, G. R. and Bergman, J. G.,** Violations of Kleinman symmetry in nonlinear optics — the forbidden coefficient of α-quartz, *J. Chem. Phys.*, 64, 27, 1975.
287. **Gray, D. E., Ed.,** *American Institute of Physics Handbook*, 2nd ed., McGraw-Hill, New York, 1963.
288. **Patel, C. K. N.,** Optical harmonic generation in the infrared using a CO_2 laser, *Phys. Rev. Lett.*, 16, 613, 1966.
289. **Jerphagnon, J., Batifol, E., and Sourbe, M.,** Generation de second harmonique d'um rayonnement laser dans le selenium et dans un alliage tellure-seleneum, *C. R. Acad. Sci. Paris*, 265, 400, 1967.
290. **Sherman, G. H. and Coleman, P. D.,** Measurement of the second harmonic generation nonlinear susceptibilities of Se, CdTe, and InSb at 28.0μm by comparison with Te, *J. Appl. Phys.*, 44, 238, 1973.
291. **Gampel, L. and Johnson, F. M.,** Index of refraction of single-crystal selenium, *J. Opt. Soc. Am.*, 59, 72, 1969.
292. **Patel, C. K. N.,** Efficient-phase-matched harmonic generation in Telluruim with a CO_2 laser at 10.6μm, *Phys. Rev. Lett.*, 15, 1027, 1965.
293. **Jerphagnon, J., Sourbe, M., and Batifol, E.,** Addition dans la tellure de deux rayonnements produits par un laser CO_2, *C. R. Acad. Sci. Paris*, 263, 1067, 1966.
294. **Jerphagnon, J.,** *Ann. Telecommun.*, 23, 203, 1968.
295. **Van Tran, N.,** Properties optiques nonlineares de tellure *L'Onde Electronique*, 48, 965, 1967.
296. **McFee, J. H., Boyd, G. D., and Schmidt, P. H.,** Redetermination of the nonlinear optical coefficients of Te and GaAs by comparison with Ag_3SbS_3, *Appl. Phys. Lett.*, 17, 57, 1970.
297. **Sherman, G. H. and Coleman, P. D.,** Absolute measurement of the second harmonic generation nonlinear susceptibility of tellurium at 28.0μm, *IEEE J. Quantum Electron.*, QE-9, 403, 1973.
298. **Panyakeow, Y., Tanigaki, Y., Shirafuji, J., and Inuishi, Y.,** Temperature dependence of second harmonic generation in tellurium with a CO_2 laser at 10.6μ, *J. Appl. Phys.*, 43, 4268, 1972.
299. **Taynai, J. D., Targ, R., and Tiffany, W. B.,** Investigation of tellurium for frequency doubling with CO_2 lasers, *IEEE J. Quantum Electron.*, 7, 1971.
300. **Sherman, G. H., Coleman, P. D., and Keezer, R. C.,** *Proc. Symp.* Submillimeter Waves, Polytechnic, Brooklyn, 1971, 165.
301. **Caldwell, R. S.,** Special Report, Contract DA36-039-SC-71131, Department of Physics, Purdue University, West Lafayette, In., January 1958.
302. **Miller, R. C. and Nordland, W. A., Jr.,** Optical second harmonic generation coefficients for guanidinium aluminum sulfate hexahydrate, *J. Appl. Phys.*, 45, 898, 1974.
303. **Smith, A. W.,** Optical harmonic generation in two ferroelectric crystals, *Appl. Opt.*, 3, 147, 1964.
304. **Boyd, G. D., Miller, R. C., Nassau, K., Bond, W. L., and Savage, A.,** $LiNbO_3$: an efficient phase-matachable nonlinear optical material, *Appl. Phys. Lett.*, 5, 1964.

305. **Miller, R. C. and Savage, A.**, Temperature dependence of the optical properties of ferroelectric $LiNbO_3$ and $LiTaO_3$, *Appl. Phys. Lett.*, 9, 169, 1966.
306. **Miller, R. C., Nordland, W. A., and Bridenbough, P. M.**, Dependence of second harmonic generation coefficients of $LiNbO_3$ on melt composition, *J. Appl. Phys.*, 42, 4145, 1971.
307. **Smith, D. S., Riccius, H. D., and Edwin, R. P.**, Refractive indices of lithium niobate, *Opt. Commun.*, 17, 332, 1976.
308. **Nelson, D. F. and Mikulyak, R. M.**, Refractive indices of congruently melting lithium niobate, *J. Appl. Phys.*, 45, 3688, 1974.
309. **Midwinder, J. E.**, Lithium niobate: effects of composition on the refractive indices and optical second harmonic generation, *J. Appl. Phys.*, 32, 3033, 1968.
310. **Bond, W. L.**, Measurement of the refractive indices of several crystals, *J. Appl. Phys.*, 36, 1674, 1965.
311. **Feichtner, J. D., Johannes, R., and Roland, G. W.**, Growth and optical properties of single crystal pyragyrite (Ag_3SbS_3), *Appl. Opt.*, 9, 1716, 1970.
312. **Hulme, K. F., Jones, O., Davies, P. H., and Hobden, M. V.**, Synthetic proustile (Ag_3AsS_3): a new crystal for optical mixing, *Appl. Phys. Lett.*, 10, 133, 1967.
313. **Boggett, D. M. and Gibson, A. F.**, Second harmonic generation in proustile, *Phys. Lett.*, 28A, 33, 1968.
314. **Chemla, D. S., Kupecek, Ph. J., and Schwartz, C. A.**, Redetermination of the nonlinear optical coefficients of proustile by comparison with pyrargyrite and gallium arsenide, *Opt. Commun.*, 7, 225, 1973.
315. **Hobden, M. V.**, *Opt. Electron.*, 1, 159, 1969.
316. **Ernst, G. J. and Willerman, W. J.**, Second harmonic generation in proustile with a CW CO_2 laser, *IEEE J. Quantum Electron.*, QE-8, 382, 1972.
317. **Singh, S., Potopowicz, J. R., and Pawelek, R.**, Second harmonic properties of tourmaline, unpublished.
318. **Bergman, J. G.**, Structural Aspects of Nonlinear Optics: Structure and Hyperpolarizability of $TlIO_3$, BTL TM 75-1527-75, Bell Laboratories, Murray Hill, N.J., 1975.
319. **Feichtner, J. D. and Roland, G. W.**, Optical properties of a new nonlinear optical material: Tl_3AsSe_3, *Appl. Opt.*, 11, 993, 1972.
320. **Nath, G. and Haussühl, S.**, Large nonlinear optical coefficient and phase matched second harmonic generation in $LiIO_3$, *Appl. Phys. Lett.*, 14, 154, 1969.
321. **Nash, F. R., Bergman, J. G., Boyd, G. D., and Turner, E. H.**, Optical nonlinearities in $LiIO_3$, *J. Appl. Phys.*, 40, 5201, 1969.
322. **Campillo, A. J. and Tang, C. L.**, Spontaneous parametric scattering of light in $LiIO_3$, *Appl. Phys. Lett.*, 16, 424, 1970.
323. **Singh, S., Potopowicz, J. R., Bonner, W. A., and Van Uitert, L. G.**, Second harmonic coefficient and temperature dependence of refractive indices of $LiIO_3$, unpublished.
324. **Okada, M. and Ieiri, S.**, Kleinman's symmetry relation in nonlinear optical coefficient of $LiIO_3$, *Phys. Lett.*, 34A, 63, 1971.
325. **Umegaki, S., Tanaka, S. I., Uchiyama, T., and Yabumoto, S.**, Refractive indices of lithium iodate between 0.4 and 2.2μ, *Opt. Commun.*, 3, 244, 1971.
326. **Crettez, J. M., Comte, J., and Coquet, E.**, Optical properties of α and β lithium iodate in the visible range, *Opt. Commun.*, 6, 26, 1972.
327. **Abdullaev, G. B., Kulevskii, L. A., Prokhorov, A. M., Savel'ev, A. D., Salaev, E. Yu., and Smirnov, V. V.**, A new effective material for nonlinear optics, *JETP Lett.*, 16, 90, 1972.
328. **Kupecek, Ph., Batifol, E., and Kuhn, A.**, Conversion de frequences optiques dans le selenuire de galluim (GaSe), *Opt. Commun.*, 11, 291, 1974.
329. **Akundov, G. A., Agaeve, A. A., Salmonov, V. M., Sharorov, Yu. P., and Yaroshetskii, I. D.**, Second harmonic generation in III-VI compounds, *Sov. Phys. Semicond.*, 7, 826, 1973.
330. **Jerphagnon, J. and Newkirk, H. W.**, Optical nonlinear susceptibilities of berylluim oxide, *Appl. Phys. Lett.*, 18, 245, 1971.
331. **Newkirk, H. W., Smith, D. K., and Kahn, J. S.**, Synthetic bromellite III — some optical properties, *Am. Mineral.*, 51, 141, 1966.
332. **Soref, R. A. and Moos, H. W.**, Optical second harmonic generation in ZnS-Cds and CdS-CdSe alloys, *J. Appl. Phys.*, 35, 2152, 1964.
333. **Herbst, R. L. and Byer, R. L.**, Efficient mixing in CdSe, *Appl. Phys. Lett.*, 19, 527, 1971.
334. **Bieniewski, T. M. and Czyzak, S. J.**, Refractive indexes of single hexagonal ZnS and CdS crystals, *J. Opt. Soc. Am.*, 53, 496, 1963.
335. **Fujii, Y., Yoshida, S., Misawa, S., Maekawa, S., and Sakudo, T.**, Nonlinear optical susceptibilities of AlN film, *Appl. Phys. Lett.*, 31, 815, 1977.
336. **Bergman, J. G., Williams, D., Crane, G. R., and Storey, R. N.**, Linear and nonlinear optical properties of lithium perchlorate trihydrate $LiClO_4{\cdot}3H_2O$, *Appl. Phys. Lett.*, 26, 571, 1975.
337. **Singh, S., Potopowicz, J. R., Van Uitert, L. G., and Wemple, S. H.**, Nonlinear optical properties of hexagonal silicon carbide, *Appl. Phys. Lett.*, 19, 53, 1971; **Miller, R. C.**, private communication.

338. **Levine, B. F., Nordland, W. A., and Shiever, J. W.,** Nonlinear optical susceptibility of AgI, *IEEE J. Quantum Electron.*, QE-9, 468, 1973.
339. **Levine, B. F., Bethea, C. G., Lambrecht, V. G., Jr., and Robbins, M.,** Nonlinear optical properties of Zn_3AgInS_5 and Zn_5AgInS_7, *IEEE J. Quantum Electron.*, QE-9, 258, 1973.
340. **Miller, R. C., Abrahams, S. C., Barns, R. L., Bernstein, J. L., and Nordland, W. A.,** Absolute signs of the second harmonic generation, electro-optic, and piezoelectric coeffients of CuCl and ZnS, *Solid State Commun.*, 9, 1463, 1971.
341. **Singh, S.,** unpublished.
342. **Chandrasekhar, S. and Madhav, M. S.,** Optical rotary dispersion of crystals of sodium chlorate and sodium bromate, *Acta Crystallogr.*, 23, 911, 1967.
343. **Chandrasekhar, S.,** Optical rotary dispersion of crystals, *Proc. R. Soc.*, A259, 531, 1961.
344. **Braunstein, R. and Ockman, N.,** Interactions of Coherent Optical Radiation with Solids, Final report prepared for the Office of Naval Research, Contract #N ONR-4128100 Arpa order #306-62, Department of the Navy, Washington, D.C., 1964.
345. **Singh, S. and Pawelek, R.,** unpublished.
346. **Marple, D. T. F.,** Refractive index of ZnSe, ZnTe and CdTe, *J. Appl. Phys.*, 35, 539, 1964.
347. **Harvey, J. E. and Wolfe, W. E.,** Refractive index of irtran 6 (hot pressed cadmium telluride) as a function of wavelength and temperature, *J. Opt. Soc. Am.*, 65, 1267, 1975.
348. **Barnes, N. P. and Plitch, M. S.,** Temperature-dependent sellmeier coefficients and coherence length for cadmium telluride, *J. Opt. Soc. Am.*, 67, 628, 1977.
349. **Jerphagnon, J., Schwab, C., and Chemla, D.,** Generation de second harmonique dans le chlorure cuivreux, *C. R. Acad. Sci. (Paris)*, 265B, 1032, 1967.
350. **Chemla, D., Kupecek, P., Schwartz, C., Schwab, C., and Goltzene, A.,** Nonlinear properties of cuprous halides, *IEEE J. Quantum Electron.*, QE-7, 126, 1971.
351. **Miller, R. C., Nordland, W. A., Abrahams, S. C., and Bernstein, J. L.,** Nonlinear optical properties of the cuprous halides, *J. Appl. Phys.*, 44, 3700, 1973.
352. **Marple, D. T. F.,** Refractive index of GaAs, *J. Appl. Phys.*, 35, 1941, 1964.
353. **Kachare, A. H., Spitzer, W. G., and Fredrickson, J. E.,** Refractive index of ion-implanted GaAs, *J. Appl. Phys.*, 47, 4209, 1976.
354. **Mooradian, A. and McWhorter, A. L.,** *International Conference on Light Scattering Solids*, Wright, G. B., Ed., Springer, New York, 1969.
355. **Garfinkel, M. and Engeler, W. F.,** Sum frequencies and harmonic generation in GaAs laser, *Appl. Phys. Lett.*, 3, 178, 1963.
356. **Chang, T. Y., Van Tran, N., and Patel, C. K. N.,** Absolute measurement of second order nonlinear coefficient for optical generation of millimeter wave difference frequencies in GaAs, *Appl. Phys. Lett.*, 13, 357, 1968.
357. **Nelson, D. F. and Turner, E. H.,** Electro-optic and piezoelectric coefficients and refractive index of gallium phosphide, *J. Appl. Phys.*, 39, 3337, 1968.
358. **Faust, W. L., Henry, C. H., and Eick, R. H.,** *Phys. Rev.*, 173, 781, 1968.
359. **Heilmeier, G. H., Ockman, N., Braunstein, R., and Kramer, D. A.,** Relationship between optical second harmonic generation and the electro-optic effect in the molecular crystal hexamine, *Appl. Phys. Lett.*, 5, 229, 1964.
360. **Landölt-Börnstein,** *Optische Konstanten*, II Band, 8 Teil, Springer-Verlag, Berlin, 1962, 2-43 to 2-397.
361. **Wynne, J. J.,** Dispersion of the lowest-order optical nonlinearily in InSb, *Phys. Rev. Lett.*, 27, 17, 1971.
362. **Slicker, T. R. and Jost, J. M.,** Linear electro-optic effect and refractive indices of cubic ZnTe, *J. Opt. Soc. Am.*, 56, 130, 1966.
363. **Barnes, N. P. and Piltch, M. S.,** Temperature-dependent sellmeier coefficients and nonlinear optics average power limit for germanium, *J. Opt. Soc. Am.*, 69, 178, 1979.
364. **Schultz, L. G. and Tangherlini, F. R.,** Optical constants of silver, gold, copper and aluminum. II. The index of refraction, *J. Opt. Soc. Am.*, 44, 362, 1954.
365. **Irani, G. B., Huen, T., and Wooten, F.,** Optical constants of silver and gold in the visible and ultraviolet, *J. Opt. Soc. Am.*, 61, 128, 1971.
366. **Jeppesen, M. A.,** Some optical thermo-optical, and piezo-optical properties of synthetic sapphire, *J. Opt. Soc. Am.*, 48, 629, 195?
367. **Kohlrausch, F.,** *Praktische Pysik*, Vol. 3, 22nd ed., Teubner, Leipzig, 1968, 74.
368. **Coblentz, W. W.,** *J. Opt. Soc. Am.*, 4, 441, 1914.
369. **Driscoll, W. G. and Vaughan, W., Eds.,** *Handbook of Optics*, Optical Society of America, McGraw Hill, New York, 1978, 7.
370. **Edwards, D. F. and Ochoa, E.,** Infrared refractive index of diamond, *J. Opt. Soc. Am.*, 71, 607, 1981.
371. **Icenogle, H. W., Platt, B. C., and Wolfe, W. L.,** Refractive indexes and temperature coefficients of germanium and silicon, *Appl. Opt.*, 15, 2348, 1976.
372. **Salzberg, C. D. and Villa, J. J.,** *J. Opt. Soc. Am.*, 43, 579, 1953.

373. **Seraphin, B. O. and Bennett, H. E.,** Optical constants, in *Semiconductors and Semimetals,* Vol. 3, Willardson, R. K. and Beer, A. C., Eds., Academic Press, New York, 535.
374. **Herzberger, M. and Salzberg, C. D.,** Refractive indices of infrared optical materials and color correction of infrared lenses, *J. Opt. Soc. Am.,* 52, 420, 1962.
375. **Laporte P. and Subtil, J. L.,** Refractive index of LiF from 105 to 200 nm, *J. Opt. Soc. Am.,* 72, 1558, 1982.
376. **Stephens, R. F. and Malitson, I. H.,** *Natl. Bur. Stand. J. Res.,* 49, 249, 1952.
377. **Stephens, R. E., Plyler, E. K., Rodney, W. S., and Spindler, R. J.,** *J. Opt. Soc. Am.,* 43, 100, 1953.
378. **Barnes, N. P., Gettemy, D. J., and Adhav, R. S.,** Variation of the refractive index with temperature and the tuning rate for KDP isomorphs, *J. Opt. Soc. Am.,* 72, 895, 1982.
379. **Salzberg, C. D. and Villa, J. J.,** Infrared refractive indexes of silicon germanium and modified solemium glass, *J. Opt. Soc. Am.,* 47, 244, 1957.
380. **Edwards, D. F. and Ochoa, E.,** Infrared refractive index of silicon, *Appl. Opt.,* 19, 4130, 1980.
381. **Malitson, I. H.,** *J. Opt. Soc. Am.,* 55, 1205, 1965.
382. **Singh, S., and Pawelek, R.,** unpublished.
383. **Koch, J.,** On the refraction and dispersion of the noble gases krypton and xenon, *Kungl. Fysiografiska Sallskapets I Lund Forhandlingar,* 19, 173, 1949.
384. **Czyzak, S. J., Baker, W. M., Crane, R. C., and Howe, J. B.,** Refractive indexes of single synthetic zinc sulfide and cadnium sulfide crystals, *J. Opt. Soc. Am.,* 47, 240, 1957.
385. **Deshotels, W. J.,** Ultraviolet transmission of dihydrogen arsenate and phosphate crystals, *J. Opt. Soc. Am.,* 50, 865, 1960.
386. **Bolognesi, G. P., Mezzetti, S., and Pandarese, F.,** Behaviour of some nonlinear optical powdered organic materials, *Opt. Commun.,* 8, 267, 1973.
387. **Delfino, M. and Gentile, P. S.,** Approximate nonlinear optical susceptibility of cubic boracites, *J. Appl. Phys.,* 51, 2264, 1980.
388. **Habout, J. M., Sarhangi, A., and Tang, C. L.,** Nonlinear optical properties of N,N[1] dimethyl urea, *Appl. Phys. Lett.,* 37, 864, 1980.
389. **Jain, K., Hewig, G. H., Cheng, Y. Y., and Crowley, J. I.,** New organic materials with large optical nonlinearities, *IEEE J. Quantum Electron.,* QE-17, 1593, 1981.
390. **Twieg, R., Azema, A., Jain, K., and Cheng, Y. Y.,** Organic materials for nonlinear optics, nitropyridine derivatives, *Chem. Phys. Lett.,* 92, 208, 1982.
391. **Jain, K., Crowley, J. I., Hewig, G. H., Cheng, Y. Y., and Twieg, R. J.,** Optically nonlinear organic materials, *Opt. Laser Technol.,* 297, 1981.
392. **Vizert, R. V., Davydov, B. L., Kotovshchikow, S. G., and Starodubtseva, M. P.,** Generation of the second harmonic of a neodymium laser in powder of noncentrosymmetric organic compounds, *Sov. J. Quantum Electron.,* 12, 214, 1982.
393. **Mollitson, I. H., Murphy, F. V., and Rodney, W. S.,** Refractive index of synthetic sapphire, *J. Opt. Soc. Am.,* 48, 72, 1958.
394. **Feldman, A. and Horowitz, D.,** Refractive index of cuprous chloride, *J. Opt. Soc. Am.,* 61, 89, 1971.
395. **Edlen, B.,** *Metrologia,* 2, 71, 1966.
396. **Peck, E. R. and Reeder, K.,** Dispersion of air, *J. Opt. Soc. Am.,* 62, 958, 1972.
397. **Peck, E. R. and Fisher, D. J.,** Dispersion of argon, *J. Opt. Soc. Am.,* 54, 1362, 1964.
398. **Old, J. G., Gentili, K. L., and Peck, E. R.,** Dispersion of carbon dioxide, *J. Opt. Soc. Am.,* 61, 89, 1971.
399. **Mansfield, C. R. and Peck, E. R.,** Dispersion of helium, *J. Opt. Soc. Am.,* 59, 199, 1969.
400. **Huber, M. C. E. and Tondello, G.,** Refractive index of he in the region 920-1910Å, *J. Opt. Soc. Am.,* 64, 390, 1974.
401. **Peck, E. R. and Huang, S.,** Refractivity and dispersion of hydrogen in the visible and near infrared, *J. Opt. Soc. Am.,* 67, 1550, 1977.
402. **Wilkenson, P. G.,** Refractive dispersion of nitrogen in the vacuum ultraviolet, *J. Opt. Soc. Am.,* 50, 1002, 1960.
403. **Hellwarth, R. W.,** Third order optical susceptibilities of liquids and solids, *Prog. Quantum Electron.,* 5, 1, 1977.
404. **Boling, N. L., Glass, A. J., and Owyoung, A.,** Empirical relationships for predicting nonlinear refractive index changes in optical solids, *IEEE J. Quantum Electron.,* QE-14, 601, 1978.
405. **Veduta, A. P. and Kirsanov, B. P.,** Variation of the refractive index of liquids and glasses in a high intensity field of a ruby laser, *Sov. Phys. JETP,* 27, 736, 1968.
406. **Hellwarth, R. W., Owyoung, A., and George, N.,** *Phys. Rev.,* A4, 2342, 1971.
407. **Owyoung, A.,** Ellipse rotation studies in laser host materials, *IEEE J. Quantum Electron.,* QE-9, 1064, 1973.
408. **Owyoung, A., Hellwarth, R. W., and George, N.,** Intensity induced changes in optical polarizations in glasses, *Phys. Rev.,* B5, 628, 1972.

409. **Hellwarth, R. W., Cherlow, J., and Yang, T. T.,** *Phys. Rev.,* B11, 964, 1975.

410. (a) **Levine, B. F., and Bethea, C. G.,** *Appl. Phys. Lett.,* 24, 445, 1974; (b) **Levine, B. F. and Bethea, C. G.,** Second and third order hyperpolarizabilities of organic molecules, *J. Chem. Phys.,* 63, 2666, 1975; (c) **Levine, B. F., and Bethea, C. G.,** Second order hyperpolarizability of a polypeptide α-helix: poly-α-benzyl-L-glutamate, *J. Chem. Phys.,* 65, 1989, 1976; (d) **Levine, B. F. and Bethea, C. G.,** Hyperpolarizability of the pyridine-iodine charge transfer complex, *J. Chem. Phys.,* 65, 2439, 1976; (e) **Levine, B. F. and Bethea, C. G.,** Effects on hyperpolarizabilities of molecular interactions in associating liquid mixtures, *J. Chem. Phys.,* 65, 2429, 1976; (f) **Levine, B. F.,** Donor-acceptor charge transfer contributions to the second order hyperpolarizability, *Chem. Phys. Lett.,* 37, 516, 1976; (g) **Levine, B. F. and Bethea, C. G.,** Charge transfer complexes and hyperpolarizabilities, *J. Chem. Phys.,* 66, 1074, 1977; (h) **Levine, B. F.,** Ultraviolet dispersion of the donor-acceptor charge-transfer contribution to the second order hyperpolarizability, *J. Chem. Phys.,* 69, 5240, 1978.

411. **Levenson, M. D. and Bloembergen, N.,** Dispersion of the nonlinear optical susceptibility tensor in centrosymmetric media, *Phys. Rev.,* B10, 4447, 1974.

412. **Finn, R. S. and Ward, J. F.,** D.C. Induced optical second harmonic generation in the inert gases, *Phys. Rev. Lett.,* 26, 285, 1971.

413. **Finn, R. S. and Ward, J. F.,** Measurements of hyperpolarizabilities for some halogenated methanes, *J. Chem. Phys.,* 60, 454, 1974.

414. **Ward, J. F. and Bigio, I. J.,** Molecular second and third order polarizabilities from measurements of second harmonic generation in gases, *Phys. Rev.,* A11, 60, 1975.

415. **Hauchecorne, G., Kerherve, F., and Mayer, G.,** *J. Phys. (Paris),* 32, 47, 1971.

416. **Buckingham, A. D. and Orr, B. J.,** *Trans. Faraday Soc.,* 65, 673, 1969.

1.2 TWO-PHOTON ABSORPTION IN CONDENSED MEDIA

W. Lee Smith*

INTRODUCTION

Two-photon absorption is a nonlinear optical phenomenon in which two quanta are simultaneously absorbed in an electronic transition by a material system that is transparent to either quantum alone. Such an absorption requires that both photons be present simultaneously in the region of the absorbing atom or molecule. Hence, large photon fluence — 10^{20} to 10^{30}/cm^2-sec — is the general requirement for the observation of two-photon absorption.

The study of two-photon absorption began with the theoretical exposition of Göppert-Mayer over 50 years ago.[1] Confirmation in the laboratory, first by Hughes and Grabner[2] in 1950, came out of atomic and molecular beam spectroscopy in the radio and microwave frequency regions. By 1960, papers discussing two-photon absorption[3] in the scientific literature numbered approximately ten.

The laser entered in 1960, allowing experimenters to begin probing matter coherently and intensely in the optical region of the spectrum. Months after its invention, the pulsed ruby laser[4,5] was used by Kaiser and Garrett[6] to excite two-photon absorption in CaF_2:Eu^{3+}, evidenced by fluorescent de-excitation in the blue spectral region. This was the second occurrence of nonlinear optics, following closely the demonstration of second-harmonic generation.[7]

With lasers, the study of two-photon absorption grew rapidly, with seminal theoretical and experimental works occurring in the 1960s. Early theoretical treatments were given by Kleinman[8] and Braunstein.[9] Abella[10] observed two-photon absorption in Cs vapor in 1962. Peticolas et al.[11] reported in 1963 two-photon-induced fluorescence in several organic crystals. In the same year Giordmaine and Howe[12] measured in liquid CS_2 the first intensity-dependent transparency and extracted the first two-photon absorption coefficient. Hopfield et al.[13] and Hopfield and Worlock[14] combined light waves from a ruby laser and a broadband lamp to perform in crystalline potassium iodide the first two-photon absorption spectroscopy. Braunstein and Ockman[15] in 1964 measured the strength of two-photon absorption in a semiconductor.

By the late 1960s, mode-controlled, Q-switched, and mode-locked ruby and Nd-doped lasers supplanted the burst-mode ruby laser in studies of two-photon absorption. The newer lasers enabled optical power in the 10^6 to 10^9 W range to be generated and used to measure even smaller two-photon absorption coefficients, and gave better precision in the measurements as well. Dye lasers have gradually replaced the flashlamp as the source of tunable-wavelength photons for two-photon spectroscopy. With these advances and the development of a number of sensitive detection schemes, the study of two-photon absorption is now widespread. The scientific literature of two-photon absorption now numbers approximately 1000 papers.

The significance of two-photon absorption for the optical study of matter was realized at the outset. Optical spectroscopy in the many decades prior to 1961 was primarily the study of transitions by which the electric dipole operator connected the initial and final states. For "allowed" transitions, these states had to be of opposite parity. Thus, "half" of the possible transitions — those between like-parity states — were inaccessible to spectroscopists. Two-photon absorption, depending quadratically on the momentum operator, removes this lim-

* Work performed under the auspices of the U.S. Department of Energy by Lawrence Livermore National Laboratory under Contract No. W-7405-ENG-48.

itation. Linear (one-photon) and two-photon spectroscopies are complementary, and between them all states of matter in the optical region may, in principle, be examined.

A second useful difference between linear and nonlinear spectroscopies is the characteristic dependence of the nonlinear absorption strength on the intensity of the applied light wave. Although this feature has the operational cost of supplying intense illumination, the inducible nature of two-photon absorption allows the spectroscopist to "pull out" weak spectral signatures, limited only by optical damage of the sample. Laser pulses in the femtosecond-duration range are now possible,[16,17] enabling extraordinary intensity to be established in a sample, yet with minimal pulse energy and lessened likelihood of damage. Proper diagnostics allow the separation of direct two-photon absorption from fluence-dependent effects, such as absorption by the two-photon created states[18,19] or high-density charge carrier dynamics in semiconductors.[20,21,22]

Two-photon excitation is uniquely useful for investigations into the (Urbach) spectral region of strong one-photon electronic transitions for two reasons. Relatively uniform bulk excitation of the sample material can be achieved rather than deposition in a thin, possibly extrinsic surface layer, and the excitation can be obtained with sources in the convenient 0.2- to 2-μm spectral region rather than in the vacuum UV.

Two-photon absorption impacts a considerable range of scientific and technological investigations. In fundamental physics, experiments involving two-photon pumping and decay have been used to argue the validity of quantum mechanics against hidden-variable theory.[23] High-density biexcitons produced with "giant" two-photon absorption in CuCl are being examined as a possible third system exhibiting Bose-Einstein condensation,[24,25] along with superfluidity and superconductivity. Laser action has been produced by two-photon absorption pumping in several semiconductors,[26,27,28] vapors,[29,30] and gases.[31] Diffraction gratings have been written in transparent crystals by two-photon absorption,[32] optical memory systems based on two-photon absorption have been proposed,[33] and two-photon photorefractivity has been reported.[34] Resistance to laser-induced breakdown in optical materials is related to two-photon (and higher-order) production of charge carriers,[35] and window materials that are usable in giant fusion lasers operating in the UV are restricted by two-photon absorption.[19] Two-photon absorption has been used for the temporal shaping of pulses[36] and for the measurement of the duration of pico- and femtosecond pulses.[37] Two-photon absorption has been examined in biological systems including the visual chromophore[38] and chlorophyll in vivo.[39,40]

SCOPE

The goal of this chapter is to provide a practical and useful compendium of the measurements of two-photon absorption in condensed media made over the past 20-odd years. Effort has been made to be complete, although omissions are inevitable in a compilation of this size. Works which report only the observations of two-photon absorption behavior, without reporting the quantitative strength or relative spectral dependence, were not included in the tables given here. Even with this selection rule, data from over 200 works are compiled in the tables.

As the title indicates, gaseous and atomic-vapor systems are not included in this review. Multiphoton photoelectron emission, multiphoton ionization, and multiphoton dissociation involves additional physical parameters (electron affinity, etc.) or absorption steps beyond two-photon absorption and are not addressed here.

OTHER REVIEWS

Reviews of two-photon absorption and spectroscopy have been given by Mahr,[41] Bredikhin et al.,[42] Worlock,[43] McClain and Harris,[44] and Frölich and Sondergeld.[45] Work on general nonlinear optics which are valuable for understanding two-photon absorption have been given by Flytzanis[46] and Levenson.[47]

DEFINITIONS

Attenuation of the intensity of a light wave as a function of distance inside a material exhibiting one- and two-photon absorption is described by

$$\frac{dI(z)}{dz} = -[\alpha + \beta I(z)]\, I(z) \tag{1}$$

Here α and β are the one- and two-photon absorption coefficients, with units of cm^{-1} and cm/W, respectively. Let I_{inc} and I(L) be the intensity incident on, and transmitted by, the sample of length, L, respectively. Then the transmission, T, is $I(L)/I_{inc}$. For conditions which allow the effect of the rear-surface reflectivity, R, on the two-photon absorption to be neglected,[48,49] the solution for the transmission is

$$T = T_L T_{NL} \tag{2}$$

where the linear transmission factor is

$$T_L = (1 - R)^2\, e^{-\alpha L} \tag{3}$$

the nonlinear transmission factor is

$$T_{NL} = \frac{1}{1 + Q} \tag{4}$$

and the nonlinear parameter, Q, is

$$Q = \frac{\beta(1 - R)\, I_{inc}\, (1 - e^{-\alpha L})}{\alpha} \tag{5}$$

The above solution is for a single value of intensity. Typical laser waveforms vary in space and time. The solutions to Equation 1 (in terms of the transmission) for the more common laser waveforms are listed in Table 1.2.1.

In Figure 1.2.1 we plot useful universal curves of the nonlinear transmission factor, T_{NL}, vs. Q for the four types of waveform in Table 1.2.1. Using otherwise known values of R and α with the values for β from the tables in this chapter, the reader may calculate Q, read T_{NL} from the curves in Figure 1.2.1, and arrive at the overall transmission, T, for any specific sample.

It is often convenient to discuss the two-photon absorption strength in terms of a cross-section per molecule and per photon fluence, σ_2, related to the two-photon coefficient, β, by:

$$\sigma_2 = \frac{h\omega}{N}\beta \tag{6}$$

Here $h\omega$ is the energy of a single photon of the incident laser wave and N is the number per unit volume of the absorbing molecule. The units of σ_2 are cm^4-sec/photon-molecule, and typical magnitudes are in the range 10^{-51} to 10^{-49}.

The relationship between two-photon absorption and other nonlinear optical phenomena,

Table 1.2.1
SOLUTIONS[a] FOR THE TRANSMISSION, T, THROUGH A ONE- AND TWO-PHOTON ABSORBING SAMPLE OF LENGTH, L

$$T = \frac{\int_{-\infty}^{\infty} dt \int_0^{\infty} 2\pi r dr \; I\,(r,L,t)}{\int_{-\infty}^{\infty} dt \int_0^{\infty} 2\pi r dr \; I\,(r,0,t)} = T_L T_{NL}$$

	Laser pulse profile: Space	Laser pulse profile: Time	
I	Flat	Flat	$T_L \frac{1}{1 + Q}$
II	Flat	Gaus	$T_L \frac{1}{\sqrt{\pi}} \int_{-\infty}^{\infty} e^{-x^2}/[1 + Qe^{-x^2}]dx$
III	Gaus	Flat	$T_L \frac{\ell n(1 + Q)}{Q}$
IV	Gaus	Gaus	$T_L \frac{2}{\sqrt{\pi}\, Q} \int_0^{\infty} \ell n[1 + Qe^{-x^2}]dx$ $= T_L \sum_{n=1}^{\infty} (-Q)^{n-1} n^{-3/2}$ for $Q < 1$

[a] Rear-surface reflection neglected.

such as stimulated Raman scattering or second-harmonic generation, can be established through the formalism of nonlinear optical susceptibilities.[50,46] As a particular example, for a two-photon absorption experiment involving a single beam of frequency, ω, linearly polarized along the [100] direction of a cubic crystal, the two-photon absorption coefficient, β, is related to the imaginary part of the third-order nonlinear susceptibility, $\chi^{(3)}$, by:

$$\beta = \frac{32\pi^2\omega}{n^2c^2} \; 3\chi^{(3)''}_{xxxx}\,(-\omega, \omega, \omega, -\omega) \tag{7}$$

Here the subscripts "x" denote the polarization direction of the involved waves (all along [100] in this example), and n is the refractive index of the material at the frequency, ω. The bracketed term is the effective third-order susceptibility. For configurations involving beams of different frequency and other polarization directions, the effective nonlinear susceptibility may involve other elements of the susceptibility tensor and different numerical prefactors. Further specific cases have been worked out in the literature.[51,52]

The real part of the third-order susceptibility, $\chi^{(3)}$, is related to the nonlinear refractive index and is the major parameter describing self-focusing (see the following chapter).

The electrostatic or cgs units of $\chi^{(3)}$ are cm^3/erg, and the mks or SI units are m^2/V^2. Conversion between the two systems is made with

$$\chi^{(3)}[\text{mks}] = \frac{4\pi}{9 \times 10^8}\, \chi^{(3)}[\text{esu}] \tag{8}$$

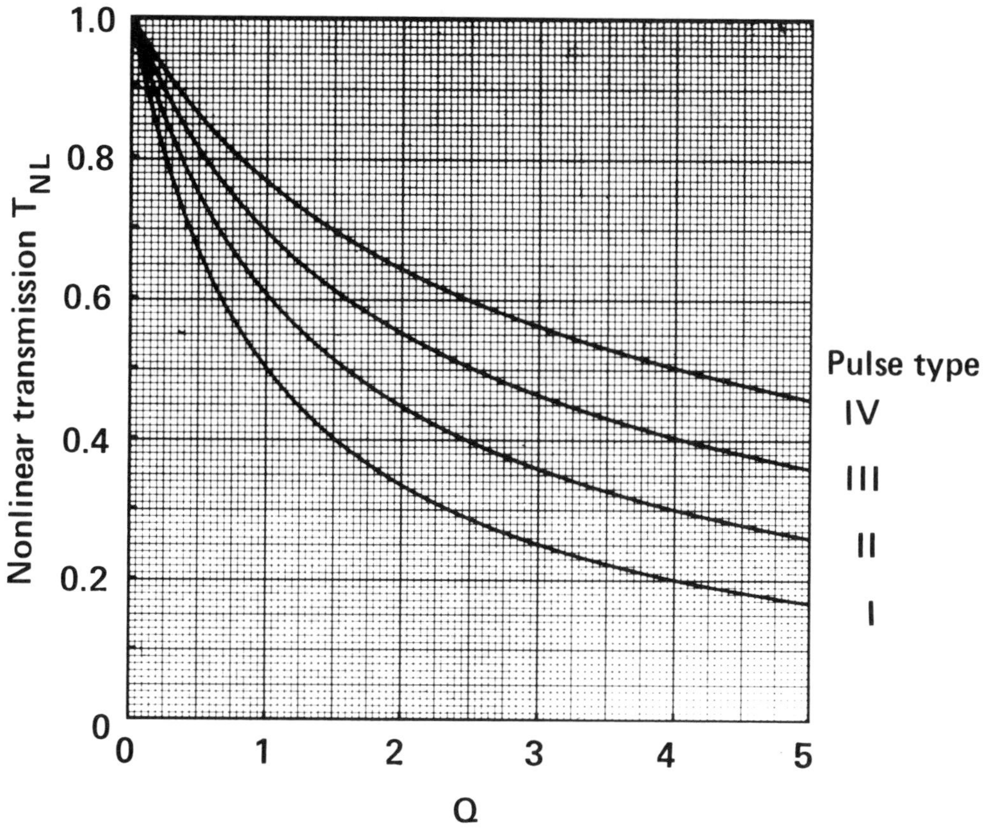

FIGURE 1.2.1. Universal curves for the nonlinear transmission factor, T_{NL}, vs. the dimensionless parameter, Q. Laser pulse types I to IV are defined in Table 1.2.1.

MEASUREMENT TECHNIQUES

Experimental methods that have been used to measure the two-photon absorption results compiled in this chapter are listed in Table 1.2.2. We also list the abbreviations used in the subsequent data tables, and a primary or recent reference to which the reader is directed for a description of each method.

CAVEAT

The two-photon absorption literature serves several scientific communities having widely differing requirements on absolute accuracy of data. For much spectroscopic research, the primary value of two-photon absorption data is for elucidation of energy states of matter. Hence, transition energies, linewidths, and relative strengths are more informative than absolute two-photon absorption strengths. Yet, for technological applications, the magnitude of the two-photon absorption coefficient may be required (and much effort spent on obtaining accurate absolute strengths) at a few selected wavelengths.

It is well known that two-photon absorption coefficients published for the same material may span a considerable range, sometimes over one order of magnitude. Self-focusing and other stimulated effects may proceed undetected and thereby influence the accuracy of two-photon absorption coefficients. In semiconductors, dopant concentrations may be unknown or unreported, yet affect the two-photon absorption coefficient. More recently, appreciation

Table 1.2.2
TECHNIQUES FOR MEASURING TWO-PHOTON ABSORPTION

		Ref.
AI1	Attenuation vs. intensity for single beam	51
AI2	Attenuation vs. intensity; 2 beams, usually 1 scanned in frequency	45
AIS	Attenuation vs. intensity; broad spectrum treated simultaneously	53
C	Calorimetry	49
CR	Comparison of two-photon loss and Raman gain	54,55
EPR	Elliptical polarization rotation	56
FCC	F-center coloration	57
GTA	Gain measurement in a two-photon-pumped amplifier	58
IA	Intracavity absorption	59
L	Luminescence	60,61
PA	Photoacoustics	62
PC	Photoconductivity	63,64
TCN	Two-channel normalization	65
TL	Thermal lensing	66
TRT	Time-resolved transmission	19

is being made of the impact of excited-state absorption, following the nonlinear absorption, on the apparent magnitude of the two-photon absorption coefficient.

It is intended that the reader refer to the original papers and, against his own requirements, assess the credibility of the data and the analysis with which the two-photon absorption coefficients were extracted.

THE DATA TABLES

The experimental data on two-photon absorption are organized into eight tables (Tables 1.2.3 to 1.2.10): small bandgap (<4 eV) crystals, large bandgap crystals, glasses (amorphous solids), organic liquids and crystals, organic dyes, polymers, biological materials (including H_2O), and rare-earth and color-center systems. Materials are listed alphabetically within each table.

The method of measurement (using the abbreviations defined in Table 1.2.2), the duration of the pulsed laser excitation, and the approximate one-photon absorption edge (if given in the reference) are included. The two-photon excitation energy is listed either as a single value or, if a spectrum was reported, as a range of energy.

The two-photon absorption coefficient, β, or two-photon absorption cross-section, σ_2, is tabulated as an absolute value, a relative value, an absolute spectrum, or a relative spectrum. For each absolute spectrum (abbreviated abs. spectr. in the tables), we include in parentheses a value of β or σ_2 at an arbitrarily chosen two-photon energy within the spectrum. The reader is directed to the original literature for the spectrum. The units of β in the tables are cm/GW*, and of σ_2, cm^4-sec/photon-molecule. Additional information (temperature other than room temperature, crystal orientation, etc.) is included in some cases.

ACKNOWLEDGMENT

The author is indebted to Mrs. T. Mauch for her skill and sustained patience in preparing this manuscript, and to Ms. G. Martinez for extensive library services.

* GW = 10^9 W.

Table 1.2.3
TWO-PHOTON ABSORPTION DATA FOR SMALL BANDGAP CRYSTALS

Material	Method	Excitation duration (nsec)	Approximate one-photon absorption edge (eV)	Applied two-photon energy (eV)	Two-photon absorption coefficient β (cm/GW)	Ref.	Additional information
$AgAsS_3$	AI2	10	2.1	2.7—3.2	Abs. spectr. (10 @ 3.1 eV)	67,68	
$AgAsS_3$	AI1	25		3.56	20	69	
$AgAsS_3$	AI1	20		2.34	3	69	
AgCl	L		~4.0	4.0—4.4	Abs. spectr. (0.5 @ 4.3 eV)	70	Indirect gap
$AgGaSe_2$	AI1	~0.01	1.8	2.33	1.4	71	
AgI	AI2		3.0	3.0—3.06	Relative spectrum	72	1.6 K
AlAs-GaAs	L		1.7	1.65—1.8	Relative spectrum	73	
As_2S_3	AI2	30	2.5	2.4—3.6	Abs. spectr. (25. @ 3.4 eV)	74	
CdP_2	AI2	15		2.4—3.2	Abs. spectr. (11. @ 2.9 eV)	75	
CdP_2	TRT			2.34	160	76	
CdP_2	TRT			3.56	800	76	
CdS	L	$\sim 3 \times 10^5$	2.5	3.56	~12	15	123 K
CdS	AI2	300		2.5—2.6	Abs. spectr. (4000 @ 2.54 eV)	77	
CdS	AI2			2.54—3.6	Abs. spectr. (30 @ 3.0 eV)	78	77 K
CdS	AI2	30		2.5—3.5	Abs. spectr. (20 @ 2.8 eV)	79	
CdS	AI1	30		3.56	30	80	
CdS	AI2	0.006		2.5—3.5	Abs. spectr. (18 @ 3.35 eV)	81	
CdS	L			2.5503—2.5540	Relative spectrum	82	1.6 K
CdS	AI1	5		2.54—2.55	Relative spectrum	83	
CdS	AI1	30		3.56	100	48	β vs. thickness
CdS	AI1	45	2.5	3.56	120; 56	84	E ∥ c; E ⊥ c
CdS	AI1	20	2.4	3.56	56	85	Self-focusing considered
CdS	L	~25	2.5	2.56—2.62	Relative spectrum	86	
CdS	AI1	0.024		3.56	19	87	
CdS	CR			3.92	56	88	

Table 1.2.3 (continued)
TWO-PHOTON ABSORPTION DATA FOR SMALL BANDGAP CRYSTALS

Material	Method	Excitation duration (nsec)	Approximate one-photon absorption edge (eV)	Applied two-photon energy (eV)	Two-photon absorption coefficient β (cm/GW)	Ref.	Additional information
CdS	CR		2.4	3.18	110	89	
CdS	AI2	40		2.5—3.4	Abs. spectr. (14.7 @ 3.4 eV)	90	
CdS	AI2	30		2.65—3.45	Abs. spectr. (35 @ 4 eV)	91	
CdS	L	~2 × 10^5		3.56	12	15	
CdS	PC	20	2.4	3.56	14	63	
CdS	AI2	30		2.5—3.5	Relative spectrum	92	Anisotropy
$CdS_{0.5}Se_{0.5}$	C	11; 26		2.33	32; 135	49	
$CdS_{0.9}Se_{0.1}$	AI1	30		3.56	70	80	
$CdS_{0.8}Se_{0.2}$	AI1	30		3.56	130	80	
CdSe	AI1			2.33	900; 390	93	E ⊥ z; E ∥ z
CdSe	AI1	15		2.33	40	94	
CdSe	PA	0.040		2.33	35	95	
CdSe	C	11; 26		2.33	25; 38	49	
CdSe	AI1	20		2.33	60—140	96	
CdSe	C	79		1.88	67	49	
CdSe	AI1			2.33	950	97	
CdSe	AI1	0.030		2.33	30	98	
CdSe	C	16		2.33	50	99	
CdSe	AI1			2.33	200	100	
CdSe	TRT	~20		1.88; 2.33	2; 20	69	
CdS_xSe_{1-x}	AI1			3.56	200—1700	101	300 K, 77 K, Anisotropy
$CdS_xSe_{1=x}$	AI1			2.33, 3.56	Relative spectrum	102	
CdTe	AI1			2.33	200	100	
CdTe	AI1			2.33	180	103	300 K, 85 K, Anisotropy
CdTe	PA	0.040		2.33	50	95	
CdTe	TCN	20	1.5	2.33	170	104	

CdTe	C	11; 38		2.33	53; 78	49	
CdTe	C	79		1.88	120	49	
CdTe	AI1	0.030		2.33	25	98	E ∥ z
CdTe	C	16		2.33	13	99	
Cu_2O	AI2	40	2.1	2.0—2.8	Relative spectrum	105	
Cu_2O	AI2	0.0005		3.4—3.9	Relative spectrum	106	20 K
CuBr	AI1	45	3.1	3.56	200	84	
CuBr	AI2	30		2.97—3.10	Relative spectrum	107	
CuCl	AI2		~3.3	3.21—3.56	Relative spectrum	108	4.2 K
CuCl	AIS	~5		3.16—3.18	$\sim 3 \times 10^6$	53	4.2 K
CuCl	AI1	16		3.16—3.20	1×10^6	109	4 K, 77 K
CuCl	AI2	~10	3.4	3.2—4.25	Relative spectrum	110	20 K
CuCl	AI1	45	3.4	3.56	45	84	
CuCl	AI2	~10	3.4	3.2—5.4	Relative spectrum	111	
CuCl	EPR	5		3.18—3.19	Relative spectrum	56	4 K
CuI	AI1	45	3.1	3.56	89	84	
GaAs	AI1			2.33	20	100	
GaAs	AI1	30		2.33	230	112	
GaAs	AI1			2.33	800	80	n-type; 10^{16} cm^{-3}
GaAs	AI1	~10		2.33	35; 78	113	
GaAs	AI1	~50	1.42	1.5—2.33	Abs. spectr. (1100 @ 1.5 eV)	114	
GaAs	AI1			2.33	80	103	300 K, 85 K; Anisotropy
GaAs	AI1	15		2.33	60	94	
GaAs	C	0.030		2.33	30	49	
GaAs	L	5		2.33	70	115	
GaAs	AI1	0.030		2.33	28	98	E ∥ z
GaAs	AI1	0.008		2.33	15	116	
GaAs	AI1	~125		1.88	33	117	
GaAs	AI1, PC	22		2.33	360	126	
GaAs	L		1.52	1.4—1.8	Abs. spectr. (5.1 @ 1.6 eV)	119	
GaAs	AI1			2.33	300	97	
GaP	AI1	50	2.9	2.33	1.7	120	
GaP	AI1	0.030		2.33	0.2	98	E ∥ [110]
GaP	TCN			3.92	250	88	
GaP	CR		2.8	3.18	250	89	
GaS	AI1	20	2.8	3.56	100	121	Direct gap

Table 1.2.3 (continued)
TWO-PHOTON ABSORPTION DATA FOR SMALL BANDGAP CRYSTALS

Material	Method	Excitation duration (nsec)	Approximate one-photon absorption edge (eV)	Applied two-photon energy (eV)	Two-photon absorption coefficient β (cm/GW)	Ref.	Additional information
GaS	PC	20	2.3	2.33	0.05	121	Indirect gap
GaSe	AI1	20	2.0	2.33	110	121	
Ge	AI1	100	0.7	0.8—1.0	2500	122	
Ge	AI1	80		1.06	50	123	
Ge	AI1; PC	~480		0.916	160	64	
$Hg_{0.78}Cd_{0.22}Te$	TRT	~200	0.17	0.233	14000	124	
$Hg_{0.78}Cd_{0.22}Te$	TRT	300		0.233	1×10^4; 3×10^4	125	300 K; 150 K
InP	TRT	200		2.33	210	112	
InP	AI1, PC	22		2.33	260	126	
InSb	PC	~200	0.228	0.233	59.6—119	127	77 K
InSb	PC	~200	0.236	0.233	0	127	2 K
InSb	PC	~200	0.228	0.257	1151—1419	127	77 K
InSb	PC	~200	0.236	0.257	946—1850	127	2 K
InSb	TRT	300		0.233	4800	125	
InSb	TRT	300	0.228	0.233	220	125	77 K
InSb	AI1	30		2.33	15000	112	
InSb	AI1; PC	~150		0.223	220	64	
InSb	AI1	~10		0.233	16000	128	
InSb	AI2	130		0.233	2000—5600	129	
InSb	TRT	~200	0.18	0.233; 0.258	8000; 14000	124	
InSb	PC	CW		0.26	2900	130	In magnetic field
KRS-V	C	38		2.33	1.6	49	
PbI_2	AI1	20	2.4	3.56	250	121	
PbI_2	AI2	~10	2.5	2.46—2.55	Relative spectrum	131	1.6—300 K
Si	AI1			2.33	7300	100	Also T dependence
Si	AI1	200	1.1	2.33	40	132	
Si	C	79		1.88	21	49	
Si	AI2	25		1.62—2.2	Relative spectrum	133	

Si	AI1	0.020		2.33	1.9; 1.5	134	20 K; 100 K
SiC	AI1	30		3.56	200	80	
SiC	L	20	3.1	3.0—4.6	Relative spectrum	135	E ∥ c; E ⊥ c
SnO_2	AI1	45	3.4	3.56	300; 34	84	E ∥ c; E ⊥ c
SnO_2	AI2	~10	~3.5	3.555—3.573	Relative spectrum	136	
Te	AI2	200		0.342	800	137	
TiO_2	AI1	0.004	3.75	4.66	14	138	
TiO_2	AI1	0.004	3.75	3.96	6.5	138	
TiO_2	CR			3.92	23	88	
TiO_2	AI2		~3.7	3.3—4.1	Abs. spectr. (300 @ 4 eV)	139	
TiO_2	CR			4.1; 4.7; 5.0	12.7; 87; 170.	140	
TiO_2	AI1	45	2.9	3.56	150; 120	84	E ∥ c; E ⊥ c
TlCl	AI2	20	3.5	3.4—4.4	Abs. spectr. (4.7 @ 4.0 eV)	141	
TlCl	AI2	~10	3.55	3.39—3.56	Relative spectrum	142	4, 20, 77 K
TlCl	AI2		3.55	3.55—4.33	Relative spectrum	143	77 K
TlCl	AI2	20	~3.4	3.4—4.4	Abs. spectr. (0.45 @ 3.8 eV)	144	
$Zn_{0.5}Cd_{0.5}S$	AI1	45	~3	3.56	60	84	
$Zn_{0.85}Cd_{0.15}Se$	AI1	10	2.65	3.56	56	145	
$Zn_{0.12}Cd_{0.88}Se$	AI1	10	1.92	3.56	620.; 260.	145	E ⊥ z; E ∥ z
$Zn_xCd_{1-x}S$	AI1		2.45—3.55	3.56	50—1	146	Anistropy
ZnO	CR			4.1; 4.7; 5.0	19.8; 20.9; 47.8	140	
ZnO	AI2	20		3.42—3.45	Relative spectrum	147	
ZnO	AI2		3.5	3.42—3.43	Abs. spectr. (2 @ 3.4 eV)	148	4.2 K in 42 kG magnetic field
ZnO	AI1	45	3.5	3.56	34; 16.	84	E ∥ c; E ⊥ c
ZnO	AI2	30		3.42—3.48	Relative spectrum	149	
ZnO	AI2	~10	~3.4	3.42—3.48	Relative spectrum	150	1.6 K
ZnP_2	AI2	15		2.4—3.2	Abs. spectr. (10 @ 2.9 eV)	75	
ZnP_2	TRT			2.34	120	76	
ZnP_2	TRT			3.56	650	76	
ZnS	PC	20	3.7	3.56	4.3	63	Cu-doped
ZnS	AI1	45	3.8	3.56	20; 0	84	E ∥ c; E ⊥ c
ZnS	PA	5	3.6	3.65—5.5	Relative spectrum	62	E ∥ z
ZnS	AI2			3.7—4.2	Abs. spectr. (2.6 @ 4.0 eV)	151	
ZnS	AI2			3.6—4.0	Relative spectrum	152	
ZnSe	AI1	10		3.56	45	145	

Table 1.2.3 (continued)
TWO-PHOTON ABSORPTION DATA FOR SMALL BANDGAP CRYSTALS

Material	Method	Excitation duration (nsec)	Approximate one-photon absorption edge (eV)	Applied two-photon energy (eV)	Two-photon absorption coefficient β (cm/GW)	Ref.	Additional information
ZnSe	AI1	30		3.56	40	80	
ZnSe	AI2	15		2.7—3.75	Abs. spectr. (10 @ 3.5 eV)	75	
ZnSe	AI2	40		2.6—3.6	Abs. spectr. (4 @ 3.0 eV)	153	
ZnSe	TCN	20	2.6	3.56	80	104	
ZnSe	CR		2.6	3.18	60	89	
ZnSe	AI2	40		2.75—3.45	Abs. spectr. (13.0 @ 3.45 eV)	234	Impurity resonances
ZnTe	AI1	30		2.33	34	112	
ZnTe	AI1, TCN	20	2.0—2.3	2.34, 3.56	20—300	154	Temperature-tuned bandgap
ZnTe	AI1	0.030		2.33	8.0	98	E ∥ z

Table 1.2.4
TWO-PHOTON ABSORPTION DATA FOR LARGE BANDGAP CRYSTALS

Material	Method	Excitation duration (nsec)	Approximate one-photon absorption edge (eV)	Applied two-photon energy (eV)	Two-photon absorption coefficient β (cm/GW)	Ref.	Additional information
ADA($NH_4H_2AsO_4$)	AI1	0.017		6.99	0.035	51	
ADP($NH_4H_2PO_4$)	AI1	0.030		9.32	0.11	155	
ADP($NH_4H_2PO_4$)	AI1	0.015	6.8	9.32	0.24	51	
ADP($NH_4H_2PO_4$)	AI1	0.017	6.8	6.99	0.0068	51	
Al_2O_3	AI1	0.017	7.3	6.99	<0.0016	51	
Al_2O_3	AI1	0.015	7.3	9.32	0.27	51	
BaF_2	AI1	0.017	9.0	6.99	<0.0036	156	

BaF_2	AI1	0.015	9.0	9.32	<0.0040	156	
$CaCO_3$	AI1	0.015	5.9	9.32	0.24	51	
CaF_2	AI1	0.015	10.0	9.32	<0.02	51	
CaF_2	CR	4.0		4.31	<0.004	54	
CD*A(CsD_2AsO_4)	AI1	0.017		6.99	0.051; 0.080	51	E ∥ z; E ⊥ z
CDA(CsH_2AsO_4)	AI1	0.017		6.99	0.028	51	
CdF_2	CR	4.0		4.31	<0.03	54	
CdF_2	AI1	0.017	9.0	6.99	<0.042	156	
CdF_2	AI1	0.015	~6.0	9.32	1.6	156	
CsBr	AI2	~10	~6.7	7.2—8.0	Abs. spectr. (10 @ 7.5 eV)	157	20 K
CsI	AI2	~10	~5.7	6.2—6.8	Abs. spectr. (40 @ 6.5 eV)	157	20 K
CsI	AI2	$\sim 10^6$	~5.5	5.7—6.8	Relative spectrum	13,14	
CsI	AI2			6.0—6.8	Relative spectrum	150	20 K
Diamond	AI1	0.021	~4.3	4.66	<0.26	158	
Diamond	CR	4.0		4.55	<0.003	54	
KBr	AI1	15		7.12	3.3	159	
KBr	CR	15		6.7	8.0	52	
KBr	AI2	~10	7.3	7.0—8.0	Relative spectrum	160	20 K, 80 K
KBr	AI1	0.015	6.0	9.32	2.0	51	
KBr	AI1	10		7.18	Relative value	161	
KCl	AI1	0.015	6.5	9.32	2.2	51	
KCl	FCC	8		9.32	1.5	57	
KDA(KH_2AsO_4)	AI1	0.017		6.99	0.048	51	
KD*A(KD_2AsO_4)	AI1	0.017		6.99	0.027	51	
KD*P(KD_2PO_4)	AI1	0.030		9.32	0.027	155	
KD*P(KD_2PO_4)	AI1	0.017		6.99	0.0054	51	
KDP(KH_2PO_4)	AI1	0.017	7.0	6.99	0.0059	51	
KDP(KH_2PO_4)	AI1	0.015	7.0	9.32	0.27	51	
KI	CR	15		6.7	18	52	
KI	AI1	15		7.12	4.4	159	
KI	AI2			6.23—6.36	Relative spectrum	162	
KI	AI2		~6.0	6.0—7.5	Relative spectrum	163	
KI	AI2			6.1—7.7	Relative spectrum	150	20 K
KI	AI2	$\sim 10^6$	~5.7	5.7—6.8	Relative spectrum	13,14	
KI	AI1	0.024		7.12	8	87	
KI	AI1	0.017	5.1	6.99	7.3	51	
KI	AI1	0.015	5.1	9.32	3.7	51	

Table 1.2.4 (continued)
TWO-PHOTON ABSORPTION DATA FOR LARGE BANDGAP CRYSTALS

Material	Method	Excitation duration (nsec)	Approximate one-photon absorption edge (eV)	Applied two-photon energy (eV)	Two-photon absorption coefficient β (cm/GW)	Ref.	Additional information
KI	AI1	10		7.12	Relative value	161	
KI	PC	20	~5.7	7.12	10	164	
$KTa_{0.65}Nb_{0.35}O_3$	AI1	0.010		4.66	14	165	
LiF	AI1	0.015	11.6	9.32	<0.02	51	
$LiIO_3$	AI2	10	4.0	4.4—4.8	<0.4	166	
$LiNbO_3$	AI1	30		3.56	10.	80	
$LiNbO_3$	AI1	0.025	4.0	4.66	3.4	167	
$LiNbO_3$	AI2	10	3.5	4.4—4.8	Abs. spectr. (1.5 @ 4.66 eV)	166	
$LiYF_4$	AI1	0.015	~11	9.32	<0.004	156	
MgF_2	AI1	0.017	11.8	6.99	<0.0062	156	
MgF_2	AI1	0.015	11.8	9.32	<0.0028	156	
NaBr	AI1	0.015	7.7	9.320	2.5	51	
NaCl	AI1	0.015	6.4	9.320	3.5	51	
RbBr	AI2	~10	7.2	7.0—8.0	Relative spectrum	160	20 K, 80 K
RbBr	CR	15		6.7	11.	52	
RbBr	AI1	0.017	5.4	6.99	2.43	51	
RbBr	AI1	0.015	5.4	9.32	2.18	51	
RbBr	AI1	10		7.12	Relative value	161	
RbCl	AI1	0.015	7.3	9.32	1.1	51	
RbI	AI2	~10	6.3	6.1—7.0	Relative spectrum	160	20 K
RbI	AI1	0.017	5.0	6.99	5.1	51	
RbI	AI1	0.015	5.0	9.32	2.5	51	
RbI	AI1	10		7.12	Relative value	161	
RDA(RbH_2AsO_4)	AI1	0.017		6.99	0.050	51	
RDP(RbH_2PO_4)	AI1	0.017		6.99	0.059	51	
SiO_2(quartz)	AI1	0.015	7.8	9.32	<0.045	51	
SrF_2	AI1	0.017		6.99	<0.0057	156	
SrF_2	AI1	0.015		9.32	<0.0054	156	

Table 1.2.4 (continued)
TWO-PHOTON ABSORPTION DATA FOR LARGE BANDGAP CRYSTALS

Material	Method	Excitation duration (nsec)	Approximate one-photon absorption edge (eV)	Applied two-photon energy (eV)	Two-photon absorption coefficient β (cm/GW)	Ref.	Additional information
$SrTiO_3$	AI1	~30		3.56	2.9	168	
$SrTiO_3$	AI2			4.1; 4.7; 5.0	1.3; 4.1; 10.2	140	
$SrTiO_3$	CR			3.92	3.	88	
$SrTiO_3$	AI2	17	4.1	3.2—4.4	Abs. spectr. (4 @ 4.2 eV)	169	
$SrTiO_3$	PA, T	5		3.2—5.5	Abs. spectr. (5 @ 5.0 eV)	170	

Table 1.2.5
TWO-PHOTON ABSORPTION DATA FOR GLASSES

Material	Method	Excitation duration (nsec)	Approximate one-photon absorption edge (eV)	Applied two-photon energy (eV)	Two-photon absorption coefficient β (cm/GW)	Ref.
As_2S_3	AI2	30	2.3	2.4—3.6	Abs. spectr. (30 @ 3.1 eV)	74
As_2S_3	AI1	~30		3.56	14	168
BK7 borosilicate	AI1	1.1, 7	3.9	7.07	0.0060	19
BK10 borosilicate	AI1	1.1, 7	4.1	7.07	0.0045	19
BK10 antisolarant	AI1	1.1, 7	4.0	7.07	0.0100	19
Holmium oxide	PA			4.26—4.32	Relative spectrum	171
LG630:Nd (Schott)	AI1	0.006		2.33	0.004	172
Silica (Corning 7940)	AI1	1.1, 7	7.8	7.07	<0.0005	19
Silica (Suprasil)	AI1	0.017	7.8	6.99	<0.0012	51
Silica (Suprasil)	AI1	0.015	7.8	9.32	0.017	156
Silica (Suprasil)	AI1	0.015	7.8	9.32	<0.045	51

Table 1.2.6
TWO-PHOTON ABSORPTION DATA FOR ORGANIC LIQUIDS AND CRYSTALS

Material	Method	Excitation duration (nsec)	Applied two-photon energy (eV)	Two-photon cross-section σ_2 10^{-50}cm^4-sec/photon-molecule	Ref.	Additional information
Aminophthalimide	L		3.56	2.	173	In ethyl alcohol
Anthracene	L		3.56	25.	173	Crystal
Anthracene	L		3.56	3.5	173	In ethyl alcohol
Anthracene	L		3.24—3.42	Relative spectrum	174	In hexane
Anthracene	L, PC	40	4.14—4.70	Relative spectrum	175	Crystal
Anthracene	L	8	3.8—4.6	Relative spectrum	176	
Anthracene	L	25	3.60—5.1	Abs. spectr. (70 @ 4.6 eV)	177	Crystal
Anthracene	AI2	20	4.4	70	178	In chloroform
Anthracene	L	25	3.60—5.1	Abs. spectr. (~1 @ 4.6 eV)	177	In cyclohexane, Benzene
Anthracene	L		3.56	Relative values	179	
Anthracene	L	20—25	3.10—3.56	Abs. spectr. (0.01 @ 3.2 eV)	61	Crystal
Anthracene	L	20—25	3.10—3.56	Abs. spectr. (0.0005 @ 3.2 eV)	61	In benzene
Anthracene	AI2	60	3.45—4.25	Abs. spectr. (20 @ 3.8 eV)	180	10 K, 77 K
Anthracene	L		3.56	0.16	181	In cyclohexane
Anthracene	L	40	3.56	0.23	60	In cyclohexane
Anthracene	L	5	3.9—5.0	Abs. spectr. (~2 @ 4 eV)	182	Crystal
Azulene	L	20	4.0—5.7	Relative spectrum	183	
Benzanthracene	L		3.56	Relative values	179	
Benzene	CR	15	6.7	2.5	52	
Benzene	L	5	4.6—5.2	Relative spectrum	184	
Benzene	TL	~5	4.6—5.3	Relative spectrum	66	
Benzene	L	6	5.8—6.8	Relative spectrum	185	4.2 K, Crystal
Benzene	L	20	4.4—6.0	Relative spectrum	186	
Benzene	L	~5	5.4—6.8	Relative spectrum	187	
Benzene	PA	1000	4.8—5.44	Abs. spectr. (10 @ 4.7 eV)	188	
Benzene	CR	~5	4.34	8.7	189	2 K
Benzene	L		3.4—4.4	Relative spectrum	190	2 K, Crystal
Benzene, deuterated	L	5	4.7—5.1	Relative spectrum	191	2 K, Crystal
Benzene	L	5	4.6—5.1	Relative spectrum	191	2 K, Crystal
Benzene	CR	4	4.72	<0.003	192	

Benzopyrene	AI1	10	3.31—4.68	Abs. spectr. (10 @ 3.6 eV)	193	
Benzopyrene	L		3.56	Relative values	179	
Binaphthyl	L		4.1—5.6	Relative spectrum	194	
Biphenyl-h_{10}	L		4.1—4.3	Relative spectrum	195	2 K
Biphenyl-d_{10}	L		4.1—4.3	Relative spectrum	195	2 K
Biphenyl	AI2, L	20	4.1—5.8	Abs. spectr. (1 @ 4.8 eV)	196	
Bipyridine	L	~5	3.7—4.7	Relative spectrum	197	In CCl_4
Bypyridine	AI2, L	20	4.1—4.6	Relative spectrum	196	
Carbon disulfide	CR	4	4.72	<0.057	192	
Carbon disulfide	CR		4.34—5.58	Abs. spectr. (1.1 @ 5.58 eV)	55	
Carbon disulfide	AI1	40	3.56	0.5	12	
Carbon disulfide	AI1	~30	3.56	0.26	168	
Carbon tetrachloride	CR	4	4.72	<0.014	192	
Chloronaphthalene	AI2	0.020	3.97—4.34	Relative spectrum	198	
Chloronaphthalene	AI2		4.2—5.3	Abs. spectr. (1 @ 4.8 eV)	199	
Chloronaphthalene	AI2		3.7	Abs. spectr. (1 @ 5.2 eV)	200	
Chrysene	L		3.56	Relative values	179	
Cyclohexane	CR		3.92	1.9	88	
Dibenzanthracene	L		3.56	Relative values	179	
Dichlorobiphenyl	AI2, L	20	4.1—5.6	Relative spectrum	196	
Difluorenyl	L		4.1—4.7	Relative spectrum	194	
Difluorobenzene	L	5	4.4—4.5	Relative spectrum	184	
Difluorobiphenyl	AI2, L	20	4.1—5.6	Relative spectrum	196	
Diphenyl	AI2	0.020	3.97—4.34	Relative spectrum	198	
Diphenylbutadiene	CR	CW	4.08	14.4	201	In benzene
Diphenylbutadiene	L		3.4—4.0	Relative spectrum	202	77 K, in EPA glass
Diphenylbutadiene	AI2	~30	3.50	1.0	203	
Diphenylhexatriene	CR	CW	4.08	43.3	201	In benzene
Diphenylhexatriene	L		2.9—3.5	Relative spectrum	204	77 K, in EPA glass
Diphenyloctatetraene	CR	CW	4.08	61.0	201	In benzene
Diphenyloxazole	CR	4	4.72	0.41	192	
Fluoranthene	L		3.56	Relative values	179	
Fluorene	L		4.1—4.7	Relative spectrum	194	
Fluorene	L		3.56	Relative values	179	
Fluorobenzene	L	5	4.5—5.1	Relative spectrum	205	Vapor
Fluorobenzene	L	5	4.6—5.0	Relative spectrum	184	
Hexatriene	TL		4.1—6.0	Abs. spectr. (2 @ 5.2 eV)	206	
Methylene iodide	AI1	0.004	4.66	2.1	138	

Table 1.2.6 (continued)
TWO-PHOTON ABSORPTION DATA FOR ORGANIC LIQUIDS AND CRYSTALS

Material	Method	Excitation duration (nsec)	Applied two-photon energy (eV)	Two-photon cross-section σ_2 10^{-50}cm^4-sec/photon-molecule	Ref.	Additional information
Naphthacene	L		3.56	Relative values	179	
Naphthalene	L		4.1—5.8	Relative spectrum	194	In cyclohexane
Naphthalene	AI2	20	3.8—3.95	Abs. spectr. (0.2 @ 3.9 eV)	207	
Naphthalene	CR		5.46	2.0	208	In CS_2
Naphthalene	L	6	3.84—6.19	Relative spectrum	243	In cyclohexane
Naphthalene-d_8	L	6	3.84—6.19	Relative spectrum	243	In cyclohexane
Naphthalene	L	6	3.84—6.19	Relative spectrum	243	Crystal
Naphthalene-d_8	L	6	3.84—6.19	Relative spectrum	243	Crystal
Nitrobenzene	CR	4	4.72	0.11	192	
Nitrobenzene	AI2		4.1—4.5	Relative spectrum	210	
Paracyclophane	L	~5	3.87—6.03	Relative spectrum	211	Crystal
Perylene	L		3.56	Relative values	179	
Phenanthrene	L		3.56	Relative values	179	
Pyrene	L		3.56	Relative values	179	
Pyrene	AI1	~30	3.56	0.22	168	
Pyrimidine	L	~5	3.8—6.0	Relative spectrum	212	In hexane
Stilbene	AI2	20	4.4	80.	178	In chloroform
Stilbene	L		4.3—5.5	Relative spectrum	213	In methylcyclohexane
Stilbene	AI2	200	3.56	80.	214	In chloroform
Stilbene	L, TL	~5	4.6—5.4	Relative spectrum	215	In hexane
Stilbene	CR	CW	4.82	12.1	201	In cyclohexane
Terphenyl	AI2, L	20	4.1—5.6	Relative spectrum	196	
Thianthrene	PC	10	3.88—5.52	Relative spectrum	216	Crystal
Toluene	L	20	4.4—6.0	Relative spectrum	186	
Toluene	L	5	4.5—4.7	Relative spectrum	184	
Triphenylene	L		3.56	Relative values	179	

Table 1.2.7
TWO-PHOTON ABSORPTION DATA FOR ORGANIC DYES

Material	Method	Excitation duration (nsec)	Applied two-photon energy (eV)	Two-photon cross-section σ_2 10^{-50}cm^4-sec/photon-molecule	Ref.	Additional information
Acridine	L		3.56	2.0	173	In ethanol
Acridine red	L	~200, ~40	2.33	Relative values	217—220	
Acridine orange	L	~200, ~40	2.33	Relative values	217—220	
Acridine yellow	L	~200, ~40	2.33	Relative values	217—220	
Acridine red	L	60	2.33	2.8	221	In ethanol
BBOT	L	40	3.56	4	60	In methylcyclohexane
Cyanine I, II, III, IV	L	~200, ~40	2.33	Relative values	217—220	
DCM	L	110	2.33	9.0	222	
Dimethyl-POPOP	L	40	3.56	11.5	60	In methylcyclohexane
Dimethyl-POPOP	GTA	0.005	3.56	600	58	In methylcyclohexane
Disodium fluorescein	L	60	2.33	0.18	221	In ethanol
DODCI	L	60	2.33	22	221	In methanol
DODCI I	L	110	2.33	38	222	
Eosin	L	30	2.33	18	223	
Eosine B, Y	L	~200, ~40	2.33	Relative values	217—220	
Hydroxycoumarin	L	4	3.8—5.0	Abs. spectr. (7.9 @ 4.1 eV)	224	
Imidocarbocyanine	L		2.23—3.29	Relative spectrum	225	In diethyleneglycol
Methyldiethyl-aminocoumarin	L; AI1	40	3.56	14.5; 50	60	In ethanol
Oxazine 725	L	110	2.33	0.41	222	
Phloxine B	L	~200, ~40	2.33	Relative values	217—220	
Rhodamine B, 6G	L	~200, ~40	2.33	Relative values	217—220	
Rhodamine B	L	CW	2.33	13	241	

Table 1.2.7 (continued)
TWO-PHOTON ABSORPTION DATA FOR ORGANIC DYES

Material	Method	Excitation duration (nsec)	Applied two-photon energy (eV)	Two-photon cross-section σ_2 10^{-50}cm^4-sec/photon-molecule	Ref.	Additional information
Rhodamine B	AI1		2.33	14.3	226	
Rhodamine B	AI1		3.56	148	226	
Rhodamine 6G	L	30	2.33	26	223	
Rhodamine 6G	AI1		2.33	12.9	226	
Rhodamine 6G	AI1		3.56	355	226	
Rhodamine 6G	L	0.025	2.1—3.5	Relative spectrum	227	
Rhodamine C	L	30	2.33	26	223	
Rhodamine 110	L	110	2.33	0.19	222	
Rhodamine 6G	L	110	2.33	11	222	
Rhodamine B	L	110	2.33	12	222	
Rhodamine 101	L	110	2.33	20	222	
Rhodamine 6G	L		2.23—3.29	Relative spectrum	256	In ethanol
Rhodamine 6G	L	60	2.33	5.5	221	In ethanol
Rhodamine B	L	60	2.33	14	221	In dichloroethane
Rubicene	L	~200, ~40	2.33	Relative values	217—220	
Safranine O	L	~200, ~40	2.33	Relative values	217—220	
Uranine	L	~200, ~40	2.33	Relative values	217—220	

Table 1.2.8
TWO-PHOTON ABSORPTION DATA FOR POLYMERS

Material	Method	Excitation duration (nsec)	Applied two-photon energy (eV)	Two-photon cross-section σ_2 $10^{-50}cm^4$-sec/photon-molecule	Ref.	Additional information
4BCMU (4 butoxycarbonylmethylurethane)	CR	~5	3.2—4.2	Relative spectrum	228	In $CHCl_3$
3BCMU (3 butoxycarbonylmethylurethane)	CR	~5	4.0	6000	229	In $CHCl_3$
Poly-chlorotri-fluoroethylene	AI1	1.1, 7	7.07	(B = 0.012 cm/GW)	230	
Polydiacetelyne-bis (toluene sulfonate)	AI2	0.015	2.3—3.0	[10^3 cm/GW @ 3 eV]	118	Crystal
Polydiacetelyne-bis (phenylurethane)	AI2	0.015	2.3—3.0	[10^2 cm/GW @ 3 eV]	118	Crystal

Table 1.2.9
TWO-PHOTON ABSORPTION DATA FOR BIOLOGICAL MATERIALS

Material	Method	Excitation duration (nsec)	Applied two-photon energy (eV)	Two-photon cross-section σ_2 $10^{-50}cm^4$-sec/photon-molecule	Ref.	Additional information
Chlorophyll-a	L	30	2.33	2.0	223	In ethanol
Chlorophyll-a	AI1	2	3.6	—	39,40	Transmission vs. Intensity
H_2O	PC	0.03	9.32	0.081	231	
NADH (enzyme)	AI1, TCN	20	3.56	340	232	
Trans-retinal	L		2.8—3.0	Relative spectrum	233	77 K
Trans-retinol (visual chromophore)	L		3.4—4.0	Relative spectrum	38	77 K

Table 1.2.10
TWO-PHOTON ABSORPTION DATA FOR RARE EARTH AND COLOR-CENTER SYSTEMS

Material	Method	Excitation duration (nsec)	Applied two-photon energy (eV)	Two-photon cross-section σ_2 10^{-50}cm^4-sec/photon-molecule	Ref.
Eu^{2+} in H_2O	L	10	7.36	40	242
Eu^{3+} in CaF_2	L	15	3.1—3.4	Relative spectrum	235
Eu^{3+} chelate	L		3.5—5.5	Abs. spectr. (3000 @ 4.5 eV)	236
Eu^{2+} in CaF_2 and SrF_2	L	30	3.4—3.46	Relative spectrum	237
F-centers in KCl	L	0.010	2.33	2.0	238
Gd^{3+} in LaF_3	L	~5	3.84—5.08	Relative spectrum	239
NaF:$F_3{}^+$ centers	L		2.33	65	240
NaF:N_1 centers	L		2.33	55	240
Tb^{3+} in CaF_2	L	15	3.1—3.4	Relative spectrum	235

REFERENCES

1. **Göppert-Mayer, M.,** Über Elementarakte mit zwei Quantensprüngen, *Ann. Phys.*, 9, 273, 1931.
2. **Hughes, V. and Grabner, L.,** The radiofrequency spectrum of $Rb^{85}F$ and $Rb^{87}F$ by the electric resonance method, *Phys. Rev.*, 79(2), 314, 1950.
3. **Kusch, P. and Hughes, V. W.,** Atomic and molecular beam spectroscopy, in *Encyclopedia of Physics*, Vol. 37, Flügge, S., Ed., Springer-Verlag, Berlin, 1959, 67.
4. **Maiman, T.,** Stimulated optical radiation in ruby, *Nature (London)*, 187, 493, 1960.
5. **Collins, R. J., Nelson, D. F., Schawlow, A. L., Bond, W. L., Garrett, C. G. B., and Kaiser, W. K.,** Coherence, narrowing, directionality and relaxation oscillations in the light emissions from ruby, *Phys. Rev. Lett.*, 5, 303, 1960.
6. **Kaiser, W. and Garrett, C. G. B.,** Two-photon excitation in CaF_2:Eu^{2+}, *Phys. Rev. Lett.*, 7(6), 229, 1961.
7. **Franken, P. A., Hill, A. E., Peters, C. W., and Weinreich, G.,** Generation of optical harmonics, *Phys. Rev. Lett.*, 7, 118, 1961.
8. **Kleinman, D. A.,** Laser and two-photon processes, *Phys. Rev.*, 125(1), 87, 1962.
9. **Braunstein, R.,** Nonlinear optical effects, *Phys. Rev.*, 125(2), 475, 1962.
10. **Abella, I. D.,** Optical double-photon absorption in cesium vapor, *Phys. Rev. Lett.*, 9(11), 453, 1962.
11. **Peticolas, W. L., Goldsborough, J. P., and Rieckhoff, K. E.,** Double photon excitation in organic crystals, *Phys. Rev. Lett.*, 10, 43, 1963.
12. **Giordmaine, J. A. and Howe, J. A.,** Intensity-induced optical absorption cross section in CS_2, *Phys. Rev. Lett.*, 11(5), 207, 1963.
13. **Hopfield, J. J., Worlock, J. M., and Park, K.,** Two-quantum absorption spectrum of KI, *Phys. Rev. Lett.*, 11(9), 414, 1963.
14. **Hopfield, J. J. and Worlock, J. M.,** Two-quantum absorption spectrum of KI and CsI, *Phys. Rev.*, 137(5A), A1455, 1965.
15. **Braunstein, R. and Ockman, N.,** Optical double-photon absorption in CdS, *Phys. Rev.*, 134(2A), A499, 1964.
16. **Fork, R. L., Greene, B. I., and Shank, C. V.,** Generation of optical pulses shorter than 0.1 psec by colliding pulse mode locking, *Appl. Phys. Lett.*, 38, 671, 1981.
17. **Shank, C. V., Fork, R. L., Yen, R., Stolen, R. H., and Tomlinson, W. J.,** Compression of femtosecond optical pulses, *Appl. Phys. Lett.*, 40, 761, 1982.
18. **James, R. B. and Smith, D. L.,** Theory of nonlinear optical absorption associated with free carriers in semiconductors, *IEEE J. Quantum Electron.*, QE-18, 1841, 1982.
19. **Smith, W. L.,** Lawrence Livermore National Laboratory,1981 Laser Program Annual Report, U.C.R.L. - 50021-81, p. 7-23.
20. **Fossum, H. J., Chen, W. S., and Ancker-Johnson, B.,** Excess carriers induced in Indium Antimonide with a carbon-dioxide laser, *Phys. Rev.*, B8, 2857, 1973.
21. **Seiler, D. G. and Hanes, L. K.,** Absorption processes near the bandgap of InSb: laser-induced hot electron and photoconductivity studies, *Opt. Commun.*, 28, 326, 1979.

22. **McManus, J. B., People, R., Aggarwal, R. L., and Wolff, P. A.,** Nonlinear absorption due to shallow donors in germanium at 10.6μm, *J. Appl. Phys.*, 52, 4748, 1981.
23. **Aspect, A., Dalibard, J., and Roger, G.,** Experimental test of Bell's inequalities using time-varying analyzers, *Phys. Rev. Lett.*, 49(25), 1804, 1982; see also *Science,* 219, 40, 1983.
24. **Hanamura, E.,** Giant two-photon absorption due to excitonic molecule, *Solid State Commun.*, 12, 951, 1973.
25. **Chase, L. L., Peyghambarian, N., Grynberg, G., and Mysyrowicz, A.,** Evidence for Bose-Einstein condensation of biexcitons in CuCl, *Phys. Rev. Lett.*, 42(18), 1231, 1979.
26. **Basov, N. G., Grasyuk, A. Z., Zubarev, I. G., Katulin, V. A., and Krokhin, O. N.,** Semiconductor quantum generator with two-photon optical excitation, *Sov. Phys. JETP,* 23(3), 366, 1966.
27. **Patel, C. K. N., Fleury, P. A., Slusher, R. E., and Frisch, H. L.,** Multiphoton plasma production and stimulated recombination radiation in semiconductors, *Phys. Rev. Lett.*, 16(22), 971, 1966.
28. **Vaucher, A. M., Cao, W., Ling, J., and Lee, C. H.,** Generation of tunable picosecond pulses from a bulk GaAs laser, *IEEE J. Quantum Electron.*, QE-18(2), 187, 1982.
29. **Jacobs, R. R., Prosnitz, D., Bischel, W. K., and Rhodes, C. K.,** Laser generation from 6 to 35μm following two-photon excitation of ammonia, *Appl. Phys. Lett.*, 29(11), 710, 1976.
30. **Lüthy, W., Burkhard, P., Gerber, T. E., and Weber, H. P.,** Laser emission at 535 nm by two-photon-dissociation of thallium-bromide, *Opt. Commun.*, 38(5,6), 413, 1981.
31. **Kligler, D., Pummer, H., Bischel, W. K., and Rhodes, C. K.,** Photolytic and energy transfer studies of laser media using two-quantum excitation, Paper THB3-1, in *Digest of Topical Meeting on Inertial Confinement Fusion,* February 7-9, San Diego, Optical Society of America, Washington, D.C., 1977.
32. **Park, H. and Steier, W. H.,** Two-photon generated color-center gratings in KBr: proposed picosecond pulsewidth measuring technique for the ultraviolet, *IEEE J. Quantum Electron.*, QE-17(5), 581, 1981.
33. **Geller, M. and Taylor, H. F.,** Three-Dimensional Information Storage in Alkali Halides by Two-Photon Absorption, Naval Electronics Laboratory Center Tech. Rep. 1749, San Diego, 1971; available from NTIS, Springfield, Va.
34. **Wood, V. E., Hartman, N. F., and Verber, C. M.,** Two-photon photorefractivity in pure and doped $LiNbO_3$, *Ferroelectrics,* 27, 237, 1980.
35. **Smith, W. L.,** Laser-induced breakdown in optical materials, *Opt. Eng.*, 17, 489, 1978.
36. **Dempsey, J., Smith, J., Holah, G. D., and Miller, A.,** Nonlinear absorption and pulse shaping in InSb, *Opt. Commun.*, 26(2), 265, 1978.
37. **Mak, P. S., Davis, C. C., Forster, B. J., and Lee, C. H.,** Measurements of picosecond light pulses with a two-photon photoconductivity detector, *Rev. Sci. Instrum.*, 51, 647, 1980.
38. **Birge, R. R., Bennett, J. A., Pierce, B. M., and Thomas, T. M.,** Two-photon spectroscopy of the visual chromophores. Evidence for a lowest excited $^1A_g^-$-like $\pi\pi^*$ state in all-trans-Retinol (Vitamin A), *J. Am. Chem. Soc.*, 78, 1533, 1978.
39. **Leupold, D., Mory, S., and Hoffmann, O.,** Nonlinear absorption of chlorophyll-a, *Acta Phys. Chem.*, 23(1), 33, 1977.
40. **Leupold, D., Voigt, B., Höxtermann, E., Ehlert, J., Hoffman, P., and Hieke, B.,** Nonlinear absorption of chlorophyll-A oligomers in dry n-hexane in comparison with chlorophyll in vivo, *Stud. Biophys.*, 83, 41, 1981.
41. **Mahr, H.,** Two-photon absorption spectroscopy, in *Quantum Electronics,* Vol. 1 (Part A), Rubin, H. and Tang, C. L., Eds., Academic Press, New York, 1975, chap. 4.
42. **Bredikhin, V. I., Galenin, M. D., and Genkin, V. N.,** Two-photon absorption and spectroscopy, *Sov. Phys. Usp.*, 16(3), 1973.
43. **Worlock, J. M.,** Two-photon spectroscopy, in *Laser Handbook,* Arecchi, F. T. and Schulz-Dubois, E. O., Eds., North-Holland, Amsterdam, 1972, chap. E6.
44. **McClain, W. M. and Harris, R. A.,** Two-photon molecular spectroscopy in liquids and gases, in *Excited States,* Vol. 3, Lim, E. C., Ed., Academic Press, New York, 1977, 1.
45. **Fröhlich, D. and Sondergeld, M.,** Experimental techniques in two-photon spectroscopy, *J. Phys. E,* 10, 761, 1977.
46. **Flytzanis, C.,** Theory of nonlinear optical susceptibilities, in *Quantum Electronics,* Vol. 1 (Part A), Rabin, H. and Tang, C. L., Eds., Academic Press, New York, 1975, chap. 1.
47. **Levenson, M. D.,** *Introduction to Nonlinear Laser Spectroscopy,* Academic Press, New York, 1982.
48. **Baltrameyunas, R., Vaitkus, Y. Y., Vishchakas, Y. K., and Gavryushin, V. I.,** Dependence of the two-photon absorption coefficient on the thickness of CdS single crystals, *Opt. Spectrosc.*, 36(6), 714, 1974.
49. **Stewart, A. F. and Bass, M.,** Intensity-dependent absorption in semiconductors, *Appl. Phys. Lett.*, 37(11), 1040, 1980.
50. **Bloembergen, N.,** *Nonlinear Optics,* W. A. Benjamin, New York, 1965; references therein.
51. **Liu, P., Smith, W. L., Lotem, H., Bechtel, J. H., Bloembergen, N., and Adhav, R. S.,** Absolute two-photon absorption coefficients at 355 and 266 nm, *Phys. Rev. B,* 17(12), 4620, 1978.

52. **Prior, Y. and Vogt, H.,** Measurements of uv two-photon absorption relative to known Raman cross sections, *Phys. Rev. B,* 19, 5388, 1979.
53. **Bivas, A., Levy, R., Phach, V. D., and Grun, J. B.,** Biexciton two-photon absorption in the nanosecond and picosecond range in copper halides, in *Physics of Semiconductors 1978,* Inst. Phys. Conf. Ser. No. 43, AIP, New York, 1979.
54. **Levenson, M. D. and Bloembergen, N.,** Dispersion of the nonlinear optical susceptibility tensor in centrosymmetric media, *Phys. Rev. B,* 10, 4447, 1974.
55. **Lotem, H. and Lynch, R. T., Jr.,** Destructive interference of imaginary resonant contributions to $\chi^{(3)}$, *Phys. Rev. Lett.,* 37(6), 334, 1976.
56. **Kuwata, M. and Nagasawa, N.,** Self-induced polarization rotation effect of an elliptically polarized beam in CuCl, *J. Phys. Soc. Jpn.,* 51, 2591, 1982.
57. **Mollenauer, L. F., Bjorklund, G. C., and Tomlinson, W. J.,** Production of stabilized coloration in alkali halides by a two-photon absorption process, *Phys. Rev. Lett.,* 35(24), 1662, 1975.
58. **Topp, M. R. and Rentzepis, P. M.,** Picosecond stimulated emission in a fluorescent solution following two-photon absorption, *Phys. Rev. A,* 3(1), 358, 1971.
59. **Reilly, J. P. and Clark, J. H.,** Broadband detection of two-photon absorption by intracavity dye laser spectroscopy, in *Advances in Laser Chemistry,* Zewail, A. H., Ed., Springer-Verlag, Berlin, 1978, 354.
60. **Hermann, J. P. and Ducuing, J.,** Absolute measurement of two-photon cross sections, *Phys. Rev. A,* 5(6), 2557, 1972.
61. **Webman, I. and Jortner, J.,** Energy dependence of two-photon-absorption cross sections in anthracene, *J. Chem. Phys.,* 50(6), 2706, 1969.
62. **Bae, Y., Song, J. J., and Kim, Y. B.,** Photoacoustic study of two-photon absorption in hexagonal ZnS, *J. Appl. Phys.,* 53(1), 615, 1982.
63. **Cingolani, A., Ferrero, F., Minafra, A., and Trigiante, D.,** Two-photon conductivity in ZnS and CdS, *Il Nuovo Cimento,* 4B, 217, 1971.
64. **Gibson, A. F., Hatch, C. B., Maggs, P. N. D., Tilley, D. R., and Walker, A. C.,** Two-photon absorption in indium antimonide and germanium, *J. Phys. C,* 9, 3259, 1976.
65. **Lotem, H., Bechtel, J. H., and Smith, W. L.,** Normalization technique for accurate measurements of two-photon absorption coefficients, *Appl. Phys. Lett.,* 28, 389, 1976.
66. **Twarowski, A. J. and Kliger, D. S.,** Multiphoton absorption spectra using thermal blooming, *Chem. Phys.,* 20, 259, 1977.
67. **Bredikhin, V. I., Genkin, V. N., and Soustov, L. V.,** Two-photon spectroscopy of proustite (Ag_3AsS_3), *Sov. J. Quantum Electron .* 6(4), 409, 1976.
68. **Aleksandrov, A. P., Bredikhin, V. I., and Genkin, V. N.,** Two-photon absorption spectrum of proustite, *Sov. Phys. Solid State,* 17(8), 1645, 1976.
69. **Hanna, D. C. and Turner, A. J.,** Nonlinear absorption measurements in proustite (Ag_3AsS_3) and CdSe, *Opt. Quantum Electron.,* 8, 213, 1976.
70. **Casalboni, M., Crisanti, F., Francini, R., and Grassano, U. M.,** Two-photon spectroscopy in AgCl, *Solid State Commun.,* 35, 833, 1980.
71. **Miller, A. and Ash, G. S.,** Two-photon absorption and short pulse stimulated recombination in $AgGaSe_2$, *Opt. Commun.,* 33(3), 297, 1980.
72. **Dinges, R., Fröhlich, D., and Uihlein, Ch.,** Two-photon absorption of anisotropic excitons in β-AgI, *Phys. Status Solidi (B),* 76, 613, 1976; **Dinges, R. and Frölich, D.,** Polariton tuning by two-photon Stark effect in Wurtzite AgI, *Solid State Commun.,* 19, 61, 1976.
73. **van der Ziel, J. P. and Gossard, A. C.,** Two-photon absorption spectrum of AlAs-GaAs monolayer crystals, *Phys. Rev. B,* 17(2), 765, 1978.
74. **Nasyrov, U.,** Two-photon absorption spectrum of crystalline and glassy As_2S_3, *Sov. Phys. Semicond.,* 12(6), 720, 1978.
75. **Mozol, P. E., Patskun, I. I., Sal'kov, E. A., and Fekeshgazi, I. V.,** Two-photon and two-step transitions in the wide-band semiconductors CdP_2, ZnP_2, and ZnSe, *Bull. Acad. Sci. U.S.S.R. Phys. Ser.,* 45(6), 1092, 1981.
76. **Lisitsa, M. P., Mozol', P. E., and Fekeshgazi, I. V.,** Laser pulse lengthening with the aid of nonlinearly absorbing semiconductors CdP_2 and ZnP_2, *Sov. J. Quantum Electron.,* 4(1), 110, 1974.
77. **Nguyen, V. T., Damen, T. C., and Gornik, E.,** Two-photon absorption spectroscopy of the intrinsic exciton fine structure in CdS, *Appl. Phys. Lett.,* 30(1), 33, 1977.
78. **Pradere, F. and Mysyrowicz, A.,** Two-photon absorption spectrum by excitons in CdS, *Proc. Tenth Int. Conf. on Physics of Semiconductors,* Keller, S. P., Ed., U.S. Atomic Energy Commission, 1970, 101.
79. **Konyukhov, V. K., Kulevskii, L. A., and Prokhorov, A. M.,** Two-photon absorption spectrum in CdS near the fundamental absorption edge, *Sov. Phys. Doklady,* 12(4), 1967.
80. **Arsen'ev, V. V., Dneprovskii, V. S., Klyshko, D. N., and Penin, A. N.,** Nonlinear absorption and limitation of light intensity in semiconductors, *Sov. Phys. JETP,* 29(3), 413, 1969.

81. **Penzkofer, A., Falkenstein, W., and Kaiser, W.,** Two-photon spectroscopy using picosecond light continua, *Appl. Phys. Lett.*, 28(6), 319, 1976.
82. **Itoh, T., Nozue, Y., and Ueta, M.,** Giant two-photon absorption and two-photon resonance Raman scattering in CdS, *J. Phys. Soc. Jpn.*, 40(6), 1791, 1976.
83. **Svorec, R. W. and Chase, L. L.,** Two-photon absorption in the vicinity of the exciton absorption in CdS, *Solid State Commun.*, 17, 803, 1975.
84. **Kobbe, G. and Klingshirn, C.,** Quantitative investigation of the two-photon absorption of ruby-laser-light in various semiconductors, *Z. Phys.*, B37, 9, 1980.
85. **Catalano, I. M. and Cingolani, A.,** Non-linear self-action effect and absolute two-photon absorption coefficient in CdS, *Solid State Commun.*, 34, 761, 1980; **Catalano, I. M., Cingolani, A., and Minafra, A.,** Transmittance, luminescence, and photocurrent in CdS under two-photon excitation, *Phys. Rev. B*, 9(2), 707, 1974.
86. **Jackel, J. and Mahr, H.,** Nonlinear optical measurements in the excitonic region of CdS at 4.2 K, *Phys. Rev. B*, 17(8), 3387, 1978.
87. **Blau, W. and Penzkofer, A.,** Intensity detection of picosecond ruby laser pulses by two photon absorption, *Opt. Commun.*, 36(5), 419, 1981.
88. **Lotem, H. and de Araujo, C. B.,** Absolute determination of the two-photon absorption coefficient relative to the inverse Raman cross section, *Phys. Rev. B*, 16(4), 1711, 1977.
89. **Araujo, C. B. de and Lotem, H.,** New measurements of the two-photon absorption in GaP, CdS, and ZnSe relative to Raman cross sections, *Phys. Rev. B*, 18(1), 1978.
90. **Regensburger, P. J. and Panizza, E.,** Two-photon absorption spectrum of CdS, *Phys. Rev. Lett.*, 18(4), 113, 1967.
91. **Bredikhin, V. I. and Genkin, V. N.,** Two-photon absorption spectrum of CdS, *Sov. Phys. Solid State*, 16(5), 860, 1974.
92. **Bespalov, M. S., Kulevskii, L. A., Makarov, V. P., Prokhorov, M. A., and Tikhonov, A. A.,** Anistropy in the two-photon absorption spectrum in CdS, *Sov. Phys. JETP*, 28, 77, 1969.
93. **Grasyuk, A. Z., Zubarev, I. G., and Mentser, A. N.,** Anisotropy of two-photon absorption in optical excitation of CdSe semiconductor lasers, *Sov. Phys. Solid State*, 10(2), 427, 1968.
94. **Dneprovskii, V. S. and Ok, S. M.,** Role of absorption by nonequilibrium carriers in determination of two-photon absorption coefficient of CdSe and GaAs crystals, *Sov. J. Quantum Electron.*, 6(3), 298, 1976.
95. **Van Stryland, E. W. and Woodall, M. A.,** Photoacoustic measurement of nonlinear absorption in solids, *NBS Spec. Publ.*, 620, 50, 1980.
96. **Bryukner, F., Dneprovskii, V. S., and Khattatov, Z. Q.,** Two-photon absorption in cadmium selenide, *Sov. J. Quantum Electron.*, 4, 749, 1974.
97. **Basov, N. G., Grasyuk, A. Z., Efimov, V. F., Zubarev, I. G., Katulin, V. A., and Popov, J. M.,** Semiconductor lasers using optical pumping, *J. Phys. Soc. Jpn. Suppl.*, 21, 276, 1966.
98. **Bechtel, J. H. and Smith, W. L.,** Two-photon absorption in semiconductors with picosecond laser pulses, *Phys. Rev. B*, 13(8), 3515, 1976.
99. **Bass, M., Van Stryland, E. W., and Stewart, A. F.,** Laser calorimetric measurement of two-photon absorption, *Appl. Phys. Lett.*, 34(2), 142, 1979.
100. **Ralston, J. M. and Chang, R. K.,** Optical limiting in semiconductors, *Appl. Phys. Lett.*, 15(6), 164, 1969; Nd:Laser induced absorption in semiconductors and aqueous $PrCl_3$ and $NdCl_3$, *Opto-Electron.*, 1, 182, 1969.
101. **Brodin, M. S., Demidenko, Z. A., Dmitrenko, K. A., and Rexnichenko, V. Y.,** Temperature dependence of the two-photon absorption coefficient of CdS_xSe_{1-x} crystals near resonance, *Sov. J. Quantum Electron.*, 5(7), 861, 1975; **Brodin, M. S., Dmitrenko, K. A., and Rexnichenko, V. Y.,** Two-photon absorption in CdS_xSe_{1-x} mixed crystals at ruby laser frequencies, *Sov. Phys. Solid State*, 13(6), 1328, 1971.
102. **Brodin, M. S., Demidenko, Z. A., Dmitrenko, K. A., Reznichenko, V. Y., and Strashnikova, M. I.,** Frequency dependences of the one- and two-photon absorption of crystals with nonparabolic conduction bands similar to CdS, *Sov. Phys. Semicond.*, 8(1), 43, 1974.
103. **Bepko, S. J.,** Anistropy of two-photon absorption in GaAs and CdTe, *Phys. Rev. B*, 12(2), 669, 1975.
104. **Catalano, I. M. and Cingolani, A.,** Non-parabolic band effect on two-photon absorption in ZnSe and CdTe, *Solid State Commun.*, 43(3), 213, 1982.
105. **Rustagi, K. C., Pradère, F., and Mysyrowicz, A.,** Two-photon absorption in Cu_2O, *Phys. Rev. B*, 8(6), 2721, 1973.
106. **Antonetti, A., Astier, R., Martin, J. L., Migus, A., Hulin, D., and Mysyrowicz, A.,** Picosecond spectroscopy in Cu_2O: exciton dynamics and two-photon absorption, *Opt. Commun.*, 38, 431, 1981.
107. **Marange, C., Bivas, A., Levy, R., Grun, J. B., and Schwab, C.,** Two-photon absorption in CuBr, *Opt. Commun.*, 6(2), 138, 1972.
108. **Bivas, A., Marange, C., Grun, J. B., and Schwab, C.,** Two-photon absorption in CuCl, *Opt. Commun.*, 6(2), 142, 1972.

109. **Gale, G. M. and Mysyrowicz, A.,** Direct creation of excitonic molecules in CuCl by giant two-photon absorption, *Phys. Rev. Lett.*, 32(13), 727, 1974.
110. **Fröhlich, D., Staginnus, B., and Schönherr, E.,** Two-photon absorption spectrum of CuCl, *Phys. Rev. Lett.*, 19(18), 1032, 1967.
111. **Fröhlich, D. and Volkenandt, H.,** Determination of Γ_3 valence band in CuCl by two-photon absorption, *Solid State Commun.*, 43(3), 189, 1982.
112. **Lee, C. C. and Fan, H. Y.,** Two-photon absorption with exciton effect for degenerate valence bands, *Phys. Rev. B*, 9(8), 3502, 1974.
113. **Zubarev, I. G., Mironov, A. B., and Mikhailov, S. I.,** Influence of deep impurity levels on nonlinear absorption of light in GaAs, *Sov. Phys. Semicond.*, 11(2), 239, 1977.
114. **Grasyuk, A. Z., Zubarev, I. G., Mironov, A. B., and Poluektov, I. A.,** Spectrum of two-photon interband absorption of laser radiation in semiconducting GaAs, *Sov. J. Quantum Electron.*, 5(8), 997, 1976.
115. **Saissy, A., Azema, A., Botineau, J., and Gires, F.,** Absolute measurement of the 1.06 μm two-photon absorption coefficient in GaAs, *Appl. Phys.*, 15, 99, 1978.
116. **Bosacchi, B., Bessey, J. S., and Jain, F. C.,** Two-photon absorption of neodymium laser radiation in gallium arsenide, *J. Appl. Phys.*, 49(8), 4609, 1978.
117. **Kleinman, D. A., Miller, R. C., and Nordland, W. A.,** Two-photon absorption of Nd laser radiation in GaAs, *Appl. Phys. Lett.*, 23(5), 243, 1973.
118. **Lequime, M. and Herman, J. P.,** Reversible creation of defects by light in one dimensional conjugated polymers, *Chem. Phys.*, 26, 431, 1977.
119. **van der Ziel, J. P.,** Two-photon absorption spectra of GaAs with 2 ω near the direct band gap, *Phys. Rev. B*, 16(6), 2775, 1977.
120. **Yee, J. H. and Chan, H. H. M.,** Two-photon indirect transitions in GaP crystal, *Opt. Commun.*, 10, 56, 1974.
121. **Catalano, I. M., Cingolani, A., and Minafra, A.,** Multiphoton processes in layered structures, *Il Nuovo Cimento*, 38B(2), 579, 1977; **Adduci, F., Catalano, I. M., Cingolani, A., and Minafra, A.,** Direct and indirect two-photon process in layered semiconductors, *Phys. Rev. B*, 15(2), 926, 1977.
122. **Wenzel, R. G., Arnold, G. P., and Greiner, N. R.,** Nonlinear loss in Ge in the 2.5-4 μm range, *Appl. Opt.*, 12(10), 2245, 1973.
123. **Zubov, B. V., Murina, T. M., Olovyagin, B. R., and Prokhorov, A. M.,** Investigation of nonlinear absorption in germanium, *Sov. Phys. Semicond.*, 5(4), 559, 1971.
124. **Miller, A., Johnston, A., Dempsey, J., Smith, J., Pidgeon, C. R., and Holah, G. D.,** Two-photon absorption in InSb and $Hg_{1-x}Cd_xTe$, *J. Phys. C*, 12, 4839, 1979.
125. **Johnston, A. M., Pidgeon, C. R., and Dempsey, J.,** Frequency dependence of two-photon absorption in InSb and $Hg_{1-x}Cd_xTe$, *Phys. Rev. B*, 22(2), 825, 1980.
126. **Lee, C. C. and Fan, H. Y.,** Two-photon absorption and photoconductivity in GaAs and InP, *Appl. Phys. Lett.*, 20, 18, 1972.
127. **Fossom, H. J. and Chang, D. B.,** Two-photon excitation rate in indium antimonide, *Phys. Rev. B*, 8(6), 2842, 1973.
128. **Doviak, J. M., Gibson, A. F., Kimmitt, M. F., and Walker, A. C.,** Two-photon absorption in indium antimonide at 10.6 μm, *J. Phys. C*, 6, 593, 1973.
129. **Areshev, I. P., Guseinaliev, M. G., Danishevskii, A. M., Kochegarov, S. F., and Subashiev, V. K.,** Investigation of nonlinear absorption of light in indium antimonide by the two-beam method, *Sov. Phys. Solid State*, 22(5), 849, 1980; Double-beam linear dichroism investigation of the two-photon absorption in InSb, *Phys. Status Solidi (B)*, 102, 383, 1980.
130. **Seiler, D. G., Goodwin, M. W., and Weiler, M. H.,** High-resolution two-photon spectroscopy in InSb at milliwatt cw powers in a magnetic field, *Phys. Rev. B*, 23(12), 6806, 1981.
131. **Fröhlich, D. and Kenklies, R.,** Two-photon absorption in PbI_2, *Il Nuovo Cimento*, 38B(2), 433, 1977.
132. **Geusic, J. E., Singh, S., Tipping, D. W., and Rich, T. C.,** Three-photon stepwise optical limiting in silicon, *Phys. Rev. Lett.*, 19(19), 1126, 1967.
133. **Gauster, W. B. and Bushnell, J. C.,** Laser-induced infrared absorption in silicon, *J. Appl. Phys.*, 41(9), 3850, 1970.
134. **Reintjes, J. F. and McGroddy, J. C.,** Indirect two-photon transitions in Si at 1.06 μm, *Phys. Rev. Lett.*, 30(19), 901, 1973.
135. **Lisitsa, M. P., Kulish, N. R., and Stolyarenko, A. V.,** Two-photon absorption spectrum of α-SiC(6H), *Sov. Phys. Semicond.*, 14(10), 1208, 1980.
136. **Fröhlich, D. and Kenklies, R.,** Band-gap assignment in SnO_2 by two-photon spectroscopy, *Phys. Rev. Lett.*, 41(25), 1750, 1978.
137. **Oudar, J., Schwartz, C. A., and Batifol, E. M.,** Influence of two-photon absorption on second-harmonic generation in semiconductors. II. Measurement of two-photon absorption in tellurium, *IEEE J. Quantum Electron.*, QE-11(8), 623, 1975.

138. **Penzkofer, A. and Falkenstein, W.,** Direct determination of the intensity of picosecond light pulses by two-photon absorption, *Opt. Commun.*, 17(1), 1976.
139. **Waff, H. S. and Park, K.,** Structure in the two-photon absorption spectrum of TiO_2 (Rutile), *Phys. Lett.*, 32A(2), 109, 1970.
140. **Lee, J. H., Scarparo, M. A. F., and Song, J. J.,** Two-photon absorption measurements of crystals relative to the Raman cross section, Proc. VIIth Int. Conf. on Raman Spectroscopy, Ottawa, Canada, 1980, 684.
141. **Fröhlich, D., Staginnus, B., and Thurm, S.,** Symmetry assignments of two-photon transitions in TlCl, *Phys. Status Solidi*, 40, 287, 1970.
142. **Fröhlich, D., Treusch, J., and Kottler, W.,** Multiphonon processes in the two-photon absorption of TlCl and the temperature dependence of the band edge, *Phys. Rev. Lett.*, 29(24), 1603, 1972.
143. **Matsuoka, M.,** Angular dependence of two photon absorption in thallous chloride, *J. Phys. Soc. Jpn.*, 23(5), 1028, 1967.
144. **Matsuoka, M. and Yajima, T.,** Two-photon absorption spectrum in thallous chloride, *Phys. Lett.*, 23(1), 54, 1966.
145. **Brodin, M. S. and Goer, D. B.,** Two-photon absorption of ruby laser radiation by semiconductor crystals of ZnSe and $Zn_xCd_{1-x}Se$, *Sov. Phys. Semicond.*, 5(2), 219, 1971.
146. **Brodin, M. S., Shevel', S. G., Kodzhespirov, F. F., and Mozharovskii, L. A.,** Two-photon absorption of ruby laser radiation in mixed $Zn_xCd_{1-x}S$ crystals, *Sov. Phys. Semicond.*, 5(12), 2047, 1972.
147. **Staginnus, B., Fröhlich, D., and Caps, T.,** Automatic 2-photon spectrophotometer, *Rev. Sci. Instrum.*, 39(8), 1129, 1968.
148. **Dinges, R., Fröhlich, D., Staginnus, B., and Staude, W.,** Two-photon magnetoabsorption in ZnO, *Phys. Rev. Lett.*, 25(14), 922, 1970.
149. **Kaule, W.,** Polarization dependence of the two quantum absorption spectrum of intrinsic excitons in ZnO, *Solid State Commun.*, 9, 17, 1971.
150. **Fröhlich, D. M.,** Two-photon spectroscopy in solids, in *Proc. Tenth Int. Conf. on Physics of Semiconductors,* Keller, S. P., Ed., U.S. Atomic Energy Commission, 1970, 95.
151. **Panizza, E.,** Two-photon absorption in ZnS, *Appl. Phys. Lett.*, 10(10), 265, 1967.
152. **Park, K. and Waff, H. S.,** Two-photon absorption spectrum of ZnS, *Phys. Lett.*, 28A(4), 305, 1968.
153. **Valtrameyunas, R., Gavryushin, V., and Vaitkus, Y.,** Frequency dependence of the coefficient of two-photon absorption in ZnSe, *Sov. Phys. Solid State*, 17(10), 2020, 1976.
154. **Catalano, I. M. and Cingolani, A.,** Absolute two-photon absorption line shape in ZnTe, *Phys. Rev. B*, 19(2), 1049, 1979.
155. **Reintjes, J. F. and Eckardt, R. C.,** Two-photon absorption on ADP and KD*P at 266.1 nm, *IEEE J. Quantum Electron.*, QE-13(9), 791, 1977; Efficient harmonic generation from 532 to 266 nm in ADP and KD*P, *Appl. Phys. Lett.*, 30(2), 91, 1977.
156. **Liu, P., Yen, R., and Bloembergen, N.,** Two-photon absorption coefficients in UV window and coating materials, *Appl. Opt.*, 18(7), 1015, 1979.
157. **Fröhlich, D., Staginnus, B., and Onodera, Y.,** Two-photon spectroscopy in CsI and CsBr, *Phys. Status Solidi*, 40, 547, 1970.
158. **Liu, P., Yen, R., and Bloembergen, N.,** Dielectric breakdown threshold, two-photon absorption, and other optical damage mechanisms in diamond, *IEEE J. Quantum Electron.*, QE-14(8), 574, 1978.
159. **Geller, M., DeTemple, T. A., and Taylor, H. F.,** Quantum efficiency for F-center production by two-photon absorption, *Solid State Commun.*, 7, 1019, 1969.
160. **Fröhlich, D. and Staginnus, B.,** New assignment of the band gap in the alkali bromides by two-photon spectroscopy, *Phys. Rev. Lett.*, 19(9), 496, 1967.
161. **Araujo, C. B. de and Lotem, H.,** Ultraviolet two-photon absorption in alkali-halides, *Phys. Rev. B*, 26, 1044, 1982.
162. **Stafford, R. G. and Park, K.,** LO-phonon-assisted two-photon absorption in KI, *Phys. Rev. Lett.*, 25(24), 1652, 1970; LO-phonon-assisted transitions in the two-photon absorption spectrum of KI, *Phys. Rev. B*, 4(6), 2006, 1971.
163. **Park, K. and Stafford, R. G.,** Evidence for an optical transition at a noncentrosymmetric point of the Brillouin zone in KI, *Phys. Rev. Lett.*, 22(26), 1426, 1969.
164. **Catalano, I. M., Cingolani, A., and Minafra, A.,** Multiphoton transitions in ionic crystals, *Phys. Rev. B*, 5, 1629, 1972.
165. **von der Linde, D., Glass, A. M., and Rodgers, K. F.,** High-sensitivity optical recording in KTN by two-photon absorption, *Appl. Phys. Lett.*, 26(1), 22, 1975.
166. **Bityurin, N. M., Bredikhin, V. I., and Genkin, V. N.,** Nonlinear optical absorption and energy structure of $LiNbO_3$ and α-$LiIO_3$ crystals, *Sov. J. Quantum Electron.*, 8(11), 1377, 1978; Two-photon absorption and the characteristics of the energy spectrum of $LiNbO_3$ and α-$LiIO_3$ crystals, *Bull. Acad. Sci. U.S.S.R. Phys. Ser. U.S.A.*, 43(2), 1979.

167. **Kurz, H. and Von der Linde, D.,** Nonlinear optical excitation of photovoltaic $LiNbO_3$, *Ferroelectrics*, 21, 621, 1978.
168. **Maker, P. D. and Terhune, R. W.,** Study of optical effects due to an induced polarization third order in the electric field strength, *Phys. Rev.*, 137(3A), A801, 1965.
169. **Shablaev, S. I., Danishevskii, A. M., Subashiev, V. K., and Babashkin, A. A.,** Investigation of the energy band structure of $SrTiO_3$ by the two-photon spectroscopy method, *Sov. Phys. Solid State*, 21(4), 662, 1979.
170. **Bae, Y., Song, J. J., and Kim, Y. B.,** Photoacoustic detection of nanosecond-pulse-induced optical absorption in solids, *Appl. Opt.*, 21(1), 35, 1982.
171. **Munir, Q., Wintner, E., and Schmidt, A. J.,** Optoacoustic detection of nonlinear absorption in glasses, *Opt. Commun.*, 36(6), 467, 1981.
172. **Penzkofer, A. and Kaiser, W.,** Nonlinear loss in Nd-doped laser glass, *Appl. Phys. Lett.*, 21(9), 427, 1972.
173. **Galanin, M. D. and Chizhikova, Z. A.,** Effective cross sections of two-photon absorption in organic molecules, *Sov. Phys. JETP Lett.*, 4(2), 27, 1966.
174. **Aleksandrov, A. P., Bredikhin, V. I., and Genkin, V. N.,** Electron-vibrational nature of two-photon absorption in centrally-symmetrical organic molecules, *Sov. Phys. JETP Lett.*, 10, 117, 1969.
175. **Strome, F. C., Jr.,** Direct, two-photon photocarrier generation in anthracene, *Phys. Rev. Lett.*, 20(1), 3, 1968.
176. **Nikumb, S. K., Hari, V., Inamdar, A. S., and Mehendale, N.,** Two photon absorption in anthracene crystal with a nitrogen laser pumped dye laser, *Ind. J. Phys.*, 54B, 100, 1980.
177. **Bergman, A. and Jortner, J.,** Two-photon spectroscopy utilizing dye lasers, *Chem. Phys. Lett.*, 15(3), 309, 1972.
178. **Kleinschmidt, J., Tottleben, W., and Rentsch, S.,** Measurements of two-photon-absorption in the cavity of an organic dye laser, *Exp. Tech. Phys.*, 22, 191, 1974.
179. **McMahon, D. H., Soref, R. A., and Franklin, A. R.,** Quantitative measurements of double-photon absorption in the polycyclic benzene ring compounds, *Phys. Rev. Lett.*, 14(26), 1060, 1965.
180. **Fröhlich, D. and Mahr, H.,** Two-photon spectroscopy in anthracene, *Phys. Rev. Lett.*, 16(20), 895, 1966.
181. **Aleksandrov, A. P. and Bredikhin, V. I.,** Measurement of the absolute value of the cross-section for two-photon absorption in anthracene molecules, *Sov. Phys. Opt. Spect.*, 30(1), 37, 1971.
182. **Kepler, R. G.,** Two-photon transitions to highly excited states in anthracene crystals, *Phys. Rev. B*, 9, 4468, 1974.
183. **Bergman, A. and Jortner, J.,** Consecutive two-photon absorption of azulene in solution utilizing dye lasers, *Chem. Phys. Lett.*, 20(1), 1973.
184. **Friedrich, D. M., Van Alsten, J., and Walters, M. A.,** Polarization and concentration dependence of two-photon forbidden electronic origins: the $^1B_{2\mu}$ of benzene in solution, *Chem. Phys. Lett.*, 76(3), 504, 1980.
185. **Bree, A., Taliani, C., and Thirunamachandran, T.,** The polarized two-photon excitation spectrum of benzene monocrystals at 6 eV, *Chem. Phys.*, 56, 123, 1981.
186. **Razumova, T. K. and Starobogatov, I. O.,** Two-photon spectroscopy of liquid benzene and toluene, *Opt. Spectrosc. (U.S.S.R.)*, 49(6), 654, 1980.
187. **Scott, T. W., Braun, C. L., and Albrecht, A. C.,** Multiphoton ionization in the organic condensed phase: a three-photon study of liquid benzene, *J. Chem. Phys.*, 76(11), 5195, 1982.
188. **Tam, A. C. and Patel, C. K. N.,** Two-photon absorption spectra and cross-section measurements in liquids, *Nature (London)*, 280, 304, 1979.
189. **Hochstrasser, R. M., Meredith, G. R., and Trommsdorff, H. P.,** Resonant four wave mixing in molecular crystals, *J. Chem. Phys.*, 73(3), 1009, 1980.
190. **Hochstrasser, R. M., Wessel, J. E., and Sung, H. N.,** Two-photon excitation spectrum of benzene in the gas phase and the crystal, *J. Chem. Phys.*, 60(1), 317, 1974.
191. **Hochstrasser, R. M., Sung, H. N., and Wessel, J. E.,** Moderate-resolution study of the two-photon spectrum of the $^1B_{2u}$ state of benzene-h_6 and benzene-d_6, *Chem. Phys. Lett.*, 24(1), 7, 1974.
192. **Levenson, M. D. and Bloembergen, N.,** Dispersion of the nonlinear optical susceptibilities of organic liquids and solutions, *J. Chem. Phys.*, 60, 1323, 1974.
193. **Krasinski, J., Majewski, W., and Glodz, M.,** Investigation of the two-photon absorption spectrum of 3,4-benzopyrene by means of a tunable dye laser, *Opt. Commun.*, 14(2), 187, 1975.
194. **Drucker, R. P. and McClain, W. M.,** Two-photon absorption in "double molecules", *Chem. Phys. Lett.*, 28(2), 255, 1974.
195. **Hochstrasser, R. M., Sung, H. N., and Wessel, J. E.,** High resolution two-photon excitation spectra, *J. Chem. Phys.*, 58(10), 4694, 1973.
196. **Drucker, R. P. and McClain, W. M.,** Polarized two-photon studies of biphenyl and several derivatives, *J. Chem. Phys.*, 61(7), 2609, 1974.

197. **Castellucci, E., Salvi, P. R., and Foggi, P.,** Two-photon excitation spectra of the lowest electronic states of 2,2′-bipyridine, *Chem. Phys.*, 66, 281, 1982.
198. **Heumann, E., Kapp, I., Renstch, S., and Triebel, W.,** Polarization-dependent picosecond two-photon absorption spectra of diphenyl and chloronaphthalene in solution, *Zh. Prikl. Spektros.*, 34(3), 481, 1981.
199. **Monson, P. R. and McClain, W. M.,** Polarization dependence of the two-photon absorption of tumbling molecules with application to liquid 1-chloronaphthalene and benzene, *J. Chem. Phys.*, 53(1), 29, 1970; Complete polarization study of the two-photon absorption of liquid 1-chloronaphthalene, *J. Chem. Phys.*, 56(10), 4817, 1972.
200. **Eisenthal, K. B., Dowley, M. W., and Peticolas, W. L.,** Two-photon spectrum of a liquid, *Phys. Rev. Lett.*, 20(3), 93, 1968.
201. **Anderson, R. J. M., Holtom, G. R., and McClain, W. M.,** Two-photon absorptivities of the all trans α,ω-diphenylpolyenes from stilbene to diphenyloctatetraene via three wave mixing, *J. Chem. Phys.*, 70(9), 4310, 1979.
202. **Bennett, J. A. and Birge, R. R.,** Two-photon spectroscopy of diphenylbutadiene. The nature of the lowest-lying $^1A_g{}^*$-$\pi\pi^*$ state, *J. Chem. Phys.*, 73(9), 4234, 1980.
203. **Swofford, R. L. and McClain, W. M.,** New two-photon absorption spectrometer and its application to diphenylbutadiene, *Rev. Sci. Instrum.*, 46(3), 246, 1975.
204. **Fang, H. L.-B., Thrash, R. J., and Leroi, G. E.,** Observation of the low-energy ^1A_g state of diphenylhexatriene by two-photon excitation spectroscopy, *Chem. Phys. Lett.*, 57(1), 59, 1978.
205. **Vasudev, R. and Brand, J. C. D.,** The two-photon excitation spectrum of fluorobenzene vapor, *J. Mol. Spectrosc.*, 75, 288, 1979.
206. **Twarowski, A. J. and Kliger, D. S.,** A search for a low-lying excited ^1A state in 1,3,5-hexatriene, *Chem. Phys. Lett.*, 50(1), 36, 1977.
207. **Kleinschmidt, J. and Torpatschow, P.,** Experimental investigation of two photon absorption in naphthalene, *Wiss. Z.*, 24, 439, 1975.
208. **Logothetis, E. M.,** Two-photon photoelectric spectroscopy in CsI, *Phys. Rev. Lett.*, 19(26), 1470, 1967.
209. **Lynch, R. T., Jr. and Lotem, H.,** Two-photon absorption measurements in organic liquids via nonlinear light mixing spectroscopy, *Chem. Phys.*, 66(5), 1905, 1977.
210. **Galanin, M. D. and Chizhikova, Z. A.,** Two-photon dichroism in nitrobenzene, *Sov. Phys. JETP Lett.*, 5, 300, 1967.
211. **Fuke, K., Nagakura, S., and Kobayashi, T.,** Two-photon absorption spectrum of (2,2)-paracyclophane, *Chem. Phys. Lett.*, 31(2), 205, 1975.
212. **Callis, P. R., Scott, T. W., and Albrecht, A. C.,** Polarized two-photon fluorescence excitation studies of pyrimidine, *J. Chem. Phys.*, 75(12), 5640, 1981.
213. **Stachelek, T. M., Pazoha, T. A., and McClain, W. M.,** Detection and assignment of the "phantom" photochemical singlet of trans-stilbene by two-photon excitation, *J. Chem. Phys.*, 66(1), 4540, 1977.
214. **Kleinschmidt, J., Rentsch, S., Tottleben, W., and Wilhelmi, B.,** Measurement of strong nonlinear absorption in stilbene-chloroform solutions, explained by the superposition of two-photon absorption and one-photon absorption from the excited state, *Chem. Phys. Lett.*, 24(1), 133, 1974.
215. **Fuke, K., Sakamoto, S., Ueda, M., and Itoh, M.,** Two-photon absorption spectrum of trans-stilbene. Trans-cis photoisomerization via upper ^1A_g state, *Chem. Phys. Lett.*, 74(3), 546, 1980.
216. **Samoc, M., Samoc, A., and Williams, D. F.,** Multiphoton processes of charge carrier generation in thianthrene crystals, *J. Chem. Phys.*, 76(7), 3768, 1982.
217. **Schäfer, F. P. and Schmidt, W.,** Geometrical model and experimental verification of two-photon absorption in organic dye solutions, *IEEE J. Quantum Electron.*, QE-2(9), 357, 1966.
218. **Schäfer, F. P. and Schmidt, W.,** Comments on two-photon absorption in organic dyes — relation with the symmetry of the levels, *Opt. Commun.*, 17(1), 1976.
219. **Foucault, B. and Hermann, J. P.,** Two-photon absorption in organic dyes — relation with the symmetry of the levels, *Opt. Commun.*, 15(3), 412, 1975.
220. **Ducuing, J., Foucault, B., and Hermann, J. P.,** Two-photon absorption in organic dyes—reply to a comment, *Opt. Commun.*, 17(3), 267, 1976.
221. **Bradley, D. J., Hutchinson, M. H. R., and Koetser, H.,** Interactions of picosecond laser pulses with organic molecules. II. Two-photon absorption cross-sections, *Proc. R. Soc. London* A, 329, 105, 1972.
222. **Li, S. and She, C. Y.,** Two-photon absorption cross-section measurements in common laser dyes at 1.06μm, *Opt. Acta*, 29(3), 281, 1982.
223. **Vsevolodov, N. N., Kostikov, L. P., Kayushin, L. P., and Gorbatenkov, V. I.,** Two-photon absorption of laser radiation by chlorophyll-a and certain organic dyes, *Biofizika (U.S.S.R.)*, 18, 755, 1973; *Biophysics* (GB), 18, 807, 1973.
224. **Parma, L. and Omenetto, N.,** Two-photon absorption of 7-hydroxycoumarine, *Chem. Phys. Lett.*, 54(3), 541, 1978.
225. **Aslanidi, E. B. and Tikhonov, E. A.,** Two-photon absorption spectra of organic dye molecules, *Opt. Spectrosc.*, 37(4), 446, 1974.

226. **Hermann, J. P. and Ducuing, J.,** Dispersion of the two-photon cross section in rhodamine dyes, *Opt. Commun.*, 6(2), 101, 1972.
227. **Rulliere, C. and Kottis, P.,** Two-photon spectroscopy of organic molecules recorded by the fluorescence induced by a picosecond light continuum generated in D_2O, *Chem. Phys. Lett.*, 75(3), 478, 1980.
228. **Chance, R. R., Shand, M. L., Hogg, C., and Silbey, R.,** Three-wave mixing in conjugated polymer solutions: two-photon absorption in polydiacetylenes, *Phys. Rev. B*, 22(8), 3540, 1980.
229. **Shand, M. L. and Chance, R. R.,** Third-order nonlinear mixing in polydiacetylene solutions, *J. Chem. Phys.*, 69(10), 4482, 1978.
230. **Smith, W. L.,** Lawrence Livermore National Laboratory, unpublished.
231. **Nikogosyan, D. N. and Angelov, D. A.,** Production of free radicals in water by high-power laser UV radiation, *Doklady Akademii Nauk SSSR*, 253(3), 733, 1980.
232. **Catalano, I. M. and Cingolani, A.,** Absolute determination of the two photon cross section in NADH, *Opt. Commun.*, 32(1), 156, 1980.
233. **Birge, R. R., Bennett, J. A., Fang, H. L.-B., and Leroi, G. E.,** The two-photon spectroscopy of all-trans retinal and related polyenes, in *Advances in Laser Chemistry*, Springer-Verlag, Berlin, 1978, 347.
234. **Baltrameyunas, R., Vaitkus, Y., and Gavryushin, V.,** Influence of impurities on the two-photon absorption spectrum of ZnSe single crystals, *Sov. Phys. Solid State*, 18(10), 1723, 1976.
235. **Makhanek, A. G. and Skripko, G. A.,** Application of two-photon spectroscopy in the study of trivalent rare-earth ions in crystals, *Phys. Status Solidi (A)*, 53, 243, 1979.
236. **Speiser, S.,** Resonance enhanced two-photon absorption spectrum of an europium chelate, *Opt. Commun.*, 43(3), 221, 1982.
237. **Fritzler, U.,** Investigations on the excited states of Eu^{2+} in CaF_2- and SrF_2-crystals by two-photon absorption spectroscopy, *Z. Phys. B*, 27, 289, 1977.
238. **Vogler, K.,** Two-photon absorption of F-centers in KCl, *Phys. Status Solidi (B)*, 107, 195, 1981.
239. **Downer, M. C., Bivas, A., and Bloembergen, N.,** Selection rule violation, anisotropy, and anomalous intensity of two-photon absorption lines in Gd^{3+}:LaF_3, *Opt. Commun.*, 41(5), 335, 1982.
240. **Casalboni, M., Grassano, U. M., and Tanga, A.,** Two-photon absorption of color centers in NaF, *Phys. Rev. B*, 19(6), 3306, 1979.
241. **Catalano, I. M. and Cingolani, A.,** Absolute two-photon fluorescence with low-power cw lasers, *Appl. Phys. Lett.*, 38(10), 745, 1981.
242. **Guiliani, J. F.,** Multiphoton excitation and emission from europium ions in aqueous solution, *Opt. Lett.*, 3(4), 149, 1978.
243. **Mikami, N. and Ito, M.,** Two-photon excitation spectra of naphthalene and naphthalene-d_8, *Chem. Phys. Lett.*, 31(3), 472, 1975.

1.3 NONLINEAR REFRACTIVE INDEX

W. Lee Smith

INTRODUCTION

In nonlinear refraction, the phase profile of a laser wave traveling through a material other than vacuum acquires a perturbation that replicates the intensity profile of the laser wave. This coupling of amplitude and phase comes about through higher-order terms in the basic relations describing electromagnetic fields in a material system. As the laser wave propagates farther through the material, the phase perturbation leads by diffraction to intensity perturbation, which generates additional phase perturbation, and so on. This intensification is self-focusing. Self-induced phase perturbation can also be used as the basis for optical switching in Fabry-Perot interferometric devices. The active field of optical bistability began with this effect. The parameter that governs the strength of these and other effects is the nonlinear refractive index, the subject of this section.

As the available intensity from pulsed lasers reached the multimegawatt level in the late 1960s, the study of nonlinear refraction grew and maintained a high level of interest through the 1970s. Much of the emphasis in the 1970s was on suppression of nonlinear refraction in high-power glass lasers. In the latter 1970s the fields of optical bistability and phase-conjugation developed, broadening the interest in nonlinear refraction. For these topics, a major goal has been the identification of materials having a very large nonlinear refractive index, thereby opening up the potential for practical utilization of these processes with low-power lasers in optical data processing and other systems. This search for a large nonlinear refractive index has been carried out in many different types of materials, including recently semiconductor quantum-well structures[1] and high-polarizability liquid crystals.[2-4] The span of nonlinear refractive index thus far reported in the literature is roughly 12 orders of magnitude.

Theoretical calculations of the nonlinear refractive index have been presented by Langhoff et al.[5] (isolated molecules), Hellwarth[6] (liquids), Sparks and Duthler[7] (dispersion of n_2 in LiF), Mollmann and Bartolotti[8] (atomic hydrogen), Ratajska-Gadomska[9] (crystals, evaluated for diamond and benzene), Bokov[10] (liquids), Chang[11] (near-resonant effects), Ackerhalt et al.[12] (SF_6 gas), Jha and Bloembergen[13] (semiconductors), and Yuen[14] (semiconductor heterostructures). Hellwarth and colleagues[15,16] have calculated for a number of glasses the frequency dependence and the relative contributions of electronic and nuclear response to the nonlinear refractive index. Numerous other calculations have been performed for the third-order susceptibility appropriate for third-harmonic generation (see Section 3.1).

Beginning with Wang,[17] several authors[18-20] have proposed empirical relationships between the linear and nonlinear refractive indexes. The range of validity and utility of these expressions has been discussed by Boling et al.[19] Additional values of n_2 calculated from these models have been tabulated by Weber et al.[21] for glasses and crystals, and Brown et al.[22] for liquids.

The nonlinear refractive index is related to a variety of other nonlinear optical effects through the formalism of nonlinear optical susceptibilities.[23-25] It is through these related effects (ellipse rotation, optical Kerr effect, self-phase modulation, optical mixing, etc.) that much of the available data on nonlinear refraction has been obtained.

The frequency-domain manifestation of nonlinear refraction — self-phase modulation — has proved useful in extending the spectral bandwidth of laser pulses across the visible region.[26] These picosecond, ''white-light'' pulses have been employed in fast, time-resolved spectroscopy. In another application of nonlinear refraction, soliton behavior has been demonstrated and studied using laser pulses propagating through single-mode glass fibers.[27] This effect is of potential significance for future fiberoptic communications systems. Here the natural dispersive group-velocity spreading of a laser pulse propagating in an optical fiber is counterbalanced by the frequency "chirp" resulting from the refractive index nonlinearity. The result is that, for properly chosen conditions, the laser pulse waveform may travel many kilometers without temporal spreading, thus making possible the desired data rate of gigabits per second. Furthermore, actual temporal narrowing may be achieved. With this effect, picosecond pulses have been compressed into the femtosecond regime.[28,29]

Self-defocusing, originating from a negative nonlinear refractive index, has been observed in atomic vapors by using a laser wavelength that is within the anomalous dispersion region of an atomic transition. This effect has been observed for both high-power laser pulses[30] and low-power, continuous laser beams.[31]

Artificial self-focusing media, composed of submicron dielectric spheres suspended in a liquid, have been employed by Smith and Ashkin and colleagues[32,33] to study self-focusing with low-power, CW laser beams. The effective nonlinear refractive index exhibited by these media can exceed the value in carbon disulfide by up to 10^5.

SCOPE AND REVIEWS

The objective of this chapter is to provide a useful tabulation of measurements of nonlinear refractive parameters made through 1982.

General treatment of the subject of third-order nonlinear susceptibilities has been given in the monograph by Hellwarth,[34] the dissertation of Owyoung,[35] and the chapter by Flytzanis.[36] Review articles on nonlinear refractive index have been written by Chang,[11] and by Weber et al.[21] Reviews of self-focusing have been given by Shen,[37] Marburger,[38] Akhmanov et al.,[39] and Svelto.[40] An article detailing the impact of self-focusing in high-power laser propagation has been given by Simmons et al.[41] Nonlinear optical phase conjugation has recently been reviewed in a special journal issue.[42] Optical bistability has been reviewed by in a similar issue[43] and has been the subject of a recent conference.[44] Nonlinear optical effects, including self-focusing, in liquid crystals have been reviewed by Arakelyan et al.[45] and Tabiryan and Zel'dovich.[46]

DEFINITIONS

The nonlinear refractive index, n_2, and the nonlinear refractive coefficient, γ, for an optical material are defined by:

$$n = n_o + n_2 < E^2 > \quad (1a)$$

and

$$n = n_o + \gamma I \quad (1b)$$

Here n is the total refractive index of the material and n_o is the familiar linear refractive index; $<E^2>$ and I are the time average of the squared optical electric field and the optical intensity (or irradiance), respectively, applied with the laser wave. Note that, whereas the intensity has the same value inside a medium as outside it, $<E^2>$ does not; the correct $<E^2>$ to be used in Equation 1 is the value inside the material. The units of n_2 are cm^3/erg (esu) or m^2/V^2 (mks); the units of γ are m^2/W or, more customarily, cm^2/W. Conversion between n_2 and γ is made via:

$$n_2[esu] = (cn_o/40\pi)\ \gamma\ [m^2/W] \quad (2)$$

where c is the light speed (m/sec) in vacuum. Numerically,

$$n_2\ [\mathrm{cm^3/erg}] = (238.7) \cdot n_o\gamma\ [\mathrm{cm^2/W}] \tag{3a}$$

$$n_2\ [\mathrm{m^2/V^2}] = (3.333 \times 10^{-6}) \cdot n_o\gamma\ [\mathrm{cm^2/W}] \tag{3b}$$

and

$$n_2\ [\mathrm{cm^3/erg}] = (7.162 \times 10^{7}) \cdot n_2\ [\mathrm{m^2/V^2}] \tag{3c}$$

Terms commonly used in the study of nonlinear refraction pertaining to self-focusing are the critical power for self-focusing, P_1, and the B-integral, B. These parameters describe, respectively, whole-beam and small-scale self-focusing.

The critical power for whole-beam self-focusing[38] is that power, contained in the laser pulse, for which self-focusing exactly counterbalances the natural diffractive spreading of the pulse. For a pulse having Gaussian transverse distribution of intensity, the critical power is given by

$$P_1 = \frac{c\lambda^2}{32\pi^2 n_2} \tag{4}$$

As an example, P_1 is ~ 870 kW for BK-7 glass, at a wavelength of 1064 nm.

For laser pulses that contain many times more power than P_1, rather than self-focusing as a whole as just mentioned, regions of the pulse self-focus individually[47] in what is known as "small-scale" self-focusing. Inevitable hot spots on the intense laser pulse are intensified as the pulse travels through matter. The maximum rate of intensification is exponential:[48]

$$I(z) = I_N(0)e^B \tag{5}$$

Here $I_N(O)$ represents the intensity characteristic of the initial hot spot on the pulse. B, the so-called "B-integral", is defined by:

$$B = \frac{2\pi\gamma}{\lambda}\int_0^L I(z)dz \tag{6}$$

and is the phase retardation (in radians) accumulated by the hot spot. Here λ is the wavelength of the light in vacuum. It is often found that serious self-focusing effects begin when B is allowed to exceed ~ π radians. For very powerful laser pulses, such as the terawatt pulses employed in fusion research, uncontrolled self-focusing can cause the pulse to "break up" into a cluster of micron-diameter spikes.[41] Laser damage to the optical material often results from small-scale self-focusing.

The fundamental physical origin of the nonlinear refractive index is made clear via the formalism of the nonlinear optical susceptibilities.[23] The central expression in this formalism is the expansion of the total optical polarization density as a power series in the electric field:

$$\begin{aligned}\vec{P} &= \vec{P}^{(1)} + \vec{P}^{(2)} + \vec{P}^{(3)} + \ldots \\ &= \chi^{(1)} \cdot \vec{E} + \chi^{(2)} : \vec{E}\vec{E} + \chi^{(3)} \vdots\, \vec{E}\vec{E}\vec{E} + \ldots\end{aligned} \tag{7}$$

Here $\chi^{(n)}$ are the nth-order complex optical susceptibilities. The real and imaginary parts of $\chi^{(1)}$ account, of course, for linear refraction and absorption of light. Recalling that the work performed on a material system by a light wave is the integral over many optical cycles of E·(dP/dt), we see that the term in the work involving $\chi^{(2)}$ would be cubic in the electric field, or 3/2-power in the optical intensity. Hence, the $\chi^{(2)}$ term cannot describe additional absorptive or refractive work. Rather, it describes in noncentrosymmetric optical media the transfer of energy from one light wave, or pair of waves, to a different wave in second-harmonic generation, sum- and difference-frequency generation, and electrooptic phenomena. These effects are addressed in the chapters by Singh in Section 1.1 of this volume and by Kaminow in Volume IV.

The real and imaginary parts of $\chi^{(3)}$ give rise to nonlinear refraction and nonlinear absorption, respectively. The preceding section was devoted to two-photon absorption. Furthermore, third-harmonic generation, stimulated Raman scattering, optical Kerr effect, self-induced ellipse rotation, and optical phase-conjugation are described by the third-order nonlinear susceptibility. As these effects are related to nonlinear refraction, information about it may be gained from their study as well.

The third-order contribution to the total polarization density is more fully written as:

$$P_i^{(3)}(\omega_4) = \sum_{j,k,\ell} \chi_{ijk\ell}^{(3)}(-\omega_4;\omega_1,\omega_2,\omega_3)E_j(\omega_1)E_k(\omega_2)E_\ell(\omega_3) \quad (8)$$

Here E_j, E_k, and E_ℓ are, in general, three separately applied electric fields, each having its own frequency (ω_1, ω_2, or ω_3) and polarization direction. By conservation of energy, $\omega_4 = \omega_1 + \omega_2 + \omega_3$. The subscripts, i, j, k, and ℓ represent directions of symmetry in the natural coordinate system of the optical material, along which the applied electric fields are aligned or projected. For isotropic media, they indicate whether the electric field components being considered are parallel or orthogonal to each other. The fourth-rank tensor $\chi^{(3)}$ contains, in general, a unique element for each of the 81 possible combinations of 4 electric field or polarization vector components, each in 3 possible spatial directions. Each element is the strength parameter for that particular geometrical combination. Fortunately, in the majority of situations of current practical interest, the number of nonzero elements is reduced by symmetry properties from 81 to a rather small number (<6) of independent elements.

For isotropic materials such as liquids, gases, and glasses, there are three nonzero elements of $\chi^{(3)}$: $\chi_{1111}^{(3)}$, $\chi_{1122}^{(3)}$, and $\chi_{1221}^{(3)}$. These are related by the isotropy relation:

$$\chi_{1111}^{(3)} = 2\chi_{1122}^{(3)} + \chi_{1221}^{(3)} \quad (9)$$

leaving two independent $\chi^{(3)}$ elements for isotropic materials. For a linearly polarized laser beam, n_2 and $\chi^{(3)}$ are related simply by:

$$n_2(LP) = \frac{12\pi}{n_o}\chi_{1111}^{(3)}(-\omega;\omega,\omega,-\omega) \quad (10)$$

for isotropic materials. For a circularly polarized laser beam in isotropic materials, this relation becomes:

$$n_2(CP) = \frac{24\pi}{n_o}\chi_{1122}^{(3)}(-\omega;\omega,\omega,-\omega) \quad (11)$$

For cubic crystalline materials, the same three $\chi^{(3)}$ elements are nonzero, and all three are independent. For a beam linearly polarized along a [100] axis, we again have

$$n_2(\mathrm{LP},100) = \frac{12\pi}{n_o} \chi^{(3)}_{1111}(-\omega, \omega, \omega, -\omega) \tag{12}$$

For a beam circularly polarized and propagating along a [100] axis, we have

$$n_2(\mathrm{CP},100) = \frac{6\pi}{n_o} (\chi^{(3)}_{1111} + 2\chi^{(3)}_{1122} - \chi^{(3)}_{1221}) \tag{13}$$

and for a circularly polarized beam propagating along a [111] axis, we have

$$n_2(\mathrm{CP},111) = \frac{4\pi}{n_o} (\chi^{(3)}_{1111} + 4\chi^{(3)}_{1122} - \chi^{(3)}_{1221}) \tag{14}$$

In materials for which the origin of n_2 is predominantly electronic rather than ''nuclear'' (orientational, vibrational, etc.) — a transparent glass, for example — we may obtain an estimate of the ratio $n_2(\mathrm{CP})/n_2(\mathrm{LP})$. For such materials, one may approximate $\chi^{(3)}_{1122} \simeq \chi^{(3)}_{1221}$ and further, with Equation 9, $\chi^{(3)}_{1111} \simeq 3\chi^{(3)}_{1122}$. Hence, from Equations 10 and 11, we have $n_2(\mathrm{CP})/n_2(\mathrm{LP}) \simeq 2/3$. This effect has been put to use in high-power laser chains limited by self-focusing.[49] For the same output power, 1/3 less B-integral is accumulated and consequently, a more uniform output energy distribution is obtained.

The response times of the several sources of refractive index nonlinearity vary from $\sim 10^{-15}$ to ~ 1 sec. The fastest source is the electron cloud distribution, responding in a few optical cycles (several femtoseconds). Molecular vibrational and other ''nuclear'' contributions also follow in a few oscillatory periods, generally a few picoseconds.

On the other hand, electrostrictive and thermal contributions to n_2 arise from macroscopic, collective response of the medium. In the electrostrictive nonlinearity,[50-52] an intense laser pulse induces a strain field proportional to the gradient of the laser intensity. This strain field results in position-dependent alteration of the density of the material, the latter giving rise to focusing of the light pulse. Because the density adjustment propagates relatively slowly at the longitudinal sound velocity (v, typically 5 km/sec), the build-up time of electrostrictive self-focusing is characteristically in the nanosecond range. The response time is given by $t = r/v$, where r is the distance over which the intensity gradient is established by the laser pulse. For pulse durations shorter than ~ 1 nsec, the transient electrostrictive contribution remains weak. The reader is referred to the work of Kerr and other authors[50,52] for treatment of electrostrictive self-focusing.

Thermal self-focusing[12] results from absorptive heating of a material sample by a laser pulse. The induced temperature rise causes a change in refractive index (dn/dT) at constant material density, and a change in refractive index due to thermoelastic strain. The latter is similar in several respects to the electrostrictive nonlinearity, and its response time is the same. However, the dn/dT term follows the instantaneous temperature of the material and so is proportional to the integrated energy of the laser pulse. The dn/dT term is positive in most glasses, but negative in most liquids.

As discussed by Feldman et al.,[52] for a general situation all of the sources of refractive index change must be considered, giving rise to the following composite expression:

$$\Delta n = \{n_2[e] + n_2[m] \cdot f(t) + n_2[ES] \cdot g(t) + n_2(T) \cdot h(t)\} < E^2 > \tag{15}$$

Here $n_2[e]$ is the instantaneously responding electronic source. Next, $n_2[m] \cdot f(t)$ represents the contributions from molecular sources. Although only one n_2 and one response function are written for this term, there are several molecular sources of nonlinear polarization (due to molecular vibration, reorientation, and redistribution or libration, etc.), each having its own response time. The build-up of electrostrictive and thermal n_2 is included by the terms involving the response functions g(t) and h(t).

MEASUREMENT TECHNIQUES

Experimental methods that have been used to measure the nonlinear refractive parameters compiled in this chapter are listed in Table 1.3.1. Also listed there are the abbreviations for the methods used in the data tables and a primary or recent reference for each method. The reader is directed to these references for descriptions of the methods. Values of $\chi^{(3)}$ measured with the techniques of third-harmonic generation (THG) and DC-field-induced second-harmonic generation (DCFISH) have not been tabulated here, as they each involve a frequency (3ω in THG, and 0 in DCFISH) significantly different from the usual single-applied-frequency situation in investigations of nonlinear refraction.

DATA TABLES

The measured data on nonlinear refraction are organized into five tables (Tables 1.3.2 to 1.3.6): liquids, large bandgap crystals, semiconductor crystals, glasses, and a final table including gaseous, polymeric, liquid crystalline, and biological materials. Materials are listed alphabetically within each table.

The method of measurement (using the abbreviations defined in Table 1.3.1), the duration and wavelength(s) of the laser excitation, and the linear refractive index are tabulated. In many references, the linear refractive index n_o is not given. Because it is needed to convert between $\chi^{(3)}$, n_2 and γ, the linear refractive index has been supplied by this author where possible. Generally, nonlinear refractive indexes are measured only to within ±15% or more; hence, linear refractive indexes are entered only to two decimal places. In some cases, the refractive index for the Fraunhofer d wavelength (587.6 nm) has been used by this author in lieu of the unavailable value of n_o at the actual wavelength(s) employed. All n_o values supplied by this author are entered in italics.

The primary tabulated data are $\chi^{(3)}_{1111}$, $\chi^{(3)}_{1122}$, $\chi^{(3)}_{1221}$, and $n_{2,LP}$ for linearly polarized light (as defined by Equations 1a, 10, or 12), and γ_{LP} for linearly polarized light (Equation 1b). Conversion of the parameter given in the original work to the other appropriate parameters listed (e.g., $\chi^{(3)}_{1111}$ to γ) has been done by the author for the convenience of the reader. Converted forms are listed in italics in the tables.

ACKNOWLEDGMENT

The author is again indebted to T. Mauch for the preparation of this manuscript, and to the Laser Fusion Program of Lawrence Livermore National Laboratory for support during its writing.

Table 1.3.1
TECHNIQUES FOR MEASURING THE NONLINEAR REFRACTIVE INDEX

	Method	Ref.
DHG	Dynamic holographic grating	2
DTLC	Damage threshold for linear vs. circular polarization	52
ER	Ellipse rotation	53
KE	"DC" Kerr effect	54
OKE	Optical Kerr effect	55
PDF	Power-dependent focus	56
PST	Power for self-trapping	57
RSS	Raman scattering spectroscopy	58
SFL	Self-focal length	59
SPA	Spatial profile analysis	31
SPM	Self-phase modulation	60
SSMG	Small-scale modulation growth	61
TBI	Two-beam interferometry	62
TII	Time-integrated interferometry	63
TRI	Time-resolved interferometry	64
TWM	Three-wave mixing	65
TWR	Temporal waveform reshaping	66
WFC	Wavefront conjugation	67

Table 1.3.2
MEASURED NONLINEAR REFRACTIVE PARAMETERS FOR LIQUIDS

Liquid	Formula	Method	Pulse duration (nsec)	Wavelength (nm)	Linear refractive index	χ_{1111} (10^{-13}cm^3/erg)	χ_{1122} (10^{-13}cm^3/erg)	χ_{1221} (10^{-13}cm^3/erg)	$n_{1,LP}$	γ_{LP} (10^{-16}cm^2/W)	Ref.
Acetic acid	CH_3CO_2H	OKE	40	694	1.37	0.27			7.4	23	68
Acetone	$(CH_3)_2CO$	TII	10	694	1.35	1.8			50	156	69
Acetone	$(CH_3)_2CO$	OKE	0.010	1064, 532	1.35	0.28			7.7	24	55
Acetone	$(CH_3)_2CO$	OKE	40	694	1.35	0.16			4.3	13	68
Acetonitrile	CH_3CN	OKE	0.010	1064, 532	1.34	0.34			9.5	30	55
Aniline	$C_6H_5NH_2$	TWM	5	581, 549	1.59	$\chi^{(3)}/\chi^{(3)}benzene = 1.95$					70
Argon (77 K)	Ar	OKE		10640	1.50	$\chi^{(3)}1111 - \chi^{(3)}1122 = 0.015$					71
Benzene	C_6H_6	OKE	55	694	1.50	$\chi^{(3)}1122 + \chi^{(3)}1221 = 0.97$					72
Benzene	C_6H_6	TBI	5	570, 532	1.50	0.55			13.8	39	62
Benzene	C_6H_6	PDF		694	1.50	0.056			~1.4	3.9	73
Benzene	C_6H_6	TWM	5	613, 577	1.50	$\chi^{(3)}(benzene)/\chi^{(3)}(CS - 3.41_2)$					74
Benzene	C_6H_6	TII	5	570, 532	1.50	0.20			5.0	14	75
Benzene	C_6H_6	ER	30	694	1.50			0.65			57
Benzene	C_6H_6	PST	30	694	1.50	$\chi^{(3)}1122 + 2\chi^{(3)}1221 = 0.9$					57
Benzene	C_6H_6	TWM	3	590, 560	1.50	0.17			4.3	12	76, 65
Benzene	C_6H_6	ER	10	1064	1.50			0.50			77
Benzene	C_6H_6	OKE	40	694	1.50	0.50			12.5	35	68
Benzene	C_6H_6	TWM		580, 550	1.50	1.14			28.6	80	78
Benzene	C_6H_6	TII	10	694	1.50	3.7	3.4		93	260	69
Benzene	C_6H_6	OKE	0.010	1064, 532	1.50	2.4			60	168	55
Benzonitrile	C_6H_5CN	TWM	5	581, 549	1.53	$\chi^{(3)}/\chi^{(3)}benzene = 1.2$					70
Bromobenzene	C_6H_5Br	TII	0.7	1320	1.53	2.1			51	140	63
Bromobenzene	C_6H_5Br	TWM	2	452, 434	1.56	$\chi^{(3)}/n_o$ relative to $C_6H_6 = 1.2$					79
Bromobenzene	C_6H_5Br	TWM	5	581, 549	1.56	$\chi^{(3)}/\chi^{(3)}benzene = 1.7$					70
Bromoform	$CHBr_3$	ER	30	694	1.58		0.07				25
Bromoform	$CHBr_3$	TWM	2	452, 434	1.60	$\chi^{(3)}/n_o$ relative to $C_6H_6 = 1.7$					79
Carbon disulfide	CS_2	ER	30	694	1.59		0.45				25
Carbon disulfide	CS_2	OKE	400	10640	1.63	0.72			17	43	80
Carbon disulfide	CS_2	OKE	40	694	1.59	5.1			120	316	68

Carbon disulfide	CS_2	OKE	55	694	1.59	$\chi^{(3)}1122 + \chi^{(3)}1221 = 5.7$					72
Carbon disulfide	CS_2	ER	20	694	*1.59*			3.18			81
Carbon disulfide	CS_2	TRI	1	1064	*1.59*	*4.6*			110	*290*	82
Carbon disulfide	CS_2	TII	0.7	1320	1.59	*5.3*			125	330	63
Carbon disulfide	CS_2	TWM	3	590, 560	*1.63*	0.87			*20*	*52*	83
Carbon disulfide	CS_2	ER	30	694	1.59			3.4			57
Carbon disulfide	CS_2	TII	5	570, 532	*1.63*	0.86			*20*	*52*	75
Carbon disulfide	CS_2	PST	30	694	1.59	$\chi^{(3)}1122 + 2\chi^{(3)}1221 = 4.7$					57
Carbon disulfide	CS_2	KE		546	*1.63*	*5.4*			124	*319*	54
Carbon disulfide	CS_2	KE		633	1.623	*5.0*			117	*302*	54
Carbon disulfide	CS_2	TWM	2	452, 434	*1.63*	$\chi^{(3)}/n_o$ *relative to* $C_6H_6 = 6.6$					79
Carbon disulfide	CS_2	TWM		580, 550	*1.63*	*6.3*			145	*373*	18
Carbon disulfide	CS_2	OKE	0.010	1064, 532	*1.63*	*8.6*			200	*514*	55
Carbon disulfide	CS_2	OKE	~250	10640	*1.63*	*~0.3*			~7	*~18*	84
Carbon monoxide (77 K)	CO	OKE		10640		$\chi^{(3)}1111 - \chi^{(3)}1122 = 0.67$					71
Carbon tetrachloride	CCl_4	ER	30	694	1.45		0.012				25
Carbon tetrachloride	CCl_4	ER	20	694	*1.45*	$\chi_{1122} + \chi_{1221} = 0.083$					85
Carbon tetrachloride	CCl_4	ER	30	694	1.45			0.09			57
Carbon tetrachloride	CCl_4	TWM	3	590, 560	*1.46*	0.11			*2.8*	*8.2*	83
Carbon tetrachloride	CCl_4	PST	30	694	1.45	$\chi^{(3)}1122 + 2\chi^{(3)}1221 = 0.12$					57
Carbon tetrachloride	CCl_4	ER	10	1064	1.45			0.041			77
Carbon tetrachloride	CCl_4	OKE	40	694	1.45	*0.08*			2	*5.8*	68
Carbon tetrachloride	CCl_4	OKE	0.010	1064, 532	*1.45*	*0.20*			5.3	*15*	55
Chlorobenzene	C_6H_5Cl	TII	0.7	1320	1.50	*1.6*			40	110	63
Chlorobenzene	C_6H_5Cl	ER	10	1064	1.51			0.98			77
Chlorobenzene	C_6H_5Cl	TWM	2	452, 434	*1.52*	$\chi^{(3)}/n_o$ *relative to* $C_6H_6 = 0.93$					79
Chlorobenzene	C_6H_5Cl	TWM	5	581, 549	*1.52*	$\chi^{(3)}/\chi^{(3)}_{benzene} = 1.4$					70
Chlorobenzene	C_6H_5Cl	TWM		580, 550	*1.52*	*2.04*			50.7	*140*	78
Chlorobenzene	C_6H_5Cl	TWM	3	590, 560	1.52	0.30			*7.4*	*20*	83

Table 1.3.2 (Continued)
MEASURED NONLINEAR REFRACTIVE PARAMETERS FOR LIQUIDS

Liquid	Formula	Method	Pulse duration (nsec)	Wavelength (nm)	Linear refractive index	χ_{1111} (10^{-13} cm^3/erg)	χ_{1122} (10^{-13} cm^3/erg)	χ_{1221} (10^{-13} cm^3/erg)	$n_{1,LP}$	γ_{LP} (10^{-16} cm^2/W)	Ref.
Chloroform	$CHCl_3$	ER	30	694	1.43		0.05				25
Chloroform	$CHCl_3$	ER	10	1064	1.43			0.15			77
Chloroform	$CHCl_3$	OKE	40	694	1.44	0.22			5.8	17	68
Chloroform	$CHCl_3$	OKE	0.010	1064, 532	1.45	0.41			10.6	30.6	55
Chloro-naphthalene	$C_{10}H_7Cl$	OKE	40	694	1.63	3.7			86	220	68
Chloro-naphthalene	$C_{10}H_7Cl$	TWM	2	452, 434	1.63	$\chi^{(3)}/n_o$ relative to C_6H_6 = 2.0					79
Cyclohexane	C_6H_{12}	TWM	0.005	539, 532	1.43			0.15			86
Cyclohexane	C_6H_{12}	ER	10	1064	1.43			0.037			77
Cyclohexane	C_6H_{12}	TWM		580, 550	1.43	0.16			4.2	12	78
Cyclohexane	C_6H_{12}	OKE	40	694	1.44	0.050			1.3	3.8	68
Cyclohexane	C_6H_{12}	OKE	0.010	1064, 532	1.43	0.15			3.9	11	55
Decane	$C_{10}H_{22}$	OKE	0.010	1064, 532	1.41	0.57			15.3	45.5	55
Decanol	$C_{10}H_{21}OH$	OKE	0.010	1064, 532	1.44	0.15			3.9	11	55
Dibromohexane	$C_6H_{12}Br_2$	TII	0.7	1320	1.49	<0.40			<10	<30	63
Dichlorobenzene	$C_6H_4Cl_2$	TII	0.7	1320	1.53	1.7			41	110	63
Dichloroethane	$C_2H_4Cl_2$	TII	0.7	1320	1.43	<0.4			<10	<30	63
Dichloroethane	$C_2H_4Cl_2$	OKE	0.010	1064, 532	1.44	0.48			12.5	36	55
Dichloromethane	CH_2Cl_2	TII	0.7	1320	1.41	<0.37			<10	<30	63
Diiodomethane	CH_2I_2	TII	0.7	1320	1.69	2.6			59	150	63
Dimethylaniline	$C_6H_5N(CH_3)_2$	TWM	5	581, 549	1.56	$\chi^{(3)}/\chi^{(3)}_{benzene}$ = 2.1					70
Dimethylsulfoxide	C_2H_6SO	ER	10	1064	1.46			0.082			77
Dimethylsulfoxide	C_2H_6SO	TWM	2	452, 434	1.48	$\chi^{(3)}/n_o$ relative to C_6H_6 = 0.80					79
Ethanol	C_2H_5OH	OKE	0.010	1064, 532	1.35	0.90			2.5	7.6	55
Ethanol	C_2H_5OH	TII	10	694	1.35	1.4			39	120	69
Ethyl iodide	C_2H_5I	OKE	0.010	1064, 532					22.6		55
Ethyl iodide	C_2H_5I	TWM	2	452, 434		$\chi^{(3)}/n_o$ relative to C_6H_6 = 1.9					79
Ethylene bromide	$C_2H_4Br_2$	OKE	0.010	1064, 532	1.53	1.4			34.3	94	55
Ethylene chloride	$C_2H_4Cl_2$	ER	10	1064	1.41			0.21			77

Fluorobenzene	C_6H_5F	TWM	5	581, 549	*1.47*	$\chi^{(3)}/\chi^{(3)}_{benzene} = 1.24$					70
Formic acid	CH_2O_2	OKE	0.010	1064, 532	*1.36*	*0.36*			10	*31*	55
Glycerin	$C_3H_8O_3$	TII	10	694	*1.47*	1.4	1.4		*36*	*100*	69
Glycerin	$C_3H_8O_3$	OKE	0.010	1064, 532	*1.47*	*0.10*			2.5	*7.1*	55
Hexadecane	$C_{16}H_{34}$	OKE	0.010	1064, 532	*1.43*	*0.59*			15.5	*45.4*	55
Hexane	C_6H_{14}	ER	30	694	1.37		0.009				25
Iodoaniline	C_6H_6IN	TII	0.7	1320	1.64	*<0.44*			<10	<30	63
Iodobenzene	C_6H_5I	TII	0.7	1320	1.59	*2.9*			69	180	63
Iodobenzene	C_6H_5I	TWM	5	581, 549	*1.61*	$\chi^{(3)}/\chi^{(3)}_{benzene} = 2.1$					70
Methanol	CH_3OH	OKE	0.010	1064, 532	*1.32*	*0.077*			2.2	*7.0*	55
Methanol	CH_3OH	ER	30	694	1.32		0.003				25
Methyl iodide	CH_3I	TII	0.7	1320	1.50	*<0.40*			<10	<30	63
Nitrobenzene	$C_6H_5NO_2$	OKE	55	694	*1.53*	$\chi^{(3)}_{1122} + \chi^{(3)}_{1221} = 4.6$					72
Nitrobenzene	$C_6H_5NO_2$	OKE	0.006	1054, 527	*1.54*	*8.6*[a]			210[a]	*570*[a]	87
Nitrobenzene	$C_6H_5NO_2$	ER	30	694	1.53			2.3			57
Nitrobenzene	$C_6H_5NO_2$	TWM	3	590, 560	*1.54*	0.40			*9.8*	*27*	83
Nitrobenzene	$C_6H_5NO_2$	PST	30	694	1.53	$\chi^{(3)}_{1122} + 2\chi^{(3)}_{1221} = 3.1$					57
Nitrobenzene	$C_6H_5NO_2$	ER	10	1064	1.53			2.6			77
Nitrobenzene	$C_6H_5NO_2$	TWM	5	581, 549	*1.54*	$\chi^{(3)}/\chi^{(3)}_{benzene} = 2.26$					70
Nitrobenzene	$C_6H_5NO_2$	TII	10	694	*1.53*	8.2	5.8		*202*	*553*	69
Nitrobenzene	$C_6H_5NO_2$	OKE	0.010	1064, 532	*1.54*	*10.2*			250	*685*	55
Nitrobenzene	$C_6H_5NO_2$	OKE	40	694	*1.53*	*3.5*			86	*240*	68
Nitrogen (77 K)	N_2	OKE		10640		$\chi^{(3)}_{1111} - \chi^{(3)}_{1122} = 0.11$					71
Nitrotoluene	$C_7H_7NO_2$	OKE	0.010	1064, 532	*1.53*	*11.0*			270	*739*	55
Octane	C_8H_{18}	OKE	0.010	1064, 532	*1.40*	*0.20*			5.5	*16*	55
Octanol	$C_8H_{17}OH$	OKE	0.010	1064, 532	*1.43*	*0.16*			4.1	*12*	55
Oxygen (77 K)	O_2	OKE		10640		$\chi^{(3)}_{1111} - \chi^{(3)}_{1122} = 0.48$					71
Pentane	C_5H_{12}	OKE	0.010	1064, 532	*1.36*	*0.15*			4.2	*13*	55
Phenol	C_6H_5OH	TWM	5	581, 549	*1.54*	$\chi^{(3)}/\chi^{(3)}_{benzene} = 1.24$					70
Phosphorus oxychloride	$POCl_3$	SFL	0.015	1052	1.47	*0.52*			13.3	*37.9*	59
Propanol	C_3H_7OH	OKE	0.010	1064, 532	*1.37*	*0.098*			2.7	*8.3*	55
Pyridine	C_5H_5N	ER	30	694	1.50		0.10				25
Pyridine	C_5H_5N	TWM		580, 550	*1.50*	*1.2*			30	*84*	78
Quinoline	C_9H_7N	TII	0.7	1320	1.59	*4.6*			110	280	63
Quinoline	C_9H_7N	TII	10	694	*1.61*	6.9	5.4		*160*	*420*	69
Salol	$C_{13}H_{10}O_3$	OKE	0.010	1064, 532	*1.25*	*15*			460	*1540*	55

Table 1.3.2 (Continued)
MEASURED NONLINEAR REFRACTIVE PARAMETERS FOR LIQUIDS

Liquid	Formula	Method	Pulse duration (nsec)	Wavelength (nm)	Linear refractive index	χ_{1111} (10^{-13}cm^3/erg)	χ_{1122} (10^{-13}cm^3/erg)	χ_{1221} (10^{-13}cm^3/erg)	$n_{1,LP}$ (10^{-13}cm^3/erg)	γ_{LP} (10^{-16}cm^2/W)	Ref.
Succinonitrile	$C_4H_4N_2$	OKE	0.010	1064, 532	*1.41*	*0.94*			25	*74*	55
Toluene	$C_6H_5CH_3$	ER	30	694	*1.48*			0.8			57
Toluene	$C_6H_5CH_3$	OKE	40	694	*1.48*	*1.2*			30	*85*	68
Toluene	$C_6H_5CH_3$	PST	30	694	*1.48*	$\chi^{(3)}_{1122} + 2\chi^{(3)}_{1221} = 1.1$					57
Toluene	$C_6H_5CH_3$	ER	10	1064	1.48			0.73			77
Toluene	$C_6H_5CH_3$	TWM	2	452, 434	*1.50*	$\chi^{(3)}/n_o$ *relative to* $C_6H_6 = 0.92$					79
Toluene	$C_6H_5CH_3$	TII	10	694	*1.48*	4.8	3.4		*120*	*350*	69
Toluene	$C_6H_5CH_3$	OKE	0.010	1064, 532	*1.48*	*2.4*			60	*170*	55
Trichloroethylene	C_2HCl_3	OKE	0.010	1064, 532	*1.47*	*1.41*			36.1	*103*	55
Trichloromethane	$CHCl_3$	TII	0.7	1320	1.43	*<0.38*			<10	<30	63
Trichloroethane	$C_2H_3Cl_3$	OKE	0.010	1064, 532	*1.43*	*0.31*			8.2	*24*	55
Water	H_2O	ER	30	694	1.33		0.003				25
Water	H_2O	PDF	0.021	532	*1.33*	*0.060*			1.7	*5.4*	88
Water	H_2O	OKE	40	694	1.33	*0.032*			0.9	*2.8*	68
Water	H_2O	TII	10	694	*1.33*	0.7			*20*	*62*	69
Water	H_2O	OKE	0.010	1064, 532	*1.33*	*0.046*			1.3	*4.1*	55
Water-heavy	D_2O	PDF	0.021	532	*1.33*	*0.071*			2.0	*6.3*	88
Xylene	$C_6H_4(CH_3)_2$	TII	10	694	*1.49*	4.4	3.6		*110*	*310*	69

[a] Orientational only.

Table 1.3.3
MEASURED NONLINEAR REFRACTIVE PARAMETERS FOR LARGE BANDGAP CRYSTALS

Crystal	Method	Pulse duration (nsec)	Wavelength (nm)	Linear refractive index	χ_{1111} (10^{-13}cm^{3}erg)	χ_{1122} (10^{-13}cm^{3}erg)	χ_{1221} (10^{-13}cm^{3}erg)	$n_{2,LP}$ (10^{-13}cm^{3}erg)	γ_{LP} (10^{-16} cm^2/W)	Ref.	Additional information
Al_2O_3:Cr	TRI	~1	1064	1.76	*0.069*			1.48	*3.52*	82	
Al_2O_3	TWM	3	560, 590	*1.76*	*0.11*			2.4	*5.7*	65	
Al_2O_3	PDF	0.030	1064	*1.76*	*0.060*			1.3	*3.1*	89	
BaF_2	TWM	4	592, 575	*1.47*	0.069	0.040	0.036	*1.8*	*5.0*	90	
BaF_2	TWM	4	575, 575-ϵ	*1.47*	0.083		0.034	*2.1*	*6.1*	76	
BaF_2	TRI	0.125	1064	1.47	*0.39*			*1.00*	2.85	91	
C(diamond)	TWM	4	545, 545-ϵ	*2.42*	0.46	0.18	0.17	*7.2*	*12.6*	76	Dispersion also given
$CaCO_3$	TWM	3	560, 590	*1.66*	*0.14*			3.2	*8.1*	65, 76	$\tilde{E} \perp$ optic axis
CaF_2	TWM	4	592, 575	*1.43*	0.04	0.020	0.017	*1.1*	*3.1*	90	
CaF_2	TWM	4	575, 575-ϵ	*1.43*	0.043		0.019	*1.1*	*3.3*	76	
CaF_2	TWM	3	560, 590	*1.43*	*0.055*			1.46	*4.3*	65	
CaF_2	TRI	0.125	1064	1.43	*0.025*			0.65	1.90	91, 92	
CaF_2	PDF	0.030	1064	1.43	0.105			2.8	*8.1*	56	
CdF_2	TWM	4	592, 575	*1.57*	0.149	0.048	0.041	*3.58*	*9.55*	90	
CdF_2	TWM	4	575, 575-ϵ	*1.57*	0.145		0.047	*3.48*	*9.29*	76	
CdF_2	TRI	0.125	1064	1.57	*0.061*			*1.46*	3.87	91	
CeF_3	TRI	0.125	1064	~1.6	*0.066*			*1.55*	4.06	91	
CsCl	TWM	0.006	1064, 532	*1.64*	0.086			*2.0*	*5.1*	93	
CsCl	TWM	0.006	1064, 532		0.029					93	In 5 mol/ℓ aqueous sol.
KBr	PDF	0.030	1064	1.544	0.58			14.2	*38.5*	56	
KCl	PDF	0.030	1064	1.479	0.13			3.3	*9.3*	56	
KF	TWM	0.006	1064, 532	*1.36*	0.014			*0.39*	*1.2*	93	
KF	TWM	0.006	1064, 532		0.020					93	In 5 mol/ℓ aqueous sol.
KH_2PO_4	TRI	0.10	1064	*1.49*	*0.040*			1.0	*2.8*	92	
KH_2PO_4	PDF	0.030	1064	1.49	0.14			3.6	*10*	56	
KI	TWM	0.006	1064, 532	*1.7*	0.38			*8.4*	*20*	93	

Table 1.3.3 (continued)
MEASURED NONLINEAR REFRACTIVE PARAMETERS FOR LARGE BANDGAP CRYSTALS

Crystal	Method	Pulse duration (nsec)	Wavelength (nm)	Linear refractive index	χ_{1111} (10⁻¹³cm³erg)	χ_{1122} (10⁻¹³cm³erg)	χ_{1221} (10⁻¹³cm³erg)	$n_{2,LP}$ (10⁻¹³cm³erg)	γ_{LP} (10^{-16} cm²/W)	Ref.	Additional information
KI	TWM	0.006	1064, 532		0.13					93	In 5 mol/ℓ aqueous sol.
KI	PDF	0.030	1064	1.638	0.49			11.2	*29*	56	
$La_2Be_2O_5$:Nd	PDF	0.030	1064	*1.98*	*0.11*			2.1	*4.4*	89	
LaF_3	TRI	0.125	1064	1.60(o)	*0.064*			1.51	3.95	91	
LiCl	TWM	0.006	1064, 532	*1.67*	0.069			*1.56*	*3.9*	93	
LiCl	TWM	0.006	1064, 532		0.027					93	In 5 mol/ℓ aqueous sol.
LiF	TWM	3	560, 590	*1.39*	*0.034*			0.92	*2.8*	65	
LiF	TRI	0.125	1064	1.39	*0.013*			0.35	1.05	91, 92	
LiF	PDF	0.030	1064	1.39	0.09			2.4	*7.2*	56	
$LiNbO_3$	TWM	5	577	*2.31*(o)			1.6			94	
$LiYF_4$	TRI	0.125	1064	1.45(o)	*0.023*			0.60	1.72	91	
MgF_2	TRI	0.125	1064	1.37(o)	*0.011*			0.30	0.92	91, 92	
NaBr	PDF	0.030	1064	1.62	0.41			9.6	*25*	56	
NaCl	PDF	0.030	1064	1.532	0.26			6.5	*18*	56	
NaF	TRI	0.125	1064	1.32	*0.015*			0.43	1.37	91	
NaF	PDF	0.030	1064	1.321	0.03			0.9	*2.9*	56	
PbF_2	TRI	0.125	1064	1.76	*0.23*			4.94	11.7	91	
SrF_2	TWM	4	592, 575	*1.43*	0.052	0.027	0.025	*1.4*	*4.0*	90	
SrF_2	TWM	4	575, 575-ϵ	*1.43*	*0.044*		0.030	*1.2*	*3.4*	76	
SrF_2	TRI	0.125	1064	1.43	*0.023*			0.60	1.76	91	
$Y_3Al_5O_{12}$:Nd	PDF	0.030	1064	1.82	0.17			3.5	*8.1*	56	
$Y_3Al_5O_{12}$	TRI	0.15	1064	*1.83*	*0.15*			3.16	*7.2*	95	$\vec{k}$ ∥ [111]
$Y_3Al_5O_{12}$	TWM	3	560, 590	*1.83*	*0.22*			4.5	*10*	65	
$Y_3Al_5O_{12}$	TRI	~1	1064	1.83	*0.17*			3.47	*7.9*	82	
$Y_3Al_5O_{12}$	ER	13	964	1.829	*0.21*			4.27	*9.8*	53	$\vec{E}$ ∥ [100]

Table 1.3.4
MEASURED NONLINEAR REFRACTIVE PARAMETERS FOR SEMICONDUCTOR CRYSTALS

Crystal	Method	Pulse duration (nsec)	Wavelength (nm)	Linear refractive index	χ_{1111} (10⁻¹³cm³erg)	χ_{1122}	χ_{1221}	$n_{2,LP}$	γ_{LP} (10^{-16}cm²/W)	Ref.	Additional information
CdS	SPA	20	694	2.42	130			2×10^3	3.5×10^3	96	
$CdS_{0.5}Se_{0.5}$	SPA	20	694	2.5	230			3.5×10^3	5.9×10^3	96	
$CdS_{0.18}Se_{0.82}$	SPA	20	694	2.6	1500			2.2×10^4	3.5×10^4	96	
CdTe	WFC	15	1064	~3	2.5×10^5			3.1×10^6	4.4×10^6	97	
CuCl	TWM		773, 694	1.94	33			640	1400	98	15 K
GaAs	TWM	~200	9200—11800	3.3	120	30		1.4×10^3	1.7×10^3	99	
Ge	TWM	~200	9200—11800	4	1000	610		9.4×10^3	9.9×10^3	99	
Ge	ER	2.3	10590	4	250	130		2.3×10^3	2.5×10^3	100	
Ge	TWM		10600	4						101	Rel. spectr.; impurity res.
Ge	WFC	300	38000	4.0	400			3.8×10^3	3.9×10^3	102	
HgCdTe	SPA	CW	10640	4.25	$\Delta n = -7 \times 10^{-3} I^{1/3}$					103	175 K; I in W/cm²
InSb	SPA	CW	5313	4	-6×10^{10}			-6×10^{11}	-6×10^{11}	104	5 K
InSb	SPA	CW	5405—5714	4	Abs. spectr., e.g., $\gamma = 10^{-4}$ cm³/W at 5555 nm					105	77 K
InSb	TWM		10600	4	$\sim 2 \times 10^6$			$\sim 2 \times 10^7$	$\sim 2 \times 10^7$	106	4 K
Si	TWM	~200	9200—11800	3.4	60	29		660	820	99	
SiC	SPA, TWR	20	694	2.68	36			510	800	107	
SiC	SPA	20	694	2.68	1.4×10^3			2×10^4	3×10^4	96	

Table 1.3.5
MEASURED NONLINEAR REFRACTIVE PARAMETERS FOR GLASSES

Glass	Method	Pulse duration (nsec)	Wavelength (nm)	Linear refractive index	χ_{1111} (10^{-13}cm^{3}erg)	χ_{1122} (10^{-13}cm^{3}erg)	χ_{1221} (10^{-13}cm^{3}erg)	$n_{2,LP}$ (10^{-13}cm^{3}erg)	γ_{LP} (10^{-16}cm^2/W)	Ref.	Additional information
Beryllium fluoride	TRI	0.15	1064	1.28	*0.0078*			0.23	*0.75*	108	
Borosilicate 517	DTLC	20	1064	1.51	*0.050*			1.24	*3.44*	52	Total n_2
Borosilicate BK-7	ER	20	694	1.52	*0.056*			1.4	*3.86*	81	"Electronic" assumption
BorosilicateBK-7	TRI	0.125	1064	1.52	*0.050*			1.24	3.43	64	
Borosilicate BK-10	TRI	0.17	355	1.50	*0.024*			0.6	1.7	109	
Borosilicate BSC	TWM	3	560,590	1.51	*0.092*			2.3	*6.4*	65	
Borosilicate BSC-2	TWR	12	694	*1.50*	*0.080*			2.0	*5.6*	110	
Flint SF-55	DTLC	20	1064	1.73	*0.38*			8.3	*20*	52	Total n_2
Fluoroberyllate:Nd	TRI	0.15	1064	1.34	*0.012*			0.33	*1.0*	108	
Fluorophosphate A86-82	TRI	0.125	1064	1.49	*0.028*			0.71	2.0	113	
Fluorophosphate FK-51	TRI	0.125	1064	1.49	*0.027*			0.69	1.94	64	
Fluorosilicate FC-5	TRI	0.125	1064	1.49	*0.042*			1.07	3.01	64	
Phosphate:Ce FR-4	TRI	0.15	1064	*1.56*	*0.081*			1.95	*5.2*	95	
Phosphate EV-1	TRI	0.125	1064	1.51	*0.036*			0.91	2.53	113	
Phosphate:Nd LHG-5	TRI	0.125	1064	1.54	*0.047*			1.16	3.15	113	
Phosphate:Nd LHG-6	TRI	0.125	1064	1.53	*0.040*			1.01	2.76	113	
Phosphate:Nd LHG-5	PDF	0.030	1064	1.54	*0.061*			1.5	*4.1*	89	
Phosphate:Nd LHG-6	PDF	0.030	1064	1.53	*0.061*			1.5	*4.1*	89	
Silica(Dynasil 4000)	TRI	0.125	1064	1.46	*0.037*			0.95	2.73	64	
Silica (fiber)	SPM	~0.15	514	*1.47*	*0.044*			1.14	*3.2*	60	
Silica(Suprasil II)	TRI	0.17	355	1.50	*0.036*			0.9	2.5	109	
Silica(Suprasil II)	SSMG	1.1	351	1.50	*0.024*			0.6	1.7	61	
Silica, $S10_2$	TII	20.	1064	*1.46*	0.044		0.010	*1.1*	*3.3*	121	
Silica, $S10_2$	ER	13	694	1.45	*0.039*			1.00	*2.88*	53	"Electronic" assumption
Silica, $S10_2$	TWM	3	560,590	1.46	*0.070*			1.8	*5.2*	65	
Silica, $S10_2$	DTLC	20	1064	1.45	*0.036*			0.93	*2.7*	52	Total n_2
Silicate C835	TRI	~1	1064	1.50	*0.073*			1.83	*5.1*	82	
Silicate C1020	TRI	~1	1064	1.50	*0.073*			1.83	*5.1*	82	
Silicate C1020	RSS		647	1.51	*0.060*			1.5	*4.2*	58	Also nuclear/elect. ratio
Silicate C2828	TRI	~1	1064	1.53	*0.084*			2.08	*5.7*	82	

Silicate ED-2	TRI	~1	1064	*1.57*	*0.064*		1.53	*4.1*	82	
Silicate ED-2	TRI	0.125	1064	1.57	*0.059*		1.41	3.77	113	
Silicate ED-2:Nd	TRI	0.125	1064	1.57	*0.059*		1.41	3.77	64	
Silicate ED-2:Nd	RSS		647	*1.57*	*0.075*		1.8	*4.8*	58	Also nuclear/elect. ratio
Silicate ED-2:Nd	TRI	0.15	1064	*1.57*	*0.063*		1.52	*4.1*	95	
Silicate ED-4	TWM	3	560,590	1.55	*0.011*		2.6	*7.0*	65	
Silicate ED-4	PDF	0.030	1064	1.55	*0.086*		2.1	*5.7*	89	
Silicate ED-4	ER	13	694	1.56	*0.072*		1.73	*4.6*	53	''Electronic'' assumption
Silicate EY-1	ER	13	694	1.61	*0.088*		2.06	*5.4*	111	''Electronic'' assumption
Silicate EY-1	TRI	0.15	1064	*1.61*	*0.076*		1.77	*4.6*	95	
Silicate GLS-1	PDF	~1	1064				1.16		114	
Silicate GLS-1	ER	29	1064			0.017			115	
Silicate GLS-2	ER	29	1064			0.019			115	
Silicate GLS-6	ER	29	1064			0.026			115	
Silicate K-108	ER	29	1064			0.018			115	
Silicate K-8	TII	10	694		1.5				69	
Silicate KGSS-1621	PDF	~1	1064				1.07		114	
Silicate LGS-1	ER	29	1064			0.023			115	
Silicate LGS-247	PDF	~1	1064				1.17		114	
Silicate LSO	ER	13	694	1.51	*0.058*		1.44	*4.0*	53	''Electronic'' assumption
Silicate SF6	TRI	~1	1064	1.77	*0.42*		9.0	*21*	82	
Silicate SF-7	ER	20	694	1.63	*0.26*		5.9	*15*	81	''Electronic'' assumption
Silicate:Tb FR-5	TRI	0.125	1064	1.67	*0.093*		2.1	5.2	64	
Silicate TF-7	TII	10	694		3.0				69	
Silicate TF-7	ER	29	1064			0.025			115	
Silicate ZF-7	TII		532				0.7		112	

Table 1.3.6
MEASURED NONLINEAR REFRACTIVE PARAMETERS FOR OTHER MATERIALS

Material	Method	Pulse duration (nsec)	Wavelength (nm)	Linear refractive index	χ_{1111} (cm³/erg)	χ_{1122} (cm³/erg)	χ_{1221} (cm³/erg)	$n_{2,LP}$ (cm³/erg)	γ_{LP} (cm²/W)	Ref.	Additional information
Gases and vapors											
Air	PDF	30	1064	1.0	$\sim 6.6 \times 10^{-18}$			$\sim 2.5 \times 10^{-16}$	1.0×10^{-15}	116	1 Atm.
Na vapor	SPA	CW	589	1.0	$7.2 \times 10^{-19} \times N$			$2.7 \times 10^{-17} \times N$	$1.1 \times 10^{-16} \times N$	31	N = Na density (cm^{-3})
C_s vapor	SPA	0.030	1064	1.0	$-3.7 \times 10^{-32} \times N$			$-1.4 \times 10^{-30} \times N$	$-5.9 \times 10^{-30} \times N$	30	N = Cs
Rb vapor	WFC	6	795	1.0	$-2.9 \times 10^{-23} \times N$			$-1.1 \times 10^{-21} \times N$	$-4.6 \times 10^{-21} \times N$	67	N = Rb
I_2 vapor	PDF	10	694	1.0	1.5×10^{-14}			5.7×10^{-13}	2.4×10^{-12}	117	
Polymeric media											
Lucite	TRI	~1	1064	1.49	1.1×10^{-14}			2.74×10^{-13}	7.7×10^{-16}	82	
Liquid crystals											
MBBA	DHG	CW	450					0.05		2	
MBBA	PST	20	694					4.4×10^{-10}		118	321 K
MBBA	OKE, ER	10	694					$\chi_{1221} = 2.2 \times 10^{-10}/T - T^*$		119	T* = 314.7 K
EBBA	SPA, TWR	15	694					$(1.2 - 24) \times 10^{-10}$		66	Phase transition
EBBA	OKE, ER	10	694					$\chi_{1221} = 1.3 \times 10^{-10}/T - T^*$		119	T* = 350.6 K
Nematic	PDF	CW	633					0.14		3	
Biological systems											
Retinol	TWM	10	1064						3.0×10^{-12}	120	Supercooled liquid
Retinol acetate	TWM	10	1064						8.2×10^{-12}	120	Supercooled liquid
Retinol acetate	TWM	10	1064						2.5×10^{-12}	120	6×10^{20} cm^{-3} in CS_2
Retinal	TWM	10	1064						1.1×10^{-11}	120	Supercooled liquid
Retinoic acid	TWM	10	1064						3.0×10^{-12}	120	3.8×10^{19} cm^{-3} in CS_2
β-Carotene	TWM	10	1064						8.8×10^{-12}	120	5.7×10^{19} cm^{-3} in CS_2

[a] MBBA: *p*-methoxybenzylidene-*p*-butylaniline.
[b] EBBA: *p*-ethoxybenzylidene-*p*-butylaniline.

REFERENCES

1. **Miller, D. A. B., Chemla, D. S., Smith, P. W., and Gossard, A. C.,** Resonant room temperature nonlinear optical processes in GaAs/GaAlAs multiple quantum well structures. Paper WQ3, in *Digest of Conference on Lasers and Electro Optics 1983,* Optical Society of America, Washington, D.C., 1983.
2. **Odulov, S. G., Reznikov, Y. A., Soskin, M. S., and Khizhnyak, A. I.,** Photostimulated transformation of molecules — a new type of ''giant'' optical nonlinearity in liquid crystals, *Sov. Phys. JETP,* 55(5), 854, 1982.
3. **Zel'dovich, B. Y., Pilipetskii, N. F., Sukhov, A. V., and Tabiryan, N. V.,** Giant optical nonlinearity in the mesophase of a nematic liquid crystal (NCL), *JETP Lett.,* 31(5), 263, 1980.
4. **Khoo, I. C., Zhuang, S. L., and Shepard, S.,** Self-focusing of a low power cw laser beam via optically induced birefringence in a nematic liquid-crystal film, *Appl. Phys. Lett.,* 39(12), 937, 1981.
5. **Langhoff, P. W., Epstein, S. T., and Karplus, M.,** Aspects of time-dependent perturbation theory, *Rev. Mod. Phys.,* 44, 602, 1972.
6. **Hellwarth, R. W.,** Effect of molecular redistribution on the nonlinear refractive index of liquids, *Phys. Rev.,* 152(1), 156, 1966.
7. **Sparks, M. and Duthler, C. J.,** Theoretical Studies of High-Power Ultraviolet and Infrared Materials, Fifth Technical Report on ARPA Contract DAHC 15-73-C-0127, Xonics, Inc., Van Nuys, Calif., 1975, 169.
8. **Mollmann, J. C. and Bartolotti, L. J.,** Third-harmonic generation and intensity-dependent refractive index of atomic hydrogen, *Int. J. Quantum Chem: Quantum Chem. Symp.,* 14, 31, 1980.
9. **Ratajska-Gadomska, B.,** Influence of the interaction between dipoles, optically induced in a crystal lattice, on the nonlinear refractive index of crystals, *Phys. Rev. B,* 26(4), 1942, 1982.
10. **Bokov, O. G.,** Theory of the nonlinear refractive index of liquids, *Sov. Phys. JETP,* 40(5), 923, 1974.
11. **Chang, T. Y.,** Fast self-induced refractive index changes in optical media: a survey, *Opt. Eng.,* 20(2), 220, 1981.
12. **Ackerhalt, J. R., Galbraith, H. W., and Goldstein, J. C.,** Self-focusing in SF_6, *Opt. Lett.,* 6(8), 377, 1981.
13. **Jha, S. S. and Bloembergen, N.,** Nonlinear optical susceptibilities in Group-IV and III-V semiconductors, *Phys. Rev.,* 171(3), 891, 1968.
14. **Yuen, S. Y.,** Third-order optical nonlinearity induced by effective mass gradient in heterostructures, *Appl. Phys. Lett.,* 42(4), 331, 1983.
15. **Hellwarth, R. W., Cherlow, J., and Yang, T.-T.,** Origin and frequency dependence of nonlinear optical susceptibilities of glasses, *Phys. Rev. B,* 11(2), 964, 1975.
16. **Heiman, D., Hellwarth, R. W., and Hamilton, D. S.,** Raman scattering and nonlinear refractive index measurements of optical glasses, *J. Non-Cryst. Solids,* 34, 63, 1979.
17. **Wang, C. C.,** Empirical relation between the linear and the third-order nonlinear optical susceptibilities, *Phys. Rev.,* 2B, 2045, 1970.
18. **Fournier, J. T. and Snitzer, E.,** The nonlinear refractive index of glass, *IEEE J. Quantum Electron.,* QE-10(5), 473, 1974.
19. **Boling, N. L., Glass, A. J., and Owyoung, A.,** Empirical relationships for predicting nonlinear refractive index changes in optical solids, *IEEE J. Quantum Electron.,* QE-14(8), 601, 1978.
20. **Garg, R.,** Empirical relationship for a nonlinear index coefficient, *Appl. Opt.,* 19(8), 1219, 1980.
21. **Weber, M. J., Milam, D., and Smith, W. L.,** Nonlinear refractive index of glasses and crystals, *Opt. Eng.,* 17(5), 463, 1978.
22. **Brown, D. C., Rinefierd, J. M., Jacobs, S. D., and Abate, J. A.,** Electronic, nuclear and total nonlinear indices of liquids, *NBS Spec. Publ.,* 568, 91, 1979.
23. **Armstrong, J. A., Bloembergen, N., Ducuing, J., and Pershan, P. S.,** Interactions between light waves in a nonlinear dielectric, *Phys. Rev.,* 127, 1918, 1962.
24. **Bloembergen, N.,** *Nonlinear Optics,* W. A. Benjamin, New York, 1965.
25. **Maker, P. D. and Terhune, R. W.,** Study of optical effects due to an induced polarization third order in the electric field strength, *Phys. Rev.,* 137(3A), A801, 1965; **Maker, P. D., Terhune, R. W., and Savage, C. M.,** Intensity-dependent changes in the refractive index of liquids, *Phys. Rev. Lett.,* 12(18), 507, 1964.
26. **Alfano, R. R., Hope, L. L., and Shapiro, S. L.,** Electronic mechanism for production of self-phase modulation, *Phys. Rev. A,* 6(1), 433, 1972.
27. **Mollenauer, L. F., Stolen, R. H., and Gordon, J. P.,** Experimental observation of picosecond pulse narrowing and solitons in optical fibers, *Phys. Rev. Lett.,* 45(13), 1095, 1980.
28. **Mollenauer, L. F. and Stolen, R. H.,** Solitons in optical fibers, *Fiberopt. Technol.,* 193, 1982.
29. **Nikolaus, B. and Grischkowsky, D.,** 90-fs Tunable optical pulses obtained by two-stage pulse compression, *Appl. Phys. Lett.,* 43(3), 228, 1983.

30. **Lehmberg, R. H., Reintjes, J., and Eckardt, R. C.,** Negative nonlinear susceptibility of cesium vapor around 1.06 μm, *Phys. Rev. A*, 13(3), 1095, 1976; Compensation of self-phase modulation by Cesium vapor, *Opt. Commun.*, 18(1), 174, 1976; Two-photon resonantly enhanced self-defocusing in Cs vapor at 1.06 μ, *Appl. Phys. Lett.*, 25(7), 374, 1974; Complete compensation of Self-phase modulation in cesium vapor at 1.06 μ, *Appl. Phys. Lett.*, 30(9), 487, 1977.
31. **Bjorkholm, J. E. and Ashkin, A.,** cw Self-focusing and self-trapping of light in sodium vapor, *Phys. Rev. Lett.*, 32(4), 129, 1974.
32. **Smith, P. W., Ashkin, A., and Tomlinson, W. J.,** Four-wave mixing in an artificial Kerr medium, *Opt. Lett.*, 6, 284, 1981.
33. **Ashkin, A., Dziedzic, J. M., and Smith, P. W.,** Continuous-wave self-focusing and self-trapping of light in artificial Kerr media, *Opt. Lett.*, 7(6), 276, 1982.
34. **Hellwarth, R. W.,** Third-order optical susceptibilities of liquids and solids, in *Progress in Quantum Electronics*, Vol. 5 (Part 1), Pergamon Press, Oxford, England, 1977.
35. **Owyoung, A.,** The Origins of the Nonlinear Refractive Indices of Liquids and Glasses, M.S. thesis, California Institute of Technology, Pasadena, 1971.
36. **Flytzanis, C.,** Theory of nonlinear optical susceptibilities, in *Quantum Electronics*, Vol. 1 (Part A), Rabin, H. and Tang, C. L., Eds., Academic Press, New York, 1975, chap. 1.
37. **Shen, Y. R.,** Self-focusing: experimental, in *Progress in Quantum Electronics*, Vol. 4 (Part 1), Sanders, J. H. and Stenholm, S., Eds., Pergamon Press, New York, 1975, 1.
38. **Marburger, J. H.,** Self-focusing: theory, in *Progress in Quantum Electronics*, Vol. 4 (Part 1), Sanders, J. H. and Stenholm, S., Eds., Pergamon Press, New York, 1975, 35.
39. **Akhmanov, S. A., Khokhlov, R. V., and Sukhorukov, A. P.,** Self-focusing, self-defocusing, and self-modulation of laser beams, in *Laser Handbook*, Arecchi, F. T. and Schulz-Dubois, E. O., Eds., North-Holland, Amsterdam, 1972, 1151.
40. **Svelto, O.,** Self-focusing, self-trapping, and self-phase modulation of laser beams, in *Progress in Optics*, Vol. 12, Wolf, E., Ed., North-Holland, Amsterdam, 1974.
41. **Simmons, W. W., Hunt, J. T., and Warren, W. E.,** Light propagation through large laser systems, *IEEE J. Quantum Electron.*, QE-17(9), 1727, 1981.
42. See the articles collected in the special issue of *Opt. Eng.*, 21(2), 1982.
43. See the articles in *Opt. Eng.*, 19(4), 1980.
44. *Technical Digest of the Topical Meeting on Optical Bistability*, Optical Society of America, Washington, D.C., 1983.
45. **Arakelyan, S. M., Lyakhov, G. A., and Chilingaryan, Yu. S.,** Nonlinear optics of liquid crystals, *Sov. Phys. Usp.*, 23(5), 245, 1980.
46. **Tabiryan, N. V. and Zel'dovich, B. Ya.,** The orientational optical non-linearity of liquid crystals. I. Nematics, *Mol. Cryst. Liq. Cryst.*, 62, 237, 1980; II. Cholesterics, *Mol. Cryst. Liq. Cryst.*, 69, 19, 1981; III. Smectics, *Mol. Cryst. Liq. Cryst.*, 69, 31, 1981.
47. **Campillo, A. J., Shapiro, S. L., and Suydam, B. R.,** Periodic breakup of optical beams due to self-focusing, *Appl. Phys. Lett.*, 23(11), 628, 1973; Relationship of self-focusing to spatial instability modes, *Appl. Phys. Lett.*, 24(4), 178, 1974.
48. **Bespalov, V. I. and Talanov, V. I.,** Filamentary structure of light beams in nonlinear liquids, *JETP Lett. (Sov. Phys.)*, 3, 307, 1966.
49. **Vlasov, S. N., Kryzhanovskii, V. I., and Yashin, V. E.,** Use of circularly polarized optical beams to suppress self-focusing instability in a nonlinear cubic medium with repeaters, *Sov. J. Quantum Electron.*, 12(1), 7, 1982.
50. **Shen, Y. R.,** Electrostriction, optical Kerr effect and self-focusing of laser beams, *Phys. Lett.*, 20(4), 378, 1966.
51. **Kerr, E. R.,** Transient and steady-state electrostrictive laser beam trapping, *IEEE J. Quantum Electron.*, QE-6(10), 616, 1970.
52. **Feldman, A., Horowitz, D., and Waxler, R. M.,** Mechanisms for self-focusing in optical glasses, *IEEE J. Quantum Electron.*, QE-9, 1054, 1973.
53. **Owyoung, A.,** Ellipse rotation studies in laser host materials, *IEEE J. Quantum Electron.*, QE-9(11), 1064, 1973.
54. **Hellwarth, R. W. and George, N.,** Nonlinear refractive indices of CS_2-CCl_4 mixtures, *Opt-Electronics*, 1, 213, 1969.
55. **Ho, P. P. and Alfano, R. R.,** Optical Kerr effect in liquids, *Phys. Rev. A*, 20(5), 2170, 1979.
56. **Smith, W. L., Bechtel, J. H., and Bloembergen, N.,** Dielectric-breakdown threshold and nonlineaer-refractive-index measurements with picosecond laser pulses, *Phys. Rev. B*, 12, 706, 1975.
57. **Wang, C. C.,** Nonlinear susceptibility constants and self-focusing of optical beams in liquids, *Phys. Rev.*, 152(1), 149, 1966.
58. **Yang, T. T.,** Raman scattering and optical susceptibilities of Nd-doped glasses, *Appl. Phys.*, 11, 167, 1976.

59. **Hongyo, M., Sasaki, T., and Yamanaka, C.**, Nonlinear effects of $POCl_3$ liquid laser, *Technol. Rep. Osaka Univ.*, 23(1121-1154), 455, 1973.
60. **Stolen, R. H. and Lin, C.**, Self-phase-modulation in silica optical fibers, *Phys. Rev. A*, 17(4), 1448, 1978.
61. **Smith, W. L., Warren, W. E., Vercimak, C. L., and White, W. T., III,** Nonlinear refractive index at 351 nm by direct measurement of small-scale self-focusing. Paper FB4, in *Digest of Conference on Lasers and Electro Optics*, Optical Society of America, Washington, D.C., 1983, 17.
62. **Owyoung, A. and Peercy, P. S.**, Precise characterization of the Raman nonlinearity in benzene using nonlinear interferometry, *J. Appl. Phys.*, 48(2), 674, 1977.
63. **Witte, K. J., Galanti, M., and Volk, R.**, n_2-Measurements at 1.32 μm of some organic compounds usable as solvents in a saturable absorber for an atomic iodine laser, *Opt. Commun.*, 34(2), 278, 1980.
64. **Milam, D. and Weber, M. J.**, Measurement of nonlinear refractive-index coefficients using time-resolved interferometry: application to optical materials for high-power neodymium laser, *J. Appl. Phys.*, 47(6), 2497, 1976.
65. **Levenson, M. D.**, Feasibility of measuring the nonlinear index of refraction by third-order frequency mixing, *IEEE J. Quantum Electron.*, QE-10(2), 110, 1974.
66. **Hanson, E. G., Shen, Y. R., and Wong, G. K. L.**, Experimental study of self-focusing in a liquid crystalline medium, *Appl. Phys.*, 14, 65, 1977; Self-focusing: from transient to quasi-steady-state, *Opt. Commun.*, 20(1), 45, 1977; **Wong, G. K. L. and Shen, Y. R.**, Transient self-focusing in a nematic liquid crystal in the isotropic phase, *Phys. Rev. Lett.*, 32(10), 527, 1974.
67. **Grischkowsky, D., Shiren, N. S., and Bennett, R. J.**, Generation of time-reversed wave fronts using a resonantly enhanced electronic nonlinearity, *Appl. Phys. Lett.*, 33(9), 805, 1978.
68. **Paillette, M.**, Recherches expérimentales sur les effets Kerr induits par une onde lumineuse, *Ann. Phys. (Paris)*, 4, 671, 1969.
69. **Veduta, A. P. and Kirsanov, B. P.**, Variation of the refractive index of liquids and glasses in a high intensity field of a ruby laser, *Sov. Phys. JETP*, 27(5), 736, 1968.
70. **Oudar, J. L., Chemla, D. S., and Batifol, E.**, Optical nonlinearities of various substituted benzene molecules in the liquid state, and comparison with solid state nonlinear susceptibilities, *J. Chem. Phys.*, 67(4), 1626, 1977.
71. **Kildal, H. and Brueck, S. R. J.**, Orientational and electronic contributions to the third-order susceptibilities of cryogenic liquids, *J. Chem. Phys.*, 73(10), 4951, 1980.
72. **Mayer, M. M. G. and Gires, F.**, Action d'une onde lumineuse intense sur l'indice de réfraction des liquides, *C. R. Acad. Sci. Paris*, 256, 2039, 1964.
73. **Owyoung, A.**, Induced focusing and defocusing of optical beams via a Raman-type nonlinearity, *Appl. Phys. Lett.*, 26(4), 168, 1975.
74. **Wiener-Avnear, E., Chandra, S., and Compaan, A.**, Third-order nonlinear susceptibility ratio by CARS of mixtures: CS_2 in C_6H_6, *Appl. Phys. Lett.*, 32(5), 286, 1978.
75. **Owyoung, A.**, Absolute determination of the nonlinear susceptibility χ_3 via two-beam nonlinear interferometry, *Opt. Commun.*, 16(2), 266, 1976.
76. **Levenson, M. D. and Bloembergen, N.**, Dispersion of the nonlinear optical susceptibility tensor in centrosymmetric media, *Phys. Rev. B.*, 10(10), 4447, 1974.
77. **Cherlow, J. M., Yang, T. T., and Hellwarth, R. W.**, Nonlinear optical susceptibilities of solvents, *IEEE J. Quantum Electron.*, 644, 1976.
78. **Song, J. J. and Levenson, M. D.**, Electronic and orientational contributions to the optical Kerr constant determined by coherent Raman techniques, *J. Appl. Phys.*, 48(8), 3496, 1977.
79. **Saikan, S. and Marowsky, G.**, Measurement of the third-order nonlinear susceptibility ratio by the CARS Maker fringe technique, *Opt. Commun.*, 26(3), 466, 1978.
80. **Thomas, P., Jares, A., and Stoicheff, B. P.**, Nonlinear refractive index and "DC" Kerr constant of liquid CS_2 at 10.6 μm, *IEEE J. Quantum Electron.*, QE-10, 493, 1974.
81. **Owyoung, A., Hellwarth, R. W., and George, N.**, Intensity-induced changes in optical polarizations in glasses, *Phys. Rev. B*, 5(2), 628, 1972.
82. **Moran, M. J., She, C.-Y., and Carman, R. L.**, Interferometric measurements of the nonlinear refractive-index coefficient relative to CS_2 in laser-system-related materials, *J. Quantum Electron.*, QE-11(6), 259, 1975.
83. **Levenson, M. D. and Bloembergen, N.**, Dispersion of the nonlinear optical susceptibilities of organic liquids and solutions, *J. Chem. Phys.*, 60(4), 1323, 1974.
84. **Owen, T. C., Coleman, L. W., and Burgess, T. J.**, Ultrafast optical Kerr effect in CS_2 at 10.6 μm, *Appl. Phys. Lett.*, 22(6), 272, 1973.
85. **Hellwarth, R. W., Owyoung, A., and George, N.**, Origin of the nonlinear refractive index of liquid CCl_4, *Phys. Rev. A*, 4(6), 2342, 1971.
86. **Zinth, W., Laubereau, A., and Kaiser, W.**, Time resolved observation of resonant and non-resonant contributions to the nonlinear susceptibility $\chi^{(3)}$, *Opt. Commun.*, 26(3), 457, 1978.

87. **Ho, P. P., Lu, P. Y., and Alfano, R. R.,** Oscillatory optically induced Kerr kinetics in nitrobenzene, *Opt. Commun.*, 30(3), 426, 1979.
88. **Smith, W. L., Liu, P., and Bloembergen, N.,** Superbroadening in H_2O and D_2O by self-focused picosecond pulses from a YAlG:Nd laser, *Phys. Rev. A*, 15(6), 2396, 1977.
89. **Smith, W. L. and Bechtel, J. H.,** Laser-induced breakdown and nonlinear refractive index measurements in phosphate glasses, lanthanum beryllate, and Al_2O_3, *Appl. Phys. Lett.*, 28, 606, 1976.
90. **Lynch, R. T., Jr., Levenson, M. D., and Bloembergen, N.,** Experimental test for deviation from Kleinman's symmetry in the third order susceptibility tensor, *Phys. Lett.*, 50A(1), 61, 1974.
91. **Milam, D., Weber, M. J., and Glass, A. J.,** Nonlinear refractive index of fluoride crystals, *Appl. Phys. Lett.*, 31(12), 822, 1977.
92. **Milam, D. and Weber, M. J.,** Time-resolved interferometric measurements of the nonlinear refractive index in laser materials, *Opt. Commun.*, 18(1), 172, 1976.
93. **Penzkofer, A., Schmailzl, J., and Glas, H.,** Four-wave mixing in alkali halide crystals and aqueous solutions, *Appl. Phys. B*, 29, 37, 1982.
94. **Wynne, J. J.,** Nonlinear optical spectroscopy of $\chi^{(3)}$ in $LiNbO_3$, *Phys. Rev. Lett.*, 29(10), 650, 1972.
95. **Bliss, E. S., Speck, D. R., and Simmons, W. W.,** Direct interferometric measurements of the nonlinear refractive index coefficient n_2 in laser materials, *Appl. Phys. Lett.*, 25(12), 728, 1974.
96. **Borshch, A. A. and Brodin, M. S.,** Nonlinear polarizability of some binary and mixed semiconductors, *Bull. Acad. Sci. USSR, Phys. Ser. (USA)*, 43(2), 98, 1978; **Borshch, A. A., Brodin, M. S., Krupa, N. N., Lukomskii, V. P., Pisarenko, V. G., Petropaviovskii, A. I., and Chernyi, V. V.,** Determination of the coefficients of the nonlinear refractive index of a CdS crystal by the nonlinear refraction method, *Sov. Phys. JETP*, 48(1), 41, 1978.
97. **Kremenitskii, V., Odulov, S., and Soskin, M.,** Backward degenerate four-wave mixing in cadmium telluride, *Phys. Status Solidi (A)*, 57, K71, 1980.
98. **Kramer, S. D., Parson, F. G., and Bloembergen, N.,** Interference of third-order light mixing and second-harmonic exciton generation in CuCl, *Phys. Rev. B*, 9(4), 1853, 1974.
99. **Wynne, J. J.,** Optical third-order mixing in GaAs, Ge, Si, and InAs, *Phys. Rev.*, 178(3), 1295, 1969.
100. **Watkins, D. E., Phipps, C. R., and Thomas, S. J.,** Determination of the third-order nonlinear optical coefficients of germanium through ellipse rotation, *Opt. Lett.*, 5(6), 248, 1980.
101. **Wood, R. A., Kahn, M. A., Wolff, P. A., and Aggarwal, R. L.,** Dispersion of the nonlinear optical susceptibility of n-type germanium, *Opt. Commun.*, 21(1), 154, 1977.
102. **Depatie, D. and Haueisen, D.,** Multiline phase conjugation at 4 μm in germanium, *Opt. Lett.*, 5(6), 252, 1980.
103. **Hill, J. R., Parry, G., and Miller, A.,** Nonlinear refractive index changes in CdHgTe at 175K with 10.6 μm radiation, *Opt. Commun.*, 43(2), 151, 1982.
104. **Weaire, D., Wherrett, B. S., Miller, D. A. B., and Smith, S. D.,** Effect of low-power nonlinear refraction on laser-beam propagation in InSb, *Opt. Lett.*, 4(10), 331, 1979.
105. **Miller, D. A. B., Seaton, C. T., Prise, M. E., and Smith, S. D.,** Band-gap-resonant nonlinear refraction in III-V semiconductors, *Phys. Rev. Lett.*, 47(3), 197, 1981.
106. **Yuen, S. Y. and Wolff, P. A.,** Difference-frequency variation of the free-carrier-induced, third-order nonlinear susceptibility in n-InSb, *Appl. Phys. Lett.*, 40(6), 457, 1982.
107. **Borshch, A. A., Brodin, M. S., and Volkov, V. I.,** Self-focusing of ruby-laser radiation in single-crystal silicon carbide, *Sov. Phys. JETP*, 45(3), 490, 1977.
108. **Weber, M. J., Cline, C. F., Smith, W. L., Milam, D., Heiman, D., and Hellwarth, R. W.,** Measurements of the electronic and nuclear contributions to the nonlinear refractive index of beryllium fluoride glasses, *Appl. Phys. Lett.*, 32(7), 403, 1978.
109. **White, W. T., III, Smith, W. L., and Milam, D.,** Direct measurement of the nonlinear refractive index coefficient γ at 355 nm in fused silica and in BK-10 glass, *Opt. Lett.*, 9, 10, 1984.
110. **Newnam, B. E. and DeShazer, L. B.,** Direct nondestructive measurement of self-focusing in laser glass, *NBS Spec. Publ.*, 356, 113, 1971.
111. **Owyoung, A.,** Nonlinear refractive index measurements in laser media, *NBS Spec. Publ.*, 387, 12, 1973.
112. **Chi, K.,** Interferometric measurement of nonlinear refractive index of ZF-7 glass, *Laser J. (China)*, 8(4), 48, 1981.
113. **Milam, D. and Weber, M. J.,** Nonlinear refractive index coefficient for Nd phosphate laser glasses, *IEEE J. Quantum Electron.*, QE-12, 512, 1976.
114. **Bondarenko, N. G., Evemina, I. V., and Makarov, A. I.,** Measurement of the coefficient of electronic nonlinearity in optical and laser glass, *Sov. J. Quantum Electron.*, 8(4), 482, 1978.
115. **Garaev, R. A., Vlasov, D. V., and Korobkin, V. V.,** Need to allow for slow nonlinearity in measurements of n_2, *Sov. J. Quantum Electron.*, 12(1), 100, 1982.
116. **Vlasov, D. V., Garaev, R. A., Korobkin, V. V., and Serov, R. V.,** Measurement of nonlinear polarizability of air, *Sov. Phys. JETP*, 49(6), 1033, 1979.

117. **Villaverde, A. B.,** Saturable absorption and self-focusing effects in molecular iodine vapour, *J. Phys. B*, 13, 1817, 1980.
118. **Narasinha Rao, D. V. G. L. and Jayaraman, S.,** Self-focusing of laser light in the isotropic phase of a nematic liquid crystal, *Appl. Phys. Lett.*, 23(10), 539, 1973.
119. **Wong, G. K. L. and Shen, Y. R.,** Study of pretransitional behavior of laser-field-induced molecular alignment in isotropic nematic substances, *Phys. Rev. A*, 10(4), 1277, 1974.
120. **Smith, P. W., Tomlinson, W. J., Eilenberger, D. J., and Maloney, P. J.,** Measurement of electronic optical Kerr coefficients, *Opt. Lett.*, 6(12), 581, 1981; **Smith, P.,** Measurement of electronic optical Kerr coefficients, Paper WF5 at CLEO '81, June 10-12, 1981, Optical Society of America, Washington, D.C.
121. **Al'tshuler, G. B., Barbashev, A. I., Karasev, V. B., Krylov, K. I., Ovchinnikov, V. M., and Sharlai, S. F.,** Direct measurement of the tensor elements of the nonlinear optical susceptibility of optical materials, *Sov. Tech. Phys. Lett.*, 3(6), 213, 1977.

DISCLAIMER

1.4 STIMULATED RAMAN SCATTERING

Fred P. Milanovich

INTRODUCTION

Stimulated Raman scattering (SRS) was predicted by Javan[1] in 1958 and observed by Woodbury and Ng[2] some 4 years later in experiments with a high-power pulsed laser. In the 2 decades that have followed this observation, over 1000 articles have been written on SRS and related topics. The field has grown from an intellectual curiosity to an applied science. For instance, SRS has been observed in glass optical fibers,[3] made ultratrace determinations of compounds,[4,5] elucidated fundamentals of molecular dynamics,[6] and generated frequency conversions for a variety of applications.[7] Numerous reviews of varying detail on SRS and closely related topics are now available.[8-15]

SPONTANEOUS RAMAN SCATTERING

The spontaneous Raman effect was discovered over 50 years ago and was rigorously described shortly thereafter by the polarizability theory of Placzek.[16] The effect saw little application until the advent of the laser in the early 1960s. With these high-intensity monochromatic sources, spontaneous Raman scattering has become a more routine spectroscopic technique.

In the Raman effect, a photon of frequency ω_ℓ can excite a material energy level (typically vibration or rotation) of frequency ω_R even though ω_ℓ is not close to resonance (i.e., $\omega_\ell \gg \omega_R$). This results, by virtue of energy conservation, in a scattered photon of frequency $\omega_s = \omega_\ell - \omega_R$. Here the subscript "s" stands for Stokes shift. If a molecule is already in an excited energy state it can return a scattered photon of greater energy (anti-Stokes) than the incident as $\omega_{as} = \omega_\ell + \omega_R$.

The cross section for the Raman effect is usually defined in an expression of the form:

$$I_s = N \frac{\partial\sigma}{\partial\Omega} I_\ell \, dz d\Omega \tag{1}$$

Here I_s is the intensity scattered at the Stokes frequency, ω_s, in the length interval dz and within solid angle $d\Omega$. The N is the number density of scatterers and I_ℓ is the incident intensity at the source frequency. The units of $\partial\sigma/\partial\Omega$ are typically cm^2/molecule · steradian with typical values of approximately 10^{-30}.

The absolute Raman cross section has been measured to high precision in relatively few compounds. Cross sections for many other molecules have been determined by relative measurements to these well-substantiated, absolute cross sections, and the status of this field has been critically reviewed recently.[17] In addition, it was shown that for N_2 gas, $\partial\sigma/\partial\Omega \cdot \omega_s^{-4}$ is a constant (independent of excitation frequency in the visible).

The Raman cross section can be expressed in terms of fundamental parameters as (see Reference 8):

$$\frac{\partial\sigma}{\partial\Omega} = \frac{\hbar\omega_s^4\eta_s}{2m\,\omega_R\,c^4\eta_\ell}\left(\frac{\partial\alpha}{\partial q}\right)^2 \tag{2}$$

Here ω_s and ω_R are the Raman Stokes and material vibration radial frequencies, respectively, m is the reduced mass of the vibration, η_ℓ and η_s are the refractive indexes of the material at the laser and Stokes frequency, respectively, and $\partial\alpha/\partial q$ is the normal mode derivative of the polarizability tensor. Raman active materials typically have values of $\partial\alpha/\partial q > 10^{-16}$ cm^2.

For incident and scattered waves polarized in the same direction the cross section becomes:

$$\frac{\partial\sigma}{\partial\Omega_{\parallel}} = \frac{\partial\sigma}{\partial\Omega}\frac{1}{1 + \rho} \tag{3}$$

where ρ is the depolarization ratio of the scattered intensity $I_{\perp}/I_{\parallel}$.

Equation 1 shows that the intensity of scattered photons at the Stokes frequency is proportional to the incident (or excitation) intensity. This is an approximate expression that is valid at low-to-moderate incident light intensity. Under this condition, the scattered photon occupies a state in the radiation field with few or no photons of the same frequency (energy). However, at high-incident intensities ($>10^7$ W/cm^2), it is possible to achieve considerable population in a given frequency mode ($n_s \gg 1$). It then follows from Bose-Einstein statistics that the probability of scattering additional photons into the mode is proportional to n_s or $I_s \propto I_\ell \cdot I_s$. This is the condition for SRS.

The above condition results in an amplified intensity at the Stokes frequency that is exponential. In fact:

$$I_s = I_s^{(0)} \exp(gI_\ell L) \tag{4}$$

Here $I_s^{(O)}$ is the initial intensity at the Stokes frequency, g is the SRS gain coefficient, and L is the length of amplifying medium.

If the accumulation of photons in the Stokes mode is greater than the losses, significant conversion of intensity from the laser frequency to the Stokes frequency can occur. For typical systems, appreciable conversion begins when the product $gI_\ell L \geqslant 25$.

A fundamental difference between SRS and basic laser action is that in SRS there is no need to achieve a population inversion to achieve amplification. The process begins with a material in the ground or unexcited state.

THIRD-ORDER SUSCEPTIBILITY AND THE SRS GAIN COEFFICIENT

The electric fields associated with high-intensity laser light can be quite large. For instance, an incident intensity of 10^7 W/cm^2 corresponds to a maximum field amplitude of $\sim 10^5$ V/cm. Under these conditions, the polarization induced in the material is inadequately described by a linear expansion of the electric field. The response to the electric field becomes nonlinear and is more correctly expressed as:

$$P = \chi^{(1)}E + \chi^{(2)}E^2 + \chi^{(3)}E^3 + \ldots \tag{5}$$

Here, e.g., $\chi^{(1)}$ is responsible for linear optical properties such as reflection and refraction, $\chi^{(2)}$ is responsible for sum and difference frequency generation, and $\chi^{(3)}$ is responsible for two-photon absorption and SRS.

The SRS gain coefficient, g, can be derived rigorously from the semiclassical theory of the interaction of light and matter. We present here a brief outline of this derivation and the resulting expression for g.

The propagation of light is described by classical electromagnetic theory. In the presence of nonlinear sources the wave equation becomes:

$$\nabla^2 E - \left(\frac{\eta}{c}\right)^2 \frac{\partial^2}{\partial t^2} E = \frac{4\pi}{c^2} \frac{\partial^2}{\partial t^2} P^{NL} \tag{6}$$

Here $P^{NL} = \chi^{(3)}E^3$, η is the refractive index, and c is the vacuum light speed.

If we consider a two-level system and a highly polarized Raman transition we can write:

$$P^{NL} = N \frac{\partial\alpha}{\partial q} \langle q \rangle E \tag{7}$$

where $\partial\alpha/\partial q$ may be treated as a frequency-independent constant, N is the number density of molecules, and $\langle q \rangle$ is the quantum-mechanical expectation value of the vibrational coordinate.

We can also write a differential equation describing the amplitude $\langle q \rangle$ of the Raman active vibration at frequency ω_R:

$$\frac{\partial^2 \langle q \rangle}{\partial t^2} + \Gamma \frac{\partial \langle q \rangle}{\partial t} + \omega_R^2 \langle q \rangle = \frac{1}{2m} \frac{\partial\alpha}{\partial q} E^2 \tag{8}$$

Here Γ is the damping constant of the vibration (inverse phase relaxation time), ω_R is the frequency of the Raman active vibration, m is the reduced mass of the vibrating moiety, and E is the electric field described below.

$$E = \frac{1}{2} E_\ell \exp [i(k_\ell z - \omega_\ell t)] + \frac{1}{2} E_s \exp [i(k_s z - \omega_s t)] + c\ c \tag{9}$$

Equations 6 to 9 allow us to express a set of coupled differential equations describing the propagation of light at frequencies ω_ℓ and ω_s in the medium. These equations can be solved subject to at least the following assumptions: the vibrations are highly damped ($\partial^2\langle q \rangle/\partial t^2 = \partial\langle q \rangle/\partial t = 0$) and the propagation of light is steady state ($\partial I_{\omega_\ell}/\partial t = \partial I_{\omega_s}/\partial t = 0$).

We can thus write an expression for the loss of intensity at ω_ℓ as:

$$g_\ell = \frac{8\pi^2\omega_\ell N}{\eta^2 c^2 m\, \omega_R \Gamma} \left(\frac{\partial\alpha}{\partial q}\right)^2 \tag{10}$$

and for the gain in intensity at ω_s as:

$$g_s = \frac{8\pi^2\omega_s N}{\eta^2 c^2 m\, \omega_R \Gamma} \left(\frac{\partial\alpha}{\partial q}\right)^2 \tag{11}$$

Here frequencies are radial and Γ is full width at half-maximum (FWHM).

Equation 11 can be put in a more convenient form by the substitution:

$$\left(\frac{\partial\alpha}{\partial q}\right)^2 = \frac{2c^4 m\, \omega_R}{\omega_s^4 \hbar} \frac{\partial\sigma}{\partial\Omega_\parallel} \tag{12}$$

Again the frequencies are radial and Equation 11 becomes:

$$g_s = \frac{16\pi^2 c^2 N}{\hbar\omega_s^3 \eta^2 \Gamma} \frac{\partial\sigma}{\partial\Omega_\parallel} \tag{13}$$

For example, take $\partial\sigma/\partial\Omega_{\parallel} = 10^{-30}$ cm^2, N = 10^{22} cm^{-3}, $\Gamma/2\pi c$ = 2 cm^{-1}, $\omega_s/2\pi c$ = 2 × 10^4 cm^{-1}, and η = 1.5. This yields $g_s \simeq 3 \times 10^{-17}$ cm · s/erg = 3×10^{-10} cm/W.

TABLES OF EXPERIMENTAL DATA

Quantitative SRS information (such as gain coefficient and Stokes wavelength) is now available for many gases, liquids, and solids. These are presented in Tables 1.4.1 to 1.4.7. In many instances, the experimental conditions necessarily violate the assumptions underlying the derivation of Equation 13. To help the reader completely evaluate the data for his own purposes, we elaborate on some of the many pitfalls confronting the data analysis:

1. Steady-state vs. transient excitation — transient effects become important when the duration of the laser pulse is comparable to or shorter than the vibrational dephasing time $1/\Gamma$. Specifically, $\tau_\ell\Gamma < gI_\ell L$ is the transient regime.[18] As a rule of thumb, transient effects can occur in liquids with $\tau_\ell < 10^{-10}$ sec and $\tau_\ell < 10^{-8}$ sec in gases.
2. Measured gain coefficients — many experiments are performed by focusing high-energy laser pulses into an SRS medium. In many instances, especially in liquids, the data reduction can be in error because of contributions from overlooked effects such as self-focusing and stimulated Brillouin scattering.
3. Calculated gain coefficients — calculations are subject to the accuracy with which the Raman cross section $\partial\sigma/\partial\Omega$ and line width, Γ, are known.[19,20] Also, the cross section in Equation 13 is frequency integrated. (*Note:* $\partial\sigma/\partial\Omega \cdot \Gamma^{-1}$ is the peak cross section for a Lorentzian lineshape.)
4. Frequency shifts — it has been shown that Stokes emission in the transient stimulated case can be shifted in frequency relative to data obtained from spontaneous Raman scattering.[21]
5. Linewidth considerations — Equation 13 is valid for a single frequency pump laser of linewidth small compared to that of the Raman transition. As the pump laser linewidth approaches or exceeds that of the Raman transition the gain dependence on linewidth becomes dependent on scattering geometry and measurement technique.[22]
6. Frequency dependence of g — subject to the conditions for which $\partial\alpha/\partial q$ is frequency independent, g is proportional to Stokes frequency ω_s. However, close to a resonance, $\partial\sigma/\partial\Omega$ varies as $\omega_s^4 \cdot (\omega_a - \omega_\ell)^{-2}$ where ω_a is the frequency of the resonance.

ACKNOWLEDGMENTS

The author wishes to thank W. K. Bischel, R. L. Byer, H. Komine, and W. L. Smith for helpful discussions.

Table 1.4.1
OBSERVED SRS TRANSITIONS IN LIQUIDS

Substance	ω_R (cm^{-1})	Ref.
Bromoform	222	23
Tetrachloroethylene	448	24
Carbon tetrachloride[a]	460	25
Ethyl iodide	497	26
Hexafluorobenzene[a]	515	25
Bromoform	539	23
Chlorine	552	13
Methylene bromide	580	25
Trichloroethylene	640	23
Carbon disulfide	655	28
Ethylene bromide	660	27
Chloroform	667	23
o-Xylene	730	29
FC104[b]	757	30
Sulfur hexafluoride	775	31
α-Dimethylphenethylamine	836	32
Dioxane	836	23
Morpholine[a]	841	25
Thiophenol[a]	916	25
Nitromethane[a]	927	25
Deuterated benzene	944	33
Potassium dihydrogen phosphate	980	34
Cumene[a]	990	25
Pyridine	991	33
1,3-Dibromobenzene	992	24
Benzene	992	33
Aniline	997	35
Styrene	998	36
m-Toluidine[a]	999	25
Acetophenone	999	37
Bromobenzene	1000	35
Chlorobenzene[a]	1001	25
tert-Butylbenzene	1000	24
Benzaldehyde[c]	1001	24
Ethylbenzoate	1001	37
Benzonitrile	1002	35
Ethylbenzene	1002	29
Toluene	1004	33
Fluorobenzene	1012	38
γ-Picoline	1016	25
m-Cresol[a]	1029	25
m-Dichlorobenzene[a]	1034	25
1-Fluoro-2-chlorobenzene[d]	1034	24
Iodo-Benzene[a]	1070	25
Benzoyl chloride[a]	1086	25
Benzaldehyde[a]	1086	25
Anisole[a]	1097	25
Pyrrole[a]	1178	25
Furan[a]	1180	25
Nitrous oxide	1289	31
Styrene	1315	36
Nitrobenzene	1344	33
1-Bromonaphthalene	1363	33
1-Chloronaphthalene	1374	39

Table 1.4.1 (continued)
OBSERVED SRS TRANSITIONS IN LIQUIDS

Substance	ω_R (cm^{-1})	Ref.
2-Ethylnaphthalene	1382	24
m-Nitrotoluene[a]	1389	25
Carbon dioxide	1392	31
Quinoline[a]	1427	25
Bromocyclohexane	1438	26
Furan[a]	1522	25
Methyl salicylate[a]	1612	25
Cinnamaldehyde	1624	39
Styrene	1631	36
3-Methylbutadiene	1638	40
Pentadiene	1655	40
Isoprene	1792	32
1-Hexyne	2116	24
Dimethyl sulfoxide[c]	2128	41
o-Dichlorobenzene[a]	2202	25
Benzonitrile	2229	39
Acetonitrile	2250	26
1,2-Dimethylaniline[a]	2292	25
Nitrogen	2327	42
Hydrobromic acid	2493	30
Hydrochloric acid	2814	43
Methylcyclohexane[a]	2817	25
Methanol	2831	23
cis, trans, 1,3-Dimethylcyclohexane	2844	24
Tetrahydrofuran	2849	39
Cyclohexane	2852	33
cis-1,2-Dimethylcyclohexane	2853	24
α-Dimethylphenethylamine	2856	32
Dioxane	2856	23
Decahydronaphthalene	2860	30
Cyclohexane	2863	23
Cyclohexanone	2863	29
cis, trans-1,3-Dimethylcyclohexane	2866	24
cis, 1,4-Dimethylcyclohexane	2866	24
Cyclohexane	2884	23
Dichloromethane[a]	2902	25
Dimethyl sulfoxide	2916	41
Morpholine[a]	2902	25
Cargille 5610[f]	2908	30
2,3-Dimethyl-1,5-hexadiene	2910	24
Limonene	2910	32
o-Xylene	2913	29
1-Hexyne	2916	24
cis-2-Heptene	2916	24
2-Octene	2918	24
Acetonitrile	2920	30
Mesitylene	2920	32
2-Bromopropane	2920	24
Acetone	2921	29
Ethanol	2921	23
cis-1,2-Dimethylcyclohexane	2921	24
Carvone	2922	32
2-Chloro-2-methylbutane	2927	24
Dimethylformamide	2930	23

Table 1.4.1 (continued) OBSERVED SRS TRANSITIONS IN LIQUIDS

Substance	ω_R (cm^{-1})	Ref.
cis, trans-1,3-Dimethylcyclohexane	2926	24
m-Xylene	2933	29
1,2-Diethyl tartrate	2933	32
o-Xylene	2933	29
Piperidine	2933	29
1,2-Diethylbenzene	2934	24
1-Bromopropane	2935	24
Piperidine	2936	29
Tetrahydrofuran	2939	39
Decahydronaphthalene	2940	30
Piperidine	2940	29
Cyclohexanone	2945	29
2-Nitropropane	2945	24
1,2 Diethyl carbonate[a]	2955	25
1,2 Dichloroethane[a]	2956	25
trans-Dichloroethylene	2956	23
Methyl fluoride	2960	31
1-Bromopropane	2962	24
2-Chloro-2-methylbutane	2962	24
α-Dimethylphenethylamine	2967	32
Dioxane	2967	23
Methyl chloride	2970	31
Cyclohexanol[a]	2982	25
Cyclopentane[a]	2982	25
Cyclopentanol[a]	2982	25
Bromocyclopentane[a]	2982	25
o-Dichlorobenzene[a]	2982	25
p-Chlorotoluene[a]	2982	25
α-Picoline[a]	2982	25
p-Xylene	2988	29
o-Xylene	2992	29
Dibutyl-phthalate[a]	2992	25
1,1,1-Trichloroethane	3018	23
Ethylene chlorohydrin[a]	3022	25
Isophorone[a]	3022	25
Nitrosodimethylamine[a]	3022	25
Propylene glycol[a]	3022	25
Cyclohexane[a]	3038	25
Styrene	3056	36
Pyridine	3058	24
Benzene	3064	33
tert-Butylbenzene	3065	24
1-Fluoro-2-chlorobenzene	3082	24
Turpentine[a]	3090	25
Pseudocumene[a]	3093	25
Acetic acid[a]	3162	25
Acetonylacetone[a]	3162	25
Methyl methacrylate[a]	3162	25
γ-Picoline[a]	3182	25
Aniline	3300	35
Water[a]	3651	25

[a] Observed at low resolution.
[b] Product of 3M Co., St. Paul, Minn.
[c] 1:1 mixture with tetrachloroethylene.
[d] Very weak and diffuse.
[e] Deuterated.
[f] Product of Cargille Laboratories, Cedar Falls, N.J.

Table 1.4.2
OBSERVED SRS TRANSITIONS IN SOLIDS

Substance	ω_R (cm^{-1})	Ref.
Quartz	128	44
Lithium niobate	152	45
α-Sulfur	216	46
Lithium niobate	248	45
Bromine	295	46
Bromine	303	46
Quartz	466	44
α-Sulfur	470	46
Chlorine	543	47
Lithium niobate	628	45
Carbon disulfide	656	47
Potassium iodate	746	48
Potassium bromate[a]	780	48
Potassium periodate	790	48
Potassium bromate	798	48
Potassium chromate	844	48
Sodium molybdate[a]	884	48
Potassium dichromate	906	48
Calcium tungstate	911	39
Potassium dihydrogen phosphate	915	34
Ammonium vanadate	915	48
Sodium tungstate	915	48
Potassium perchlorate[a]	936	48
Potassium chlorate[a]	938	48
Ammonium sulfate[a]	975	48
Potassium sulfate[a]	985	48
Stilbene	997	49
Polystyrene	1001	23
Calcium nitrate	1050	48
Calcium nitrate tetrahydrate[a]	1052	48
Potassium nitrate	1060	48
Magnesium nitrate dihydrate[a]	1060	48
Ammonium nitrate[a]	1062	48
Magnesium nitrate hexahydrate[a]	1063	48
Sodium nitrate	1075	48
Calcite	1086	46
Diamond	1332	46
Naphthalene	1380	39
Anthracene	1403	50
Stilbene	1591	49
Potassium thiocyanate	1040	48
Potassium ferricyanide[a]	2100	48
Triglycine sulfate	2422	25
Triglycine sulfate	2702	25
Triglycine sulfate	3022	25
Polystyrene	3054	23

[a] 77 K (not observed at 293 K).

Table 1.4.3
OBSERVED SRS TRANSITIONS IN GASES

Substance	ω_R (cm^{-1})	Ref.
Barium vapor[a]	IR[b]	51
Cesium vapor[a]	IR[b]	52,53
Hydrogen fluoride	FIR[b]	54
Potassium vapor[a]	IR[b]	53,55
Rubidium vapor[a]	IR[b]	56
p-H_2	354	57,58
Carbon tetrafluoride	980	59
Oxygen	1552	29
Nitrogen	2331	60
Potassium vapor	2721	61
Methane	2916	62
Deuterium	2991	62
Hydrogen deuteride	3628	63
Hydrogen	4155	62

[a] Stimulated electronic Raman scattering (SERS).

[b] Generally tunable transitions in the infrared (IR) and far infrared (FIR).

Table 1.4.4
CALCULATED[a] AND MEASURED GAIN (g_s) FACTOR FOR STIMULATED RAMAN TRANSITIONS IN LIQUIDS (*l*), GASES (g), AND SOLIDS (s)

Substance	Pump wavelength (nm)	Frequency (cm^{-1})	Linewidth (cm^{-1})	Gain (g_s) × 10^9 (cm/W)	g_s calc. × 10^9 (cm/W)	g_s at 532 nm[b]	g_s Relative to $C_6H_6(l)$[c]	Ref.
Benzene (*l*)	532	992	2.15	5.5[d]		5.5	1.0	64
	532	992	2.15	5.5[e]		5.5		65
	532	992	2.15	4.3 ± 0.9[f]		4.3 ± 0.9		66
	694.3	992	2.15		2.8	5.9[g,h]		67
Oxygen (*l*)	694.3	1552	0.117	16.0 ± 0.5	14.5 ± 0.4	21.5	3.9	67
Nitrogen (*l*)	532	2327	0.067	30.0	24.0 ± 7.0	30.0	5.4	68
	694.3	2327	0.067	16.0 ± 0.55	17.0 ± 0.5	21.5	3.9	67
	1060	2327	0.067	10.0	9.0 ± 3.0	23.2	4.2	69
	1315	2327	0.067	5.0 ± 2.0	6.0 ± 2.0	15.6 ± 6.2	2.9	70
Carbon disulfide (*l*)	694.3	655.6	0.50	24.0		32.2	5.9	67
Methanol (*l*)	694.3	2837	18	0.4		0.53	0.10	71
Carbon tetrachloride (*l*)	597.6	458		1.3		1.5	0.27	72
	532						0.12[i]	73
Acetone (*l*)	532						0.20[i]	73
Cyclohexane (*l*)	532						0.25[i]	73
Hydrogen[j] (g)	694.3	4155			1.5[k]	2.14		74
	694.3	4155		1.9 ± 0.3[k]	1.5[k]	2.7		74
	694.3	4155		1.5[k]		2.1		75
	353	4155		5.0[k]	5.7[k]			63
	308	4155			6.7[k]			63
	10600	354			0.07	0.09		76
Hydrogen deuteride (g)	353	3628		0.2				63
Nitrogen (g, 1 atm)	694.3	2330.7		0.0022		0.0030		60
	694.3	2330.7	0.075		0.0027[l]	0.0038		77
Carbon tetrafluoride (g, 500 atm)	532	980	0.17	24.6 ± 2[m]		2.46 ± 2[m]		59
Calcite (s)	532	1086	1.2	5.5		5.5		78
Fused quartz (s)	495.4	420		0.017		0.016		79

Potassium dihydrogen phosphate (s)	532	918	18	0.21 ± 0.05[f]	0.21 ± 0.05	66

[a] Only those supporting measured values or of special interest are given.

[b] Except where noted $g_s/532 = \frac{\omega_s/532}{\omega_s/\text{meas}} g_s/\text{meas}$.

[c] For qualitative use only; see above.
[d] Estimated from stimulated threshold.
[e] Estimated from stimulated conversion.
[f] Direct measurements with single-frequency lasers.
[g] Extrapolated using $I \alpha \omega_s^4 (\omega_a - \omega_\ell)^{-2}$; $\omega_a = 39{,}000\ \text{cm}^{-1}$.
[h] See Reference 80 for discussion of peak cross-section measurements.
[i] Relative threshold measurement.
[j] For detailed analysis of H_2 cross-section pump frequency dependence see References 81 and 82.
[k] Pressure independent gain; Q(1) transition.
[l] Corrected using linewidth of Reference 60; Q(6) transition.
[m] Transient gain.

REFERENCES

1. **Javan, A.,** Transitions a plusieurs quanta et amplification maser dans les systemes a deux niveaux, *J. Phys. Radium*, 19, 806, 1958.
2. **Woodbury, E. J. and Ng, W. K.,** Ruby laser operations in the near IR, *Proc. IRE*, 50, 2367, 1962.
3. **Stolen, R. H., Ippen, E. P., and Tynes, A. R.,** Raman oscillation in glass optical waveguide, *Appl. Phys. Lett.*, 20, 62, 1972.
4. **Owyoung, A.,** Coherent Raman gain spectroscopy using CW laser sources, *IEEE J. Quantum Electron.*, QE-14, 192, 1978.
5. **Levine, B. F. and Bethea, C. G.,** Ultrahigh sensitivity stimulated Raman gain spectroscopy, *IEEE J. Quantum Electron.*, QE-16(1), 85, 1980.
6. **Laubereau, A. and Kaiser, W.,** Vibrational dynamics of liquids and solids investigated by picosecond light pulses, *Rev. Mod. Phys.*, 50, 607, 1978.
7. **Byer, R. L. and Herbst, R. L.,** Parametric oscillation and mixing, in *Topics in Applied Physics Series*, Vol. 16, Shen, Y. R., Ed., Springer-Verlag, Berlin, 1977, 81.
8. **Penzkofer, A., Laubereau, A., and Kaiser, W.,** High intensity Raman interactions, *Prog. Quantum Electron.*, 6, 55, 1979.
9. **Levenson, M. D. and Song, J. J.,** Coherent Raman spectroscopy, in *Coherent Nonlinear Optics, Recent Advances*, Springer-Verlag, Berlin, 1980, 293.
10. **Bloembergen, N.,** Stimulated Raman and two-photon absorption processes in molecules, *J. Mol. Struct. (Netherlands)*, 59, 331, 1980.
11. **Maier, M.,** Applications of stimulated Raman scattering, *Appl. Phys.*, 11(3), 209, 1976.
12. **Shen, Y. R.,** Recent advances in nonlinear optics, *Rev. Mod. Phys.*, 48, 1, 1976.
13. **Kaiser, W. and Maier, M.,** Stimulated Rayleigh, Brillouin and Raman spectroscopy, in *Laser Handbook*, Vol. 2, Arrecchi, F. T. and Schultz-Dubois, E. O., Eds., North-Holland, Amsterdam, 1972, 1078.
14. **Grasyuk, A. Z.,** Raman lasers (review), *Sov. J. Quantum Electron. U.S.A.*, 4(3), 269, 1974.
15. **Bloembergen, N.,** The stimulated Raman effect, *Am. J. Phys.*, 35, 989, 1967.
16. **Placzek, G.,** *Handbuch der Radioilogie*, 2nd ed., Maric, E., Ed. Academische Verlagsgesellschaft, Leipzig, 1934, 206.
17. **Schrötter, H. W. and Klöckner, H. W.,** Raman scattering cross sections in gases and liquids, in *Topics in Applied Physics Series*, Vol. 2, Weber, A., Ed., Springer-Verlag, Berlin, 1979, 123.
18. **Carman, R. L., Mack, M. E., Shimizu, F., and Bloembergen, N.,** Forward picosecond stokes-pulse generation in transient stimulated Raman scattering, *Phys. Rev. Lett.*, 23, 1327, 1969.
19. **Skinner, J. G. and Nilsen, W. G.,** Absolute Raman scattering cross-section measurement of the 992 cm^{-1} line of benzene, *J. Opt. Soc. Am.*, 58, 113, 1968.
20. **Colles, M. J. and Griffiths, J. E.,** Relative and absolute Raman scattering cross sections in liquids, *J. Chem. Phys.*, 56, 3383, 1972.
21. **Zinth, W. and Kaiser, W.,** Frequency shifts in stimulated Raman scattering, *Opt. Commun.*, 32, 507, 1980.
22. **Akmanov, S. A., D'yakov, Yu. E., and Pavlov, L. I.,** Statistical phenomena in Raman scattering stimulated by a broad-band pump, *Sov. Phys. JETP*, 39, 349, 1974.
23. **Kern, S. and Feldman, B.,** *Stimulated Raman Emission*, Vol. 3, Massachusetts Institute of Technology, Lincoln Laboratory, Bedford, Mass., 1974, 18.
24. **Barrett, J. J. and Tobin, M. C.,** Stimulated Raman emission frequencies in 21 organic liquids, *J. Opt. Soc. Am.*, 56, 129, 1966.
25. **Martin, M. D. and Thomas, E. L.,** Infrared difference frequency generation, *IEEE J. Quantum Electron.*, QE-2, 196, 1966.
26. **El-Sayed, M. A., Johnson, F. M., and Duardo, J.,** A comparative study of the coherent Raman processes using the ruby and the second harmonic neodymium giant-pulsed lasers, *J. Chim. Phys.*, 1, 227, 1967.
27. **Prasada Rao, T. A. and Seetharaman, N.,** Amplification of stimulated Raman scattering by a dye, *Ind. J. Pure Appl. Phys.*, 13, 207, 1975.
28. **Giordmaine, J. A. and Howe, J. A.,** Intensity-induced optical absorption cross section in CS_2, *Phys. Rev. Lett.*, 11, 207, 1963.
29. **Geller, M., Bortfeld, D. P., and Sooy, W. R.,** New Woodbury-Raman laser materials, *Appl. Phys. Lett.*, 3, 36, 1963.
30. **Smith, W. L. and Milanovich, F. P.,** Lawrence Livermore National Laboratory, Livermore, CA, private communication, 1983.
31. **Maple, J. R. and Knudtson, J. T.,** Transient stimulated vibrational Raman scattering in small molecule liquids, *Chem. Phys. Lett.*, 56, 241, 1978.
32. **Wright, J. K., Carmichael, C. H. H., and Brown, B. J.,** Narrow linewidth output from a Q-switched Nd^{3+}/glass laser, *Phys. Lett.*, 16, 264, 1965.

33. **Eckardt, G., Hellwarth, R. W., McClung, F. J., Schwarz, S. E., and Weiner, D.,** Stimulated Raman scattering from organic liquids, *Phys. Rev. Lett.*, 9, 455, 1962.
34. **Srivastava, M. K. and Crow, R. W.,** Raman susceptibility measurements and stimulated Raman effect in KDP, *Opt. Commun.*, 8, 82, 1973.
35. **Maker, P. D. and Terhune, R. W.,** Study of optical effects due to an induced polarization third order in the electric field strength, *Phys. Rev.*, 137, A801, 1965.
36. **Bortfeld, D. P., Geiller, M., and Eckhardt, G.,** Combination lines in the stimulated Raman spectrum of styrene, *J. Chem. Phys.*, 40, 1770, 1964.
37. **Orlovich, V. A.,** Measurement of the coefficient of stimulated Raman scattering in organic liquids with the aid of an amplifier with transverse pumping, *Zh. Prikl. Spektrosk.*, 23, 224, 1975.
38. **Calvieilo, J. A. and Heller, Z. H.,** Raman laser action in mixed liquids, *Appl. Phys. Lett.*, 5, 112, 1964.
39. **Eckhardt, G.,** Selection of Raman laser materials, *IEEE J. Quantum Electron.*, QE-2, 1, 1966.
40. **Subov, V. A., Sushchinskii, M. M., and Shuvalton, I. K.,** Investigation of the excitation threshold of induced Raman scattering, *J. Exp. Theor. Phys. U.S.S.R.*, 47, 784, 1964.
41. **Decker, C. D.,** High-efficiency stimulated Raman scattering/dye radiation source, *Appl. Phys. Lett.*, 33, 323, 1978.
42. **Stoicheff, B. P.,** Characteristics of stimulated Raman radiation generated by coherent light, *Phys. Lett.*, 7, 186, 1963.
43. **Cotter, D., Hanna, D. C., and Wyatt, R.,** A high power, widely tunable infrared source based on stimulated electronic Raman scattering in caesium vapour, *Opt. Commun.*, 16, 256, 1976.
44. **Tannenwald, P. E. and Thaxter, J. B.,** Stimulated Brillouin and Raman scattering in quartz at 2.1 to 293 Kelvin, *Science*, 134, 1319, 1966.
45. **Gelbwachs, J., Pantell, R. H., Puthoff, H. E., and Yarborough, J. M.,** A tunable stimulated Raman oscillator, *Appl. Phys. Lett.*, 14, 1969.
46. **Eckhardt, G., Bortfeld, D. P., and Geller, M.,** Stimulated emission of stokes and anti-stokes Raman lines from diamond, calcite and α-sulfur single crystals, *Appl. Phys. Lett.*, 3, 137, 1963.
47. **Aleksandrov, I. V., Bobovitch, Ya. S., Bortkevich, A. V., and Tsenter, M. Ya.,** Raman scattering in crystalline chlorine and bromine, *Opt. Spektrosk.*, 36, 150, 1974.
48. **Kondilenko, I. I., Korotkov, P. A., and Maly, V. I.,** Temperature dependence of SRS thresholds for some salts of inorganic acids in the crystalline phase, *Opt. Commun.*, 10, 50, 1974.
49. **Weinberg, D. L.,** Stimulated Raman emission in crystals and organic liquids, *Mass. Inst. Technol. Lincoln Lab. Solid-State Res. Rep.*, 3, 31, 1965.
50. **Avanesyan, O. S., Benderskii, V. A., Brikenshtein, V. K. H., Broude, V. L., Lavrushko, A. G., Tartakovskii, I. I., and Filippov, P. V.,** Characteristics of stimulated light generation and stimulated Raman scattering in anthracene crystals, *Sov. J. Quantum Electron.*, 7, 403, 1977.
51. **Carlsten, J. L. and Dunn, P. C.,** Stimulated stokes emission with a dye laser: intense tunable radiation in the infrared, *Opt. Commun.*, 14, 8, 1975.
52. **Cotter, D., Hanna, D. C., Kärkkäinen, P. A., and Wyatt, R.,** Stimulated electronic Raman scattering as a tunable infrared source, *Opt. Commun.*, 15, 143, 1975.
53. **Sorokin, P. P., Wynne, J. J., and Landkard, J. R.,** Tunable coherent IR source based upon four-wave parametric conversion in alkali metal vapors, *Appl. Phys. Lett.*, 22, 342, 1973.
54. **DeMartino, A., Frey, R., and Pradere, F.,** Tunable far infrared generation in hydrogen fluoride, *Opt. Commun.*, 27, 262, 1978.
55. **Cotter, D., Hanna, D. C., Kärkkäinen, P. A., and Wyatt, R.,** Stimulated electronic Raman scattering as a tunable infrared source, *Opt. Commun.*, 15, 143, 1975.
56. **May, P., Bernage, P., and Bocquet, H.,** Stimulated electronic Raman scattering in rubidium vapour, *Opt. Commun.*, 29, 369, 1979.
57. **Byer, R. L. and Trutna, W. R.,** 16-μm Generation by CO_2-pumped rotational Raman scattering in H_2, *Opt. Lett.*, 3, 144, 1978.
58. **Rabinowitz, P., Stein, A., Brickman, R., and Kaldor, A.,** Efficient tunable H_2 Raman laser, *Appl. Phys. Lett.*, 35, 739, 1979.
59. **Pochon, E.,** Determination of the spontaneous Raman linewidth of CF_4 by measurements of stimulated Raman scattering in both transient and steady states, *Chem. Phys. Lett.*, 77, 500, 1981.
60. **Kinkaid, B. E. and Fontana, J. R.,** Raman cross-section determination by direction stimulated Raman gain measurements, *Appl. Phys. Lett.*, 28, 12, 1975.
61. **Rokni, M. and Yatsiv, S.,** Resonance Raman effects in free atoms of potassium, *Phys. Lett.*, 24, 277, 1967.
62. **Minck, R. W., Terhune, R. W., and Rado, W. G.,** Laser-stimulated Raman effect and resonant four-photon interactions in gases H_2, D_2, and CH_4, *Appl. Phys. Lett.*, 3, 181, 1963.
63. **Komine, H.,** Northrop Corp., Palos Verdes, Calif., private communication, 1983.
64. **Ausenegg, F. and Deserno, V.,** Stimulated Raman scattering excited by light of 5300 Å, *Opt. Commun.*, 2, 295, 1970.

65. **Colles, M. J.,** Efficient stimulated Raman scattering from picosecond pulses, *Opt. Commun.*, 1, 169, 1969.
66. **Smith, W. L., Milanovich, F. P., and Henesian, M.,** Lawrence Livermore National Laboratory, private communication, 1983.
67. **Grun, J. B., McQuillan, A. K., and Stoicheff, B. P.,** Intensity and gain measurements on the stimulated Raman emission in liquid O_2 and N_2, *Phys. Rev.*, 181, 61, 1969.
68. **Akmanov, S. A., Zhdanov, B. V., Kovrigin, A. I., and Pershin, S. A.,** Effective stimulated scattering in the ultraviolet and dispersion of gain in the 1.06-0.26 μ band, *JETP Lett.*, 15, 185, 1972.
69. **Grasyuk, A. Z., Efinkov, V. F., Zubarev, I. G., Mishin, V. I., and Smirnov, V. G.,** Laser based on Raman scattering in liquid nitrogen, *JETP Lett.*, 8, 291, 1968.
70. **Vakhonev, M. B., Volkov, V. N., Grasyuk, A. Z., and Kirkin, A. N.,** Determination of the gain in stimulated Raman scattering under spatially inhomogeneous pumping conditions, *Sov. J. Quantum Electron.*, 6, 1369, 1976.
71. **Reinhold, I. and Maier, M.,** Gain measurements of stimulated Raman scattering using a tunable dye laser, *Opt. Commun.*, 5, 31, 1972.
72. **Görner, H., Maier, M., and Kaiser, W.,** Raman gain in liquid core fibers, *J. Raman Spectrosc.*, 2, 363, 1974.
73. **Gazengel, J., Xuan, N. P., and Rivoire, G.,** Stimulated Raman scattering thresholds for ultra-short excitation, *Opt. Acta*, 26, 1245, 1979.
74. **Bloembergen, N., Lallemand, B. P., Pine, A., and Simova, P.,** Controlled stimulated Raman amplification and oscillation in hydrogen gas, *IEEE J. Quantum Electron.*, QE-3, 197, 1967.
75. **Hagenlocker, E. E., Minck, R. W., and Rado, W. G.,** Effects of phonon lifetime on stimulated optical scattering in gases, *Phys. Rev.*, 154, 226, 1967.
76. **Byer, R. L.,** A 16-μm source for laser isotope enrichment, *IEEE J. Quantum Electron.*, YE-12, 732, 1976.
77. **Minck, R. W., Hagenlocker, E. E., and Rado, W. G.,** *Consideration and Evaluation of Factors Influencing the Stimulated Optical Scattering in Gases*, Scientific Laboratory, Ford Motor Company, Dearborn, Mich., SC66-24, 1966.
78. **Bisson, G. and Mayer, G.,** Effects Raman stimule's dans la cakite, *Cr. Acad. Sci. (Paris)*, 265, 397, 1967.
79. **Ippen, E. P.,** Low-power quasi-cw Raman oscillator, *Appl. Phys. Lett.*, 16, 303, 1970.
80. **Owyoung, A. and Percy, P. S.,** Precise characterization of the Raman nonlinearity in benzene using nonlinear interferometry, *J. Appl. Phys.*, 48, 674, 1977.
81. **Bischel, W. K. and Black, G.,** Wavelength dependence of Raman scattering cross sections from 200-600 Nm, *Proc. Top. Conf. Eximer Lasers*, in press.
82. **Bischel, W. K. and Dyer, M. J.,** Temperature dependence of the Raman linewidth in H_2, *Proc. Top. Conf. Eximer Lasers*, in press.

Section 2
Radiation Damage

2. RADIATION DAMAGE IN OPTICALLY TRANSMITTING CRYSTALS AND GLASSES

R. T. Williams* and E. Joseph Friebele

2.1. INTRODUCTION

This section summarizes effects of energetic particles and short wavelength photons on the optical properties of transparent materials. The damage to be considered is mostly atomic in scale, as distinguished from dielectric breakdown, thermal fracture, or other processes of macroscopic damage depending on intensity or fluence of laser radiation. In the sense defined, radiation damage to optical materials in laser systems can be caused by a variety of factors, including radiation in the application environment, electron-beam pump sources, or short wavelength laser radiation itself. On the atomic scale, radiation damage may involve a change in the charge state of a defect in the material, atomic or ionic displacement within the solid, retention of impinging ions, sputtering of the surface, or nuclear transmutation of constituent atoms. Electronic (charge rearrangement) damage is normally associated with ionizing radiation, which includes X-rays, gamma rays, and energetic electrons or ions. In some materials, UV or even visible radiation can have the same effect, which is to produce mobile electrons or holes in a material containing deep traps. Atomic displacement can be caused either by direct momentum transfer from an impinging particle or through essentially photochemical utilization of electronic excitation energy imparted by the radiation.

Displacement damage by elastic collision is the more general of the two mechanisms, occurring in any material if the impinging particle carries sufficient momentum. The criterion of sufficient momentum follows from the requirement that, subject to conservation of energy and momentum in the collision, the recoil atom must have received enough energy to break bonds as required and move to occupy a defect (e.g., interstitial) position in a strained lattice. Photochemical mechanisms of displacement damage occur less generally and can usually be identified with particular classes of materials. When a photochemical channel is operative, however, it is often far more effective in terms of defect yield for a given radiation dose than is the pure collisional mechanism. Furthermore, the energy threshold for typical photochemical channels is of order 10 eV, as opposed to photon or electron energy thresholds of order 10^5 eV for collisional displacement.

During irradiation, charge carriers and a variety of atomic defects are typically produced, and these will be mobile to varying degrees. During their periods of mobility, the intermediate products may annihilate by recombination, combine to form a new defect or charge state, or become associated with impurities, dislocations, and other defects. Both the type and amount of damage may depend dramatically on the type of radiation, ambient temperature, the thermal history of the material, impurity content, mechanical strain, total previous dose and dose rate, and time after irradiation. The size of this parameter space renders a comprehensive tabulation of radiation effects in arbitrary optical materials for arbitrary conditions virtually impossible to compile or use, even if such comprehensive data were available. Therefore, we have tried to present the available data in the physical context of atomic and electronic defects, accompanied by discussions that we hope will facilitate interpolations and extrapolations where there are gaps in the data. On the other hand, this summary is not intended as a review of the physics of defects. For detailed information, the original literature cited herein or reviews such as References 1 to 9 should be consulted.

* This work was completed while Dr. Williams was affiliated with the Naval Research Laboratory, Washington, D.C.

This chapter is primarily concerned with the transparency of optical materials in a radiation environment. Thus, optical absorption and related luminescence are the principal defect characteristics that are cataloged in the following summaries. Since most applications can be expected to involve operation near room temperature, we have chosen to present room-temperature data where possible. When behavior at other temperatures is significantly different or better understood, those properties are described as well. In many cases there is no choice but to use low-temperature data, since researchers measuring defect properties have usually found it advantageous to work at cryogenic temperatures. Data on annealing and bleaching of defects are presented. Damage thresholds and efficiencies of defect production are given when available. Data on short-lived defects causing transient optical absorption after pulsed irradiation are included.

Crystalline materials are presented approximately in descending order of ionicity, grouped as halides, oxides, other chalcogenide and III-V compounds, and Group IV elements. Glasses are then presented in the order of silicon dioxide, doped SiO_2, silicate, borate, phosphate, and fluoride glasses. Because of its close relationship to SiO_2-based glasses, quartz is discussed at the start of the glass section rather than with the other oxide crystals. The most important or otherwise illustrative materials for which data are available are described in text with accompanying figures.

Data bases that were used in compiling this summary included INSPEC, NTIS, ''Radiation Damage of Laser Materials: 1964-July, 1981 (Citations from the NTIS Data Base),'' and bibliographies given in the reviews listed as References 1 to 9. An especially valuable resource was consultation with colleagues doing research on radiation effects in solids.

BASIC TERMS AND FORMULAS

Smakula Equation

The integrated optical absorption coefficient for transitions between two electronic states, i and j, of a defect is related to the number, N, of defects per unit volume by the generalized Smakula equation:[3,10]

$$Nf_{ij} = \frac{nmc}{2\pi^2 e^2 \hbar} \left[\frac{E_0}{E_{eff}}\right]^2 \int_0^\infty \alpha_{ij}(E)\, dE \qquad (1)$$

where E is the transition energy, n is the refractive index (assumed approximately constant) of the solid medium at the transition frequency, and m, e, and c are the electron mass, electron charge, and velocity of light, respectively. E_{eff}/E_0 is the effective field ratio, which takes into account the effect of a polarizable medium on the optical electric field at the defect position. The quantity f_{ij} is the oscillator strength of the i,j electronic transition. The oscillator strength has a value between 0 and 1 if a single electron participates, but can be significantly larger than 1 for a given defect transition if more electrons can participate. Trapped-hole centers and defect-perturbed exciton creation in alkali halides are examples in which f can exceed unity. The effective field ratio appropriate to a particular defect is usually known only approximately. If the defect wavefunctions are very diffuse, E_{eff}/E_0 is close to unity, whereas for compact centers the ratio can be reasonably approximated by the classical Lorentz expression:

$$\frac{E_{eff}}{E_0} = \frac{n^2 + 2}{3} \qquad (2)$$

Substituting the Lorentz local field approximation and assuming a Gaussian band shape, one obtains:

$$Nf_{ij} = 0.87 \times 10^{17} \frac{n}{(n^2 + 2)^2} \alpha_{ij} (\max) W_{ij} \tag{3}$$

where α_{ij} (max) is the absorption coefficient (cm^{-1}) at the maximum of the band, W_{ij} is the full width at half-maximum in electron volts, and N is the number of defects per cubic centimeter.

When it is possible to obtain an independent measure of defect concentration, e.g., by electron paramagnetic resonance (EPR) or chemical methods, Equation 3 allows an experimental determination of oscillator strength. More often, the oscillator strength is known or can be estimated, and Equation 3 gives a direct (but typically 50% uncertain) relation between defect concentration and the area under an absorption band.

Radiation Dosage and Units

Among the reproduced figures and literature citations in this section, the units roentgen, rad, MeV/cm^3, gray, neutrons/cm^2, and nvt can all be found in reference to radiation exposure or dose. In consulting the literature on radiation damage, it is clearly useful to be able to make interconversions among units where appropriate, and a summary here may be helpful.

The roentgen (R) is defined as the unit of X-ray or gamma-ray exposure which would produce 1 electrostatic unit of ionic charge in 1 cm^3 of dry air at 0°C and standard atmospheric pressure. This can equivalently be expressed as 1.61×10^{11} ion pairs produced per gram of dry air. Since an average of 33.5 eV is expended in producing each ion pair in air, the energy deposited on absorption of 1 R of X-rays in 1 g of air is about 86 erg.[11] For perspective with relation to personnel dosimetry, we note that exposure of soft biological tissue to 1 R of X-rays results in an energy deposition of approximately 97 erg/g of material. Clearly, in other materials the absorbed energy density will be somewhat different.

The rad is a unit of dosage defined directly as the absorbed dose of any ionizing radiation which deposits 100 erg/g of absorbing material.

$$1 \text{ rad} = 100 \text{ erg/g (absorbed)} \tag{4}$$

Note that exposure characterizes the incident radiation, while dose characterizes absorbed radiation. In SI units, the unit of absorbed dose is the gray (Gy):

$$1 \text{ Gy} = 100 \text{ rad} = 1 \text{ j/kg (absorbed)} \tag{5}$$

Another unit of absorbed dose which is encountered is MeV/cm^3.

If radiation interacts primarily by photoproduction or Compton scattering from electrons in the material, the absorbed dose (energy/mass unit) received from a given exposure will scale approximately as the ratio of the mean atomic number to mean atomic weight for the material considered. Thus, the relation:

$$\text{dose (rads)} \approx \frac{(\bar{Z}/\bar{A})}{(\bar{Z}/\bar{A})_{air}} \times 0.86 \times \text{exposure (R)} \tag{6}$$

provides a rough rule for conversion of X-ray exposure to absorbed dose as long as the photon energy is much higher than characteristic binding energies in the solid. Otherwise, explicit X-ray or UV absorption coefficients characteristic of bound electrons must be used to compute dose. For particles, the appropriate range-energy relation must be used.

The rad or gray can be used to specify radiation exposure, but then the detector material must be stated. The rad (Si) is an exposure unit that would produce a dose of exactly 1 rad in silicon, but a somewhat different dose in a material with different Z/A (different absorption coefficient or stopping power). It is worth noting that Z/A does not change radically from material to material, so as a *rough* approximation the exposure in roentgens will be close to the dose in rads for high-energy X-rays, gamma rays, and charged particles.

Neutron exposure is not particularly well described by the units defined for ionizing radiation. Neutrons can produce some ionization by secondary processes, but interact predominantly with nuclei. Furthermore, the neutron reaction cross section may vary dramatically with target nucleus type and neutron energy. Thus, the correspondence of dosage and exposure is not as smooth a function of atomic number as it is for high-energy ionizing radiation. It has been customary in much of the literature on radiation damage to specify integrated neutron flux, neutron energy (spectrum), and target material separately, rather than a single absorbed dose. The integrated neutron flux is stated most simply as neutrons per square centimeter. One often sees the notation ''nvt'', which simply signifies that an integrated flux is stated, implying neutrons per square centimeter.

Heavy charged particles interact significantly with both nuclei and electrons. However, interactions with the electrons dominate in most cases, and the radiation dosage units defined above for X-rays and electrons serve as well for most proton or ion irradiations. Note that the dose received from a given integrated particle flux is much higher for ions than for electrons of equivalent kinetic energy. The stopping ranges of charged particles of different energy, mass, and charge are discussed in References 1 and 4, and must be taken into account when converting from integrated charged particle flux to ionizing radiation dose.

Displacement Threshold

Atomic (or ionic) displacement damage will occur in any material if the impinging particle can transfer kinetic energy to a lattice atom in excess of the threshold displacement energy, T_d, characteristic of the particular atom and the surrounding solid. The energy which can be transferred depends on the incident particle kinetic energy, E, incident and target particle rest masses, m and M, and on collision angle, as dictated by conservation of energy and momentum.[1,3,4] The maximum energy transfer, in a relativistic head-on collision, is

$$T_{max} = 2(E + 2mc^2)E/Mc^2 \qquad (7)$$

The incident-particle threshold energy for damage is that value of E for which $T_{max} = T_d$. A collision cascade yielding locally high defect densities can occur if the displaced atoms themselves have enough kinetic energy to initiate subsequent displacements. The mechanisms of atomic displacement by energetic particles have been reviewed by Seitz and Koehler,[4] Cobert and Bourgoin,[2] and Sonder and Sibley,[1] to name a few.

The displacement energy, T_d, includes the energies involved in breaking bonds between the struck atom and its neighbors, in overcoming potential barriers due to ion-core repulsion between the moving ion and neighbors encountered along the path of motion, and strain energy of the lattice in accommodating the new defect. The displacement energy can be different for motion along different crystallographic directions. Since roughly the same bonds are broken in displacement of either element in a binary compound, values of T_d for cations and anions in a given solid are usually about the same. However, due to different efficiencies of energy transfer from an incident particle to atoms of different masses (Equation 7), the incident-particle threshold energies for displacement of cations and anions in a given material may be quite different. For a number of materials, T_d is about twice the bond energy of the solid.[2]

Electron-Hole Pair Creation Energies

Photochemical lattice damage and charge-rearrangement reactions generally occur in proportion to the number of electron-hole pairs created by passage of an energetic charged particle or photon. The discussion of alkali halides to follow illustrates one such photochemical mechanism of lattice damage. Clearly, the bandgap or lowest exciton energy constitutes the threshold for generation of an electron-hole (e-h) pair. (Release of charge from trap levels in the gap is considered separately.) When the incident charged particle or photon has energy much larger than the bandgap, E_g, multiple e-h pairs are produced. Empirically, the average energy which is expended in creation of each e-h pair by high-energy ionizing radiation is

$$E_{e\text{-}h} \approx 2.8\ E_g \tag{8}$$

This rule arose originally out of observations on semiconductors, but its applicability to a large number of insulators has also been demonstrated.[12] It allows the estimation of e-h pair density corresponding to a given radiation dose.

Defect Nomenclature

Table 2.2.1 lists labels which have been assigned to various point defects in insulator and semiconductor crystals. Tables 2.3.2 and 2.3.5, presented later, deal with SiO_2 and silicate glasses. The left and right sides of Table 2.2.1 reflect the different terminologies which developed early in the studies of halide crystals and elemental semiconductors, respectively. Some ambiguity arises in the materials of medium ionicity, particularly II-VI compounds, which are the meeting ground of the two nomenclature systems.

Some labels have changed over the years to fit a more codified scheme as the relations to other defects have become more clearly appreciated. The system proposed for polar crystals by Sonder and Sibley[1] has been adopted here. The greatest possibility for confusion may center on the letter V, used in reference to vacancies in nonpolar or moderately polar semiconductors, and also used initially as a generic label for hole traps in alkali halides, many of which turn out not to involve vacancies. It has been recommended that in ionic crystals the term "V center" should refer specifically to cation vacancy centers.[1] Thus, for example, the halogen di-interstitial, formerly termed the V_4 center, will be referred to here as the I_2 center. Charge states relative to the normal charge of the lattice site occupied by a defect are shown by superscripts. Defect aggregates are generally indicated with subscripts. For example, a nearest-neighbor pair of F centers, formerly termed an M center, is referred to here as an F_2 center in keeping with Reference 1.

2.2 CRYSTALS

HALIDE CRYSTALS

Alkali Halides

Radiation damage in the alkali halides is principally a low-energy photochemical process initiated by the excitation of valence electrons across the optical bandgap, whether the excitation is by high-energy particles or UV photons. (See References 1 and 13 to 16 for reviews of photochemical defect production.) The primary photochemical defect pair in pure alkali halides is a halide ion vacancy (with trapped electron) and an interstitial halogen atom resulting from the vacancy formation. Halogen interstitials are significantly more mobile than the vacancies and do not persist as isolated defects at room temperature. Recombination of vacancies and interstitials exerts a dominant influence on the observed efficiency of producing stable defects.

A negative-ion vacancy on which one electron is trapped is termed an ''F center'', with optical absorption bands corresponding to transitions of the bound electron. The lowest-energy transition, denoted 1s — 2p in an effective mass model — has an oscillator strength near unity and, along with the β transition of the F center discussed below, typically dominates the spectrum of radiation-induced defects at room temperature (Figures 2.2.1 to 2.2.4).* Because defect optical transitions are strongly coupled to lattice vibrations in the alkali halides, the lowest electronic transition of the F center is broadened into a smooth, nearly Gaussian absorption band a few tenths of an electronvolt wide at room temperature. Higher-energy transitions terminating in final states degenerate with the conduction band comprise a weak shoulder on the short-wavelength side of the F band.[3] A localized excitonic transition on the ion shell surrounding the F center gives rise to β band absorption (Figure 2.2.2) situated close to the fundamental absorption edge.[18,20,21] A β band should accompany the F band in all alkali halides, and their respective absorption strengths should be comparable ($f_\beta \approx f_F \approx 1$), though measurements have been made in only a few cases. The bare ionized negative-ion vacancy, an F^+ center, exhibits a similar perturbed-exciton transition (α band) at slightly lower energy.[18] Almost any defect can be expected to induce such perturbed-exciton transitions near the UV absorption edge. The F center may trap a second electron at low temperature, giving rise to the F^- center which has a broad visible/IR absorption band but is not stable at room temperature.[3]

Most of the optical absorption observed on the long-wavelength side of the F band in pure crystals at room temperature is attributable to F-aggregate defects.[22] The spectra of KCl and LiF in Figures 2.2.1 and 2.2.3 serve as examples. Aggregates of F centers may occur by the initial formation of vacancies on adjacent lattice sites, or by subsequent association of vacancies through thermally or optically activated migration. The symbols F_2 (formerly M) and F_3 (formerly R) denote aggregates of two and three F centers, respectively, located on adjacent lattice sites. The aggregates are optically anisotropic and exhibit two or more absorption bands with distinct polarization properties. If defect migration is prevented, as at low temperature in darkness, the relation of F-center concentration, [F], and F-aggregate concentration, $[F_n]$, is determined statistically, so that

$$[F_2] \propto [F]^2, \qquad [F_3] \propto [F]^3, \quad \text{etc.}$$

The growth of F and F-aggregate absorption with radiation dose is illustrated in Figures 2.2.3 and 2.2.5. Exposure to light absorbed in the F band or heating of the crystal can cause migration and aggregation of the F centers (Figure 2.2.1), substantially increasing the num-

* Figures 2.2.1 to 2.2.96 may be found on pp. 319 to 380.

bers of F_2, F_3, and larger aggregate centers. At very high dose with simultaneous or subsequent heating above room temperature, large aggregates amounting to voids containing metal colloids or halogen fluid are formed.[24] These have characteristic optical attenuation bands resulting from scattering and absorption, as well as affecting mechanical properties of the crystal.

The optical absorption spectra of LiF irradiated with X-rays, protons, electrons, and deuterons at room temperature (Figure 2.2.3) share a close similarity, differing mainly in relative magnitude of the F-aggregate bands. Vacuum UV transmittance of electron-irradiated LiF is shown in Figure 2.2.4. The F band at 248 nm and accompanying β band at 111.7 nm appear first at low dose. With increasing dose, the spectrum is filled in sequentially by F_2, F_3, and F_4 bands at 450, 313 to 380, and 518 to 540 nm, respectively. The results of neutron irradiation are similar, but represent a higher local density of damage, as in ion bombardment. Most of the slow-neutron-induced radiation damage in LiF is attributable to 4.8-MeV alpha particles created in the nuclear reaction ^{6}Li(n,α) ^{3}H + 4.8 MeV.[25] The resulting high damage density along the alpha particle tracks favors aggregate formation at a lower average dose. Fast neutrons produce energetic recoil ions with similar effect.

The broad near-UV absorption from 3 to 5 eV in Figures 2.2.1 and 2.2.2 is attributable primarily to trapped-hole centers, generically termed "V center" in the older literature. It has been recommended[1] that the term "V center" be reserved for a positive-ion vacancy trapping a hole in the adjacent halide shell, i.e., the antimorph of an F center for the alkali sublattice. Note, however, that alkali vacancies are probably not produced photochemically. If irradiation and measurement were conducted at low-temperature, broad absorption bands due to self-trapped holes (V_k centers) and interstitial halogen atoms (molecular ions, H centers) would be prominent in this spectral region. (Hole centers are reviewed in References 26 and 27.) However, since these defects are very mobile at room temperature, one finds on warming the crystal a variety of impurity-stabilized interstitial hole traps and other defects with absorption bands in the 3- to 5-eV range. H centers are stabilized to about 200 K by substitutional monovalent cations, producing H_A centers, by divalent cations, producing H_Z centers,[28] and by other H centers, forming interstitial halogen pairs (I_2, formerly V_4).[29] The I_2 absorption band is near 4.7 eV, while the H, H_A, and H_Z bands are all near 3.5 eV. Larger aggregates tend to be stable at higher temperatures. The UV absorption bands historically termed V_2 and V_3 have been associated with dislocation loops formed by interstitial halogen.[30] The radiation-enhanced absorption overlapping the OH band in Figure 2.2.1 may have this origin.

Although Figures 2.2.1 to 2.2.4 illustrate the particular examples of KCl, KI, and LiF, the principal features of the absorption spectra are common to most of the alkali halides, and many of the transition energies vary in a systematic way from material to material. The wavelength of the peak of the F band follows the empirical Ivey relation:[1,3,31]

$$\lambda_F = 703\ d^{1.84} \tag{9}$$

where d is the lattice constant of the alkali halide in angstroms and λ_F is the wavelength of the F-band peak in angstroms. Similar relations hold for a number of other trapped electron centers.[22] The α- and β-band transitions should scale in energy approximately with the fundamental band edge.

The dependence of the F-center absorption band shape on temperature is shown for KBr in Figure 2.2.6. The full width at half-maximum (FWHM) is given as a function of temperature by:

$$\mathrm{FWHM} = W_0\ [\coth\ (h\nu_g/2kT)]^{1/2} \tag{10}$$

where W_0 is the limiting FWHM approached at zero temperature, ν_g is the local-mode vibrational frequency which interacts most strongly with the defect in its electronic ground state, h is Planck's constant, k is Boltzmann's constant, and T is the absolute temperature. This is illustrated for LiF and KCl in Figure 2.2.7. As temperature increases, the F-band peak shifts to lower energy, with the energy of the band maximum, E_m, given by

$$E_m = E_m(0) - A\,[\coth(h\nu_g/2kT) - 1] \qquad (11)$$

Radiation damage is found to increase the absorption coefficient at 10.6 μm in pure and doped KCl. In pure monocrystalline KCl at room temperature, the change in absorption coefficient with X-ray dose is approximately 4×10^{-5} $cm^{-1}/10^6$ R. In deformed (pressed) or doped KCl the absorption change is larger (Figure 2.2.8) and in all cases appears to correlate with F-aggregate concentration, specifically F_3 centers.[33,34]

The growth of F-center absorption in KCl as a function of radiation dose at room temperature is shown in Figures 2.2.9 to 2.2.11 for different dose rates and sample pretreatments. It is common to refer to "easy" Stage I coloration, "linear" Stage II coloration, and late-stage saturation regions in such plots of defect absorption coefficient vs. radiation dose. The steep slope of Stage I coloration is usually indicative of simple charge-transfer mechanisms whereby an electron is captured by a halide vacancy already in existence, although such behavior can also occur when saturable interstitial sinks control the vacancy-interstitial recombination rate. For example, divalent metal impurities are charge-compensated by generation of alkali vacancies, both of which act as traps for interstitial halogen. The magnitude of Stage I coloration is very dependent on history of the materials, including thermal treatment (Figure 2.2.10), mechanical strain (Figure 2.2.11), or prior radiation damage. The approximate linear range of dose dependence comprises Stage II, in which new halide vacancies (and interstitials) are produced by the radiation and achieve stable separations. In this linear range, approximately 4×10^3 eV of ionizing energy must be absorbed on average for each stable F center produced at room temperature in KCl. At higher doses, the defects start to interact to a significant degree, and the yield of isolated F centers falls below the linear extrapolation. Obviously, tabulated values of defect production efficiency must be used with caution, taking account of dose rate, total dose, impurity content, deformation, temperature, thermal history, etc. The major influences on stable F-center yield are generally those factors controlling the mobility of interstitial halogen atoms and thus the recombination probability, or those factors offering alternative channels for electron-hole recombination. Certain impurities can inhibit the late-stage F-center production at room temperature,[36] possibly because the traps that initially looked like interstitial sinks eventually saturate and act as sources of recombining interstitials.

Radiation-induced coloration can often be removed by appropriate heating or optical bleaching, although the perfect crystal is not necessarily restored. Thermal annealing can cause the release of trapped electrons or the migration of interstitials and vacancies. If the annealing or bleaching results from electronic charge redistribution, the coloration will likely return with a relatively small subsequent radiation exposure (i.e., Stage I coloration). Recombination of vacancies and interstitials restores the perfect crystal lattice, but defect mobility also implies that larger, more stable aggregates can be formed. F centers in alkali halides acquire significant mobility at temperatures in the range of 200 to 400°C. Annealing of damage in gamma-irradiated LiF is illustrated in Figure 2.2.12. At the high defect densities achieved in neutron-irradiated samples, higher temperatures may be required for annealing (Figure 2.2.13), and optical attenuation extending to long wavelengths may be introduced by colloid formation (Figure 2.2.14).

Alkali halides are subject to strong transient radiation-induced optical absorption at low temperature. The major contributors are self-trapped excitons and recombining vacancy-interstitial pairs. Valence holes in alkali halides spontaneously and quickly ($\sim 10^{-13}$ sec) become localized (self-trapped) in an antibonding electron orbital on two adjacent halide ions, which relax toward each other in response to the molecular bonding that results.[27] This self-trapped hole (V_k center) becomes a self-trapped exciton (STE) when it captures an electron in a bound effective mass orbital. Radiative decay of STEs restores the perfect lattice and accounts for the strongly Stokes-shifted intrinsic luminescence in alkali halides (Figure 2.2.15). Nonradiative decay of STEs is the principal channel for photochemical vacancy-interstitial pair production.[13-16,38,39] STE triplet states can have long lifetimes at low temperature, e.g., 5 msec in KCl, but undergo rapid nonradiative decay above characteristic quenching temperatures ranging from 12 to ~80 K in various alkali halides. While in existence, the STEs exhibit strong optical absorption bands in the visible and UV spectrum due to excitation of the self-trapped holes, and in the visible or IR spectrum due to excitation of the bound electrons.[42,43] The STE absorption bands in nine alkali halides at low temperature are shown in Figure 2.2.16. Transient and stable components of F, H, and V_k bands are also shown. The transient absorption is comparatively quite strong, with roughly one STE (or F center) formed for each electron-hole recombination event at low temperature. The oscillator strengths in each of the two main STE bands, as well as in the F band, are of order unity. As temperature increases, the decay time of STE absorption and luminescence decreases rapidly above a characteristic quenching temperature, illustrated for KBr in Figure 2.2.17. At room temperature, the decay time of STE absorption is typically in the 10- to 100-psec range.

Figure 2.2.18 shows the development and decay of STE and F absorption bands in NaCl from 17 psec to 2 min after bandgap excitation, at temperatures of 80 and 300 K. The bandgap excitation in this case was two-photon absorption of 266-nm laser radiation. Other experiments with short laser pulses have shown that F centers are formed in the ground state within 10 psec of electron-hole pair generation in KCl (Figure 2.2.19) and comparably short formation times are found in other alkali halides.[44-47] Furthermore, the primary photochemical efficiency (i.e., measured before vacancy-interstitial recombination can occur) may be of order unity at high temperature, as shown for KCl in Figure 2.2.20. Figure 2.2.21 illustrates the dramatic role of defect stabilization in determining the temperature-dependent yield of stable F centers measured 10 sec after the pulses which gave the primary efficiencies shown in Figure 2.2.20.

MgF_2

Radiation damage in pure MgF_2 is dominated by photochemical generation of fluoride ion vacancies under irradiation conditions which create electron-hole pairs in the crystal.[48,49] The prominent defect-associated optical absorption bands at room temperature are at 117, 260 and 370 nm (Figures 2.2.22 and 2.2.23). All of these are associated in some way with the F center, an electron trapped at a fluoride ion vacancy. The 260-nm band is the lowest-energy electronic transition of an isolated F center. In the MgF_2 crystal structure (cassiterite), the point symmetry of the F center is C_{2v}. The F absorption band is thus slightly anisotropic for light polarized parallel and perpendicular to the c axis of the crystal, as indicated in Figure 2.2.22. The 370-nm band arises from F_2 (di-vacancy) centers having C_{2h} site symmetry. A weaker absorption band at 400 nm is attributed to one or more of the other three site symmetries: C_1, C_{2v}, D_{2h}, of F_2 centers in MgF_2. The β band of the F center at 117 nm (Figures 2.2.23 and 2.2.24) has the same localized excitonic origin as described for the alkali halides. In the spectrum of a neutron-irradiated sample (Figures 2.2.24 and 2.2.25), a featureless background of radiation-induced optical absorption fills in the spectral region between 260 and 117 nm. This absorption seems to be particular to neutron irradiation or other treatment[52] giving locally high densities of defects. There are no indications of strong IR absorption induced by radiation in MgF_2. Measurements out to 7 μm have been reported,[53] in which weak absorption features of unknown origin, but suspected of involving water or OH^- ions, were found at 3, 6, and 7 μm.

If the crystal is exposed to light in the F absorption band, motion and consequent aggregation of the F centers occur at room temperature, enhancing the 370-nm F_2 band as shown in Figure 2.2.26. Luminescence spectra resulting from excitation of the F and F_2 bands, respectively, are shown in Figures 2.2.27 and 2.2.28. The luminescence is observable at room temperature, as well as lower temperatures. The effect of thermal annealing of the defects in an electron-irradiated sample is shown in Figure 2.2.29. Above about 700 K, the vacancies become mobile and the F band starts to decrease, while the F_2 bands at 370 and 400 nm experience a dramatic rise due to aggregation of the vacancies. Above 800 K, virtually all the defects are annihilated. In neutron-irradiated material, somewhat higher temperatures are typically needed for annealing. Figure 2.2.30 shows absorption spectra after annealing neutron-irradiated MgF_2 at various temperatures.

Figure 2.2.31 shows the radiation dose dependence of F center absorption in different samples of MgF_2 whose content of certain impurities is summarized in the caption. As in the case of alkali halides discussed above, trace impurities typically enhance the Stage I coloration but can have the effect of suppressing later stages of F center production at high dose. The approximate slope of the linear dose dependence range in Figure 2.2.31 indicates that at least 4×10^5 eV must be deposited for each stable F center formed at room temperature. This is roughly 100 times the energy per stable F center in KCl.

The F-center production yield in MgF_2 has an unusual dependence on temperature as indicated in Figure 2.2.32. At temperatures in the range 80 to 150 K, the radiation resistance becomes extremely high. It is believed that in this temperature range, newly formed close pairs of vacancies and interstitials recombine, whereas at higher temperature the interstitial fluorine atoms escape to immobilizing trap sites or dislocations. At very low temperature, vacancy-interstitial recombination is inhibited at all but the closest separations. This is supported by observations of transient absorption produced by pulsed electron irradiation. At low temperature, a transient absorption band very similar to the F band but shifted 0.4 eV to lower energy (Figure 2.2.33) is formed with an efficiency 10^4 times that of stable F centers. One such center is formed for each 28 eV deposited in the crystal.[56] The decay of the absorption band is accompanied by the intrinsic 385-nm recombination luminescence of MgF_2, as shown in Figure 2.2.34. The temperature dependences of the absorption and luminescence decay times are plotted in Figure 2.2.35. The transient defect responsible for this absorption and luminescence can be described as a self-trapped exciton (STE) roughly analogous to the STE in alkali halides discussed above. In MgF_2 the STE is equivalently describable as a close vacancy-interstitial pair capable of undergoing radiative recombination to the ground state of the perfect lattice.[56]

CaF_2, SrF_2, BaF_2

At room temperature, the radiation sensitivity of CaF_2 is strongly influenced by impurities, including rare earth ions, oxygen, and particularly yttrium.[57-66] Yttrium is a common and persistent impurity in CaF_2 since the ionic radii of Y^{3+} and Ca^{2+} are the same. Gorlich et al.[62] have cataloged radiation-induced absorption spectra associated with a wide array of rare earths and other impurities in CaF_2, SrF_2, and BaF_2. Irradiated CaF_2 containing yttrium exhibits a characteristic 4-band absorption spectrum with peaks at 225, 335, 400, and 580 nm, as shown in Figure 2.2.36.

If CaF_2 can be prepared free of yttrium and other sensitizing impurities, its radiation resistance at room temperature is quite good compared to most other halide materials. Indeed, the radiation resistance of CaF_2 is frequently cited as a sensitive indicator of material purity. Figure 2.2.37 shows the absorption spectrum produced in a relatively pure synthetic crystal by an X-ray exposure of 1.7×10^6 R. The spectrum is rather featureless, and cannot be unambiguously ascribed to F centers or other intrinsic defects. At low temperature, F centers are formed by irradiation of CaF_2 and can be clearly identified,[67-71] as in Figure 2.2.38.

In the terms of first- and second-stage regions of a damage vs. dose curve, the absorption of Figure 2.2.37 produced by a 1.7×10^6 R exposure must be regarded as Stage I (early) coloration. There are as yet no complete data specifying the yield of F centers in pure CaF_2 at room temperature in the Stage II (linear dose dependence) region. Judging from the Stage I yield, however, very pure CaF_2 could be comparable to or better than MgF_2 in radiation resistance at room temperature. Figure 2.2.39 shows a spectrum of radiation-induced vacuum UV absorption in CaF_2, though it appears that the material employed had yttrium impurity relating it more to Figure 2.2.36 than Figure 2.2.37. The spectrum for BaF_2 from the same work is shown in Figure 2.2.40. Gorlich et al[63] have commented that their observations on pure material at room temperature indicate that radiation resistance improves in the sequence BaF_2, SrF_2, CaF_2, in contrast to the suggestion based on Figures 2.2.39 and 2.2.40. Radiation-induced absorption spectra in Gorlich's pure synthetic crystals of SrF_2 and BaF_2 are shown in Figures 2.2.41 and 2.2.42. Colloids can be produced by high doses of irradiation at elevated temperature.

Transient absorption bands lasting milliseconds after pulsed irradiation at low temperature are shown in Figure 2.2.43. The absorbing centers are termed self-trapped excitons STEs in analogy to the alkali halides and MgF_2. The STE in fluorite materials resembles a nearest-neighbor pair of an F center and interstitial atom, capable of radiative recombination to restore the perfect crystal ground state. The oscillator strength in each of the two main transient absorption bands is approximately unity, and roughly one such STE is formed for each electron-hole pair recombination at low temperature. The transient absorption decay time in CaF_2 ranges from about 10^{-3} sec at 10 K to 5×10^{-5} sec at 77 K, but has not been measured directly at room temperature. The corresponding decay time of luminescence (Figure 2.2.44) indicates an STE lifetime of about 10^{-6} sec at room temperature.

Alkali/Alkaline-Earth Fluorides, ABF_3

Alkali and alkaline earth metals combine with fluorine in the proportion ABF_3 to form wide-gap transparent crystals, typically having the Perovskite structure. Perovskite-structured $KMgF_3$ can be visualized as composed of cubes having K^+ ions at the corners, Mg^{2+} ions at the body centers, and F^- ions at the face centers. Radiation damage occurs by a photochemical mechanism, resulting in defects very similar to those in alkali or alkaline earth halides.[73-78] The room-temperature absorption spectrum is dominated by F and F-aggregate centers. Self-trapped holes (V_k centers) and isolated interstitial fluorine hole centers are observable at temperatures up to about 100 K.

Figure 2.2.45 shows the absorption spectrum of $KMgF_3$ which has been irradiated at room temperature with 2-MeV electrons. The principal F, F_2, and F_3 absorption bands are at 270, 445, and 395 nm, respectively. The F-center concentration as a function of absorbed 2-MeV electron dose is shown for a series of irradiation temperatures in Figure 2.2.46. The defect yield increases with increasing irradiation temperature up to about 233 K and then decreases. It should be noted that the quantity plotted is fn_F, where n_F is the F-center concentration and f is the oscillator strength of the principal F-center transition. Although not explicitly determined, $f \approx 1$ is a reasonable estimate based on other halide materials.

Absorption spectra of $NaMgF_3$ irradiated with 2-MeV electrons at various temperatures are shown in Figure 2.2.47. The broad 350-nm absorption band evident at 10 and 80 K is due to self-trapped holes that anneal at about 120 K. The 290- and 405-nm bands result from F and F_2 centers, while the 230-nm band is suggested to be associated with stabilized interstitial aggregates. The growth of 290-nm absorption coefficient is shown as a function of 2-MeV electron dose at several irradiation temperatures in Figure 2.2.48. Thermal stability of the F band and 350-nm absorption is indicated by the isochronal annealing curves in Figure 2.2.49. In $RbMgF_3$, F and F-aggregate centers persist up to temperatures in the range of 600 K. Absorption spectra of $RbCaF_3$ irradiated with 2-MeV electrons at various temperatures are shown in Figure 2.2.50.

Thallium Halides

Thallium halides, TlCl, TlBr, TlI, and mixtures such as TlBrI (KRS-5), are used primarily for IR-transmitting optics. An absorption band at 530 nm, produced in TlCl by exposure to UV light of unspecified spectrum and dose, was reported in 1930[79] (Figure 2.2.51). The optical spectrum behaved very similarly to that of UV-irradiated AgCl, and was therefore attributed to thallium colloid formation. To the extent that the silver-halide analogy is valid, one might expect that electrons and holes produced by bandgap excitation ($h\nu > 3.4$ eV in TlCl) would convert the constituent ions to thallium atoms and halogen atoms (holes). Recombination is prevented if the holes migrate to a free surface or grain boundary, where they are trapped and possibly evolve halogen molecules. Aggregates of thallium atoms could then form as precipitated metal particles. In silver halides, these processes are inhibited in the interior of a well-annealed single crystal, but can be enhanced by mechanical deformation or by the presence of certain impurities. The analogy to silver halides remains somewhat conjectural, however, since there appears to have been little additional work on photochemical damage in thallium halides. An attempt to produce coloration of TlCl by X-ray irradiation at both room temperature and 100 K showed that the material is remarkably resistant to visible discoloration under those conditions.[80] Specifically, exposure of pure TlCl at room temperature for 64 hr at the exit window of a tungsten-target X-ray tube operating at 50 kV and 40 mA produced no evident coloration. For comparison, it was noted that in KCl irradiated under similar conditions, an F-center concentration of 10^{17} cm^{-3} would be produced in 15 min. Thus, TlCl would appear to be at least 10^3 times more resistant to visible discoloration by X-rays than is KCl. The reports of UV photolysis and resistance to X-ray irradiation in the same material may at first seem contradictory, but are reconcilable in terms of total dose and dose rate. That is, the small absorption depth of UV radiation above the bandgap of TlCl means that the excitation density may be extremely high. This is known in general to favor colloid formation, whereas the isolated point defects produced by X-ray irradiation may recombine more readily than form colloids, analogous to reciprocity failure of silver halides. Observation of a broad transient absorption band centered near 600 nm in TlCl irradiated by a pulse of 500-keV electrons has been reported in connection with measurements of recombination luminescence, but no quantitative transient absorption measurements are available.[81]

The thallium halides have large dielectric constants, so that trapped electrons at defects should be only weakly bound and thus may be expected to have transitions only in the IR. On the other hand, such shallow electron traps would likely be thermally ionized at room temperature. There is yet no direct experimental evidence concerning such states in thallium halide crystals. A 3.0-eV absorption band in TlCl has been attributed to a localized excitonic transition in the vicinity of a chloride ion vacancy complexed with a chalcogen impurity, roughly analogous to the α band in alkali halides.[82] Although the vacancies seen in that study were introduced chemically rather than by radiation, one might expect a similar band in neutron-irradiated TlCl, for example.

In summary, it appears that if the surfaces are protected from UV radiation, TlCl (and, by implication, other thallium halides) are quite resistant to ionizing radiation damage.

OXIDE CRYSTALS

The photochemical mechanisms of lattice defect formation that are so prevalent in halide crystals are apparently not important in oxide and other chalcogenide crystals. Direct collisional energy transfer from primary or secondary energetic particles is the main mechanism for production of lattice defects. Electronic processes are effective in altering the charge states of impurities and other lattice defects already existing in the material.

MgO

The alkaline earth oxides are the simplest oxide materials and probably the best understood in regard to radiation damage. Although many of the alkaline earth oxides are hygroscopic and thus not widely used in optical systems, MgO is relatively insoluble, hard, and durable. A discussion of radiation damage in MgO will serve to illustrate many of the concepts applicable to other oxides as well. The accepted nomenclature of defects in alkaline earth oxides is very similar to that for halides, and has been discussed systematically in Reference 1. A negative-ion vacancy containing the same number of electrons as the charge of the normal lattice ion is defined to be the F center. Thus, in oxide crystals, an oxygen vacancy with two trapped electrons is an F center, a vacancy with one electron is an F^+ center, and the bare oxygen vacancy is denoted F^{2+}. In oxides of divalent metals the complementary positive-ion vacancy defects are V, V^-, and V^{2-}, trapping 2, 1, and 0 holes, respectively (see Table 2.2.1).

Transition metals are common multivalent impurities in MgO. Each valence state of the impurity can have one or more characteristic absorption bands, and changes of valence state are often accompanied by luminescence.[83,84] Thus, radiation-induced valence changes of impurity ions can have a significant effect on the optical properties. Absorption bands due to Fe^{3+} are found at 210 and 285 nm in MgO.

The F and F^+ absorption bands overlap in MgO, the F peak occurring at 5.01 eV and the F^+ peak at 4.92 eV (Figure 2.2.53).[85-87] The F and F^+ luminescence bands are observed at 2.4 and 3.13 eV, respectively. A number of defect cluster centers in irradiated MgO have optical absorption bands in the visible spectrum, as also shown in Figure 2.2.53. The dose dependences of F-band absorption for neutron and electron irradiation are shown in Figures 2.2.54 and 2.2.55. The defect aggregate bands are formed more readily by fast neutron irradiation than by electron irradiation, and are approximately proportional to the F-center concentration over a wide range. In samples that are heavily damaged by ion bombardment, a broad absorption feature appears between the 5-eV F band and the fundamental absorption edge. The growth of this feature with dose is shown in Figure 2.2.56. There is also evidence that the radiation-induced disorder produces a broadened tail of absorption at the fundamental band edge.

The annealing of F centers in MgO is plotted in Figure 2.2.57. The 400°C annealing temperature in electron- and neutron-irradiated samples indicates that the annealing occurs principally through recombination of mobile oxygen interstitials with relatively immobile vacancies. The oxygen vacancies do not become mobile until the temperature reaches 1000°C, as shown in the curve for the additively colored crystal. Additive coloration, i.e., heating the crystal in magnesium vapor, produces a crystal with many oxygen vacancies but few oxygen interstitials. The annealing of the defect cluster absorption bands is illustrated in Figure 2.2.58. Doping with lithium appears to suppress radiation damage in MgO.[89]

The V^- center, a magnesium ion vacancy with one hole trapped on the surrounding shell of oxygen ions, exhibits an absorption band at 2.3 eV and has an oscillator strength of about 0.1.[90-93] Typically, the magnesium ion vacancies form complexes with partially charge-compensating impurity ions, such as Al^{3+} substituting for Mg^{2+}, and OH^- or F^- substituting for O^{2-}. Upon capturing a hole, the resulting V_{Al}, V_{OH} and V_F centers all have optical absorption bands overlapping the V^- band at 2.3 eV. From the optical standpoint, the main differences in these constituents of the composite V band are in thermal stability. The V-center absorption anneals in the temperature range from 320 to 450 K, as shown by the curve labeled ''electronic'' in Figure 2.2.59. However, this annealing process involves only the untrapping of the hole from the magnesium vacancy, and a brief exposure to ionizing

Table 2.2.1
NOMENCLATURE OF POINT DEFECTS IN REPRESENTATIVE ELEMENTAL AND BINARY COMPOUND CRYSTALS

		I—VII	II—VII$_2$	II—VI		IV	
Defect description	**Sublattice**	**LiF**	**MgF$_2$**	**MgO**	**ZnSe**	**Si**	**C**
Vacancy	−	F^+, (α)	F^+	F^{2+}	V_{Se}^{2+}	V^0	GR, V^0
Vacancy	+	V^-	V^{2-}	V^{2-}	V_{Zn}^{2-}		
Vacancy + electron(e)	−	F	F	F^+	V_{Se}^+	V^-	
Vacancy + hole (h)	+	V, (V_1)	V^-	V^-, (V′)	V_{Zn}^-	V^+	
Vacancy + 2 e	−	F^-, (F′)	F^-	F	V_{Se}	V^{2-}	
Vacancy + 2 h	+	V^+	V	V	V_{Zn}	V^{2+}	
2-Vacancy neutral cluster	−	F_2, (M)	F_2	F_2	$(V_{Se})_2$	V_2	
3-Vacancy neutral cluster	−	F_3, (R)	F_3	F_3	$(V_{Se})_3$	V_3	
n-Vacancy neutral cluster	−	F_n, (N, n = 4)		F_n			
Vacancy + e + subst. cation of host valence	−	F_A	F_A	F_A			
Vacancy + h + subst. anion	+	V_A (e.g., V_{OH})		(e.g., V_{OH}, V_F)	A (V_{Zn} + donor)	A (V + O_{sub})	
Interstitial atom		I, I_F^0, I_{Li}^0 I_F^0, I_{Mg}^0		I_O^0, I_{Mg}^0	I_{Se}^0, I_{Zn}^0	I^0	
Interstitial atom pair		I_2^0, (V_4)					
Interstitial atom cluster		I_n^0, (V_2, V_3)		I_n^0			
Interstitial ion		I^-, I^+		I^-, I^+		I^-, I^+	
"Split interstitial" bonded atom pair on subst. site		H	H			I^0	
H Center + monovalent cation		H_A					
H Center + divalent cation		H_Z					
Self-trapped hole		V_k, $[X_2^-]$	$[X_2^-]$				
Self-trapped exciton		STE, $[X_2^{2-}]$*	STE				

Note: For nomenclature in crystalline SiO_2 and silicate glasses, see Tables 2.2.2 and 2.2.5. Older or less preferred symbols are given in parentheses. It should be recognized that the defect descriptions are brief and do not necessarily convey nuances of defect structure in different materials. For example, although the F center in LiF and a V_{Se}^+ defect in ZnSe are both anion vacancies trapping one electron, basic properties such as the degree of electron localization may be quite different. Roman numeral subscripts are used as a general representation for chemical symbols from the appropriate group of the periodic table. The designation of (− or +) sublattice clearly refers only to the binary compounds.

radiation will restore the 2.3-eV absorption band. If the crystal is heated to still higher temperature and exposed to ionizing radiation to repopulate the hole traps after each heating stage, the curve labeled "intrinsic" in Figure 2.2.59 results, corresponding to annihilation of the magnesium ion vacancies.

Although ionizing radiation produces the V^- absorption band, the requisite magnesium vacancies must already have been present. Typical MgO samples contain magnesium vacancies at a concentration in the range of 10^{16} to 10^{18} cm^{-3}, depending on impurity compensation, thermal history, and mechanical deformation. The saturation value of gamma-radiation-induced V^- absorption is shown as a function of temperature from which the crystal has been quenched in Figure 2.2.60. Hydrogen impurity in MgO is found to enhance formation of V^- centers by radiation.[95]

The IR stretching band for an OH^- ion adjacent to a magnesium vacancy in MgO occurs at 3296 cm^{-1}. Capture of a hole to form the V_{OH} center has the effect of shifting the OH stretching band to 3323 cm^{-1}. This is a radiation-induced IR band shift that is reversible by thermal annealing or optical bleaching of the 2.3-eV V band.

Under ionizing irradiation, nominally pure MgO emits luminescence typically at 4.9, 3.3, and 1.7 eV. The 4.9-eV band is attributable to recombination of an electron with a hole trapped at a V-type center as discussed above. The 1.7-eV luminescence occurs in radiative hole capture by Cr^{2+} impurity ions. The 3.3-eV luminescence also appears to arise from radiative hole capture at an impurity, possibly Fe^{2+}. Time-resolved measurements of luminescence in MgO following pulsed 500-keV electron excitation indicate that the recombination kinetics are dominated by initial hole capture at V centers. Subsequent recombination with electrons trapped at impurity sites occurs by pairwise tunneling at temperatures below roughly 400 K, in combination with thermally activated hole release above 400 K.[97] Luminescence decay curves are shown in Figure 2.2.61. Transient optical absorption in the V-band region near 2.3 eV has been observed following pulsed 500-keV electron irradiation, but no quantitative data are available.[97]

Al_2O_3

Ionizing radiation produces two prominent absorption bands in nominally pure Al_2O_3, shown in Figure 2.2.62 at 3.08 and 5.45 eV. The 3.08-eV (410-nm) band is a composite V band analogous to the V bands in MgO. In Al_2O_3, a hole trapped on an oxygen ion adjacent to an aluminum vacancy comprises a V^{2-} center. Similarly, an aluminum vacancy with two adjacent O^- ions (trapped holes) is a V^- center, or $V^-{}_{OH}$ if one of the O^- ions is replaced by OH^-.[91-99] The aluminum vacancies are not formed by ionizing radiation in general, but would be present as a result of thermal history, impurity compensation, deformation, or prior particle irradiation. The 5.56-eV (227-nm) band is due to Cr^{2+} trace impurity, almost universally present, which is converted from Cr^{3+} to Cr^{2+} by ionizing radiation. In the case of intentional chromium doping (ruby), UV light reduces Cr^{3+} to Cr^{2+}, and the corresponding holes are trapped at aluminum vacancies to form V centers (V^{2-}, $V_{OH}{}^-$, etc.) absorbing light in a broad band peaked near 410 nm. This absorption imparts a characteristic brown color to the solarized laser rod and competes with the chromium ions for pump light.[100] Thermal annealing of the composite V band at 410 nm and the $V_{OH}{}^-$ IR absorption at 3316 cm^{-1} are shown in Figure 2.2.63, along with the thermoluminescence that results when holes released from V centers are captured at impurities, principally Cr^{2+}. Radiation-induced hole capture at $V_{OH}{}^{2-}$, converting it to $V_{OH}{}^-$, shifts the OH stretch band in IR absorption as shown in Figure 2.2.64.

Irradiation by particles capable of displacing lattice ions by collisional energy transfer yields additional absorption bands. As shown for the case of fast fission neutron irradiation in Figure 2.2.65, the absorption is a continuum of overlapping bands, falling off almost monotonically from the UV (6.1-eV band) to the IR. The 6.1-eV absorption band is due to the F center, an oxygen vacancy occupied by two electrons. Bands at 4.8 and 5.4 eV arise from 1A → 1B and 1A → 2A transitions, respectively, of the F^+ center.[104] The 1A → 2A transition is symmetry forbidden for light with electric vector parallel to the crystal c axis. Figure 2.2.66 shows absorption spectra measured with polarized light at 295 and 77 K in an Al_2O_3 crystal bombarded with 14-MeV (fusion) neutrons to a dose of 3.9×10^{16} neutrons per square centimeter. Also shown is the F^+ luminescence band at 3.8 eV, and its excitation spectrum.

The growth of the F and F^+ absorption bands as a function of fission neutron dose is shown in Figure 2.2.67. Thermal annealing of the bands induced by both fission-neutron and 14-MeV neutron irradiation is shown in Figure 2.2.68.

ZnO

Energetic particle irradiation of ZnO produces a yellow coloration due to the absorption spectrum shown in Figure 2.2.70, as induced by 1.7-MeV electrons and fast (> 1-MeV) neutrons. X-ray and ^{60}Co gamma irradiation produce no observable change in absorption. For comparison, the optical absorption produced by additive coloration, i.e., heating above 825°C in zinc vapor, is also shown. It should be noted that Figure 2.2.70 displays "induced absorption" obtained by subtracting the spectrum of an uncolored crystal. The rather featureless spectrum is attributed at least partially to the F^+ center (oxygen vacancy trapping one electron) having an absorption band peaked near 3 eV.[109] The very broad low-energy tail of absorption probably involves other defects as well. The failure to resolve a distinct peak in the spectra for irradiated crystals is likely due to disorder-induced broadening of the fundamental absorption edge (normally occurring at about 393 nm). The apparent threshold of incident electron energy for production of F^+ absorption in as-grown ZnO has been found to be above 600 keV, too high for a reasonable oxygen displacement energy.[108] Subsequent work has revealed that the threshold electron energy for oxygen displacement is 310 keV, corresponding to 57 eV displacement energy.[110] However, oxygen vacancies so produced can apparently not achieve the observable F^+ charge state unless damage is also produced on the zinc sublattice, providing electron traps. The threshold energy for zinc displacement is about 900 keV, corresponding to $T_d(Z_n) = 57$ eV.

The growth of 410-nm absorption coefficient with 1.7-MeV electron irradiation dose at 60°C is shown in Figure 2.2.71 for pure and lithium-doped ZnO. Annealing of the defect absorption is shown in Figure 2.2.72. In the additively colored crystal containing oxygen vacancies but few oxygen interstitials, the F^+ centers are stable to 800°C. The F^+ centers in irradiated crystals are unstable above about 250°C, presumably annihilated by diffusing interstitial oxygen. The well-defined single annealing stage in lithium-doped electron-irradiated material, along with the enhanced coloration efficiency shown in Figure 2.2.71, imply that the lithium impurity is an effective trap for interstitial oxygen, further indicating the importance of oxygen interstitial stabilization in determining defect yield.

TiO_2

Titanium dioxide (rutile) is reported to be unusually resistant to damage by particle irradiation as well as ionizing radiation.[102,111,112] Neutron irradiation to 4.8×10^{18} neutrons per square centimeter fast and 1.7×10^{19} neutrons per square centimeter thermal produced a general loss of transparency of 5% in a sample approximately 2 mm thick ($\alpha \approx 0.25$ cm^{-1}), but no distinct absorption bands. For comparison, note that an exposure of MgO to 4.8×10^{18} neutrons per square centimeter results in an F band with $\alpha \approx 1300$ cm^{-1}. Neutron-irradiated TiO_2 exhibits a structured spectrum of photoconductivity, as shown in Figure 2.2.73, indicating at least the introduction of some defects. Gamma radiation at room temperature is said to make no significant changes in the photoconductivity spectrum.

OTHER CHALCOGENIDE AND III-V CRYSTALS

ZnS

High-energy particle irradiation of ZnS produces optically absorbing defects by collisional displacement on both the zinc and sulfur sublattices.[113-115] The transmission spectra of single-crystal ZnS before and after neutron irradiation are shown in Figure 2.2.74. Irradiations were conducted at or above room temperature, while the measurements shown were made at liquid helium temperature. Also shown is the effect of UV ($\lambda < 380$ nm) illumination at low temperature. Four irradiation-induced features are evident: increased absorption near the fundamental edge and discrete bands near 540, 430, and 390 nm. The 540- and 430-nm bands have been attributed to transitions of the F^+ center (sulfur vacancy + e^-). Absorption at 390 nm has been identified with the V^- center (zinc vacancy + h^+). Close pairs of donors and zinc vacancies (A centers) absorb light near 360 nm (Figure 2.2.75). It is well known that radiation damage decreases the efficiency of activated ZnS phosphors. As monitored by phosphor efficiency, damage produced by neutrons can be annealed by heating to about 700°C.[116]

ZnSe

The most evident radiation effects in ZnSe involve zinc vacancies. The displacement energy on the zinc sublattice is 10 eV, corresponding to an incident electron energy of 240 keV (Figure 2.2.76).[117] The isolated zinc vacancy, denoted V^{2-}, absorbs light in a band near 500 nm (Figure 2.2.77). In the resulting photoionized state, V^-, a hole is trapped on a selenium atom adjacent to the vacancy.

The trapped hole induces a trigonal distortion of the vacancy site, with a <111> symmetry axis which can be aligned by applied stress at low temperature.[117] The V^- center exhibits weak optical absorption bands at 1.4 and 2.65 eV. As shown in Figure 2.2.78 by means of stress-induced dichroism, the absorption bands are linearly polarized with respect to the symmetry axis of the V^- defect. Illumination with IR light near 1.03 μm releases electrons from donor levels to recombine with the holes at V^- defects, emitting 720-nm luminescence as the V^{2-} population is restored. If the vacancy is in close association with a donor, the absorption and emission bands are shifted slightly, to approximately 465 and 625 nm, respectively.[117] This vacancy-donor close pair is termed an "A center". The thermal evolution of zinc vacancies and their complexes is shown in Figure 2.2.79, as deduced from electron paramagnetic resonance studies.[119,120] Irradiation at low temperature produces isolated zinc vacancies (V) and close vacancy-interstitial pairs in configurations labeled I to IV, each having a characteristic temperature of annealing. The isolated zinc vacancy becomes mobile above 400 K, forming complexes with various impurities that persist to about 600 K.

GaAs

IR absorption induced in GaAs by irradiation with fission neutrons is shown in Figure 2.2.80 for doses ranging from 2×10^{16} to 1.4×10^{18} neutrons per square centimeter. The irradiation temperature was about 100°C and the measurements were made at room temperature.[121] The most noticeable feature is the smooth background of irradiation-induced absorption increasing approximately as the square of the photon energy. Superimposed on this background is a band at about 0.47 eV. A weak feature is observable at about 0.25 eV for the highest doses. It has been suggested that the E^2 dependence of absorption coefficient in neutron-irradiated GaAs indicates the existence of small metallic inhomogeneities about 10 nm in radius.[121-123] Such metallic regions could result from the disorder and high pressure associated with a collision cascade or thermal spike due to irradiation with fast neutrons or ions. It seems reasonable to attribute the 0.47-eV optical band to transitions out of a defect level found in electrical measurements to lie about 0.5 eV below the conduction band.[121]

Figure 2.2.81 shows the effectiveness of annealing at 480 and 540°C in restoring most of the original transparency of a neutron-irradiated crystal.[124-125]

Irradiation with 1.5-MeV electrons produces lattice damage in GaAs, but the resulting IR spectrum is qualitatively different from that of neutron-irradiated material.[121] This behavior is consistent with the idea of a thermal spike in neutron-irradiated material as discussed above, since 1.5-MeV electrons would not produce such a spike. As measured at room temperature (Figure 2.2.82), the absorption coefficient increases exponentially with photon energy below the fundamental band edge. The exponentially broadened absorption edge in the presence of radiation-induced defects is attributable to electric microfield effects.[126] Measurements made at 85 K reveal a broad absorption band centered at 1 eV in electron-irradiated GaAs (Figure 2.2.83). Isochronal annealing of the 1-eV absorption and the exponentially broadened absorption near the fundamental edge are shown in Figure 2.2.84.

By serving as recombination centers and scattering centers, radiation-induced defects reduce both the number of free carriers and their mobilities. Annealing stages in the recovery of conductivity in electron- and neutron-irradiated GaAs are indicative of defect stabilities. Annealing is almost complete at 250°C in electron-irradiated GaAs,[125,127-128] but in the neutron-irradiated material there are stages at 250, 450, and 650°C. Through its effect on free-carrier concentration, irradiation of heavily doped semiconductors can have a pronounced influence on IR reflectivity, shown in Reference 129.

GROUP IV ELEMENTAL CRYSTALS

Diamond

High-purity insulating (Type IIa) diamond is normally transparent from the intrinsic IR 2-phonon vibrational band to the fundamental absorption edge at 5.4 eV.[130-134] X-ray and UV radiation have little effect on the optical spectrum. Energetic-particle or gamma-ray irradiation capable of inducing atomic displacements produces a characteristic optical absorption spectrum as shown in Figure 2.2.85. The 2-eV absorption band and associated 1.673-eV zero-phonon line are labeled GR1. The GR defect is thought to be an isolated neutral vacancy.[134-137] The broad UV absorption band around 4 eV is closely correlated with the GR1 band in production and annealing, and is thought to arise principally from the GR defect, along with the zero-phonon lines GR2-7 that are superimposed on the UV band.

The GR defect absorption is linear in radiation dose for a given particle energy, as illustrated in Figure 2.2.86. The slope, or defect production efficiency, depends on the incident particle energy as shown in Figure 2.2.87, and exhibits a threshold below which no damage occurs. These data indicate that the carbon displacement energy in diamond is 80 eV. Ion bombardment produces a conducting surface layer due to formation of amorphous or graphitic carbon.[139] Isochronal annealing of the GR1 band is shown in Figure 2.2.88. Annealing starts near 300°C, but some GR defects persist up to 1000°C. It is concluded that vacancies are not mobile below 1000°C, and that the early stages of annealing are due to interstitials released from trap sites, including interstitial clusters.[140] Changes in the absorption spectrum of an electron-irradiated Type IIa diamond after annealing at 600°C for periods of 4 and 28 hr are shown in Figure 2.2.89. The resulting absorption band near 2.6 eV is labeled TH5, and appears to be formed at the expense of the GR (vacancy) defect. After additional heating at 850°C for 21 hr, the irradiated sample approaches the pre-irradiation Type IIa spectrum (Figure 2.2.90), but retains a featureless residual absorption that suggests formation of highly disordered or graphitic regions. In annealing diamond, the phase transition to graphite at 1800°C must be kept in mind. Furthermore, heating to 1000°C or above must be done in vacuum to prevent oxidation.

Type I diamonds are less pure than Type II, the principal contaminant being nitrogen.[130] Type Ia material contains as much as 0.5% nitrogen, much of it in precipitated platelets on

<100> planes. In Type Ib diamonds, the nitrogen impurity is more evenly dispersed. Before irradiation, Type I diamonds typically exhibit a prominent absorption band centered near 3.25 eV, on which fine line structure (N2-N4 series) can be resolved. Type I material is opaque above 4 eV. After irradiation causing atomic displacement, a spectrum such as that shown by curve A in Figure 2.2.91 is typically found. In addition to the familiar GR band at 2 eV, a band at 3.4 eV, labeled ND, is observed, including sharp lines ND1-6.[142] Subsequent heat treatment depletes the GR and ND defect absorption bands, and a new absorption band and line series labeled H3,4 grows in near 2.6 eV (Figure 2.2.91). These are probably due to the association of radiation-produced defects with nitrogen.

Luminescence is described in References 143 to 145.

Silicon

The atomic displacement energy in crystalline silicon at room temperature is about 13 eV, corresponding to 145 keV incident electron energy at the threshold for damage. Silicon vacancies and interstitials are the primary defects produced, but both are extremely mobile and do not survive as isolated defects at room temperature.[146-148] For example, silicon interstitials apparently can move through the lattice even at 4 K under irradiation conditions. On encountering one of a number of substitutional impurities including carbon, aluminum, and boron, a silicon interstitial may exchange places with it, so that the end result is an interstitial impurity produced by silicon displacement.[148-151] The vacancy diffuses readily above 150 K until stabilized by association with impurities or another vacancy. Thus, room-temperature irradiation produces a variety of defects depending crucially on impurities in the crystal. The defects have characteristic IR absorption bands due either to electronic transitions or local vibrational modes. IR bands at 1.8, 3.3, and 3.9 μm are associated with divacancies (Figure 2.2.92). Unlike the isolated vacancy, divacancies are stable at room temperature as indicated by the annealing curves in Figure 2.2.93. Radiation-induced defects associate with oxygen impurity atoms to produce a number of IR local-mode bands (Figure 2.2.94). A prominent band at 12 μm (830 cm^{-1}) is attributed to an oxygen + vacancy complex (A center).[153-157] A survey of radiation-induced IR absorption in silicon containing oxygen (dashed lines) and carbon + oxygen (full lines) is given in Figure 2.2.95.[151]

Radiation damage dramatically alters the recombination luminescence in silicon, quenching the intrinsic excitonic luminescence and replacing it with various defect bands.[158-159] A luminescence line at 1.28 μm and associated vibronic structure are commonly found in silicon irradiated or subsequently annealed in the temperature range 20 to 200°C. It has been shown by optically detected magnetic resonance that the responsible defect is a complex of two substitutional carbon atoms and a silicon interstitial.[160]

Germanium

Germanium is similar to silicon in radiation damage properties as manifested in the optical spectrum. Isolated vacancies become mobile above 65 K, and isolated interstitials are very mobile at low temperature under irradiation conditions.[161] Thus, radiation damage in germanium is very sensitive to the presence of any impurities that can stabilize the primary radiation-induced defects. IR absorption in irradiated oxygen-doped germanium is shown in Figure 2.2.96 for different annealing conditions. The 1.6-μm (618 cm^{-1}) band has been attributed to the germanium A center, a vacancy + oxygen complex.[162,163] The A center anneals at about 150°C, while other oxygen-defect complexes may persist to 400°C. It is reported that no radiation-induced absorption bands are observed in the 9- to 16-μm range in oxygen-free ($< 10^{16}$ cm^{-3}) germanium.[163] Divacancies in germanium have been suggested to account for a 120 K annealing stage.[161] No IR optical transitions have been ascribed to divacancies in germanium.

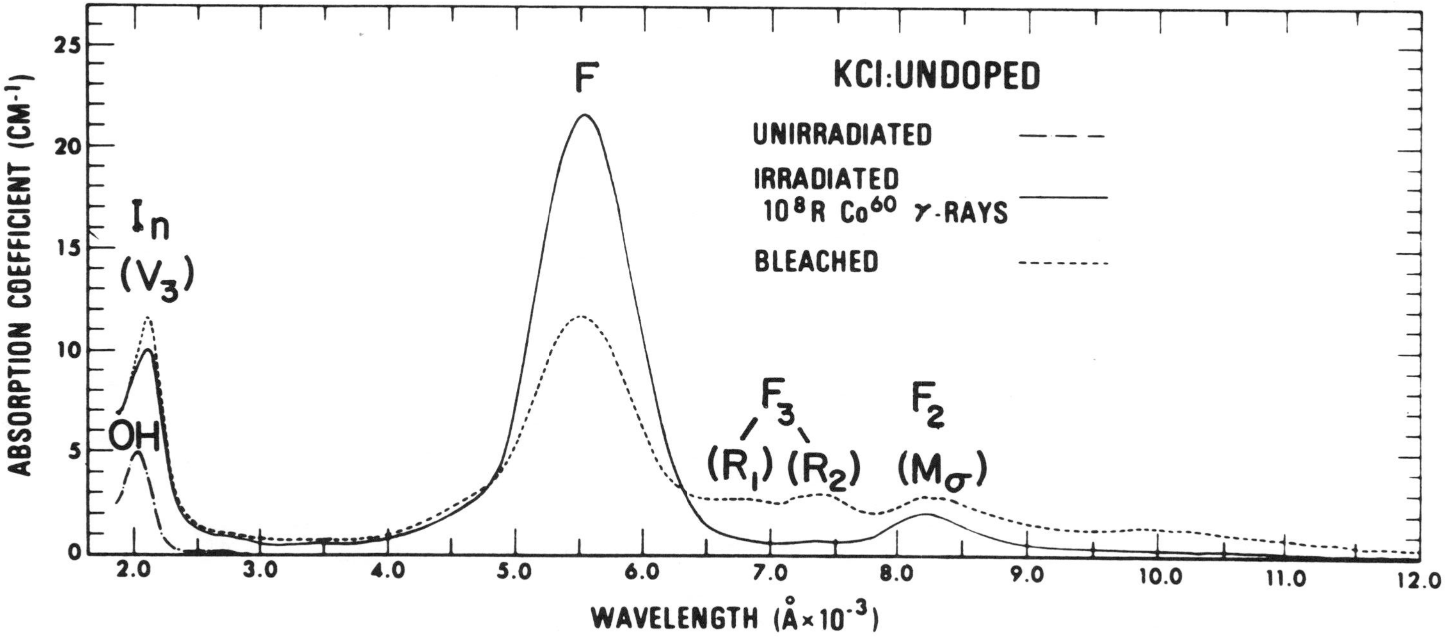

FIGURE 2.2.1. Optical absorption spectra of KCl single crystal before and after gamma irradiation, and after bleaching with a broadband light source (glow bar). Defect absorption bands are labeled using the nomenclature summarized in Table 2.2.1. Irradiation and measurements were made at room temperature. (From Stoebe, T. G., Spry, R. J., and Lewis, J., *Proc. 5th Conf. Infrared Laser Window Mater.*, Andrews, C. F. and Strecker, C. L., Eds., NTIS No. AD-A026363/2ST, 1976, 266. With permission.)

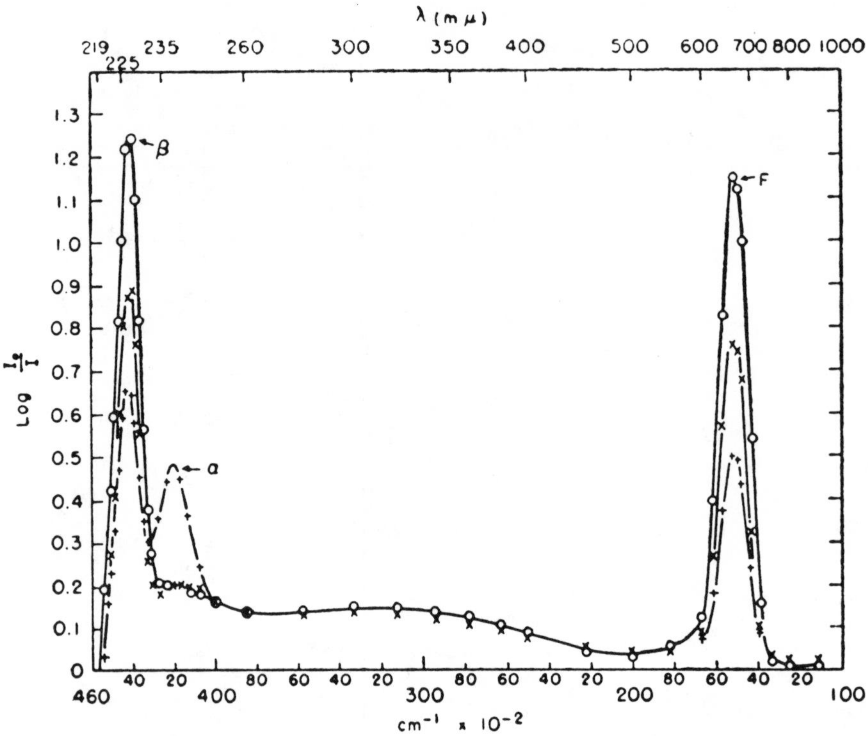

FIGURE 2.2.2. Optical absorption spectra of KI single crystal following treatment as indicated, with all measurements made at 78 K: 0-0-0 exposure to soft X-rays at room temperature; x-x-x 75 hr in dark at room temperature; + − + − + 35 min bleach with F-band light at 78 K. (From Delbecq, C. J., Pringsheim, P., and Yuster, P., *J. Chem. Phys.*, 19, 574, 1951. With permission.)

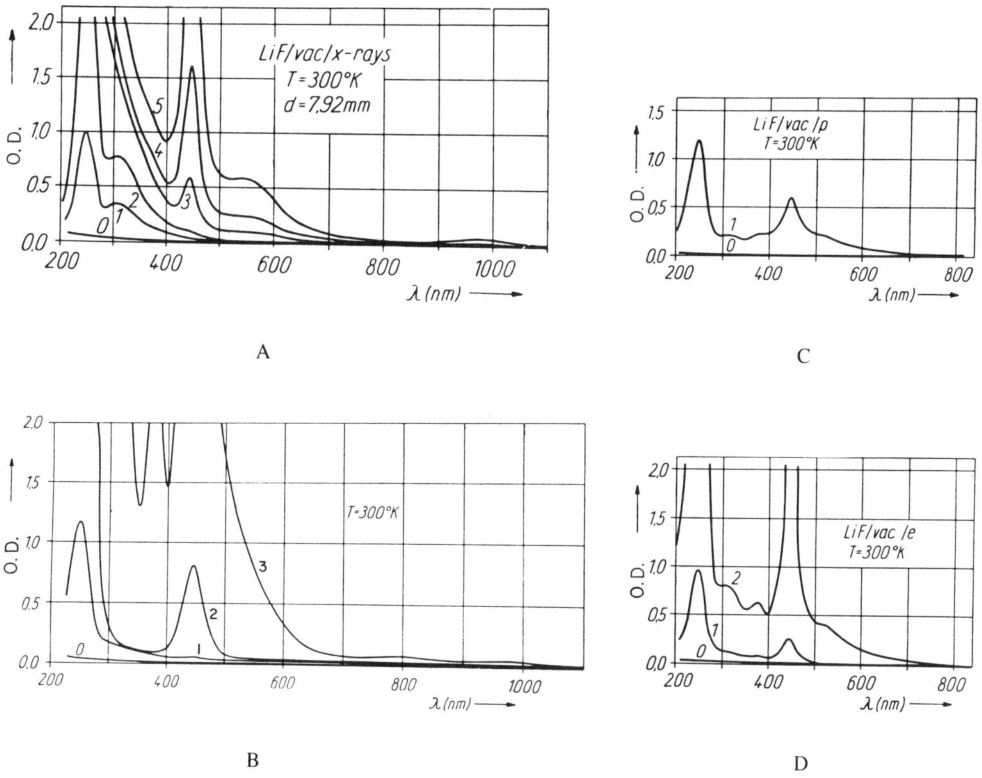

FIGURE 2.2.3. Optical absorption spectra of vacuum-grown single-crystal LiF following room-temperature irradiation as stated. Optical density (O.D. = log I_0/I) is plotted. (A) X-ray irradiation, 0) nonirradiated, 1) 5.37 × 10^3 rad, 2) 4.29 × 10^4 rad, 3) 5.26 × 10^5 rad, 4) 1.08 × 10^6 rad, 5) 2.68 × 10^6 rad; (B) deuteron irradiation, 0) nonirradiated, 1) 10^{12} deuterons/cm^2, 2) 10^{15} deuterons/cm^2, 3) 10^{19} deuterons/cm^2; (C) proton irradiation, 0) nonirradiated, 1) 4.0 × 10^{17} protons/cm^2; (D) 50-keV electron irradiation, 0) nonirradiated, 1) 1.8 × 10^{18} electrons/cm^2, 2) 5.4 × 10^{18} electrons/cm^2. (From Görlich, P., Karras, H., and Kotitz, G., *Phys. Status Solidi*, 3, 1629, 1963. With permission.)

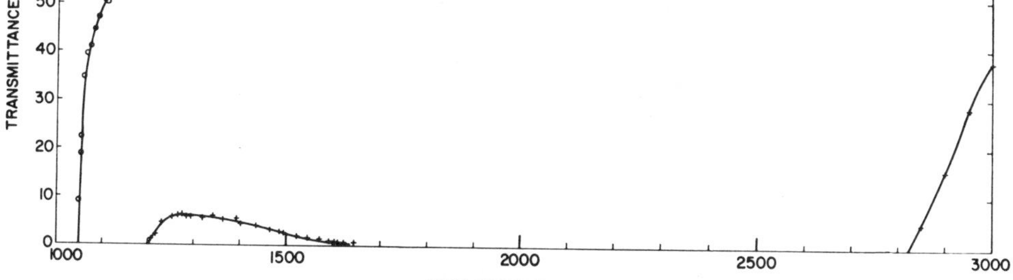

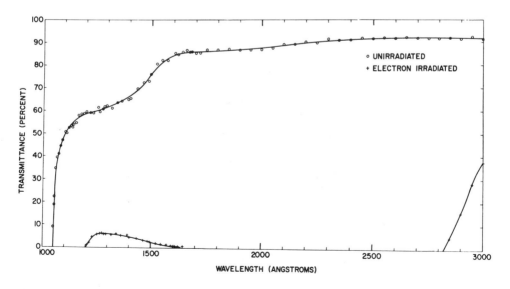

FIGURE 2.2.4. Transmittance of single-crystal LiF before and after irradiation by 1 to 2 MeV electrons to an integrated flux of 10^{14} electrons/cm^2, incident along the direction of the optical probe beam. Irradiation and measurement at room temperature. (From Heath, D. and Sacher, P. A., *Appl. Opt.*, 5, 937, 1966. With permission.)

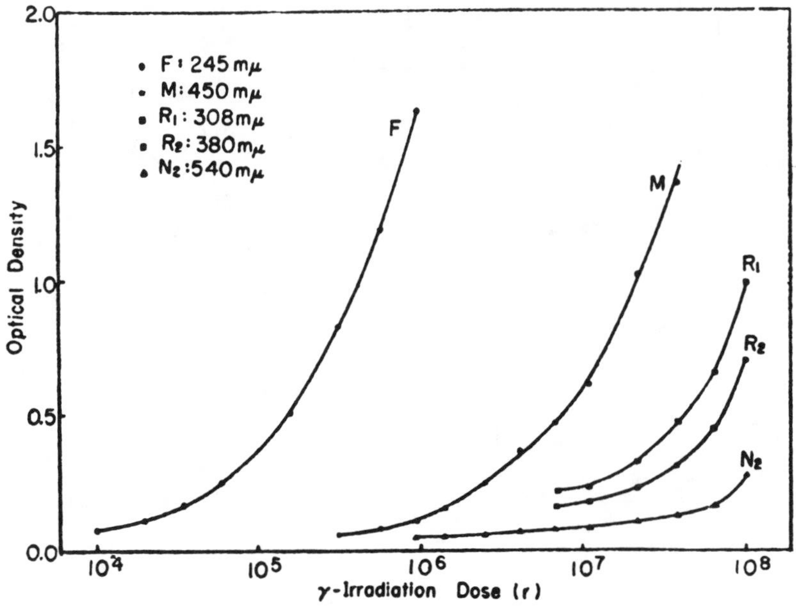

FIGURE 2.2.5. Variation in the optical density of F and F-aggregate defect absorption bands in LiF as a function of increasing gamma radiation dose. The temperature during irradiation was less than 50°C. (From Kubo, K., *J. Phys. Soc. Jpn.*, 16, 2294, 1961. With permission.)

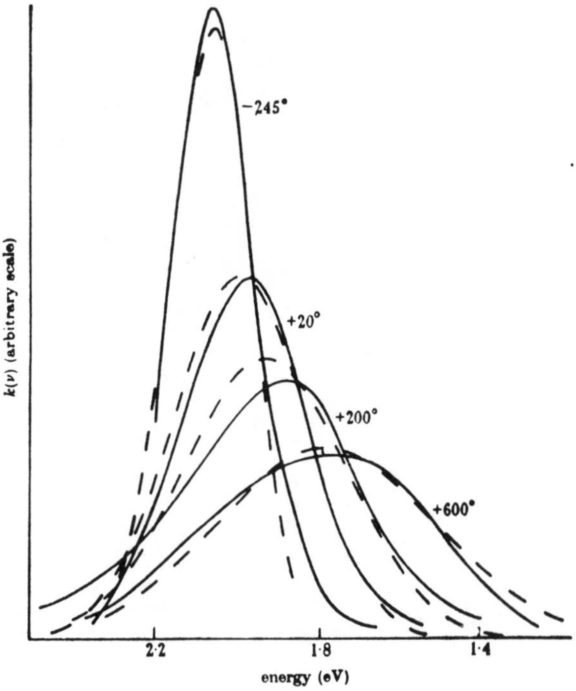

FIGURE 2.2.6. Variation with temperature of the absorption band shape for the lowest-energy F-center transition in KBr. Solid line, experimental; dashed line, theoretical band shape using the model of Huang and Rhys. (Reprinted with permission from Schulman, J. H. and Compton, W. D., *Color Centers in Solids*, Copyright 1962, Pergamon Press, New York.)

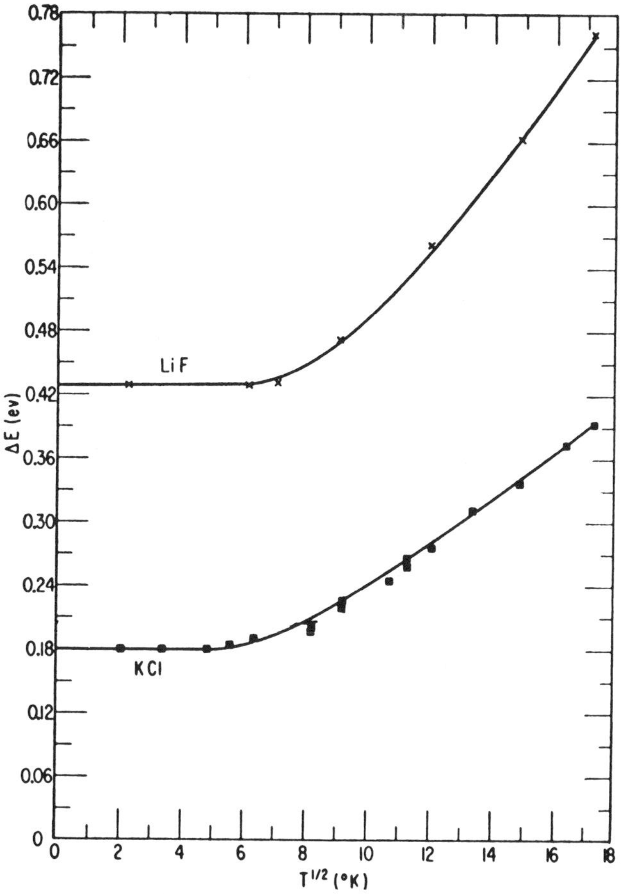

A

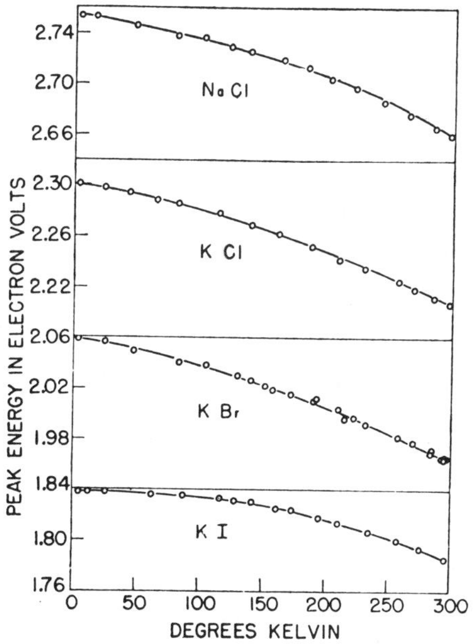

B

FIGURE 2.2.7. (A) Variation of the width of the F band (FWHM) as a function of temperature. The curves are the best fits to the experimental points using Equation 10 and choosing $\nu_g = 2.6 \times 10^{12}$ sec^{-1} for KCl and $\nu_g = 4.1 \times 10^{12}$ sec^{-1} for LiF; (B) variation of the peak of the F-center absorption band with temperature for NaCl, KCl, KBr, and KI. (From Russell, G. A. and Klick, C. C., *Phys. Rev.*, 101, 1473, 1956. With permission.)

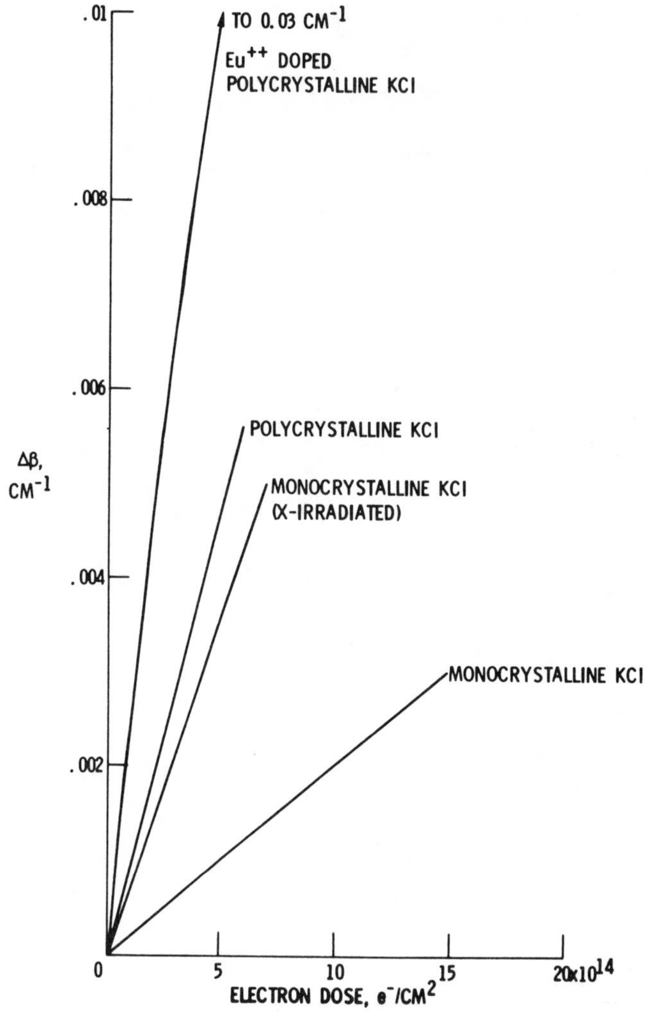

FIGURE 2.2.8. Change in optical absorption coefficient, $\Delta\beta$, at 10.6 μm as a function of 2-MeV electron exposure for electron-irradiated monocrystalline, pure polycrystalline, and Eu^{++}-doped polycrystalline KCl, at room temperature. The X-irradiation dose is represented here as an equivalent electron exposure, where $\Delta\beta = 0.005\ cm^{-1}$ is achieved for a 300-kV X-ray exposure of 1.1×10^8 R. (From Grimes, H. H., Maisel, J. E., and Hartford, R. H., *Proc. Fifth Conf. Infrared Laser Window Materials*, Andrews, C. R. and Strecker, C. L., Eds., NTIS No. N76-1329-68ST, 1976, 280. With permission.)

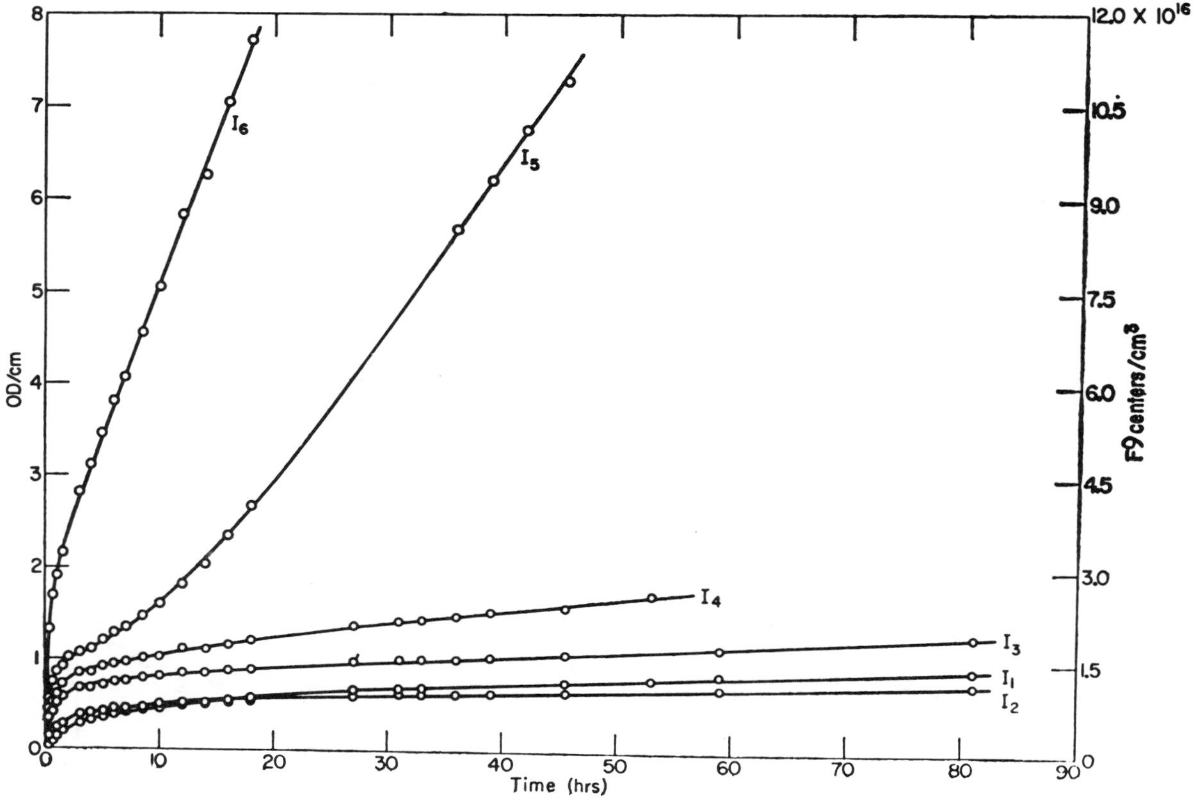

FIGURE 2.2.9. Growth of F-center optical absorption in KCl at room temperature as a function of X-ray exposure time, for six different exposure rates, I_1 to I_6, of 300, 1200, 2700, 4800, 14700, and 30000 R/hr, respectively. (From Mitchell, P. V., Wiegand, D. A., and Smoluchowski, R., *Phys. Rev.*, 121, 484, 1961. With permission.)

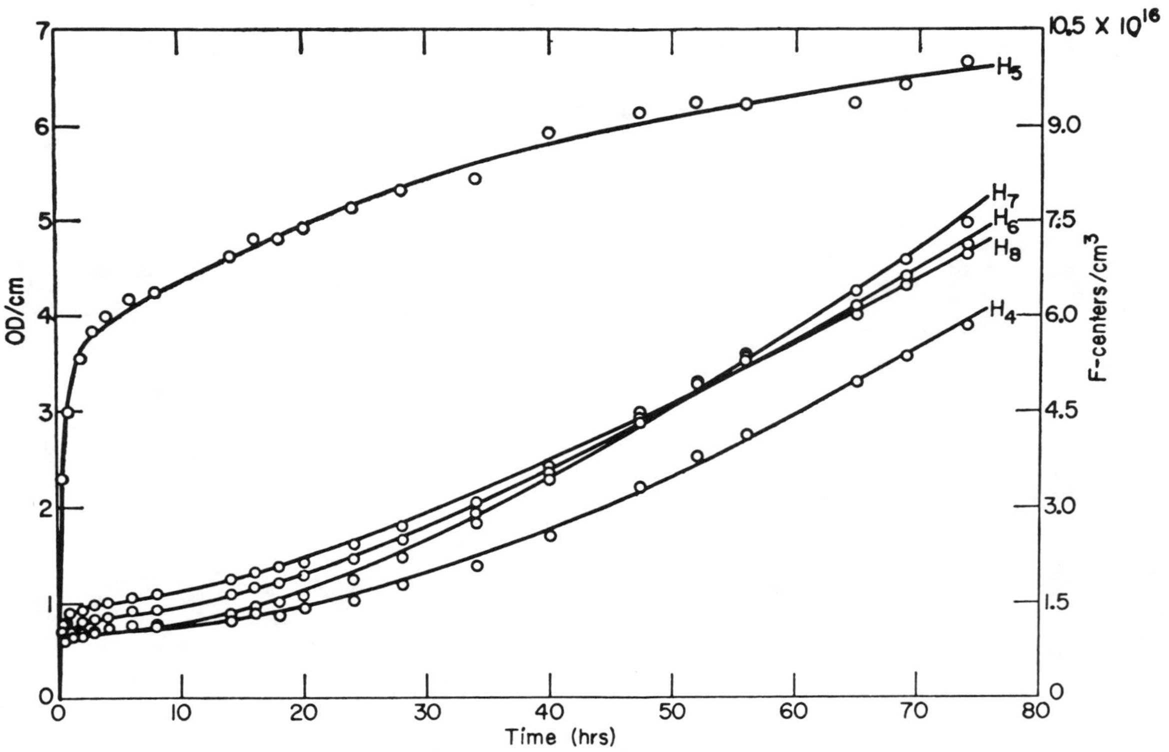

FIGURE 2.2.10. Growth of F-center optical absorption in KCl at room temperature as a function of time of X-ray exposure at 11500 R/hr, for 5 different thermal pretreatments, H4: as cleaved, H5: 500°C, 10 hr, air quenched, H6: 450°C, 10 hr, air quenched, H7: 450°C, 10 hr, cooled 1°C/min, H8: 500°C, 10 hr, cooled 1°C/min. (From Mitchell, P. V., Wiegand, D. A., and Smoluchowski, R., *Phys. Rev.*, 121, 484, 1961. With permission.)

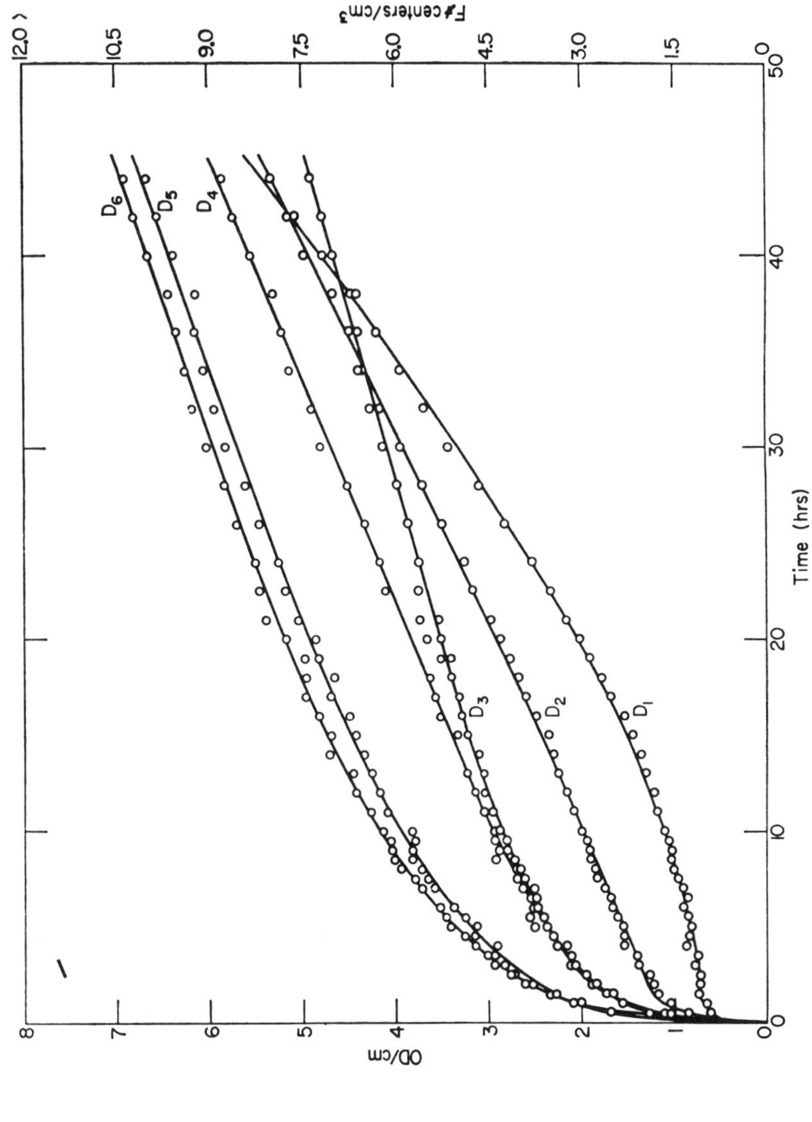

FIGURE 2.2.11. Growth of F-center optical absorption in KCl at room temperature as a function of time of X-ray exposure at 13900 R/hr, for 6 different treatments of mechanical deformation, D1-D6: as cleaved, 0.6% deformed, 1.32, 1.56, 2.02, and 3.04% deformed, respectively. (From Mitchell, P. V., Wiegand, D. A., and Smoluchowski, R., *Phys. Rev.*, 121, 484, 1961. With permission.)

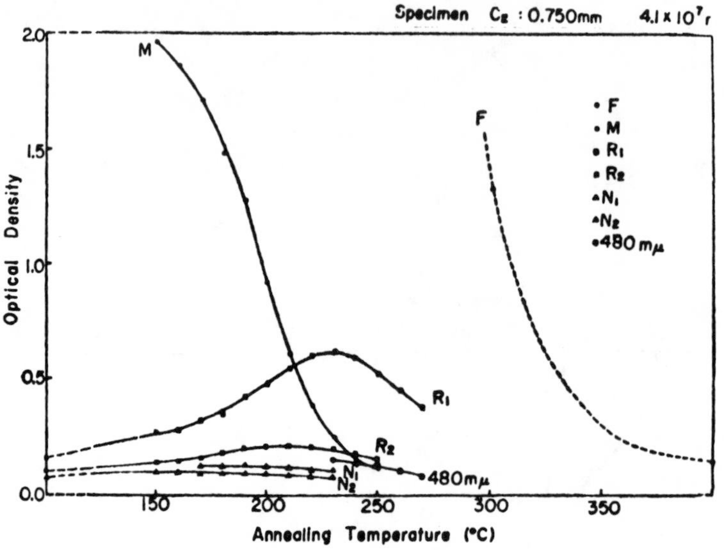

FIGURE 2.2.12. Variation of F and F-aggregate optical absorption bands with annealing in LiF gamma irradiated to 4.1 × 10^7 R. (From Kubo, K., *J. Phys. Soc. Jpn.*, 16, 2294, 1961. With permission.)

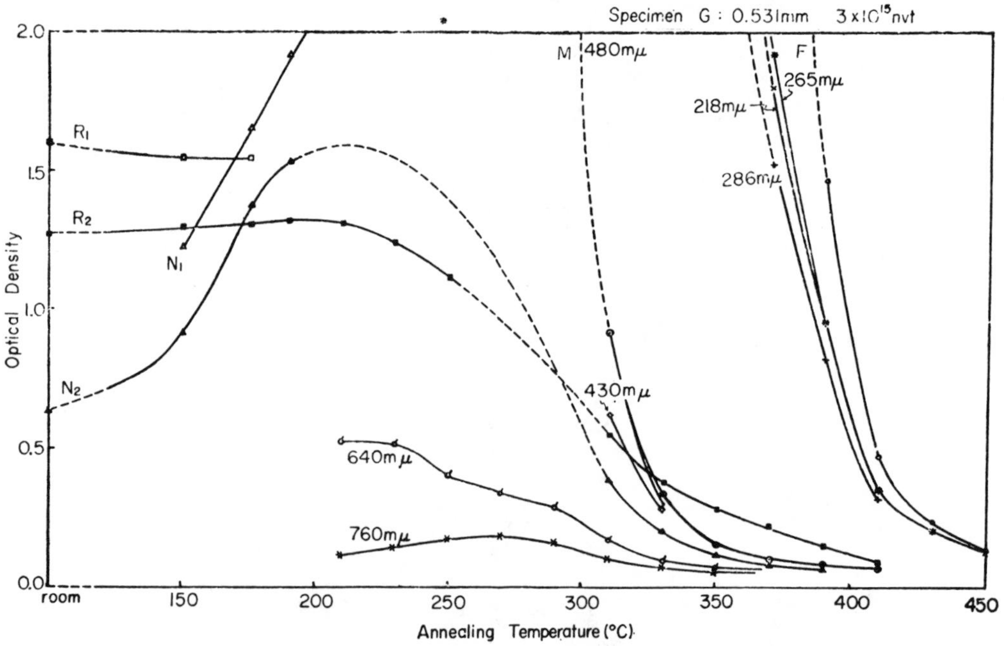

FIGURE 2.2.13. Variation of optical absorption bands, labeled by F-aggregate defect type or wavelength of the band peak, with annealing in LiF irradiated by fission neutrons for an integrated flux of 3 × 10^{15} neutrons/cm^2. (From Kubo, K., *J. Phys. Soc. Jpn.*, 16, 2294, 1961. With permission.)

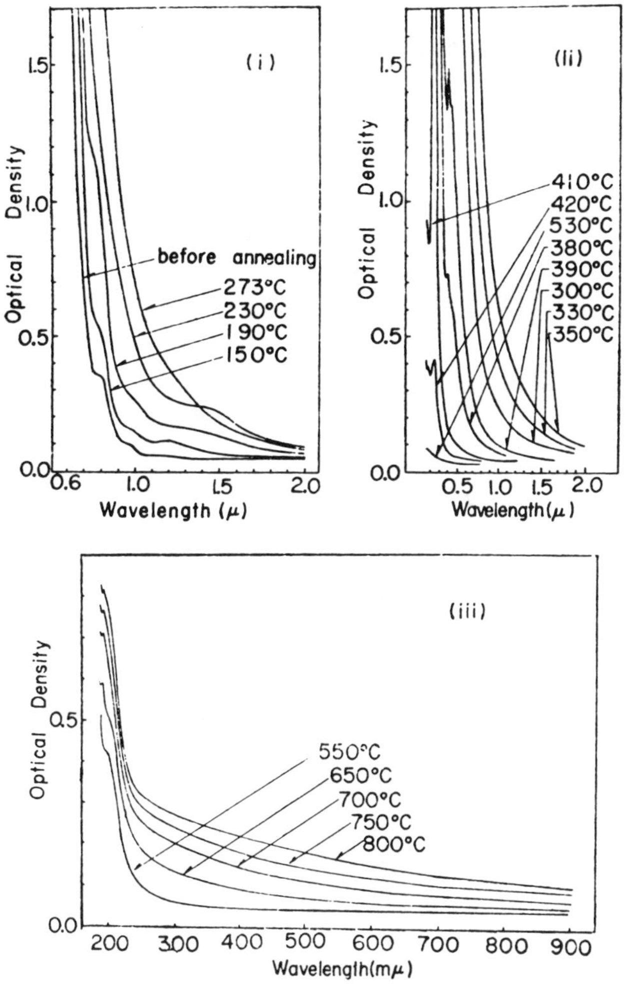

FIGURE 2.2.14. Change in the optical absorption of LiF, 0.5 mm thick, irradiated with 2×10^{17} neutrons/cm^2, with successive annealing at temperatures from 150 to 800°C. (i) From room temperature up to 280°C; (ii) from 300 to 530°C; (iii) from 550 to 800°C. (From Kubo, K., *J. Phys. Soc. Jpn.*, 16, 2294, 1961. With permission.)

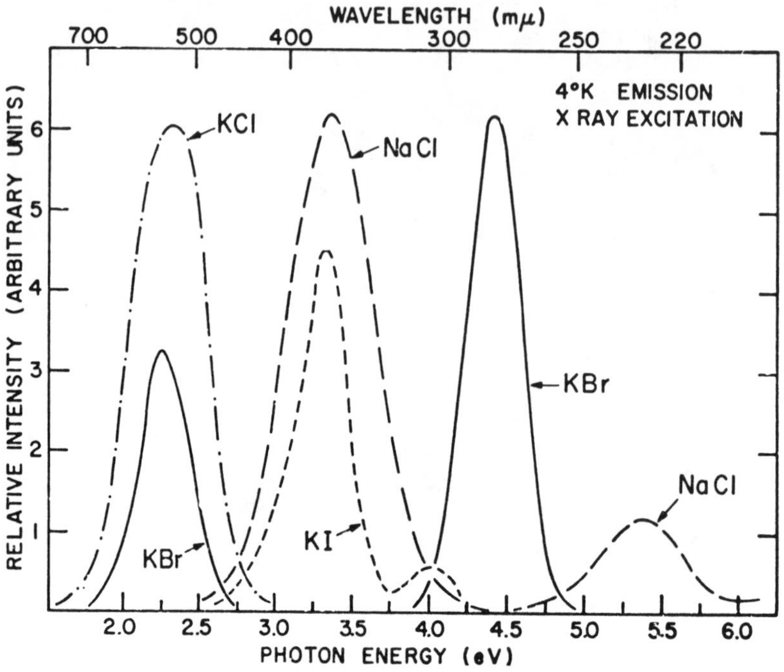

FIGURE 2.2.15. Intrinsic luminescence from recombination of electrons and holes (as self-trapped excitons) in several alkali halides at low temperature. (From Kabler, M. N., *Phys. Rev.*, 136, A1296, 1964. With permission.)

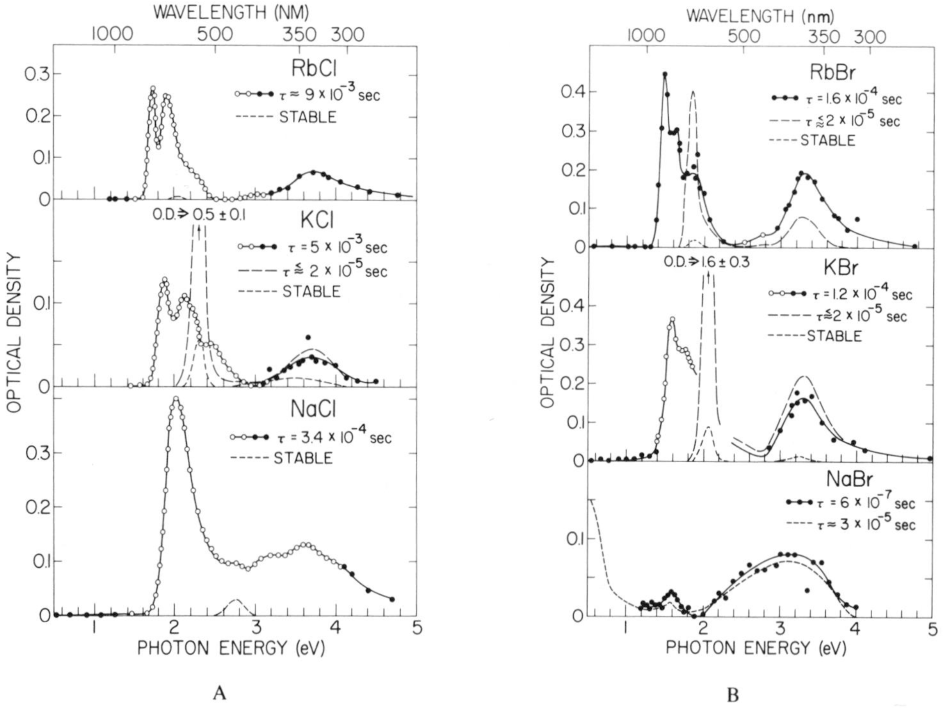

FIGURE 2.2.16. Time-resolved spectra of optical absorption induced in alkali halide crystals at low temperature by a pulse of 500-keV electrons. The spectra shown in solid lines with experimental points share the decay time of the emission from the lowest triplet state of the self-trapped exciton. (From Williams, R. T. and Kabler, M. N., *Phys. Rev. B*, 9, 1897, 1974. With permission.)

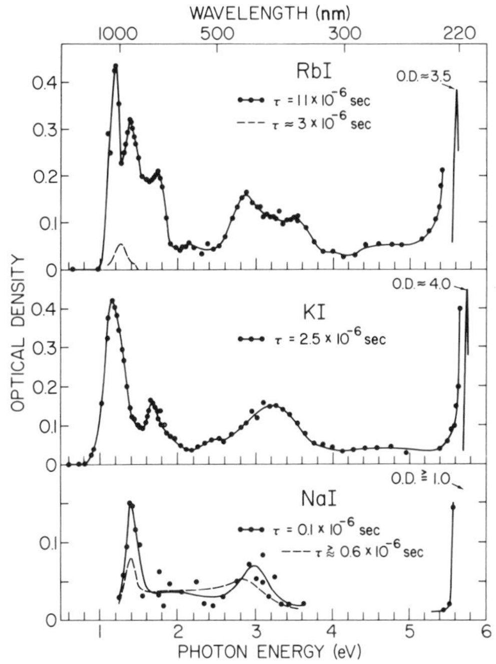

FIGURE 2.2.16C.

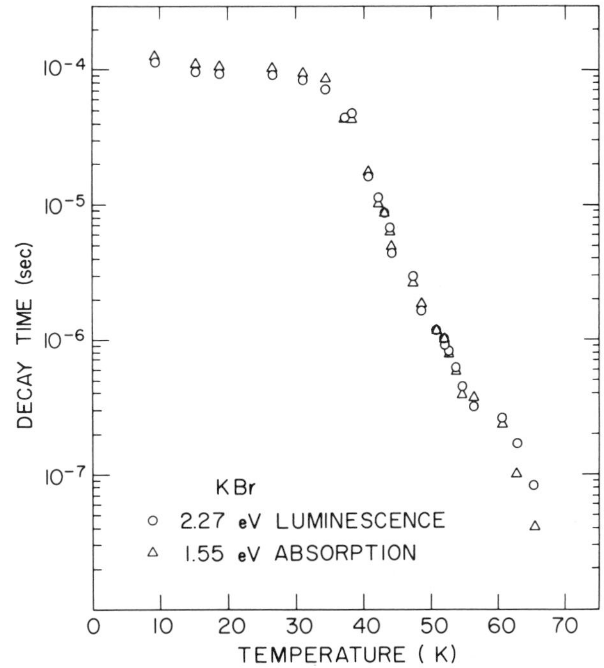

FIGURE 2.2.17. Temperature dependence of the decay time of self-trapped exciton (triplet state) luminescence and the 1.5-eV absorption band in KBr. (From Williams, R. T. and Kabler, M. N., *Phys. Rev. B*, 9, 1897, 1974. With permission.)

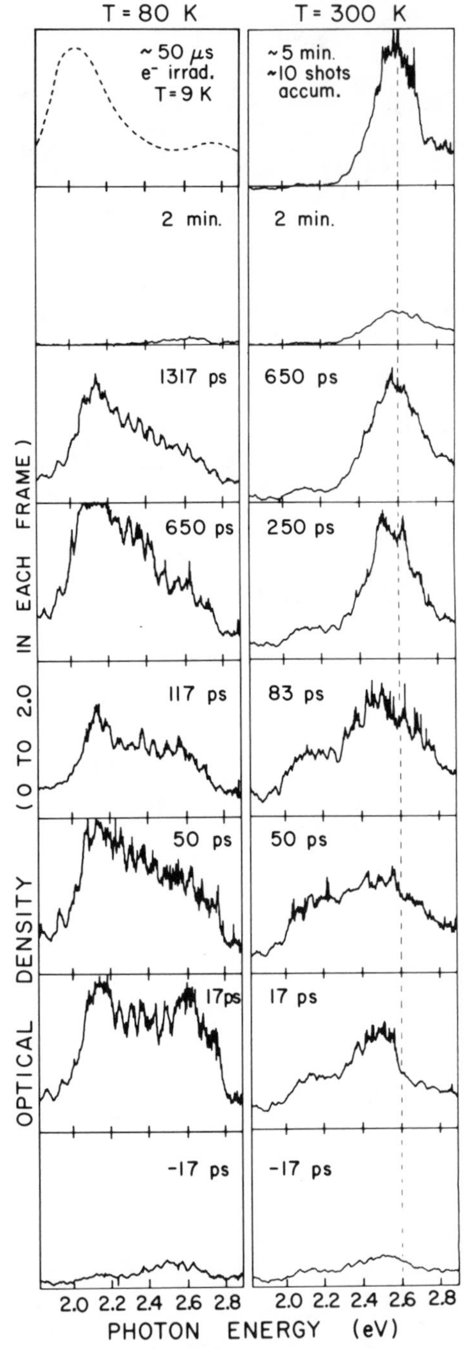

FIGURE 2.2.18. Time series of absorption spectra for NaCl at 80 K (left) and 300 K (right). The probe-pulse delay after the 266-nm (two photon absorption) excitation pulse is given in each frame. The top left frame is taken from Reference 43. STE absorption (peak near 2.1 eV) is dominant at low temperature, while F absorption (peak near 2.6 eV) is dominant at room temperature. (From Williams, R. T., Craig, B. B., and Faust, W. L., *Phys. Rev. Lett.*, 52, 1709, 1984. With permission.)

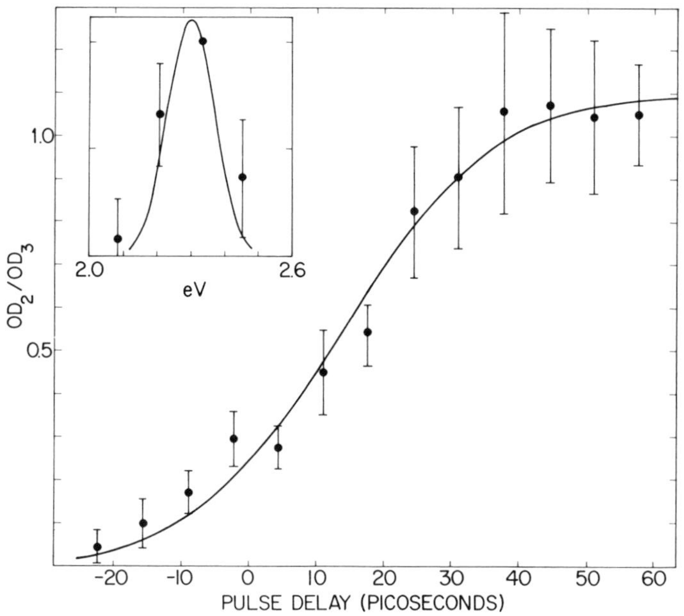

FIGURE 2.2.19. Fractional optical density ($^0D_2/^0D_3$) at 532 nm vs. delay after bandgap excitation by a 266-nm pulse in KCl at T ≈ 25 K. The density $^0D_3$ is measured at 10 nsec. The curve is a best-fit convolution of pulse shapes with an exponential form for defect production, yielding $\tau \approx 11$ psec. Inset: a four-point absorption spectrum measured about 85 psec after bandgap excitation; the curve is the normal F band at 25 K. (From Bradford, J. N., Williams, R. T., and Faust, W. L., *Phys. Rev. Lett.*, 35, 300, 1975. With permission.)

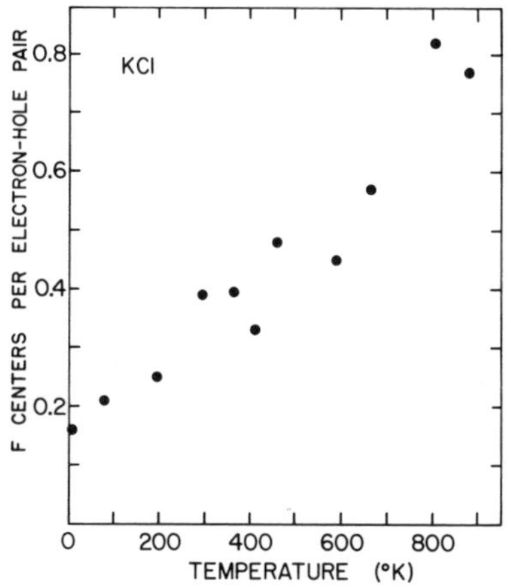

FIGURE 2.2.20. From measurements of 532-nm absorption made 46 psec after bandgap excitation of KCl by a 266-nm pulse, the number of F centers produced per two-photon absorption event has been computed and is plotted as a function of temperature. The production efficiency has been corrected for the temperature dependence of the F-band spectrum and approximately corrected for a small component of 532-nm absorption due to self-trapped excitons. (From Williams, R. T., Bradford, J. N., and Faust, W. L., *Phys. Rev. B*, 18, 7038, 1978. With permission.)

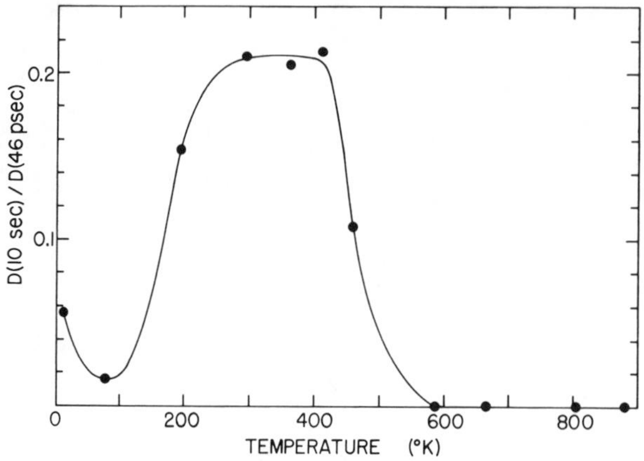

FIGURE 2.2.21. Ratio of optical densities at 532 nm measured 10 sec and 46 psec after bandgap excitation of KCl is plotted as a function of crystal temperature. (From Williams, R. T., Bradford, J. N., and Faust, W. L., *Phys. Rev. B*, 18, 7038, 1978. With permission.)

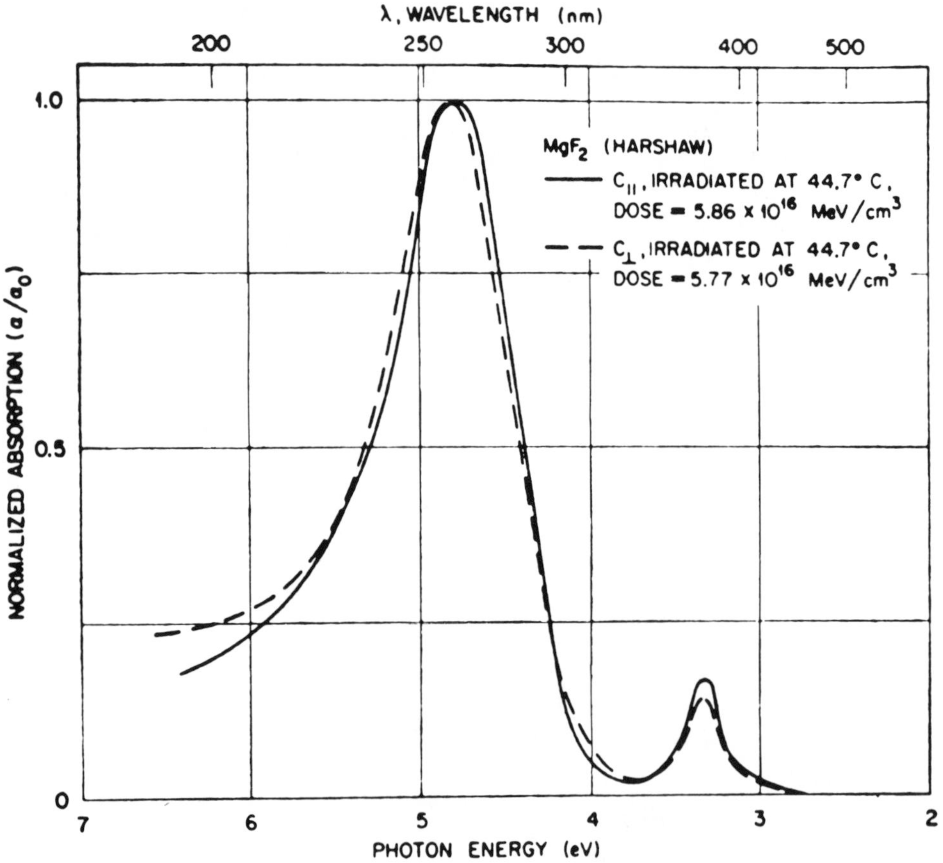

FIGURE 2.2.22. Normalized absorption coefficient vs. photon energy for two samples of MgF_2 cut with different orientations and irradiated to about the same gamma dose. (From Sibley, W. A. and Facey, O. E., *Phys. Rev.*, 174, 1076, 1978. With permission.)

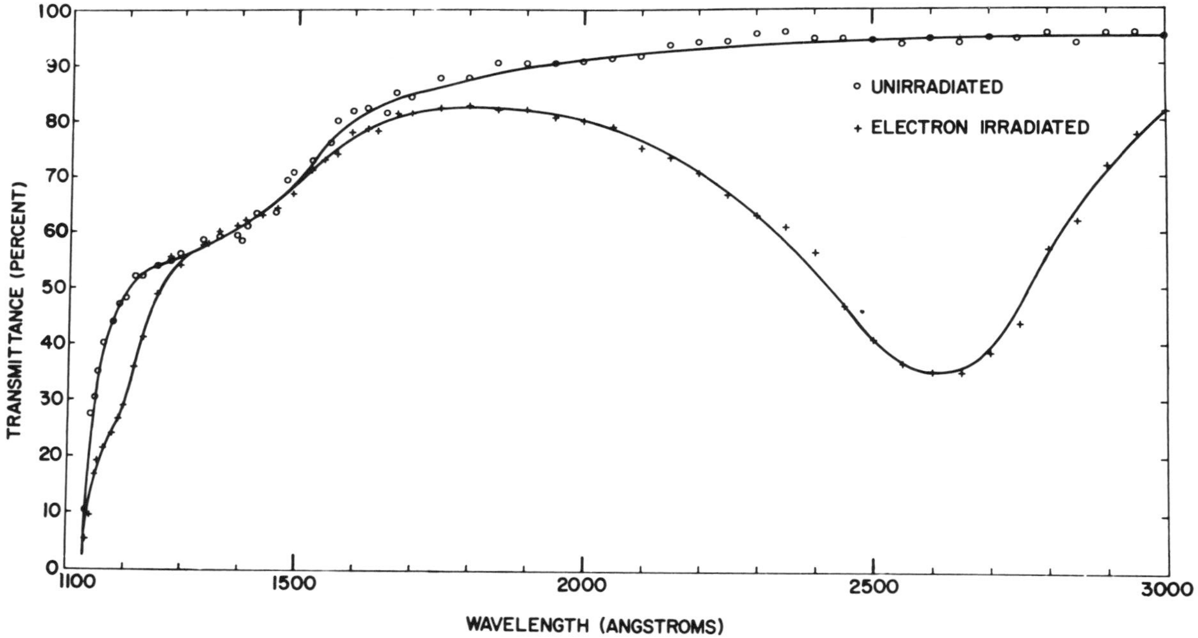

FIGURE 2.2.23. Transmittance of MgF_2 before and after irradiation by 1 to 2 MeV electrons to an integrated flux of 10^{14} electrons/cm^2, incident along the direction of the optical probe beam. Irradiation and measurement at room temperature. (From Heath, D. and Sacher, P. A., *Appl. Opt.*, 5, 937, 1966. With permission.)

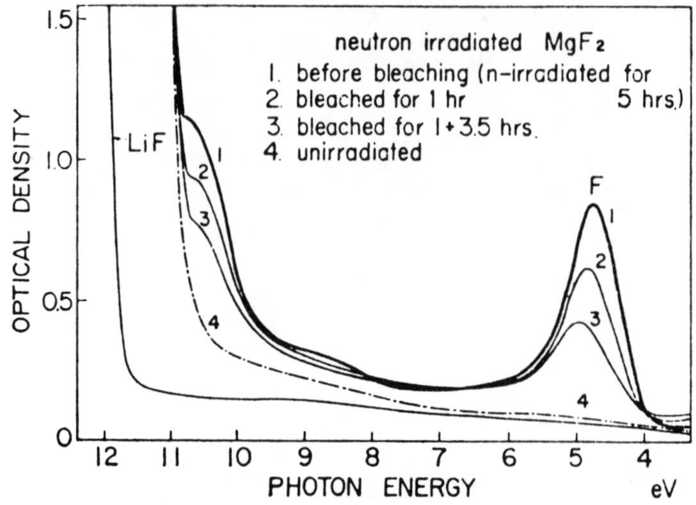

FIGURE 2.2.24. Optical absorption spectra of MgF_2 crystal neutron irradiated (average thermal neutron flux 2.2×10^{13} cm^{-2} sec^{-1}) for 5 hr and subsequently bleached at the F-band wavelength. (From Tsuboi, T., Kato, R., and Nakagawa, M., *J. Phys. Soc. Jpn.*, 16(1961) Fig. 1 on p. 645. With permission.)

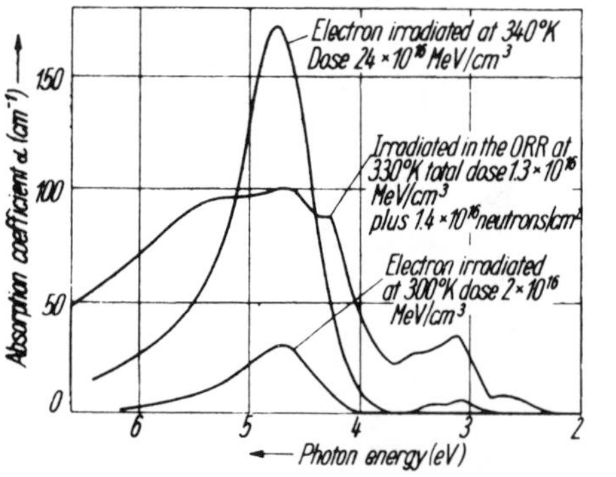

FIGURE 2.2.25. Comparison of optical absorption spectra for electron- and neutron-irradiated MgF_2. (From Facey, O. E., Lewis, D. L., and Sibley, W. A., *Phys. Status Solidi*, 32, 831, 1969. With permission.)

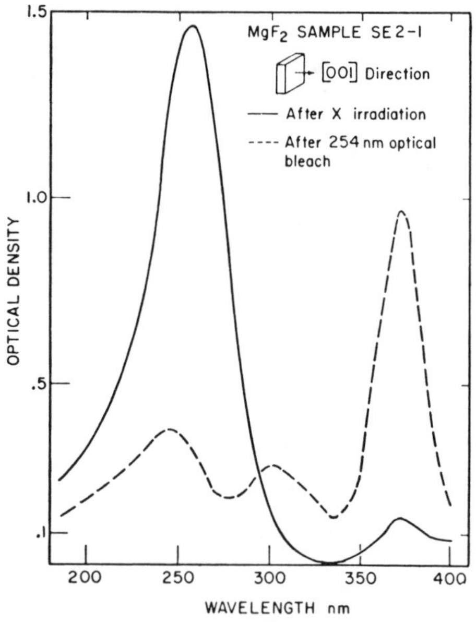

FIGURE 2.2.26. Optical absorption spectra of an MgF_2 crystal (0.05 cm thick) after irradiation at room temperature and after a subsequent 254-nm optical bleach. (From Blunt, R. F. and Cohen, M. I., *Phys. Rev.*, 153, 1031, 1967. With permission.)

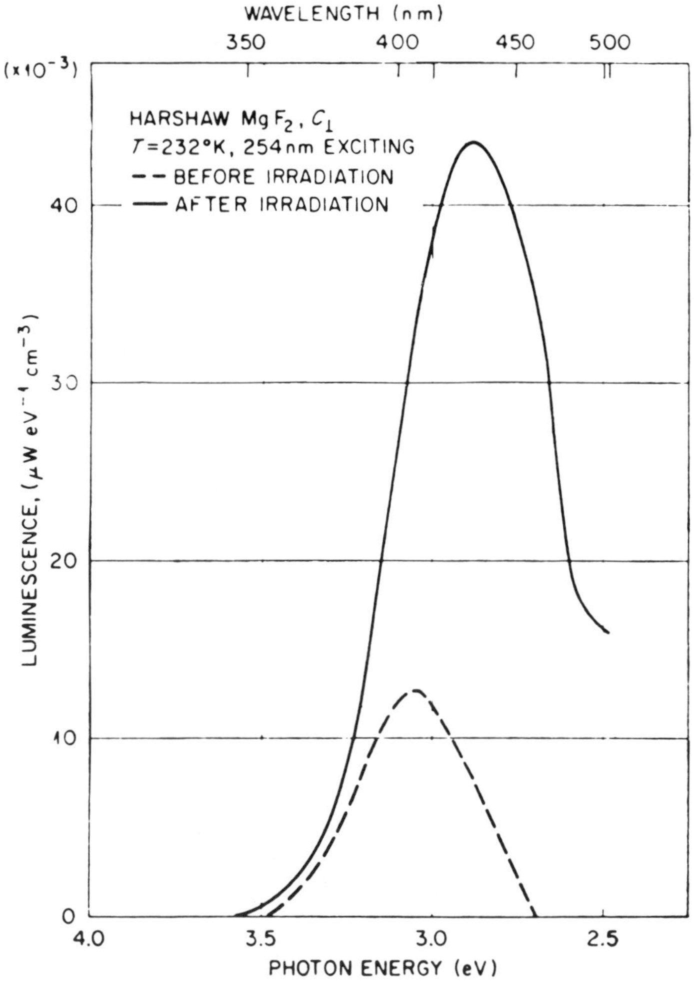

FIGURE 2.2.27. Luminescence excited by 254-nm light in unirradiated and electron-irradiated MgF_2. (From Facey, O. E. and Sibley, W. A., *Phys. Rev.*, 186, 926, 1969. With permission.)

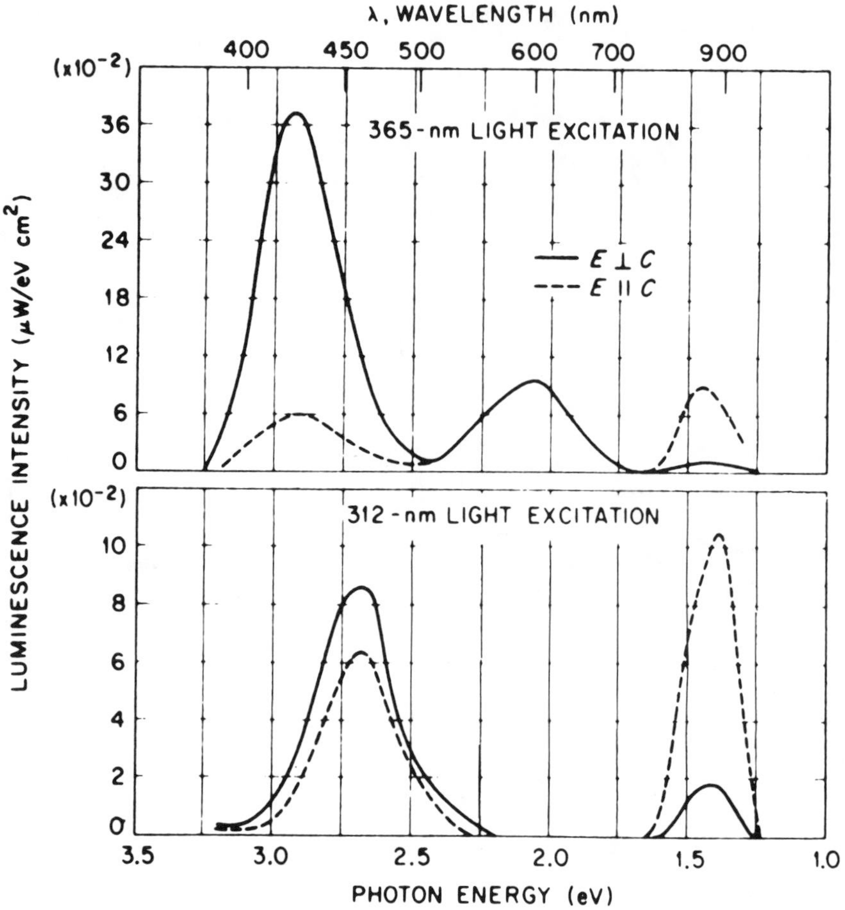

FIGURE 2.2.28. Luminescence at 7 K from neutron-irradiated MgF_2 excited by either 365- or 312-nm light. The solid line depicts data for the electric vector of the emitted light perpendicular to the c axis of the crystal. The dashed line shows data for the electric vector of the emitted light parallel to the crystalline c axis. (From Facey, O. E. and Sibley, W. A., *Phys. Rev. B*, 2, 1111, 1970. With permission.)

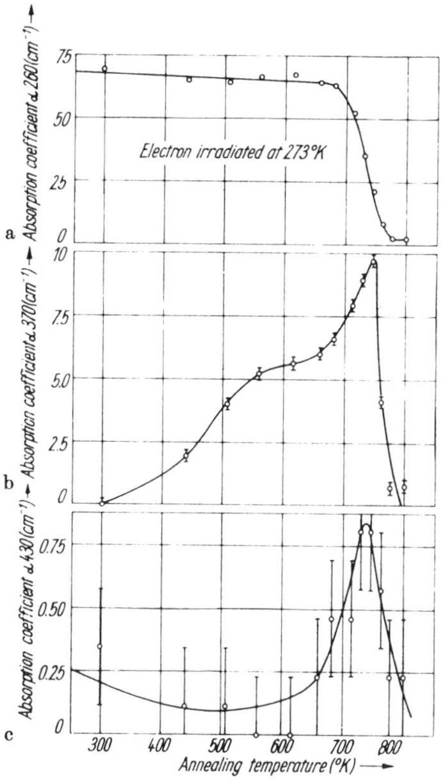

FIGURE 2.2.29. Absorption coefficients at (A) 260 nm; (B) 370 nm, and (C) 430 nm vs. annealing temperature for electron-irradiated MgF_2. (From Facey, O. E., Lewis, D. L., and Sibley, W. A., *Phys. Status Solidi*, 32, 831, 1969. With permission.)

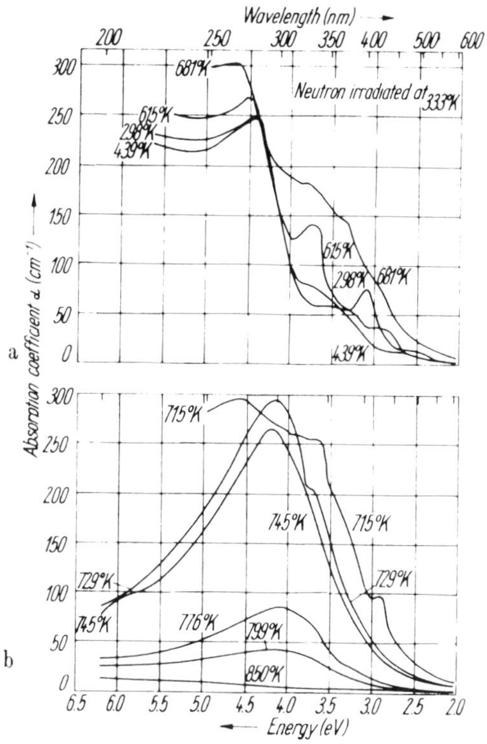

FIGURE 2.2.30. Absorption spectra after different annealing runs for reactor irradiated MgF_2. (From Facey, O. E., Lewis, D. L., and Sibley, W. A., *Phys. Status Solidi*, 32, 831, 1969. With permission.)

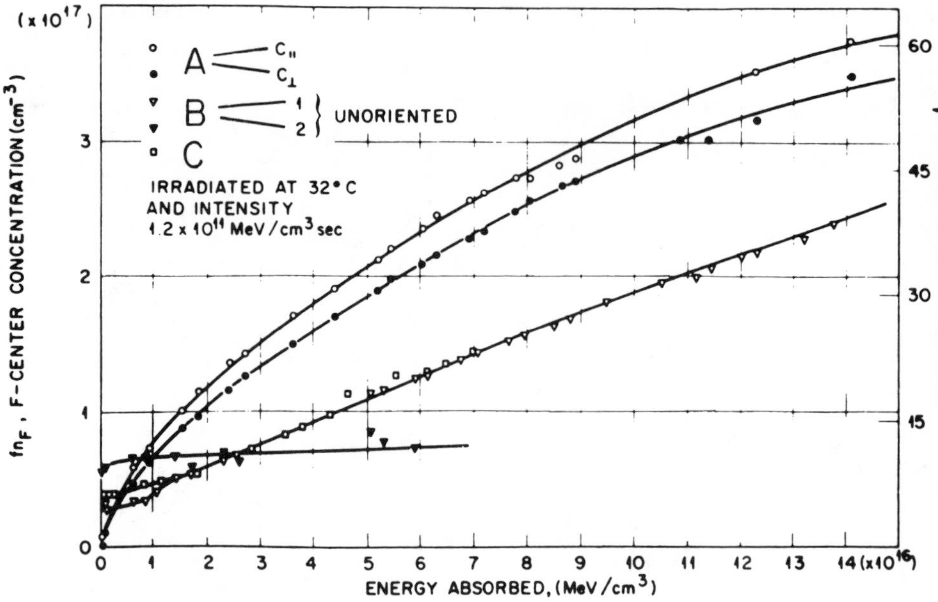

FIGURE 2.2.31. F-center concentration and absorption coefficient vs. radiation dose of 1.7-MeV electrons, for MgF_2 crystals from three commercial sources, characterized by the following impurity analyses (μg/g):

Element	Source A	Source B	Source C
Al	2—10	5—10	—
Ba	<1	<1—10	—
Ca	≤20	8	—
Co	≤10	4	—
Cr	<1—9	<1—50	45
Cu	10—40	5,20	—
Fe	<4	<4—10	150—180
K	3	8	—
Mn	1	3—50	—
Na	<10	<10	69
Mo	<10—60	<10—55	—
Pb	<10	<10	—
Si	100	<10	—
Zn	<100	<100	—

(From Sibley, W. A. and Facey, O. E., *Phys. Rev.*, 174, 1076, 1968. With permission.)

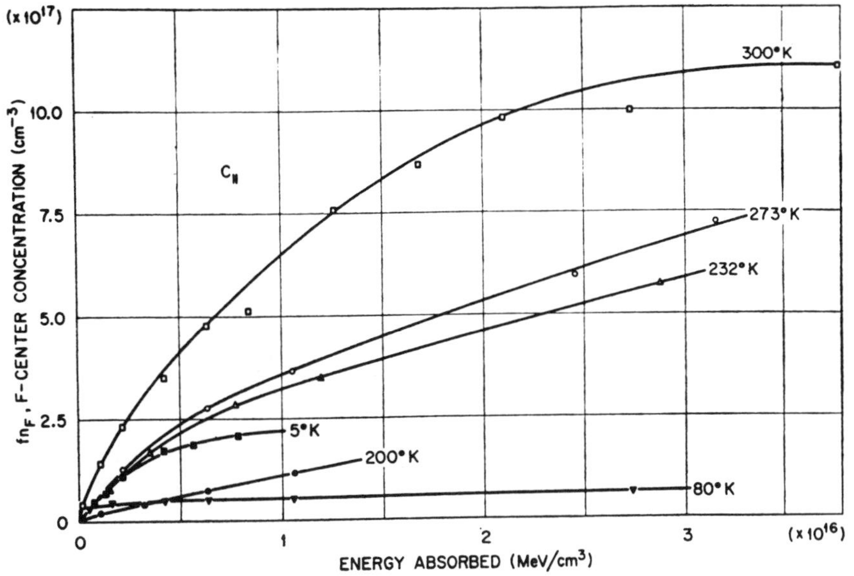

FIGURE 2.2.32. Plot of the F-center concentration vs. energy absorbed as a function of irradiation temperature in MgF_2 at a fixed radiation intensity of 1.2×10^{12} MeV/cm^3-sec. (From Sibley, W. A. and Facey, O. E., *Phys. Rev.*, 174, 1076, 1968. With permission.)

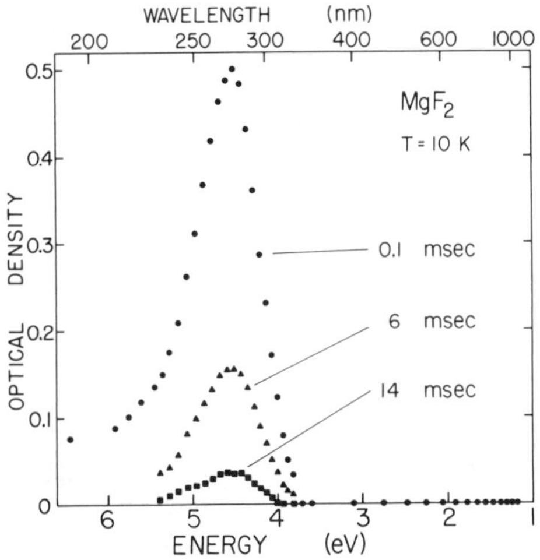

FIGURE 2.2.33. Optical absorption in MgF_2 at 10 K is shown for delays of 0.1, 6, and 14 msec after irradiation by a pulse of 500-keV electrons. Approximately one absorbing center of assumed unit oscillator strength is formed for each 28 eV of radiation energy absorbed in the crystal. The stable absorption remaining after several minutes is about 10^4 times lower than the maximum transient absorption shown here. (From Williams, R. T., Marquardt, C. L., Williams, J. W., and Kabler, M. N., *Phys. Rev. B*, 15, 5003, 1977. With permission.)

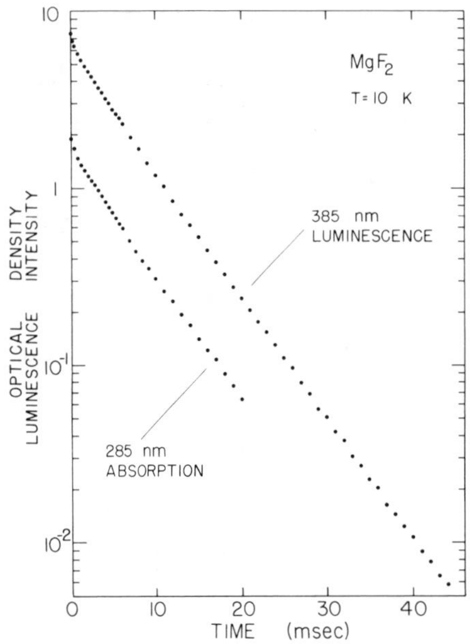

FIGURE 2.2.34.

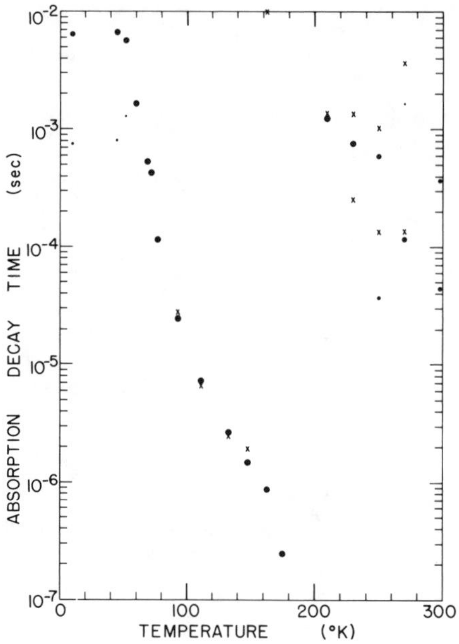

FIGURE 2.2.35.

FIGURE 2.2.34. Decay of optical density at 285 nm and luminescence at 385 nm following electron-pulse irradiation is shown for MgF_2 at 10 K. The initial values are arbitrary. FIGURE 2.2.35. Decay-time constants of 285-nm absorption (circular points) and 255-nm absorption (crosses) are plotted as functions of temperature for MgF_2 irradiated by an electron pulse. The area of each circular point-indicator is proportional to the fractional contribution of the corresponding decay time component to the total 285-nm absorption immediately after excitation. (From Williams, R. T., Marquardt, C. L., Williams, J. W., and Kabler, M. N., *Phys. Rev. B*, 15, 5003, 1977. With permission.)

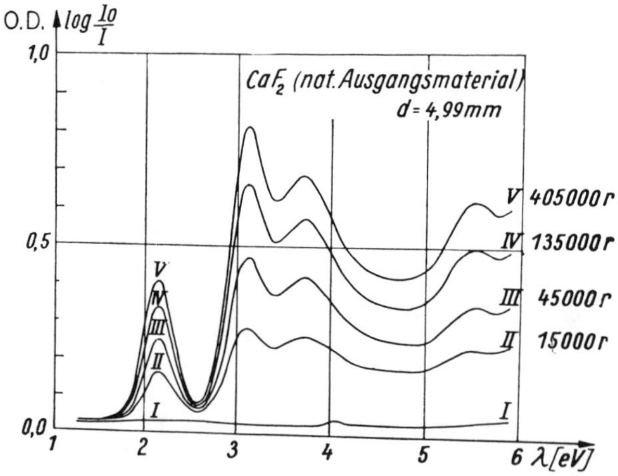

FIGURE 2.2.36. Spectra of optical density for several X-ray irradiation exposures of a CaF_2 crystal containing yttrium as an impurity. Irradiation and measurement at room temperature. (From Gorlich, P., Karras, H., and Lehmann, R., *Phys. Status Solidi*, 1, 389, 1961. With permission.)

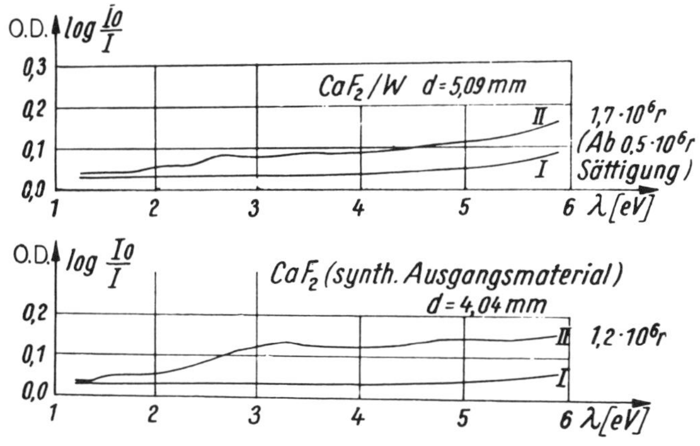

FIGURE 2.2.37. Spectra of optical density in unirradiated (I) and X-ray irradiated (II) CaF_2 of higher purity than in Figure 2.2.36. Irradiation and measurement at room temperature. (From Gorlich, P., Karras, H., and Lehmann, R., *Phys. Status Solidi,* 1, 389, 1961. With permission.)

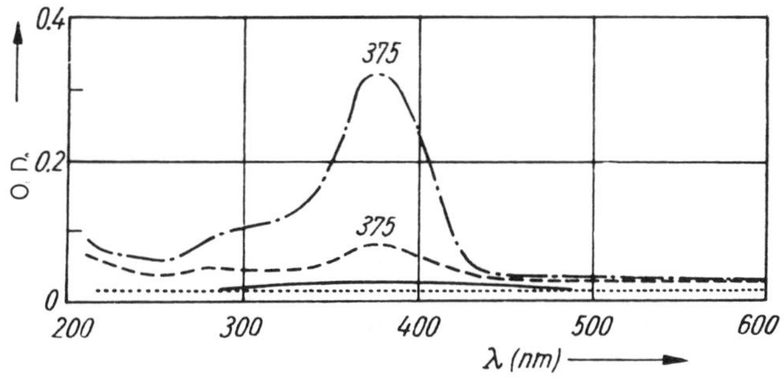

FIGURE 2.2.38. Optical density spectra of pure CaF_2, ... before irradiation; ———— irradiated and measured at room temperature; ---- irradiated and measured at 84 K; —.—.—. irradiated and measured at 37 K. Radiation exposure = 2.1 × 10^5 R. Sample thickness = 5.32 nm. (From Görlich, P., Karras, H., Symanoski, Ch., and Ullmann, P., *Phys. Status Solidi,* 25, 93, 1968. With permission.)

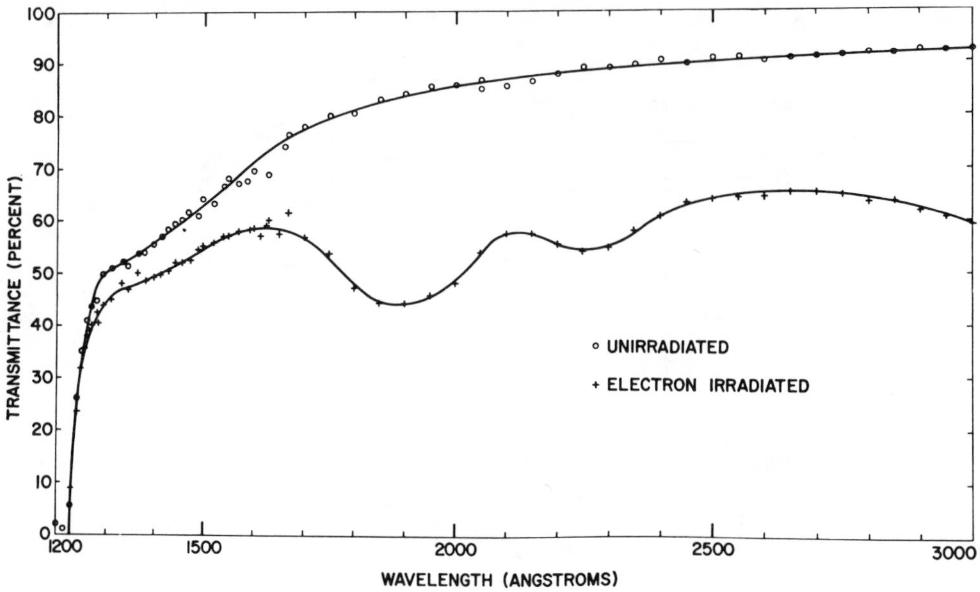

FIGURE 2.2.39. Transmittance of CaF_2 before and after irradiation by 1 to 2 MeV electrons to an integrated flux of 10^{14} electrons/cm^2 at room temperature. (From Heath, D. and Sacher, P. A., *Appl. Opt.*, 5, 937, 1966. With permission.)

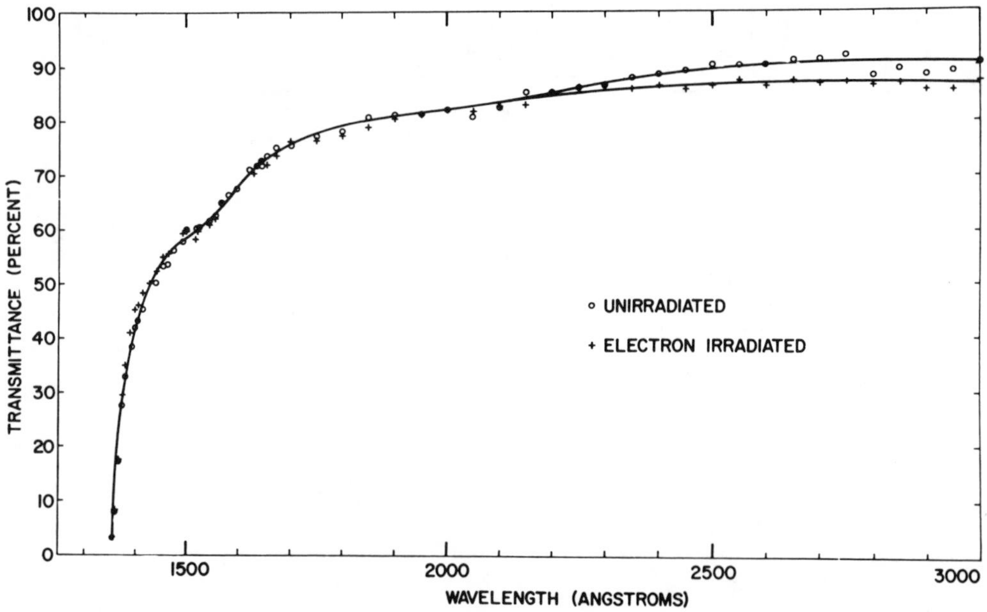

FIGURE 2.2.40. Transmittance of BaF_2 before and after irradiation by 1 to 2 MeV electrons to an integrated flux of 10^{14} electrons/cm^2, at room temperature. (From Heath, D. and Sacher, P. A., *Appl. Opt.*, 5, 937, 1966. With permission.)

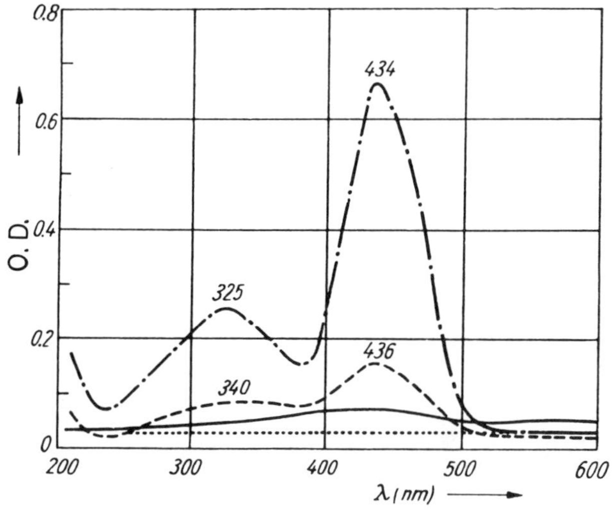

FIGURE 2.2.41. Optical density spectra of pure SrF_2, ... before irradiation; ——— irradiated and measured at room temperature; ---- irradiated and measured at 84 K; —.—.—. irradiated and measured at 37 K. Radiation exposure = 1 × 10^5 R. Sample thickness = 4.85 mm. (From Görlich, P., Karras, H., Symanoski, Ch., and Ullmann, P., *Phys. Status Solidi,* 25, 93, 1968. With permission.)

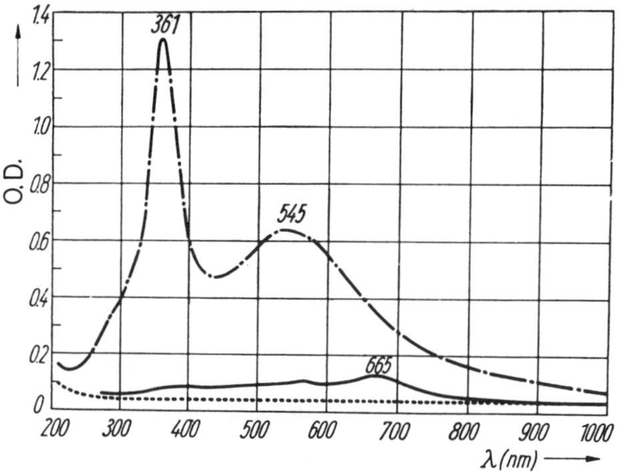

FIGURE 2.2.42. Optical density spectra of pure BaF_2, ... before irradiation; ---- irradiated and measured at room temperature; —.—.—. irradiated and measured at 84 and 37 K, respectively. Radiation exposure = 3.1 × 10^5 R (300 K) and 0.015 × 10^5 R (84 and 37 K). Sample thickness = 4.72 nm. (From Görlich, P., Karras, H., Symanowski, Ch., and Ullmann, P., *Phys. Status Solidi,* 25, 93, 1968. With permission.)

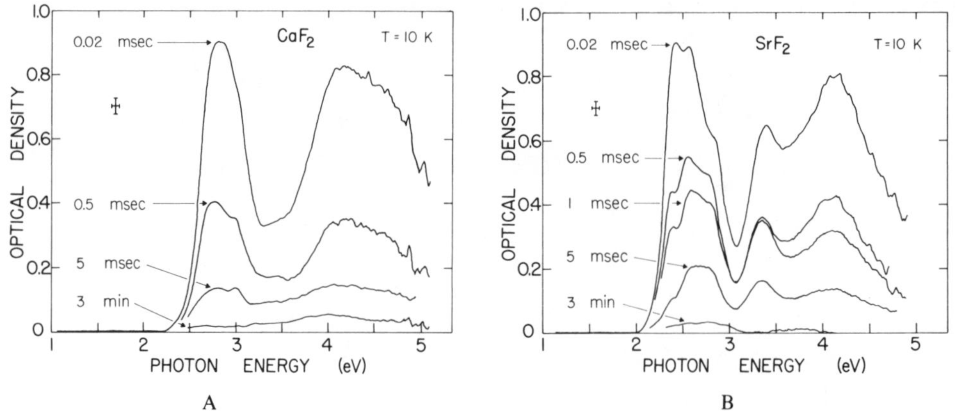

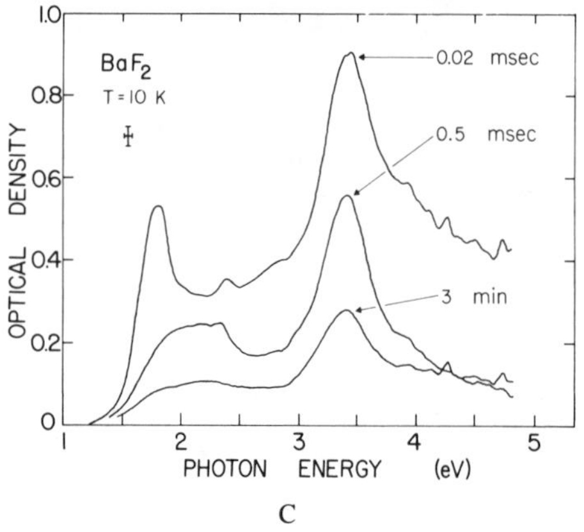

FIGURE 2.2.43. Transient optical absorption in CaF_2, SrF_2, and BaF_2 at T = 10 K is shown for the indicated delays after irradiation by a pulse of 500-keV electrons. (From Williams, R. T., Kabler, M. N., Hayes, W., and Stott, J. P., *Phys. Rev. B*, 14, 725, 1976. With permission.)

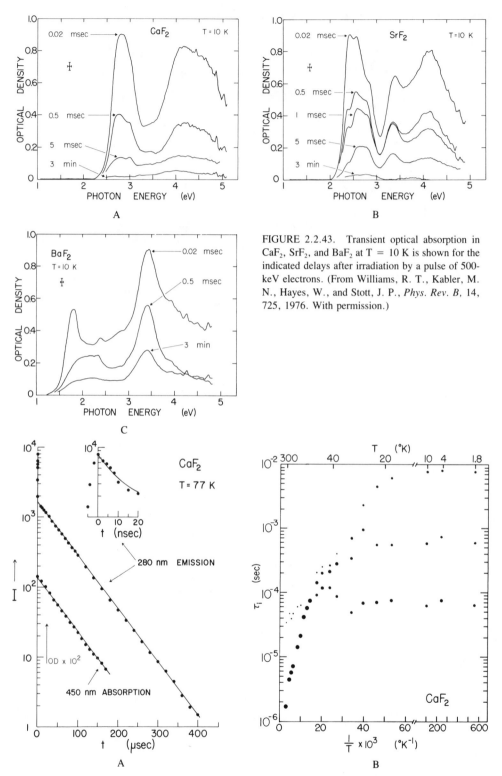

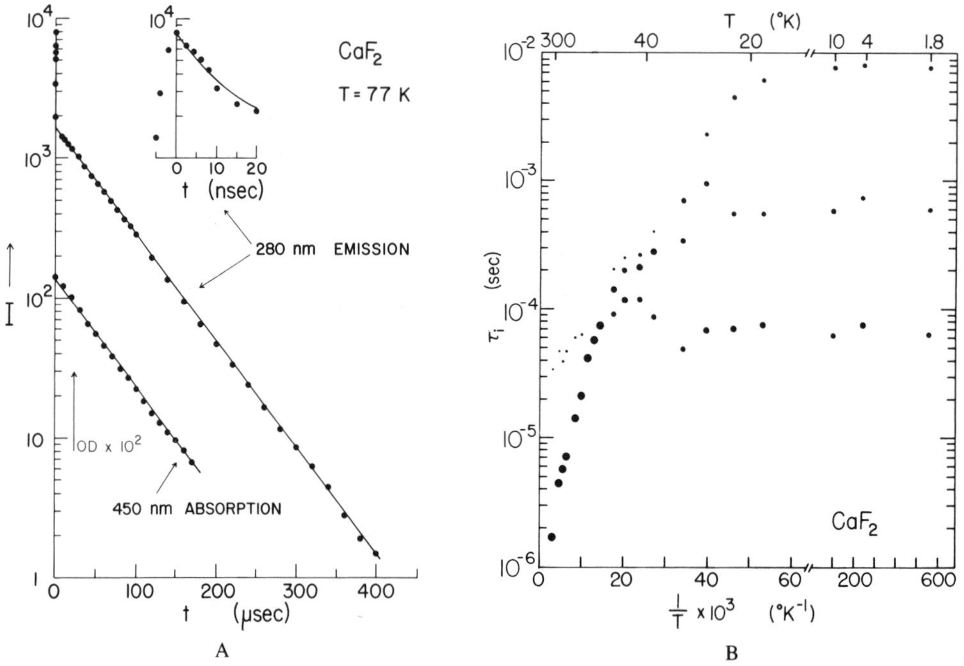

FIGURE 2.2.44. (A) Decay of 280-nm emission and 450-nm absorption in CaF_2 at T = 77 K following pulsed electron irradiation. (B) time constants, τ_i, determined by fitting the decay of 280-nm emission with a sum of exponential components, are plotted as functions of temperature. The area of each point indicator for τ_i is proportional to the time-integrated intensity of that component, $A_i\tau_i$, where $\Sigma_i A_i\tau_i = 1$ at a given temperature. (From Williams, R. T., Kabler, M. N., Hayes, W., and Stott, J. P., *Phys. Rev. B*, 14, 725, 1976. With permission.)

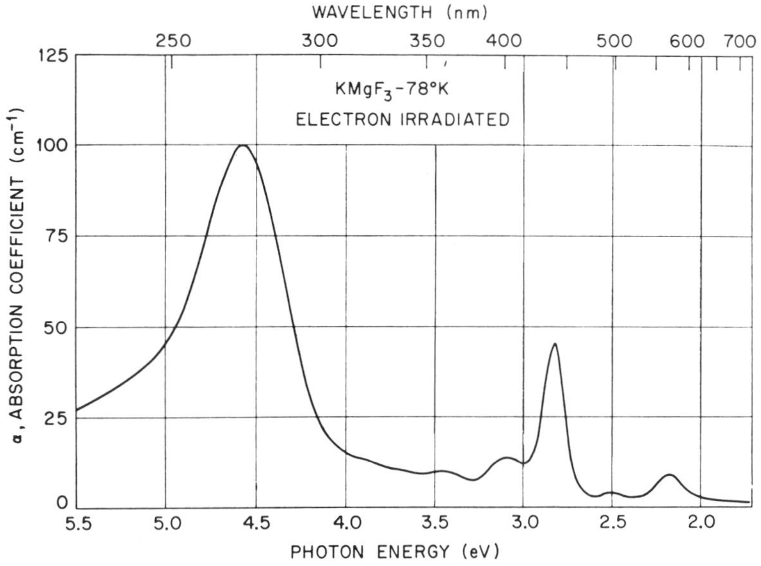

FIGURE 2.2.45. Optical absorption spectrum for $KMgF_3$ crystal irradiated by 2-MeV electrons at 78 K. (From Riley, C. R. and Sibley, W. A., *Phys. Rev. B*, 1, 2789, 1970. With permission.)

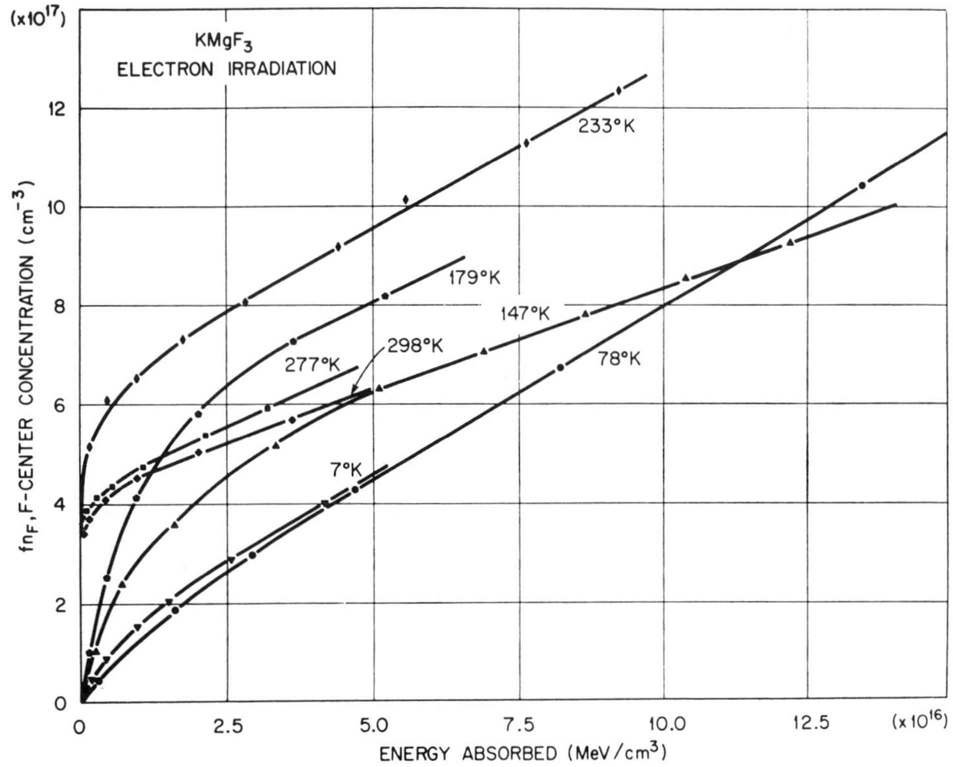

FIGURE 2.2.46. F-center concentration vs. 2-MeV electron radiation dose in $KMgF_3$ at different temperatures as indicated. (From Riley, C. R. and Sibley, W. A., *Phys. Rev. B*, 1, 2797, 1970. With permission.)

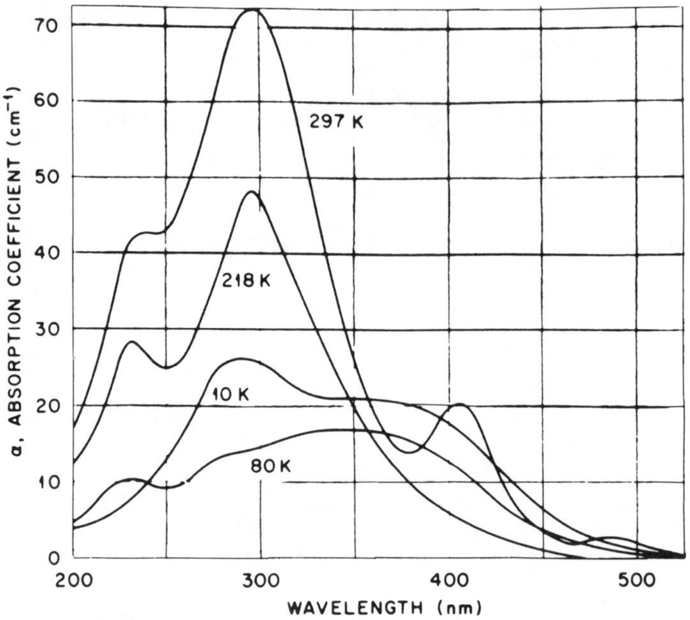

FIGURE 2.2.47. Absorption spectra of $NaMgF_3$ irradiated at temperatures indicated. The irradiation doses for the four samples were, for ascending order of irradiation temperatures, 1.7, 1.7, 1.6, and 2.1×10^{14} MeV/cm³. (From Seretlo, J. R., Martin, J. J., and Sonder, E., *Phys. Rev. B,* 14, 5404, 1976. With permission.)

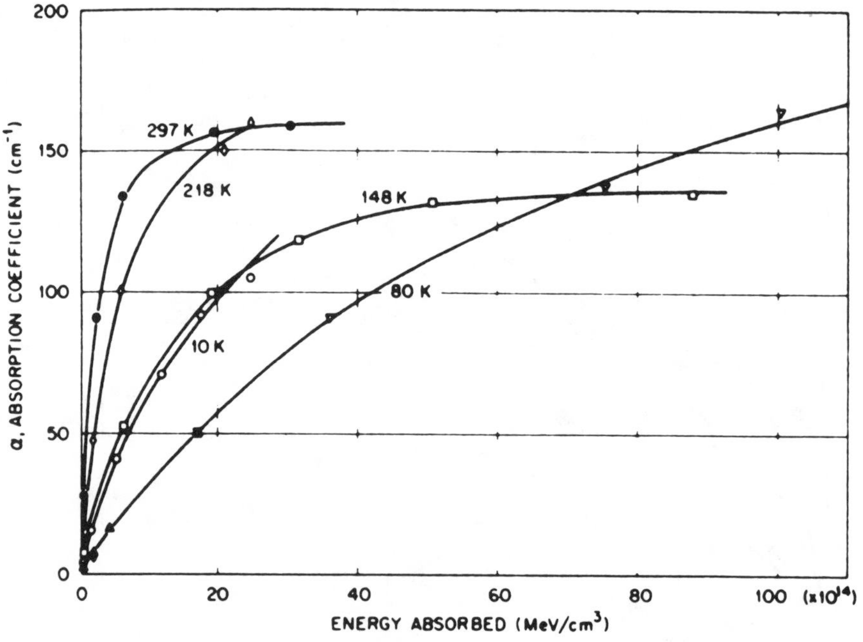

FIGURE 2.2.48. Growth of the 290-nm absorption band in $NaMgF_3$. The temperatures labeling the curves indicate sample temperature during irradiation. (From Seretlo, J. R., Martin, J. J., and Sonder, E., *Phys. Rev. B,* 14, 5404, 1976. With permission.)

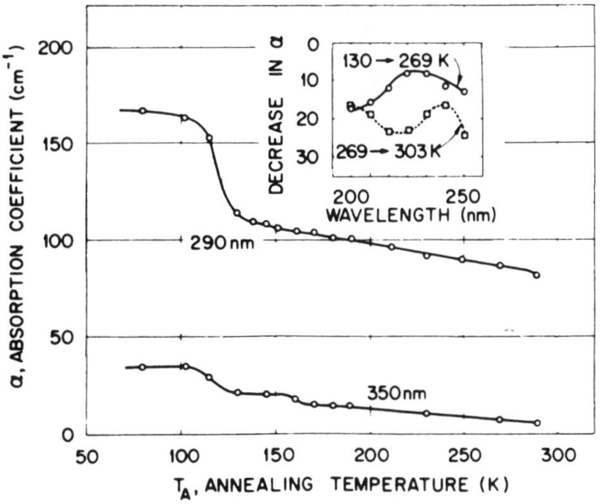

FIGURE 2.2.49. Stability of radiation-induced absorption in Na-MgF_3. The inset depicts spectra in the range 200 to 260 nm, obtained by taking the difference in absorption coefficient between curves obtained after 130 and 296 K annealing (circles) and 269 and 303 K annealing (squares). (From Seretlo, J. R., Martin, J. J., and Sonder, E., *Phys. Rev. B,* 14, 5404, 1976. With permission.)

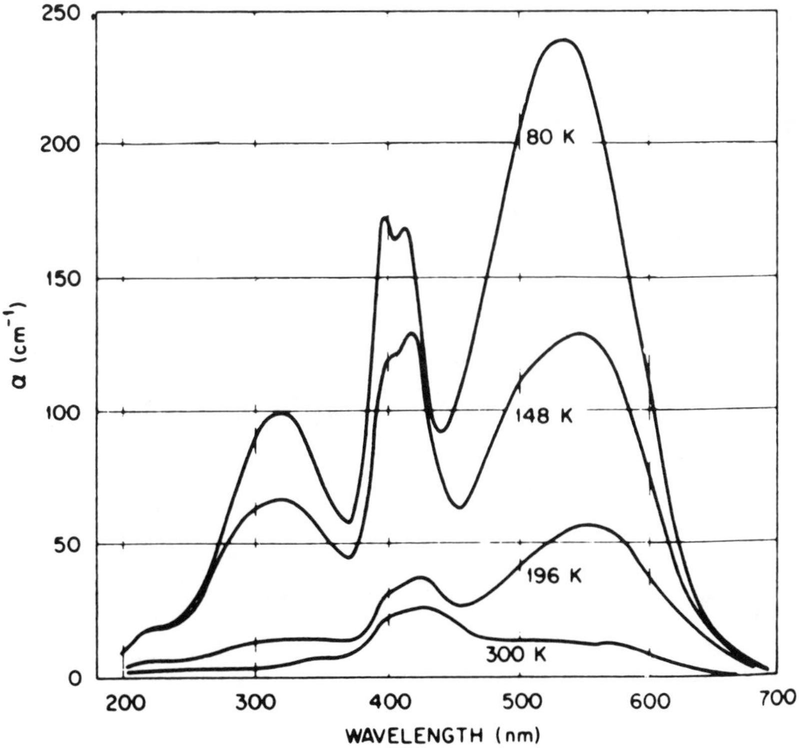

FIGURE 2.2.50. Absorption spectra of $RbCaF_3$ irradiated at temperatures indicated, and measured at 80 K. The irradiation doses for the four samples were, for ascending irradiation temperatures, 10, 8.5, 11, and 10 × 10^{14} MeV/cm^3. (From Seretlo, J. R., Martin, J. J., and Sonder, E., *Phys. Rev. B,* 14, 5404, 1976. With permission.)

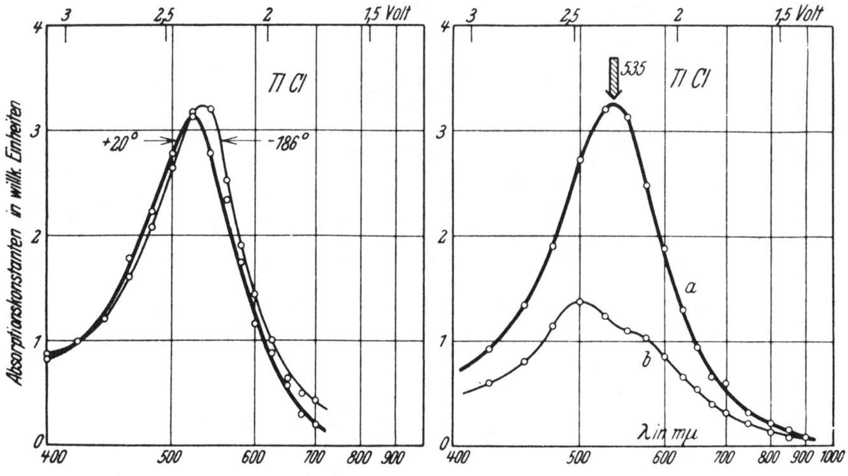

A B

FIGURE 2.2.51. (A) Optical absorption spectra of UV-induced color centers in crystalline TlCl at room temperature and at 87 K; (B) effect of bleaching the UV-induced coloration with 535-nm light. (From Hilsch, R. and Pohl, R. W., *Z. Physik.*, 64, 606, 1930. With permission.)

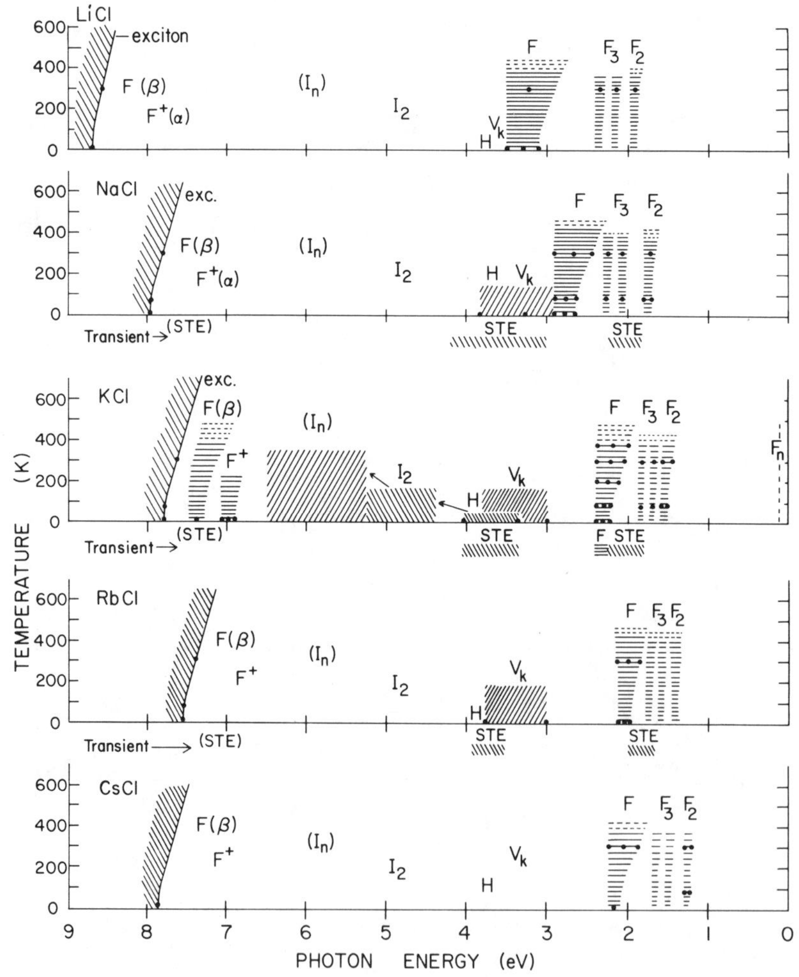

FIGURE 2.2.52A.

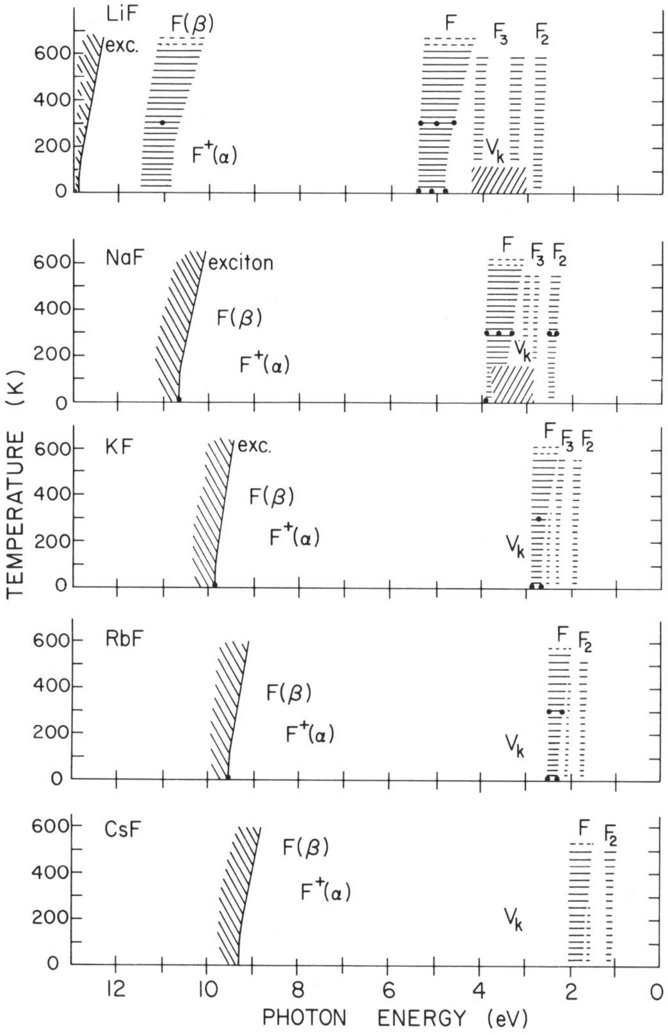

B

FIGURE 2.2.52. Schematic figure summarizing peak energies and widths (FWHM) of radiation-induced defect absorption bands in alkali halide crystals as a function of temperature. Shaded regions represent the extent, between half-maximum points, of the indicated absorption bands as estimated or computed from Equations 10 and 11. Experimental measurements of band peaks or widths are indicated by points at the corresponding temperatures. Approximate annealing temperature is indicated by the high-temperature termination of the band representation. A defect band whose observation in a given material is not yet reported, but whose existence can be anticipated based on observations in other materials, is indicated by the defect label alone at the anticipated energy. The energy of the lowest exciton peak (exc.) is indicated for reference as the ultimate limit of UV transparency. In addition to the temperature-dependent shift of the exciton peak as indicated, the excitonic (Urbach) absorption edge becomes broader with increasing temperature (not shown). The reader is cautioned that this figure is intended primarily to illustrate trends from material to material and qualitative temperature dependences. It is necessarily approximate except where measurements are actually indicated, and is speculative in regard to anticipated defect band locations. Tabulations of parameters for F, F_2, and F_3 centers in alkali halides are given in Reference 3.

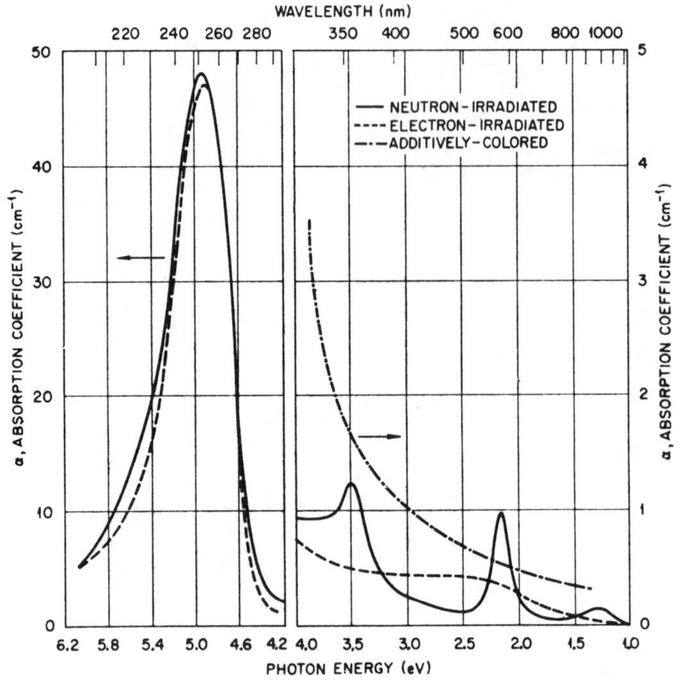

FIGURE 2.2.53. Comparison of absorption spectra of MgO crystals produced by neutron irradiation, electron irradiation, and additive coloration. The 250-nm band for the additively colored crystal (α_{250} = 400 cm^{-1}) is not shown here. (From Chen, Y., Williams, R. T., and Sibley, W. A., *Phys. Rev.*, 182, 960, 1969. With permission.)

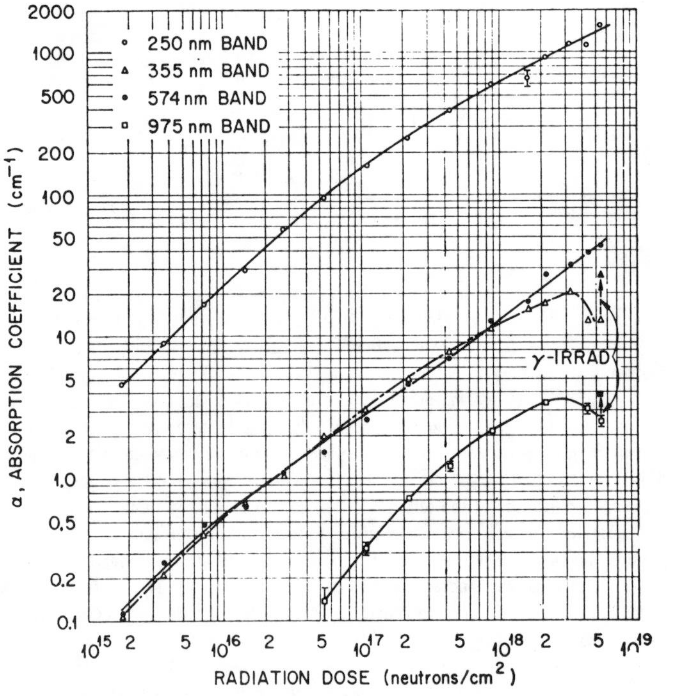

FIGURE 2.2.54. Absorption coefficient vs. neutron dose for the bands at 250, 355, 574, and 975 nm in MgO. (From Chen, Y., Williams, R. T., and Sibley, W. A., *Phys. Rev.*, 182, 960, 1969. With permission.)

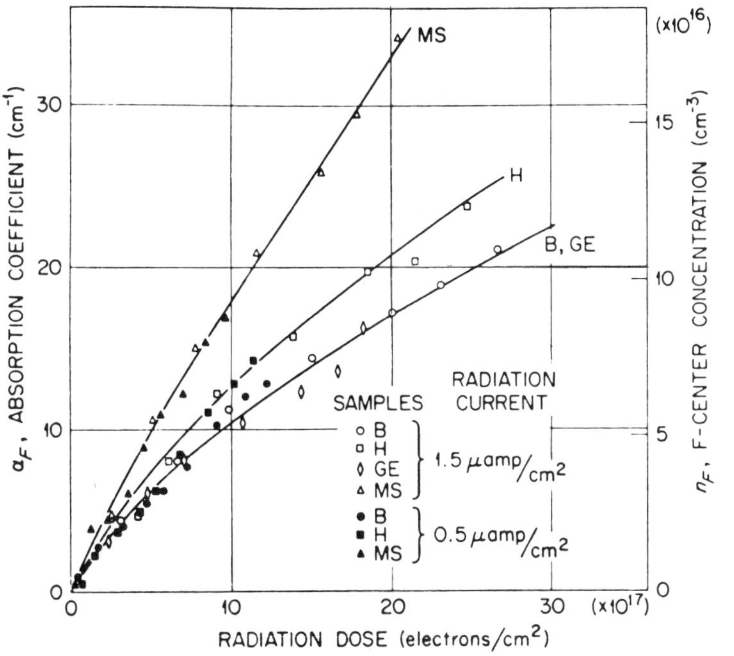

FIGURE 2.2.55. Absorption coefficient at the peak of the F band vs. electron dose for MgO crystals. Samples H, B, and GE contain less than 10 ppm Fe, whereas the crystals labeled MS contain 60 ppm Fe. (From Sibley, W. A. and Chen, Y., *Phys. Rev.*, 160, 712, 1967. With permission.)

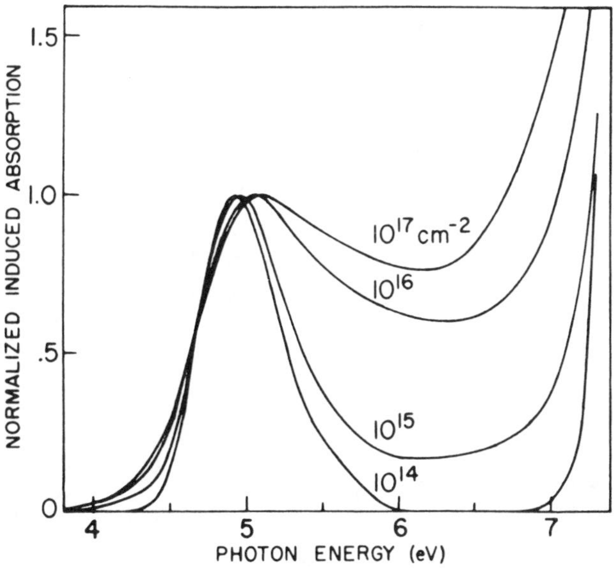

FIGURE 2.2.56. Room-temperature absorption spectra normalized to the peak of the F-type band of induced absorption resulting from room-temperature irradiation to 3-MeV Ne^+ fluences of 10^{14}, 10^{15}, 10^{16}, and 10^{17} cm^{-2}. (From Evans, B. D., *Phys. Rev. B*, 9, 5222, 1974. With permission.)

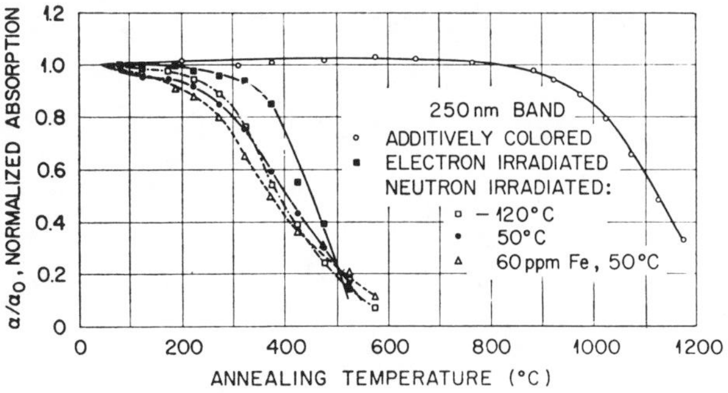

FIGURE 2.2.57. Normalized absorption coefficient α/α_0 of the 250-nm band vs. annealing temperature for electron-irradiated, neutron-irradiated, and additively colored crystals of MgO. (From Chen, Y., Williams, R. T., and Sibley, W. A., *Phys. Rev.*, 182, 960, 1969. With permission.)

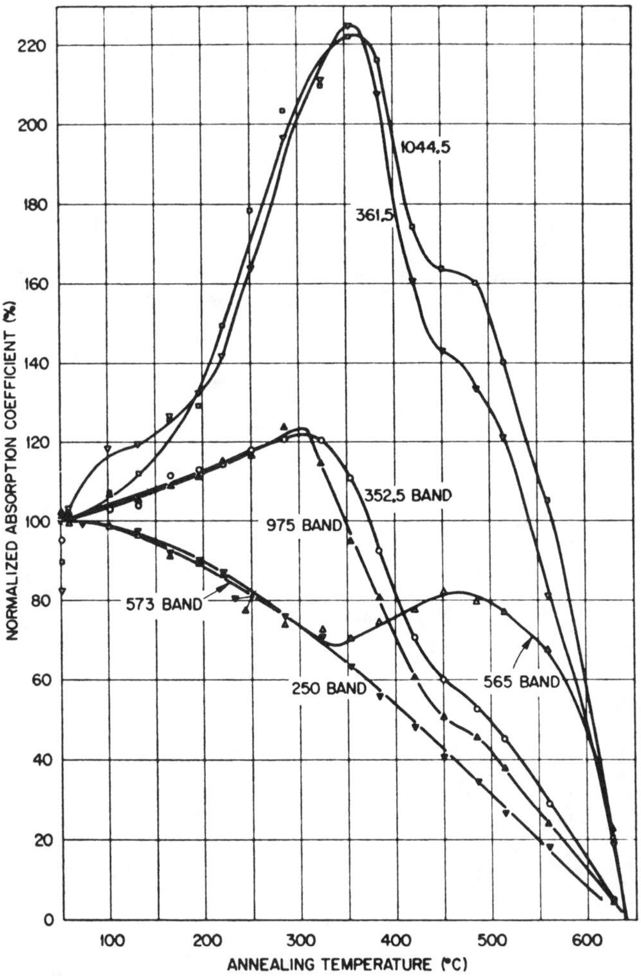

FIGURE 2.2.58. Normalized absorption coefficients α/α_0 of absorption bands and two zero-phonon lines in MgO irradiated with 2.2×10^{18} neutrons/cm^2 vs. annealing temperature. The initial absorption coefficients for the 250-, 352.5-, 573-, and 975-nm bands are 920, 22, 34, and 4 cm^{-1}, respectively, and those for the zero-phonon lines at 361.5 and 1044.5 nm are 6 and 10 cm^{-1}, respectively. (From Chen, Y., Williams, R. T., and Sibley, W. A., *Phys. Rev.*, 182, 960, 1969. With permission.)

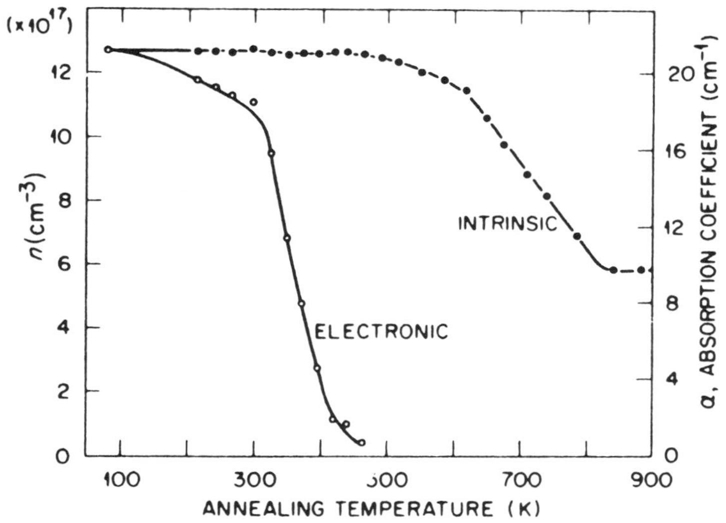

FIGURE 2.2.59. Isochronal annealing of V-type centers in MgO electron-irradiated to a dose in excess of 10^{18} e/cm^2 and gamma irradiated to produce V^-, V_{Al}, and V_{OH} centers. The lower curve (open circles) which describes the thermal stability of the hole trapped at these centers, was obtained by optical measurement after each isochronal anneal. The upper curve (solid circles) was obtained by repopulating the centers with holes by a short ionizing irradiation following each isochronal anneal and therefore portrays the thermal stability of vacancies which can trap holes. (From Abraham, M. M., Chen, Y., and Unruh, W. P., *Phys. Rev. B*, 9, 1974. With permission.)

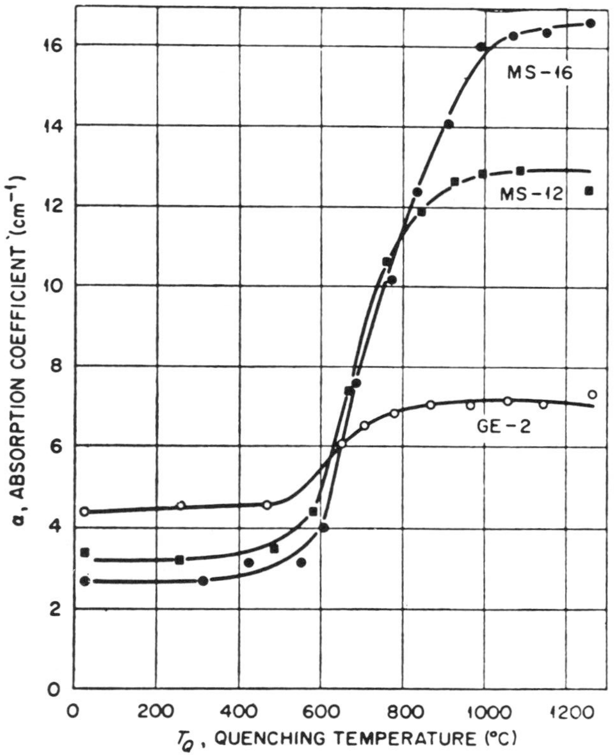

FIGURE 2.2.60. Absorption coefficient of the 2.3-eV band in MgO (due to V-type centers) vs. quenching temperature. Each measurement is preceded by a 20-min gamma irradiation. (From Chen, Y. and Sibley, W. A., *Phys. Rev.*, 154, 842, 1967. With permission.)

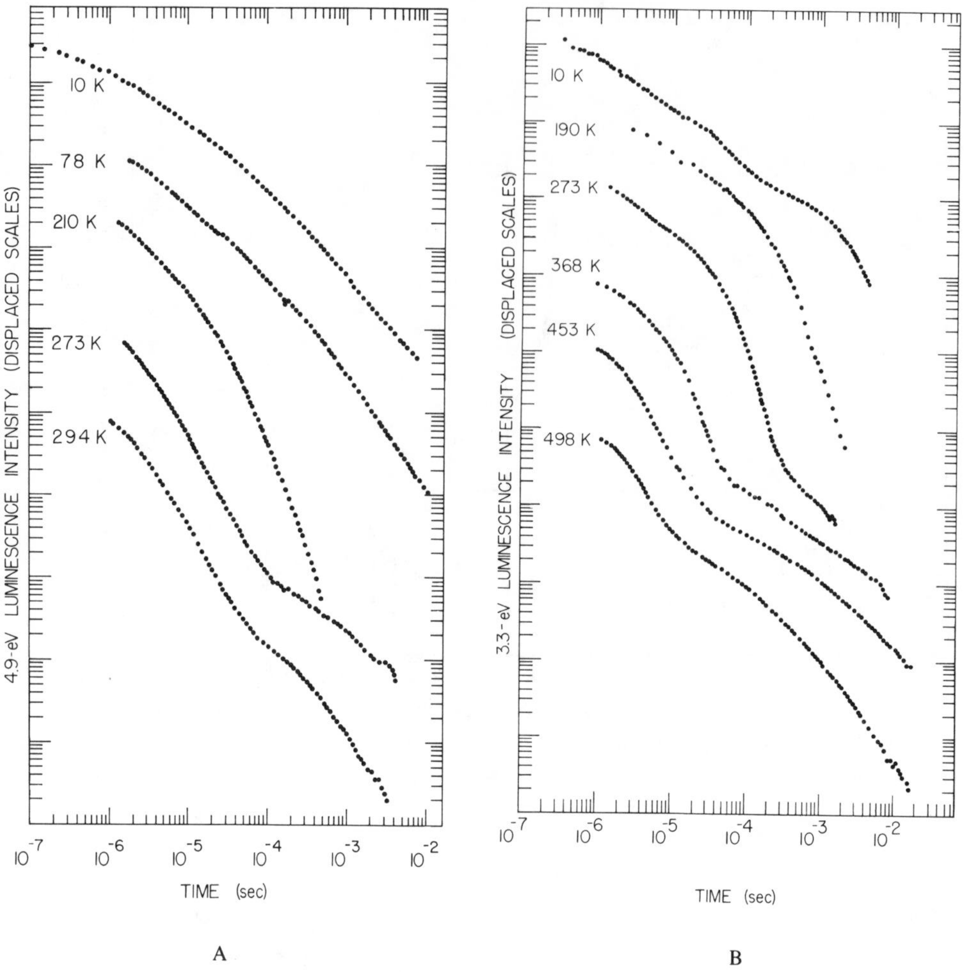

FIGURE 2.2.61. Luminescence intensity at 4.9 eV (A) and 3.3 eV (B) is shown as a function of time after electron pulse excitation at the temperatures indicated. Both axes are logarithmic; the data for different temperatures are displaced arbitrarily along the vertical (intensity) axis for convenient graphical presentation. (From Williams, R. T., Williams, J. W., Turner, T. J., and Lee, K. H., *Phys. Rev. B*, 20, 1687, 1979. With permission.)

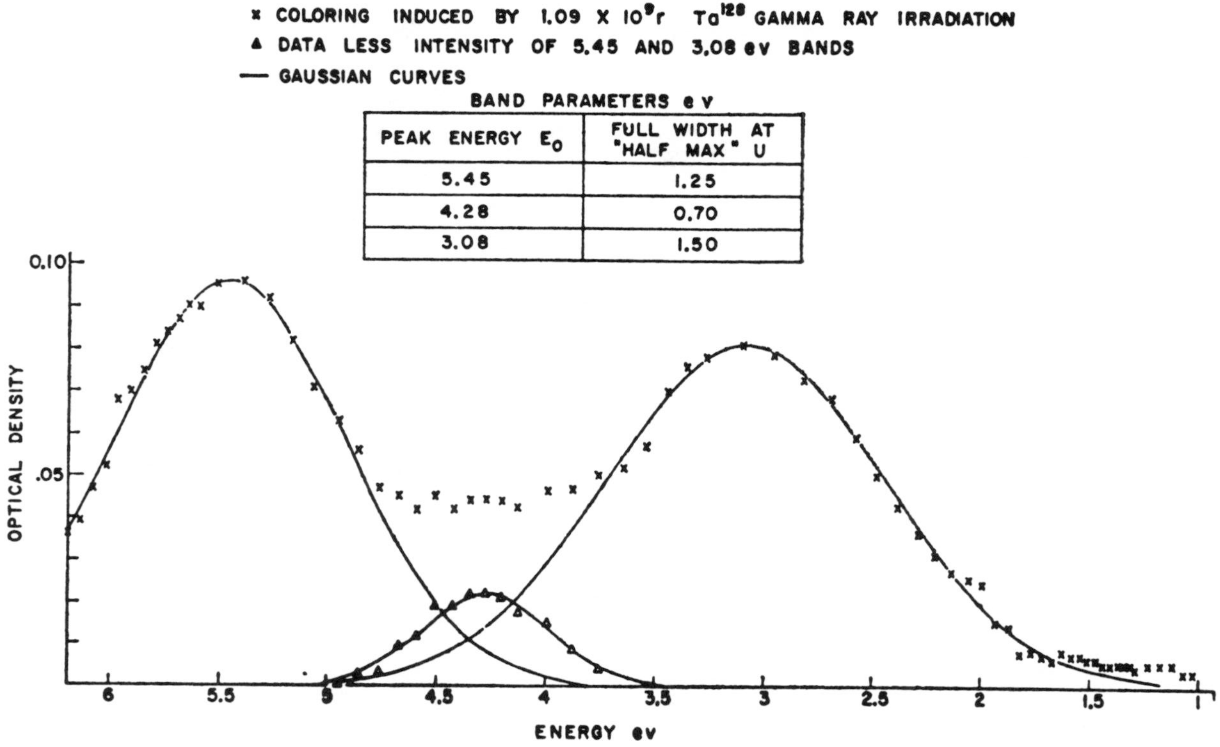

FIGURE 2.2.62. Gamma-ray-induced absorption spectrum of Al_2O_3. The spectrum obtained by an irradiation of only 3×10^4 R remains unchanged by additional gamma-ray irradiation even up to 10^9 R. (From Levy, P. W., *Phys. Rev.*, 123, 1226, 1961. With permission.)

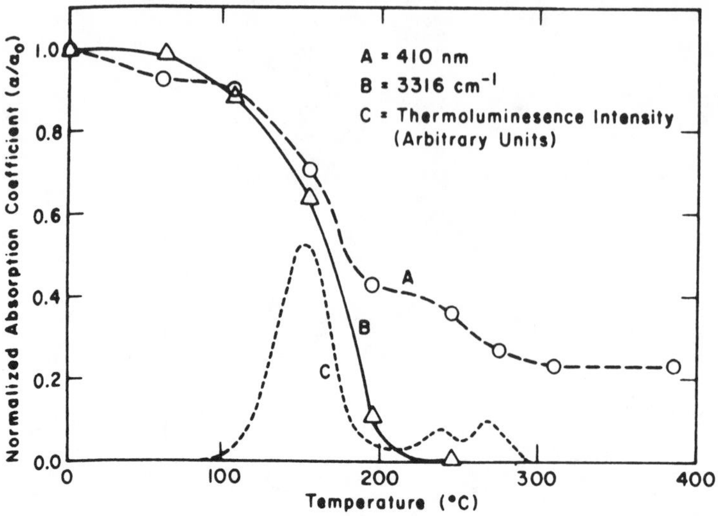

FIGURE 2.2.63. Isochronal annealing data for (A) the 410-nm and (B) the 1316 cm^{-1} absorption band in gamma-irradiated Al_2O_3. Curve C shows thermoluminescence corresponding to annealing of the V centers. (Reprinted with permission from Turner, T. J. and Crawford, J. H., Jr., *Solid State Commun.*, 17, 167, 1975. Copyright 1975, Pergamon Press, Ltd.)

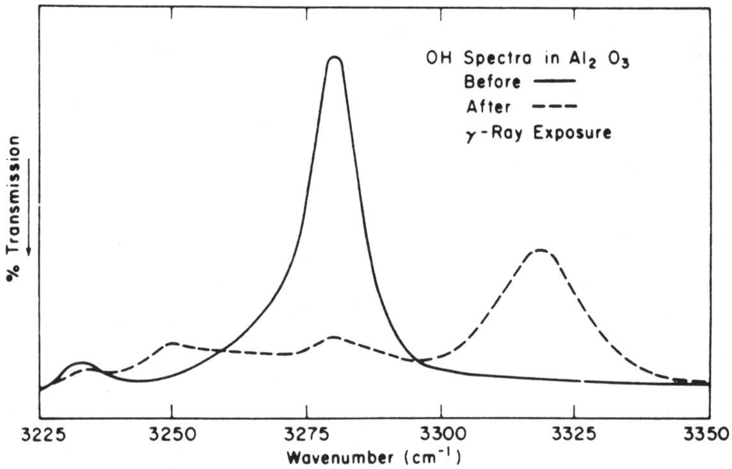

FIGURE 2.2.64. OH^- stretching bands in Al_2O_3 before and after gamma irradiation. (Reprinted with permission from Turner, T. J. and Crawford, J. H., Jr., *Solid State Commun.*, 17, 167, 1975. Copyright 1975, Pergamon Press, Ltd.)

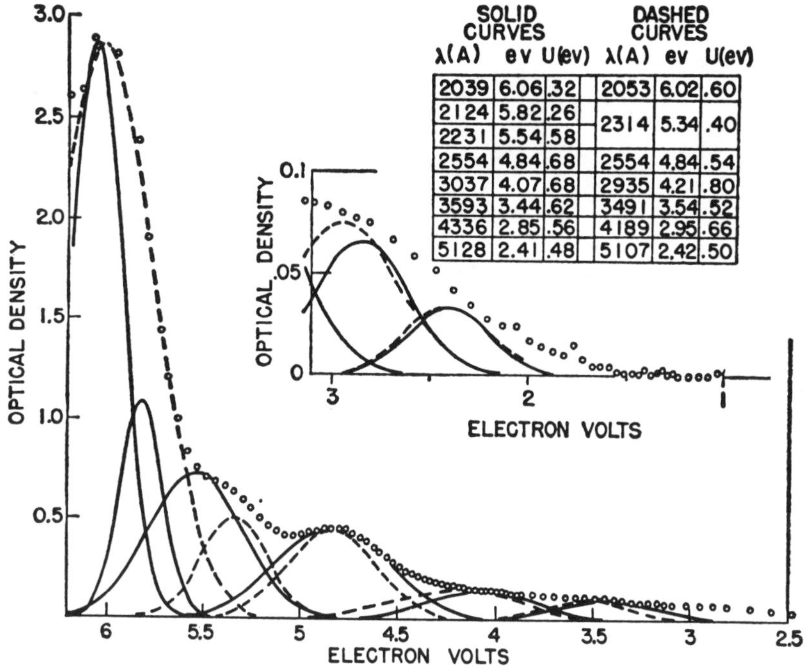

FIGURE 2.2.65. Optical absorption spectrum of Al_2O_3 induced by reactor-neutron irradiation. Results of curve-fitting with Gaussian bands are shown by solid and dashed curves. (From Levy, P. W., *Phys. Rev.*, 123, 1226, 1961. With permission.)

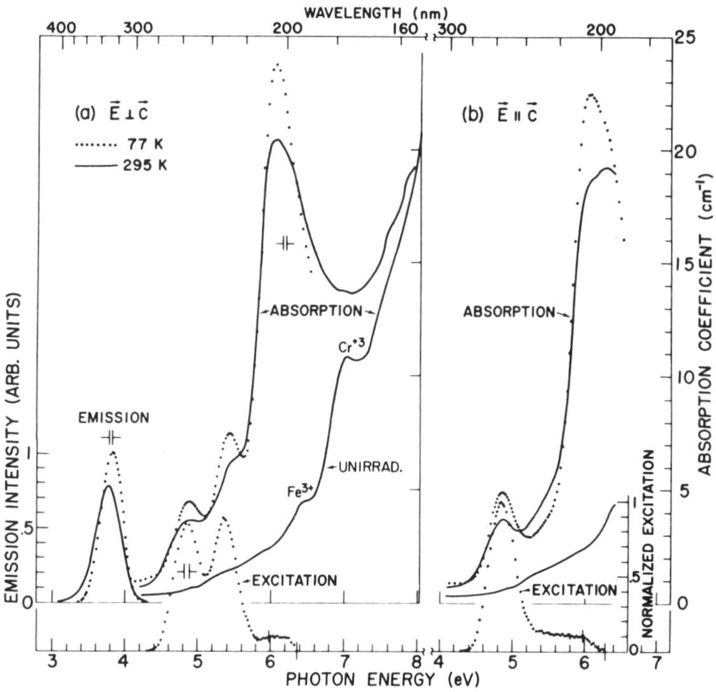

FIGURE 2.2.66. Polarized optical absorption from high-purity Czochralski sapphire irradiated to 3.9×10^{16} 14-MeV neutrons/cm^2 followed by a 20-min exposure to 300°C. Emission and excitation from other samples bombarded to 1.7×10^{16} neutrons/cm^2. (a) E $\perp$ c; (b) E $\parallel$ c. (From Evans, B. D. and Stapelbroek, M., *Phys. Rev. B*, 18, 7089, 1978. With permission.)

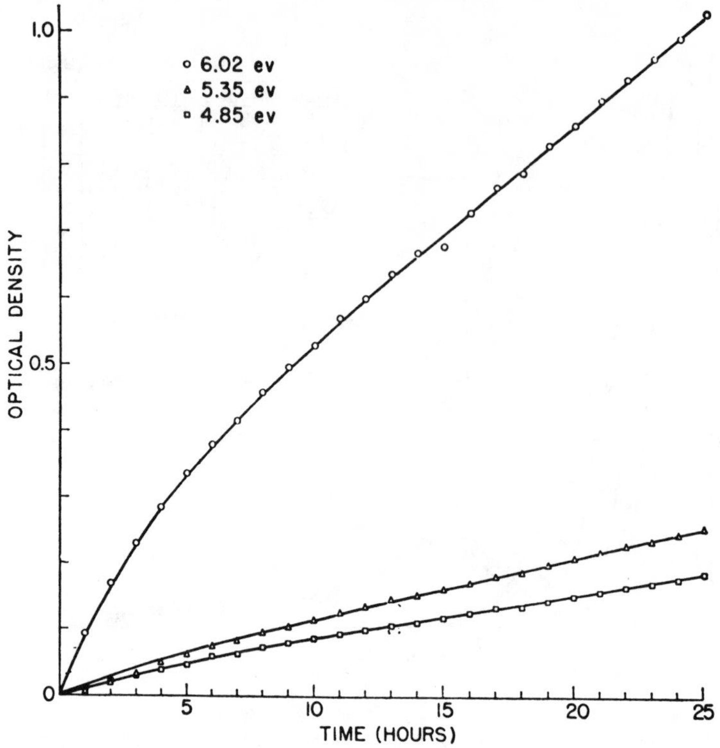

FIGURE 2.2.67. Growth of the three most prominent neutron-induced bands in Al_2O_3, as a function of irradiation time at a total (fast and slow) flux of about 2.9 × 10^{12} neutrons cm^{-2} sec^{-1}. (From Levy, P. W., *Phys. Rev.*, 123, 1226, 1961. With permission.)

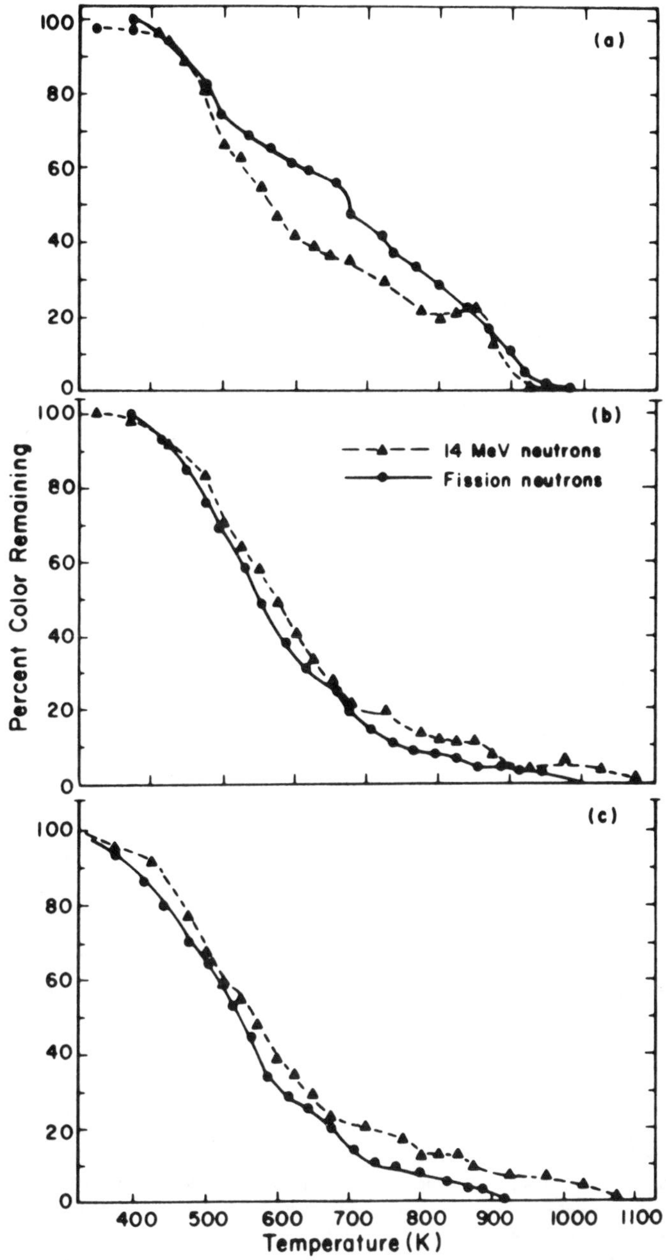

FIGURE 2.2.68. Isochronal annealing of optical absorption bands at (A) 6.02 eV, (B) 5.34 eV, and (C) 4.85 eV in single-crystal Al_2O_3 after irradiation with 14-MeV neutrons and fission neutrons. (From Bunch, J. M. and Clinard, F. W., Jr., *J. Am. Ceram. Soc.*, 57, 279, 1974. With permission.)

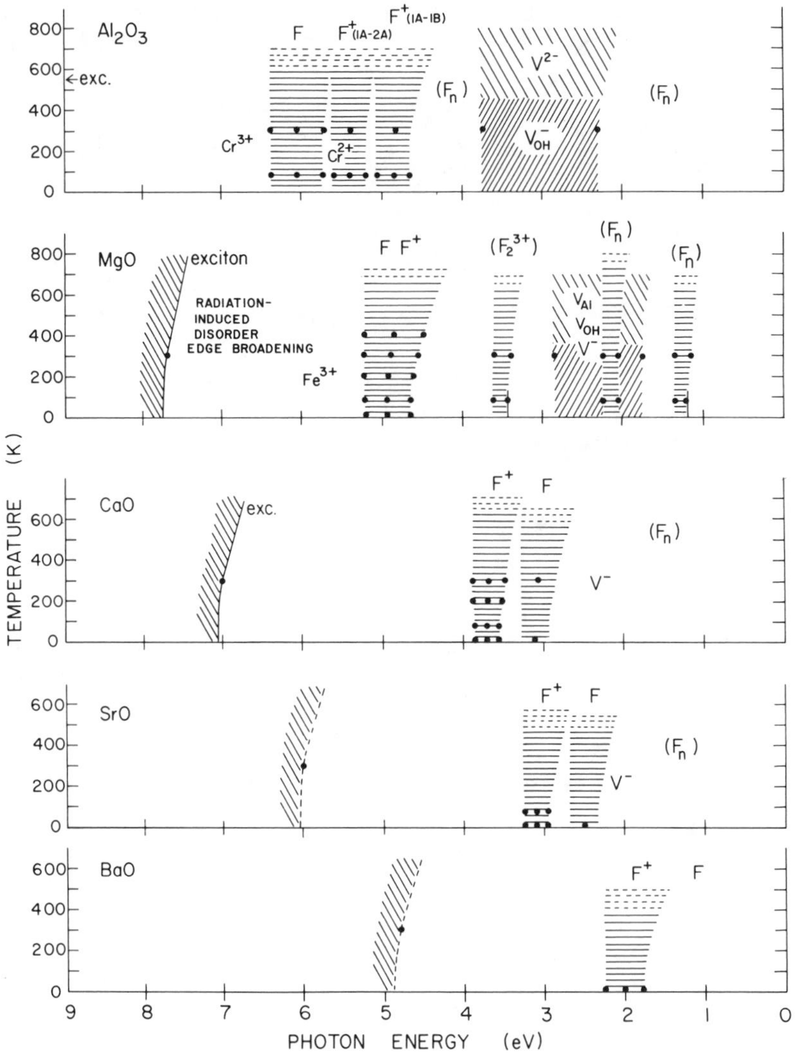

FIGURE 2.2.69. Schematic figure summarizing peak energies and widths (FWHM) of radiation-induced defect absorption bands in several oxide crystals as a function of temperature. Shaded regions represent the extent, between half-maximum points, of the indicated absorption bands as estimated or computed from Equations 10 and 11. Experimental measurements of band peaks or widths are indicated by points at the corresponding temperatures. Approximate annealing temperature is indicated by the high-temperature termination of the band representation. A defect band whose observation in a given material is not yet reported, but whose existence can be anticipated based on observations in other materials, is indicated by the defect label alone at the anticipated energy. The energy of the lowest exciton peak (exc.) is indicated for reference as the ultimate limit of the UV transparency. The reader is cautioned that this figure is intended primarily to illustrate trends from material to material and qualitative temperature dependences. It is necessarily approximate except where measurements are actually indicated, and is speculative in regard to anticipated defect band locations.

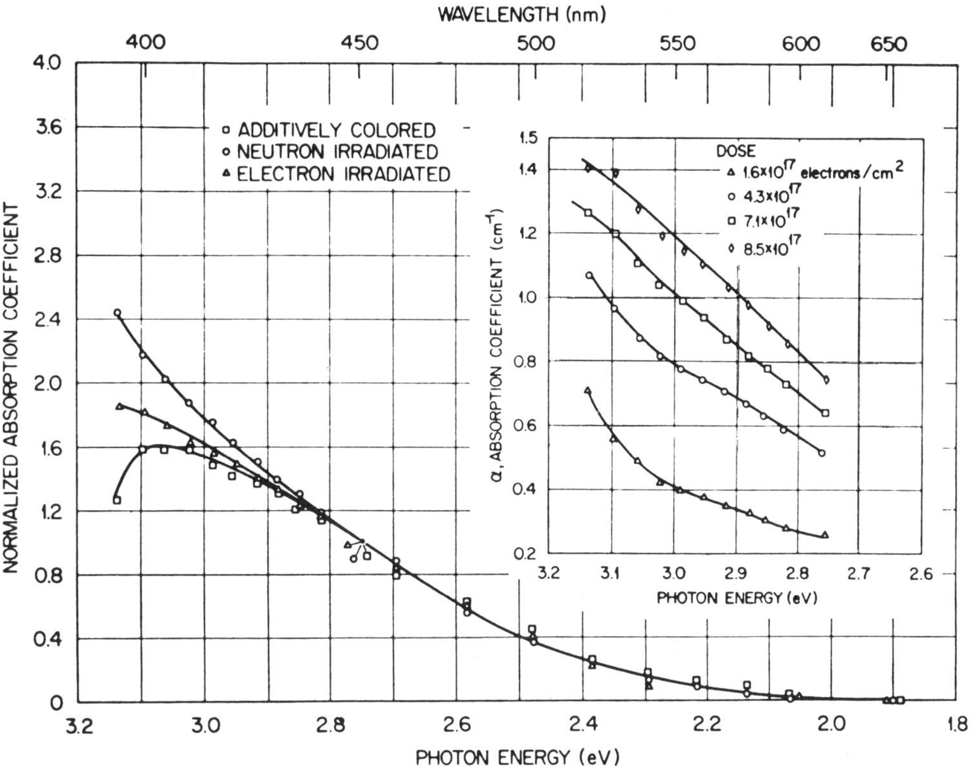

FIGURE 2.2.70. Spectra of absorption induced by additive coloration, neutron irradiation, and electron irradiation of ZnO. The pre-irradiation spectrum has been subtracted in each case. The inset shows the change in character of the absorption with dose for an electron-irradiated specimen. (From Vehse, W. E., Sibley, W. A., Keller, F. J., and Chen, Y., *Phys. Rev.*, 167, 828, 1968. With permission.)

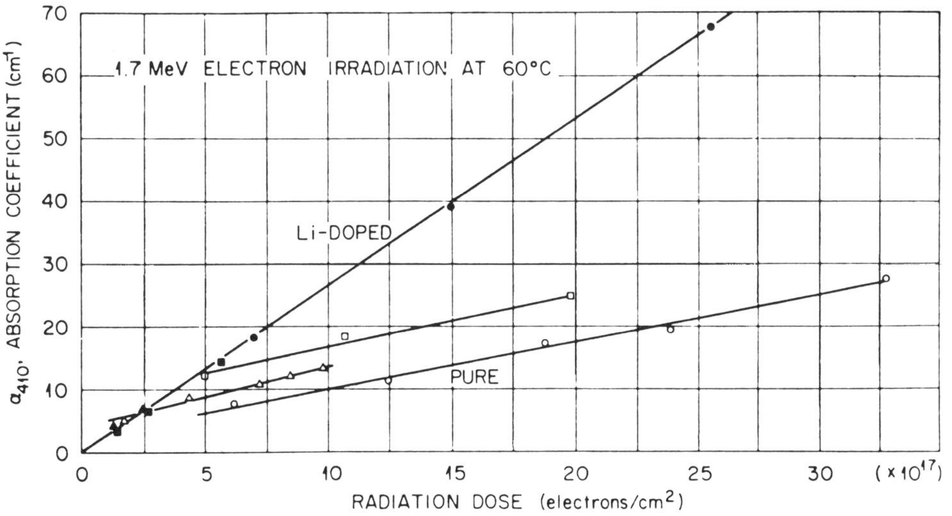

FIGURE 2.2.71. A plot of the absorption coefficient at 410 nm vs. electron-radiation dose for several "pure" and doped samples of ZnO. The radiation energy was 1.7 MeV and the sample temperature was 60°C during irradiation. (From Vehse, W. E., Sibley, W. A., Keller, F. J., and Chen, Y., *Phys. Rev.*, 167, 828, 1968. With permission.)

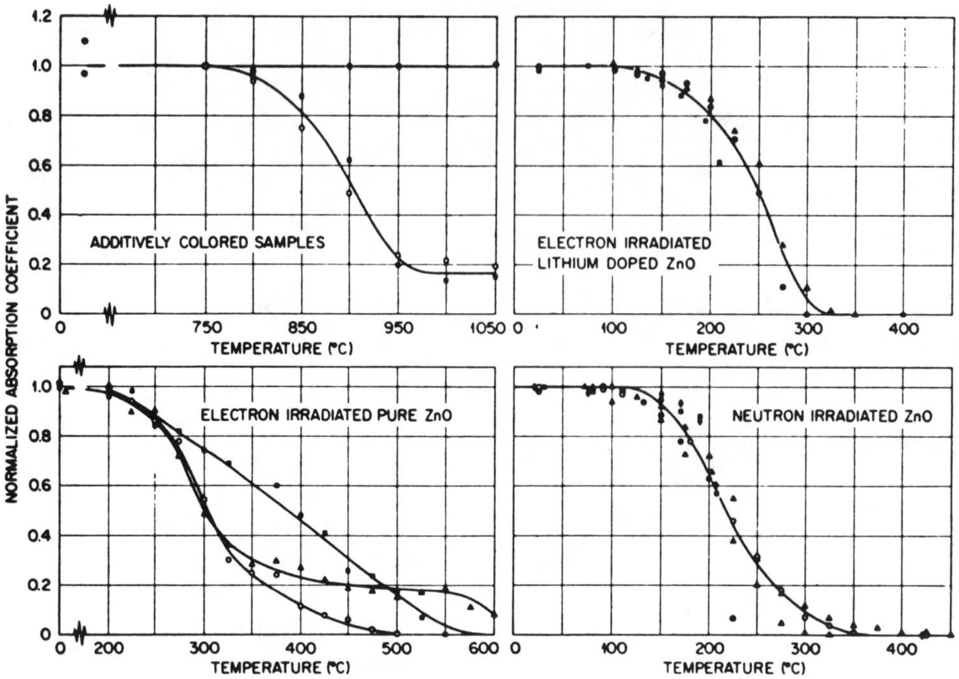

FIGURE 2.2.72. Annealing data for pure and lithium-doped ZnO colored by irradiation and heat treatment in Zn vapor. The data have all been normalized such that before annealing the induced absorption coefficient at 410 nm is one. The open symbols represent data on pure crystals while the full symbols are for data on Li-doped samples. (A) Additively colored samples; (B) electron irradiated doped ZnO; (D) neutron irradiated ZnO. (From Vehse, W. E., Sibley, W. A., Keller, F. J., and Chen, Y., *Phys. Rev.*, 167, 828, 1968. With permission.)

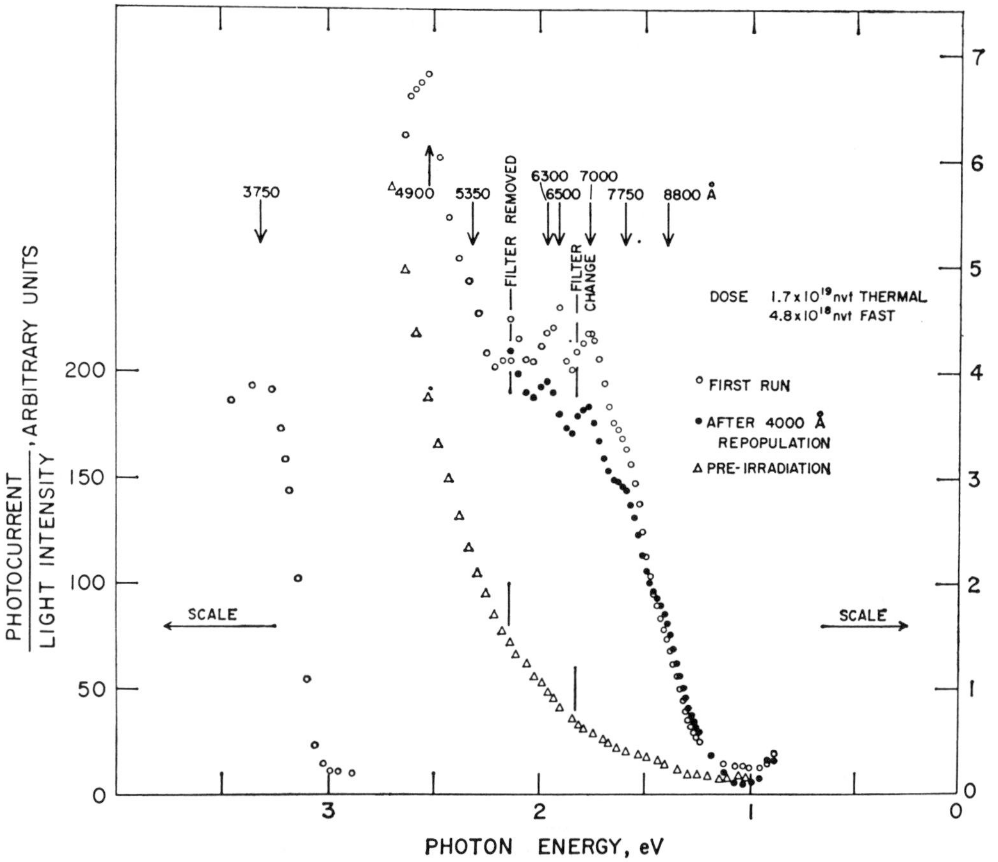

FIGURE 2.2.73. Photoconductivity spectrum of TiO_2 before and after neutron irradiation. Measurements made at 77 K. (From Townsend, P. D., Kan, H. K. A., and Levy, P. W., *Proc. Br. Ceram. Soc.*, 1, 71, 1964. With permission.)

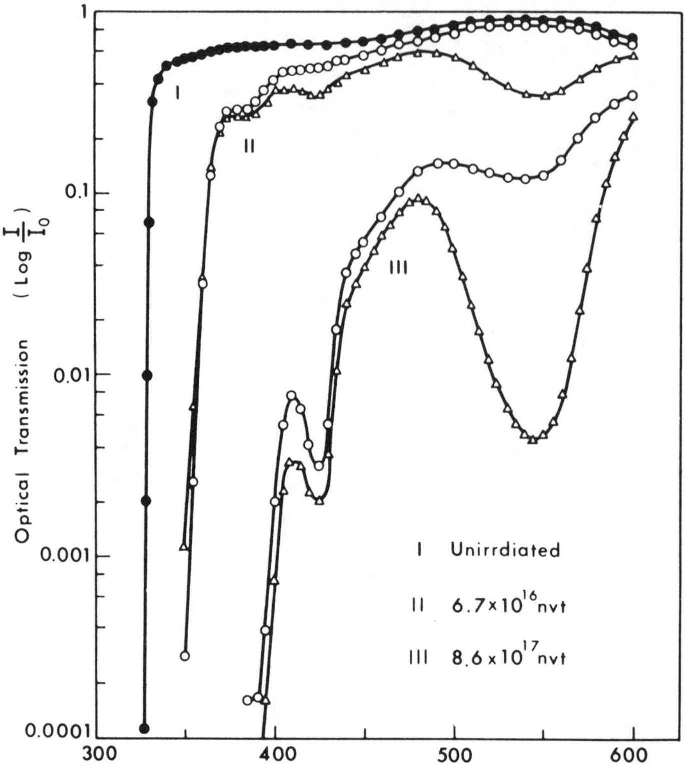

FIGURE 2.2.74. Optical transmission spectra of neutron-irradiated ZnS single cyrstals measured at liquid helium temperature. Circles: after the samples had been kept in the dark at room temperature; triangles: after UV illumination at liquid helium temperature. (From Yoshida, T., Seiyama, T., Shono, Y., and Kitagawa, M., *Appl. Phys. Lett.*, 9, 26, 1966. With permission.)

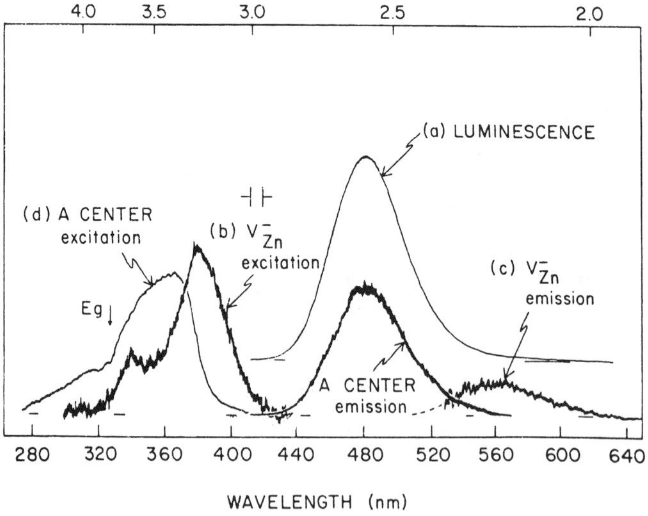

FIGURE 2.2.75. Spectral dependence of (A) luminescence; (B) excitation; (C) emission as monitored by V_{Zn} ODMR signal; (D) excitation; and (E) emission as monitored by A center ODMR signal in electron-irradiated ZnS. (Reprinted with permission from Lee, K. M., O'Donnell, K. P., and Watkins, G. D., *Solid State Commun.*, 41, 881, 1982. Copyright 1982, Pergamon Press, Ltd.)

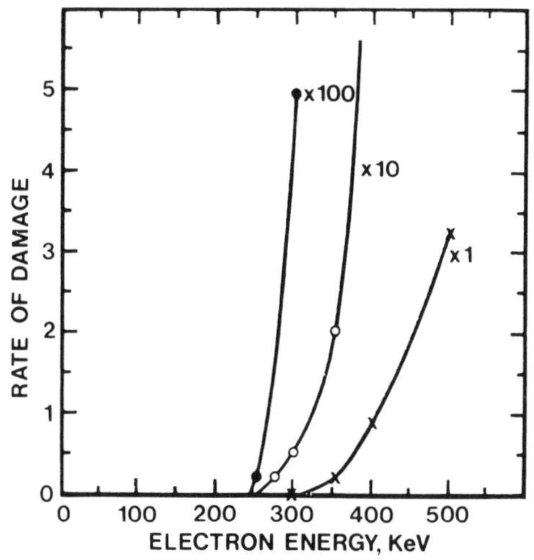

FIGURE 2.2.76. Damage in ZnSe as monitored by fluorescence as a function of bombarding electron energy at 85 K. (From Kulp and Detweiler and Reference 117.)

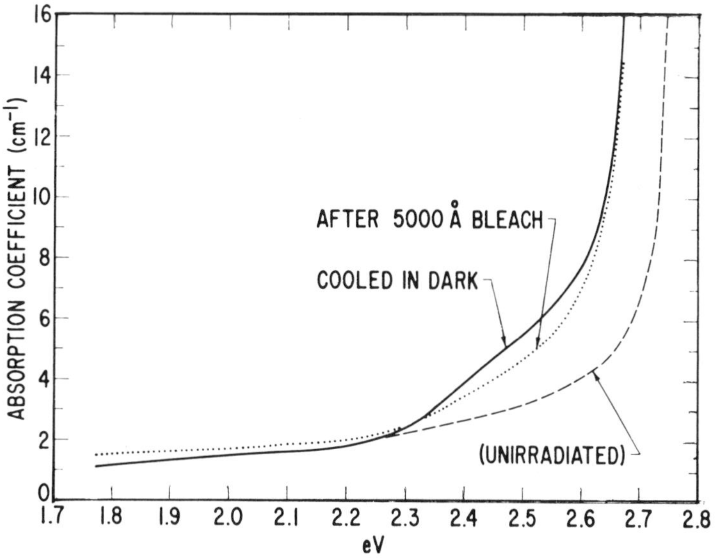

FIGURE 2.2.77. Optical absorption in ZnSe measured at 2.0 K before and after 1.5-MeV electron irradiation at room temperature. The material was originally n-type ($\sim 10^{17}$ donors/cm^3) and the irradiation dose was $\sim 2 \times 10^{18}$ e/cm^2. Bleaching at 5000 A reveals a band tentatively identified with the double minus charge state of the zinc vacancy. (From Watkins, G. D., *Rad. Effects*, 9, 105, 1971. With permission.)

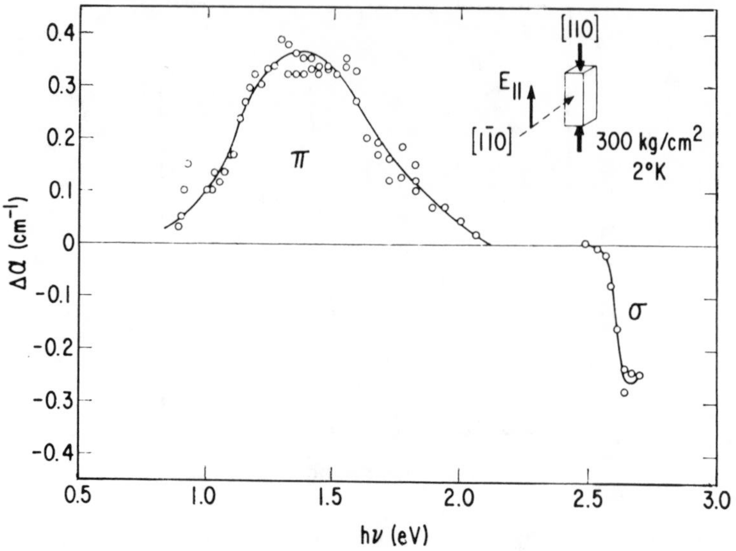

FIGURE 2.2.78. Absorption bands associated with the single minus charge state of the zinc vacancy in ZnSe, as revealed by stress-induced dichroism at 2.0 K. (From Watkins, G. D., *Rad. Effects,* 9, 105, 1971. With permission.)

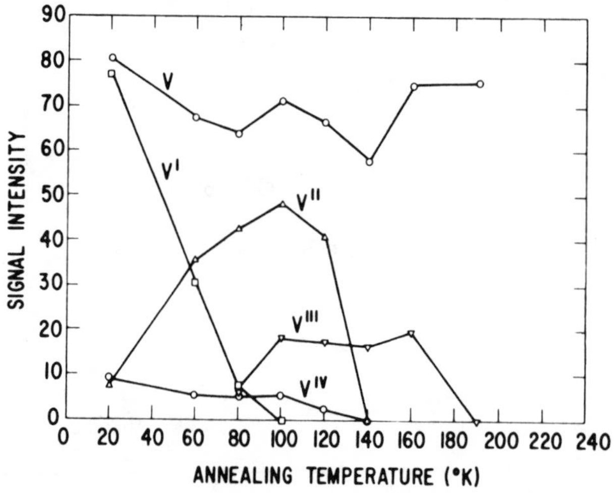

FIGURE 2.2.79. Intensities of EPR spectra observed at 20.4 K in irradiated ZnSe after 15-min isochronal anneals. The sample was irradiated at 20.4 K with 1.5-MeV electrons to a fluence of 4×10^{17} electrons/cm^2. The curve V arises from isolated zinc vacancies, while V^I, V^{II}, V^{III}, and V^{IV} are zinc lattice vacancies perturbed by zinc interstitials displaced to a near-neighbor site in the primary radiation event. (From Watkins, G. D., *Phys. Rev. Lett.*, 33, 223, 1974. With permission.)

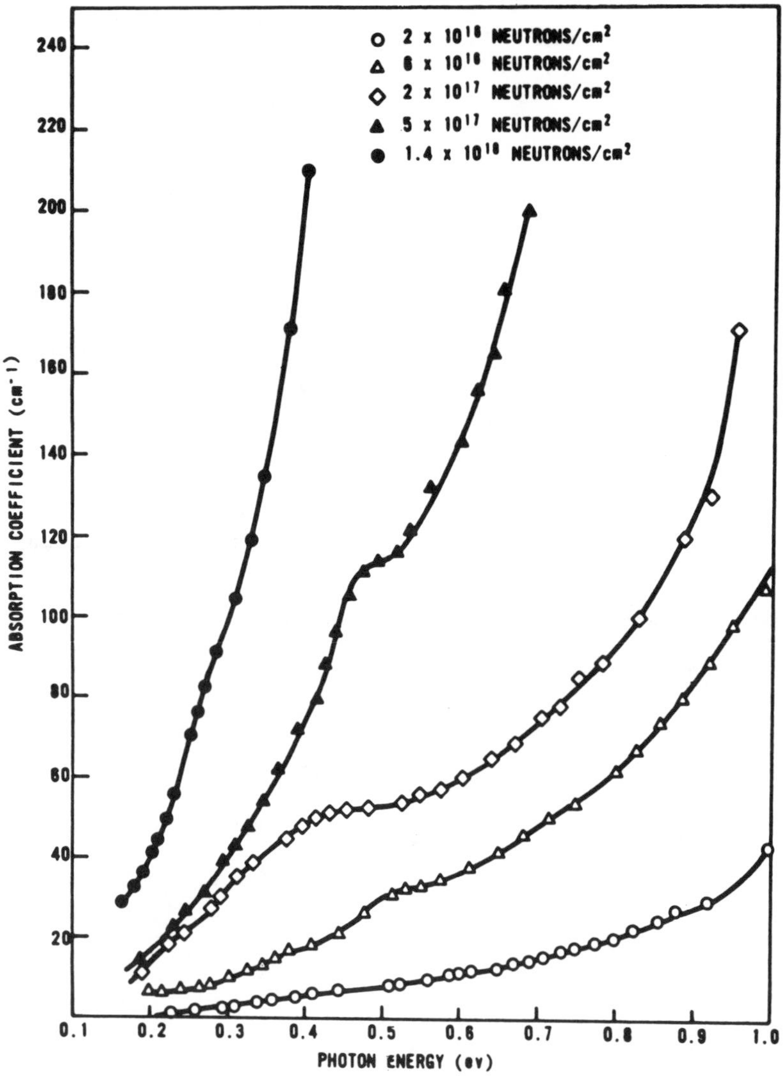

FIGURE 2.2.80. Optical absorption spectra in neutron-irradiated GaAs at room temperature. (From Vaidyanathan, K. V. and Watt, L. A., *Rad. Effects*, 10, 99, 1971. With permission.)

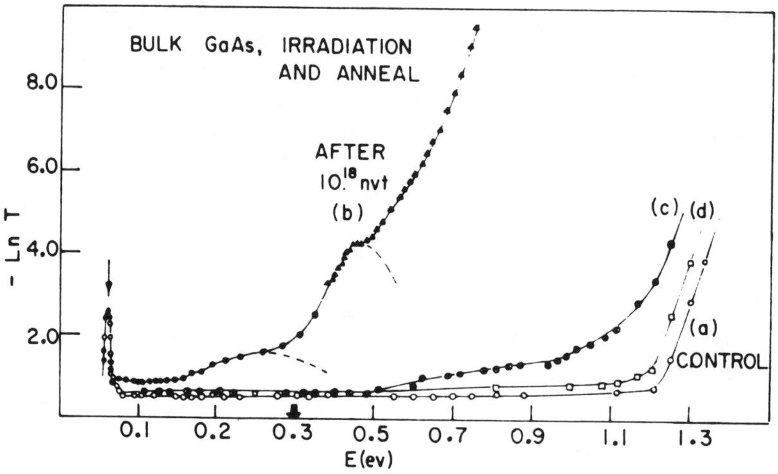

FIGURE 2.2.81. The effect of irradiation and annealing on the absorption edge of bulk GaAs. Curve (a) is the unirradiated or control absorption curve. Curve (b) is after irradiation and shows two absorption bands in the free carrier region — one at 0.25 eV and one at 0.50 eV. The dotted lines are drawn in to emphasize these bands. The lattice band at 0.061 eV is given as a reference point. Curve (c) is after a 480°C anneal for 5 hr and (d) is after a further anneal for 1.5 hr at 540°C. (From Pankey, T., Jr. and Davey, J. E., *J. Appl. Phys.*, 41, 697, 1970. With permission.)

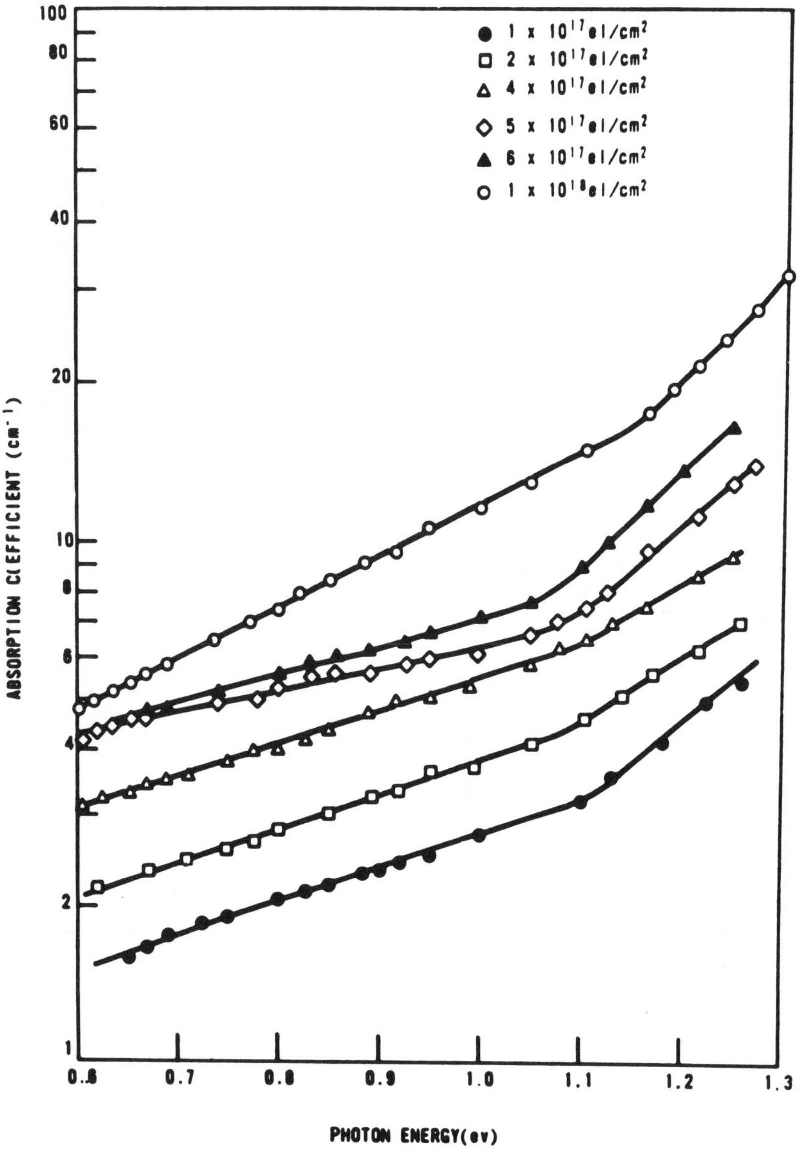

FIGURE 2.2.82. Absorption spectra of electron-irradiated GaAs at room temperature. (From Vaidyanathan, K. V. and Watt, L. A., *Rad. Effects*, 10, 99, 1971. With permission.)

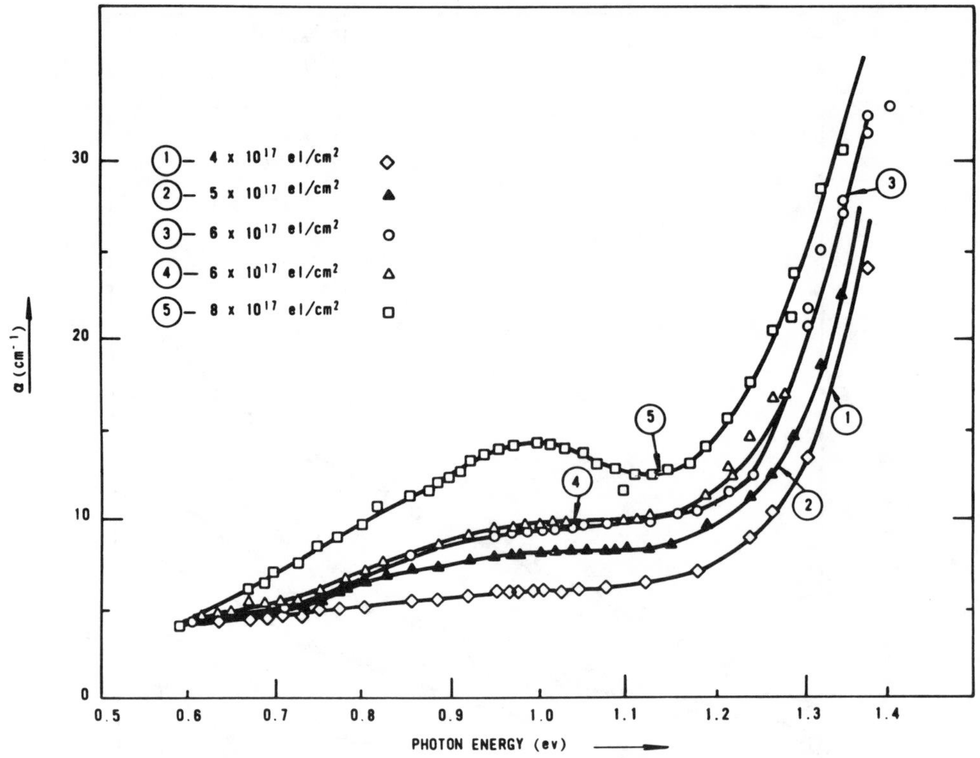

FIGURE 2.2.83. Absorption spectra in GaAs electron irradiated and measured at 85 K. (From Vaidyanathan, K. V. and Watt, L. A., *Rad. Effects*, 10, 99, 1971. With permission.)

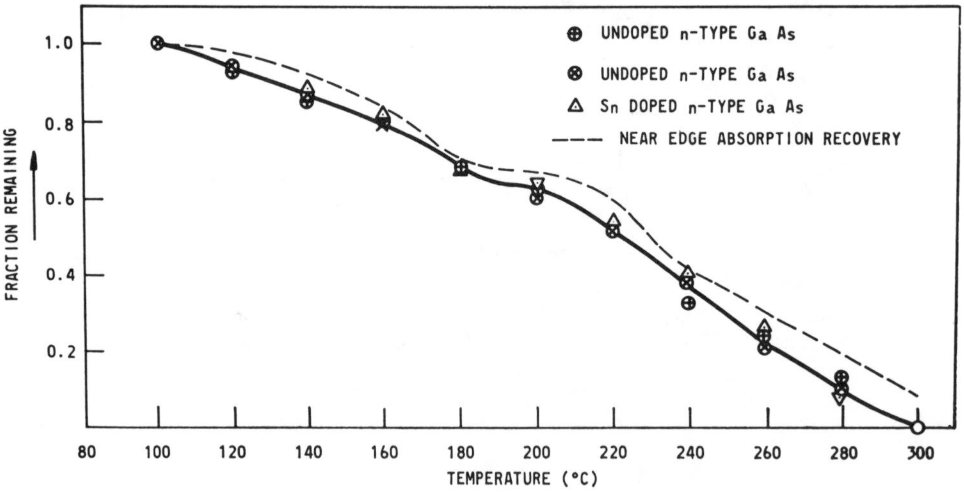

FIGURE 2.2.84. Isochronal recovery of the 1.0-eV absorption band after 30 min anneal at each temperature in electron-irradiated GaAs. (From Vaidyanathan, K. V. and Watt, L. A., *Rad. Effects*, 10, 99, 1971. With permission.)

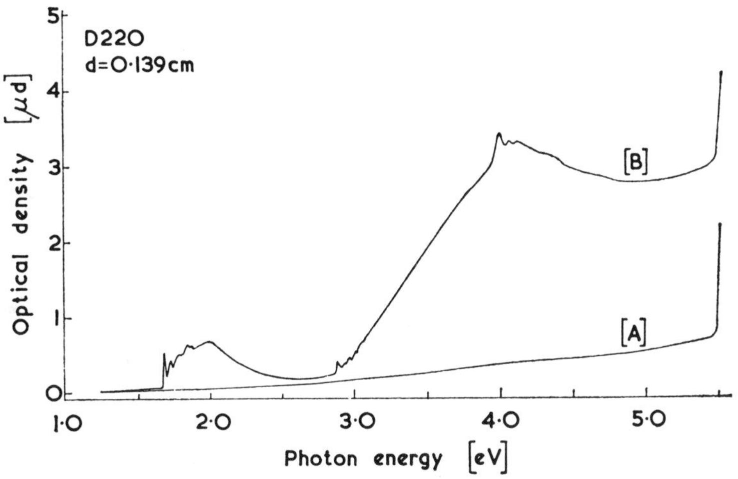

FIGURE 2.2.85. Absorption spectra of type IIa diamond measured at 85 K. (A) Before and (B) after a 1-MeV electron irradiation of 2.26 × 10^{18} electrons/cm^2 at 2 μA/cm^2. (From Clark, C. D., Duncan, I., Lomer, J. N., and Whippey, P. W., *Proc. Br. Ceram. Soc.*, 1, 85, 1964. With permission.)

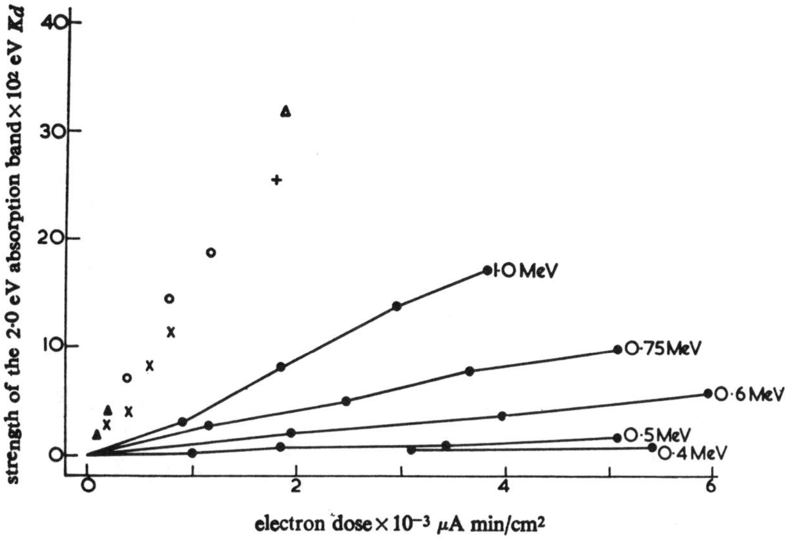

FIGURE 2.2.86. Rates of increase of the absorption band produced at 2.0 eV during electron irradiation at a constant flux of 2 μA/cm^2. Results for a range of electron energies are shown: O, results obtained at 1.5 MeV; ▲, 1.75 MeV; △, 2.0 MeV. (From Clark, C. D., Kimmey, P. J., and Mitchell, E. W. J., *Disc. Faraday Soc.*, 31, 96, 1961. With permission.)

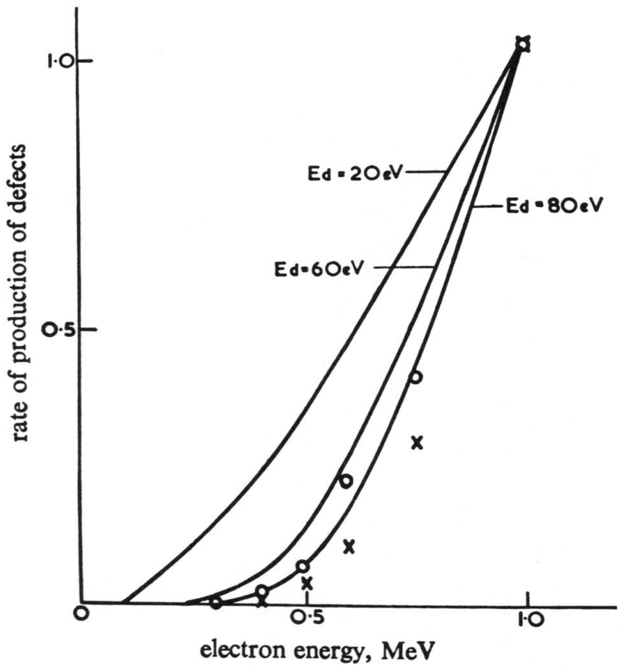

FIGURE 2.2.87. The full line curves show the theoretical results for the energy dependence of the rate of production of vacancy interstitial pairs in diamond, normalized at 1.0 MeV, for assumed displacement threshold energies of 20, 60, and 80 eV. Experimental results for the rates of production of the absorption band at 2.0 eV during irradiation are shown by 0. Experimental results for the rates of change of resistivity of semiconducting diamond during irradiation are shown by X. (From Clark, C. D., Kemmey, P. J., and Mitchell, E. W. J., *Disc. Faraday Soc.*, 31, 96, 1961. With permission.)

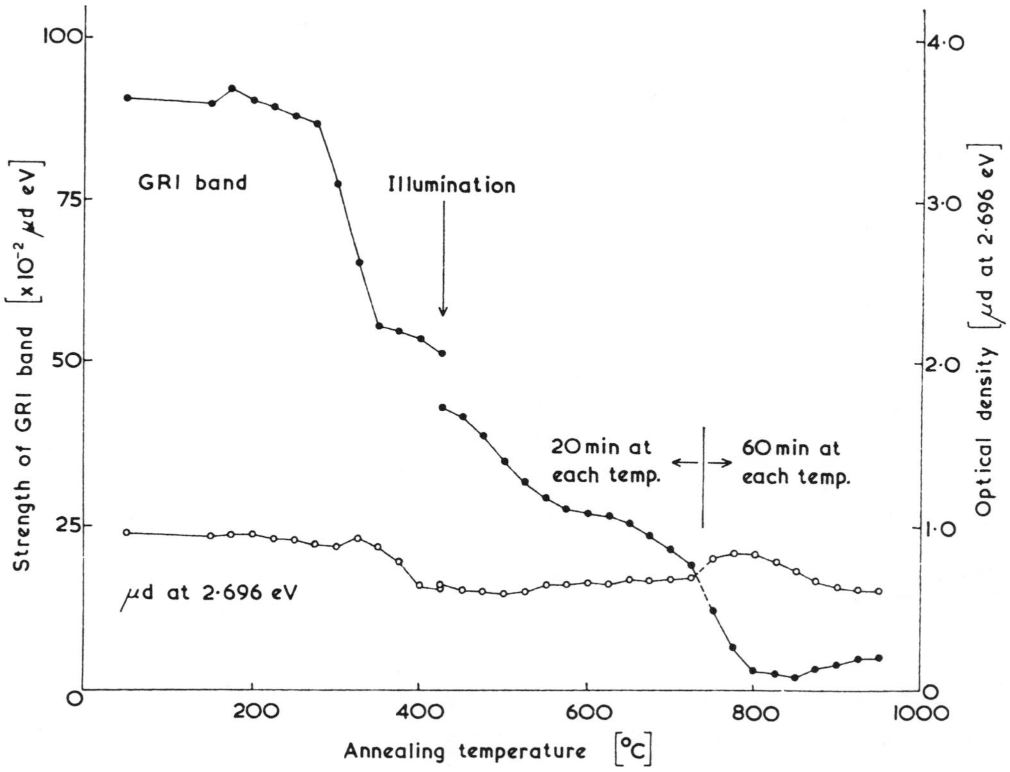

FIGURE 2.2.88. Isochronal annealing of the GRl band and the optical density at 2.695 eV, in a crystal irradiated with 2.0 MeV electrons. 20-min anneals were given at temperatures up to 750°C, and 60-min anneals above 750°C. The break in the curves at 425°C was the result of a single optical bleaching experiment using a high-pressure mercury lamp. (From Clark, C. D., Duncan, I., Lomer, J. N., and Whippey, P. W., *Proc. Br. Ceram. Soc.*, 1, 85, 1964. With permission.)

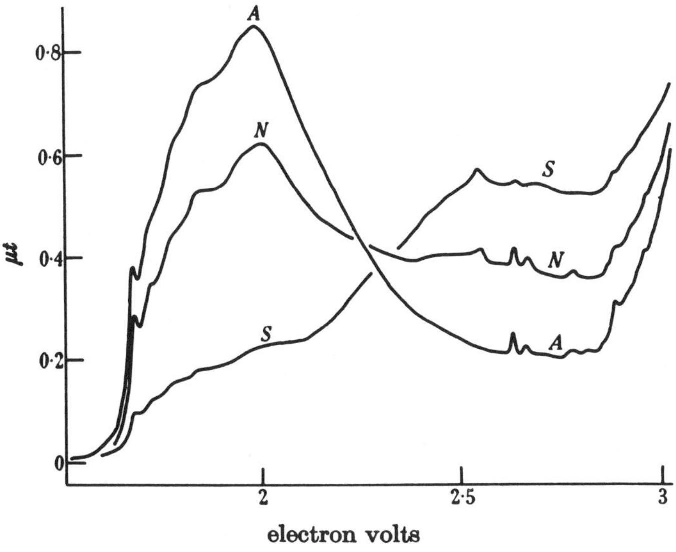

FIGURE 2.2.89. Changes in the optical absorption spectrum of a type II a diamond caused by irradiation with 1-MeV electrons (curve A) and subsequent isothermal heating at 600°C after 4 hr (curve N), 28 hr (curve S). All spectra were recorded at 290 K. (From Clark, C. D., Ditchburn, R. W., and Dyer, H. B., *Proc. R. Soc. London,* A237, 75, 1956. With permission.)

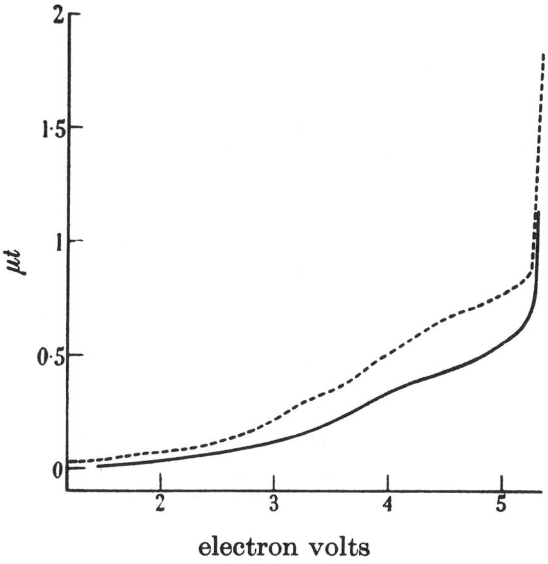

FIGURE 2.2.90. Absorption spectra of a natural type IIa diamond before (——) and after (----) 0.6 MeV electron irradiation and heating at 850°C for 21 hr. The spectra were recorded at 290 K. (From Clark, C. D., Ditchburn, R. W., and Dyer, H. B., *Proc. R. Soc. London,* A237, 75, 1956. With permission.)

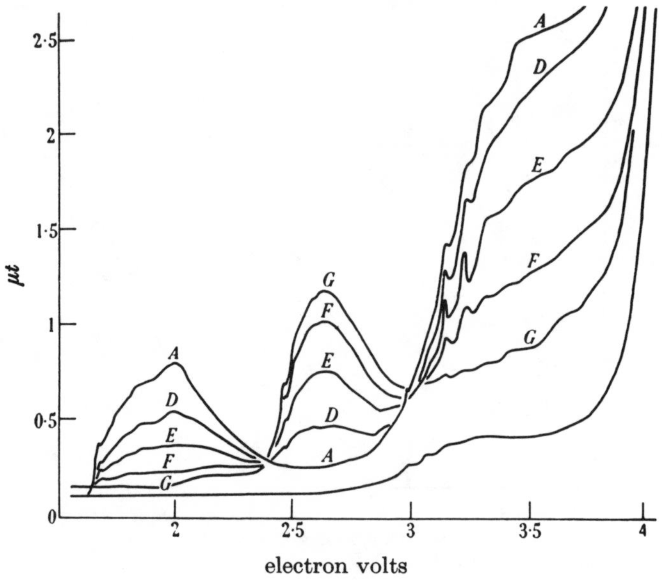

FIGURE 2.2.91. Absorption spectra of a natural type I diamond after 1-MeV electron irradiation (curve A) and after heating at 580°C for 1 hr (curve D), 630°C for 1/2 h (curve E), 680°C for 1/2 h (curve F), and 930°C for 18 hr (curve G). All spectra were recorded at 290 K. The lowest curve shows the absorption before irradiation. (From Clark, C. D., Ditchburn, R. W., and Dyer, H. B., *Proc. R. Soc. London,* A237, 75, 1956. With permission.)

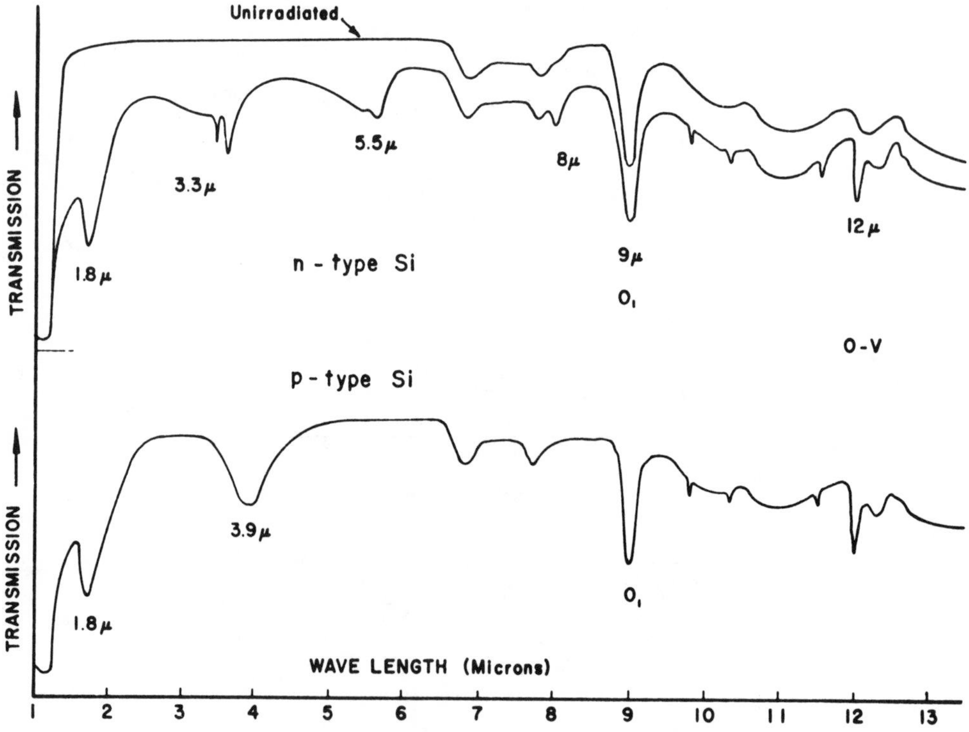

FIGURE 2.2.92. Schematic diagram of radiation-induced defect infrared absorption bands in silicon (both n- and p-type) in the wavelength region 1 to 13 μm. (From Cheng, L. J., Corelli, J. C., Corbett, J. W., and Watkins, G. D., *Phys. Rev.,* 152, 761, 1966. With permission.)

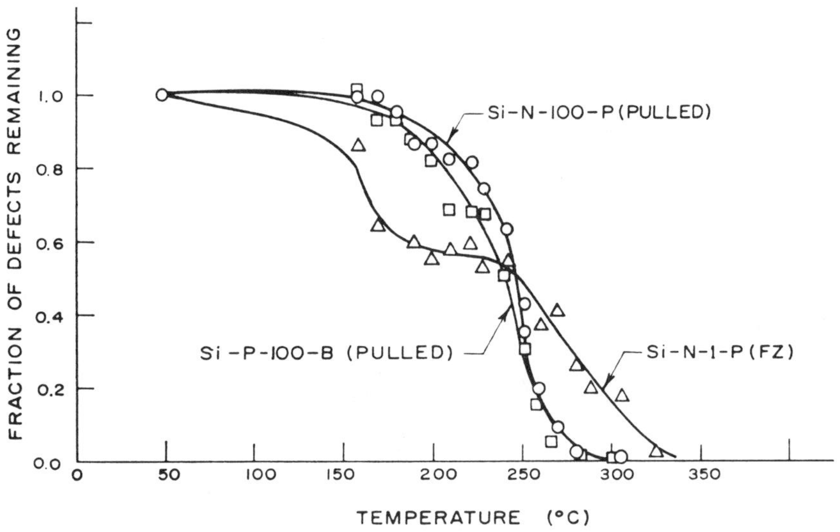

FIGURE 2.2.93. Isochronal anneal (20 min at each temperature) of the 1.8-μm band in floating-zone and pulled silicon. (From Cheng, L. J., Corelli, J. C., Corbett, J. W., and Watkins, G. D., *Phys. Rev.*, 152, 761, 1966. With permission.)

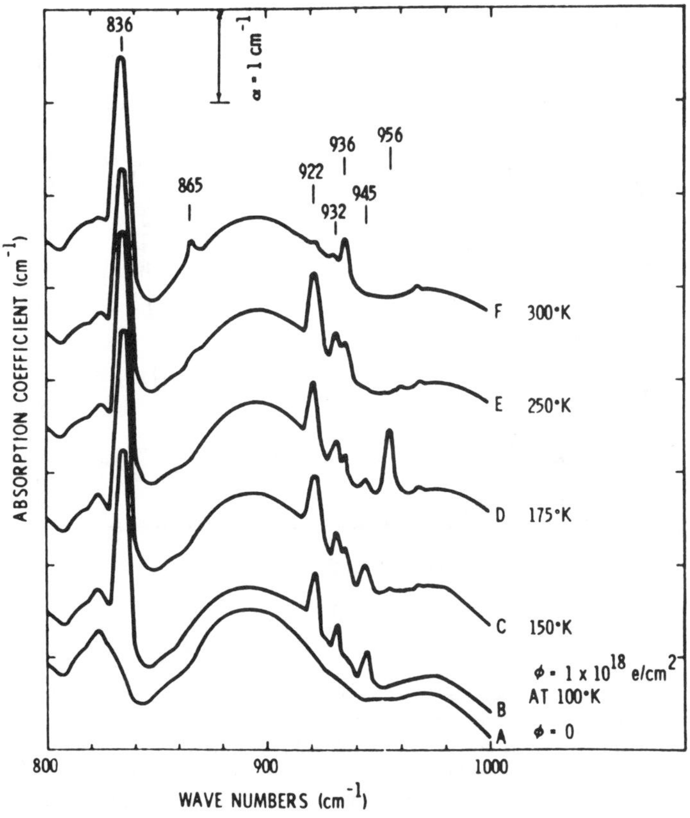

FIGURE 2.2.94. Infrared absorption spectra of n-type oxygen-doped silicon. Spectra are displaced vertically for clarity. All spectra recorded at 80 K. (A) Pre-irradiation; (B) after irradiation at 100 K to a fluence of 10^{18} e/cm^2; (C) after 20 min annealing at 150 K; (D) after 20 min annealing at 175 K; (E) after 20 min annealing at 250 K; (F) after 20 min annealing at 300 K. (From Whan, R. E. and Vook, F. L., *Phys. Rev.*, 153, 814, 1967. By permission of Sandia National Laboratories.)

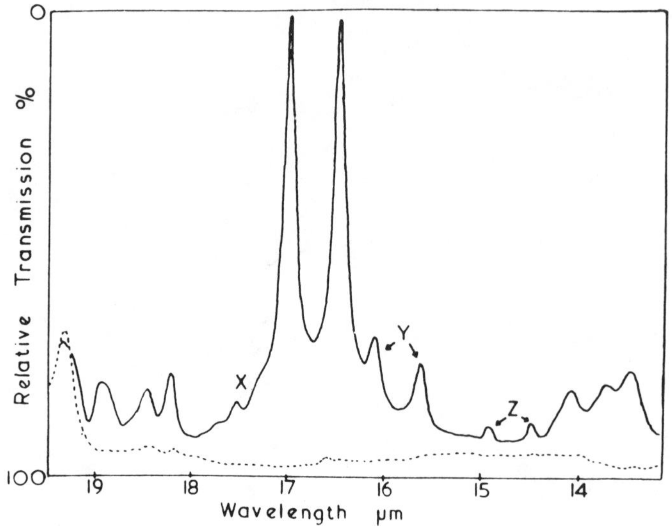

A

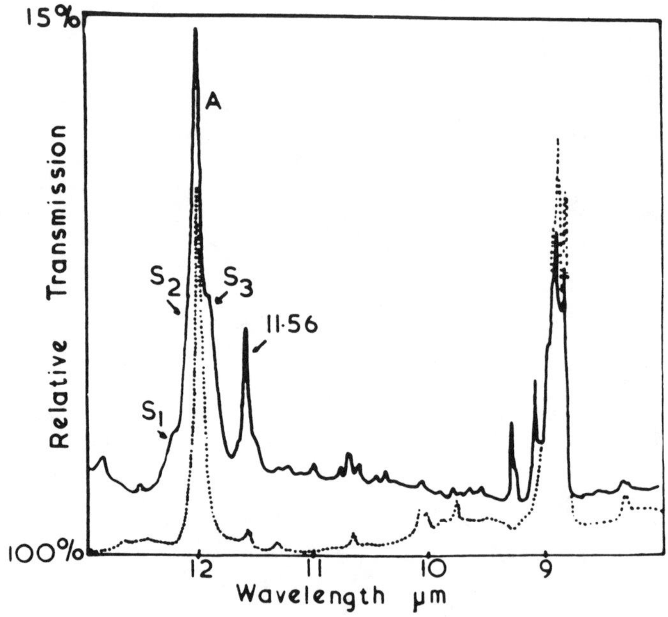

B

FIGURE 2.2.95. Transmission at 80 K of two samples containing 2×10^{18} oxygen atoms/cm^3 but differing in carbon content, after an electron dose of 5×10^{18} cm^{-2}, measured relative to an undoped floating zone crystal. Dashed trace for sample with less than 10^{17} carbon atoms/cm^3; continuous trace for sample with 2×10^{18} carbon atoms/cm^3. Data in three spectral regions are shown in (A), (B), and (C). (Reprinted with permission from Bean, A. R., Newman, R. C., and Smith, R. S., *J. Phys. Chem. Solids,* 31, 739, 1970. Copyright 1970, Pergamon Press, Ltd.)

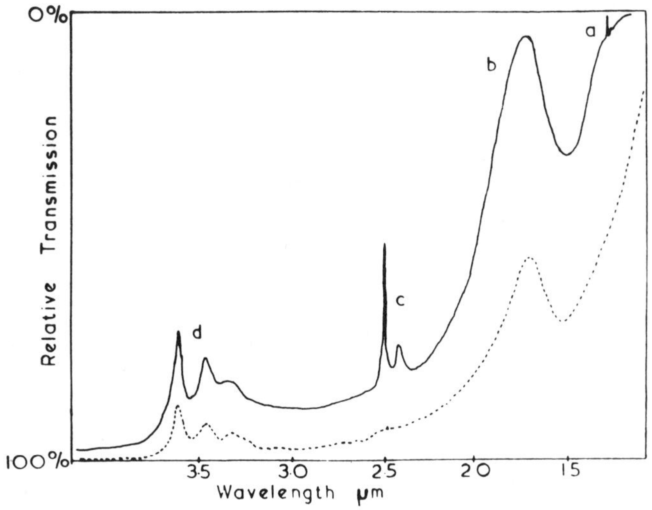

FIGURE 2.2.95C.

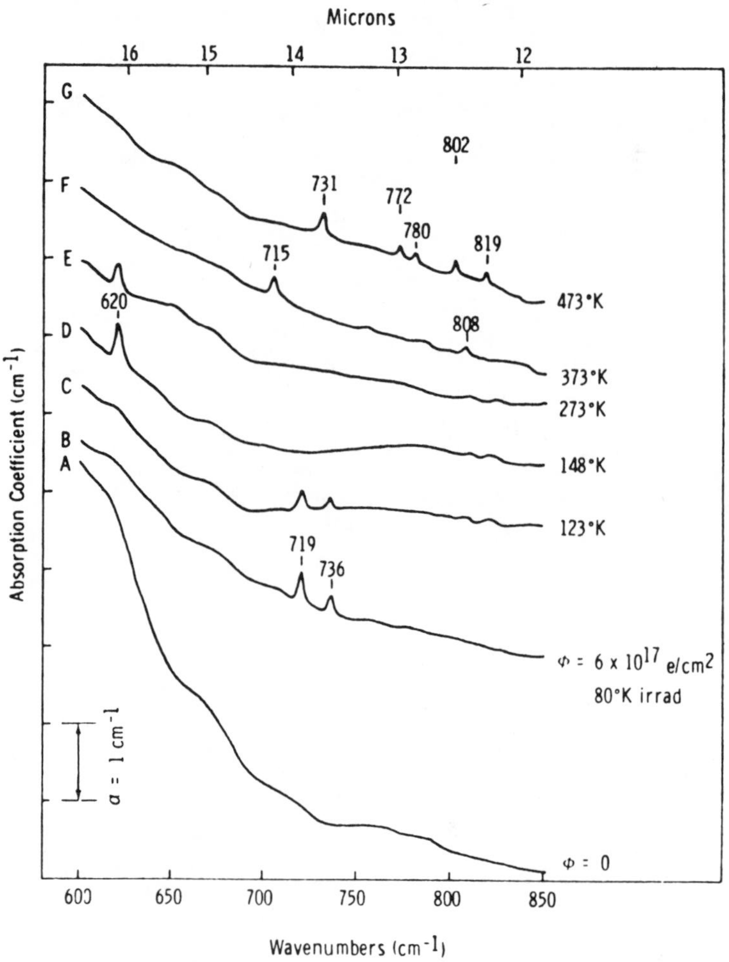

FIGURE 2.2.96. IR spectra of oxygen-doped germanium. Spectra have been vertically displaced for clarity. All spectra recorded at 80 K. (A) Pre-irradiation; (B) after 80 K irradiation to 6 × 10^{17} e/cm^2 fluence; (C) after 20-min anneal at 123 K; (D) after 20-min. anneal at 148 K; (E) after 20-min. anneal at 273 K; (F) after 20-min. anneal at 373 K; (G) after 20-min. anneal at 473 K. (From Whan, R. E., *Phys. Rev.*, 140, A690, 1965. By permission of Sandia National Laboratories.)

2.3. GLASSES

CRYSTALLINE AND AMORPHOUS SiO_2

The forms of SiO_2 that are of interest as optical materials include crystal quartz and amorphous silica:[165] quartz fused in an electric arc or by a flame in a Verneuille process (commonly referred to as Types I and II, respectively), and synthetic silica produced by flame hydrolysis (Type III) or plasma fusion (Type IV). These different forms of SiO_2 vary considerably in the nature and concentration of impurities such as aluminum, alkali metal, and OH. Crystal quartz and Types I and II silica have both alkali and aluminum as common impurities and have OH concentrations of 5 to 20 ppm for the crystalline quartz and Type I silica or 150 to 400 ppm for the Type II silica, whereas the synthetic silicas are practically free from metallic impurities; the OH content of Type III is quite large, ~1200 ppm, whereas that of Type IV is about 0.4 ppm. Since the optical properties of irradiated SiO_2 depend to a much larger extent on the impurity content than whether the material is crystalline or amorphous, and since a number of intrinsic induced absorptions are common to both forms, this section will discuss both crystalline and amorphous SiO_2 together.

Intrinsic Defect Absorption

Radiation effects in SiO_2 have been reviewed by a number of authors,[5,8] and the reader is referred to these articles for an entry point into the literature. Table 2.3.1 contains the details of the optical bands and associated defects, where determined. Studies of the radiation-induced optical absorptions in SiO_2 not associated with impurities are facilitated by the Types III and IV synthetic amorphous silicas. The transitions of "intrinsic" defect centers with significant oscillator strengths occur primarily in the UV spectral region and are identical both in these materials and in crystalline and fused quartz. Example spectra of irradiated Type III Suprasil 1 and Type IV Suprasil W-1 silica and the resolution of the experimental data into component Gaussian bands are shown in Figures 2.3.97 and 2.3.98.* Three bands centered near 4.7 eV (0.260 μm), 5.5 eV (0.230 μm), and 5.85 eV (0.212 μm) and the tail of a band near 6.5 eV are evident in both materials. A similar resolution into these four absorption bands has been performed on Type III Corning 7940 by Levy,[166] but the 4.7-eV band is absent from the radiation-induced spectrum of Corning 7943 (a synthetic silica prepared by flame hydrolysis of $SiCl_4$ and subsequently dried in H_2 to remove the OH). Because the intensity of the 4.7-eV band varies substantially from sample to sample, it is likely that it is either due to an impurity or in some way related to the stoichiometry of the silica since Corning 7943 is known to be oxygen deficient due to the drying procedure. No definitive identification of this band has yet been reported.

The absorptions at 5.5 and 5.85 eV in irradiated crystalline quartz and amorphous quartz and silica have been well characterized as arising from electrons trapped in sp^3 hybrid orbitals of Si at the site of oxygen vacancies.[167-169] Although the early literature considered this "E′" center to be formed by an electron being trapped by a Si, recent theoretical and experimental evidence confirm that the "E′" defect center is generated when a hole is trapped in a Si-Si bond, allowing one of the SiO_3 groups to relax into a planar configuration. The 5.85-eV band has been correlated with the E'_1 defect center identified by ESR in both quartz and silica by Weeks and Sonder, who determined the oscillator strength to be 0.14 ± 0.04.[170] Unfortunately, because of interference between the E'_1 and E'_2 ESR lines in irradiated amorphous SiO_2, the E'_2 center, a variant of the E′ defect in which a proton interacts weakly with the defect electron, could be resolved only in crystalline quartz. The correlation between the 5.5-eV band and the E′ center was thus established for this material, and an oscillator[2] strength of 0.28 was calculated.[170] Nevertheless, the induced optical spectra of the amorphous

* Figures 2.3.97 to 2.3.154 may be found on pp. 400 to 441.

Table 2.3.1
SUMMARY OF RADIATION-INDUCED OPTICAL BANDS IN SiO_2

Label	E_o	U	f	Occurrence	Defect	Sample ref.
	8.2			γ-Irradiation		176
	8.0			γ-Irradiation		176
E	7.6	0.5	0.65	Pure SiO_2 Alkali-doped SiO_2	Peroxy radical Hole trap	173, 175—177
D	7.15	0.8		Heavy ion irrad.		173, 177
C	6.1			Impure SiO_2		5
E_1'	5.85	0.6	0.14	Pure and lightly doped SiO_2	Hole trapped in neutral 0 vacancy	167, 170
E_2'	5.5	0.4	0.28	Pure SiO_2	E_1' + proton	169
	5.3	0.4		Alkali-doped SiO_2	E_1' + alkali	213
B_2	5.1	0.4		Heavy ion irradiation	Neutral oxygen vacancy	173
D_0	4.7			Pure SiO_2		173
B	4.0	1.2		Impure SiO_2	Electron trap	191
A_3	2.9	1.5		Impure SiO_2	Al center	191, 195
A_2	2.5	0.8		Impure SiO_2	Al center variant	191, 194
	2.0	0.26		Pure SiO_2 - low OH	Nonparamagnetic e′ trap	183
	2.0	0.33		Pure SiO_2 - high OH	Nonbridging oxygen hole center	183
A_1	1.9	0.7		Impure SiO_2	Al center variant	191

Note: E_0 = energy at the center of the band (eV); U = width of the band at half maximum (eV); f = oscillator strength.

silica and crystalline quartz are so similar that it is likely that such a correlation exists between the 5.5.-eV band and the E'_2 center in the glass, as well.

A word of caution is in order concerning the confusion that exists in the literature between the E'_1 center absorption and the "C" band which absorbs at the same energy in impure silica and in crystal and fused quartz (see Table 2.3.1). The C band, which develops strongly at low doses in these less pure materials, has been associated with aluminum impurities, whereas the E'_1 band is definitely an intrinsic defect.[5] Nevertheless, the labels are used almost interchangeably in the literature.

The transient absorption spectra of crystal quartz and amorphous silica measured after pulsed electron irradiation (Figure 2.3.99) are quite similar to the "permanent" spectra measured after γ irradiation (Figures 2.3.97 to 2.3.98).[171] The same three component bands at 4.7, 5.5., and 5.85 eV are evident in both cases, indicating the identical natures of the damage produced by pulsed and steady-state exposures.

There are additional radiation-induced absorptions reported at higher energies in quartz and silica. The dominant band following ionizing and particle irradiation, which has been labeled the E band, occurs at 7.6 eV (0.160 μm) as shown in Figure 2.3.100. The band is quite intense in both neutron and γ-irradiated samples.[172,173] The absorption has been correlated by Stapelbroek et al., with the peroxy radical defect Si–O–O characterized by ESR;[174] the oscillator strength for this band has been estimated to be 0.65 ± 0.3.[175] A band at 7.1 eV of width 0.8 eV is observed in silica after heavy ion irradiation (Figure 2.3.100) by Antonini et al.[173] Additional bands at 8 and 8.2 eV has been reported,[176] and the existence of weak bands at lower energies has been suggested. However, the principal absorptions attributed to intrinsic defect centers induced by ionizing radiation are those at 4.7, 5.5, 5.85, and 7.6 eV. Neutron or heavy ion irradiation has been found to induce an additional band at 5.1 eV (0.245 μm) called the B_2 band (Figures 2.3.100 and 2.3.101), which has been tentatively attributed to a doubly charged oxygen vacancy, i.e., Si □ Si.[173,177]

The growth behaviors of the E'_1 and E'_2 center bands in Corning 7940 irradiated by various energy electron beams at 77 K are shown in Figure 2.3.102.[178] Arnold and Compton[178] attributed the initial slope of these curves to population of existing defects in the glass by the radiolytic charges, and the linear behavior at higher fluxes to concurrent defect production and trapping. For comparison, the growth behavior of the E'_1 band similarly electron-irradiated at room temperature is shown in Figure 2.3.103.[179] Palma and Gagosz[179] observed identical behavior when the silica sample was irradiated with γ rays from a ^{60}Co source. They found good agreement between their experimental growth behavior and a theoretical model based on simultaneous trapping and defect generation by the ionizing radiation. It is significant to note that the specific growth behavior of the E'_1 center appears to depend somewhat upon the type of silica irradiated. Korneienko et al.[180] observed the ESR intensity of the E' center to be proportional to $D^{0.2}$ between 10^4 and 10^6 rad and proportional to $D^{1.5}$ above 10^7 rad in a Soviet Type III KSG silica (where D is the dose in rads), and Griscom and Friebele[181] reported growth of the ESR intensity approximately proportional to $D^{0.1}$ between 10^3 and 10^7 rad and to $D^{0.8}$ between 10^7 and 10^8 rad in (Type III) Suprasil synthetic silica. The behavior in the low-dose region is attributed to metastable E' centers that anneal during the irradiation, whereas that at higher doses is ascribed to the creation of additional oxygen vacancies via direct knock-on processes. Hence, the damage and the exponent will be dose-rate dependent in the low-dose-rate region.

The dependence of the absorption coefficient of the E'_1 optical band in Corning 7940 on the energy of the electron beam during irradiation at 77 K is shown in Figure 2.3.104.[178] The much greater efficiency of coloration in the oxygen-deficient Corning 7943 is readily apparent. It is also interesting to note that the absorption decreases with increasing electron energy for a constant flux.

At very large dose rates, the additional mechanism of radiation annealing, i.e., actual removal of defect sites and healing of the glass network during the irradiation, was reported by Palma and Gagosz.[179] Figure 2.3.105 illustrates this effect in Corning 7940 irradiated with 1.5 MeV electrons at various dose rates. Note also that the peak in the absorption coefficient vs. dose rate at 0.4 Mrad/sec becomes more pronounced with increasing irradiation temperature. Similar annealing is observed when Corning 7940 is pulse-reactor-irradiated at 61 Mrad/sec, and this effect has been confirmed by Korneienko et al.[180] in their experiments on KSG silica.

An absorption band that is commonly observed in the visible spectral region in irradiated high-purity synthetic silica is located near 2.0 eV (0.630 μm), as shown in Figure 2.3.106.[181] Because it is a rather weak absorption, it has been reported only in bulk samples irradiated to high doses ($\gtrsim 10^6$ rad) or at lower doses in optical fibers with pure silica cores, which provide much longer path lengths. The band is more intense and more clearly resolved in Suprasil than in Suprasil W, and interestingly, it may be induced without irradiation by drawing silica with a low OH content (such as the Type IV Suprasil W) into an optical fiber.[182] The growth behavior of the band in a Suprasil core fiber is shown in Figure 2.3.107, and the annealing behaviors in bulk Suprasil and Suprasil W are shown in Figure 2.3.108. The precise origin of this absorption is difficult to determine since unambiguous correlations with paramagnetic defect centers have not been forthcoming. However, annealing results, such as those shown in Figure 2.3.108a, are evidence that the 2.0 eV band is due to the nonbridging oxygen hole center (NBOHC), i.e., Si–O· in high OH content silica such as Suprasil.[183] In contrast, the band in low OH content Suprasil W arises from a center which is nonparamagnetic since it is not observed by ESR, and the center appears to be an electron trap since its band height correlates with the sum of the ESR intensities of all the paramagnetic hole traps, as shown in Figure 2.3.108b.[183] A possible model for this defect center is Si_3^-, i.e., a 3-coordinated silicon that has trapped a *pair* of electrons.

Emission

The luminescence emission from crystalline and amorphous SiO_2 is quite complex, consisting of a number of bands which vary in center position and halfwidth depending upon the type of irradiation, the temperature, and whether the sample is crystalline or glassy.[171] Although some of the principal bands that have been observed are influenced by the impurities present, they all are directly related to intrinsic defects in the SiO_2. The centers of the emission bands are near 1.91 eV (0.65 μm), 2.82 eV (0.440 μm), 4.28 eV (0.290 μm), and 6.70 eV (0.185 μm). Figure 2.3.109 depicts the luminescence band at 1.91 eV in neutron-irradiated Type IV Suprasil W silica;[184] Figure 2.3.110 contains a summary of cathodoluminescence, radioluminescence, and transient luminescence of the 2.82- and 4.28-eV bands in crystalline and amorphous samples;[171] Figure 2.3.111 shows the spectrum of the 6.70-eV band in α quartz.[185]

The luminescence at 1.91 eV shown in Figure 2.3.109 has been reported in vitreous silica samples following both ionizing and neutron irradiation.[183,184,186,187] The photoluminescence is approximately three times more efficient at 90 than at 300 K,[184] and the excitation band has been determined to be centered at 4.7 eV.[184,186] Interestingly, similar to the 2.0-eV absorption, this luminescence can be induced in low OH content synthetic silica, such as Suprasil W, by drawing it into an optical fiber[182] or by irradiation of either low or high OH content bulk silica (Figure 2.3.112)[183] or pure or doped silica core fibers.[188] The decay of the photoluminescence in such a fiber is shown in Figure 2.3.113, where Sigel and Marrone[188] showed that the data can be resolved into a two-stage exponential recovery with decay times of 2.2 and 18 μsec at 300 K. In contrast, a single exponential decay was measured by Skuja et al.[187] in neutron-irradiated silica with a decay time of 15 μsec at 100 K and 10 μsec at 300 K. There is a substantial body of evidence to associate the 2.0 eV luminescence in high OH content silica such as Suprasil with the radiation-induced nonbridging oxygen hole center,[183,188] and that in low OH content silica such as Suprasil W with a nonparamagnetic electron trap, e.g., Si_3^-.[183]

It is apparent that the 2.82-eV luminescence is common to all amorphous and crystalline samples shown in Figure 2.3.110, although the center and width of the band depends on the temperature and impurity content. As shown in Figure 2.3.114, a marked temperature dependence of the quantum efficiency of the band is noted by comparing the data of the Sawyer quartz sample at 4.2, 77, and 300 K. It is also apparent that the quantum efficiencies of the emission in both the fused quartz (Infrasil) and synthetic silica glasses are low compared to crystalline quartz, even at 4.2 K.[171]

In contrast to the exponential decay behavior of the 2.82-eV radioluminescence band in crystalline quartz following pulsed electron irradiation, the luminescence decay in the amorphous SiO_2 samples is quite complex, as evidenced by the substantial curvature in the data shown in Figure 2.3.114 due to a wide range of decay times.[171] At higher temperatures, Skuja et al.[187] report simple exponential decay of the photoluminescence of this band in neutron-irradiated Type III KSG and Corning 7940 silicas with a decay time of 60 nsec at a temperature of 100 K and <50 nsec at 300 K. Thus, it appears that those sites characterized by the longer-lived luminescence are the most susceptible to thermal quenching. The origin of the 2.82-eV "blue" luminescence in SiO_2 is not yet fully understood, although various authors have been able to draw successful correlations between the presence of the emission and behavior of E'_1 centers (see discussion in Reference 171).

As seen in Figure 2.3.110, the 4.28-eV emission is present only in Corning 7943 and in samples that have been neutron irradiated. The decay of the 2.82- and 4.28-eV emissions in Corning 7940 and 7943 following pulsed electron irradiation are shown in Figure 2.3.115, plotted on an expanded time scale relative to Figure 2.3.114. As shown by Griscom,[171] and the 4.28-eV emission decays much more rapidly than the 2.82-eV blue emission, and significantly, the decay matches that of the time derivative of the transient E′ optical absorption at 5.85 eV. This result establishes that the 4.28-eV luminescence results from the electron-hole recombination at the sites of oxygen vacancies.[171] Before the arrival of the electron, the holes trapped at these sites are manifested as E′ centers. That the emission is not observed in Corning 7940 or other SiO_2 samples unless they have been neutron irradiated is consistent with the fact that there is a much larger concentration of oxygen vacancies in Corning 7943 and in the neutron-irradiated materials. It is interesting to note the difference in decay times for the 4.28-eV radioluminescence of ~100 μsec at 4.2 K evident in Figure 2.3.114 with that of the <10 nsec reported for photoluminescence at 100 K.[187]

The 6.70-eV luminescence shown in Figure 2.3.111 was observed by Treadway et al. during steady-state irradiation of α quartz and various high-purity silicas with 2.5-MeV electrons.[185] The intensity was found to scale with dose rate, and the conversion efficiency of ionizing radiation into radioluminescence was found to be 3×10^{-5} for crystalline quartz and 9×10^{-6} for Suprasil. The origin of this luminescence is still debated.

For comparison, the emission spectrum of Corning 7940 under electron bombardment is compared in Figure 2.3.116 with those of two alkaline earth fluorides and sapphire. The emission of the glass is significantly less than that of the fluorides over the whole spectral region, and an advantage obtains for sapphire only in the 450 to 550 nm range. Reference 189, from which this figure was taken, also contains an extensive table of luminescence efficiencies of these and other potential photomultiplier window materials.

Impurity-Related Absorption

When crystalline or fused quartz is irradiated, strong optical absorptions in the visible spectral region are generated in addition to the intrinsic bands described above.[5,8] In fact, one can easily determine whether a sample of SiO_2 is quartz or high-purity synthetic (Type III or IV) silica by examining an irradiated specimen with the eye. The induced absorption spectrum is quite complex with a number of broad overlapping bands. Although Lell et al.[5] and Arnold[190] tabulate peak positions in a variety of samples, Gaussian resolutions of the spectra into component bands, such as shown in Figure 2.3.117, are required to fully characterize the absorption.[191] The bands shown in Figure 2.3.117 are tabulated in Table 2.3.1. [In addition, a B_2 band at 5.1 eV (0.245 μm) has been reported in ion-bombarded silica.[173,177]] It is significant to note that the B band at 3.9 eV, which had been previously thought to exist only in fused quartz, is clearly evident in the deconvoluted spectrum of the irradiated crystal in Figure 2.3.117, as shown by Nassau and Prescott.[191] The position of the peak in the amorphous material seems to depend upon the alkali impurity present (Figure 2.3.118).[192]

The actual color of the irradiated quartz depends on the impurities present, the irradiation history, the temperature of irradiation, and the thermal history of the sample. Nassau and Prescott[191] have shown that it is the A_3 band which gives rise to the smoky color; this band is also correlated with the aluminum center, delineated by ESR to be a hole trapped on an oxygen bridging between a substitutional aluminum and silicon.[193] Koumvakalis[194] assumed a correlation between the ESR intensity of the aluminum center and the peak height of the induced optical spectra at 2.5 eV (Figure 2.3.119), identified as the A_2 band, and derived an oscillator strength of 0.07. However, Gaussian resolutions of the spectrum were not performed in this study, so this result is open to question. The observed variations in peak heights in Figure 2.3.119 are apparently due to variations in both A_2 and A_3 intensity. The A_3 band alone has been firmly correlated with the substitutional aluminum center;[191] however, the other A bands and the B and C bands also seem to be related to aluminum and/or alkali impurities.[5,8,171]

Although the alkali is not directly involved in the substitutional aluminum center, it must be present for charge compensation prior to irradiation;[195] as a result, the intensities of the induced bands are dramatically affected by irradiation temperature and whether or not the crystal has been field-swept. Figure 2.3.119A shows that irradiation of an unswept crystal at 77 K results in much less damage than irradiation at 300 K because the alkali are hindered from diffusing away at the lower temperature. In Figure 2.3.119B the field sweeping has removed the alkali and replaced them with hydrogen atoms, which are mobile under irradiation at 77 K. The lower induced absorption at room temperature in the latter case is attributed to the lack of stable electron traps.[194] These figures serve to illustrate the complex dependence of the damage in crystalline quartz upon temperature, sample history, etc. as well as careful characterization of the optical spectra required by techniques such as Gaussian resolution into component bands.

Mitchell and Paige[172] have shown that the radiation-induced absorption in crystalline and fused quartz may be optically or thermally bleached. Figure 2.3.120 shows that bleaching of a γ-irradiated quartz sample with UV light results in a uniform decrease in the various absorption bands between 2 and 6 eV.[172] In contrast, a similar bleach of a neutron-irradiated crystal preferentially bleaches the A and C bands.[172] Thermal bleaching can occur at temperatures as low as 240°C in both X- and neutron-irradiated crystalline quartz (Figure 2.3.121), and annealing at 400°C is sufficient to remove all coloration.[190,191]

Corresponding to the substitutional aluminum hole trap there is an electron trap due to germanium present as an impurity in many single crystal and fused quartz samples. The two centers are intimately involved in that not only do they trap opposite charges, but the alkali ions that diffuse away from the aluminum centers subsequently charge-compensate the substitutional germanium after it has trapped an electron (see discussion in Reference 171). The resultant defect absorbs at 4.6 eV (0.27 μm) if the charge-compensating alkali is Li, and at 4.3 eV if the alkali is Na.[196] The absorptions anneal in the 450 to 550 K temperature range, depending upon the sample.[197] As might be expected, these bands do not appear if the sample is irradiated at low temperatures where the alkali cannot diffuse from the aluminum to the germanium, but two corresponding bands at 4.2 eV (0.296 μm) and 4.8 eV (0.26 μm) are evident after irradiation at 77 K.[198]

DOPED SiO_2 FOR FIBEROPTIC APPLICATIONS

Glasses used for the core and cladding of optical fiber waveguides are composed of either pure fused silica or silica doped with index-modifying elements such as Ge, B, P, or F.[199] Because extremely low metallic and OH impurity concentrations are required to minimize the attenuation of the fiber, the radiation-induced absorption is due to defects intrinsic to the host material and dopants, rather than impurities. Fiber communication systems are designed for use in the 0.8- to 1.6-μm spectral range, and the radiation-induced bands here are of much lower strength than those encountered, e.g., in the visible or near UV.

Figure 2.3.122 contains a summary of the radiation-induced optical spectra in a wide variety of pure and doped silica core fiberoptic waveguides.[200] The ordinate is in terms of dB/km, which is a common unit of attenuation in the fiberoptics literature; 1 dB/km is equivalent to an absorption coefficient of 2.3×10^{-6} cm^{-1}. Although there are few resolved absorption bands in the 0.4 to 1.8 μm range in the fibers, the amount of radiation damage is seen to depend strongly on the type and quantity of dopant.[201] For example, the Ge-doped silica core fibers have the lowest induced losses of the doped silica core group in Figure 2.3.122A, but addition of P increases the loss at short wavelengths and causes a band near 1.6 μm to appear.[202] The induced band at 2 eV (0.63 μm) described in the previous section is evident in the silica core fibers in Figure 2.3.122B, and the damage at long wavelengths is less than that in the doped silica core fibers. There is extensive literature characterizing the growth and recovery of the radiation-induced attenuation in fibers such as these, primarily at the wavelengths of interest for communication — 0.85, 1.3, and 1.5 μm — and the reader is referred to the literature.[203,204]

In spite of the obvious technological importance of high-purity doped silica glasses, there have been few comprehensive studies undertaken to characterize the defect centers and their corresponding optical absorptions that cause the radiation damage in these materials. (Although several materials, such as Ge-doped silica[204] and B-doped silica[205] have been studied by ESR, correlations with the optical absorptions are lacking.) Two cases in which such

Table 2.3.2
AVERAGE PEAK ENERGIES, WIDTHS, AND OSCILLATOR STRENGTHS (f) OF COMPONENT GAUSSIAN BANDS USED TO DECONVOLUTE INDUCED ABSORPTION SPECTRA OF P-DOPED SiO_2 AND ASSIGNMENTS BASED ON ESR-OPTICAL CORRELATIONS

λ(μm)	E(ev)	σ(eV)	f	Defect label	Radical
Bulk results					
0.570	2.2	0.35	0.5	POHC[b]	PO_4^{2-}
0.510	2.5	0.63	0.5	POHC[b]	PO_4^{2-}
0.400	3.1	0.73	0.5	Low-T POHC[b]	PO_4^{3-}
0.300[a]	4.1[a]	0.64[a]	—	—	
0.270	4.5	1.27	0.035	P_2	PO_4^{4-}
0.260	4.8	0.41	0.014	P_4	PO_2^{2-}
0.240	5.3	0.74	0.5	POHC[b]	PO_4^{2-}
0.180	7.0	2.76	—	—	
Fiber results					
2.760	0.45	0.80	—	—	
1.570	0.79	0.29	0.0007	P_1	PO_3^{2-}
1.160	1.07	0.32	—	—	

[a] Component required only for $T_A > 623$ K.
[b] POHC ESR intensity correlates to sum of intensities of all these bands.

combined studies have been successful are pure silica, as described in the Section ''Intrinsic Defect Absorption'' and P-doped silica.[202] The latter material is important because P is a common dopant in fiber waveguide cores even though, as shown in Figures 2.3.106 and 2.3.122, its addition increases radiation sensitivity of the fiber.

The induced optical spectra of bulk and fiber samples of P-doped SiO_2 in the 1.0 to 6.0 eV and 0.5 to 1.0 eV spectral ranges have been successfully resolved into component Gaussian bands (Figure 2.3.123), and these bands have been correlated by Griscom et al. with specific defect centers identified by ESR.[202] Table 2.3.2 summarizes the optical bands, their identification, and the oscillator strengths calculated from the ESR and optical data. The defect center populations have been found to grow linearly with dose up to 10^6 rad and then exhibit substantial saturation in the 10^6- to 10^9-rad dose region.[181,202] The annealing behavior of the optical bands following irradiation at 77 K is shown in Figure 2.3.124, together with the corresponding ESR data. It is significant to note that the P_1 defect center absorption increases with annealing temperature before decreasing precipitously at 750 K (Figure 2.3.124C). Since it is this absorption which primarily affects the attenuation at the operational wavelengths for optical fibers,[183,203,204] irradiation at higher temperatures will actually result in increased loss.[183]

The radiation-induced optical spectrum in Ge-doped silica core fibers has been measured as a function of dose during steady-state irradiation,[183] and a typical example is shown in Figure 2.3.125. No resolved bands are evident, and it appears that the damage in the 0.4- to 1.1-μm region results from the tail of a strong induced absorption at shorter wavelengths or a shifting of the edge to longer wavelengths.

Pulse irradiation of Ge-doped silica fibers that are not codoped with P results in a very large transient absorption, even at long wavelengths such as 0.85 or 1.3 μm.[204] Spectral measurements of this transient absorption following a 3-nsec pulsed electron irradiation are shown in Figure 2.3.126. Once again, no bands are clearly resolved, although there seems to be a weak absorption band near 0.48 μm in the spectrum taken 60 sec after irradiation. Similar spectra have been observed for periods up to 24 hr after steady γ irradiation. Certainly, more study is in order to identify the origin of both the transient and steady-state absorptions in the 0.4- to 1.1-μm region. It is not surprising that the induced band is at wavelengths shorter than 0.4 μm since the Ge-related absorptions in fused quartz are in the 0.25- to 0.30-μm range.[196-198] The lack of alkali to stabilize the Ge centers in high purity optical fiber glasses may account for the transient nature of the absorption.

Although relatively few studies of the thermal annealing of radiation-induced optical absorption in optical fibers have been reported, considerable interest has been generated in optical bleaching.[206] The recovery of the induced attenuation at 0.85 μm in binary Ge-doped silica core fibers and pure silica core fibers has been observed by Friebele and Gingerich[206] to be enhanced by optical signals carried in the fiber. The spectrum of the damage that can be optically bleached in the doped silica fiber appears to be simply a tail of a band at shorter wavelengths than 0.4 μm, similar to the absorption spectrum in Figures 2.3.125 and 2.3.126. In contrast, as shown in Figure 2.3.127, there is a distinct band that is photobleached from pure silica core fibers. The center wavelength and width of this band varies from fiber to fiber, which implies that the corresponding defect may be related to impurities in the glass.[183,207] It has been suggested that alkali clusters might account for the observed behavior, but this hypothesis remains unconfirmed at this time. Nevertheless, photobleaching is an important means for reducing the radiation damage in some optical fibers.

MULTICOMPONENT GLASSES

The behavior of multicomponent glasses exposed to radiation is important since they are used as lenses, windows, and other optical elements in a large number of optical devices. These glasses are typically much more sensitive to radiation than pure silica unless they have been intentionally doped with radiation-protecting elements.

The radiation damage in commercial glasses has been of interest since the early days of nuclear research when they were first used as windows in hot cells and as optical elements for remote viewing. In 1950 Monk surveyed the radiation-induced absorption in a large number of glasses from several manufacturers.[208,209] Unfortunately, he reported neither the dose nor the sample thickness used, so the data are of only semiqualitative usefulness. The early work was reviewed in a series of papers by Sun and Kreidl,[210] who included a bibliography of 227 references.

One of the principal difficulties encountered in attempting to understand radiation damage in commercial glasses is that their compositions are quite complex. Added to the major glass-forming constituent, which is usually SiO_2, are one or more alkali oxides: Na_2O, K_2O, or Li_2O; alkaline earth oxides: BaO, CaO, or MgO; additional glass-forming oxides such as Al_2O_3, B_2O_3, or P_2O_5; and possibly ZnO or ZrO_2. Lead oxide PbO is added for radiation shielding and raising the index of refraction of the glass, and fining agents such as As_2O_3 or Sb_2O_3 are often used. Finally, the glass may be doped with CeO_2 to ''radiation-protect'' it, as will be described in the Section ''Alkali Silicate Glasses''. The result of having so many constituents in the glass, together with the impurities that may be present in the raw materials used for the batch, is that the optical spectra after irradiation are relatively broad and featureless with no well-resolved absorption bands.

The present understanding of radiation damage processes in multicomponent glasses derives from systematic studies of binary compositions such as alkali silicate, alkali borate or alkali phosphate glasses, and some simple ternary compositions that include aluminum or an alkaline earth. The laboratory samples have been made from high-purity reagents in order to minimize the effects of multivalent impurities and dopants that compete with the intrinsic defects in the glass for radiation-produced electrons and holes. This section will first describe the results of studies of these model glasses and then present data on both standard and protected commercial glasses. It should be noted that several excellent reviews of optical and electron spin resonance spectroscopic studies of model glasses have been published,[5-8] and the reader is referred to these references for a detailed treatment of the subject.

Table 2.3.3
GAUSSIAN RESOLUTION OF THE INDUCED ABSORPTION SPECTRA OF OXIDIZED SODA-LIME-SILICATE, BORATE AND PHOSPHATE GLASSES AFTER A ROOM TEMPERATURE EXPOSURE OF 3.9 × 10⁷ R[a]

Silicate glasses $2SiO_2 \cdot 0.35Na_2O \cdot 0.30CaO$			Borate glasses $B_2O_3 \cdot 0.35Na_2O \cdot 0.30CaO$			Phosphate glasses $P_2O_5 \cdot 0.35Na_2O \cdot 0.30CaO$		
E_o	α_m	U	E_o	α_m	U	E_o	α_m	U
2.0	3.8	0.45	2.2	6.6	0.80	2.3	5.6	0.52
2.9	12.1	1.3	3.0	5.7	1.5	3.0	5.6	1.1
4.0	7.8	0.93	5.1	11.2	2.9	5.1	17.0	2.0
5.5	28.1	1.5						

Note: E_0 = energy at the center of the band (eV); α_m = intensity at the center of the band (cm^{-1}); U = width of the band at half maximum (eV).

[a] Shown in Figure 2.3.128.

Model Glasses

The optical absorption bands induced by irradiation depends on a number of parameters, including the glass former (SiO_2, B_2O_3, or P_2O_5), the type and concentration of the alkali or alkaline earth, the concentration of OH, the redox conditions during glass melting, and the temperature at which the glass is irradiated and measured. Although this presents a formidable parameter space, careful, comprehensive studies of the silicate system (and to a lesser extent, the borate and phosphate systems) have resulted in a relatively unified understanding of the radiation damage in simple glasses.

Examples of the induced optical spectra in soda lime silicate, borate, and phosphate glasses of similar composition are shown in Figure 2.3.128, and the component optical bands derived by Gaussian resolution by Bishay and Ferguson[211] are tabulated in Table 2.3.3. These glasses were all prepared under identical oxidizing conditions and irradiated at room temperature.[211] One might infer from the figure that the damage is quite different in the three glasses. However, inspection of the table reveals that the three resolved bands centered in the 2.0- to 2.3-eV range, near 2.9 to 3.0 eV, and at 5.1 to 5.5 eV, are common to all three glasses. (The center energy of the second band depends strongly on the alkali in the glass and can vary from 2.5 to 4.0 eV, *vide infra.*) Only the 4.0-eV band appears missing from the borate and phosphate glasses.

Alkali Silicate Glasses

Silicate glass has been the most thoroughly studied, and the understanding of radiation effects is much more complete than in the borate or phosphate systems. For this reason, we will describe results of studies of silicates in some detail and then summarize results for the other systems.

As shown in Table 2.3.3, the resolved absorption bands in soda-lime-silicate glass melted under oxidizing conditions and irradiated at room temperature are located near 2.0, 2.9, 4.0, and 5.5 eV. (The latter band occurs at 5.3 eV in binary alkali-silicate glasses.) Studies of irradiated lightly alkali-doped glasses of high purity have revealed similar bands[212] plus an induced band centered at 7.6 eV.[213] Although it has been proposed by Sigel that this band is due to holes trapped on oxygens in the glass,[213] its identity is unproven at this time.

Figure 2.3.129 illustrates the effect of varying the alkali content on the induced optical spectrum. As shown by van Wieringen and Kats,[214] the band near 2.7 eV in the potassium-silicate glass (which is equivalent to the 2.9-eV band in the soda-lime-silicate glass of Figure 2.3.128) is observed to increase with increasing alkali and then saturate; the band at 2.0 eV grows in at the higher concentrations, and the 4.0-eV band increases monotonically.[214] Also clearly evident in this figure is the so-called "radiation-induced decrease" in the UV absorption. The absorption of the glass after irradiation at energies higher than 4.8 eV actually becomes less than that prior to irradiation. This effect has been since explained as arising from a small concentration of Fe^{3+} present in the glass as an impurity.[215] The Fe^{3+} ion, which initially absorbs with large oscillator strength near 5.3 eV, is converted to Fe^{2+} upon irradiation. Since divalent iron does not absorb in the UV, there is a net reduction in absorption after the irradiation. The presence of the Fe^{3+} band also causes the apparent shift in energy of the 4.0 eV band, as shown in Figure 2.3.129.

As the alkali is changed, the intensities of the 2.0- and 4.0-eV bands increase in the order Li-Na-K-Rb, followed by a decrease in Cs, as shown in Figure 2.3.130. The 2.6-eV band maintains a constant absolute intensity, but it decreases in intensity relative to the other two in the order Li-Na-K-Rb.[214,216] The central energy of this band varies with the type of alkali from 2.99 eV for Li to 2.53 eV for Cs, as shown in Figures 2.3.130 and 2.3.131, where the central energy is plotted as a function of half the field strength of the alkali ion.[217] As described below, the linear dependence shown in Figure 2.3.131 is expected and consistent with the model of the defect center that is associated with this absorption. Also plotted in Figure 2.3.131 are similar data for alkali phosphate,[217] lightly doped alkali-aluminosilicate,[192] and alkali-alkaline earth-silicate glasses.[216] Note that the slopes of these curves are quite similar.

The nature of the traps giving rise to the induced absorption bands in alkali silicate glasses has been determined by a number of investigators by doping the glass with multivalent elements that compete successfully with the intrinsic traps for the radiation-produced electrons and holes (see "Glasses Containing Multivalent Metal Ions".)[7,8,218-221] Commonly used dopants include Ce, Mn, Eu, Ti, or Fe. On this basis, Stroud has shown that the 5.3- and 4.0-eV bands are electron traps and the 2.7- and 2.0-eV bands are hole traps in oxidized alkali silicate glasses.[222,223] The 5.3-eV band has been attributed to an E'_2 type of defect,[213,224] involving an electron trapped on a silicon with a neighboring alkali. The electron trap associated with the 4.0-eV band has not been identified. The hole center bands are associated with holes trapped on nonbridging oxygens created as a result of adding alkali oxide Me_2O to silica. The bands at 2.7 and 2.0 eV are tentatively correlated by Schreurs with defects consisting of one or two nonbridging oxygens, respectively, on the same SiO_4 tetrahedron.[225] Although the alkali is free to diffuse away after the oxygen ion traps a hole, it apparently remains in the vicinity since it continues to influence both the ESR and optical absorption spectra, especially in the case of the single nonbridging oxygen defect.[8,226] For example, note the effect of the alkali on the center energy of the 2.7-eV band, as shown in Figure 2.3.131. Well-founded models for the two defect centers, termed the HC_1 and HC_2 for the centers with one and two nonbridging oxygens, respectively, have been developed on the basis of ESR spectroscopy by Griscom.[226] The correlation of the defect centers with the hole center bands, although yet to be firmly established, does explain the fact that the 2.0-eV hole center band is observed only at high alkali content (Figure 2.3.129) since two nonbridging oxygens on the same SiO_4 are to be expected only when the alkali content is relatively large.

One additional observation about the 2.0- and 2.7-eV hole center and 4.0-eV electron center bands is that Mackey et al. have shown that their intensity varies linearly with the OH content in the glass, as measured by the intensity of the Si-OH stretching vibration at 2.7 μm (e.g., Figure 2.3.132).[227] This effect has been attributed to the presence of the proton increasing the oscillator strength (transition probability) of these bands.

When alkali silicate glasses are prepared under reducing conditions or irradiated at low temperatures, the absorption spectrum is markedly different from that seen in Figures 2.3.128 to 2.3.130.[218,224] An example of a Na_2O-3-SiO_2 glass irradiated at room temperature by Smith and Cohen[218] is shown in Figure 2.3.133. The resolved bands at 2.0, 2.7, and 4.0

Table 2.3.4
COLOR CENTERS IN ALKALI SILICATE GLASSES

E_0	U	Defect	Charge trapped	Concentration dependence	Thermal stability
1.7—2.1	*	E_1^-	Electron	Reduced ≈ oxidized	<77—220 K
2.0	0.5	HC_2	Hole	α OH	~450 K
2.1—2.2	*	E_2^-	Electron	α Reduction	~300—350 K
2.7	0.95	HC_1	Hole	α OH	450—500 K
4.0	1.45	E_3^-	Electron	α OH	~450 K
5.3	>1	E_2'	Electron	α Reduction α Alkali conc. (0.01—0.5 mol %)	400—500 K
5.85	0.6	E_1'	Hole	Obs. in lightly doped	
7.6	~1		Hole?	Ind. of alkali conc.	?

Note: E_0 = energy at the center of the band (eV); U = width of the band at half maximum (eV); * = asymmetric composite band.

eV in the oxidized glass have been replaced by a broad, asymmetric band near 2.2 eV, and the 5.3-eV band is more clearly resolved. The lack of the 2.0-, 2.7-, and 4.0-eV bands in the reduced glass may be explained by the fact that protons have been removed, resulting in a very low oscillator strength for these transitions.[227]

Mackey et al. have shown that the 2.2-eV band in the reduced glass is thermally unstable, as can be seen by comparing the reduced sodium disilicate glass irradiated at 350 K (Figure 2.3.134)[224] with the reduced sodium trisilicate glass irradiated at 300 K (Figure 2.3.133);[218] the remnants of the 2.0-, 2.7-, and 4.0-eV bands are observed at 350 K, and the 5.3-eV band has gained intensity. The 2.2-eV band is also susceptible to optical bleaching.[224] If the glass is irradiated at temperatures lower than 300 K, however, the 2.2-eV band dominates the spectrum of both the oxidized and reduced samples, as seen in Figure 2.3.135. Note that the band appears to shift to longer wavelength with decreasing temperature so that it is observed at 1.8 eV when the glass is irradiated at 77 K.[224]

The nature of the asymmetric band near 1.8 to 2.2 eV has been inferred by Mackey et al. from careful studies of the optical spectra of both undoped and Eu^{3+}-doped glasses,[221] together with thermoluminescence and optical and thermal bleaching data. It has been determined that the absorption actually comprises two bands, one of which is centered in the 1.7- to 2.1-eV range (E_1^-) and the other being centered near 2.1 to 2.2 eV (E_2^-). Both bands are electron traps; the former occurs in about the same concentration in both reduced and oxidized glasses but is stable only up to 220 K (Figure 2.3.135). The latter band increases with the increasing degree of reduction during melting and is stable up to 300 to 350 K. The shift of the asymmetric absorption with temperature is due to the different thermal stabilities of the two bands.[221] It is obvious that much of the confusion in the literature (e.g., Reference 224 vs. 228) concerning the nature of the radiation-induced absorption near 2.0 eV can be explained by the simultaneous occurrence of both the 2.0-eV hole center band and the E^- electron center bands in this spectral region. A summary of the[1,2] radiation-induced bands observed and characterized in alkali silicate glasses is presented in Table 2.3.4.

A significant property of the $E_{1,2}^-$ electron center absorption is that its peak energy is strongly influenced by the alkali ion in the glass. Figure 2.3.136 contains a plot of the limited peak energy data available for alkali silicate glass[224] as a function of the inverse square of the next nearest neighbor distance between the alkali and oxygen ions, calculated as the sum of the Pauli ionic radii. An electron center band of similar shape and energy was also observed in irradiated alkali borate glasses by Beekenkamp,[217] and these data are shown in Figure 2.3.136. Note that the peak positions depend on the irradiation temperature in both the silicate and borate glasses. ESR and optical absorption studies of a potassium borate glass have correlated the $E^{-}_{1,2}$ electron center band with a defect consisting of an electron delocalized on a small cluster of 4 to 8 potassium ions.[229] Such a defect is analogous to the F center in alkali halides, where the electron is shared among the ions neighboring the anion vacancy. Similar band position data for the F center in a number of alkali halide crystals are shown in Figure 2.3.136 plotted as the inverse square of the sum of the ionic radii of the cation and anion, an approximation to the empirical $d^{-1.84}$ law described in a previous section. There is an obvious difference in the slopes of the curves for the F centers and the alkali clusters in the glasses, but a 17% increase in the assumed distance between the alkali and oxygen brings the slopes of the curves for the silicate and borate glasses into agreement with that of the alkali halides. Such a lengthening of the interatomic distances in the clusters is not unreasonable since the glass is expected to have a more open structure than the crystal. The similar dependence of peak energy on interatomic distance is further evidence that the alkali cluster center giving rise to the $E^{-}_{1,2}$ optical bands[1,2] is analogous to the F center in alkali halide crystals.

Alkali Borate Glasses

As shown in Figure 2.3.128, the radiation-induced optical spectra of irradiated alkali borate glass is quite similar to the spectra of irradiated alkali silicate glass. Table 2.3.4 shows that similar bands are induced in the borate glass near 2.2, 3.0, and 5.1 eV, but there is no band in the borates that is equivalent to the 4.0-eV band found in the silicate glass.[211,217,231]

Bishay has determined that the two low-energy bands are hole traps by comparing the induced absorption spectra of borate glasses that are undoped with those doped with small concentrations of Ce;[230] the valence state of the cerium was controlled by the redox conditions during glass melting. A tentative correlation has been established between the annealing behavior of the induced optical absorption of a 20% K_2O-80% B_2O_3 glass at 2.4, 3.0, and 3.8 eV with the ESR intensity of a defect center comprising a hole trapped on an oxygen bridging between two 4-coordinated borons in a diborate structure by Griscom.[8,229] This defect is in contrast to the commonly observed boron-oxygen-hole-center, which consists of a hole trapped on an oxygen bridging between a 3- and 4-coordinated boron. It is important to note that in spite of the obvious similarities in the induced optical spectra of the silicate and borate glasses shown in Figure 2.3.128, substantial differences exist in the detailed structure of the defect centers giving rise to these absorptions.

The energy of the band maxima of the hole center absorptions depends on the type and concentration of alkali and alkaline earth present in the glass.[231] The variation is similar to that of the hole center bands in silicate glasses shown in Figures 2.3.131 and, e.g., Figures 2.3.137 and 2.3.138. Tables 2.3.3 and 2.3.5 show representative values for four borate glasses. Bishay has shown that the intensity of absorption is also strongly influenced by the type of alkali and decreases by over a factor of two as Li is replaced with Na and K as shown in Figure 2.3.139.[231] The dependence on alkaline earth is much less, but there is a slight decrease as Ca is replaced by Mg, Sr, or Ba. The dependence of the band intensity on the alkali concentration is quite complex and seems to reflect changes in boron coordination.[211]

The induced band near 5.0 eV has been tentatively identified by Bishay as an electron trap from the results of Ce-doping.[231] Similarities have been found between the annealing data of the optical absorption in the 3- to 5-eV region and the ESR intensity of the boron electron center by Griscom.[229] This defect, which comprises an electron trapped by a boron in an *sp* orbital and shared with a neighboring alkali, apparently contributes to the induced absorption in this spectral region, although the correlation with the 5.0-eV band is not firmly established.

Table 2.3.5
GAUSSIAN RESOLUTIONS OF THE INDUCED ABSORPTION IN BORATE GLASSES

1.0 Na_2O—4.5 B_2O_3 1.2 × 10^6 R			30% K_2O—70% B_2O_3 2.2 × 10^6 R			0.75—2.75 CaO—Al_2O_3—2.5 B_2O_3 1 × 10^7 R		
E_0	α_m	U	E_0	α_m	U	E_0	α_m	U
			1.9	4.75	1.04			
2.5	0.38	0.88	2.6	1.25	0.80	2.1—2.3	0.7—1.3	0.6—1.0
3.6	0.95	1.24	3.8	3.00	1.60	3.4—3.6	3.0—4.5	1.5—2.1
4.9	0.16	1.72	5.0	0.95	.60	5.0—5.2	0.5—2.4	0.7—1.4

Note: E_0 = energy at the center of the band (eV); α_m = intensity at the center of the band (cm^{-1}); U = width of the band at half maximum (eV).

If an alkali borate glass either contains more than 20% alkali or is irradiated at low temperature,[7,229] an additional band in the 1.5- to 2.4-eV range is observed (Figure 2.3.137 vs. 2.3.138 and Table 2.3.5). As discussed in a subsequent section, a correlation has been firmly established by Griscom between this absorption and the alkali electron center in which an electron is shared by a cluster of alkali ions in the glass.[229] The defect is completely analogous to the $E_{1,2}^-$ center observed in silicate glasses and the F center in alkali halides. It can be optically[232] or thermally bleached.[229] As shown in Figure 2.3.136, the center of the band is strongly dependent on the type of alkali in the glass and the temperature of irradiation and measurement.

The confusion in the literature regarding the nature of the induced visible absorption in alkali borate glasses can now be attributed to the fact that both the 2.2-eV hole center band and the alkali electron center band are located in the same spectral region, and in contrast to silicate glasses, the latter is stable at room temperature in some borate glasses.

Alkali Phosphate Glasses

The radiation-induced absorption spectra of alkali phosphate glasses consist of 3 bands located at 2.3 eV, near 2.9 eV, and at 5.1 eV.[211,217,233] An example of such a spectrum is shown in Figure 2.3.128, and the results of a Gaussian resolution are tabulated in Table 2.3.3. By comparing the induced ESR and optical spectra of glasses that were undoped with those doped with Mn^{2+}, the two low-energy bands were identified by Schreurs and Tucker as hole traps and the absorption at 5.1 eV was found to be due to an electron trap.[219,220] The energy of the 2.3-eV band was independent of the type of alkali, but as shown in Figure 2.3.131, the energy of the 2.9-eV band varied with alkali in a manner that is similar to that of the hole centers in alkali silicate (and borate) glasses.[217] The strength of both the 2.3- and 2.9-eV bands increased with increasing alkali content in a CaO-Na_2O-2P_2O_5 glasses, and as Li was replaced by Na or K in this glass, the visible absorption increased by a factor of 1.2, whereas the UV absorption decreased by a similar amount, as shown in Figure 2.3.140.[233]

When the 2.3- and 2.9-eV hole center bands were suppressed by competitive trapping in a Mn^{2+}-doped glass, a broad asymmetric band located at 2.25 eV (0.55 μm) was observed (Figure 2.3.141).[219,220] Schreurs and Tucker[220] identified this band as Mn^{2+} that had trapped a hole on the basis of the similar center energy and shape to that of Mn^{3+}. However, the center of the Mn^{3+} band occurs in other glasses at higher energies, in the range of 2.5 to 2.6 eV (0.50 to 0.48 μm),[234] as shown in Figure 2.3.141. It seems likely that the band observed in the doped phosphate glass is a composite of Mn^{3+} and the alkali electron center. Indeed, a transient electron center band has been observed by Barkatt et al. in undoped sodium metaphosphate glass at 1.8 eV following pulsed irradiation.[235]

A good correlation exists between the optical bands observed in irradiated alkali phosphate glass and in irradiated P-doped silica described in a previous section. The 2.3- and 2.9-eV bands in the former correlate with the 2.3- and 3.1-eV bands attributed, respectively, to the high and low temperature phosphorus-oxygen-hole-centers (POHC) in the doped glass. These two defects have been determined by Griscom et al.[202] to be holes trapped on two or one nonbridging oxygens of a PO_4 structural unit, in analogy to the HC_2 and HC_1 defect centers in silicate glasses. The induced band in the alkali phosphate glasses near 5.1 eV correlates with the 4.8-eV band in P-doped silica; the absorption being associated with the P_2 defect comprising an electron trapped in a antibonding orbital of a PO_2 unit.

In concluding this section on model glasses, it is useful to compare the radiation-induced absorption bands in the silicate, borate, and phosphate glass systems. Before doing so, it should be noted, however, that it is unadvisable to stretch the comparison between systems too far — the defects centers are not identical in all systems. Nevertheless, it can be said that bands associated with holes trapped on oxygens have been identified near 2.0 to 2.6 and 2.5 to 3.0 eV in all 3 glass systems, and the latter band position depends on the type of alkali in the glass. Alkali electron center bands have been identified in all glasses in the 1.5- to 2.4-eV range, and the stability of the band appears to depend on the glass former as phosphate $<$ silicate $<$ borate.[236] An electron trap has been found in all glasses near 5.1 eV, which appears to comprise an electron trapped in a bonding orbital of the glass former with a neighboring alkali ion. Finally, an electron trap at 4.1 eV has been observed only in silicate glasses; the corresponding defect center has not been identified.

Glasses Containing Multivalent Metal Ions

When a glass that contains multivalent ions is irradiated, these ions are available as potential traps for the radiolytic electrons and holes. Whether the ions trap the charges in preference to the intrinsic traps described in the previous sections depends on the relative capture cross sections and trap depths. Trapping by the multivalent ions seems to be favored in most cases, and the behavior of glasses when irradiated therefore depends primarily on the type and concentration of these dopants (or impurities). It is important to note that the presence of the dopants does not alter the intrinsic trapping sites of the glass. Rather, by providing alternate sites, they retard the formation rate and increase the recovery rate of the intrinsic color centers.

The most thoroughly studied multivalent ion dopant is cerium.[7] Depending upon the redox conditions during glass melting, the cerium will be either entirely in the trivalent state for a strongly reduced melt, or as a mixture of Ce^{3+} and Ce^{4+} for normal or slightly reduced melt conditions. Although "radiation protection" of glass by Ce doping was reported before 1950,[208-210] it was not until 1962 that the protection mechanism was elucidated in alkali silicate glass by Stroud;[222,223] subsequent experiments by Bishay confirmed the mechanism in aluminoborate glass.[230]

The absorption band of Ce^{3+} occurs at 3.95 eV (0.314 μm) and that of Ce^{4+} at 5.17 eV (0.240 μm) with a half width of 0.65 eV.[223,237] When a sodium trisilicate glass doped with 1% Ce was prepared under reducing conditions and irradiated, Stroud observed that formation of the two hole center bands at 2.0 and 2.8 eV was inhibited, and a band in the UV centered at 4.96 eV (0.250 μm) with halfwidth of 0.72 eV was formed.[223] The visible portion of the spectra are shown in Figure 2.3.142. The band at 4.96 eV was attributed to a hole trapped by a Ce^{3+}, creating a Ce^{3++} center, which is essentially a Ce^{4+} on a site normally occupied by a Ce^{3+}.[237] As shown in Figure 2.3.142, the Ce^{3+} resulting from a strongly reduced melt suppresses the visible absorption, but allows an increase in the near UV absorption due to enhancement of the electron band at 4.0 eV in silicate glass. If the glass is less strongly reduced so that there is a mixture of Ce^{3+} and Ce^{4+}, the radiation protection is enhanced — the Ce^{3+} ions trap the radiolytic holes, inhibiting the formation of the 2.0 and 2.8 eV bands in the visible, while the Ce^{4+} traps the radiolytic electrons, inhibiting the formation of the 4.0-eV band in the UV and its tail into the visible.[222]

Similar results have been obtained in potassium aluminoborate glass by Bishay.[230] Ce^{3+} suppresses the 2.36-eV hole center band, but when the glass is highly reduced, an induced band at 1.9 eV is observed. Although this band was initially attributed by Bishay to (Ce^{3+} + e), it seems more reasonable that it is the alkali electron center described in the previous sections. Its identity as an electron trap was confirmed by the fact that it was suppressed by Ce^{4+} present when the glass was prepared under normal atmosphere. A broad induced band at 3.5 eV was found to be enhanced by Ce doping, whether or not the melt was strongly reduced — a result that indicates that the band may be a composite of different types of traps with greater cross sections for radiolytic charges than Ce.

The growth and recovery of the induced absorption bands at 1.90, 2.25, and 3.0 eV in various Ce-doped barium aluminoborate glasses have been resolved into the sum of exponential and linear terms by Levy et al.[238] The absorption in the base glass was found to steadily increase with increasing dose, whereas the 3.0-eV band in the Ce-doped glass rapidly reached saturation. In contrast, the visible bands initially increased and then decreased with increasing dose. Recovery data following pulsed irradiation of doped and undoped lead flint and barium crown glasses, such as that shown in Figure 2.3.143, further elucidated the protection mechanism. Friebele found that not only does cerium inhibit the formation of the intrinsic color centers, as evidenced by the decrease in loss at short time with increased Ce content, it simultaneously enhances the recovery since the slopes of the recovery curves increase with increasing Ce content.[239]

It is important to note that the behavior of cerium is not unique; many multivalent elements compete successfully with the intrinsic traps for the radiolytic charges. Examples that were previously discussed include Fe^{3+} impurities in alkali silicate glass (see Figure 2.3.129)[214,216] and Mn^{2+} doped into calcium phosphate glass (Figure 2.3.141).[219,220] The so-called "radiation protection" afforded by cerium doping derives from the fact that both valence states absorb in the UV and that they both have larger capture cross sections for the electrons and holes than do the intrinsic traps. Furthermore, these glasses are "radiation protected" only against the induced *visible* absorption bands; their UV transmittance is quite poor, both before and after irradiation.[237]

The effect of other multivalent ion dopants on the induced optical absorption has been reported in a number of studies. Eu^{3+} has been used by Mackey et al. as a dopant to elucidate the alkali electron centers $E^{-}_{1,2}$ in soda silicate glass since this ion is an effective electron trap.[221] Ti^{4+} also serves as an electron trap, and the behavior of glasses doped with Ti has been well documented by Bishay[7] and Arafa.[240] Following irradiation, a band was observed at 2.4 eV (0.52 μm) with a halfwidth of 1.0 eV arising from the (Ti^{4+} + e) center, i.e., a Ti^{3+} ion on a site normally occupied by a Ti^{4+}. The (Ti^{4+} + e) band was compared with that of the Ti^{3+} ion whose peak energy is 2.5 eV (0.495 μm) and halfwidth is 1.3 eV.[240] The behavior of a number of glasses doped with ions such as As, Nb, Sb, Tb, Tl, Pr, Dy, Ho, Sm, Tu, Er, and Yb has also been reported, and the reader is referred to the literature (e.g., see References 231, 233, 241, and 242).

Glasses Containing Heavy Metal Ions

Heavy metal ions such as Pb and La are incorporated into glass to raise the index of refraction, and in the case of Pb to provide shielding from radiation when the glass is used as a window in a nuclear reactor, hot cell, or other radiation source. Relatively large quantities of these metals can be added without adversely affecting glass forming or enhancing crystallization; e.g., commercially available double-extra-dense flint glass contains approximately 80 mol % PbO!

There have been a number of reports of the radiation-induced optical absorption in simple lead-containing glasses, such as alkali phosphates,[233] silicates[243] and borates,[244] and more complex glasses containing alumina.[245] The principal feature of the spectrum of many of these studies shown in Figure 2.3.144 is a band at 1.59 eV (0.780 μm) with a width at half maximum of 0.57 eV.[211,233,244] However, this band is not observed in all lead-containing glasses. In general, if the glass contains less than ~25 mol % PbO, the band will be observed, but if the glass contains more than 35 mol % PbO, no resolved band is evident. No postulate has been advanced for the occurrence of the 1.5-eV band or its associated defect center that is consistent with the various observations. It is known from nuclear magnetic resonance studies by Bray and co-workers that Pb^{2+} is incorporated into glass as an ionic network modifier for concentrations of PbO <25 to 30 mol %, whereas at larger concentrations, the Pb^{2+} is bonded covalently as a glass former in PbO_4 structural units.[246-248] ESR studies by Friebele have established that Pb^{2+} traps a hole upon irradiation and becomes Pb^{3+},[249] so it appears that the 1.59-eV absorption can be associated with a Pb^{3+} on a network modifying site.

In addition to the 1.59-eV band, two bands with peak energies of 2.36 eV (0.525 μm) and 3.31 eV (0.375 μm) and widths at half maximum of 1.09 and 1.01 eV, respectively, have been observed in Pb-containing glasses by Barker et al.[243] These bands are seen clearly in the induced spectrum of an irradiated 80 PbO-20 SiO_2 glass shown in Figure 2.3.145. Although the origin of these bands has not been determined, the growth behavior with increasing dose suggests that the defect centers not only exist in the glass prior to irradiation, but that they are created by the 1- to 2-MeV gamma ray photons used in the experiment.[243] It seems reasonable to associate these bands with Pb^{3+} in network-forming PbO_4 units.

An interesting behavior that has been reported for lead-containing glasses is rapid thermal and optical bleaching of the induced absorption in the visible provided that the glass does not contain alkali.[5,243,250] The addition of a few percent alkali stabilizes the damage although some slight recovery is noted over long periods of time, as shown in Figure 2.3.143. Although the Pb^{2+} ions provide traps for the radiolytic holes, they easily recombine with the electrons unless there are alkali present to provide permanent traps for these charges. The radiation damage in lead-containing glasses can therefore be minimized and the recovery enhanced by excluding alkali and by cerium doping to provide alternate traps for the holes.

There are a number of reports of the radiation-induced optical absorption of glasses containing up to 50% La, and in general the level of damage decreases with increasing La content. These glasses include calcium phosphate,[233] soda silicate,[242] and soda silicate glass doped with cerium,[242,251] It was reported by Bishay that the addition of 0.5% La to a potassium aluminoborate glass caused a slight increase in induced absorption.[231] Although specific absorption bands associated with La have not been identified and the understanding of the radiation damage in La-containing glasses is not complete, La appears to be an attractive alternative to Pb for raising the refractive index of glasses for radiation environments. Of course, the radiation shielding properties of Pb-containing glasses will be superior.

Multicomponent Glasses for Radiation Dosimeters

The sensitivity of multicomponent glasses to radiation has led to their use as radiation dosimeters in some applications. The requirements of a good dosimeter are linearity, or at least reproducibility, response over a broad range of doses, minimum fading, high sensitivity in the dose range of interest, and minimum dependence on the energy of the radiation. Although no one glass fully satisfies these criteria over the 0- to 10^9-rad exposure range, silver-activated phosphate glass has been shown to be an attractive dosimeter by Schulman et al.[252] In the 10- to 10^4-rad range the photoluminescence intensity of the glass can be used as a dosimeter.[253] Following irradiation the glass is exposed to a calibrated UV source and the luminescence output is measured. Although the photoluminescence is linear with dose to 10^4 rads, some saturation is evident at higher doses. Therefore, in the 10^4- to 10^7-rad range the increase in optical density with exposure is more useful for dosimetry, and silver-

Table 2.3.6
APPROXIMATE COMPOSITION (wt%) OF COMMERCIAL GLASSES[250]

	SiO_2	PbO	Na_2O	K_2O	M_gO	CaO	Al_2O_3	Fe_2O_3	As_2O_3	CeO_2
EDF	36	60		4			0.05			
OW7	35	59		4			0.05		0.2	1.0
DEDF	20	80								
OW8	20	79								0.5
Plate glass	73		13		3	9	1.2	0.08		
Plate glass + Ce	73		14		3	8	1.8	0.03		1.7

Note: Shown in Figures 2.3.147 to 2.3.149.

activated phosphate glass has been successfully calibrated and used for dosimetry of both 50 kV X-ray and 2 MeV-electron sources in this range by Ritz and Cheek.[254] As shown in Figure 2.3.146, other glasses, such as the cobalt-activated borosilicate glass, have been shown to have greater sensitivity and linearity in the 10^4 to $10^5 \times 10^6$-rad range.[253] More significantly, these glasses show significantly less fading than the silver-activated glasses.

The long path length afforded by optical fibers has been employed by Evans et al. to increase the sensitivity of glass dosimeters.[255] Fiberoptic dosimeters consist simply of an LED light source, a length of fiber, and a Si PIN detector; by cutting the fiber to different lengths, a broad range of doses can be covered with good response and linearity. A fiberoptic dosimeter system has been demonstrated onboard a satellite using a number of zinc barium crown glass fibers with lengths varying from a few centimeters to several meters.[255] The system provided accurate data from 0 to >6000 rad during its lifetime. The simplicity of fabrication and readout and the very low power consumption of such a system makes it extremely attractive for many dosimeter applications.

COMMERCIAL OPTICAL GLASSES

The optical transmission of commercial glasses after exposure to irradiation can be substantially degraded, and the glasses take on a deep brown color due to the tail of UV absorptions extending through the visible spectral region.[256] The spectrum is considerably less well resolved than that of two or three component laboratory melts due to the many constituents and impurities that may be incorporated in the commercial glass. The compositions are tailored to provide glasses of the required optical properties such as index of refraction, dispersion, homogeneity, quality, etc.

Compositions of two commercial flint glasses and a commercial plate glass together with their radiation-protected equivalents are given in Table 2.3.6; these data were taken from a 1961 publication by Barker and Richardson[250] and may or may not be typical of compositions available today. Nevertheless, they serve to illustrate the types of constituents used in making optical glass.

The transmission spectra of these glasses before and after various irradiation doses are shown in Figures 2.3.147 to 2.3.149. The effects of Ce protection are immediately obvious. As shown in Figures 2.3.147A and 2.3.148A, respectively, there is a substantial reduction in the visible transmission of the extra-dense flint (EDF) and double-extra-dense flint (DEDF) glasses after exposures of only 2×10^4 rads. On the other hand, the transmissions of the cerium-protected equivalents (glasses OW7 and OW8 shown in Figures 2.3.147B and 2.3.148B, respectively), are nearly the same before irradiation and after exposure to 10^8 rads. Of course, the unirradiated transmission of the cerium-containing glass is degraded in the short wavelength region because of the intense Ce^{3+} and Ce^{4+} absorptions in the UV. Nevertheless, the yellow tint of OW7 and OW8 glasses would be quite acceptable for many window applications.

It is interesting to note that the radiation-induced transmission loss in the EDF glass is actually greater than that in the DEDF glass for doses $>10^5$ rads, in spite of the larger Pb content in the DEDF glass (Table 2.3.6). Apparently, the enhanced radiation sensitivity of the EDF glass is due to the presence of alkali (K_2O), which stabilizes the damage, as described in a previous section. The effect of alkali is also evident in the thermal bleaching behavior of the two glasses.[250] Although the absorptions of both the EDF and DEDF glasses at 0.65 μm decreased by a factor of ~2 if the glasses are maintained in the dark for 5000 min following the irradiation, the EDF glass recovers only slightly at longer times whereas the DEDF continues its fading. Analogous behavior is evident in the optical bleaching when the glasses are exposed to sunlight following irradiation; the DEDF glass recovers completely within 100 min of the irradiation, but 300 min are required to optically bleach the damage in the EDF glass.

The transmission spectra of the plate glass before and after irradiation are shown in Figure 2.3.149A. There is substantial degradation in the optical transparency of this glass, but as shown in Figure 2.3.149B, cerium is effective in ameliorating the damage to a large extent. Although the cerium-protected plate glass maintains visible transmission after exposure to radiation, it would not be useful as a radiation-shielding glass since it contains no lead.

A set of standard and protected commercial optical glasses are prepared by Kreidl and Hensler.[256] These included a barium silicate (BSC-2), two light barium crown (LBC-1 and LBC-2), a dense flint (DF-2), and EDF-1 glasses; the transmission spectra are shown in Figure 2.3.150. It is apparent that all unprotected glasses suffer severe degradation after exposure to 10^6 rad (curve 5 in each figure), while addition of cerium is effective in largely protecting the glasses against exposure to the same dose (curve 2 in each figure). In most cases there is a substantial decrease in transmission of these glasses after exposure to 5×10^8 rad (curve 3). It should be noted that the protected glasses were specifically developed to have the same optical properties as their unprotected counterparts while having resistance to radiation-induced absorption.[256]

The glass manufacturers have developed radiation-protected optical glasses whose visible transmission is, in most cases, only slightly degraded when exposed to as much as 10^8 rad. These glasses are offered in a wide variety of refractive indexes and dispersions, providing the designer of an optical system with a larger degree of flexibility. For example, Schott Glaswerke[257] has a series of 23 radiation-resistant optical glasses with indexes varying from 1.51980 to 1.80911, and values of reciprocal dispersion from 63.62 to 25.28; other glasses are available on special request. The manufacturers should be contacted to provide specific data on the glasses of interest.

A number of other glasses were examined for their potential as radiation-resistant optical fiber waveguides by Evans and Sigel,[258] and a summary of the induced spectra is shown in Figure 2.3.151; the data comprise both fiber and bulk samples. It is apparent that there is a broad range of responses and that the level of damage experienced for a given dose is very dependent on the composition of the material. Note especially the effect of cerium addition in radiation protecting the three Na-Ca-silicate glasses; the induced visible absorption of the glass with 1% Ce is virtually unmeasurable in the 1-cm thick sample. It should also be noted that the thermal (and optical) bleaching that may be evident in these glasses depends on the composition as well, so the damage in some may fade out at a faster rate than in others. Figure 2.3.152 contains recovery data of several glasses and fibers, and once again the effect of cerium in lessening the damage and in enhancing the recovery of the Schott F2 lead flint glass and the Galileo zinc crown glasses is evident when these data are compared with the Schott R1 and cerium-doped zinc crown glasses, respectively.

Table 2.3.7

GAUSSIAN RESOLUTION OF THE INDUCED ABSORPTION SPECTRUM OF ZIRCONIUM-BARIUM-LANTHANUM-ALUMINUM-LITHIUM (ZBLAL) FLUORIDE GLASS SHOWN IN FIG. 2.3. 154, LEAD-DOPED ZBLAL AND BINARY ZB FLUORIDE GLASSES X-IRRADIATED AND MEASURED AT 80 K. PEAK ENERGIES (E_0) AND WIDTHS (W) ARE IN eV; INTENSITIES (I) ARE CALCULATED AS THE PRODUCT OF THE ABSORPTION COEFFICIENT TIMES THE WIDTH AT HALF MAXIMUM (eV/cm)

Glass		Pb^{3+}	Zr^{3+}	Pb^{+}		$V_H + V_K$			
ZBLAL	Eo		2.69			4.27	5.00	5.21	6.12
	W		1.45			1.25	1.94	0.80	1.40
	I		11.6			33.4	29.6	3.5	20.7
ZB	Eo		2.67			4.23	5.00		6.10
	W		1.36			1.11	1.94		1.4
	I		13.9			59.1	47.1		4.5
ZBLAL + Pb	Eo	1.53	2.68	3.07	4.05	4.27	5.00	5.30	
	W	0.40	1.40	1.00	0.65	1.25	1.94	0.80	
	I	0.50	3.1	4.83	1.88	12.5	104.5	6.0	

After Friebele, E. J. and Tran, D. C., *J. Non-Cryst. Solids*, in press.

HALIDE GLASSES

Non-oxide glasses in which halides are the anionic species are of interest as optical materials because they offer the potential of excellent transmission from the UV to the mid-IR.[259] In spite of the fact that halide glasses have been known to exist since 1926,[260] relatively few investigations of their radiation-induced optical properties have been reported.

Halide glasses of technological interest include those based on beryllium fluoride and the complex glasses, such as the heavy metal fluorides.[261] The optical absorption typical of irradiated vitreous BeF_2 or compound alkaline earth-alumino-beryllium fluoride glasses consists of bands at 4.6 and 2.2 eV (270 and 570 nm).[262,263] The nature of the defect giving rise to the latter band has been inferred to be an electron trap by doping the glasses with rare earth ions that compete successfully for the radiation-produced electrons or holes. The 4.6 eV band has thus been assumed to be due to a hole trap by Galimov et al.[263] Although it is tempting to try to correlate these bands with the results shown in Section 2.2 for halide crystals, an understanding of the structure of the defect centers associated with the radiation-induced optical absorptions in these beryllium glasses is at best speculative at this time.

On the other hand, Cl_2^- ions analogous to V_K centers in KCl have been observed and characterized by ESR techniques in a potassium chloride-potassium borate glass,[264] and the optical spectra of the glass and crystal have been found to be similar,[265] as shown in Figure 2.3.153.

Heavy fluoride glasses such as zirconium-barium (ZB), zirconium-barium-lanthanum-aluminum-lithium (ZBLAL), and ZBLAL doped with lead were X-irradiated at low temperature by Friebele et al.,[266] and the resulting optical spectra were resolved into component Gaussian bands; an example is shown in Figure 2.3.154. By performing isochronal anneals and resolving each spectrum into bands with the constraint that the position and width of each band remain constant and only the height of the band could change, it was possible to correlate several of the optical transitions with defect centers previously identified by Cases et al.[267] using ESR. As shown in Table 2.3.7, these include Zr^{3+}, the V_H and V_K trapped hole centers (corresponding to F^0 and F_2^-, respectively), Pb^+, and Pb^{3+}. The addition of

Cl to the ZBLAL glass has been found to result in the appearance of two additional radiation-induced bands at 3.9 and 4.78 eV,[268] which were correlated with the Cl_2^- and FCl^- defect centers identified by ESR.[269] Thus, in analogy with the oxide glasses, it is apparent that many of the defect centers and optical absorption bands that are induced in fluoride glasses by irradiation are similar to those that have been identified and thoroughly studied in crystalline fluorides, as discussed in Section 2.2.

Obviously, there is still much to be learned about the effects of radiation on halide glasses, and it seems likely that interest in this area will increase as halide glasses find more applications as windows with wide transparency ranges and as infrared transmitting optical fibers. Although no measureable radiation-induced absorption has been reported in the 2.5 to 7 μm region in bulk samples of halide glasses,[270,271] losses comparable to those in doped silica fibers have been measured in the 2 to 4 μm band using the longer path length afforded by fluoride glass fibers.[272] However, the defect centers responsible for these absorptions have not been determined at this time. The understanding of the radiation properties of halide glasses will likely be forthcoming only as a result of careful studies of pure materials, using a judicious combination of spectroscopic techniques such as optical absorption and electron spin resonance.

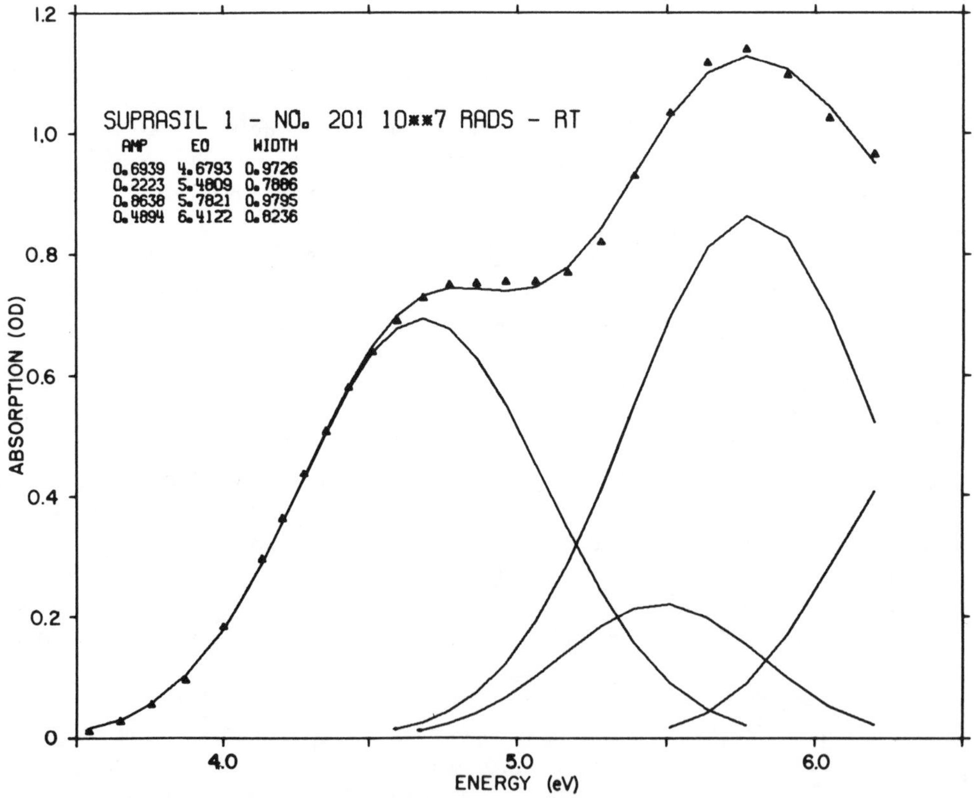

FIGURE 2.3.97. Radiation-induced optical absorption spectrum typical of high OH content Type III synthetic silica. The experimental spectrum is shown as the points; the solid lines represent the decomposition into component Gaussian bands and the fit to the data. The bands at 5.5 and 5.8 eV are the E_2' and E_1' bands, respectively. Sample length = 6 cm. (From Friebele, E. J. and Griscom, D. L., *Treatise on Materials Science and Technology,* Vol. 17, Glass II, Tomozowa, M. and Doremus, R. H., Eds., Academic Press, New York, 1979, 257. With permission.)

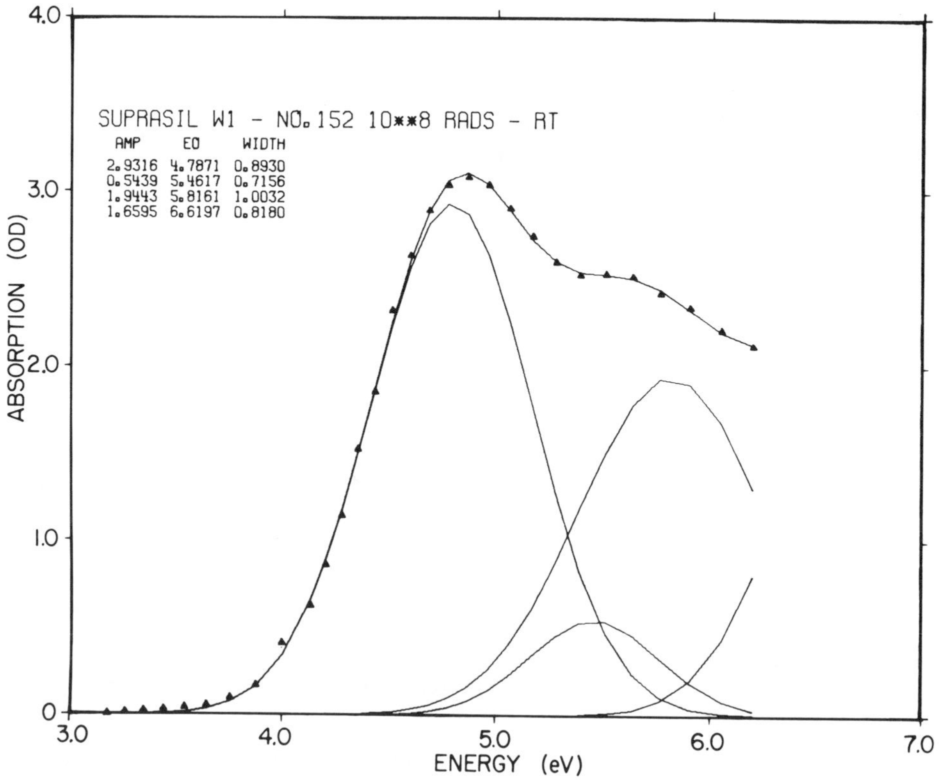

FIGURE 2.3.98. Radiation-induced optical absorption spectrum typical of low OH content Type IV synthetic silica. The experimental spectrum is shown as the points; the solid lines show the component Gaussian bands and the fit to the data. Sample length = 6 cm. (From Friebele, E. J. and Griscom, D. L., *Treatise on Materials Science and Technology,* Vol. 17, Glass II, Tomozowa, M. and Doremus, R. H., Eds., Academic Press, New York, 1979, 257. With permission.)

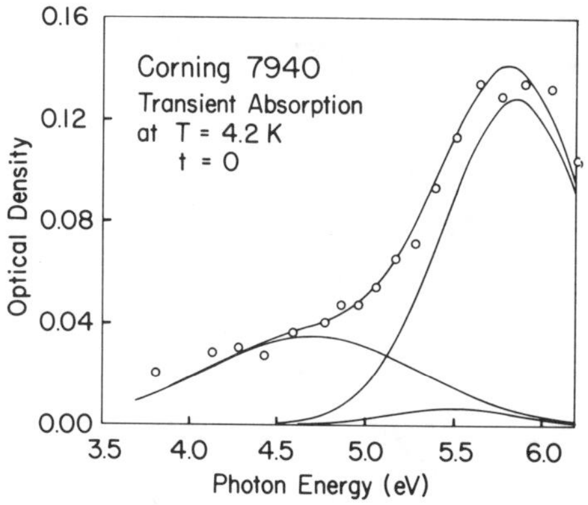

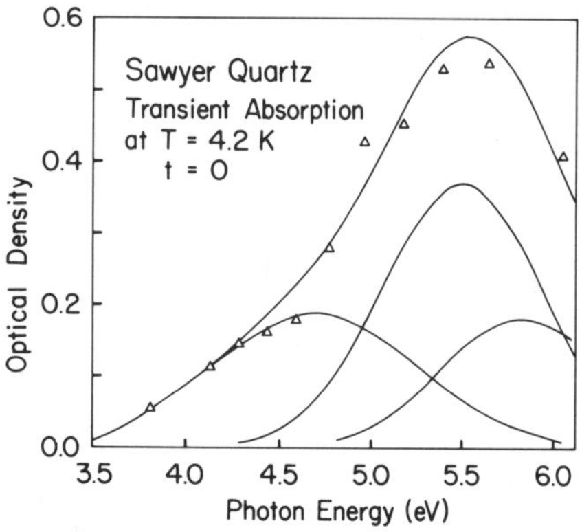

FIGURE 2.3.99. Transient radiation-induced optical absorption spectra of Type III Corning 7940 synthetic silica and field swept crystalline Sawyer quartz immediately following pulsed irradiation by 0.5-MeV electrons. (From Griscom, D. L., *Proc. 33rd Annual Symposium on Frequency Control,* Electronic Industries Association, Washington, D.C., 1979, 98. With permission.)

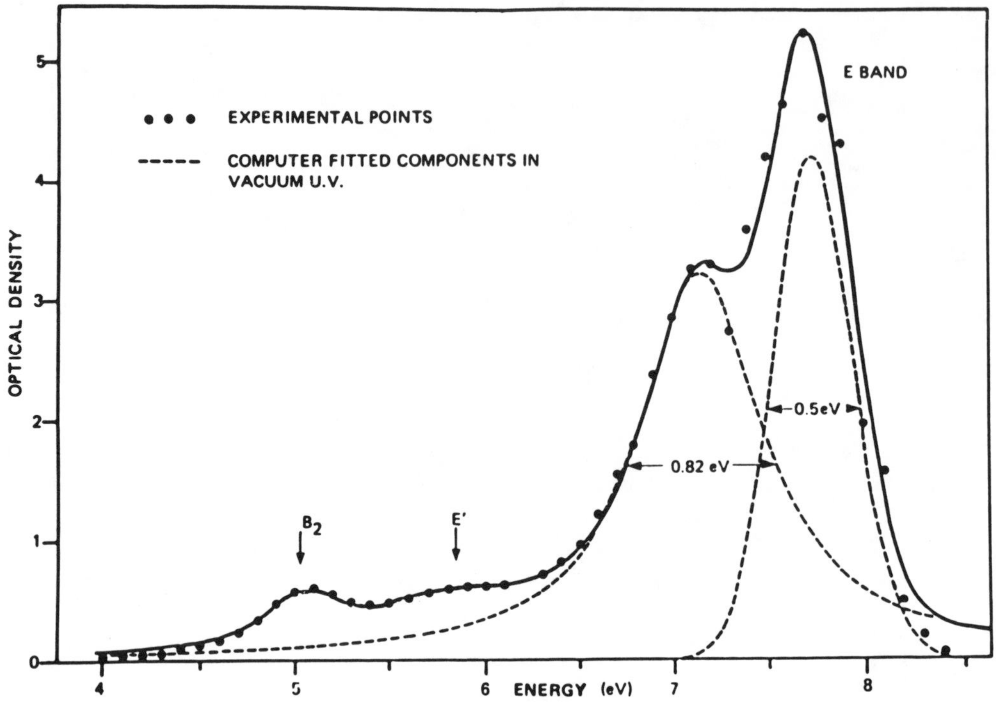

FIGURE 2.3.100. Absorption spectrum of Tetrasil B silica (OH content = 100 ppm) following a room-temperature irradiation of 10^{14} particles per square centimeter with 46.5-MeV Ni^{+6} ions. Points are experimental data and dashed lines are Gaussian bands that fit the vacuum UV data. (From Antonini, M., Camagni, P., Gibson, P. N., and Manara, A., *Rad. Effects,* 65, 41, 1982. With permission.)

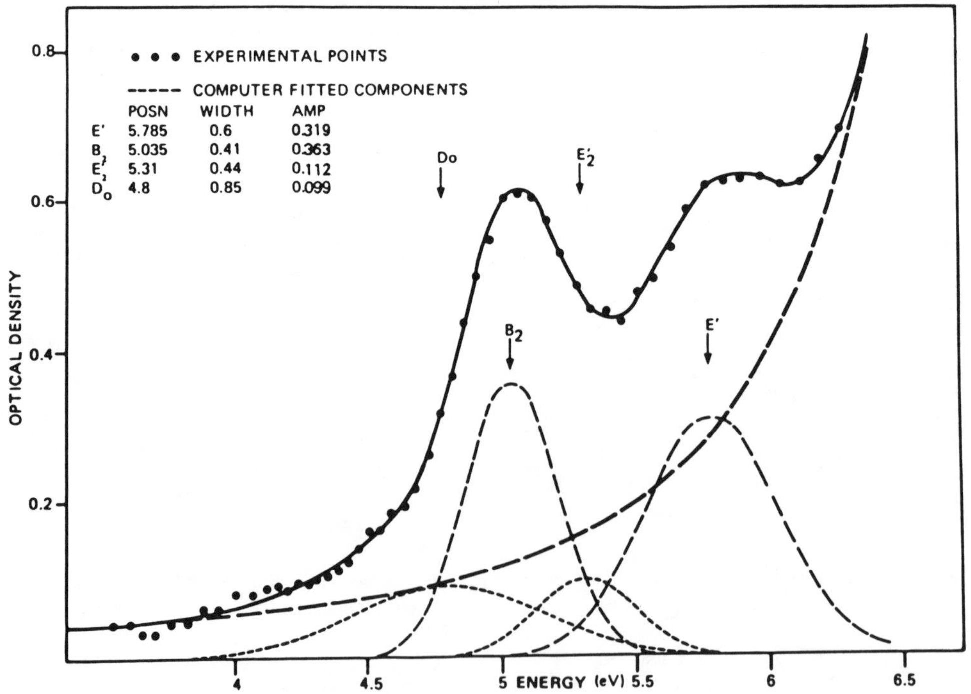

FIGURE 2.3.101. Details of the low-energy spectrum of Figure 2.3.100, showing the experimental data (points) and the component Gaussian bands (dashed lines). The thick dashed line is the tail of the band at 7.15 eV. (From Antonini, M., Camagni, P., and Gibson, P. N., and Manara, A., *Rad. Effects,* 65, 41, 1982. With permission.)

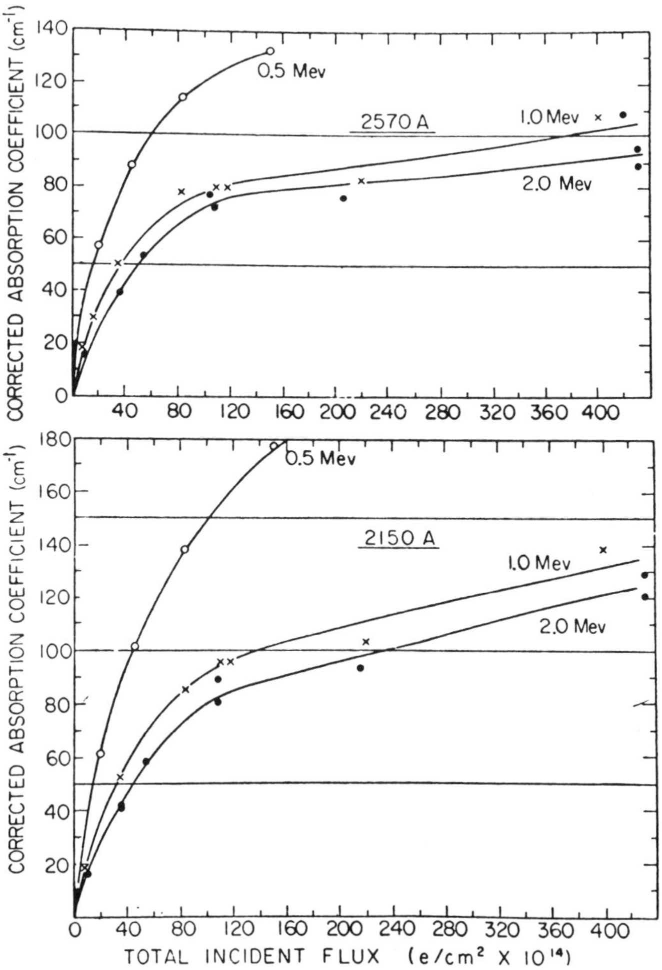

FIGURE 2.3.102. Intensity of the E_1' (0.215 μm) and E_2' (0.257 μm) absorption bands in high OH content Type III Corning 7940 synthetic silica during electron irradiation at 77 K as a function of dose and electron energy. The absorption coefficients have been corrected to take into account the effective increase in sample thickness at low energies due to multiple scattering events as the electron transverses the sample. (From Arnold, G. W. and Compton, W. D., *Phys. Rev.*, 116, 802, 1959. With permission.)

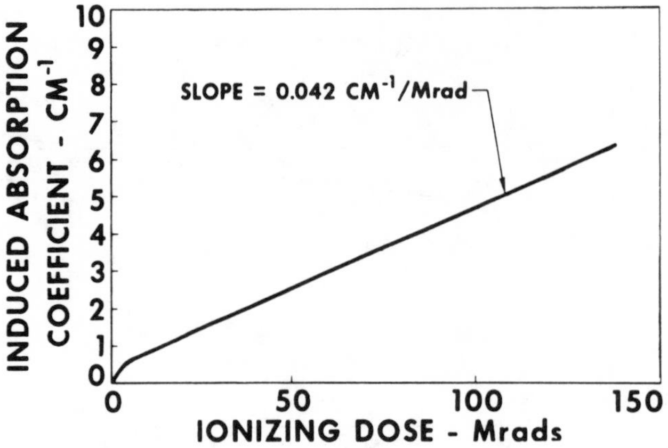

FIGURE 2.3.103. Growth of the E_1' optical absorption band at 5.8 eV in Type III Corning 7940 synthetic silica (OH content = 1200 ppm) during room temperature irradiation with 1.5-MeV electrons at a dose rate of 0.02 Mrad/sec. (From Palma, G. E. and Gagosz, R. M., *J. Phys. Chem. Solids,* 33, 177, 1972. With permission.)

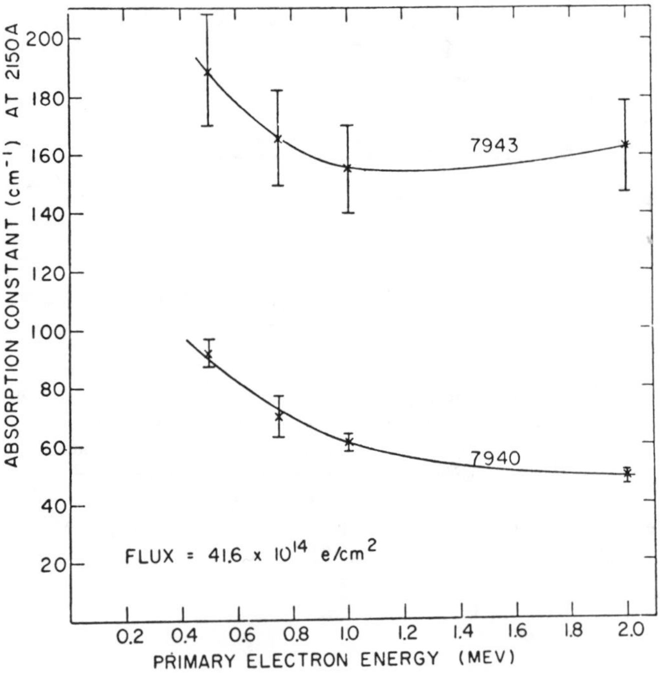

FIGURE 2.3.104. Dependence of the absorption coefficient of the E_1' band at 5.8 eV on the primary electron energy in high OH (Corning 7940) and low OH (Corning 7943) Type III synthetic silicas irradiated at 77 K with a total flux of 4.16 × 10^{15} e/cm^2. (From Arnold, G. W. and Compton, W. D., *Phys. Rev.,* 116, 802, 1959. With permission.)

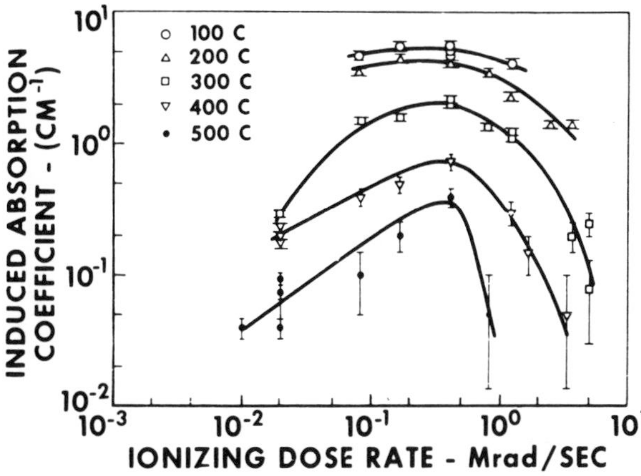

FIGURE 2.3.105. Dependence of the steady-state induced absorption coefficient of the E_1' center at 5.8 eV in high OH content Type III Corning 7940 synthetic silica on the dose rate and temperature during 1.5 MeV electron irradiation. (The data at 0.02 and 0.01 Mrad/sec were obtained from reactor irradiations.) The decrease in absorption coefficient above 0.4 Mrad/sec is due to radiation-annealing of the defect centers in the glass. (From Palma, G. E. and Gagosz, R. M., *J. Phys. Chem. Solids*, 33, 177, 1972. With permission.)

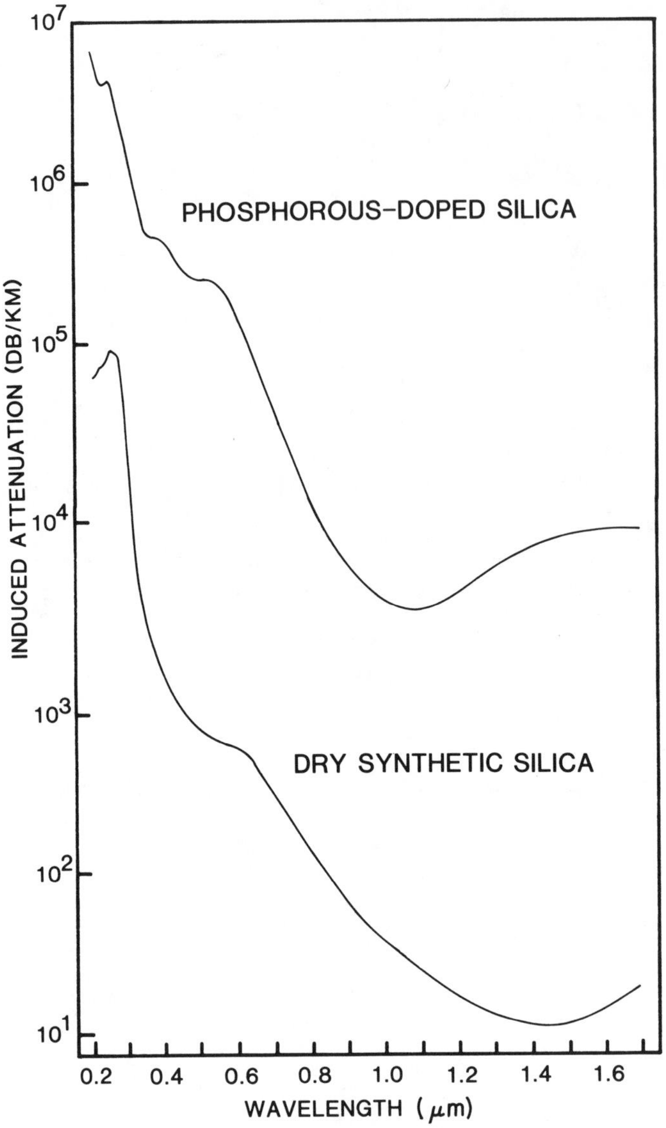

FIGURE 2.3.106. Radiation-induced optical absorption spectra of low OH content Type IV Suprasil W-1 synthetic silica and P-doped silica following room temperature ^{60}Co irradiation of 10^5 rad. Data between 0.2 and 0.8 μm were measured on bulk samples; data between 0.4 and 1.7 μm were measured on fiber samples. Note that 1 dB/km is equivalent to an absorption coefficient of 2.303×10^{-6} cm^{-1}. (From Griscom, D. L. and Friebele, E. J., *Rad. Effects,* 65, 63, 1982. With permission.)

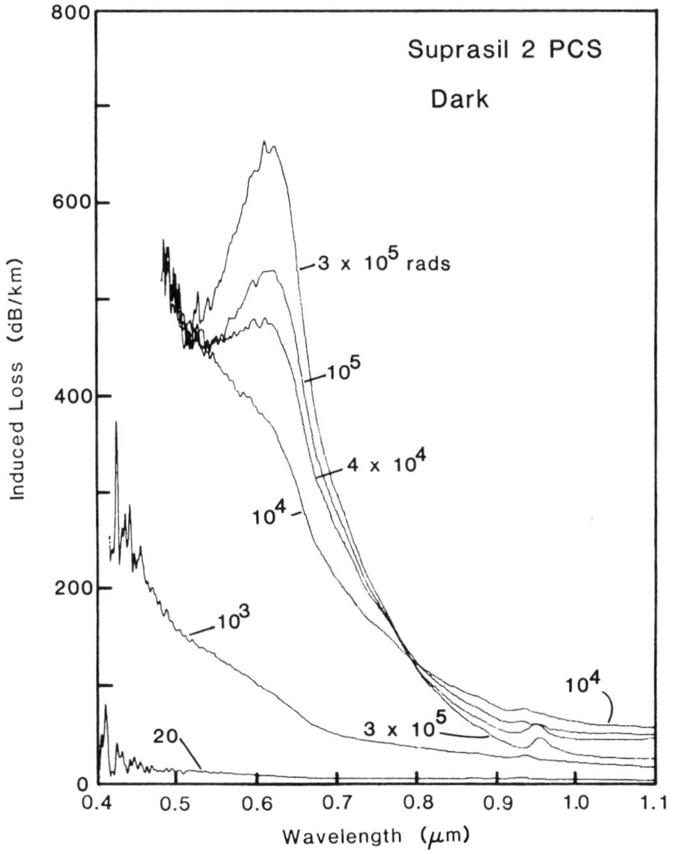

FIGURE 2.3.107. Radiation-induced optical absorption spectra of a high OH content Type III synthetic silica core fiber measured during room-temperature ^{60}Co irradiation. To prevent photobleaching of the damage,[206] the light was transmitted only during the measurement time of 1 sec. (From Griscom, D. L. and Friebele, E. J., *Rad. Effects,* 65, 63, 1982. With permission.)

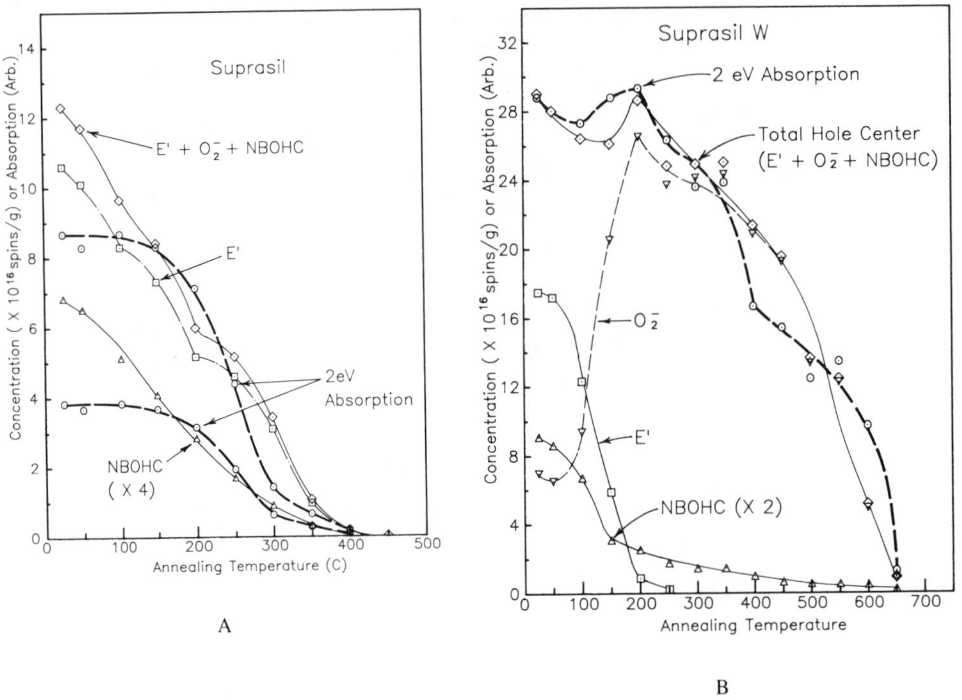

FIGURE 2.3.108. Comparison of the annealing behavior of the radiation-induced optical absorption at 2.0 eV with that of the ESR intensities of the various paramagnetic defect centers in irradiated (A) Suprasil and (B) Suprasil W, indicating the correlation of the band in Suprasil with the nonbridging oxygen hole center (NBOHC) and the band in Suprasil W with the sum of the intensities of the various hole traps. The 6 mm $\times$ 6.5 cm sample rods were exposed to 2 $\times$ 10^8 rads in a ^{60}Co source at 23 C. (After Reference 183.)

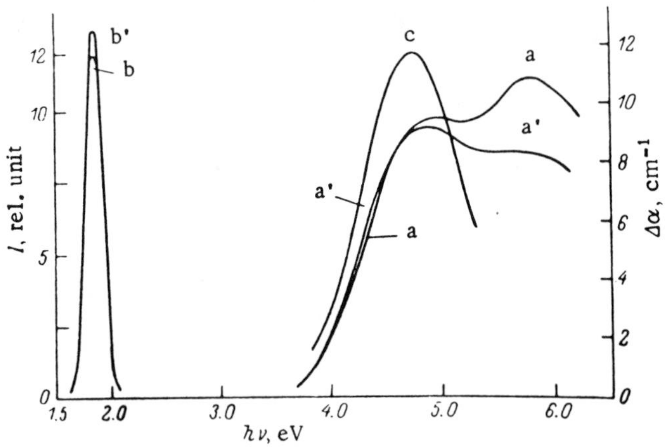

FIGURE 2.3.109. Radiation-induced optical absorption (A), photoluminescence (B), and photoluminescence excitation (C) spectra of low OH content Type IV Suprasil W-1 silica neutron-irradiated at 400 K to a dose of 10^{18} n/cm^2. Following exposure to an unfiltered Hg lamp, the absorption (a′) and the photoluminescence (b′) were measured. The temperature of measurement was 300 K. (From Silin, A. R., Skuya, L. N., and Shendrik, A. V., *Fiz. i Khim. Stek.*, 4, 352, 1978. Copyright 1978, Plenum Press. With permission.)

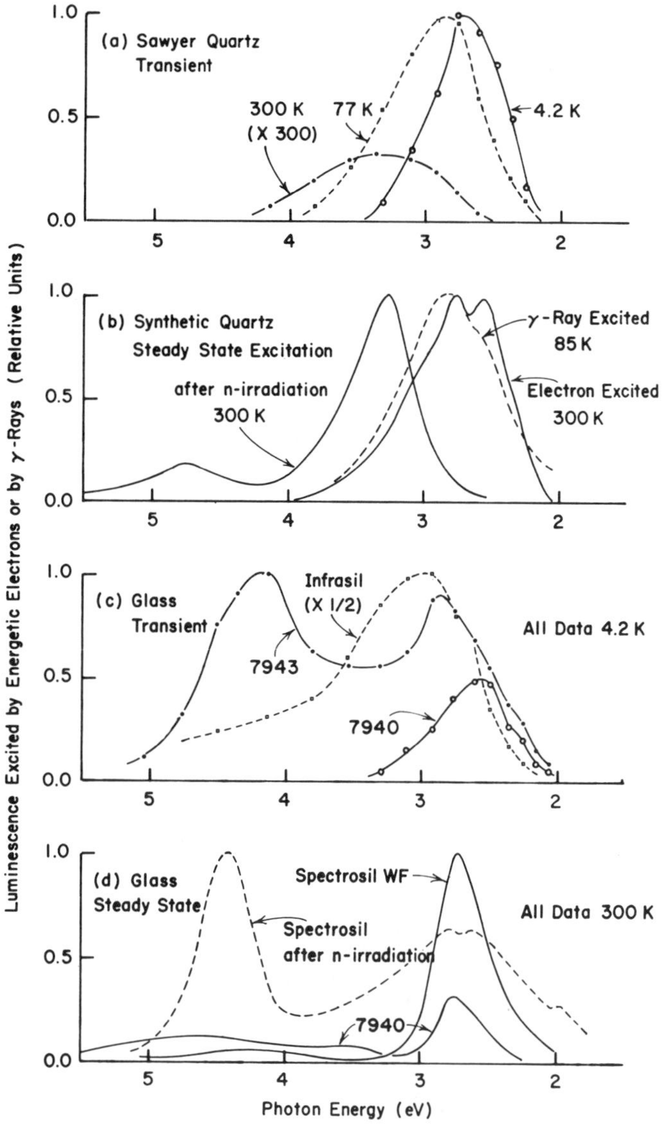

FIGURE 2.3.110. Luminescence emission in the 2 to 5.5-eV spectral region of various crystalline and amorphous SiO_2 samples stimulated by both pulsed and steady-state irradiation. (From Griscom, D. L., *Proc. 33rd Annual Symposium on Frequency Control,* 1979, 98. With permission.)

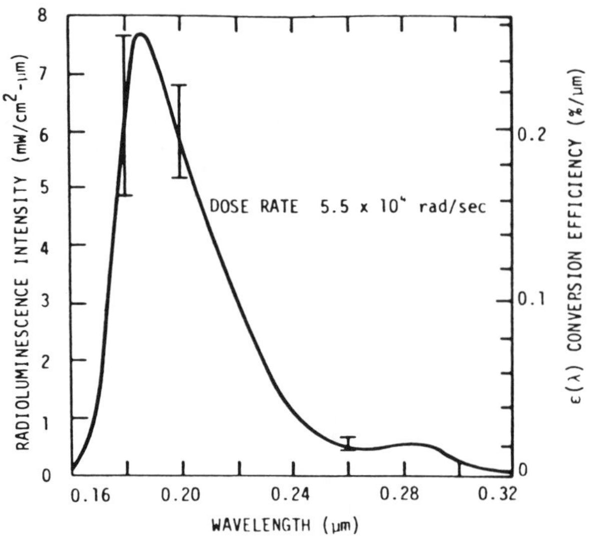

FIGURE 2.3.111. Spectrum of the radioluminescence emission of Sawyer Z growth quartz measured from 3.8 to 7.7 eV during irradiation at room temperature with 2.5 MeV electrons. Note the 7.6-eV emission band in the vacuum UV portion of the spectrum. (From Treadaway, M. J., Passenheim, B. C., and Kitterer, B. D., *IEEE Trans. Nucl. Sci.*, 22, 2253, 1975. With permission.)

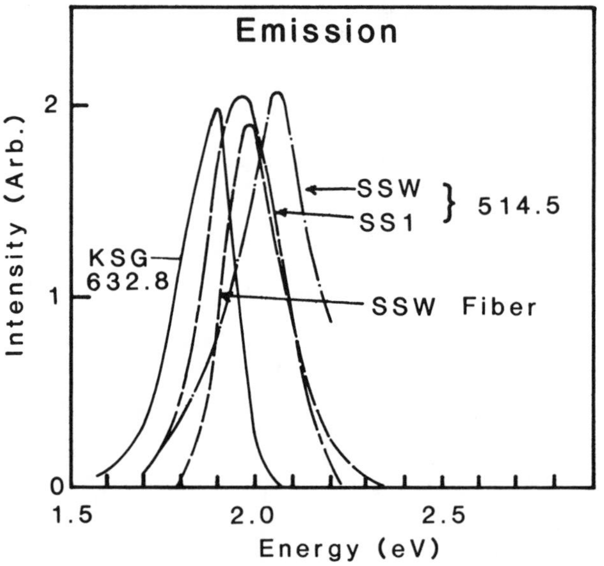

FIGURE 2.3.112. Radiation-induced photoluminescence spectra of various bulk silica samples pumped with either a HeNe (KSG) or Ar (Suprasil and Suprasil W) laser and the resonant luminescence spectrum of a Suprasil W core optical fiber. (Fiber data from Reference 182; KSG data from Reference 184. After Reference 183.)

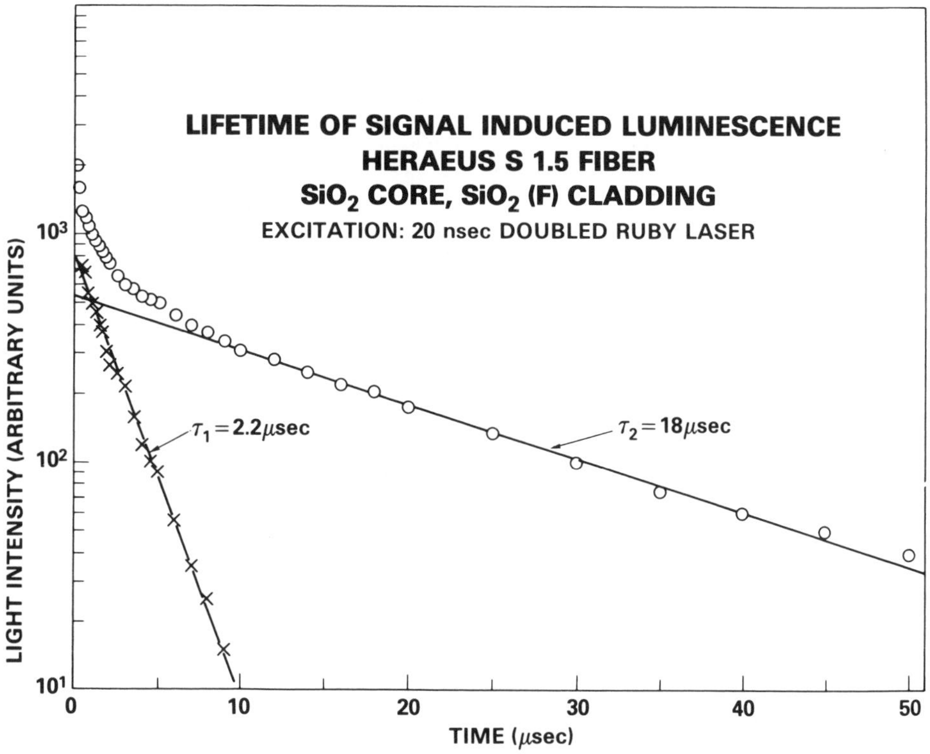

FIGURE 2.3.113. Decay of the 1.9-eV photoluminescence in an irradiated fiber with a low OH content Type IV Suprasil W synthetic silica core. The data (circles) were obtained at room temperature following excitation by a doubled ruby laser (0.3471 μm); also shown are the decomposition of the decay curve into composite exponentials. (From Sigel, G. H., Jr. and Marrone, M. J., *J. Non-Cryst. Solids*, 45, 235, 1981. With permission.)

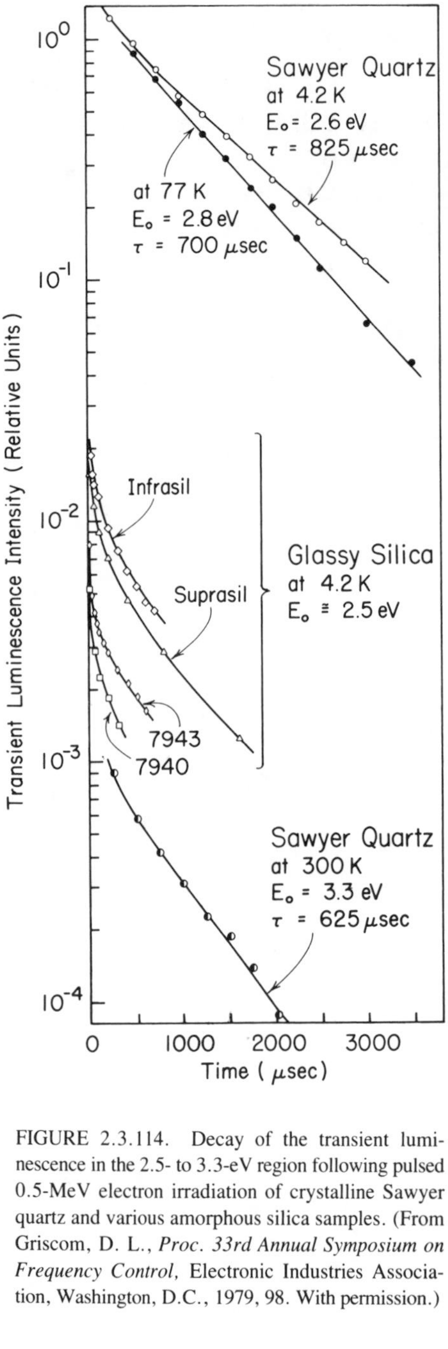

FIGURE 2.3.114. Decay of the transient luminescence in the 2.5- to 3.3-eV region following pulsed 0.5-MeV electron irradiation of crystalline Sawyer quartz and various amorphous silica samples. (From Griscom, D. L., *Proc. 33rd Annual Symposium on Frequency Control,* Electronic Industries Association, Washington, D.C., 1979, 98. With permission.)

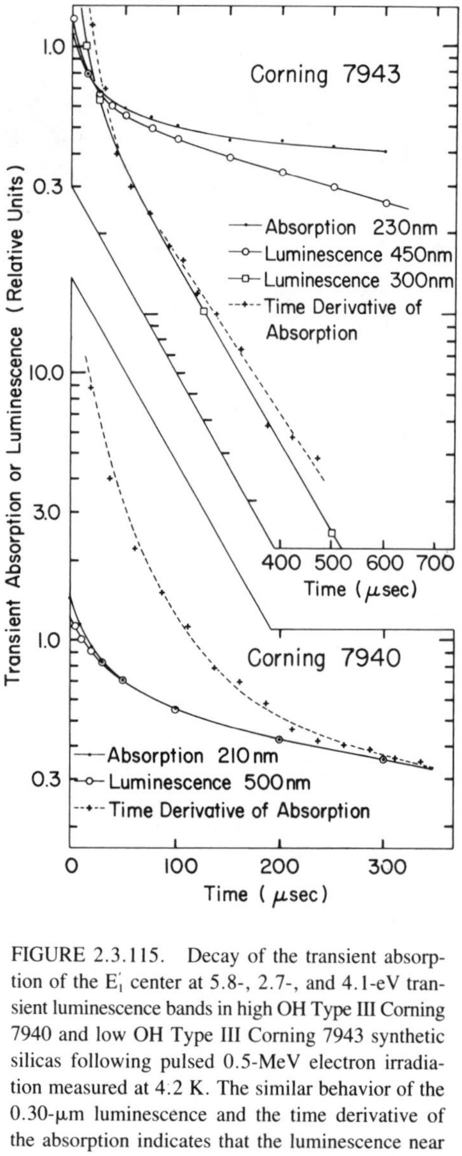

FIGURE 2.3.115. Decay of the transient absorption of the E_1' center at 5.8-, 2.7-, and 4.1-eV transient luminescence bands in high OH Type III Corning 7940 and low OH Type III Corning 7943 synthetic silicas following pulsed 0.5-MeV electron irradiation measured at 4.2 K. The similar behavior of the 0.30-μm luminescence and the time derivative of the absorption indicates that the luminescence near 4.3 eV results from the annihilation of E_1' centers at the sites of oxygen vacancies. (From Griscom, D. L., *Proc. 33rd Annual Symposium on Frequency Control,* Electronic Industries Association, Washington, D.C., 1979, 98. With permission.)

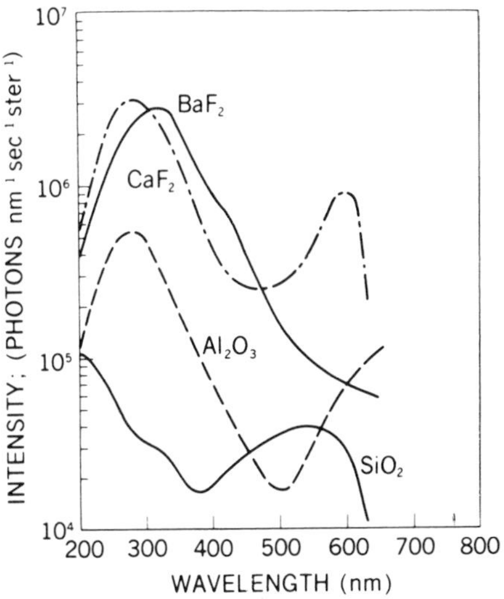

FIGURE 2.3.116. Emission spectra of several optical materials measured at room temperature during exposure to a ^{90}Sr-^{90}Y electron source. The average electron energies of the ^{90}Sr and ^{90}Y sources are 0.194 and 0.92 MeV, respectively, and the source intensity was 0.8 mCi. (From Viehmann, W., Eubands, A. G., Pieper, G. F., and Bredekamp, S. H., *Appl. Opt.*, 14, 2104, 1975. With permission.)

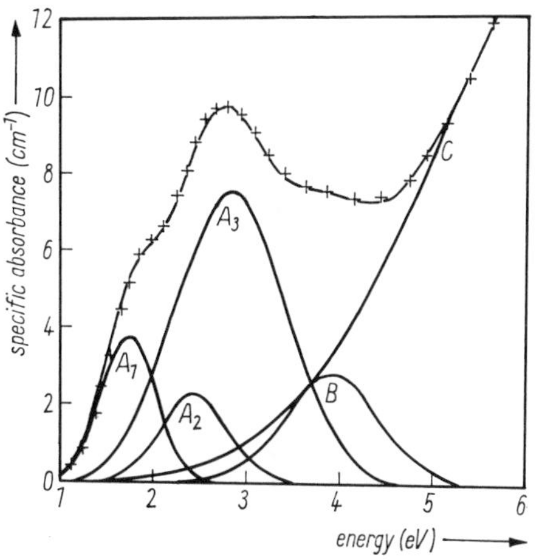

FIGURE 2.3.117. Polarized radiation-induced optical absorption spectrum of crystal quartz irradiated to a dose of 15 Mrads at room temperature in a ^{60}Co cell. Experimental spectrum is shown as crosses, and the Gaussian component bands and their sum are shown as the solid lines. The parameters of the component bands are contained in Table 2.3.1. (From Nassau, K. and Prescott, B. E., *Phys. Status Solidi*, 29, 659, 1975. With permission.)

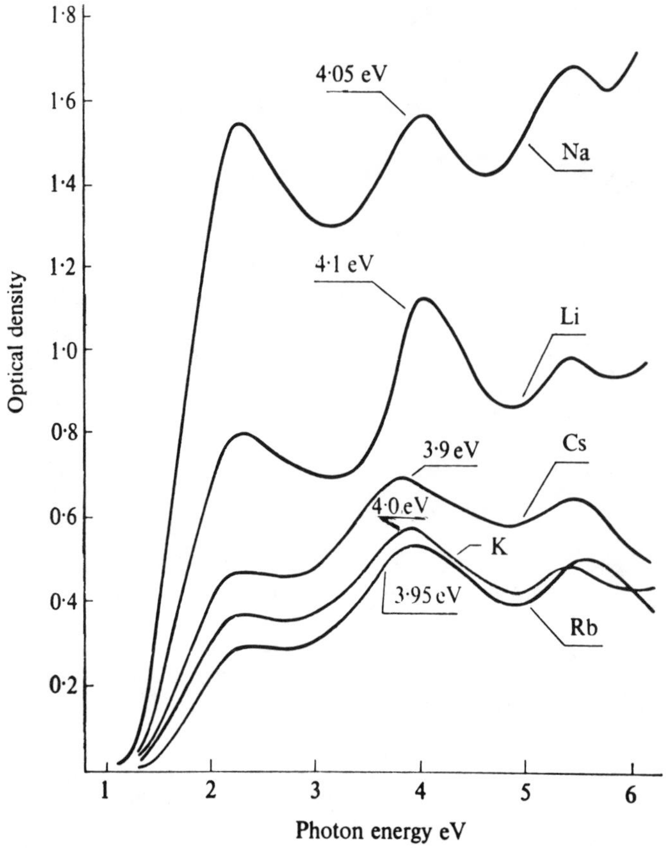

FIGURE 2.3.118. Absorption spectra of fused silica samples doped with 0.2 wt% alkali and an equal amount of Al. The samples have been exposed to a ^{60}Co dose of 10^7 rad while being maintained at 0°C. The peak position of the band near 4 eV, which depends on alkali type, is plotted in Figure 2.3.131. (From Lell, E., *Phys. Chem. Glass.*, 3, 84, 1962. Reproduced with the permission of the publisher and Bausch & Lomb Incorporated.)

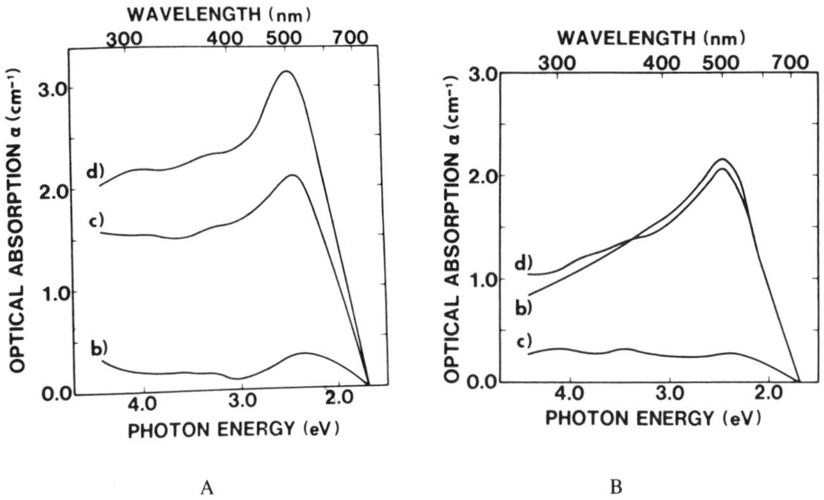

FIGURE 2.3.119. Absorption spectra of (A) unswept and (B) swept X-growth Z-plate Western Electric crystal quartz measured in the following sequence: b) after a 4-min irradiation with a 1.7-MeV electron Van de Graff accelerator at 77 K; c) after a similar irradiation at room temperature; and d) after a reirradiation at 77 K. The initial irradiation causes less damage in the unswept sample because the alkali cannot diffuse away from the aluminum center at 77 K, whereas the protons that have replaced the alkali in the process of sweeping the crystal can diffuse at 77 K and the induced loss is therefore greater. (From Koumraklis, N., *J. Appl. Phys.*, 51, 5528, 1980. With permission.)

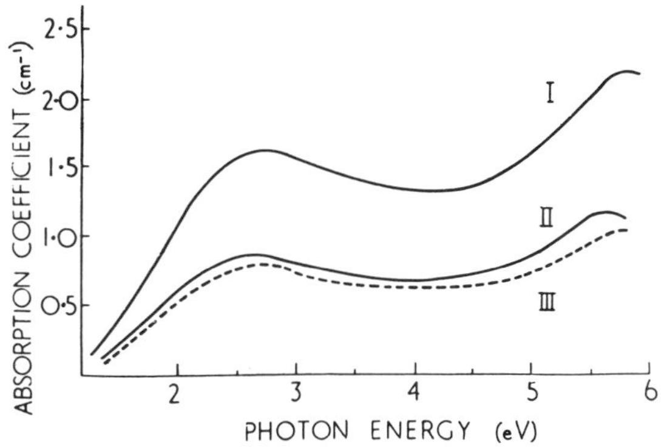

FIGURE 2.3.120. Absorption spectra of natural crystalline quartz: I) after a ^{60}Co dose of 60 MR at room temperature, II) after a subsequent UV exposure for 22 hr, and III) the difference between the two curves showing the bleached spectrum. (From Mitchell, E. W. J. and Paige, E. G. S., *Philos. Mag.*, 1, 1085, 1956. With permission.)

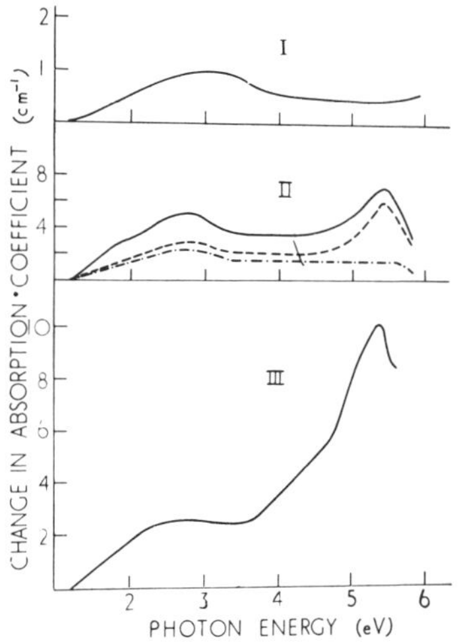

FIGURE 2.3.121. Spectra of crystal quartz showing the absorption that is bleached by a 240°C anneal for 9 hr in samples: I) irradiated with a 125-MR dose of 50 kV X-rays; II) irradiated with a neutron dose of 2.4 × 10^{17} n/cm^2 and bleached for 10 min (dashed line) and subsequently for 8 hr 50 min (dot-dashed line); and III) neutron-irradiated to 9.2 × 10^{18} n/cm^2. (From Mitchell, E. W. J. and Paige, E. G. S., *Philos. Mag.*, 1, 1085, 1956. With permission.)

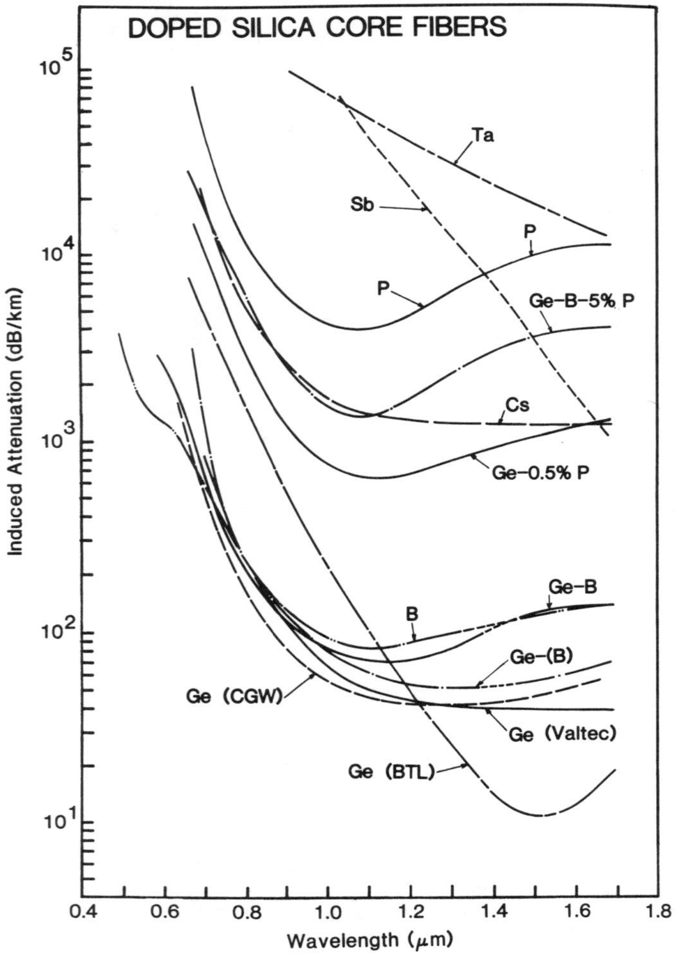

A

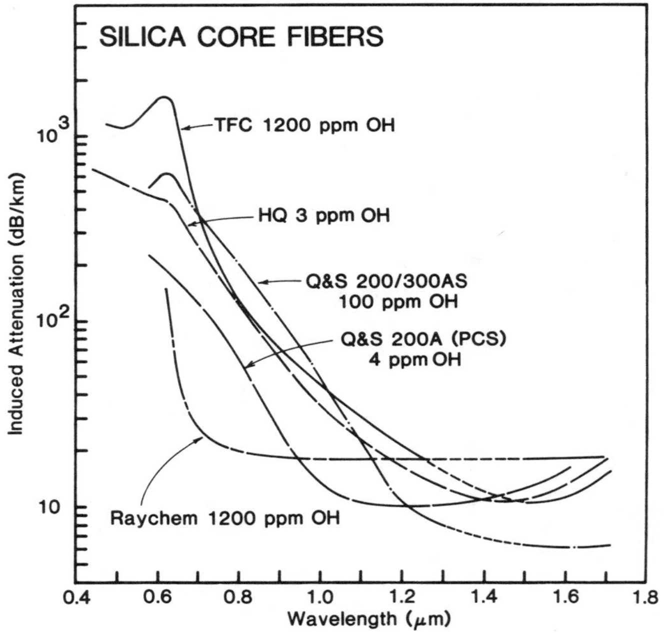

B

FIGURE 2.3.122. Induced optical absorption spectra of (A) doped silica and (B) pure silica core fibers measured 1 hr after a room-temperature irradiation of 10^5 rad in a ^{60}Co source. Sample lengths varied from 5 to 30 m. Note that 1 dB/km corresponds to an absorption coefficient of 2.303×10^{-6} cm^{-1}. (From Friebele, E. J., Gingerich, M. E., and Long, K. J., *Appl. Opt.*, 21, 547, 1982. With permission.)

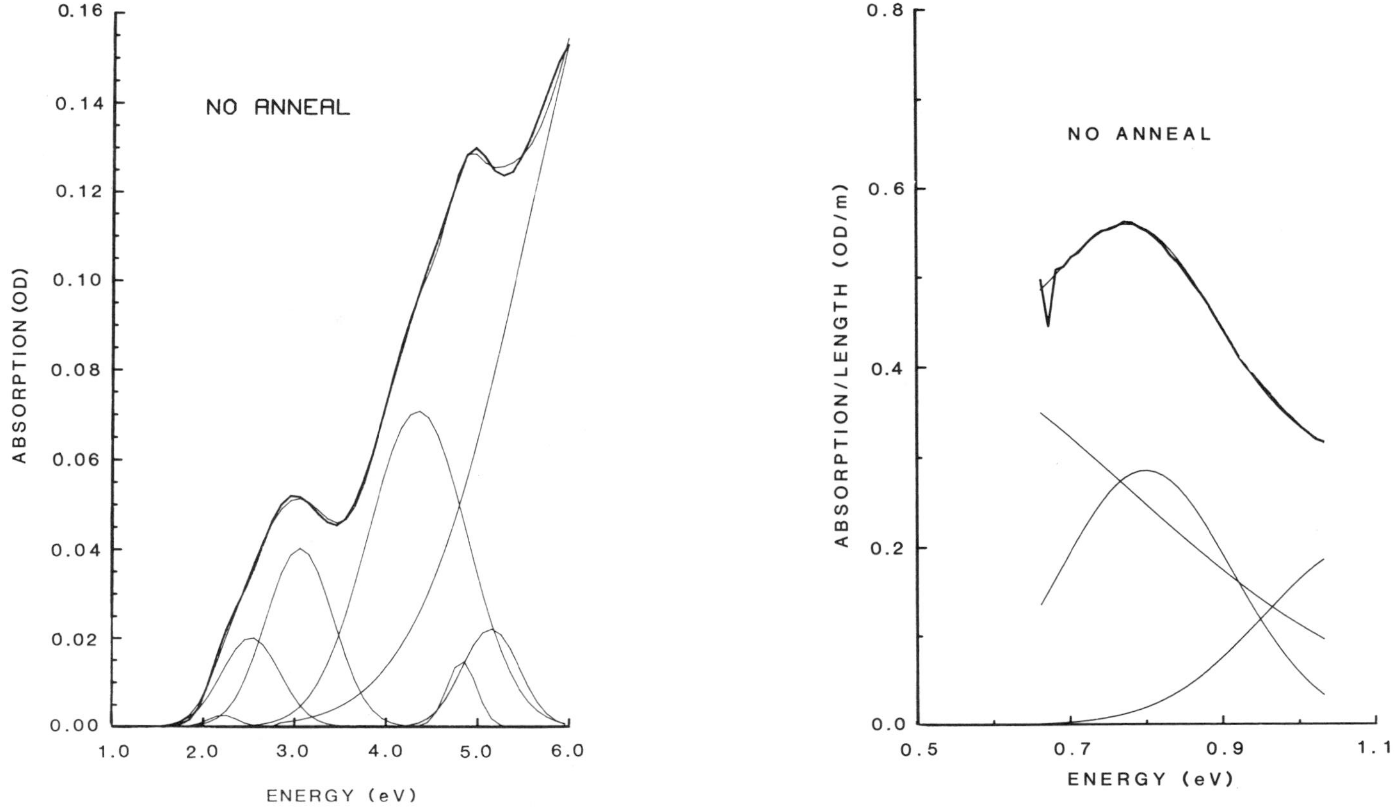

FIGURE 2.3.123. Induced absorption spectra of pure silica doped with 9.5% P_2O_5 following a 50-kV X-ray irradiation of ~10^6 rad. The irradiation and measurement were carried out at 77 K. The heavy curve is the experimental spectrum and the light curves show the decomposition into Gaussian bands and the fit to the data. Table 2.3.3 contains the parameters of the bands. A bulk sample of 0.4-cm thickness was used for the data of (A, left), and a 1-m P-doped silica core fiber sample was used for (B, right). (From Griscom, D. L., Friebele, E. J., Long, K. G., and Fleming, J. W., *J. Appl. Phys.*, in press. With permission.)

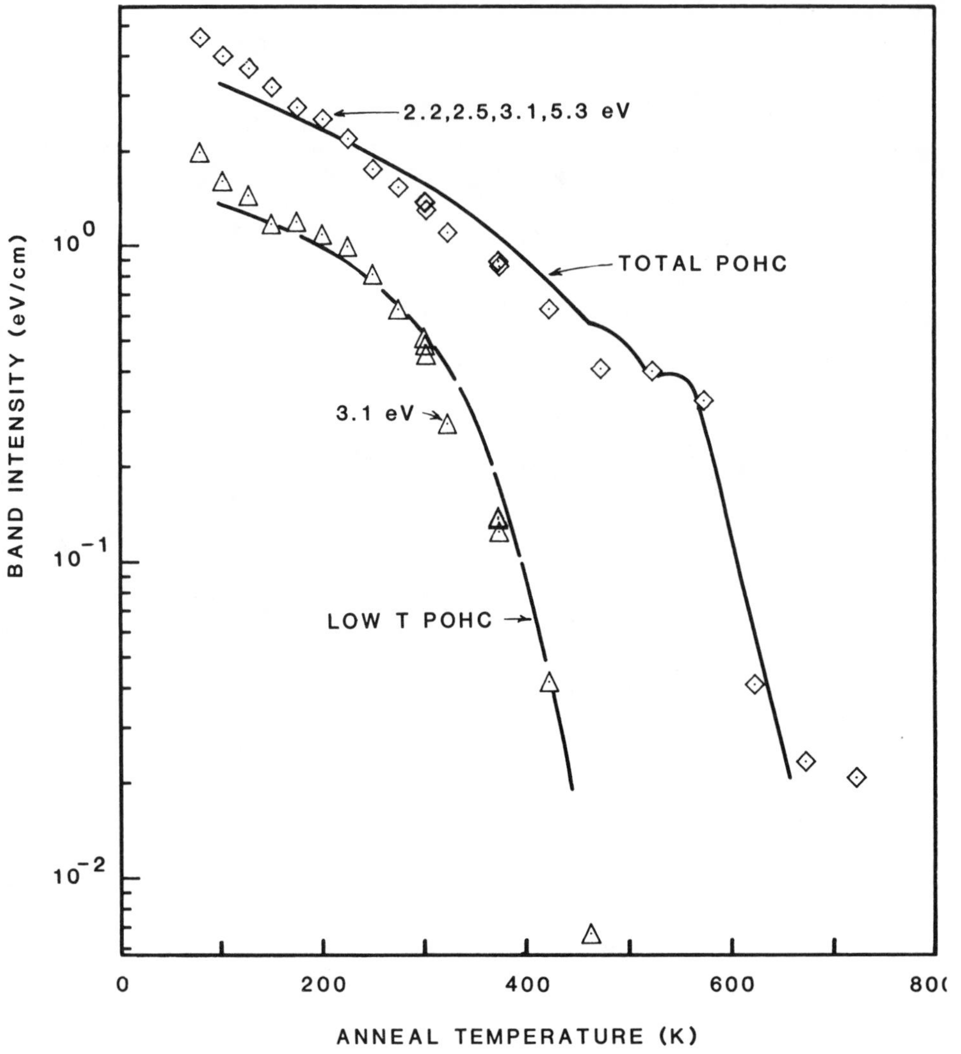

A

FIGURE 2.3.124. Change in the optical absorption band and ESR intensities as a function of annealing temperature for the various bands resolved in Figure 2.3.123 showing the correlations between defect centers and optical absorptions, which are tabulated in Table 2.3.2. (From Griscom, D. L., Friebele, E. J., Long, K. G., and Fleming, J. W., *J. Appl. Phys.*, in press. With permission.)

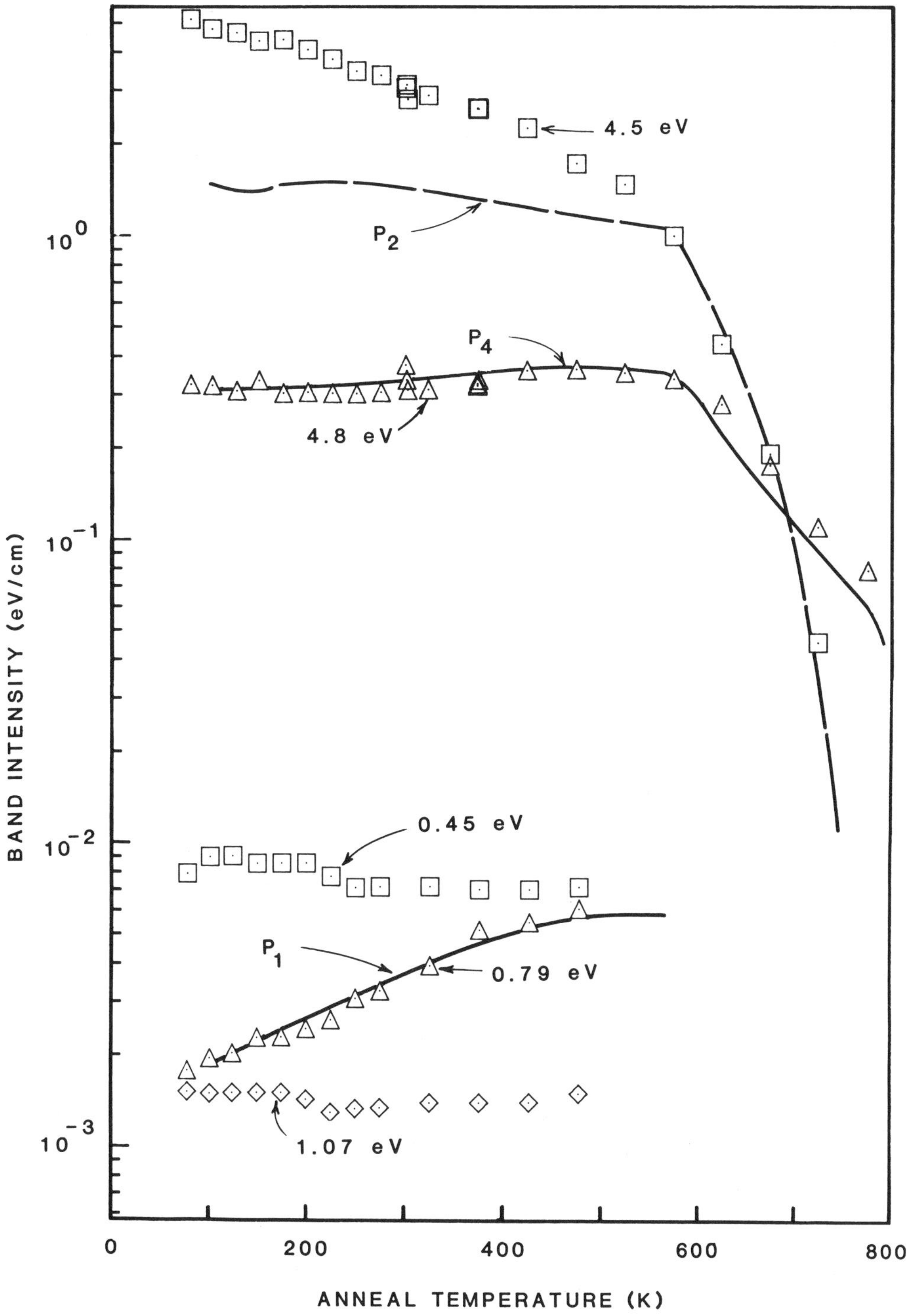

FIGURE 2.3.124B

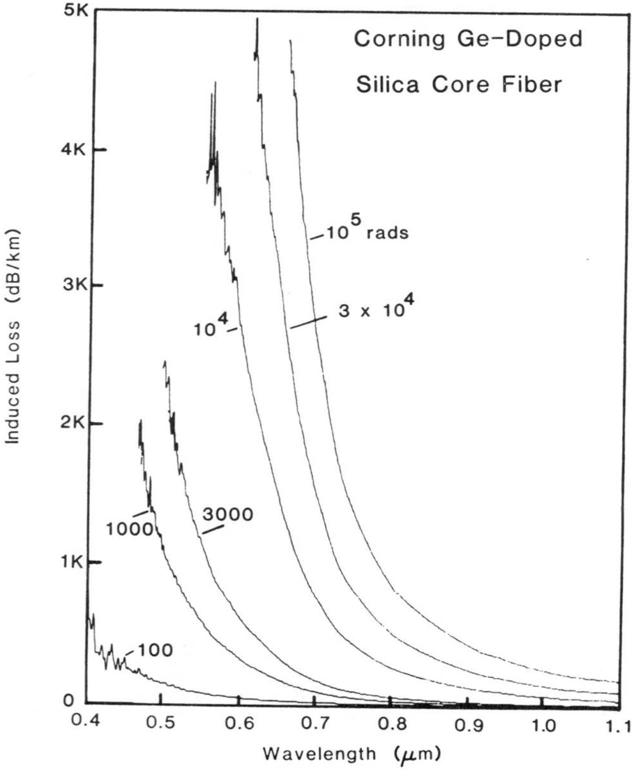

FIGURE 2.3.125. Induced optical absorption spectra of a Ge-doped silica core fiber measured *in situ* at room temperature during ^{60}Co irradiation. Sample length = 10 m. Note that 1 dB/km = 2.303 × 10^6 cm^{-1}. (From Friebele, E. J., *Proc. 2nd Int. Conf. on Rad. Effects in Insulators,* Arnold, G. W. and Borders, J. A., Eds., in press.)

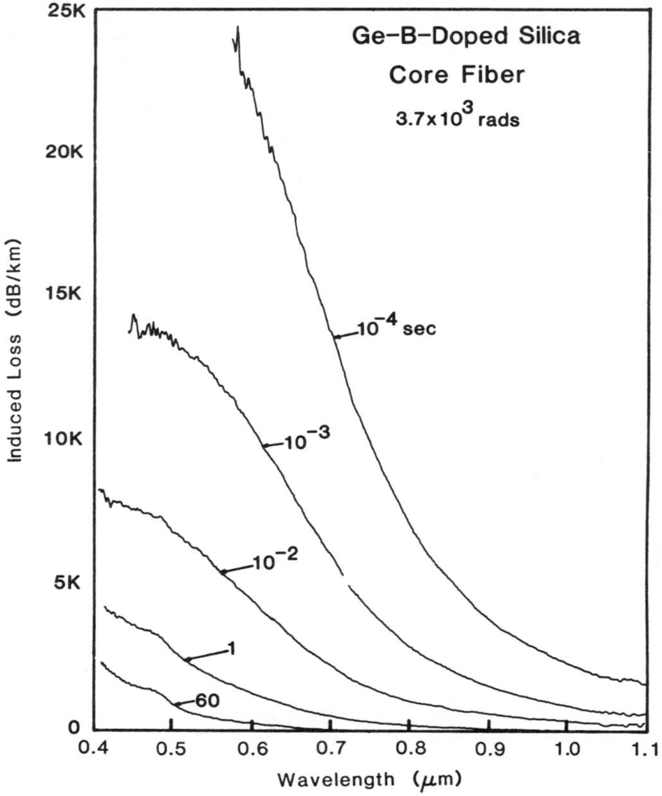

FIGURE 2.3.126. Decay of the induced optical absorption in a Ge-doped silica core fiber following a room-temperature 3-nsec pulsed irradiation with 0.5-MeV electrons. Sample length = 1 m. (From Friebele, E. J., *Proc. 2nd Int. Conf. on Rad. Effects in Insulators,* Arnold, G. W. and Borders, J. A., Eds., in press.)

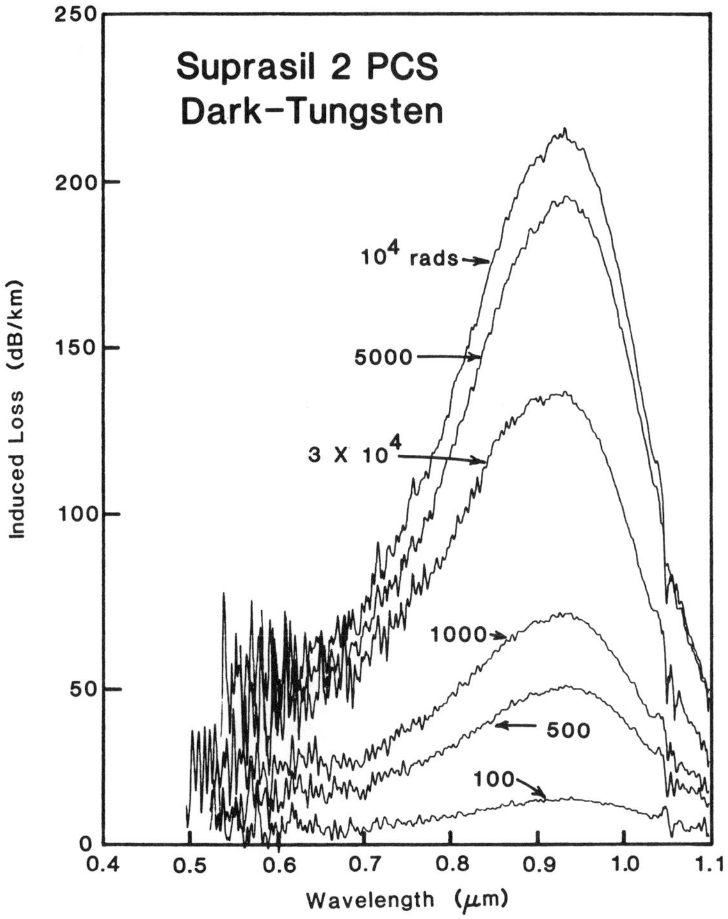

FIGURE 2.3.127. Optical spectra showing the absorption that was photobleached from a high OH content Type III Suprasil 2 silica core-polymer clad fiber by tungsten light as measured during *in situ* exposure to ^{60}Co radiation at room temperature. The position and breadth of the photobleached band has been found to vary from fiber to fiber. Sample length = 10 m. (From Friebele, E. J., *Proc. 2nd Int. Conf. on Rad. Effects in Insulators,* Arnold, G. W. and Borders, J. A., Eds., in press.)

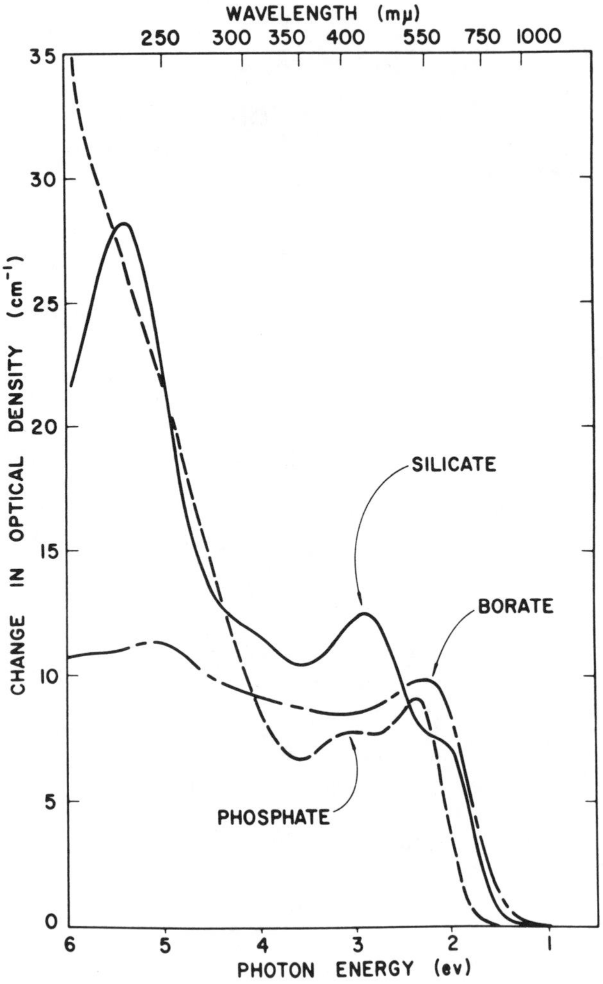

FIGURE 2.3.128. Induced optical absorption in soda lime silicate, borate and phosphate glasses following room temperature ^{60}Co irradiations of 3.9×10^7 R. The glass compositions and the results of Gaussian resolutions of the spectra into component bands are shown in Table 2.3.4. (From Bishay, A. M. and Ferguson, K. R., *Advances in Glass Technology,* Am. Ceram. Soc., Columbus, Ohio, 1962, 133. With permission.)

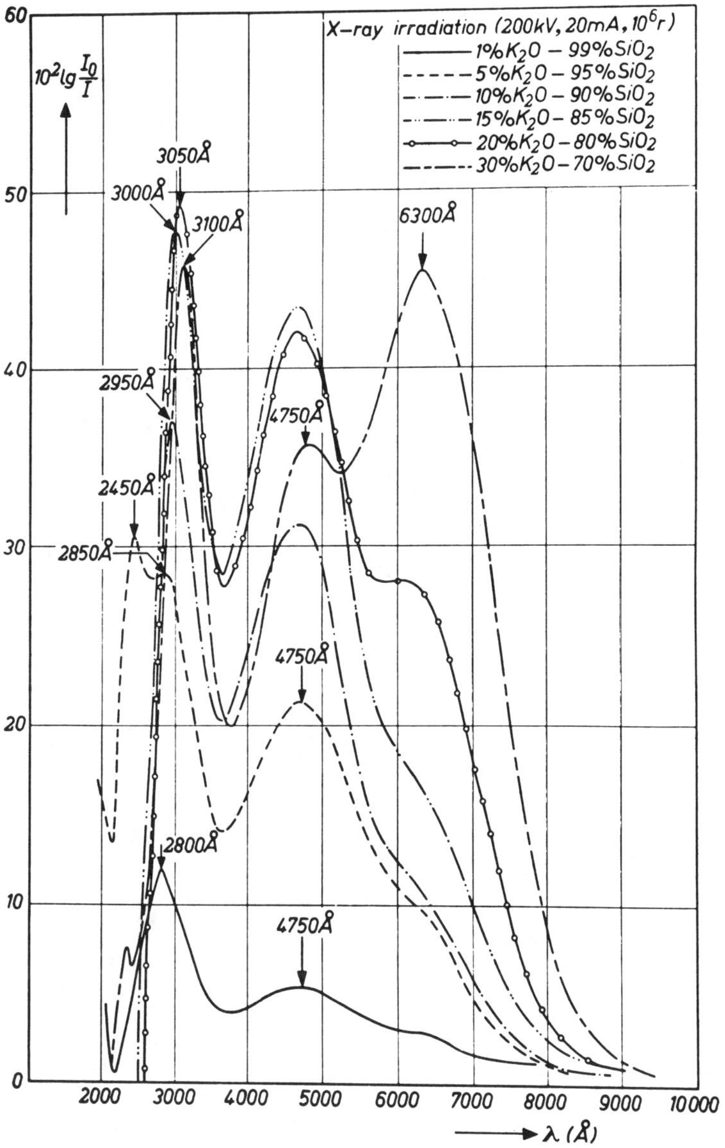

FIGURE 2.3.129. Optical absorption induced at room temperature by X-ray irradiation of potassium-silicate glasses as a function of alkali content. The apparent negative induced absorption at wavelengths shorter than 0.25 μm is an artifact due to the presence of iron impurities in the glass. (From Kats, A. and Stevels, J. M., *Philips Res. Rep.*, 11, 115, 1956. With permission.)

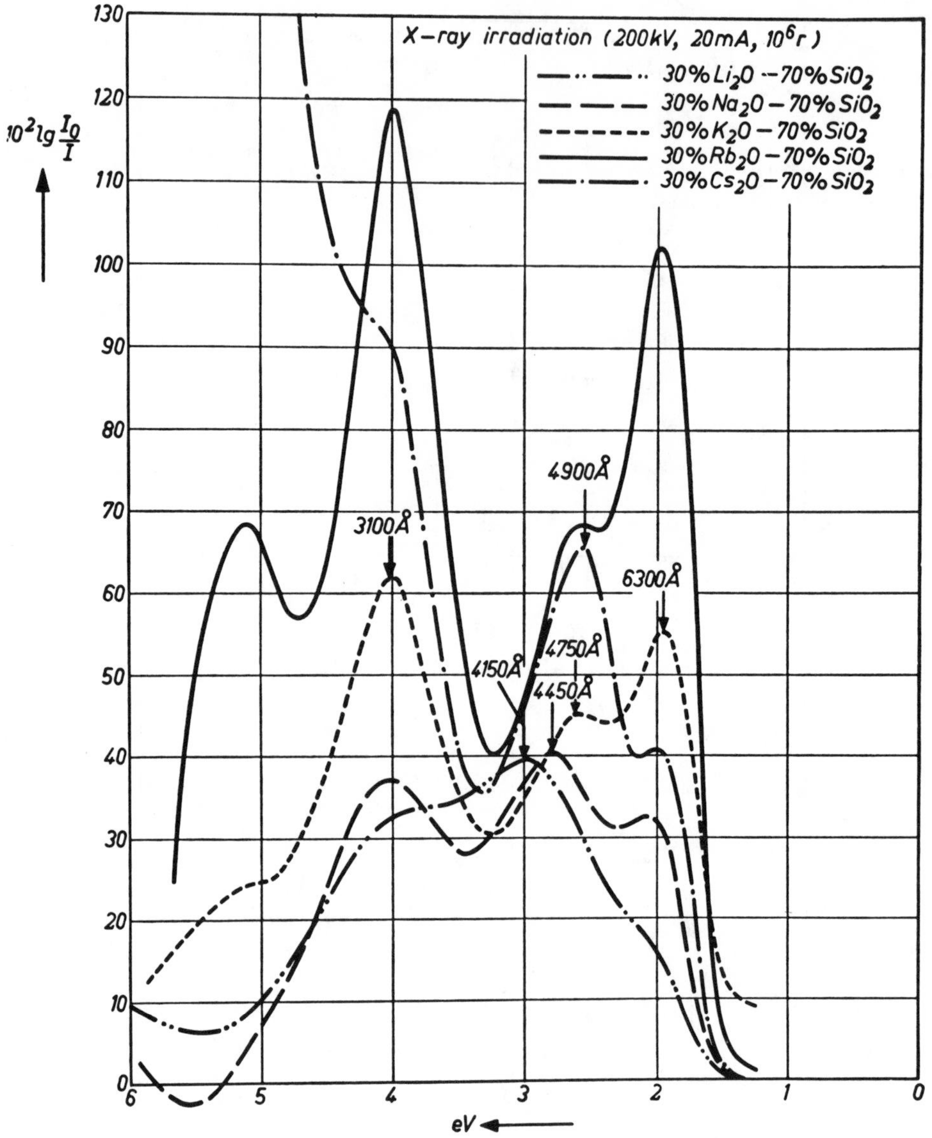

FIGURE 2.3.130. Optical absorption induced at room temperature by X-ray exposure in 30% Me_2O-70% SiO_2 glasses as a function the type of alkali ion. The peak positions of the band located near 2.7 eV are plotted in Figure 2.2.131. The negative induced absorption in the Na-silicate glass is an artifact due to iron impurities in the glass. (From van Wieringen, J. S. and Kats, A., *Philips Res. Rep.*, 12, 432, 1957. With permission.)

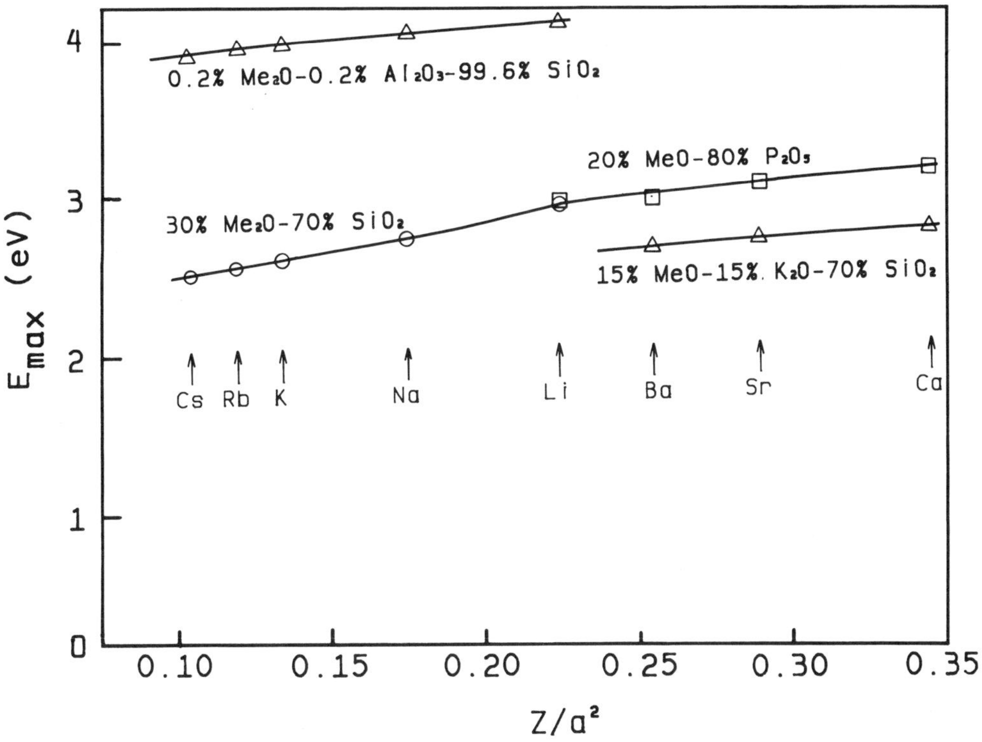

FIGURE 2.3.131. Peak energy of the hole center band located in the 2.5- to 4.0-eV region vs. half the field strength of the ion (Z/a^2) for various alkali and alkaline earth silicate and phosphate glasses. Z is the charge of the metal ion and a is ion spacing, taken as the sum of the Pauling radii of the metal and oxygen ions.

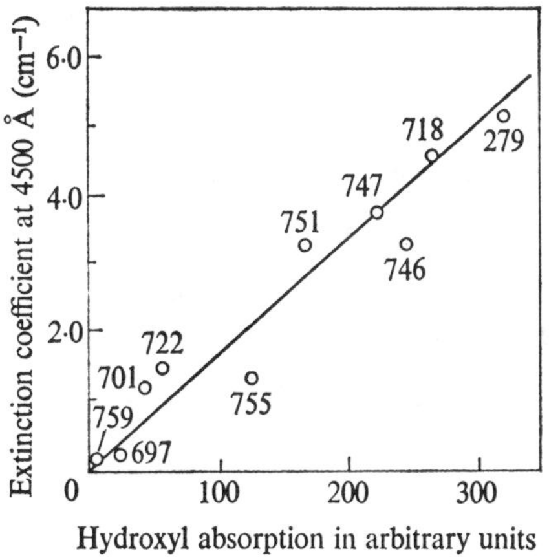

FIGURE 2.3.132. Dependence of the height of the 0.45-μm hole center absorption band on OH content in Na_2O-$3SiO_2$ glass X-irradiated at room temperature with 2×10^6 R. (From Mackey, J. H., Boss, J. W., and Kopp, M., *Phys. Chem. Glasses,* 11, 205, 1970. With permission.)

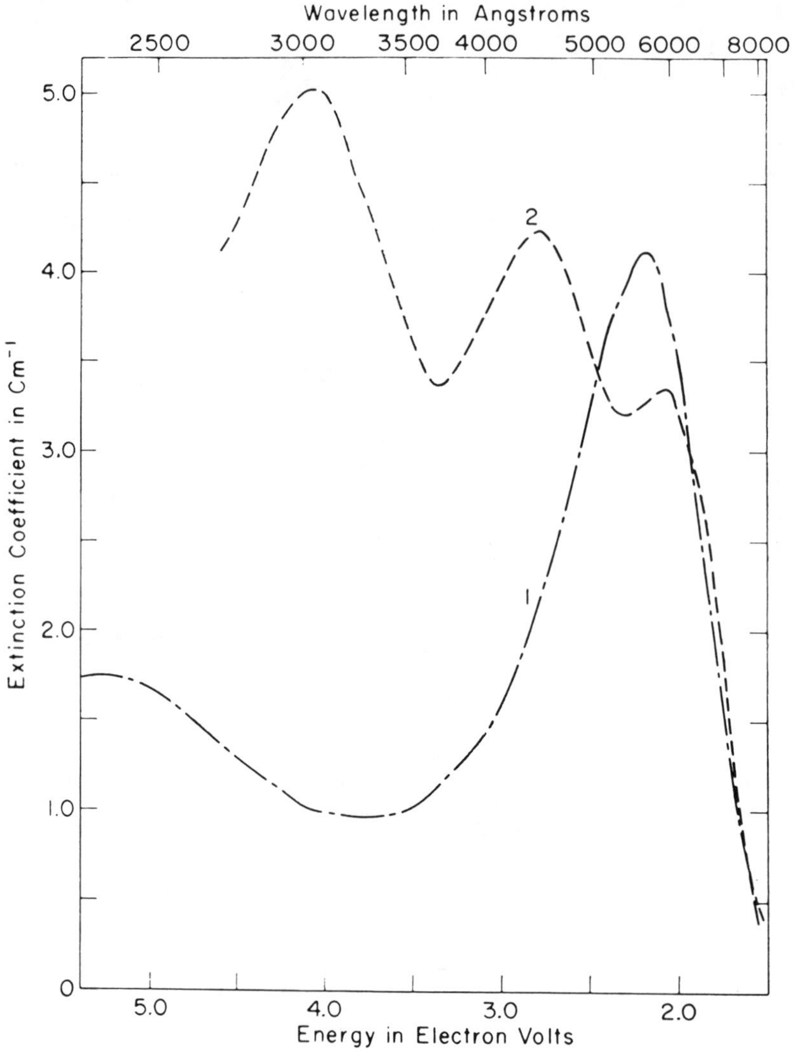

FIGURE 2.3.133. Induced absorption spectra in (1) reduced and (2) oxidized Na_2O-$3SiO_2$ glass following an X-irradiation at room temperature. The reduced glass received a dose of 1.5×10^6 R; the oxidized glass received twice this amount. (From Smith, H. L. and Cohen, A. J., *J. Am. Ceram. Soc.*, 47, 564, 1964. With permission.)

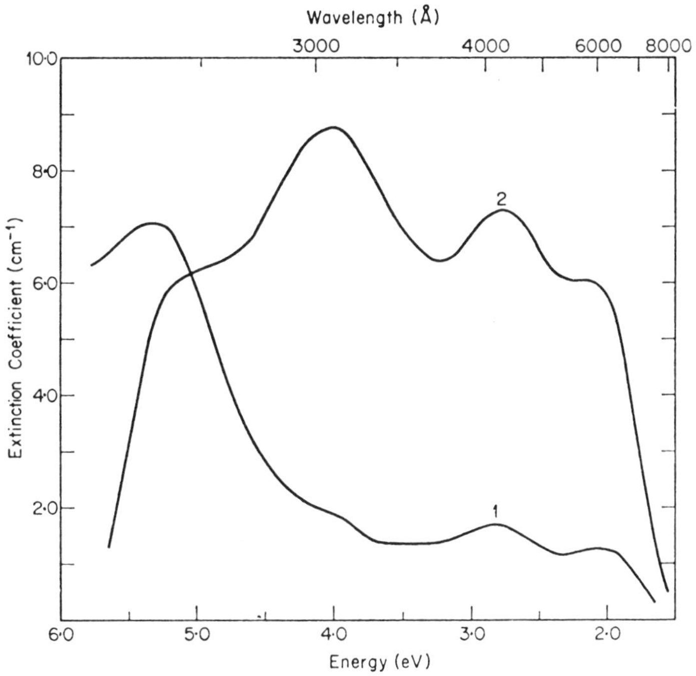

FIGURE 2.3.134. Induced absorption spectra of a (1) reduced and (2) oxidized Na_2O-$2SiO_2$ glass following a 2.5×10^6 rad X-ray exposure at 350 K. (From Mackey, J. H., Smith, H. L., and Halperin, A., *J. Phys. Chem. Solids*, 27, 1759, 1966. With permission.)

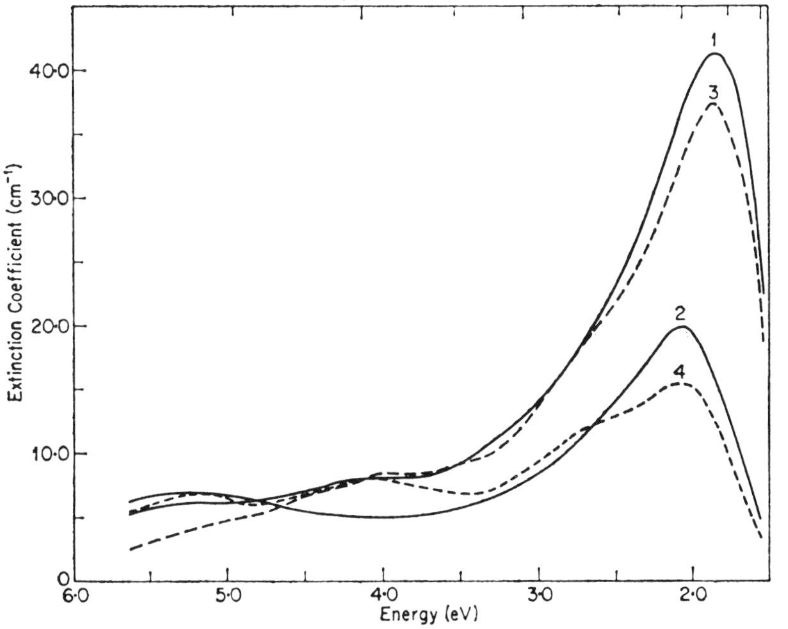

FIGURE 2.3.135. Absorption spectra induced by a 10^6 rad X-ray irradiation of (1,2) reduced and (3,4) oxidized Na_2O-$2SiO_2$ glass. The temperature during exposure and measurement was either 77 K (1,3) or 210 K (2,4). (From Mackey, J. H., Smith, H. L., and Halperin, A., *J. Phys. Chem. Solids*, 27, 1759, 1966. With permission.)

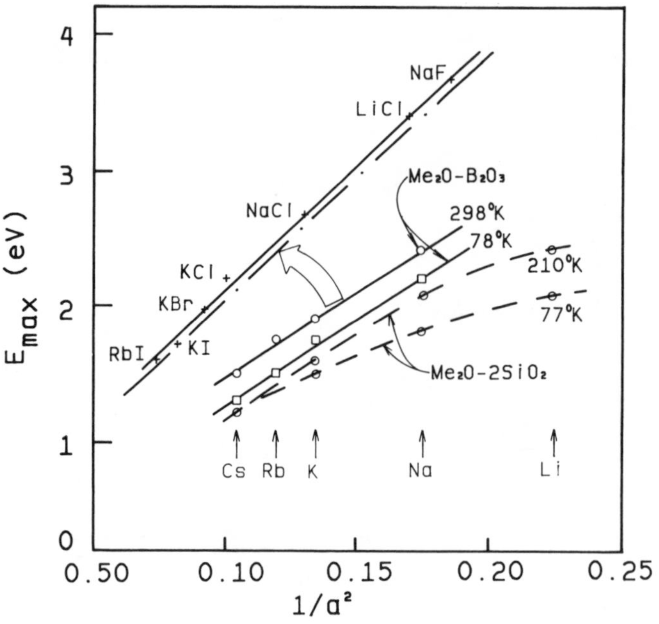

FIGURE 2.3.136. Peak energy of the F-center band in various alkali-halide crystals and peak energy of the alkali electron center band in alkali borate[217] and alkali silicate[224] glasses plotted vs. the inverse square of the sum of the Pauling radii of the alkali and halide ions or of the alkali and oxygen ions. The arrow indicates that a 17% increase in the alkali-oxygen bond distance brings the data for the alkali borate glasses into good agreement with that of the alkali halides.

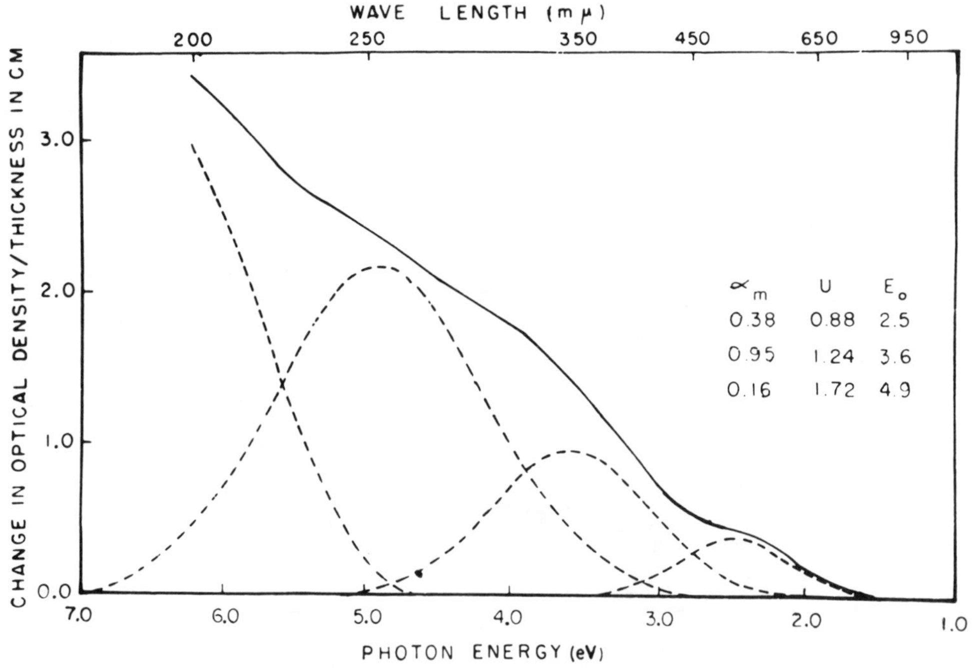

FIGURE 2.3.137. Induced optical absorption (solid line) in a 1.0 Na_2O-4.5 B_2O_3 glass following a room temperature exposure of 1.2×10^6 R. Dashed lines are resolution of the experimental spectrum into Gaussian bands. (From Bishay, A., *J. Non-Cryst. Solids*, 3, 54, 1970. With permission.)

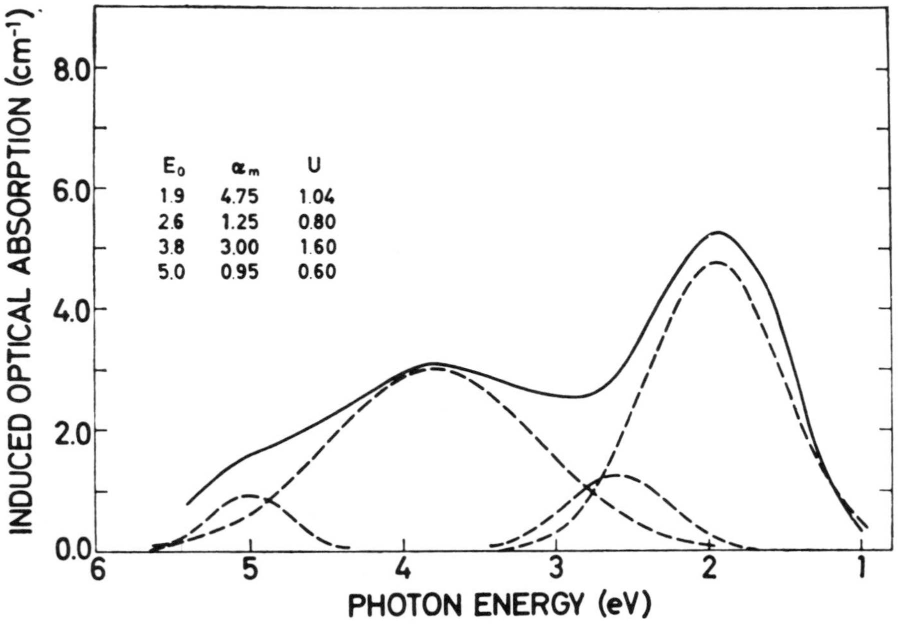

FIGURE 2.3.138. Optical absorption induced in a 3 K_2O-7 B_2O_3 glass following a room temperature exposure of 2.2×10^6 R. The experimental spectrum is the solid line and the dashed lines are the Gaussian components; the parameters are contained in Table 2.3.6. The 1.9-eV band clearly evident in this glass is the alkali electron center, which is stable at room temperature in alkali borate glasses containing more than ~20% alkali. (From Bishay, A., *J. Non-Cryst. Solids,* 3, 54, 1970. With permission.)

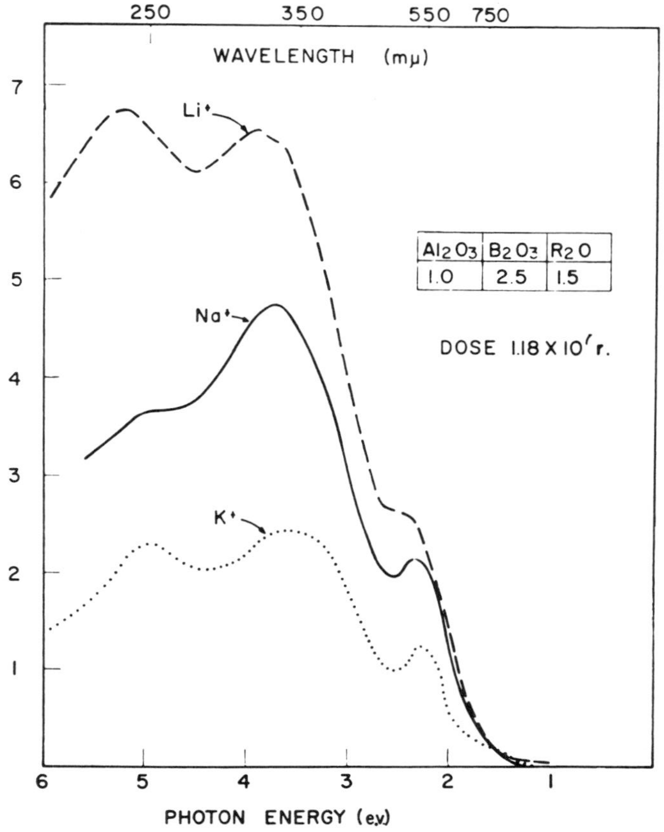

FIGURE 2.3.139. Induced optical absorption in 1.5 Me_2O-1.0 Al_2O_3-2.5 B_2O_3 glasses following a room temperature ^{60}Co exposure of 1.18×10^7 R showing the effect of varying the type of alkali in the glass. (From Bishay, A.M., *J. Am. Ceram. Soc.*, 44, 289, 1961. With permission.)

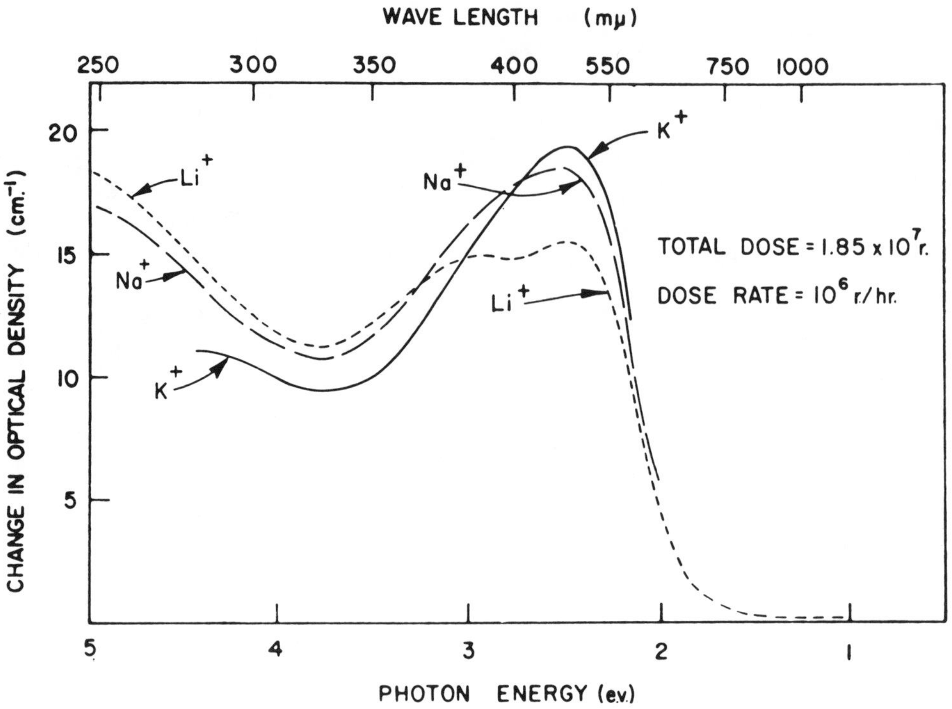

FIGURE 2.3.140. Induced optical absorption spectra of various Me_2O-CaO-P_2O_5 glasses following a 1.85×10^7 R ^{60}Co exposure at room temperature. (From Bishay, A.M., *J. Am. Ceram. Soc.*, 44, 545, 1961. With permission.)

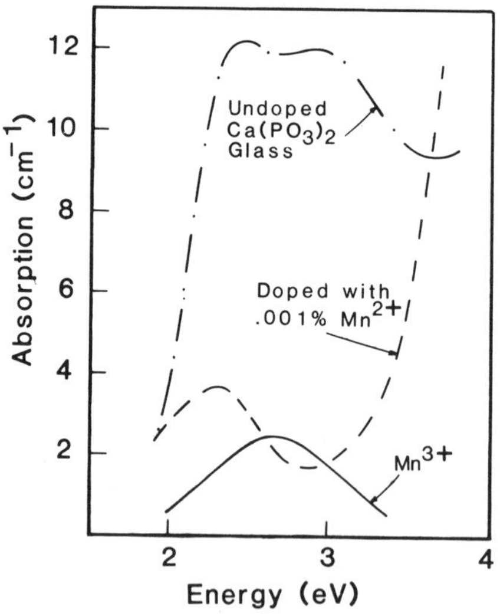

FIGURE 2.3.141. Optical absorption spectra of undoped CaO-P_2O_5 glass and a similar sample doped with 0.001% Mn^{2+} induced by an exposure to a 5×10^5 R dose of 50 kV X-rays at room temperature.[220] Also shown is the spectrum of Mn^{3+} doped into lithium borate glass.[234] The lack of agreement between the irradiated Mn^{2+}-doped phosphate glass and the Mn^{3+} spectrum indicates that a substantial portion of the induced absorption near 2.2 eV in the former glass is due to the alkali electron center.

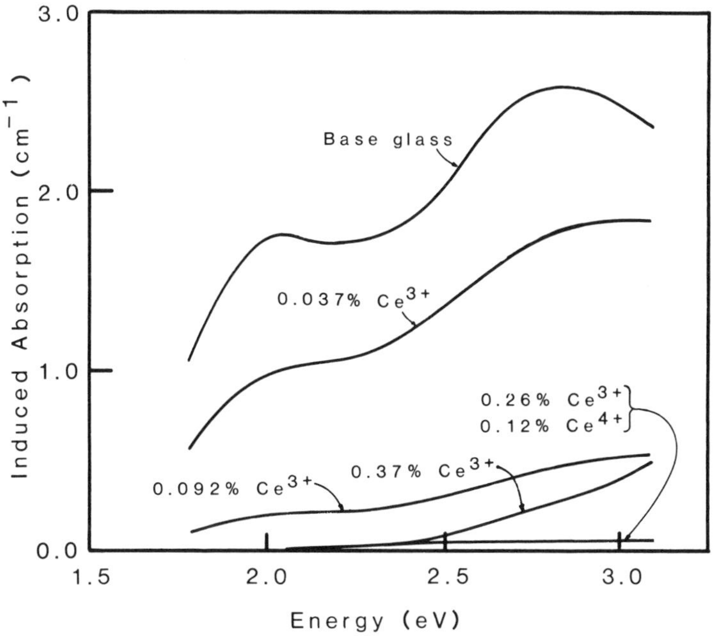

FIGURE 2.3.142. Effect of cerium doping on the optical absorption spectra induced in Na_2O-$3SiO_2$ glass by a room-temperature 250-kV X-ray dose of 2.2×10^5 R. The optimum protection by cerium is achieved in the case where both valence states are present to preferentially trap the radiolytic electrons and holes. (From Stroud, J. S., *J. Chem. Phys.*, 37, 836, 1962. With permission.)

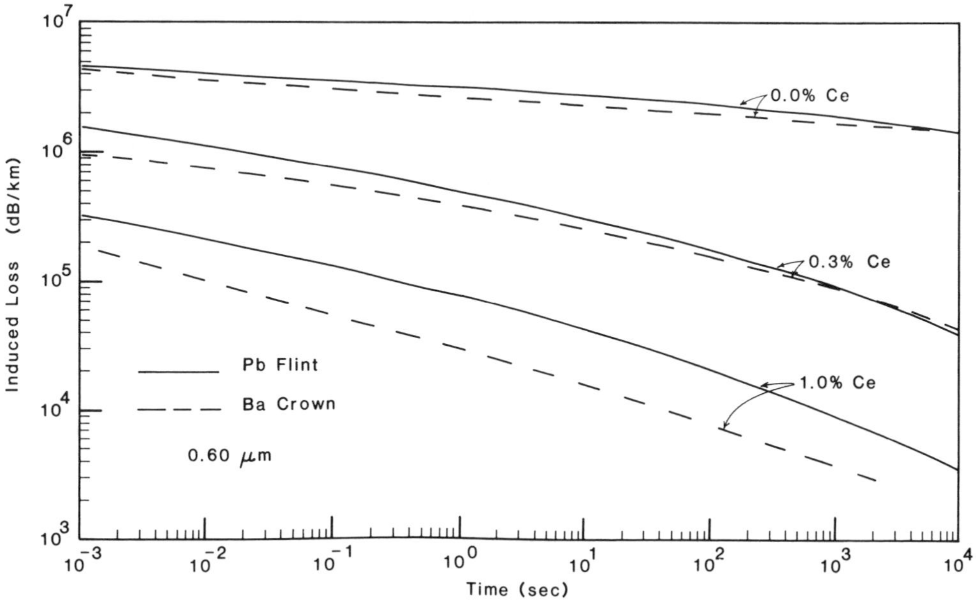

FIGURE 2.3.143. Room-temperature recovery of the induced absorption at 0.60 μm in lead flint and barium crown glasses as a function of Ce content following a 3-nsec, 10^5-rad dose of 0.5-MeV electrons. (From Friebele, E. J., *App. Phys. Lett.*, 27, 210, 1975. With permission.)

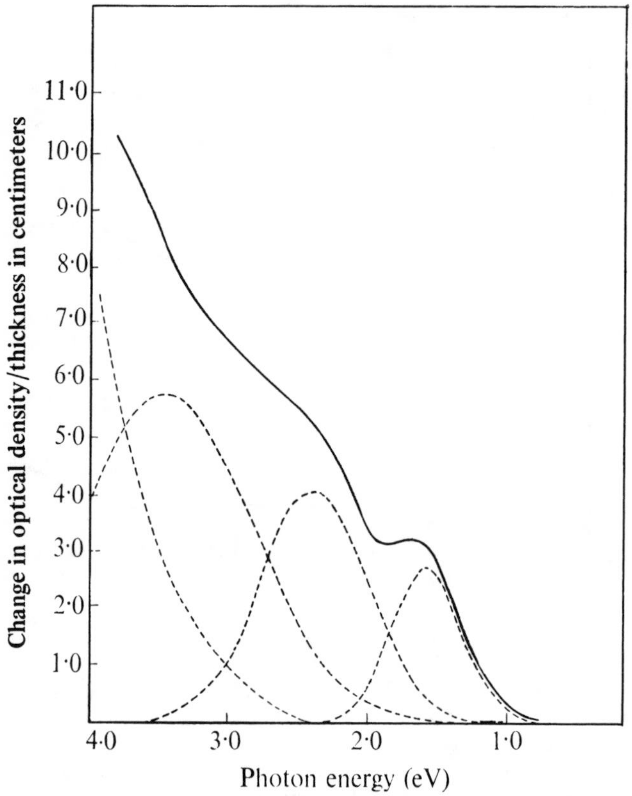

FIGURE 2.3.144. Optical absorption spectrum of a 27.4-mol% PbO-72.6% B_2O_3 glass following a 3 × 10^6 rad ^{60}Co irradiation at room temperature. The data are the heavy black line and the Gaussian components are shown as the dashed lines. Note the presence of the band at 1.59 eV, which is due to Pb^{3+} in network modifying sites. (From Bishay, A. M. and Moklad, M., *Phys. Chem. Glasses,* 6, 24, 1965. With permission.)

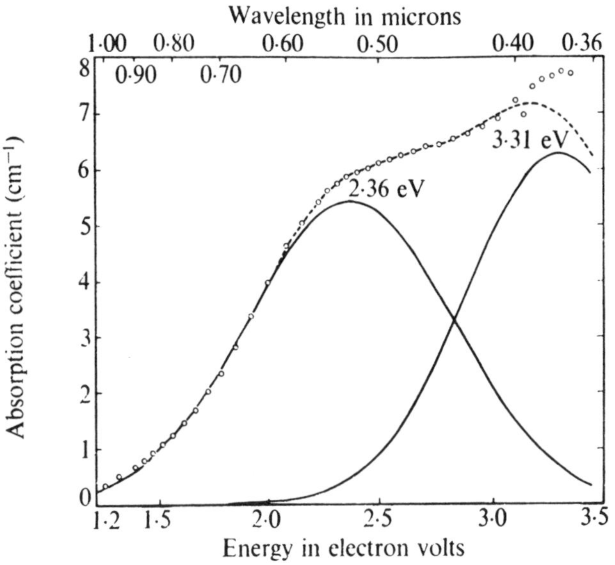

A

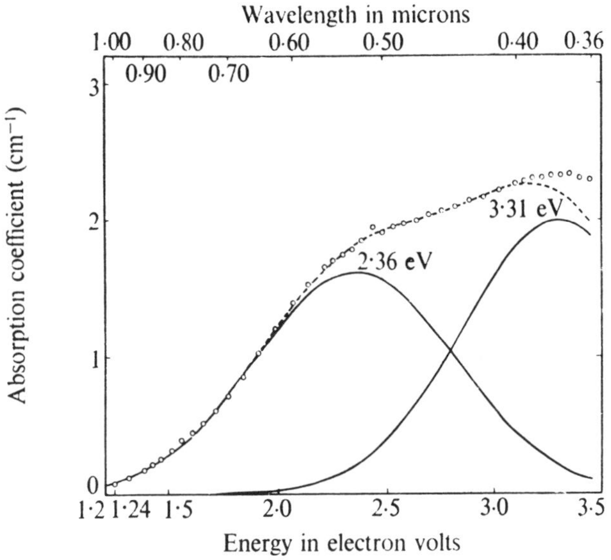

B

FIGURE 2.3.145. Optical absorption spectra of 80 PbO-20 SiO_2 glasses measured (A) 42 min after a 6 × 10^7 rad dose and (B) 33 min after a 3.2 × 10^4 rad dose of γ rays at room temperature. The data are the points, the component Gaussian bands are the solid lines, and the dashed line is the fit to the data. (From Barker, R. S., McConkey, E. A. G., and Richardson, D. A., *Phys. Chem. Glasses,* 6, 24, 1965. By permission of the publisher and Pilkington Brothers P. L. C.)

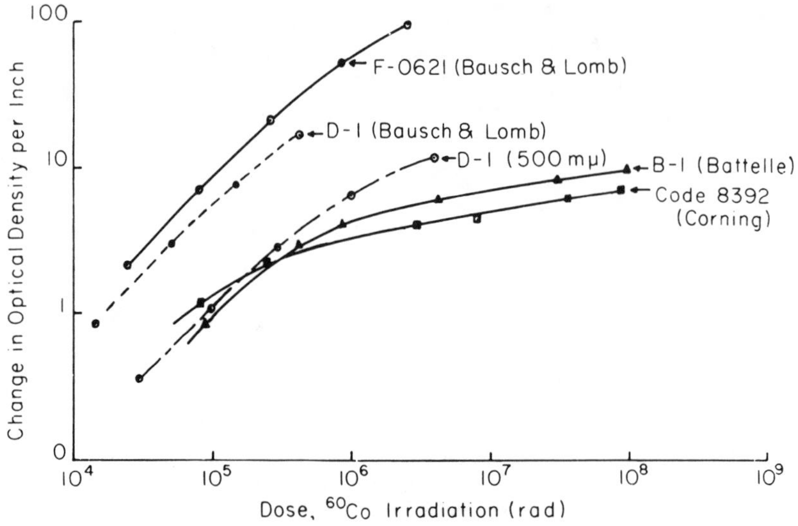

FIGURE 2.3.146. Dose dependence of induced optical absorption in a number of glasses proposed for dosimetry applications. All data shown were measured at 0.40 μm, except for the D-1 glass, which was measured at both 0.40 and 0.50 μm. B-1 is an antimonate glass, D-1 is a silver-activated phosphate glass, F-0621 is a cobalt-activated borosilicate glass, and Code 8392 is a high-lead glass. (From Blair, G. E., *J. Am. Ceram. Soc.*, 43, 426, 1960. With permission.)

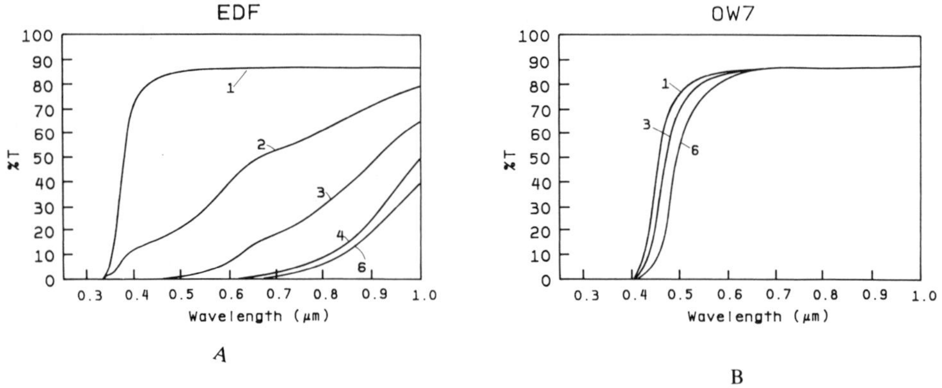

FIGURE 2.3.147. Spectral transmission curves for (A) extra-dense flint glass and (B) its cerium-protected equivalent measured ~15 min after a room temperature γ-ray dose. The approximate compositions of these glasses are contained in Table 2.3.7. The curves are: 1) unirradiated; 2) 2×10^4 rad; 3) 10^5 rad; 4) 10^6 rad; and 6) 10^8 rad. (From Barker, R. S., and Richardson, D. A., *J. Am. Ceram. Soc.*, 44, 552, 1961. With permission.)

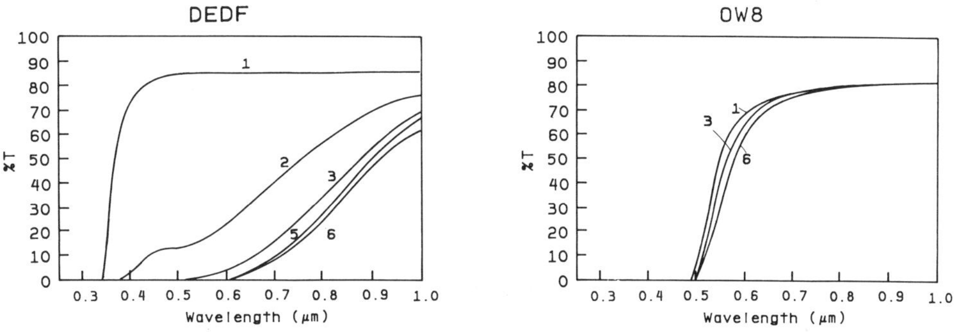

A B

FIGURE 2.3.148. Spectral transmission curves for (A) double-extra-dense flint glass and (B) its cerium-protected equivalent measured ~15 min after a room temperature γ-ray dose. The approximate compositions of these glasses are contained in Table 2.3.7. The curves are: 1) unirradiated; 2) 2 × 10^4 rad; 3) 10^5 rad; 5) 2 × 10^7 rad; and 6) 10^8 rad. (From Barker, R. S. and Richardson, D. A., *J. Am. Ceram. Soc.*, 44, 552, 1961. With permission.)

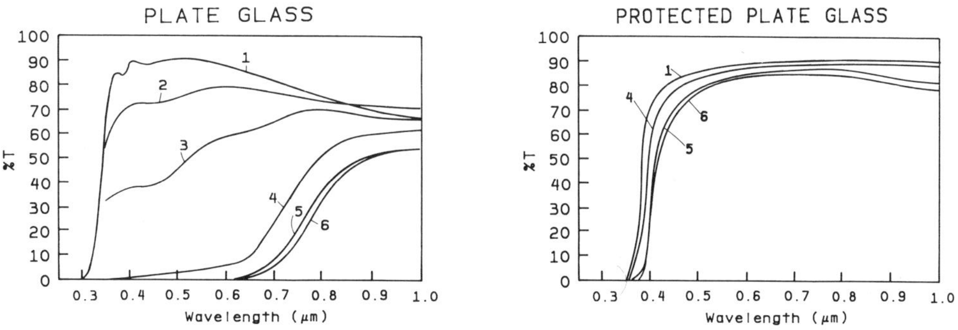

A B

FIGURE 2.3.149. Spectral transmission curves for (A) plate glass, and (B) its cerium-protected equivalent measured ~15 min after a room temperature γ-ray exposure. The approximate compositions of these glasses are contained in Table 2.3.7. The curves are: 1) unirradiated; 2) 2 × 10^4 rad; 3) 10^5 rad; 4) 10^6 rad; 5) 2 × 10^7 rad; and 6) 10^8 rad. (From Barker, R. S. and Richardson, D. A., *J. Am. Ceram. Soc.*, 44, 552, 1961. With permission.)

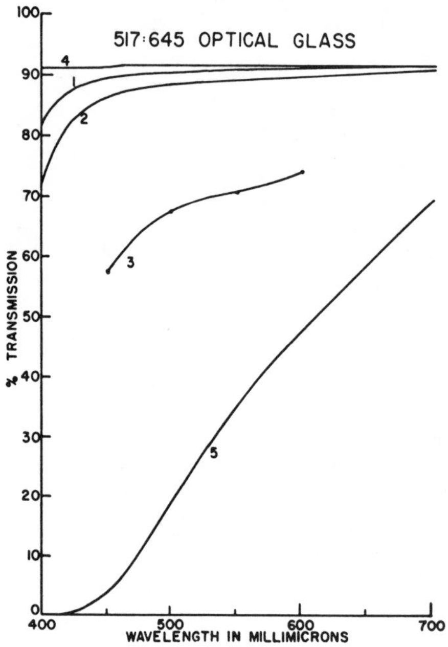

A

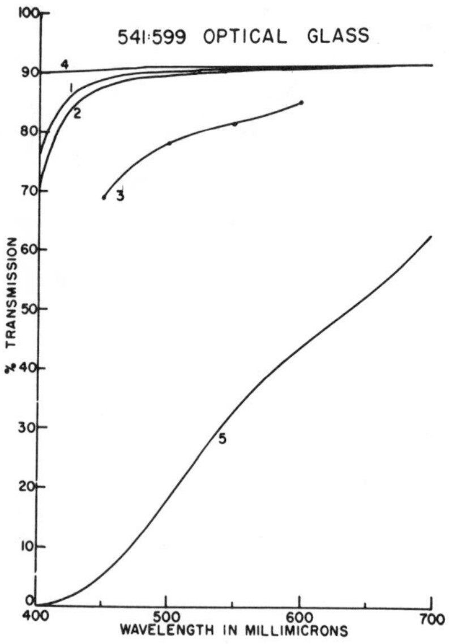

B

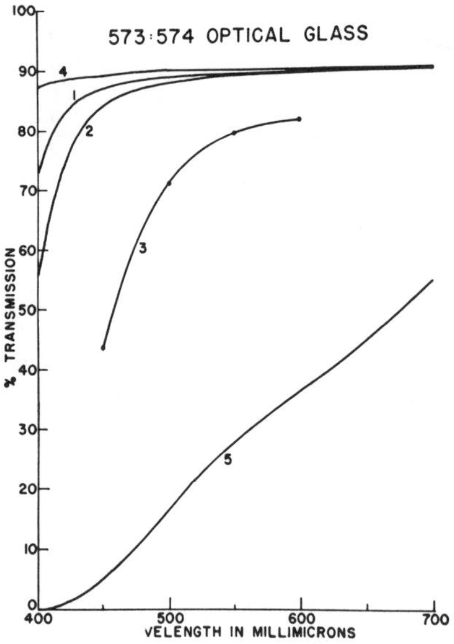

C

FIGURE 2.3.150. Transmission curves of protected and unprotected commercial optical glasses after exposure to ^{60}Co γ-rays. The numeric glass codes "A:B" are shown at the top of the figures where A = $n_D - 1$, where n_o is the refractive index and B = 10ν where (ν) is the reciprocal dispersion. The equivalent glass codes are A) BSC-2; B) LBC-1; C) LBC-2; D) DF-2; and E) EDF-1. The curves are: 1) unirradiated protected glass; 2) after 10^6 R; 3) after 5×10^8 R; 4) unirradiated unprotected glass; and 5) unprotected glass after 10^6 R. (From Kreidl, N. J. and Hensler, J. R., *J. Opt. Soc. A.*, 47, 73, 1957. With permission.)

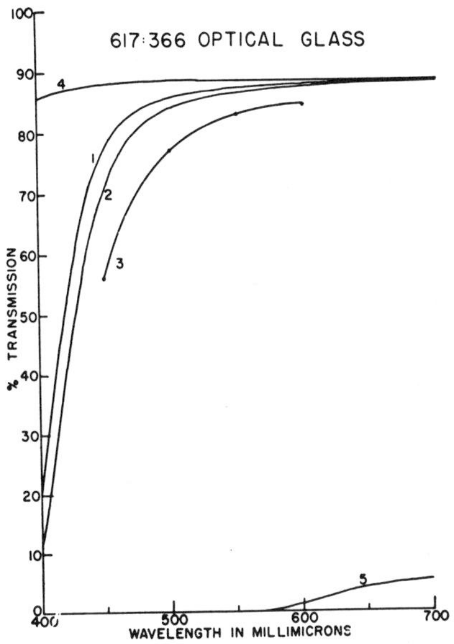

FIGURE 2.3.150D.

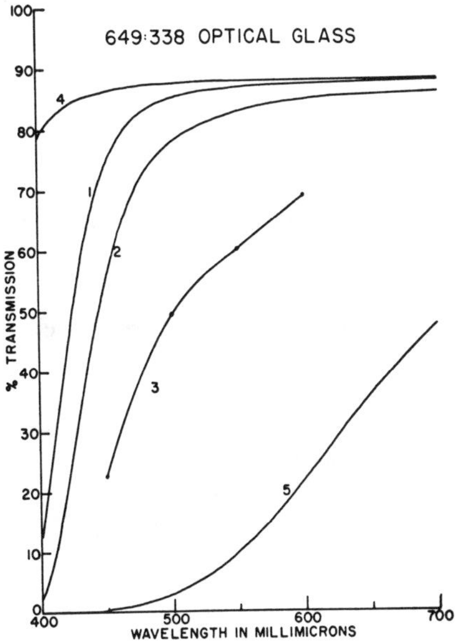

FIGURE 2.3.150E.

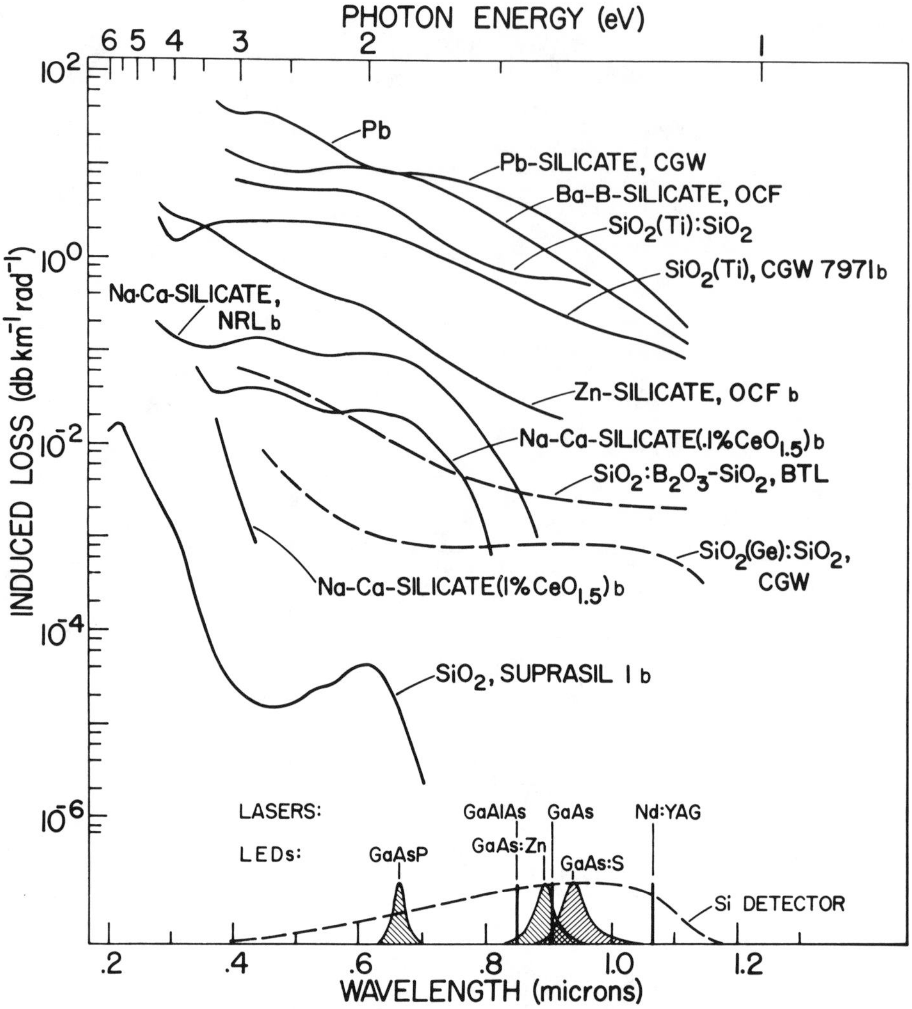

FIGURE 2.3.151. Induced absorption in a number of optical fibers and bulk glasses measured 1 hr after a room-temperature ^{60}Co irradiation. The bulk materials are labeled with the small "b". Doses ranged from 2×10^3 rad to 5×10^5 rad. The manufacturers of the samples were Corning Glass Works (CGW), Owens Corning Fiberglass (OCF), Bell Laboratories (BTL), and the Naval Research Laboratory (NRL). The Suprasil 1 sample was manufactured by Heraeus Quarzschmelze. Note that 1 dB/km is equivalent to an absorption coefficient of 2.303×10^{-6} cm^{-1}. (From Evans, B. D. and Sigel, G. H., Jr., *IEEE Trans. Nucl. Sci.*, 22, 2462, 1975. With permission.)

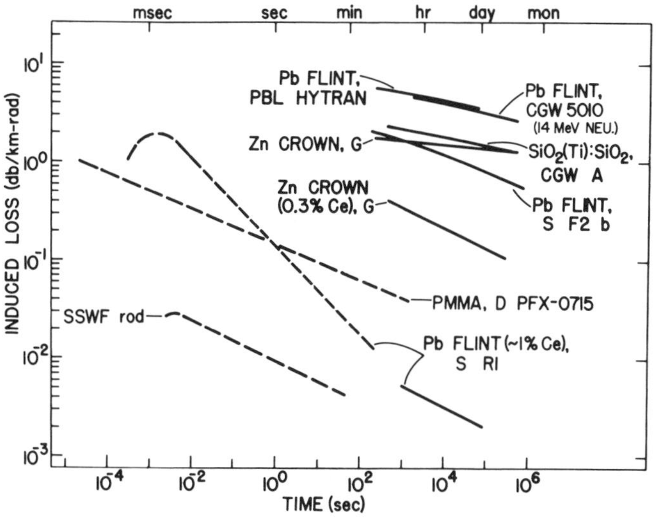

FIGURE 2.3.152. Recovery of the radiation-induced absorption in bulk glasses and fibers at 0.80 μm following either pulsed electron (dashed lines) or γ-ray (solid curves) exposures at room temperature. Also shown are recovery data for the Pb flint glass fiber following a 14-MeV neutron exposure. The manufacturers indicated in the figure are Corning Glass Works (CGW), Schott Glaswerke (S), Galileo (G), and DuPont (D). The Suprasil W rod (SSWF) was manufactured by Heraeus Quarzschmelze. Note that 1 dB/km is equivalent to an absorption coefficient of 2.303×10^{-6} cm^{-1}. (From Evans, B. D. and Sigel, G. H., Jr., *IEEE Trans. Nucl. Sci.*, 22, 2462, 1975. With permission.)

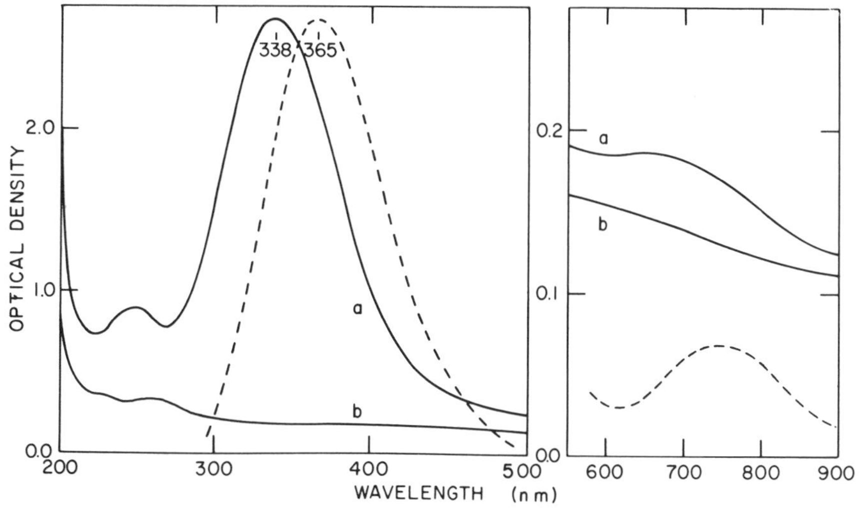

FIGURE 2.3.153. Optical absorption of a 7.5% KCl-22.5% $KO_{1/2}$-70%B_2O_3 (A) after a ray dose of 4.2×10^7 R, and (B) prior to irradiation. The dashed curves show the absorption of Cl_2^- centers in irradiated KCl. (From Griscom, D. L., *J. Chem. Phys.*,51, 5186, 1969. With permission.)

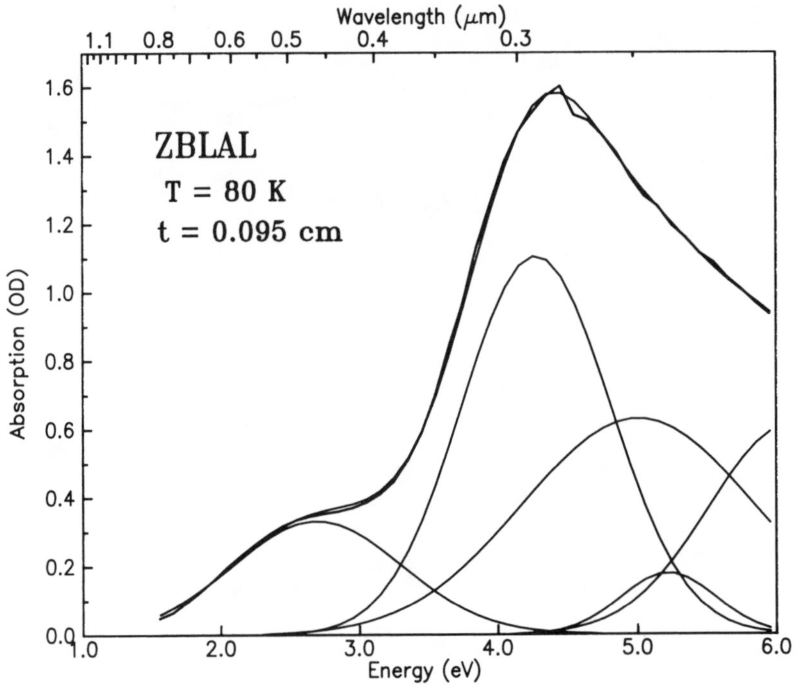

FIGURE 2.3.154. Optical absorption spectrum (bold line) induced in a ZBLAL glass by low temperature X-irradiation, together with computed spectrum based on a resolution into 5 Gaussian bands (narrow lines). The identification of the bands by correlation of annealing behaviors of the optical and ESR spectra is shown in Table 2.3.7. (After Reference 266.)

2.4. REFERENCES

1. **Sonder, E. and Sibley, W. A.,** Defect creation by radiation in polar crystals, in *Point Defects in Solids,* Vol. 1, Crawford, J. H., Jr. and Slifkin, L. M., Eds., Plenum Press, New York, 1972, chap. 4.
2. **Corbett, J. W. and Bourgoin, J. C.,** Defect creation in semiconductors, in *Point Defects in Solids,* Vol. 2, Crawford, J. H., Jr. and Slifkin, L. M., Eds., Plenum Press, New York, 1975, chap. 1.
3. **Fowler, W. B.,** Electronic states and optical properties of color centers, in *Physics of Color Centers,* Fowler, W. B., Ed., Academic Press, New York, 1968, chap. 2.
4. **Seitz, F. and Koehler, J. S.,** Displacement of atoms during irradiation, in *Solid State Physics,* Vol. 2, Seitz, F. and Turnbull, D., Eds., Academic Press, New York, 1956, 307.
5. **Lell, E., Kreidl, N. J., and Hensler, J. R.,** Radiation effects in quartz, silica and glasses, *Progr. Ceram. Sci.,* 41, 1, 1966.
6. **LeClerc, P. P.,** Action des rayonnements sur les verres, *Bull. Inf. Sci. Tech. Commiss. Energy Atom.,* 98, 7, 1968.
7. **Bishay, A.,** Radiation induced color centers in multicomponent glasses, *J. Non-Cryst. Solids,* 3, 54, 1970.
8. **Friebele, E. J. and Griscom, D. L.,** Radiation effects in glass, in *Treatise on Materials Science and Technology,* Vol. 17 (Glass II), Tomozowa, M. and Doremus, R. H., Eds., Academic Press, New York, 1979, 257.
9. **Shulman, J. H. and Compton, W. D.,** *Color Centers in Solids,* Pergamon Press, New York, 1962.
10. **Smakula, A.,** Uber erregung und entfarbung lichtelektrisch leitender alkalihalogenide, *Z. Physik,* 59, 603, 1930.
11. **Glasstone, S.,** *Sourcebook on Atomic Energy,* Van Nostrand, New York, 1958, 594.
12. **Alig, R. C. and Bloom, S.,** Electron-hole-pair creation energies in semiconductors, *Phys. Rev. Lett.,* 35, 1522, 1975; Secondary-electron-escape probabilities, *J. Appl. Phys.,* 49, 3476, 1978.
13. **Williams, R. T.,** Photochemistry of F-center formation in halide crystals, *Semiconduct. Insul.,* 3, 251, 1978.
14. **Kabler, M. N. and Williams, R. T.,** Vacancy-interstitial pair production via electron-hole recombination in halide crystals, *Phys. Rev. B,* 18, 1948, 1978.
15. **Itoh, N.,** Creation of lattice defects by electronic excitation in alkali halides, *Adv. Phys.,* 31, 491, 1982.
16. **Kabler, M. N.,** Ionization damage processes in inorganic materials, in *Proc. NATO Advanced Study Institute on Radiation Damage Processes in Materials,* Dupuy, C. H. S., Ed., Noordhoff Int., Leyden, 1975, 171.
17. **Stoebe, T. G., Spry, R. J., and Lewis, J.,** Optical properties of europium-doped potassium chloride laser window materials, in *Proc. Fifth Conf. Infrared Laser Window Materials,* Andrews, C. F. and Strecker, C. L., Eds., NTIS No. AD-AO26363/2ST, Springfield, Va., 1976, 266.
18. **Delbecq, C. J., Pringsheim, P., and Yuster, P.,** A new type of short wavelength absorption band in alkali halides containing color centers, *J. Chem. Phys.,* 19, 574, 1951.
19. **Görlich, P., Karras, H., and Kotitz, G.,** Studies on the coloration of LiF crystals, *Phys. Status Solidi,* 3, 1629, 1963.
20. **Heath, D. and Sacher, P. A.,** Effects of a simulated high-energy environment on the ultraviolet transmittance of optical materials between 1050 Å and 3000 Å, *Appl. Opt.,* 5, 937, 1966.
21. **Fuchs, R.,** Theory of the beta band in alkali halide crystals, *Phys. Rev.,* 111, 387, 1958; **Luty, F. and Zizelmann, W.,** Localized exciton absorption at point defects in KCl and KBr, *Solid State Commun.,* 2, 179, 1964.
22. **Compton, W. D. and Rabin, H.,** F-aggregate centers in alkali halide crystals, in *Solid State Physics,* Vol. 16, Seitz, F. and Turnbull, D., Eds., Academic Press, New York, 1964, 121.
23. **Kubo, K.,** Radiation effect in LiF crystals, *J. Phys. Soc. Jpn.,* 16, 2294, 1961.
24. **Hobbs, L. W.,** Point defect stabilization in ionic crystals at high defect concentrations, *J. Phys. Colloq.,* 37, C7-3, 1976.
25. **Van den Bosch, A.,** Gamma radiolysis of LiF, *Rad. Effects,* 19, 129, 1973.
26. **Itoh, N.,** Interstitial and trapped-hole centers in alkali halides, *Cryst. Lattice Def.,* 3, 115, 1972.
27. **Kabler, M. N.,** Hole centers in halide lattices, in *Point Defects in Solids,* Vol. 1, Crawford, J. H., Jr. and Slifkin, L. M., Eds., Plenum Press, New York, 1972, chap. 6.
28. **Hoshi, J., Saidoh, M., and Itoh, N.,** Stabilization of the interstitial center by divalent impurities in alkali halides, *Cryst. Lattice Def.,* 6, 15, 1975.
29. **Mitsushima, Y., Morita, K., Matsunami, N., and Itoh, N.,** Determination of the V_4-center structure in KBr crystal by means of double alignment channeling technique, *J. Phys. Colloq.,* 37, C7-95, 1976.
30. **Hobbs, L. W., Hughes, A. E., and Pooley, D.,** Nature of interstitial clusters in alkali halides, *Phys. Rev. Lett.,* 28, 234, 1972; **Ishii, T. and Rolfe, J.,** V centers in potassium bromide crystals, *Phys. Rev.,* 141, 758, 1966.
31. **Ivey, H.,** Spectral location of the absorption due to color centers in alkali halide crystals, *Phys. Rev.,* 72, 341, 1947.

32. **Russell, G. A. and Klick, C. C.,** Configuration coordinate curves for F centers in alkali halide crystals, *Phys. Rev.,* 101, 1473, 1956.
33. **Grimes, H. H., Maisel, J. E., and Hartford, R. H.,** Radiation effects on β10.6 of pure and europium doped KCl, in *Proc. Fifth Conf. Infrared Laser Window Materials,* Andrews, C. R. and Strecker, C. L., Eds., NTIS No. N76-132968ST, Springfield, Va., 1976, 180.
34. **Magee, T. J., Johnson, N., and Peng, J.,** The influence of high-energy electron irradiation on the 10.6 μm absorption of mixed KCl crystals, *Phys. Status Solidi (A),* 30, 81, 1975.
35. **Mitchell, P. V., Wiegand, D. A., and Smoluchowski, R.,** Formation of F centers in KCl by x-rays, *Phys. Rev.,* 121, 484, 1961.
36. **Sonder, E. and Templeton, L. C.,** Radiation defect production and annihilation in KCl near room temperature, *Phys. Rev.,* 164, 1106, 1967.
37. **Kabler, M. N.,** Low-temperature recombination luminescence in alkali halide crystals, *Phys. Rev.,* 136, A1296, 1964.
38. **Pooley, D.,** F-centre production in alkali halides by electron-hole recombination, and a subsequent (110) replacement sequence: a discussion of the electron-hole recombination, *Proc. Phys. Soc. (London),* 87, 245, 1966.
39. **Hersh, H. N.,** Proposed excitonic mechanism of color-center formation in alkali halides, *Phys. Rev.,* 148, 928, 1966.
40. **Kabler, M. N. and Patterson, D. A.,** Evidence for a triplet state of the self-trapped exciton in alkali halide crystals, *Phys. Rev. Lett.,* 19, 652, 1967.
41. **Ikezawa, M. and Kojima, T.,** Luminescence of alkali halide crystals induced by UV light at low temperature, *J. Phys. Soc. Jpn.,* 27, 1551, 1969.
42. **Fuller, R. G., Williams, R. T., and Kabler, M. N.,** Transient optical absorption by self-trapped excitons in alkali halide crystals, *Phys. Rev. Lett.,* 25, 446, 1970.
43. **Williams, R. T. and Kabler, M. N.,** Excited-state absorption spectroscopy of self-trapped excitons in alkali halides, *Phys. Rev. B,* 9, 1897, 1974.
44. **Williams, R. T., Craig, B. B., and Faust, W. L.,** F-center formation in NaCl: picosecond spectroscopic evidence for halogen diffusion on the lowest excitonic potential surface, *Phys. Rev. Lett.,* 52, 1709, 1984.
45. **Williams, R. T., Bradford, J. N., and Faust, W. L.,** Short-pulse optical studies of exciton self-trapping and F-center formation in NaCl, KCl, and NaBr, *Phys. Rev. B,* 18, 7038, 1978; **Bradford, J. N., Williams, R. T., and Faust, W. L.,** Study of F-center formation in KCl on a picosecond time scale, *Phys. Rev. Lett.,* 35, 300, 1975.
46. **Suzuki, Y., Okunura, M., and Hirai, M.,** Relaxation of 2p excitons and F center formation in picosecond range in RbBr, RbI, and KBr, *J. Phys. Soc. Jpn.,* 47, 184, 1979.
47. **Ortega, J. M.,** Study of the self-trapped exciton and F-center formation on a picosecond time scale in KBr, *Phys. Rev. B,* 19, 3222, 1979; **K'Hertoghe, J. and Jacobs, G.,** Self-trapped exciton and F centre formation by picosecond laser pulses in alkali bromides and iodides, *Phys. Status Solidi (B),* 95, 291, 1979.
48. **Sibley, W. A. and Facey, O. E.,** Color centers in MgF_2, *Phys. Rev.,* 174, 1076, 1968.
49. **Buckton, M. R. and Pooley, D.,** The radiation damage process in MgF_2, *J. Phys. C,* 5, 1553, 1972.
50. **Tsuboi, T., Kato, R., and Nakagawa, M.,** Irradiation-induced VUV absorption in MgF_2 crystals, *J. Phys. Soc. Jpn.,* 25, 645, 1968.
51. **Facey, O. E., Lewis, D. L., and Sibley, W. A.,** Electron and neutron damage in MgF_2 crystals, *Phys. Status Solidi,* 32, 831, 1969.
52. **Reft, C. S., Becher, J., and Kernell, R. L.,** Proton-induced degradation of VUV transmission of LiF and MgF_2, *Appl. Opt.,* 19, 4156, 1980.
53. **Blunt, R. F. and Cohen, M. I.,** Irradiation-induced color centers in magnesium fluoride, *Phys. Rev.,* 153, 1031, 1967.
54. **Facey, O. E. and Sibley, W. A.,** Optical absorption and luminescence of irradiated MgF_2, *Phys. Rev.,* 186, 926, 1969.
55. **Facey, O. E. and Sibley, W. A.,** M centers in MgF_2 crystals, *Phys. Rev. B,* 2, 1111, 1970.
56. **Williams, R. T., Marquardt, C. L., Williams, J. W., and Kabler, M. N.,** Transient absorption and luminescence in MgF_2 following electron pulse excitation, *Phys. Rev. B,* 15, 5003, 1977.
57. **Fong, F. K. and Yocom, P. N.,** Crystal growth and color centers of alkaline-earth halides, *J. Chem. Phys.,* 41, 1383, 1964.
58. **Ratnam, V. V.,** On the colour centres and X-ray luminescence of calcium fluoride crystals, *Phys. Status Solidi,* 16, 559, 1966.
59. **Kamikawa, T. and Kunio, O.,** Induced color centers in CaF_2 crystals irradiated with neutrons at liquid nitrogen temperature, *J. Phys. Soc. Jpn.,* 24, 115, 1968.
60. **Bontinck, W.,** Colour centers in synthetic fluorite crystals, *Physica,* 24, 639, 1958.
61. **Hayes, W. and Staebler, D. L.,** Oxidation-reactions and photochromism, in *Crystals with the Fluorite Structure,* Hayes, W., Ed., Clarendon Press, Oxford, 1974, chap. 7.

62. **Görlich, P., Karras, H., and Lehmann, R.**, Uber die optischen eigenschaften der erdalkalihalogenide vom flusspat-Type(I), *Phys. Status Solidi,* 1, 389, 1961.
63. **Görlich, P., Karras, H., Symanowski, Ch., and Ullmann, P.**, The colour center absorption of X-ray coloured alkaline earth fluoride crystals, *Phys. Status Solidi,* 25, 93, 1968.
64. **Patterson, D. A. and Fuller, R. G.**, F band in X- and electron-irradiated CaF_2, *Phys. Rev. Lett.,* 18, 1123, 1967.
65. **Catlow, C. R. A.**, Radiation damage and photochromism in the alkaline earth fluorides, *J. Phys. C,* 12, 969, 1979.
66. **Staebler, D. L. and Schnatterly, S. E.**, Optical studies of a photochromic color center in rare-earth-doped CaF_2, *Phys. Rev. B,* 3, 516, 1971.
67. **Hayes, W. and Stoneham, A. M.**, Colour centres, in *Crystals with the Fluorite Structure,* Hayes, W., Ed., Oxford University Press, London, 1974, chap. 4.
68. **Hayes, W. and Lambourn, R. F.**, Production of F and F-aggregate centres in CaF_2 and SrF_2 by irradiation, *Phys. Status Solidi (B),* 57, 693, 1973.
69. **Beaumont, J. H., Hayes, W., Kirk, D. L., and Summers, G. P.**, An investigation of trapped holes and trapped excitons in alkaline earth fluorides, *Proc. R. Soc. London,* A315, 69, 1970.
70. **Mukerji, A., Tanton, G. A., and Williams, J. E.**, X-ray induced F centers in CaF_2, *Phys. Status Solidi,* 22, K19, 1967.
71. **Cavenett, B. C., Hayes, W., Hunter, I. C., and Stoneham, A. M.**, Magnetooptical properties of F centres in alkaline earth fluorides, *Proc. R. Soc. London,* A309, 53, 1969.
72. **Williams, R. T., Kabler, M. N., Hayes, W., and Stott, J. P.**, Time-resolved spectroscopy of self-trapped excitons in fluorite crystals, *Phys. Rev. B,* 14, 725, 1976.
73. **Riley, C. R. and Sibley, W. A.**, Color centers in $KMgF_3$, *Phys. Rev. B,* 1, 2789, 1970.
74. **Seretlo, J. R., Martin, J. J., and Sonder, E.**, Optical absorption of radiation-produced defects in $NaMgF_3$ and $RbCaF_3$, *Phys. Rev. B,* 14, 5404, 1976.
75. **Koumvakalis, N. and Sibley, W. A.**, Radiation damage of $RbMgF_3$, *Phys. Rev. B,* 13, 4509, 1976.
76. **Hall, T. P. P. and Leggeat, A.**, Defect centers in $KMgF_3$ produced by X-irradiation at room temperature, *Solid State Commun.,* 7, 1657, 1969.
77. **Alcala, R., Koumvakalis, N., and Sibley, W. A.**, The self-trapped hole and thermoluminescence in $KMgF_3$, *Phys. Status Solidi (A),* 30, 449, 1975.
78. **Hayes, W., Owen, I. B., and Pilipenko, G. I.**, Optical detection of exciton EPR in $KMgF_3$, *J. Phys. C,* 8, L407, 1975.
79. **Hilsch, R. and Pohl, R. W.**, Zur photochemie der alkali - und silberhalogenidkristalle, *Z. Physik,* 64, 606, 1930.
80. **Christy, R. W. and Dimock, J. D.**, Color centers in TlCl, *Phys. Rev.,* 141, 806, 1966.
81. **Marquardt, C. L. and Williams, R. T.**, Recombination centers in TlBr, *J. Luminescence,* 9, 440, 1974.
82. **Brothers, A. D. and Lynch, D. W.**, Temperature and pressure dependence of the optical absorption and luminescence of a defect in TlCl, *Phys. Rev.,* 159, 687, 1967.
83. **Sibley, W. A., Kolopus, J. L., and Mallard, W. C.**, A study of the effect of deformation on the ESR, luminescence, and absorption of MgO single crystals, *Phys. Status Solidi,* 31, 223, 1969.
84. **Chao, C. C.**, Charge-transfer luminescence of Cr^{3+} in magnesium oxide, *J. Phys. Chem. Solids,* 32, 2517, 1971.
85. **Sibley, W. A. and Chen, Y.**, Radiation damage in MgO, *Phys. Rev.,* 160, 712, 1967.
86. **Chen, Y., Kolopus, J. L., and Sibley, W. A.**, Luminescence of the F^+ center in MgO, *Phys. Rev.,* 186, 865, 1969.
87. **Chen, Y., Williams, R. T., and Sibley, W. A.**, Defect cluster centers in MgO, *Phys. Rev.,* 182, 960, 1969.
88. **Evans, B. D.**, Spectral study of Ne^+-bombarded crystalline MgO, *Phys. Rev. B,* 9, 5222, 1974.
89. **Chen, Y. and Abraham, M. M.**, Evidence for suppression of radiation damage in Li-doped MgO, *J. Am. Ceram. Soc.,* 59, 101, 1976.
90. **Abraham, M. M., Chen, Y., and Unruh, W. P.**, Formation and stability of V^- and V_{Al} centers in MgO, *Phys. Rev. B,* 9, 1974.
91. **Hughes, A. E. and Henderson, B.**, Color centers in simple oxides, in *Point Defects in Solids,* Vol. 1, Crawford, J. H., Jr. and Slifkin, L. M., Eds., Plenum Press, New York, 1972, chap. 7.
92. **Norgett, M. J., Stoneham, A. M., and Pathak, A. P.**, Electronic structure of the V^- center in MgO, *J. Phys. C,* 10, 555, 1977.
93. **Searle, T. M. and Glass, A. M.**, The thermal decay of the V′ center in magnesium oxide, *J. Phys. Chem. Solids,* 29, 609, 1968.
94. **Chen, Y. and Sibley, W. A.**, Study of ionization-induced radiation damage in MgO, *Phys. Rev.,* 154, 842, 1967.
95. **Chen, Y., Abraham, M. M., Templeton, L. C., and Unruh, W. P.**, Role of hydrogen and deuterium on the V^--center formation in MgO, *Phys. Rev. B,* 11, 881, 1975.

96. **Lee, K. H. and Crawford, J. H.,** X-ray stimulated luminescence in MgO, *J. Luminescence*, 20, 9, 1979.
97. **Williams, R. T., Williams, J. W., Turner, T. J., and Lee, K. H.,** Kinetics of radiative recombination in magnesium oxide, *Phys. Rev. B*, 20, 1687, 1979.
98. **Levy, P. W.,** Color centers and radiation-induced defects in Al_2O_3, *Phys. Rev.*, 123, 1226, 1961.
99. **Turner, T. J. and Crawford, J. H., Jr.,** V centers in single crystal Al_2O_3, *Solid State Commun.*, 17, 167, 1975.
100. **Stickley, C. M., Miller, H., Hoell, E. E., Gallagher, C. C., and Bradbury, R. A.,** Color centers and ruby-laser output energy degradation, *J. Appl. Phys.*, 40, 1792, 1969.
101. **Arnold, G. W., Krefft, G. B., and Norris, C. B.,** Atomic displacement and ionization effects on the optical absorption and structural properties of ion-implanted Al_2O_3, *Appl. Phys. Lett.*, 25, 540, 1974.
102. **Compton, W. D. and Arnold, G. W., Jr.,** Radiation effects in fused silica and α-Al_2O_3, *Disc. Faraday Soc.*, 31, 130, 1961.
103. **Lee, K. H. and Crawford, J. H., Jr.,** Electron centers in single-crystal Al_2O_3, *Phys. Rev. B*, 15, 1977.
104. **Evans, B. D. and Stapelbroek, M.,** Optical properties of the F^+ center in crystalline Al_2O_3, *Phys. Rev. B*, 18, 7089, 1978.
105. **Bunch, J. M. and Clinard, F. W., Jr.,** Damage of single-crystal Al_2O_3 by 14-MeV neutrons, *J. Am. Ceram. Soc. Disc. Notes*, 57, 279, 1974.
106. **Lee, K. H. and Crawford, J. H., Jr.,** Luminescence of the F center in sapphire, *Phys. Rev. B*, 19, 3217, 1979.
107. **Levy, P. W.,** Annealing of the defects and color centers in unirradiated and in reactor irradiated Al_2O_3, *Disc. Faraday Soc.*, 31, 118, 1961.
108. **Vehse, W. E., Sibley, W. A., Keller, F. J., and Chen, Y.,** Radiation damage in ZnO single crystals, *Phys. Rev.*, 167, 828, 1968.
109. **Smith, J. M. and Vehse, W. E.,** ESR of electron irradiated ZnO confirmation of the F^+ center, *Phys. Lett.*, 31A, 147, 1970.
110. **Locker, D. R. and Meese, J. M.,** Displacement thresholds in ZnO, *IEEE Trans. Nucl. Sci.*, NS-19(6), 237, 1972.
111. **Townsend, P. D., Kan, H. K. A., and Levy, P. W.,** Effects of irradiation on the optical absorption and photoconductivity of rutile, *Proc. Br. Ceram. Soc.*, 1, 71, 1964.
112. **Purcell, T. and Weeks, R. A.,** Paramagnetic defects in TiO_2 produced by radiation, *J. Chem. Phys.*, 54, 2800, 1971.
113. **Yoshida, T., Seiyama, T., Shono, Y., and Kitagawa, M.,** Optical absorption of neutron-irradiated ZnS, *App. Phys. Lett.*, 9, 26, 1966.
114. **Lee, K. M., O'Donnell, K. P., and Watkins, G. D.,** Optically detected magnetic resonance of the zinc vacancy in ZnS, *Solid State Commun.*, 41, 881, 1982.
115. **Leutwein, K., Rauber, A., and Schneider, J.,** Optical and photoelectric properties of the F center in ZnS, *Solid State Commun.*, 5, 783, 1967.
116. **Smith, A. W. and Turkevich, J.,** Effect of neutron bombardment on a zinc sulfide phosphor, *Phys. Rev.*, 94, 857, 1954.
117. **Watkins, G. D.,** Irradiation effects in II-VI compounds, *Rad. Effects*, 9, 105, 1971.
118. **Lee, K. M., Dang, L. S., and Watkins, G. D.,** An ODMR study of the zinc vacancy in zinc selenide, *Inst. Phys. Conf. Ser. No. 59*, AIP, New York, 1981, 353.
119. **Watkins, G. D.,** Lattice defects in II-VI compounds, *Inst. Phys. Conf. Ser. No. 31*, AIP, New York, 1977, 95.
120. **Watkins, G. D.,** EPR observation of close frenkel pairs in irradiated ZnSe, *Phys. Rev. Lett.*, 33, 223, 1974.
121. **Vaidyanathan, K. V. and Watt, L. A.,** Infrared absorption studies in neutron- and electron-irradiated GaAs, *Rad. Effects*, 10, 99, 1971.
122. **McNichols, J. L. and Ginell, W. S.,** Theory of anomalous infrared attenuation in neutron-irradiated compound semiconductors, *J. Appl. Phys.*, 38, 656, 1967.
123. **Burkig, V. C., McNichols, J. L., and Ginell, W. S.,** Infrared absorption in neutron-irradiated GaAs, *J. Appl. Phys.*, 40, 3268, 1969.
124. **Pankey, T., Jr. and Davey, J. E.,** Effects of neutron irradiation on the optical properties of thin films and bulk GaAs and GaP, *J. Appl. Phys.*, 41, 697, 1970.
125. **Aukerman, L. W., Davis, P. W., Graft, R. D., and Shilliday, T. S.,** Radiation effects in GaAs, *J. Appl. Phys.*, 34, 3590, 1963.
126. **Redfield, D.,** Effect of defect fields on the optical absorption edge, *Phys. Rev.*, 130, 916, 1963.
127. **Jeong, M. U., Shirafuji, J., and Inuishi, Y.,** Annealing behavior of bulk n-GaAs irradiated by electrons at 77K, *Rad. Effects*, 10, 93, 1971.
128. **Eisen, F. H.,** III-V compound review, *Rad. Effects*, 9, 235, 1971.
129. **Kahan, A., Bouthillette, L., and DeAngelis, H. M.,** Electron irradiation of heavily doped GaAs: Si and GaAs:Te, *Rad. Effects*, 9, 99, 1971.

130. **Clark, C. D. and Mitchell, E. W. J.,** Radiation induced defects in diamond, *Rad. Effects,* 9, 219, 1971.
131. **Clark, C. D., Ditchburn, R. W., and Dyer, H. B.,** The absorption spectra of natural and irradiated diamonds, *Proc. R. Soc. London,* A234, 363, 1956.
132. **Clark, C. D., Dean, P. J., and Harris, P. V.,** Intrinsic edge absorption in diamond, *Proc. R. Soc. London,* 277, 312, 1964.
133. **Hardy, J. R. and Smith, S. D.,** Two-phonon infrared lattice absorption in diamond, *Philos. Mag.,* 6, 1163, 1961.
134. **Clark, C. D., Duncan, I., Lomer, J. N., and Whippey, P. W.,** Correlation of optical and electron spin resonance centres in electron-irradiated diamond, *Proc. Br. Ceram. Soc.,* 1, 85, 1964.
135. **Collins, A. T.,** High-resolution optical spectra of the GR defect in diamond, *J. Phys. C,* 11, 1957, 1978.
136. **Walker, J., Vermeulen, L. A., and Clark, C. D.,** Electronic transitions at the diamond vacancy, *Proc. R. Soc. London,* A341, 253, 1974.
137. **Lannoo, M. and Stoneham, A. M.,** The optical absorption of the neutral vacancy in diamond, *J. Phys. Chem. Solids,* 29, 1987, 1968.
138. **Clark, C. D., Kemmey, P. J., and Mitchell, E. W. J.,** Optical and electrical effects of radiation in diamond, *Disc. Faraday Soc.,* 31, 96, 1961.
139. **Wentorf, R. H. and Darrow, K. A.,** Semiconducting diamonds by ion bombardment, *Phys. Rev.,* 137, A1614, 1965.
140. **Horszowski, S. M.,** Isothermal annealing of electron radiation damage in natural semiconducting diamond, *Rad. Effects,* 30, 213, 1976.
141. **Clark, C. D., Ditchburn, R. W., and Dyer, H. B.,** The absorption spectra of irradiated diamonds after heat treatment, *Proc. R. Soc. London,* A237, 75, 1956.
142. **Dyer, H. B. and DuPreez, L.,** Irradiation damage in type I diamond, *J. Chem. Phys.,* 42, 1898, 1965.
143. **Dean, P. J. and Male, J. C.,** Luminescence excitation spectra and recombination radiation of diamond in the fundamental absorption region, *Proc. R. Soc. London,* 277, 330, 1964.
144. **Dean, P. J.,** Bound excitons and donor-acceptor pairs in natural and synthetic diamond, *Phys. Rev.,* 139, A588, 1965.
145. **Gerasimenko, N. N., Il'in, V. E., Lezheiko, L. V., Smirnov, L. S., Sobolev, E. V., and Yur'eva, O. P.,** *Sov. Phys. Solid State,* 9, 2898, 1968.
146. **Curtis, O. L., Jr.,** Effects of point defects on electrical and optical properties of semiconductors, in *Point Defects in Solids,* Crawford, J. H., Jr. and Slifkin, L. M., Eds., Plenum Press, New York, 1975, chap. 3.
147. **Stein, H. J.,** Defects in silicon: concepts and correlations, *Rad. Effects,* 9, 195, 1971.
148. **Watkins, G. D.,** A microscopic view of radiation damage in semiconductors using EPR as a probe, *IEEE Trans. Nucl. Sci.,* 16(6), 13, 1969.
149. **Cherki, M. and Kalma, A. H.,** Interstitial defects in p-type silicon, *IEEE Trans. Nucl. Sci.,* 16(6), 24, 1969.
150. **Devine, S. D. and Newman, R. C.,** One phonon absorption from aluminum complexes in silicon compensated by lithium or electron irradiation, *J. Phys. Chem. Solids,* 31, 685, 1970.
151. **Bean, A. R., Newman, R. C., and Smith, R. S.,** Electron irradiation damage in silicon containing carbon and oxygen, *J. Phys. Chem. Solids,* 31, 739, 1970.
152. **Cheng, L. J., Corelli, J. C., Corbett, J. W., and Watkins, G. D.,** 1.8-, 3.3- and 3.9-μ bands in irradiated silicon: correlations with the divacancy, *Phys. Rev.,* 152, 761, 1966.
153. **Whan, R. E. and Vook, F. L.,** Infrared studies of defect production in n-Type Si: irradiation-temperature dependence, *Phys. Rev.,* 153, 814, 1967.
154. **Corbett, J. W., Watkins, G. D., Chrenko, R. M., and McDonald, R. S.,** Defects in irradiated silicon. II. Infrared absorption of the Si-A center, *Phys. Rev.,* 121, 1015, 1961.
155. **Corbett, J. W., Watkins, G. D., and McDonald, R. S.,** New oxygen infrared bands in annealed irradiated silicon, *Phys. Rev.,* 135, A1381, 1964.
156. **Brelot, A. and Charlemagne, J.,** Infrared studies of low temperature electron irradiated silicon containing germanium, oxygen, carbon, *Rad. Effects,* 9, 65, 1971.
157. **Ramdas, A. K. and Rao, M. G.,** Infrared absorption spectra of oxygen-defect complexes in irradiated silicon, *Phys. Rev.,* 142, 451, 1966.
158. **Noonan, J. R., Kirkpatrick, C. G., and Streetman, B. G.,** Photoluminescence from Si irradiated with 1.5-MeV electrons at 100°K, *J. Appl. Phys.,* 47, 3010, 1976.
159. **Jones, C. E., Johnson, E. S., Compton, W. D., Noonan, J. R., and Streetman, B. G.,** Temperature, stress, and annealing effects on the luminescence from electron-irradiated silicon, *J. Appl. Phys.,* 44, 5403, 1973.
160. **Lee, K. M., O'Donnell, K. P., Weber, J., Cavanett, B. C., and Watkins, G. D.,** Optical detection of magnetic resonance for a deep-level defect in silicon, *Phys. Rev. Lett.,* 48, 38, 1982.
161. **MacKay, J. W. and Klontz, E. E.,** Review of radiation effects in germanium, *Rad. Effects,* 9, 27, 1971.

162. **Becker, J. F. and Corelli, J. C.,** Infrared properties of 40-60 MeV electron-irradiated germanium, *J. Appl. Phys.*, 36, 3606, 1965.
163. **Whan, R. E.,** Investigations of oxygen-defect interactions between 25 and 700°K in irradiated germanium, *Phys. Rev.*, 140, A690, 1965.
164. **Emtzev, V. V., Mashovets, T. V., Maximov, M., and Vitovskii, N. A.,** On the donor-vacancy type complexes in germanium, *Rad. Effects,* 9, 181, 1971.
165. **Bruckner, R.,** Properties and structure of vitreous silica. I and II, *J. Non-Cryst Solids,* 15, 123, 1970.
166. **Levy, P. W.,** Reactor and gamma-ray induced coloring of Corning fused silica, *J. Phys. Chem. Solids,* 13, 287, 1960.
167. **Nelson, C. M. and Weeks, R. A.,** Trapped electrons in irradiated quartz and silica. I. Optical absorption, *J. Am. Ceram. Soc.*, 43, 396, 1960.
168. **Weeks, R. A. and Nelson, C. M.,** Trapped electrons in irradiated quartz and silica. II. Electron spin resonance, *J. Am. Ceram. Soc.*, 43, 399, 1960; **Silsbee, R. H.,** Electron spin resonance in neutron-irradiated quartz, *J. Appl. Phys.*, 32, 1459, 1961.
169. **Weeks, R. A.,** Paramagnetic spectra of E_2' centers in crystalline quartz, *Phys. Rev.*, 130, 570, 1963.
170. **Weeks, R. A. and Sonder, E.,** The relation between the magnetic susceptibility, electron spin resonance, and the optical absorption of the E_1' center in fused silica, in *Symposium on Paramagnetic Resonance,* Vol. 2, Low, W., Ed., Academic Press, New York, 1963, 869.
171. **Griscom, D. L.,** Point defects and radiation damage processes in α-quartz, in *Proc. 33rd Annual Symposium on Frequency Control,* Electronic Industries Association, Washington, D.C., 1979, 98.
172. **Mitchell, E. W. J. and Paige, E. G. S.,** The optical effects of radiation induced atomic damage in quartz, *Philos. Mag.*, 1, 1085, 1956.
173. **Antonini, M., Camagni, P., Gibson, P. N., and Manara, A.,** Comparison of heavy-ion, proton and electron irradiation effects in vitreous silica, *Rad. Effects,* 65, 41, 1982.
174. **Friebele, E. J., Griscom, D. L., Stapelbroek, M., and Weeks, R. A.,** Fundamental defect centers in glass: the peroxy radical in irradiated high-purity, fused silica, *Phys. Rev. Lett.*, 42, 1346, 1979.
175. **Stapelbroek, M., Griscom, D. L., Friebele, E. J., and Sigel, G. H., Jr.,** Oxygen-associated trapped-hole centers in high-purity fused silicas, *J. Non-Cryst. Solids,* 32, 313, 1979.
176. **Nelson, C. M. and Weeks, R. A.,** Vacuum-ultraviolet absorption studies of irradiated silica and quartz, *J. Appl. Phys.*, 32, 883, 1961.
177. **Antonini, M., Camagni, P., Manara, A., and Moro, L.,** Heavy-ion and electron irradiation effects in vitreous silica, *J. Non-Cryst. Solids,* 44, 321, 1981.
178. **Arnold, G. W. and Compton, W. D.,** Radiation effects in silica at low temperatures, *Phys. Rev.*, 116, 802, 1959.
179. **Palma, G. E. and Gagosz, R. M.,** Optical absorption in fused silica during irradiation: radiation annealing of the C-band, *J. Phys. Chem. Solids,* 33, 177, 1972.
180. **Korneienko, L. S., Rybaltovskii, A. O., and Chernov, P. V.,** Electron paramagnetic resonance and thermoluminescence of γ-irradiated vitreous silica produced by vapor phase synthesis, *Fiz. i Khim. Stek.* 2, 396, 1976.
181. **Griscom, D. L. and Friebele, E. J.,** Effects of ionizing radiation on amorphous insulators, *Rad. Effects,* 65, 63, 1982.
182. **Kaiser, P.,** Drawing-induced coloration in vitreous silica fibers, *J. Opt. Soc. Am.*, 64, 475, 1974.
183. **Friebele, E. J., Griscom, D. L., and Marrone, M. J.,** The optical absorption and luminescence bands near 2 eV in irradiated and drawn synthetic silica, *J. Non-Cryst. Solids,* 71, 133, 1985; **Friebele, E. J., Askins, C. G., Gingerich, M. E., and Long, K. J.,** Optical fiber waveguides in radiation environments. II, *Nucl. Instrum. Meth. Phys. Res.*, B1, 355, 1984.
184. **Silin, A. R., Skuya, L. N., and Shendrik, A. V.,** Intrinsic radiation defects in vitreous silica: nonbridging oxygen, *Fiz. i Khim. Stek.*, 4, 352, 1978.
185. **Treadaway, M. J., Passenheim, B. C., and Kitterer, B. D.,** Luminescence and absorption of electron-irradiated common optical glasses, sapphire and quartz, *IEEE Trans. Nucl. Sci.*, 22, 2253, 1975.
186. **Usmanova, S. Kh., Amosov, A. V., Vakhidor, Sh. A., Sanaev, B., and Yudin, D. M.,** Effect of neutron irradiation on the spectral characteristics of vitreous silica, *Fiz. i Khim. Stek.*, 1, 15, 1975.
187. **Skuja, L. N., Silin, A. R., and Mares, J.,** Decay time and polarization properties of luminescence centers in vitreous silica, *Phys. Status Solidi,* A50, K149, 1978.
188. **Sigel, G. H., Jr. and Marrone, M. J.,** Photoluminescence in as-drawn and irradiated silica optical fibers: an assessment of the role of non-bridging oxygen defect centers, *J. Non-Cryst. Solids,* 45, 235, 1981.
189. **Viehmann, W., Eubands, A. G., Pieper, G. F., and Bredekamp, J. H.,** Photomultiplier window materials under electron irradiation: fluorescence and phosphorescence, *Appl. Opt.*, 14, 2104, 1975.
190. **Arnold, G. W.,** Defects in natural and synthetic quartz, *J. Phys. Chem. Solids,* 13, 306, 1960.
191. **Nassau, K. and Prescott, B. E.,** A reinterpretation of smoky quartz, *Phys. Status Solidi,* A29, 659, 1975.
192. **Lell, E.,** Radiation effects in doped fused silica, *Phys. Chem. Glass.*, 3, 84, 1962.

193. **O'Brien, M. C. M.,** The structure of the color centers in smoky quartz, *Proc. R. Soc. London,* A231, 404, 1955.
194. **Koumvakalis, N.,** Defects in crysalline SiO_2: optical absorption of the aluminum-associated hole center, *J. Appl. Phys.,* 51, 5528, 1980.
195. **Weeks, R. A.,** Paramagnetic resonance and optical absorption in gamma-ray irradiated alpha quartz: the "A1" center, *J. Am. Ceram. Soc.,* 53, 176, 1970.
196. **Anderson, J. H. and Weil, J. A.,** Paramagnetic resonance of color centers in germanium-doped quartz, *J. Chem. Phys.,* 31, 427, 1959.
197. **Anderson, J. H., Feigl, F. J., and Schlesinger, M.,** The effects of heating on color centers in germanium-doped quartz, *J. Phys. Chem. Solids,* 35, 1425, 1974.
198. **Halperin, A. and Ralph, J. E.,** Optical studies of anisotropic color centers in germanium-doped quartz, *J. Chem. Phys.,* 39, 63, 1963.
199. **Schultz, P. C.,** Progress in optical waveguide process and materials, *Appl. Opt.,* 18, 3684, 1979.
200. **Friebele, E. J., Gingerich, M. E., and Long, K. J.,** Radiation damage of optical fiber waveguides at long wavelengths, *Appl. Opt.,* 21, 547, 1982.
201. **Friebele, E. J., Schultz, P. C., and Gingerich, M. E.,** Compositional effects on the radiation response of Ge-doped silica-core optical fiber waveguides, *Appl. Opt.,* 19, 2910, 1980.
202. **Griscom, D. L., Friebele, E. J., Long, K. J., and Fleming, J. W.,** Fundamental defect centers in glass: electron spin resonance and optical studies of irradiated phosphorous-doped silica glass and optical fibers, *J. Appl. Phys.,* 54, 3743, 1983.
203. **Friebele, E. J.,** Optical fiber waveguides in radiation environments, *Opt. Eng.,* 18, 552, 1979.
204. **Friebele, E. J., Griscom, D. L., and Sigel, G. H., Jr.,** Defect centers in a germanium-doped silica core optical fiber, *J. Appl. Phys.,* 45, 3424, 1974; **Kawazoe, H.,** Effect of modes of glass formation on structure of intrinsic or photon induced defects centered on III, IV, or IV cations in oxide glasses, *J. Non Cryst. Solids,* 71, 231, 1985.
205. **Griscom, D. L., Sigel, G. H., Jr., and Ginther, R. J.,** Defect centers in a pure-silica-core borosilicate-clad optical fiber: esr studies, *J. Appl. Phys.,* 47, 960, 1976.
206. **Friebele, E. J. and Gingerich, M. E.,** Photobleaching effects in optical fiber waveguides, *Appl. Opt.,* 20, 3448, 1981.
207. **Friebele, E. J. and Gingerich, M. E.,** Radiation-induced optical absorption bands in low loss optical fiber waveguides, *J. Non-Cryst. Solids,* 38/39, 245, 1980.
208. **Monk, G. S.,** The Coloration of Optical Materials by High Energy Radiations, Argonne National Laboratory Report, Argonne, Ill., ANL-4536, 1950.
209. **Monk, G. S.,** Coloration of optical glass by high-energy radiation, *Nucleonics,* 1952, 52.
210. **Sun, K-H. and Kreidl, N. J.,** Coloration of glass by radiation. I. II. III, *Glass Ind.,* 33, 511, 589, 651, 1952.
211. **Bishay, A. M. and Ferguson, K. R.,** Gamma-ray-induced coloring of glasses in relation to their structure, in *Advances in Glass Technology,* American Ceramic Society, Columbus, Ohio, 1962, 133.
212. **Lell, E.,** Synthesized impurity centers in fused silica, *J. Am. Ceram. Soc.,* 43, 422, 1960.
213. **Sigel, G. H., Jr.,** Ultraviolet spectra of silicate glasses: a review of some experimental evidence, *J. Non-Cryst. Solids,* 13, 372, 1974.
214. **van Wieringen, J. S. and Kats, A.,** Paramagnetic resonance and optical investigation of silicate glasses and fused silica, colored by x-rays, *Philips Res. Rep.,* 12, 432, 1957.
215. **Sigel, G. H., Jr. and Ginther, R. J.,** The effect of iron on the ultraviolet absorption of high purity soda-silica glass, *Glass Technol.,* 9, 66, 1968.
216. **Kats, A. and Stevels, J. M.,** The effect of u.v. and x-ray radiation on silicate glasses, fused silica and quartz crystals, *Philips Res. Rep.,* 11, 115, 1956.
217. **Beekenkamp, P.,** Color centers in borate, phosphate and borophosphate glasses, *Philips Res. Rep.,* Suppl. 4, 1, 1966.
218. **Smith, H. L. and Cohen, A. J.,** Color centers in x-irradiated soda-silica glasses, *J. Am. Ceram. Soc.,* 47, 564, 1964.
219. **Stroud, J. S., Schreurs, J. W. H., and Tucker, R. F.,** Charge trapping and the electronic structure of glass, in *Proc. VII Int. Congress on Glass, Brussels,* Gordon & Breach, New York, 1966, 7.
220. **Schreurs, J. W. H. and Tucker, R. F.,** Experimental and theoretical investigation of radiation effects in glasses, in *Proc. Int. Conf. Phys. Non-Cryst. Solids, Delft,* North-Holland, Amsterdam, 1956, 616.
221. **Mackey, J. H., Smith, H. L., and Nahum, J.,** Competitive trapping in sodium disilicate glasses doped with Eu^{+3}, *J. Phys. Chem. Solids,* 27, 1773, 1966.
222. **Stroud, J. S.,** Color centers in a cerium-containing silicate glass, *J. Chem. Phys.,* 37, 836, 1962.
223. **Stroud, J. S.,** Thermal stability of colour centres in a silicate glass, *Phys. Chem. Glasses,* 5, 71, 1964.
224. **Mackey, J. H., Smith, H. L., and Halperin, A.,** Optical studies in x-irradiated high purity sodium silicate glasses, *J. Phys. Chem. Solids,* 27, 1759, 1966.

225. **Schreurs, J. W. H.,** Study of some trapped hole centers in x-irradiated alkali silicate glasses, *J. Chem. Phys.*, 47, 818, 1967.
226. **Griscom, D. L.,** Defects in amorphous insulators, *J. Non-Cryst. Solids*, 31, 241, 1978.
227. **Mackey, J. H., Boss, J. W., and Kopp, M.,** Paramegnetic centres in irradiated, oxidized, and reduced sodium silicate glasses, *Phys. Chem. Glasses*, 11, 205, 1970.
228. **Yokota, R.,** Color centers in alkali silicate and borate glasses, *Phys. Rev.*, 95, 1145, 1954.
229. **Griscom, D. L.,** ESR and optical studies of alkali-associated trapped-electron centers in alkali borate glasses irradiated at 77 K, *J. Non-Cryst. Solids*, 6, 275, 1971.
230. **Bishay, A. M.,** Role of cerium in suppression of gamma-ray induced coloring of borate glasses, *J. Am. Ceram. Soc.*, 45, 389, 1962.
231. **Bishay, A. M.,** Gamma irradiation studies of some borate glasses, *J. Am. Ceram. Soc.*, 44, 289, 1961.
232. **Bukharaev, A. A. and Yafaev, N. R.,** Laser and thermal bleaching of color centers in sodium borate glasses, *Phys. Status Solidi*, A50, 711, 1978.
233. **Bishay, A. M.,** Gamma-ray induced coloring of some phosphate glasses, *J. Am. Ceram. Soc.*, 44, 545, 1961.
234. **Paul, A.,** Optical absorption of trivalent manganese in binary alkali borate glasses, *Phys. Chem. Glasses*, 11, 168, 1970.
235. **Barkatt, A., Ottolenghi, M., and Rabani, J.,** Pulse radiolysis of sodium metaphosphate glasses, *J. Phys. Chem.*, 76, 203, 1972.
236. **Wong, J. and Angell, C. A.,** *Glass Structure by Spectroscopy*, Marcel Dekker, New York, 1976, 377ff.
237. **Stroud, J. S.,** Photoionization of Ce^{3+} in glass, *J. Chem. Phys.*, 35, 844, 1961.
238. **Levy, P. W., Mattern, P. L., and Lengweiler, K.,** Studies on nonmetals during irradiation. V. Growth and decay of color centers in barium aluminoborate glasses containing cerium, *J. Am. Ceram. Soc.*, 57, 176, 1974.
239. **Friebele, E. J.,** Radiation protection of fiber optic materials: effect of cerium doping on the radiation-induced absorption, *Appl. Phys. Lett.*, 27, 210, 1975.
240. **Arafa, S.,** The titanium centre induced in irradiated alkali diborate glasses, *Phys. Chem. Glasses*, 15, 42, 1974.
241. **Galimov, D. G., Lun'kin, S. P., and Yudin, D. M.,** The effect of γ-irradiation on calcium aluminosilicate glasses activated by rare-earth and transition metals, *Z. Prik. Spektrosk.*, 11, 1127, 1969.
242. **Byurganovskaya, G. V. and Vasil'kova, A. A.,** The radiation and optical stability of lanthanum-containing glasses, *Opt. Technol.*, 39, 36, 1972. **Byrgonovskaya, G. V., Vasil'kova, A. A., and Kasumova, L. N.,** Radiooptical properties of high-silica glasses, *Izv. Akad. Nauk SSSR, Neo. Mat.*, 6, 63, 1970.
243. **Barker, R. S., McConkey, E. A. G., and Richardson, D. A.,** Effect of gamma radiation on the optical absorption of lead silicate glass, *Phys. Chem. Glasses*, 6, 24, 1965.
244. **Bishay, A. M. and Maklad, M.,** Radiation induced optical absorption in lead borate glasses in relation to structure changes, *Phys. Chem. Glasses*, 7, 149, 1966.
245. **Bishay, A. M.,** A gamma-ray induced absorption band in some lead borate glasses, *J. Am. Ceram. Soc.*, 43, 417, 1960.
246. **Bray, P. J., Leventhal, M., and Hooper, H. O.,** Nuclear magnetic resonance investigations of the structure of lead borate glasses, *Phys. Chem. Glasses*, 4, 47, 1963.
247. **Leventhal, M. and Bray, P. J.,** Nuclear magnetic resonance investigations of compounds and glasses in the systems PbO-B_2O_3 and PbO-SiO_2, *Phys. Chem. Glasses*, 6, 113, 1965.
248. **Kim, K. S., Bray, P. J., and Merrin, S.,** Nuclear magnetic resonance studies of the glasses in the system PbO-B_2O_3-SiO_2, *J. Chem. Phys.*, 64, 4459, 1976.
249. **Friebele, E. J.,** The Pb^{3+} center in irradiated lead silicate glass, in *Proc. XII Int. Congr. on Glass, Vol. III, Prague*, CVPs-Důmtechniky Praha (Prague), 1977, 87.
250. **Barker, R. S. and Richardson, D. A.,** Effect of gamma radiation on spectral transmission of some commercial glasses, *J. Am. Ceram. Soc.*, 44, 552, 1961.
251. **Brekhovskikh, S. M. and Tyul'kin, V. A.,** Paramagnetic and optical absorption radiation centers in silicate glasses containing lanthanum, *Z. Prik. Spektrosk.*, 18, 111, 1973.
252. **Schulman, J. H., Ginther, R. J., Klick, C. C., Alger, R. S., and Levy, R. A.,** Dosimetry of x-rays and gamma rays by radiophotoluminescence, *J. Appl. Phys.*, 22, 1479, 1951.
253. **Blair, G. E.,** Applications of radiation effects in glasses in low- and high-level dosimetry, *J. Am. Ceram. Soc.*, 43, 426, 1960.
254. **Ritz, V. H. and Cheek, C. H.,** Radiation intensity measurements with silver-activated glass block dosimeters, *Rad. Res.*, 25, 537, 1965.
255. **Evans, B. D., Sigel, G. H., Jr., Langworthy, J. B., and Faraday, B. J.,** The fiber optic dosimeter on the navigational technology satellite 2, *IEEE Trans. Nucl. Sci.*, 25, 1619, 1978.
256. **Kreidl, N. J. and Hensler, J. R.,** Gamma radiation insensitive optical glasses, *J. Opt. Soc. Am.*, 47, 73, 1957.

257. Radiation Resistant Optical Glasses (Cerium Stabilized), Schott Glaswerke Bulletin, No. 7600e-1975, Schott Glaswerke, Mainz, F.R.G.
258. **Evans, B. D. and Sigel, G. H., Jr.,** Radiation resistant fiber optic materials and waveguides, *IEEE Trans. Nucl. Sci.*, 22, 2462, 1975.
259. **Baldwin, C. M., Almeida, R. M., and Mackenzie, J. D.,** Halide glasses, *J. Non-Cryst. Solids,* 43, 309, 1981.
260. **Goldschmidt, V. M.,** *Skrifter Norske Vid. Akad. (Oslo)* 8, 127, 1926.
261. **Tran, D. C., Sigel, G. H., Jr., and Bendow, B.,** Heavy metal fluoride glasses and fibers: a review, *IEEE J. Lightwave Tech.*, LT-2, 566, 1984.
262. **Tsurikova, G. A., Yudin, D. M., Petrovskii, G. T., and Galimov, D. G.,** Structural features of some beryllium fluoride glasses, *Izv. Akad. Nauk SSSR, Neo. Mat.*, 6, 498, 1970.
263. **Galimov, D. G., Karapetyan, G. O., Petrovskii, G. T., Tsurikova, G. A., and Yudin, D. M.,** Action of hard radiation on fluoroberyllate glasses activated by ions of rare-earth and transition metals, *Isv. Akad. Nauk SSSR, Neo. Mat.*, 5, 1807, 1969.
264. **Griscom, D. L., Taylor, P. C., and Bray, P. J.,** Paramagnetic resonance of room-temperature-stable V-type centers in γ-irradiated alkali halide-boron oxide glasses, *J. Chem. Phys.*, 50, 977, 1969.
265. **Griscom, D. L.,** Optical absorption of Cl_2^- hole-type centers in irradiated alkali halide-alkali borate glasses, *J. Chem. Phys.*, 51, 5186, 1969.
266. **Friebele, E. J. and Tran, D. C.,** Radiation effects in ZrF_4 based glasses. II. Optical absorption, *J. Non-Cryst. Solids,* in press.
267. **Cases, R., Griscom, D. L., and Tran, D. C.,** Radiation effects in ZrF_4 based glasses. I. Electron Spin Resonance, *J. Non-Cryst. Solids,* in press.
268. **Friebele, E. J. and Tran, D. C.,** Optical absorption studies of chlorine-associated defect centers in irradiated ZrF_4 based glasses, *J. Non-Cryst. Solids,* in press.
269. **Griscom, D. L. and Tran, D. C.,** Chlorine-associated defect centers in irradiated ZrF_4 based glasses, *J. Non-Cryst. Solids,* in press.
270. **Rosiewicz, A. and Gannon, J. R.,** Effects of γ-irradiation on mid-ir transmitting glasses, *Elect. Lett.*, 17, 184, 1981.
271. **Levin, K. H., Tran, D. C., Ginther, R. J., and Sigel, G. H., Jr.,** Optical properties of fiber and bulk zirconium fluoride glass, *J. Glass Tech.*, 24, 143, 1983.
272. **Ohishi, Y., Mitachi, S., Takahashi, S., and Miyashita, T.,** γ-ray irradiation effect on transmission loss for ZrF_4-based optical fibers, *Elect. Lett.*, 19, 830, 1983; **Ohishi, Y., Mitachi, S., Takahashi, S., and Miyashita, T.,** Effects of gamma rays on optical transmission of ZrF_4-based fluoride fibers, *IEE Proc.*, 132, 114, 1985.

Index

INDEX

A

B

C

D

E

H

I

K

L

M

N

O

P

Q

R

T

U

V

W

X

Y

Z